Signals and Systems

Signals and Systems

Simon Haykin

McMaster University

Barry Van Veen

University of Wisconsin

JOHN WILEY & SONS, INC.

New York ■ Chichester ■ Weinheim ■ Brisbane ■ Singapore ■ Toronto

To Nancy and Kathy, Emily, David, and Jonathan

EDITOR *Bill Zobrist*
MARKETING MANAGER *Katherine Hepburn*
SENIOR PRODUCTION MANAGER *Lucille Buonocore*
SENIOR PRODUCTION EDITOR *Monique Calello*
SENIOR DESIGNER *Laura Boucher*
TEXT DESIGNER *Nancy Field*
COVER DESIGNER *Laura Boucher*
COVER PHOTO *Courtesy of NASA*
ILLUSTRATION EDITOR *Sigmund Malinowski*
ILLUSTRATION *Wellington Studios*

This book was set in Times Roman by UG division of GGS Information Services and printed and bound by Quebecor Printing, Kingsport. The cover was printed by Phoenix Color Corporation.

This book is printed on acid-free paper. ∞

The paper in this book was manufactured by a mill whose forest management programs include sustained yield harvesting of its timberlands. Sustained yield harvesting principles ensure that the numbers of trees cut each year does not exceed the amount of new growth.

Library of Congress Cataloging-in-Publication Data
Haykin, Simon
    Signals and systems / Simon Haykin, Barry Van Veen.
      p.   cm.
    Includes index.
    ISBN 0-471-13820-7 (cloth : alk. paper)
    1. Signal processing.   2. System analysis.   3. Linear time
invariant systems.   4. Telecommunication systems.   I. Van Veen,
Barry.   II. Title.
    TK5102.5.H37   1999
    621.382′2—dc21   97-52090
                                                                      CIP

Printed in the United States of America

10 9 8 7 6 5 4 3 2

Preface

The study of signals and systems is basic to the discipline of electrical engineering at all levels. It is an extraordinarily rich subject with diverse applications. Indeed, a thorough understanding of signals and systems is essential for a proper appreciation and application of other parts of electrical engineering, such as *signal processing, communication systems,* and *control systems.*

This book is intended to provide a modern treatment of signals and systems at an introductory level. As such, it is intended for use in electrical engineering curricula in the sophomore or junior years and is designed to prepare students for upper-level courses in communication systems, control systems, and digital signal processing.

The book provides a balanced and integrated treatment of continuous-time and discrete-time forms of signals and systems intended to reflect their roles in engineering practice. Specifically, these two forms of signals and systems are treated side by side. This approach has the pedagogical advantage of helping the student see the fundamental similarities and differences between discrete-time and continuous-time representations. Real-world problems often involve mixtures of continuous-time and discrete-time forms, so the integrated treatment also prepares the student for practical usage of these concepts. This integrated philosophy is carried over to the chapters of the book that deal with applications of signals and systems in modulation, filtering, and feedback systems.

Abundant use is made of examples and drill problems with answers throughout the book. All of these are designed to help the student understand and master the issues under consideration. The last chapter is the only one without drill problems. Each chapter, except for the last chapter, includes a large number of end-of-chapter problems designed to test the student on the material covered in the chapter. Each chapter also includes a list of references for further reading and a collection of historical remarks.

Another feature of the book is the emphasis given to design. In particular, the chapters dealing with applications include illustrative design examples.

MATLAB, acronym for MATrix LABoratory and product of The Math Works, Inc., has emerged as a powerful environment for the experimental study of signals and systems. We have chosen to integrate MATLAB in the text by including a section entitled "Exploring Concepts with MATLAB" in every chapter, except for the concluding chapter. In making this choice, we have been guided by the conviction that MATLAB provides a computationally efficient basis for a "Software Laboratory," where concepts are explored and system designs are tested. Accordingly, we have placed the section on MATLAB before the "Summary" section, thereby relating to and building on the entire body of material discussed in the preceding sections of the pertinent chapter. This approach also offers the instructor flexibility to either formally incorporate MATLAB exploration into the classroom or leave it for the students to pursue on their own.

Each "Exploring Concepts with MATLAB" section is designed to instruct the student on the proper application of the relevant MATLAB commands and develop additional insight into the concepts introduced in the chapter. Minimal previous exposure to MATLAB is assumed. The MATLAB code for all the computations performed in the book, including the last chapter, are available on the Wiley Web Site: http://www.wiley.com/college.

There are 10 chapters in the book, organized as follows:

▶ Chapter 1 begins by motivating the reader as to what signals and systems are and how they arise in communication systems, control systems, remote sensing, biomedical signal processing, and the auditory system. It then describes the different classes of signals, defines certain elementary signals, and introduces the basic notions involved in the characterization of systems.

▶ Chapter 2 presents a detailed treatment of time-domain representations of linear time-invariant (LTI) systems. It develops convolution from the representation of an input signal as a superposition of impulses. The notions of causality, memory, stability, and invertibility that were briefly introduced in Chapter 1 are then revisited in terms of the impulse response description for LTI systems. The steady-state response of a LTI system to a sinusoidal input is used to introduce the concept of frequency response. Differential- and difference-equation representations for linear time-invariant systems are also presented. Next, block diagram representations for LTI systems are introduced. The chapter finishes with a discussion of the state-variable description of LTI systems.

▶ Chapter 3 deals with the Fourier representation of signals. In particular, the Fourier representations of four fundamental classes of signals are thoroughly discussed in a unified manner:

  ▶ Discrete-time periodic signals: the discrete-time Fourier series
  ▶ Continuous-time periodic signals: the Fourier series
  ▶ Discrete-time nonperiodic signals: the discrete-time Fourier transform
  ▶ Continuous-time nonperiodic signals: the Fourier transform

A novel feature of the chapter is the way in which similarities between these four representations are exploited and the differences between them are highlighted. The fact that complex sinusoids are eigenfunctions of LTI systems is used to motivate the representation of signals in terms of complex sinusoids. The basic form of the Fourier representation for each signal class is introduced and the four representations are developed in sequence. Next, the properties of all four representations are studied side by side. A strict separation between signal classes and the corresponding Fourier representations is maintained throughout the chapter. It is our conviction that a parallel, yet separate, treatment minimizes confusion between representations and aids later mastery of proper application for each. Mixing of Fourier representations occurs naturally in the context of analysis and computational applications and is thus deferred to Chapter 4.

▶ Chapter 4 presents a thorough treatment of the applications of Fourier representations to the study of signals and LTI systems. Links between the frequency-domain and time-domain system representations presented in Chapter 2 are established. Both analysis and computational applications are then used to motivate derivation of the relationships between the four Fourier representations and develop the student's skill in applying these tools. The continuous-time and discrete-time Fourier transform representations of periodic signals are introduced for analyzing problems in which there is a mixture of periodic and nonperiodic signals, such as application of a periodic input to a LTI system. The Fourier transform representation for discrete-time signals is then developed as a tool for analyzing situations in which there is a mixture of continuous-time and discrete-time signals. The sampling process and continuous-time signal reconstruction from samples are studied in detail within this context. Systems for discrete-time processing of continuous-time signals are also discussed, including the issues of oversampling, decimation, and interpolation. The chapter concludes by developing relationships between the discrete-time Fourier series and the discrete-time and continuous-time Fourier transforms in order to introduce the computational aspects of the Fourier analysis of signals.

▶ Chapter 5 presents an introductory treatment of linear modulation systems applied to communication systems. Practical reasons for using modulation are described. Amplitude modulation and its variants, namely, double sideband-suppressed carrier modulation, single sideband modulation, and vestigial sideband modulation, are discussed. The chapter also includes a discussion of pulse-amplitude modulation and its role in digital communications to again highlight a natural interaction between continuous-time and discrete-time signals. The chapter includes a discussion of frequency-division and time-division multiplexing techniques. It finishes with a treatment of phase and group delays that arise when a modulated signal is transmitted through a linear channel.

▶ Chapter 6 discusses the Laplace transform and its use for the complex exponential representations of continuous-time signals and the characterization of systems. The eigenfunction property of LTI systems and the existence of complex exponential representations for signals that have no Fourier representation are used to motivate the study of Laplace transforms. The unilateral Laplace transform is studied first and applied to the solution of differential equations with initial conditions to reflect the dominant role of the Laplace transform in engineering applications. The bilateral Laplace transform is introduced next and is used to study issues of causality, stability, invertibility, and the relationship between poles and zeros and frequency response. The relationships between the transfer function description of LTI systems and the time-domain descriptions introduced in Chapter 2 are developed.

▶ Chapter 7 is devoted to the z-transform and its use in the complex exponential representation of discrete-time signals and the characterization of systems. As in Chapter 6, the z-transform is motivated as a more general representation than that of the discrete-time Fourier transform. Consistent with its primary role as an analysis tool, we begin with the bilateral z-transform. The properties of the z-transform and techniques for inversion are introduced. Next, the z-transform is used for transform analysis of systems. Relationships between the transfer function and time-domain descriptions introduced in Chapter 2 are developed. Issues of invertibility, stability, causality, and the relationship between the frequency response and poles and zeros are revisited. The use of the z-transform for deriving computational structures for implementing discrete-time systems on computers is introduced. Lastly, use of the unilateral z-transform for solving difference equations is presented.

▶ Chapter 8 discusses the characterization and design of linear filters and equalizers. The approximation problem, with emphasis on Butterworth functions and brief mention of Chebyshev functions, is introduced. Direct and indirect methods for the design of analog (i.e., continuous-time) and digital (i.e., discrete-time) types of filters are presented. The window method for the design of finite-duration impulse response digital filters and the bilateral transform method for the design of infinite-duration impulse response digital filters are treated in detail. Filter design offers another opportunity to reinforce the links between continuous-time and discrete-time systems. The chapter builds on material presented in Chapter 4 in developing a method for the equalization of a linear channel using a discrete-time filter of finite impulse response. Filters and equalizers provide a natural vehicle for developing an appreciation for how to design systems required to meet prescribed frequency-domain specifications.

▶ Chapter 9 presents an introductory treatment of the many facets of linear feedback systems. The various practical advantages of feedback and the cost of its application are emphasized. The applications of feedback in the design of operational amplifiers and feedback control systems are discussed in detail. The stability problem, basic to the study of feedback systems, is treated in detail by considering the following methods:

  ▶ The root-locus method, related to the closed-loop transient response of the system
  ▶ Nyquist stability criterion, related to the open-loop frequency response of the system

The Nyquist stability criterion is studied using both the Nyquist locus and Bode diagram. The chapter also includes a discussion of sampled data systems to illustrate the natural interaction between continuous-time and discrete-time signals that occurs in control applications.

▶ Chapter 10, the final chapter in the book, takes a critical look at limitations of the representations of signals and systems presented in the previous chapters of the book. It highlights other advanced tools, namely, time–frequency analysis (the short-time Fourier transform and wavelets) and chaos, for the characterization of signals. It also highlights the notions of nonlinearity and adaptivity in the study of systems. In so doing, the student is made aware of the very broad nature of the subject of signals and systems and reminded of the limitations of the linear, time-invariance assumption.

In organizing the material as described, we have tried to follow theoretical material by appropriate applications drawn from the fields of communication systems, design of filters, and control systems. This has been done in order to provide a source of motivation for the reader.

The material in this book can be used for either a one- or two-semester course sequence on signals and systems. A two-semester course sequence would cover most, if not all, of the topics in the book. The material for a one-semester course can be arranged in a variety of ways, depending on the preference of the instructor. We have attempted to maintain maximum teaching flexibility in the selection and order of topics, subject to our philosophy of truly integrating continuous-time and discrete-time concepts. Some sections of the book include material that is considered to be of an advanced nature; these sections are marked with an asterisk. The material covered in these sections can be omitted without disrupting the continuity of the subject matter presented in the pertinent chapter.

The book finishes with the following appendices:

▶ Selected mathematical identities
▶ Partial fraction expansions
▶ Tables of Fourier representations and properties
▶ Tables of Laplace transforms and properties
▶ Tables of z-transforms and properties

A consistent set of notations is used throughout the book. Except for a few places, the derivations of all the formulas are integrated into the text.

The book is accompanied by a detailed *Solutions Manual* for all the end-of-chapter problems in the book. A copy of the *Manual* is only available to instructors adopting this book for use in classrooms and may be obtained by writing to the publisher.

Acknowledgments

In writing this book over a period of four years, we have benefited enormously from the insightful suggestions and constructive inputs received from many colleagues and reviewers:

- Professor Rajeev Agrawal, *University of Wisconsin*
- Professor Richard Baraniuk, *Rice University*
- Professor Jim Bucklew, *University of Wisconsin*
- Professor C. Sidney Burrus, *Rice University*
- Professor Dan Cobb, *University of Wisconsin*
- Professor Chris DeMarco, *University of Wisconsin*
- Professor John Gubner, *University of Wisconsin*
- Professor Yu Hu, *University of Wisconsin*
- Professor John Hung, *Auburn University*
- Professor Steve Jacobs, *University of Pittsburg*
- Dr. James F. Kaiser, *Bellcore*
- Professor Joseph Kahn, *University of California-Berkeley*
- Professor Ramdas Kumaresan, *University of Rhode Island*
- Professor Troung Nguyen, *Boston University*
- Professor Robert Nowak, *Michigan State University*
- Professor S. Pasupathy, *University of Toronto*
- Professor John Platt, *McMaster University*
- Professor Naresh K. Sinha, *McMaster University*
- Professor Mike Thomson, *University of Texas-Pan America*
- Professor Anthony Vaz, *McMaster University*

We extend our gratitude to them all for helping us in their own individual ways to shape the book into its final form.

Barry Van Veen is indebted to his colleagues at the University of Wisconsin, and Professor Willis Tompkins, Chair of the Department of Electrical and Computer Engineering, for allowing him to teach the Signals and Systems Classes repeatedly while in the process of working on this text.

We thank the many students at both McMaster and Wisconsin, whose suggestions and questions have helped us over the years to refine and in some cases rethink the presentation of the material in this book. In particular, we thank Hugh Pasika, Eko Onggo Sanusi, Dan Sebald, and Gil Raz for their invaluable help in preparing some of the computer experiments, the solutions manual, and in reviewing page proofs.

The idea of writing this book was conceived when Steve Elliott was the Editor of Electrical Engineering at Wiley. We are deeply grateful to him. We also wish to express our gratitude to Charity Robey for undertaking the many helpful reviews of the book, and Bill Zobrist, the present editor of Electrical Engineering at Wiley, for his strong support. We wish to thank Monique Calello for dextrously managing the production of the book, and Katherine Hepburn for her creative promotion of the book.

Lastly, Simon Haykin thanks his wife Nancy, and Barry Van Veen thanks his wife Kathy and children Emily and David, for their support and understanding throughout the long hours involved in writing this book.

Simon Haykin
Barry Van Veen

To Nancy and Kathy, Emily, David, and Jonathan

Contents

CHAPTER 9   *Application to Feedback Systems*   556

CHAPTER 10   *Epilogue*   648

Notation

[·]   indicates discrete-valued independent variable, for example, $x[n]$

(·)   indicates continuous-valued independent variable, for example, $x(t)$

► Lowercase functions denote time-domain quantities, for example, $x(t)$, $w[n]$

► Uppercase functions denote frequency- or transform-domain quantities

$X[k]$ discrete-time Fourier series coefficients for $x[n]$

$X[k]$ Fourier series coefficients for $x(t)$

$X(e^{j\Omega})$ discrete-time Fourier transform of $x[n]$

$X(j\omega)$ Fourier transform of $x(t)$

$X(s)$ Laplace transform of $x(t)$

$X(z)$ z-transform of $x[n]$

► Boldface lowercase symbols denote vector quantities, for example, **q**

► Boldface uppercase symbols denote matrix quantities, for example, **A**

► Subscript δ indicates continuous-time representation for a discrete-time signal

$x_\delta(t)$ continuous-time representation for $x[n]$

$X_\delta(j\omega)$ Fourier transform of $x_\delta(t)$

► Sans serif type indicates MATLAB variables or commands, for example,

`X = fft(x,n)`

$0°$ is defined as 1 for convenience

arctan   refers to the four-quadrant function and produces a value between $-\pi$ to π radians.

Symbols

$|c|$ magnitude of complex quantity c

$\arg\{c\}$ phase angle of complex quantity c

$\text{Re}\{c\}$ real part of c

$\text{Im}\{c\}$ imaginary part of c

c^* complex conjugate of c

j square root of -1

i square root of -1 used by MATLAB

$\mathcal{T}$ sampling interval in seconds

T fundamental period for continuous-time signal in seconds

N fundamental period for discrete-time signal in samples

ω (angular) frequency for continuous-time signal in radians/second

Ω	(angular) frequency for discrete-time signal in radians
ω_o	fundamental (angular) frequency for continuous-time periodic signal in radians/second
Ω_o	fundamental (angular) frequency for discrete-time periodic signal in radians
$u(t)$, $u[n]$	step function of unit amplitude
$\delta[n]$, $\delta(t)$	impulse function of unit strength
$H\{\cdot\}$	representation of a system as an operator H
$S^\tau\{\cdot\}$	time shift of τ units
H^{-1}, h^{-1}	superscript -1 denotes inverse system
$*$	denotes convolution operation
$H(e^{j\Omega})$	discrete-time system frequency response
$H(j\omega)$	continuous-time system frequency response
$h[n]$	discrete-time system impulse response
$h(t)$	continous-time system impulse response
$y^{(n)}$	superscript (n) denotes natural response
$y^{(f)}$	superscript (f) denotes forced response
$y^{(p)}$	superscript (p) denotes particular solution
$\xleftrightarrow{DTFS;\ \Omega_o}$	discrete-time Fourier series pair with fundamental frequency Ω_o
$\xleftrightarrow{FS;\ \omega_o}$	Fourier series pair with fundamental frequency ω_o
$\xleftrightarrow{DTFT}$	discrete-time Fourier transform pair
$\xleftrightarrow{FT}$	Fourier transform pair
$\xleftrightarrow{\mathscr{L}}$	Laplace transform pair
$\xleftrightarrow{\mathscr{L}_u}$	unilateral Laplace transform pair
$\xleftrightarrow{z}$	z-transform pair
$\xleftrightarrow{z_u}$	unilateral z-transform pair
$\text{sinc}(u)$	$\sin(\pi u)/\pi u$
$\circledast$	periodic convolution of two periodic signals
$\cap$	intersection
$T(s)$	closed-loop transfer function
$F(s)$	return difference
$L(s)$	loop transfer function
ϵ_{ss}	steady-state error
K_p	position error constant
K_v	velocity error constant
K_a	acceleration error constant
P.O.	percentage overshoot
T_p	peak time
T_r	rise time

T_s settling time

$X(\tau, j\omega)$ short-time Fourier transform of $x(t)$

$W_x(\tau, a)$ wavelet transform of $x(t)$

Abbreviations

A/D   analog-to-digital (converter)

AM   amplitude modulation

BIBO   bounded input bounded output

CW   continuous wave

D/A   digital-to-analog (converter)

dB   decibel

DOF   degree of freedom

DSB-SC   double sideband-suppressed carrier

DTFS   discrete-time Fourier series

DTFT   discrete-time Fourier transform

FDM   frequency-division multiplexing

FFT   fast Fourier transform

FIR   finite-duration impulse response

FM   frequency modulation

FS   Fourier series

FT   Fourier transform

Hz   hertz

IIR   infinite-duration impulse response

LTI   linear time-invariant (system)

MRI   magnetic resonance image

MSE   mean squared error

PAM   pulse-amplitude modulation

PCM   pulse-code modulation

PM   phase modulation

QAM   quadrature-amplitude modulation

ROC   region of convergence

rad   radian(s)

s   second

SSB   single sideband modulation

STFT   short-time Fourier transform

TDM   time-division multiplexing

VSB   vestigial sideband modulation

WT   wavelet transform

FIGURE 1.6 Perspective view of Mount Shasta (California) derived from a pair of stereo radar images acquired from orbit with the Shuttle Imaging Radar (SIR-B). (Courtesy of Jet Propulsion Laboratory.)

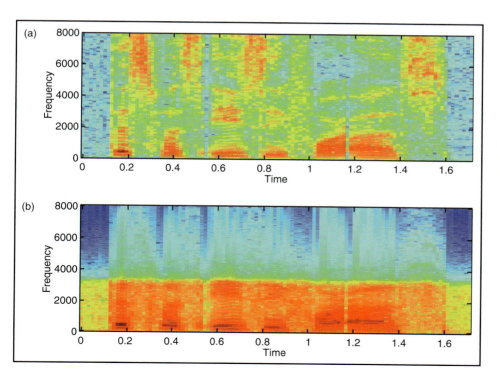

FIGURE 10.5 Spectrograms of speech signals. (a) Noisy version of the speech signal produced by a female speaker saying the phrase: "This was easy for us." (b) Filtered version of the speech signal.

1 Introduction

1.1 What Is a Signal?

Signals, in one form or another, constitute a basic ingredient of our daily lives. For example, a common form of human communication takes place through the use of speech signals, be that in a face-to-face conversation or over a telephone channel. Another common form of human communication is visual in nature, with the signals taking the form of images of people or objects around us.

Yet another form of human communication is through electronic mail over the *Internet*. In addition to mail, the Internet provides a powerful medium for searching for information of general interest, advertising, telecommuting, education, and games. All of these forms of communication over the Internet involve the use of information-bearing signals of one kind or another. Other real-life examples where signals of interest arise are discussed in what follows.

By listening to the heartbeat of a patient and monitoring his/her blood pressure and temperature, a doctor is able to diagnose the presence or absence of an illness or disease. These quantities represent signals that convey information to the doctor about the state of health of the patient.

In listening to a weather forecast over the radio, we hear references made to daily variations in temperature, humidity, and the speed and direction of prevalent winds. The signals represented by these quantities help us, for example, to form an opinion about whether to stay indoors or go out for a walk.

The daily fluctuations in the prices of stocks and commodities on world markets, in their own ways, represent signals that convey information on how the shares in a particular company or corporation are doing. On the basis of this information, decisions are made on whether to make new investments or sell old ones.

A probe exploring outer space sends valuable information about a faraway planet back to an Earth station. The information may take the form of radar images representing surface profiles of the planet, infrared images conveying information on how hot the planet is, or optical images revealing the presence of clouds around the planet. By studying these images, our knowledge of the unique characteristics of the planet in question is enhanced significantly.

Indeed, the list of what constitutes a signal is almost endless.

A signal is formally defined as a function of one or more variables, which conveys information on the nature of a physical phenomenon. When the function depends on a single variable, the signal is said to be *one-dimensional*. A speech signal is an example of a one-dimensional signal whose amplitude varies with time, depending on the spoken word and who speaks it. When the function depends on two or more variables, the signal is said to be *multidimensional*. An image is an example of a two-dimensional signal, with the horizontal and vertical coordinates of the image representing the two dimensions.

▌1.2 *What Is a System?*

In the examples of signals mentioned above, there is always a system associated with the generation of each signal and another system associated with the extraction of information from the signal. For example, in speech communication, a sound source or signal excites the vocal tract, which represents a system. The processing of speech signals usually relies on the use of our ears and auditory pathways in the brain. In the situation described here, the systems responsible for the production and reception of signals are biological in nature. They could also be performed using electronic systems that try to emulate or mimic their biological counterparts. For example, the processing of a speech signal may be performed by an automatic speech recognition system in the form of a computer program that recognizes words or phrases.

There is no unique purpose for a system. Rather, the purpose depends on the application of interest. In an automatic speaker recognition system, the function of the system is to extract information from an incoming speech signal for the purpose of *recognizing* or *identifying* the speaker. In a communication system, the function of the system is to *transport* the information content of a message signal over a communication channel and deliver it to a destination in a reliable fashion. In an *aircraft landing system*, the requirement is to keep the aircraft on the extended centerline of a runway.

A system is formally defined as an entity that manipulates one or more signals to accomplish a function, thereby yielding new signals. The interaction between a system and its associated signals is illustrated schematically in Fig. 1.1. The descriptions of the input and output signals naturally depend on the intended application of the system:

- ▶ In an automatic speaker recognition system, the input signal is a speech (voice) signal, the system is a computer, and the output signal is the identity of the speaker.
- ▶ In a communication system, the input signal could be a speech signal or computer data, the system itself is made up of the combination of a transmitter, channel, and receiver, and the output signal is an estimate of the original message signal.
- ▶ In an aircraft landing system, the input signal is the desired position of the aircraft relative to the runway, the system is the aircraft, and the output signal is a correction to the lateral position of the aircraft.

▌1.3 *Overview of Specific Systems*

In describing what we mean by signals and systems in the previous two sections, we mentioned several applications of signals and systems. In this section we will expand on five

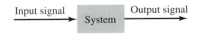

FIGURE 1.1   Block diagram representation of a system.

FIGURE 1.2   Elements of a communication system. The transmitter changes the message signal into a form suitable for transmission over the channel. The receiver processes the channel output (i.e., the received signal) to produce an estimate of the message signal.

of those applications systems, namely, communication systems, control systems, remote sensing, biomedical signal processing, and auditory systems.

■ COMMUNICATION SYSTEMS

There are three basic elements to every communication system, namely, *transmitter, channel*, and *receiver*, as depicted in Fig. 1.2. The transmitter is located at one point in space, the receiver is located at some other point separate from the transmitter, and the channel is the physical medium that connects them together. Each of these three elements may be viewed as a system with associated signals of its own. The purpose of the transmitter is to convert the message signal produced by a source of information into a form suitable for transmission over the channel. The message signal could be a speech signal, television (video) signal, or computer data. The channel may be an optical fiber, coaxial cable, satellite channel, or mobile radio channel; each of these channels has its specific area of application.

As the transmitted signal propagates over the channel, it is distorted due to the physical characteristics of the channel. Moreover, noise and interfering signals (originating from other sources) contaminate the channel output, with the result that the received signal is a corrupted version of the transmitted signal. The function of the receiver is to operate on the received signal so as to reconstruct a recognizable form (i.e., produce an estimate) of the original message signal and deliver it to the user destination. The signal-processing role of the receiver is thus the reverse of that of the transmitter; in addition, the receiver reverses the effects of the channel.

Details of the operations performed in the transmitter and receiver depend on the type of communication system being considered. The communication system can be of an analog or digital type. In signal-processing terms, the design of an *analog communication system* is relatively simple. Specifically, the transmitter consists of a *modulator* and the receiver consists of a *demodulator*. *Modulation* is the process of converting the message signal into a form that is compatible with the transmission characteristics of the channel. Ordinarily, the transmitted signal is represented as amplitude, phase, or frequency variation of a sinusoidal carrier wave. We thus speak of amplitude modulation, phase modulation, or frequency modulation, respectively. Correspondingly, through the use of amplitude demodulation, phase demodulation, or frequency demodulation, an estimate of the original message signal is produced at the receiver output. Each one of these analog modulation/demodulation techniques has its own advantages and disadvantages.

In contrast, a *digital communication system* is considerably more complex, as described here. If the message signal is of analog form, as in speech and video signals, the transmitter performs the following operations to convert it into digital form:

▶ *Sampling*, which converts the message signal into a sequence of numbers, with each number representing the amplitude of the message signal at a particular instant of time.

▶ *Quantization*, which involves representing each number produced by the sampler to the nearest level selected from a finite number of discrete amplitude levels. For example, we may represent each sample as a 16-bit binary number, in which case there are 2^{16} amplitude levels. After the combination of sampling and quantization, we have a representation of the message signal that is *discrete* in both time and amplitude.

▶ *Coding*, the purpose of which is to represent each quantized sample by a codeword made up of a finite number of symbols. For example, in a binary code the symbols may be 1's or 0's.

Unlike the operations of sampling and coding, quantization is completely irreversible: that is, a loss of information is always incurred by its application. However, this loss can be made small, and nondiscernible for all practical purposes, by using a quantizer with a sufficiently large number of discrete amplitude levels. As the number of discrete amplitude levels increases, the length of the codeword must also increase in a corresponding way.

If, however, the source of information is discrete to begin with, as in the case of a digital computer, none of the above operations would be needed.

The transmitter may involve additional operations, namely, data compression and channel encoding. The purpose of data compression is to remove redundant information from the message signal and thereby provide for efficient utilization of the channel by reducing the number of bits/sample required for transmission. Channel encoding, on the other hand, involves the insertion of redundant elements (e.g., extra symbols) into the codeword in a controlled manner; this is done to provide protection against noise and interfering signals picked up during the course of transmission through the channel. Finally, the coded signal is modulated onto a carrier wave (usually sinusoidal) for transmission over the channel.

At the receiver, the above operations are performed in reverse order. An estimate of the original message signal is thereby produced and delivered to the user destination. However, as mentioned previously, quantization is irreversible and therefore has no counterpart in the receiver.

It is apparent from this discussion that the use of digital communications may require a considerable amount of electronic circuitry. This is not a significant problem since the electronics are relatively inexpensive, due to the ever-increasing availability of very-large-scale-integrated (VLSI) circuits in the form of silicon chips. Indeed, with continuing improvements in the semiconductor industry, digital communications are often more cost effective than analog communications.

There are two basic modes of communication:

1. *Broadcasting*, which involves the use of a single powerful transmitter and numerous receivers that are relatively cheap to build. Here information-bearing signals flow only in one direction.

2. *Point-to-point communication*, in which the communication process takes place over a link between a single transmitter and a single receiver. In this case, there is usually a bidirectional flow of information-bearing signals. There is a transmitter and receiver at each end of the link.

The broadcasting mode of communication is exemplified by the radio and television that are integral parts of our daily lives. On the other hand, the ubiquitous telephone provides the means for one form of point-to-point communication. Note, however, that in this case the link is part of a highly complex telephone network designed to accommodate a large number of users on demand.

Another example of point-to-point communication is the *deep-space communications* link between an Earth station and a robot navigating the surface of a distant planet. Unlike telephonic communication, the composition of the message signal depends on the direction of the communication process. The message signal may be in the form of computer-generated instructions transmitted from an Earth station that command the robot to perform specific maneuvers, or it may contain valuable information about the chemical composition of the soil on the planet that is sent back to Earth for analysis. In order to reliably communicate over such great distances, it is necessary to use digital communications. Figure 1.3(a) shows a photograph of the robot, named *Pathfinder*, which landed on

(a)

(b)

FIGURE 1.3 (a) Snapshot of *Pathfinder* exploring the surface of Mars. (b) The 70-meter (230-foot) diameter antenna located at Canberra, Australia. The surface of the 70-meter reflector must remain accurate within a fraction of the signal wavelength. (Courtesy of Jet Propulsion Laboratory.)

Mars on July 4, 1997, a historic day in the National Aeronautics and Space Administration's (NASA's) scientific investigation of the solar system. Figure 1.3(b) shows a photograph of the high-precision, 70-meter antenna located at Canberra, Australia, which is an integral part of NASA's worldwide Deep Space Network (DSN). The DSN provides the vital two-way communications link that guides and controls (unmanned) planetary explorers and brings back images and new scientific information collected by them. The successful use of DSN for planetary exploration represents a triumph of communication theory and technology over the challenges presented by the unavoidable presence of noise.

Unfortunately, every communication system suffers from the presence of *channel noise* in the received signal. Noise places severe limits on the quality of received messages. Owing to the enormous distance between our own planet Earth and Mars, for example, the average power of the information-bearing component of the received signal, at either end of the link, is relatively small compared to the average power of the noise component. Reliable operation of the link is achieved through the combined use of (1) *large antennas* as part of the DSN and (2) *error control*. For a parabolic-reflector antenna (i.e., the type of antenna portrayed in Fig. 1.3(b)), the *effective area* is generally between 50% and 65% of the physical area of the antenna. The received power available at the terminals of the antenna is equal to the effective area times the power per unit area carried by the incident electromagnetic wave. Clearly, the larger the antenna, the larger the received signal power will be, hence the use of large antennas in DSN.

Turning next to the issue of error control, it involves the use of a *channel encoder* at the transmitter and a *channel decoder* at the receiver. The channel encoder accepts message bits and adds *redundancy* according to a prescribed rule, thereby producing encoded data at a higher bit rate. The redundant bits are added for the purpose of protection against channel noise. The channel decoder exploits the redundancy to decide which message bits were actually sent. The combined goal of the channel encoder and decoder is to minimize the effect of channel noise: that is, the number of errors between the channel encoder input (derived from the source of information) and the encoder output (delivered to the user by the receiver) is minimized on average.

■ CONTROL SYSTEMS

Control of physical systems is widespread in the application of signals and systems in our industrial society. As some specific examples where control is applied, we mention aircraft autopilots, mass-transit vehicles, automobile engines, machine tools, oil refineries, paper mills, nuclear reactors, power plants, and robots. The object to be controlled is commonly referred to as a *plant*; in this context, an aircraft is a plant.

There are many reasons for using control systems. From an engineering viewpoint, the two most important ones are the attainment of a satisfactory response and robust performance, as described here:

1. *Response.* A plant is said to produce a satisfactory response if its output follows or tracks a specified reference input. The process of holding the plant output close to the reference input is called *regulation*.

2. *Robustness.* A control system is said to be robust if it exhibits good regulation, despite the presence of external disturbances (e.g., turbulence affecting the flight of an aircraft) and in the face of changes in the plant parameters due to varying environmental conditions.

The attainment of these desirable properties usually requires the use of *feedback*, as illustrated in Fig. 1.4. The system in Fig. 1.4 contains the abstract elements of a control

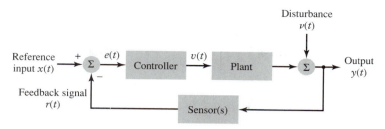

FIGURE 1.4 Block diagram of a feedback control system. The controller drives the plant, whose disturbed output drives the sensor(s). The resulting feedback signal is subtracted from the reference input to produce an error signal $e(t)$, which, in turn, drives the controller. The feedback loop is thereby closed.

system and is referred to as a *closed-loop control system* or *feedback control system*. For example, in an aircraft landing system the plant is represented by the aircraft body and actuator, the sensors are used by the pilot to determine the lateral position of the aircraft, and the controller is a digital computer.

In any event, the plant is described by mathematical operations that generate the output $y(t)$ in response to the plant input $v(t)$ and external disturbance $v(t)$. The sensor included in the feedback loop measures the plant output $y(t)$ and converts it into another form, usually electrical. The sensor output $r(t)$ constitutes the feedback signal. It is compared against the reference input $x(t)$ to produce a difference or error signal $e(t)$. This latter signal is applied to a *controller*, which, in turn, generates the actuating signal $v(t)$ that performs the controlling action on the plant. A control system with a single input and single output, as illustrated in Fig. 1.4, is referred to as a *single-input/single-output (SISO) system*. When the number of plant inputs and/or the number of plant outputs is more than one, the system is referred to as a *multiple-input/multiple-output (MIMO) system*.

In either case, the controller may be in the form of a digital computer or microprocessor, in which case we speak of a *digital control system*. The use of digital control systems is becoming more and more common because of the flexibility and high degree of accuracy afforded by the use of a digital computer as the controller. Because of its very nature, the use of a digital control system involves the operations of sampling, quantization, and coding that were described previously.

Figure 1.5 shows the photograph of a NASA (National Aeronautics and Space Administration) space shuttle launch, which relies on the use of a digital computer for its control.

■ REMOTE SENSING

Remote sensing is defined as the process of acquiring information about an object of interest without being in physical contact with it. Basically, the acquisition of information is accomplished by *detecting and measuring the changes that the object imposes on the surrounding field*. The field can be electromagnetic, acoustic, magnetic, or gravitational, depending on the application of interest. The acquisition of information can be performed in a *passive* manner, by listening to the field (signal) that is naturally emitted by the object and processing it, or in an *active* manner, by purposely illuminating the object with a well-defined field (signal) and processing the echo (i.e., signal returned) from the object.

The above definition of remote sensing is rather broad in that it applies to every possible field. In practice, however, the term "remote sensing" is commonly used in the

FIGURE 1.5   NASA space shuttle launch. (Courtesy of NASA.)

context of electromagnetic fields, with the techniques used for information acquisition covering the whole electromagnetic spectrum. It is this specialized form of remote sensing that we are concerned with here.

The scope of remote sensing has expanded enormously since the 1960s due to the advent of satellites and planetary probes as space platforms for the sensors, and the availability of sophisticated digital signal-processing techniques for extracting information from the data gathered by the sensors. In particular, sensors on Earth-orbiting satellites provide highly valuable information about global weather patterns and dynamics of clouds, surface vegetation cover and its seasonal variations, and ocean surface temperatures. Most importantly, they do so in a reliable way and on a continuing basis. In planetary studies, spaceborne sensors have provided us with high-resolution images of planetary surfaces; the images, in turn, have uncovered for us new kinds of physical phenomena, some similar to and others completely different from what we are familiar with on our planet Earth.

The electromagnetic spectrum extends from low-frequency radio waves through microwave, submillimeter, infrared, visible, ultraviolet, x-ray, and gamma-ray regions of the spectrum. Unfortunately, a single sensor by itself can cover only a small part of the electromagnetic spectrum, with the mechanism responsible for wave–matter interaction being influenced by a limited number of physical properties of the object of interest. If, therefore, we are to undertake a detailed study of a planetary surface or atmosphere, then the simultaneous use of *multiple sensors* covering a large part of the electromagnetic spectrum is required. For example, to study a planetary surface, we may require a suite of sensors covering selected bands as follows:

> ► *Radar* sensors to provide information on the surface physical properties of the planet under study (e.g., topography, roughness, moisture, and dielectric constant)
> ► *Infrared* sensors to measure the near-surface thermal properties of the planet
> ► *Visible and near-infrared* sensors to provide information about the surface chemical composition of the planet
> ► *X-ray* sensors to provide information on radioactive materials contained in the planet

The data gathered by these highly diverse sensors are then processed on a computer to generate a set of images that can be used collectively to enhance the knowledge of a scientist studying the planetary surface.

Among the electromagnetic sensors mentioned above, a special type of radar known as *synthetic aperture radar* (SAR) stands out as a unique imaging system in remote sensing. It offers the following attractive features:

> ► Satisfactory operation day and night and under all weather conditions
> ► High-resolution imaging capability that is independent of sensor altitude or wavelength

The realization of a high-resolution image with radar requires the use of an antenna with large aperture. From a practical perspective, however, there is a physical limit on the size of an antenna that can be accommodated on an airborne or spaceborne platform. In a SAR system, a large antenna aperture is synthesized by signal-processing means, hence the name "synthetic aperture radar." The key idea behind SAR is that an array of antenna elements equally spaced along a straight line is equivalent to a single antenna moving along the array line at a uniform speed. This is true provided that we satisfy the following requirement: the signals received by the single antenna at equally spaced points along the array line are *coherently* recorded; that is, amplitude and phase relationships among the received signals are maintained. Coherent recording ensures that signals received from the single antenna correspond to the signals received from the individual elements of an equivalent antenna array. In order to obtain a high-resolution image from the single-antenna signals, highly sophisticated signal-processing operations are necessary. A central operation in the signal processing is the *Fourier transform*, which is implemented efficiently on a digital computer using an algorithm known as the *fast Fourier transform (FFT) algorithm*. Fourier analysis of signals is one of the main focal points of this book.

The photograph in Fig. 1.6 shows a perspective view of Mt. Shasta (California), which was derived from a stereo pair of SAR images acquired from Earth orbit with the Shuttle Imaging Radar (SIR-B). The color version of this photograph appears on the color plate.

■ BIOMEDICAL SIGNAL PROCESSING

The goal of biomedical signal processing is to extract information from a biological signal that helps us to further improve our understanding of basic mechanisms of biological function or aids us in the diagnosis or treatment of a medical condition. The generation of many *biological signals* found in the human body is traced to the electrical activity of large groups of nerve cells or muscle cells. Nerve cells in the brain are commonly referred to as *neurons*. Figure 1.7 shows morphological types of neurons identifiable in a monkey cerebral cortex, based on studies of primary somatic sensory and motor cortex. This figure illustrates the many different shapes and sizes of neurons that exist.

Irrespective of the signal origin, biomedical signal processing begins with a temporal record of the biological event of interest. For example, the electrical activity of the heart

FIGURE 1.6   Perspective view of Mount Shasta (California) derived from a pair of stereo radar images acquired from orbit with the Shuttle Imaging Radar (SIR-B). (Courtesy of Jet Propulsion Laboratory.) See Color Plate.

is represented by a record called the *electrocardiogram* (ECG). The ECG represents changes in the potential (voltage) due to electrochemical processes involved in the formation and spatial spread of electrical excitations in the heart cells. Accordingly, detailed inferences about the heart can be made from the ECG.

Another important example of a biological signal is the *electroencephalogram* (EEG). The EEG is a record of fluctuations in the electrical activity of large groups of neurons in

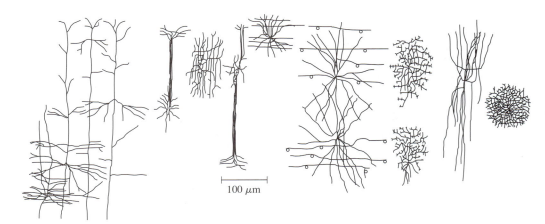

100 μm

FIGURE 1.7   Morphological types of nerve cells (neurons) identifiable in a monkey cerebral cortex, based on studies of primary somatic sensory and motor cortex. (Reproduced from E. R. Kandel, J. H. Schwartz, and T. M. Jessel, *Principles of Neural Science*, Third Edition, 1991; courtesy of Appleton and Lange.)

the brain. Specifically, the EEG measures the electrical field associated with the current flowing through a group of neurons. To record the EEG (or the ECG for that matter) at least two electrodes are needed. An active electrode is placed over the particular site of neuronal activity that is of interest, and a reference electrode is placed at some remote distance from this site; the EEG is measured as the voltage or potential difference between the active and reference electrodes. Figure 1.8 shows three examples of EEG signals recorded from the hippocampus of a rat.

A major issue of concern in biomedical signal processing—in the context of ECG, EEG, or some other biological signal—is the detection and suppression of artifacts. An *artifact* refers to that part of the signal produced by events that are extraneous to the biological event of interest. Artifacts arise in a biological signal at different stages of processing and in many different ways, as summarized here:

▶ *Instrumental artifacts*, generated by the use of an instrument. An example of an instrumental artifact is the 60-Hz interference picked up by the recording instruments from the electrical mains power supply.

▶ *Biological artifacts*, in which one biological signal contaminates or interferes with another. An example of a biological artifact is the electrical potential shift that may be observed in the EEG due to heart activity.

▶ *Analysis artifacts*, which may arise in the course of processing the biological signal to produce an estimate of the event of interest.

Analysis artifacts are, in a way, controllable. For example, roundoff errors due to quantization of signal samples, which arise from the use of digital signal processing, can be made nondiscernible for all practical purposes by making the number of discrete amplitude levels in the quantizer large enough.

What about instrumental and biological artifacts? A common method of reducing their effects is through the use of *filtering*. A *filter* is a *system* that performs a desired

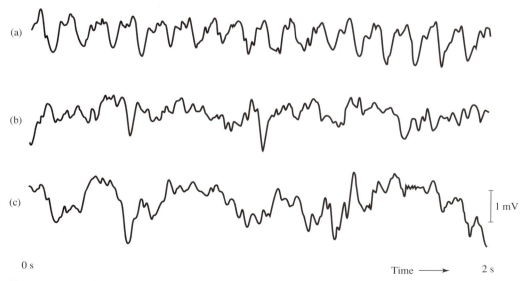

FIGURE 1.8   The traces shown in (a), (b), and (c) are three examples of EEG signals recorded from the hippocampus of a rat. Neurobiological studies suggest that the hippocampus plays a key role in certain aspects of learning or memory.

operation on a signal or signals. It passes signals containing frequencies in one frequency range, termed the filter passband, and removes signals containing frequencies in other frequency ranges. Assuming that we have *a priori* knowledge concerning the signal of interest, we may estimate the range of frequencies inside which the significant components of the desired signal are located. Then, by designing a filter whose passband corresponds to the frequencies of the desired signal, artifacts with frequency components outside this passband are removed by the filter. The assumption made here is that the desired signal and the artifacts contaminating it occupy essentially nonoverlapping frequency bands. If, however, the frequency bands overlap each other, then the filtering problem becomes more difficult and requires a solution beyond the scope of the present book.

■ AUDITORY SYSTEM

For our last example of a system, we turn to the mammalian auditory *system*, the function of which is to discriminate and recognize complex sounds on the basis of their frequency content.

Sound is produced by vibrations such as the movements of vocal cords or violin strings. These vibrations result in the compression and rarefaction (i.e., increased or reduced pressure) of the surrounding air. The disturbance so produced radiates outward from the source of sound as an *acoustical wave* with alternating highs and lows of pressure.

The ear, the organ of hearing, responds to incoming acoustical waves. It has three main parts, with their functions summarized as follows:

▶ The *outer ear* aids in the collection of sounds.

▶ The *middle ear* provides an acoustic impedance match between the air and the cochlea fluids, thereby conveying the vibrations of the *tympanic membrane* (eardrum) due to the incoming sounds to the inner ear in an efficient manner.

▶ The *inner ear* converts the mechanical vibrations from the middle ear to an "electrochemical" or "neural" signal for transmission to the brain.

The inner ear consists of a bony spiral-shaped, fluid-filled tube, called the *cochlea*. Sound-induced vibrations of the tympanic membrane are transmitted into the *oval window* of the cochlea by a chain of bones, called *ossicles*. The lever action of the ossicles provides some amplification of the mechanical vibrations of the tympanic membrane. The cochlea tapers in size like a cone toward a tip, so that there is a *base* at the oval window, and an *apex* at the tip. Through the middle of the cochlea stretches the *basilar membrane*, which gets wider as the cochlea gets narrower.

The vibratory movement of the tympanic membrane is transmitted as a *traveling wave* along the length of the basilar membrane, starting from the oval window to the apex at the far end of the cochlea. The wave propagates along the basilar membrane, much like the snapping of a rope tied at one end causes a wave to propagate along the rope from the snapped end to the fixed end. As illustrated in Fig. 1.9, the wave attains its peak amplitude at a specific location along the basilar membrane that depends on the frequency of the incoming sound. Thus, although the wave itself travels along the basilar membrane, the envelope of the wave is "stationary" for a given frequency. The peak displacements for high frequencies occur toward the base (where the basilar membrane is narrowest and stiffest). The peak displacements for low frequencies occur toward the apex (where the basilar membrane is widest and most flexible). That is, as the wave propagates along the basilar membrane, a *resonance* phenomenon takes place, with the end of the basilar membrane at the base of the cochlea resonating at about 20,000 Hz and its other end at the

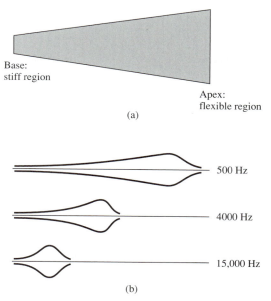

Base:
stiff region

Apex:
flexible region

(a)

500 Hz

4000 Hz

15,000 Hz

(b)

FIGURE 1.9 (a) In this diagram, the basilar membrane in the cochlea is depicted as if it were uncoiled and stretched out flat; the "base" and "apex" refer to the cochlea, but the remarks "stiff region" and "flexible region" refer to the basilar membrane. (b) This diagram illustrates the traveling waves along the basilar membrane, showing their envelopes induced by incoming sound at three different frequencies.

apex of the cochlea resonating at about 20 Hz; the resonance frequency of the basilar membrane *decreases* gradually with distance from base to apex. Consequently, the spatial axis of the cochlea is said to be *tonotopically ordered*, because each location is associated with a particular resonance frequency or tone.

The basilar membrane is a *dispersive medium*, in that higher frequencies propagate more slowly than do lower frequencies. In a dispersive medium, we distinguish two different velocities, namely, *phase velocity* and *group velocity*. The phase velocity is the velocity at which a crest or valley of the wave propagates along the basilar membrane. The group velocity is the velocity at which the envelope of the wave and its energy propagate.

The mechanical vibrations of the basilar membrane are transduced into electrochemical signals by hair cells that rest in an orderly fashion on the basilar membrane. There are two main types of hair cells: *inner hair cells* and *outer hair cells*, with the latter being by far the most numerous type. The outer hair cells are *motile* elements. That is, they are capable of altering their length, and perhaps other mechanical characteristics, which is believed to be responsible for the compressive *nonlinear* effect seen in the basilar membrane vibrations. There is also evidence that the outer hair cells contribute to the sharpening of tuning curves from the basilar membrane and on up the system. However, the inner hair cells are the main sites of *auditory transduction*. Specifically, each auditory neuron synapses with an inner hair cell at a particular location on the basilar membrane. The neurons that synapse with inner hair cells near the base of the basilar membrane are found in the periphery of the auditory nerve bundle, and there is an orderly progression toward synapsing at the apex end of the basilar membrane with movement toward the center of the bundle. The tonotopic organization of the basilar membrane is therefore anatomically preserved in the auditory nerve. The inner hair cells also perform *rectification* and *com-*

pression. The mechanical signal is approximately half-wave rectified, thereby responding to motion of the basilar membrane in one direction only. Moreover, the mechanical signal is compressed nonlinearly, such that a large range of incoming sound intensities is reduced to a manageable excursion of electrochemical potential. The electrochemical signals so produced are carried over to the brain, where they are further processed to become our hearing sensations.

In summary, in the cochlea we have a wonderful example of a biological system that operates as a *bank of filters* tuned to different frequencies and uses nonlinear processing to reduce dynamic range. It enables us to discriminate and recognize complex sounds, despite the enormous differences in intensity levels that can arise in practice.

■ ANALOG VERSUS DIGITAL SIGNAL PROCESSING

The signal processing operations involved in building communication systems, control systems, instruments for remote sensing, and instruments for the processing of biological signals, among the many applications of signal processing, can be implemented in two fundamentally different ways: (1) analog or continuous-time approach and (2) digital or discrete-time approach. The analog approach to signal processing was dominant for many years, and it remains a viable option for many applications. As the name implies, *analog signal processing* relies on the use of analog circuit elements such as resistors, capacitors, inductors, transistor amplifiers, and diodes. *Digital signal processing*, on the other hand, relies on three basic digital computer elements: adders and multipliers (for arithmetic operations) and memory (for storage).

The main attribute of the analog approach is a natural ability to solve differential equations that describe physical systems, without having to resort to approximate solutions for them. These solutions are also obtained in *real time* irrespective of the input signal's frequency range, since the underlying mechanisms responsible for the operations of the analog approach are all physical in nature. In contrast, the digital approach relies on numerical computations for its operation. The time required to perform these computations determines whether the digital approach is able to operate in real time, that is, to keep up with the changes in the input signal. In other words, the analog approach is assured of real-time operation, but there is no such guarantee for the digital approach.

However, the digital approach has the following important advantages over analog signal processing:

▶ *Flexibility*, whereby the same digital machine (hardware) can be used for implementing different versions of a signal-processing operation of interest (e.g., filtering) merely by making changes to the software (program) read into the machine. On the other hand, in the case of an analog machine, the system has to be redesigned every time the signal-processing specifications are changed.

▶ *Repeatability*, which refers to the fact that a prescribed signal-processing operation (e.g., control of a robot) can be repeated exactly over and over again when it is implemented by digital means. In contrast, analog systems suffer from parameter variations that can arise due to changes in the supply voltage or room temperature.

For a given signal-processing operation, however, we usually find that the use of a digital approach requires greater circuit complexity than an analog approach. This was an issue of major concern in years past, but this is no longer so. As remarked earlier, the ever-increasing availability of VLSI circuits in the form of silicon chips has made digital electronics relatively cheap. Consequently, we are now able to build digital signal processors that are cost competitive with respect to their analog counterparts over a wide frequency range that includes both speech and video signals. In the final analysis, however, the choice of an analog or digital approach for the solution of a signal-processing problem can only be determined by the application of interest, the resources available, and the cost involved in building the system. It should also be noted that the vast majority of systems built in practice are "mixed" in nature, combining the desirable features of both analog and digital approaches to signal processing.

1.4 *Classification of Signals*

In this book we will restrict our attention to one-dimensional signals defined as single-valued functions of time. "Single-valued" means that for every instant of time there is a unique value of the function. This value may be a real number, in which case we speak of a *real-valued signal*, or it may be a complex number, in which case we speak of a *complex-valued signal*. In either case, the independent variable, namely, time, is real valued.

The most useful method of signal representation for a given situation hinges on the particular type of signal being considered. We may identify five methods of classifying signals based on different features:

1. *Continuous-time and discrete-time signals.*
One way of classifying signals is on the basis of how they are defined as a function of time. In this context, a signal $x(t)$ is said to be a *continuous-time signal* if it is defined for all time t. Figure 1.10 represents an example of a continuous-time signal whose amplitude or value varies continuously with time. Continuous-time signals arise naturally when a physical waveform such as an acoustic wave or light wave is converted into an electrical signal. The conversion is effected by means of a *transducer*; examples include the microphone, which converts sound pressure variations into corresponding voltage or current variations, and the photocell, which does the same for light-intensity variations.

On the other hand, a *discrete-time signal* is defined only at discrete instants of time. Thus, in this case, the independent variable has discrete values only, which are usually uniformly spaced. A discrete-time signal is often derived from a continuous-time signal by *sampling* it at a uniform rate. Let $\mathcal{T}$ denote the sampling period and n denote an integer that may assume positive and negative values. Sampling a continuous-time signal $x(t)$ at time $t = n\mathcal{T}$ yields a sample of value $x(n\mathcal{T})$. For convenience of presentation, we write

$$x[n] = x(n\mathcal{T}), \qquad n = 0, \pm 1, \pm 2, \ldots \tag{1.1}$$

Thus a discrete-time signal is represented by the sequence numbers $\ldots, x[-2], x[-1], x[0], x[1], x[2], \ldots$, which can take on a continuum of values. Such a sequence of numbers is referred to as a *time series*, written as $\{x[n], n = 0, \pm 1, \pm 2, \ldots\}$ or simply $x[n]$. The

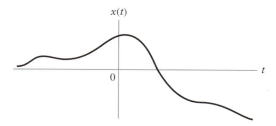

FIGURE 1.10   Continuous-time signal.

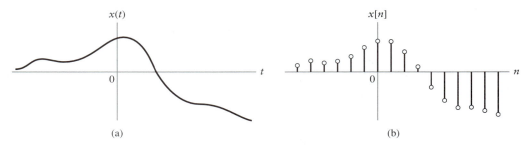

FIGURE 1.11   (a) Continuous-time signal $x(t)$. (b) Representation of $x(t)$ as a discrete-time signal $x[n]$.

latter notation is used throughout this book. Figure 1.11 illustrates the relationship be-tween a continuous-time signal $x(t)$ and discrete-time signal $x[n]$ derived from it, as de-scribed above.

Throughout this book, we use the symbol t to denote time for a continuous-time signal and the symbol n to denote time for a discrete-time signal. Similarly, parentheses $(\cdot)$ are used to denote continuous-valued quantities, while brackets $[\cdot]$ are used to denote discrete-valued quantities.

2. *Even and odd signals.*
A continuous-time signal $x(t)$ is said to be an *even signal* if it satisfies the condition

$$x(-t) = x(t) \quad \text{for all } t \tag{1.2}$$

The signal $x(t)$ is said to be an *odd signal* if it satisfies the condition

$$x(-t) = -x(t) \quad \text{for all } t \tag{1.3}$$

In other words, even signals are *symmetric* about the vertical axis or time origin, whereas odd signals are *antisymmetric* (asymmetric) about the time origin. Similar remarks apply to discrete-time signals.

EXAMPLE 1.1   Develop the even/odd decomposition of a general signal $x(t)$ by applying the definitions of Eqs. (1.2) and (1.3).

Solution: Let the signal $x(t)$ be expressed as the sum of two components $x_e(t)$ and $x_o(t)$ as follows:

$$x(t) = x_e(t) + x_o(t)$$

Define $x_e(t)$ to be even and $x_o(t)$ to be odd; that is,

$$x_e(-t) = x_e(t)$$

and

$$x_o(-t) = -x_o(t)$$

Putting $t = -t$ in the expression for $x(t)$, we may then write

$$x(-t) = x_e(-t) + x_o(-t)$$
$$= x_e(t) - x_o(t)$$

Solving for $x_e(t)$ and $x_o(t)$, we thus obtain

$$x_e(t) = \frac{1}{2}\Big(x(t) + x(-t)\Big)$$

and

$$x_o(t) = \frac{1}{2}\Big(x(t) - x(-t)\Big)$$

The above definitions of even and odd signals assume that the signals are real valued. Care has to be exercised, however, when the signal of interest is complex valued. In the case of a complex-valued signal, we may speak of conjugate symmetry. A complex-valued signal $x(t)$ is said to be *conjugate symmetric* if it satisfies the condition

$$x(-t) = x^*(t) \tag{1.4}$$

where the asterisk denotes complex conjugation. Let

$$x(t) = a(t) + jb(t)$$

where $a(t)$ is the real part of $x(t)$, $b(t)$ is the imaginary part, and j is the square root of -1. The complex conjugate of $x(t)$ is

$$x^*(t) = a(t) - jb(t)$$

From Eqs. (1.2) to (1.4), it follows therefore that a complex-valued signal $x(t)$ is conjugate symmetric if its real part is even and its imaginary part is odd. A similar remark applies to a discrete-time signal.

▶ **Drill Problem 1.1**   Consider the pair of signals shown in Fig. 1.12. Which of these two signals is even, and which one is odd?

Answer: $x_1(t)$ is even, and $x_2(t)$ is odd.   ◀

▶ **Drill Problem 1.2**   The signals $x_1(t)$ and $x_2(t)$ shown in Figs. 1.12(a) and (b) constitute the real and imaginary parts of a complex-valued signal $x(t)$. What form of symmetry does $x(t)$ have?

Answer: $x(t)$ is conjugate symmetric.   ◀

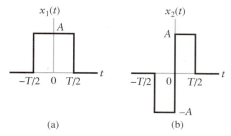

(a)   (b)

FIGURE 1.12   (a) One example of continuous-time signal. (b) Another example of continuous-time signal.

3. *Periodic signals, nonperiodic signals.*
A *periodic signal* $x(t)$ is a function that satisfies the condition

$$x(t) = x(t + T) \quad \text{for all } t \tag{1.5}$$

where T is a positive constant. Clearly, if this condition is satisfied for $T = T_0$, say, then it is also satisfied for $T = 2T_0, 3T_0, 4T_0, \ldots$. The smallest value of T that satisfies Eq. (1.5) is called the *fundamental period* of $x(t)$. Accordingly, the fundamental period T defines the duration of one complete cycle of $x(t)$. The reciprocal of the fundamental period T is called the *fundamental frequency* of the periodic signal $x(t)$; it describes how frequently the periodic signal $x(t)$ repeats itself. We thus formally write

$$f = \frac{1}{T} \tag{1.6}$$

The frequency f is measured in hertz (Hz) or cycles per second. The *angular frequency*, measured in radians per second, is defined by

$$\omega = \frac{2\pi}{T} \tag{1.7}$$

since there are 2π radians in one complete cycle. To simplify terminology, ω is often referred to simply as frequency.

Any signal $x(t)$ for which there is no value of T to satisfy the condition of Eq. (1.5) is called an *aperiodic* or *nonperiodic signal*.

Figures 1.13(a) and (b) present examples of periodic and nonperiodic signals, respectively. The periodic signal shown here represents a square wave of amplitude $A = 1$ and period T, and the nonperiodic signal represents a rectangular pulse of amplitude A and duration T_1.

▶ **Drill Problem 1.3**   Figure 1.14 shows a triangular wave. What is the fundamental frequency of this wave? Express the fundamental frequency in units of Hz or rad/s.

Answer: 5 Hz, or 10π rad/s.   ◀

The classification of signals into periodic and nonperiodic signals presented thus far applies to continuous-time signals. We next consider the case of discrete-time signals. A discrete-time signal $x[n]$ is said to be periodic if it satisfies the condition:

$$x[n] = x[n + N] \quad \text{for all integers } n \tag{1.8}$$

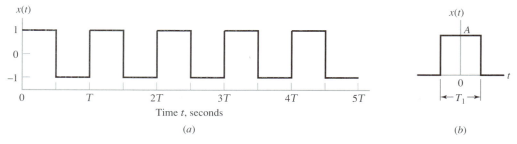

FIGURE 1.13   (a) Square wave with amplitude $A = 1$, and period $T = 0.2$ s. (b) Rectangular pulse of amplitude A and duration T_1.

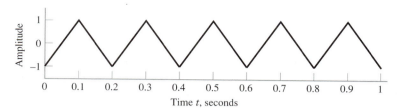

FIGURE 1.14   Triangular wave alternating between -1 and $+1$ with fundamental period of 0.2 second.

where N is a positive integer. The smallest value of integer N for which Eq. (1.8) is satisfied is called the fundamental period of the discrete-time signal $x[n]$. The fundamental angular frequency or, simply, fundamental frequency of $x[n]$ is defined by

$$\Omega = \frac{2\pi}{N} \tag{1.9}$$

which is measured in radians.

The differences between the defining equations (1.5) and (1.8) should be carefully noted. Equation (1.5) applies to a periodic continuous-time signal whose fundamental period T has any positive value. On the other hand, Eq. (1.8) applies to a periodic discrete-time signal whose fundamental period N can only assume a positive integer value.

Two examples of discrete-time signals are shown in Figs. 1.15 and 1.16; the signal of Fig. 1.15 is periodic, whereas that of Fig. 1.16 is aperiodic.

▶ **Drill Problem 1.4**   What is the fundamental frequency of the discrete-time square wave shown in Fig. 1.15?

Answer: $\pi/4$ radians.   ◀

4. Deterministic signals, random signals.
A *deterministic signal* is a signal about which there is no uncertainty with respect to its value at any time. Accordingly, we find that deterministic signals may be modeled as

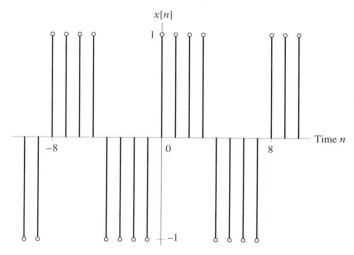

FIGURE 1.15   Discrete-time square wave alternating between -1 and $+1$.

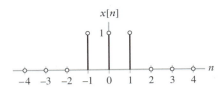

FIGURE 1.16   Aperiodic discrete-time signal consisting of three nonzero samples.

completely specified functions of time. The square wave shown in Fig. 1.13(a) and the rectangular pulse shown in Fig. 1.13(b) are examples of deterministic signals, and so are the signals shown in Figs. 1.15 and 1.16.

On the other hand, a *random signal* is a signal about which there is uncertainty before its actual occurrence. Such a signal may be viewed as belonging to an ensemble or group of signals, with each signal in the ensemble having a different waveform. Moreover, each signal within the ensemble has a certain probability of occurrence. The ensemble of such signals is referred to as a *random process*. The *noise* generated in the amplifier of a radio or television receiver is an example of a random signal. Its amplitude fluctuates between positive and negative values in a completely random fashion. The EEG signal, exemplified by the waveforms shown in Fig. 1.8, is another example of a random signal.

5. *Energy signals, power signals.*
In electrical systems, a signal may represent a voltage or a current. Consider a voltage $v(t)$ developed across a resistor R, producing a current $i(t)$. The *instantaneous power* dissipated in this resistor is defined by

$$p(t) = \frac{v^2(t)}{R} \tag{1.10}$$

or, equivalently,

$$p(t) = Ri^2(t) \tag{1.11}$$

In both cases, the instantaneous power $p(t)$ is proportional to the squared amplitude of the signal. Furthermore, for a resistance R of 1 ohm, we see that Eqs. (1.10) and (1.11) take on the same mathematical form. Accordingly, in signal analysis it is customary to define power in terms of a 1-ohm resistor, so that, regardless of whether a given signal $x(t)$ represents a voltage or a current, we may express the instantaneous power of the signal as

$$p(t) = x^2(t) \tag{1.12}$$

Based on this convention, we define the *total energy* of the continuous-time signal $x(t)$ as

$$E = \lim_{T \to \infty} \int_{-T/2}^{T/2} x^2(t)\, dt$$
$$= \int_{-\infty}^{\infty} x^2(t)\, dt \tag{1.13}$$

and its *average power* as

$$P = \lim_{T \to \infty} \frac{1}{T} \int_{-T/2}^{T/2} x^2(t)\, dt \tag{1.14}$$

From Eq. (1.14) we readily see that the average power of a periodic signal $x(t)$ of fundamental period T is given by

$$P = \frac{1}{T} \int_{-T/2}^{T/2} x^2(t)\, dt \tag{1.15}$$

The square root of the average power P is called the *root mean-square* (rms) value of the signal $x(t)$.

In the case of a discrete-time signal $x[n]$, the integrals in Eqs. (1.13) and (1.14) are replaced by corresponding sums. Thus the total energy of $x[n]$ is defined by

$$E = \sum_{n=-\infty}^{\infty} x^2[n] \tag{1.16}$$

and its average power is defined by

$$P = \lim_{N\to\infty} \frac{1}{2N} \sum_{n=-N}^{N} x^2[n] \tag{1.17}$$

Here again we see from Eq. (1.17) that the average power in a periodic signal $x[n]$ with fundamental period N is given by

$$P = \frac{1}{N} \sum_{n=0}^{N-1} x^2[n]$$

A signal is referred to as an *energy signal*, if and only if the total energy of the signal satisfies the condition

$$0 < E < \infty$$

On the other hand, it is referred to as a *power signal*, if and only if the average power of the signal satisfies the condition

$$0 < P < \infty$$

The energy and power classifications of signals are mutually exclusive. In particular, an energy signal has zero average power, whereas a power signal has infinite energy. It is also of interest to note that periodic signals and random signals are usually viewed as power signals, whereas signals that are both deterministic and nonperiodic are energy signals.

▶ **Drill Problem 1.5**

(a) What is the total energy of the rectangular pulse shown in Fig. 1.13(b)?

(b) What is the average power of the square wave shown in Fig. 1.13(a)?

Answer: (a) $A^2 T_1$. (b) 1.   ◀

▶ **Drill Problem 1.6**   What is the average power of the triangular wave shown in Fig. 1.14?

Answer: 1/3.   ◀

▶ **Drill Problem 1.7**   What is the total energy of the discrete-time signal shown in Fig. 1.16?

Answer: 3.   ◀

▶ **Drill Problem 1.8**   What is the average power of the periodic discrete-time signal shown in Fig. 1.15?

Answer: 1   ◀

▌1.5 *Basic Operations on Signals*

An issue of fundamental importance in the study of signals and systems is the use of systems to process or manipulate signals. This issue usually involves a combination of some basic operations. In particular, we may identify two classes of operations, as described here.

  1. *Operations performed on dependent variables.*

  Amplitude scaling. Let $x(t)$ denote a continuous-time signal. The signal $y(t)$ resulting from amplitude scaling applied to $x(t)$ is defined by

$$y(t) = cx(t) \tag{1.18}$$

where c is the scaling factor. According to Eq. (1.18), the value of $y(t)$ is obtained by multiplying the corresponding value of $x(t)$ by the scalar c. A physical example of a device that performs amplitude scaling is an electronic *amplifier*. A resistor also performs amplitude scaling when $x(t)$ is a current, c is the resistance, and $y(t)$ is the output voltage.

  In a manner similar to Eq. (1.18), for discrete-time signals we write

$$y[n] = cx[n]$$

  Addition. Let $x_1(t)$ and $x_2(t)$ denote a pair of continuous-time signals. The signal $y(t)$ obtained by the addition of $x_1(t)$ and $x_2(t)$ is defined by

$$y(t) = x_1(t) + x_2(t) \tag{1.19}$$

A physical example of a device that adds signals is an audio *mixer*, which combines music and voice signals.

  In a manner similar to Eq. (1.19), for discrete-time signals we write

$$y[n] = x_1[n] + x_2[n]$$

  Multiplication. Let $x_1(t)$ and $x_2(t)$ denote a pair of continuous-time signals. The signal $y(t)$ resulting from the multiplication of $x_1(t)$ by $x_2(t)$ is defined by

$$y(t) = x_1(t)x_2(t) \tag{1.20}$$

That is, for each prescribed time t the value of $y(t)$ is given by the product of the corresponding values of $x_1(t)$ and $x_2(t)$. A physical example of $y(t)$ is an *AM radio signal*, in which $x_1(t)$ consists of an audio signal plus a dc component, and $x_2(t)$ consists of a sinusoidal signal called a carrier wave.

  In a manner similar to Eq. (1.20), for discrete-time signals we write

$$y[n] = x_1[n]x_2[n]$$

FIGURE 1.17   Inductor with current $i(t)$, inducing voltage $v(t)$ across its terminals.

FIGURE 1.18   Capacitor with voltage $v(t)$ across its terminals, inducing current $i(t)$.

Differentiation. Let $x(t)$ denote a continuous-time signal. The derivative of $x(t)$ with respect to time is defined by

$$y(t) = \frac{d}{dt} x(t) \tag{1.21}$$

For example, an *inductor* performs differentiation. Let $i(t)$ denote the current flowing through an inductor of inductance L, as shown in Fig. 1.17. The voltage $v(t)$ developed across the inductor is defined by

$$v(t) = L \frac{d}{dt} i(t) \tag{1.22}$$

Integration. Let $x(t)$ denote a continuous-time signal. The integral of $x(t)$ with respect to time t is defined by

$$y(t) = \int_{-\infty}^{t} x(\tau) \, d\tau \tag{1.23}$$

where τ is the integration variable. For example, a capacitor performs integration. Let $i(t)$ denote the current flowing through a capacitor of capacitance C, as shown in Fig. 1.18. The voltage $v(t)$ developed across the capacitor is defined by

$$v(t) = \frac{1}{C} \int_{-\infty}^{t} i(\tau) \, d\tau \tag{1.24}$$

2. *Operations performed on the independent variable.*

Time scaling. Let $x(t)$ denote a continuous-time signal. The signal $y(t)$ obtained by scaling the independent variable, time t, by a factor a is defined by

$$y(t) = x(at)$$

If $a > 1$, the signal $y(t)$ is a *compressed* version of $x(t)$. If, on the other hand, $0 < a < 1$, the signal $y(t)$ is an expanded (stretched) version of $x(t)$. These two operations are illustrated in Fig. 1.19.

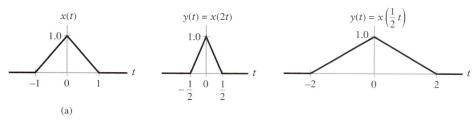

FIGURE 1.19   Time-scaling operation: (a) continuous-time signal $x(t)$, (b) compressed version of $x(t)$ by a factor of 2, and (c) expanded version of $x(t)$ by a factor of 2.

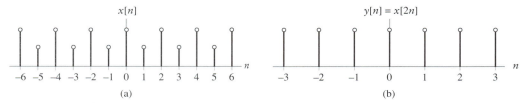

FIGURE 1.20   Effect of time scaling on a discrete-time signal: (a) discrete-time signal $x[n]$, and (b) compressed version of $x[n]$ by a factor of 2, with some values of the original $x[n]$ lost as a result of the compression.

In the discrete-time case, we write

$$y[n] = x[kn], \qquad k > 0$$

which is defined only for integer values of k. If $k > 1$, then some values of the discrete-time signal $y[n]$ are lost, as illustrated in Fig. 1.20 for $k = 2$.

Reflection. Let $x(t)$ denote a continuous-time signal. Let $y(t)$ denote the signal obtained by replacing time t with $-t$, as shown by

$$y(t) = x(-t)$$

The signal $y(t)$ represents a reflected version of $x(t)$ about the amplitude axis.

The following two cases are of special interest:

▶ Even signals, for which we have $x(-t) = x(t)$ for all t; that is, an even signal is the same as its reflected version.

▶ Odd signals, for which we have $x(-t) = -x(t)$ for all t; that is, an odd signal is the negative of its reflected version.

Similar observations apply to discrete-time signals.

EXAMPLE 1.2   Consider the triangular pulse $x(t)$ shown in Fig. 1.21(a). Find the reflected version of $x(t)$ about the amplitude axis.

Solution: Replacing the independent variable t in $x(t)$ with $-t$, we get the result $y(t) = x(-t)$ shown in Fig. 1.21(b).

Note that for this example, we have

$$x(t) = 0 \quad \text{for } t < -T_1 \text{ and } t > T_2$$

Correspondingly, we find that

$$y(t) = 0 \quad \text{for } t > T_1 \text{ and } t < -T_2$$

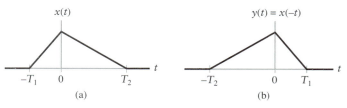

FIGURE 1.21   Operation of reflection: (a) continuous-time signal $x(t)$ and (b) reflected version of $x(t)$ about the origin.

▶ **Drill Problem 1.9**   The discrete-time signal $x[n]$ is defined by

$$x[n] = \begin{cases} 1, & n = 1 \\ -1, & n = -1 \\ 0, & n = 0 \text{ and } |n| > 1 \end{cases}$$

Find the composite signal $y[n]$ defined in terms of $x[n]$ by

$$y[n] = x[n] + x[-n]$$

Answer: $y[n] = 0$ for all integer values of n.   ◀

▶ **Drill Problem 1.10**   Repeat Drill Problem 1.9 for

$$x[n] = \begin{cases} 1, & n = -1 \text{ and } n = 1 \\ 0, & n = 0 \text{ and } |n| > 1 \end{cases}$$

Answer: $y[n] = \begin{cases} 2, & n = -1 \text{ and } n = 1 \\ 0, & n = 0 \text{ and } |n| > 1 \end{cases}$ ◀

Time shifting. Let $x(t)$ denote a continuous-time signal. The time-shifted version of $x(t)$ is defined by

$$y(t) = x(t - t_0)$$

where t_0 is the time shift. If $t_0 > 0$, the waveform representing $x(t)$ is shifted intact to the right, relative to the time axis. If $t_0 < 0$, it is shifted to the left.

EXAMPLE 1.3   Figure 1.22(a) shows a rectangular pulse $x(t)$ of unit amplitude and unit duration. Find $y(t) = x(t - 2)$.

Solution: In this example, the time shift t_0 equals 2 time units. Hence, by shifting $x(t)$ to the right by 2 time units we get the rectangular pulse $y(t)$ shown in Fig. 1.22(b). The pulse $y(t)$ has exactly the same shape as the original pulse $x(t)$; it is merely shifted along the time axis.

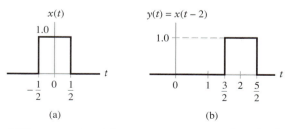

(a)   (b)

FIGURE 1.22   Time-shifting operation: (a) continuous-time signal in the form of a rectangular pulse of amplitude 1.0 and duration 1.0, symmetric about the origin; and (b) time-shifted version of $x(t)$ by 2 time units.

In the case of a discrete-time signal $x[n]$, we define its time-shifted version as follows:

$$y[n] = x[n - m]$$

where the shift m must be an integer; it can be positive or negative.

▶ **Drill Problem 1.11**   The discrete-time signal $x[n]$ is defined by

$$x[n] = \begin{cases} 1, & n = 1, 2 \\ -1, & n = -1, -2 \\ 0, & n = 0 \text{ and } |n| > 2 \end{cases}$$

Find the time-shifted signal $y[n] = x[n + 3]$.

Answer: $y[n] = \begin{cases} 1, & n = -1, -2 \\ -1, & n = -4, -5 \\ 0, & n = -3, n < -5, \text{ and } n > -1 \end{cases}$ ◀

▪ PRECEDENCE RULE FOR TIME SHIFTING AND TIME SCALING

Let $y(t)$ denote a continuous-time signal that is derived from another continuous-time signal $x(t)$ through a combination of time shifting and time scaling, as described here:

$$y(t) = x(at - b) \tag{1.25}$$

This relation between $y(t)$ and $x(t)$ satisfies the following conditions:

$$y(0) = x(-b) \tag{1.26}$$

and

$$y\left(\frac{b}{a}\right) = x(0) \tag{1.27}$$

which provide useful checks on $y(t)$ in terms of corresponding values of $x(t)$.

To correctly obtain $y(t)$ from $x(t)$, the time-shifting and time-scaling operations must be performed in the correct order. The proper order is based on the fact that the scaling operation always replaces t by at, while the time-shifting operation always replaces t by $t - b$. Hence the time-shifting operation is performed first on $x(t)$, resulting in an intermediate signal $v(t)$ defined by

$$v(t) = x(t - b)$$

The time shift has replaced t in $x(t)$ by $t - b$. Next, the time-scaling operation is performed on $v(t)$. This replaces t by at, resulting in the desired output

$$\begin{aligned} y(t) &= v(at) \\ &= x(at - b) \end{aligned}$$

To illustrate how the operation described in Eq. (1.25) can arise in a real-life situation, consider a voice signal recorded on a tape recorder. If the tape is played back at a rate faster than the original recording rate, we get compression (i.e., $a > 1$). If, on the

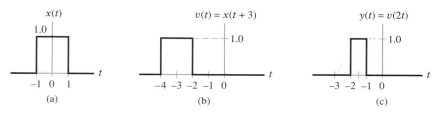

FIGURE 1.23 The proper order in which the operations of time scaling and time shifting should be applied for the case of a continuous-time signal. (a) Rectangular pulse $x(t)$ of amplitude 1.0 and duration 2.0, symmetric about the origin. (b) Intermediate pulse $v(t)$, representing time-shifted version of $x(t)$. (c) Desired signal $y(t)$, resulting from the compression of $v(t)$ by a factor of 2.

other hand, the tape is played back at a rate slower than the original recording rate, we get expansion (i.e., $a < 1$). The constant b, assumed to be positive, accounts for a delay in playing back the tape.

EXAMPLE 1.4 Consider the rectangular pulse $x(t)$ of unit amplitude and duration of 2 time units depicted in Fig. 1.23(a). Find $y(t) = x(2t + 3)$.

Solution: In this example, we have $a = 2$ and $b = -3$. Hence shifting the given pulse $x(t)$ to the left by 3 time units relative to the time axis gives the intermediate pulse $v(t)$ shown in Fig. 1.23(b). Finally, scaling the independent variable t in $v(t)$ by $a = 2$, we get the solution $y(t)$ shown in Fig. 1.23(c).

Note that the solution presented in Fig. 1.23(c) satisfies both of the conditions defined in Eqs. (1.26) and (1.27).

Suppose next that we purposely do not follow the precedence rule; that is, we first apply time scaling, followed by time shifting. For the given signal $x(t)$, shown in Fig. 1.24(a), the waveforms resulting from the application of these two operations are shown in Figs. 1.24(b) and (c), respectively. The signal $y(t)$ so obtained fails to satisfy the condition of Eq. (1.27).

This example clearly illustrates that if $y(t)$ is defined in terms of $x(t)$ by Eq. (1.25), then $y(t)$ can only be obtained from $x(t)$ correctly by adhering to the precedence rule for time shifting and time scaling.

Similar remarks apply to the case of discrete-time signals.

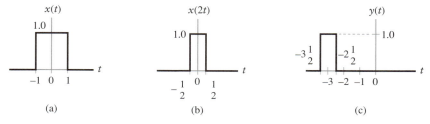

FIGURE 1.24 The incorrect way of applying the precedence rule. (a) Signal $x(t)$. (b) Time-scaled signal $x(2t)$. (c) Signal $y(t)$ obtained by shifting $x(2t)$ by 3 time units.

EXAMPLE 1.5   A discrete-time signal $x[n]$ is defined by

$$x[n] = \begin{cases} 1, & n = 1, 2 \\ -1, & n = -1, -2 \\ 0, & n = 0 \text{ and } |n| > 2 \end{cases}$$

Find $y[n] = x[2n + 3]$.

Solution: The signal $x[n]$ is displayed in Fig. 1.25(a). Time shifting $x[n]$ to the left by 3 yields the intermediate signal $v[n]$ shown in Fig. 1.25(b). Finally, scaling n in $v[n]$ by 2, we obtain the solution $y[n]$ shown in Fig. 1.25(c).

Note that as a result of the compression performed in going from $v[n]$ to $y[n] = v[2n]$, the samples of $v[n]$ at $n = -5$ and $n = -1$ (i.e., those contained in the original signal at $n = -2$ and $n = 2$) are lost.

▶ **Drill Problem 1.12**   Consider a discrete-time signal $x[n]$ defined by

$$x[n] = \begin{cases} 1, & -2 \le n \le 2 \\ 0, & |n| > 2 \end{cases}$$

Find $y[n] = x[3n - 2]$.

Answer: $y[n] = \begin{cases} 1, & n = 0, 1 \\ 0, & \text{otherwise} \end{cases}$ ◀

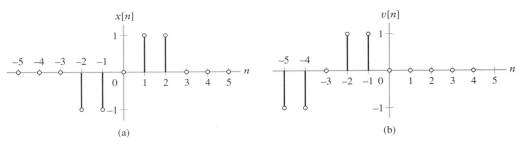

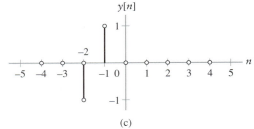

FIGURE 1.25   The proper order of applying the operations of time scaling and time shifting for the case of a discrete-time signal. (a) Discrete-time signal $x[n]$, antisymmetric about the origin. (b) Intermediate signal $v[n]$ obtained by shifting $x[n]$ to the left by 3 samples. (c) Discrete-time signal $y[n]$ resulting from the compression of $v[n]$ by a factor of 2, as a result of which two samples of the original $x[n]$ are lost.

1.6 *Elementary Signals*

There are several elementary signals that feature prominently in the study of signals and systems. The list of elementary signals includes exponential and sinusoidal signals, the step function, impulse function, and ramp function. These elementary signals serve as building blocks for the construction of more complex signals. They are also important in their own right, in that they may be used to model many physical signals that occur in nature. In what follows, we will describe the above-mentioned elementary signals, one by one.

■ EXPONENTIAL SIGNALS

A real exponential signal, in its most general form, is written as

$$x(t) = Be^{at} \tag{1.28}$$

where both B and a are real parameters. The parameter B is the amplitude of the exponential signal measured at time $t = 0$. Depending on whether the other parameter a is positive or negative, we may identify two special cases:

▶ *Decaying exponential*, for which $a < 0$
▶ *Growing exponential*, for which $a > 0$

These two forms of an exponential signal are illustrated in Fig. 1.26. Part (a) of the figure was generated using $a = -6$ and $B = 5$. Part (b) of the figure was generated using $a = 5$ and $B = 1$. If $a = 0$, the signal $x(t)$ reduces to a dc signal equal to the constant B.

For a physical example of an exponential signal, consider a "lossy" capacitor, as depicted in Fig. 1.27. The capacitor has capacitance C, and the loss is represented by shunt resistance R. The capacitor is charged by connecting a battery across it, and then the battery is removed at time $t = 0$. Let V_0 denote the initial value of the voltage developed

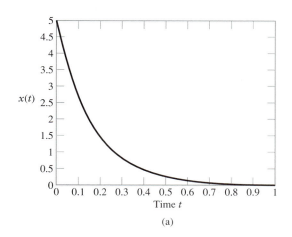

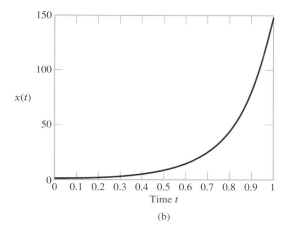

FIGURE 1.26   (a) Decaying exponential form of continuous-time signal. (b) Growing exponential form of continuous-time signal.

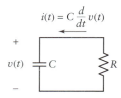

FIGURE 1.27   Lossy capacitor, with the loss represented by shunt resistance R.

across the capacitor. From Fig. 1.27 we readily see that the operation of the capacitor for $t \geq 0$ is described by

$$RC \frac{d}{dt} v(t) + v(t) = 0 \tag{1.29}$$

where $v(t)$ is the voltage measured across the capacitor at time t. Equation (1.29) is a *differential equation of order one*. Its solution is given by

$$v(t) = V_0 e^{-t/RC} \tag{1.30}$$

where the product term RC plays the role of a *time constant*. Equation (1.30) shows that the voltage across the capacitor decays exponentially with time at a rate determined by the time constant RC. The larger the resistor R (i.e., the less lossy the capacitor), the slower will be the rate of decay of $v(t)$ with time.

The discussion thus far has been in the context of continuous time. In discrete time, it is common practice to write a real exponential signal as

$$x[n] = Br^n \tag{1.31}$$

The exponential nature of this signal is readily confirmed by defining

$$r = e^{\alpha}$$

for some α. Figure 1.28 illustrates the decaying and growing forms of a discrete-time exponential signal corresponding to $0 < r < 1$ and $r > 1$, respectively. This is where the case of discrete-time exponential signals is distinctly different from continuous-time exponential signals. Note also that when $r < 0$, a discrete-time exponential signal assumes alternating signs.

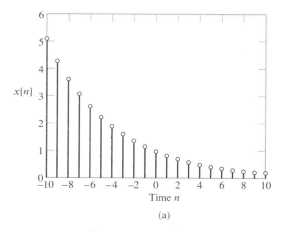

(a)

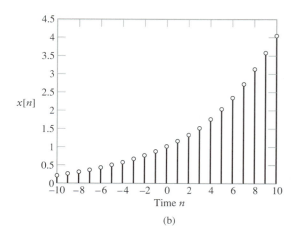

(b)

FIGURE 1.28   (a) Decaying exponential form of discrete-time signal. (b) Growing exponential form of discrete-time signal.

The exponential signals shown in Figs. 1.26 and 1.28 are all real valued. It is possible for an exponential signal to be complex valued. The mathematical forms of complex exponential signals are the same as those shown in Eqs. (1.28) and (1.31), with some differences as explained here. In the continuous-time case, the parameter B or parameter a or both in Eq. (1.28) assume complex values. Similarly, in the discrete-time case, the parameter B or parameter r or both in Eq. (1.31) assume complex values. Two commonly encountered examples of complex exponential signals are $e^{j\omega t}$ and $e^{j\Omega n}$.

■ SINUSOIDAL SIGNALS

The continuous-time version of a sinusoidal signal, in its most general form, may be written as

$$x(t) = A \cos(\omega t + \phi) \tag{1.32}$$

where A is the amplitude, ω is the frequency in radians per second, and ϕ is the phase angle in radians. Figure 1.29(a) presents the waveform of a sinusoidal signal for $A = 4$ and $\phi = +\pi/6$. A sinusoidal signal is an example of a periodic signal, the period of which is

$$T = \frac{2\pi}{\omega}$$

We may readily prove this property of a sinusoidal signal by using Eq. (1.32) to write

$$
\begin{aligned}
x(t + T) &= A \cos(\omega(t + T) + \phi) \\
&= A \cos(\omega t + \omega T + \phi) \\
&= A \cos(\omega t + 2\pi + \phi) \\
&= A \cos(\omega t + \phi) \\
&= x(t)
\end{aligned}
$$

which satisfies the defining condition of Eq. (1.5) for a periodic signal.

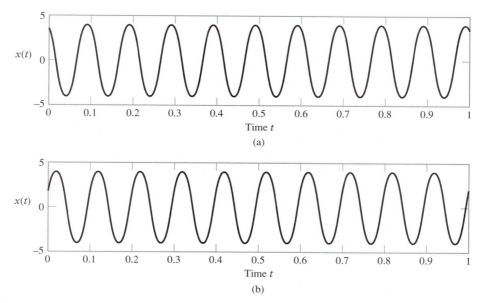

FIGURE 1.29 (a) Sinusoidal signal $A \cos(\omega t + \phi)$ with phase $\phi = +\pi/6$ radians. (b) Sinusoidal signal $A \sin(\omega t + \phi)$ with phase $\phi = +\pi/6$ radians.

To illustrate the generation of a sinusoidal signal, consider the circuit of Fig. 1.30 consisting of an inductor and capacitor connected in parallel. It is assumed that the losses in both components of the circuit are small enough for them to be considered "ideal." The voltage developed across the capacitor at time $t = 0$ is equal to V_0. The operation of the circuit in Fig. 1.30 for $t \geq 0$ is described by

$$LC \frac{d^2}{dt^2} v(t) + v(t) = 0 \tag{1.33}$$

where $v(t)$ is the voltage across the capacitor at time t, C is its capacitance, and L is the inductance of the inductor. Equation (1.33) is a *differential equation of order two*. Its solution is given by

$$v(t) = V_0 \cos(\omega_0 t), \qquad t \geq 0 \tag{1.34}$$

where ω_0 is the *natural angular frequency of oscillation* of the circuit:

$$\omega_0 = \frac{1}{\sqrt{LC}} \tag{1.35}$$

Equation (1.34) describes a sinusoidal signal of amplitude $A = V_0$, frequency $\omega = \omega_0$, and phase angle $\phi = 0$.

Consider next the discrete-time version of a sinusoidal signal, written as

$$x[n] = A \cos(\Omega n + \phi) \tag{1.36}$$

The period of a periodic discrete-time signal is measured in samples. Thus for $x[n]$ to be periodic with a period of N samples, say, it must satisfy the condition of Eq. (1.8) for all integer n and some integer N. Substituting $n + N$ for n in Eq. (1.36) yields

$$x[n + N] = A \cos(\Omega n + \Omega N + \phi)$$

For the condition of Eq. (1.8) to be satisfied, in general, we require that

$$\Omega N = 2\pi m \quad \text{radians}$$

or

$$\Omega = \frac{2\pi m}{N} \text{ radians/cycle}, \quad \text{integer } m, N \tag{1.37}$$

The important point to note here is that, unlike continuous-time sinusoidal signals, not all discrete-time sinusoidal systems with arbitrary values of Ω are periodic. Specifically, for the discrete-time sinusoidal signal described in Eq. (1.36) to be periodic, the angular frequency Ω must be a rational multiple of 2π, as indicated in Eq. (1.37). Figure 1.31 illustrates a discrete-time sinusoidal signal for $A = 1$, $\phi = 0$, and $N = 12$.

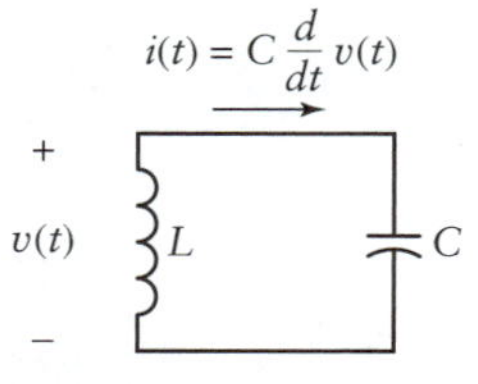

FIGURE 1.30   Parallel LC circuit, assuming that the inductor L and capacitor C are both ideal.

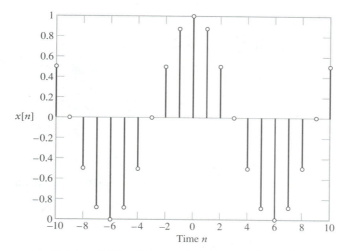

FIGURE 1.31 Discrete-time sinusoidal signal.

EXAMPLE 1.6 A pair of sinusoidal signals with a common angular frequency is defined by

$$x_1[n] = \sin[5\pi n]$$

and

$$x_2[n] = \sqrt{3}\, \cos[5\pi n]$$

(a) Specify the condition which the period N of both $x_1[n]$ and $x_2[n]$ must satisfy for them to be periodic.

(b) Evaluate the amplitude and phase angle of the composite sinusoidal signal

$$y[n] = x_1[n] + x_2[n]$$

Solution:

(a) The angular frequency of both $x_1[n]$ and $x_2[n]$ is

$$\Omega = 5\pi \text{ radians/cycle}$$

Solving Eq. (1.37) for the period N, we get

$$N = \frac{2\pi m}{\Omega}$$
$$= \frac{2\pi m}{5\pi}$$
$$= \frac{2m}{5}$$

For $x_1[n]$ and $x_2[n]$ to be periodic, their period N must be an integer. This can only be satisfied for $m = 5, 10, 15, \ldots$, which results in $N = 2, 4, 6, \ldots$.

(b) We wish to express $y[n]$ in the form

$$y[n] = A\cos(\Omega n + \phi)$$

Recall the trigonometric identity

$$A\cos(\Omega n + \phi) = A\cos(\Omega n)\cos(\phi) - A\sin(\Omega n)\sin(\phi)$$

Identifying $\Omega = 5\pi$, we see that the right-hand side of this identity is of the same form as $x_1[n] + x_2[n]$. We may therefore write

$$A \sin(\phi) = -1 \quad \text{and} \quad A \cos(\phi) = \sqrt{3}$$

Hence

$$\tan(\phi) = \frac{\sin(\phi)}{\cos(\phi)} = \frac{\text{amplitude of } x_1[n]}{\text{amplitude of } x_2[n]}$$

$$= \frac{-1}{\sqrt{3}}$$

from which we find that $\phi = -\pi/6$ radians. Similarly, the amplitude A is given by

$$A = \sqrt{(\text{amplitude of } x_1[n])^2 + (\text{amplitude of } x_2[n])^2}$$

$$= \sqrt{1 + 3} = 2$$

Accordingly, we may express $y[n]$ as

$$y[n] = 2 \cos(5\pi n - \pi/6)$$

▶ **Drill Problem 1.13**   Consider the following sinusoidal signals:

(a) $x[n] = 5 \sin[2n]$

(b) $x[n] = 5 \cos[0.2\pi n]$

(c) $x[n] = 5 \cos[6\pi n]$

(d) $x[n] = 5 \sin[6\pi n/35]$

Determine whether each $x(n)$ is periodic, and if it is, find its fundamental period.

Answer: (a) Nonperiodic. (b) Periodic, fundamental period = 10. (c) Periodic, fundamental period = 1. (d) Periodic, fundamental period = 35.   ◀

▶ **Drill Problem 1.14**   Find the smallest angular frequencies for which discrete-time sinusoidal signals with the following fundamental periods would be periodic: (a) $N = 8$, (b) $N = 32$, (c) $N = 64$, (d) $N = 128$.

Answer: (a) $\Omega = \pi/4$. (b) $\Omega = \pi/16$. (c) $\Omega = \pi/32$. (d) $\Omega = \pi/64$.   ◀

■ **RELATION BETWEEN SINUSOIDAL AND COMPLEX EXPONENTIAL SIGNALS**

Consider the complex exponential $e^{j\theta}$. Using *Euler's identity*, we may expand this term as

$$e^{j\theta} = \cos\theta + j\sin\theta \tag{1.38}$$

This result indicates that we may express the continuous-time sinusoidal signal of Eq. (1.32) as the real part of the complex exponential signal $Be^{j\omega t}$, where B is itself a complex quantity defined by

$$B = Ae^{j\phi} \tag{1.39}$$

That is, we may write

$$A \cos(\omega t + \phi) = \text{Re}\{Be^{j\omega t}\} \tag{1.40}$$

where Re{ } denotes the real part of the complex quantity enclosed inside the braces. We may readily prove this relation by noting that

$$Be^{j\omega t} = Ae^{j\phi}e^{j\omega t}$$
$$= Ae^{j(\omega t + \phi)}$$
$$= A\cos(\omega t + \phi) + jA\sin(\omega t + \phi)$$

from which Eq. (1.40) follows immediately. The sinusoidal signal of Eq. (1.32) is defined in terms of a cosine function. Of course, we may also define a continuous-time sinusoidal signal in terms of a sine function, as shown by

$$x(t) = A\sin(\omega t + \phi) \tag{1.41}$$

which is represented by the imaginary part of the complex exponential signal $Be^{j\omega t}$. That is, we may write

$$A\sin(\omega t + \phi) = \text{Im}\{Be^{j\omega t}\} \tag{1.42}$$

where B is defined by Eq. (1.39), and Im{ } denotes the imaginary part of the complex quantity enclosed inside the braces. The sinusoidal signal of Eq. (1.41) differs from that of Eq. (1.32) by a phase shift of 90°. That is, the sinusoidal signal $A\cos(\omega t + \phi)$ lags behind the sinusoidal signal $A\sin(\omega t + \phi)$, as illustrated in Fig. 1.29(b) for $\phi = \pi/6$.

Similarly, in the discrete-time case we may write

$$A\cos(\Omega n + \phi) = \text{Re}\{Be^{j\Omega n}\} \tag{1.43}$$

and

$$A\sin(\Omega n + \phi) = \text{Im}\{Be^{j\Omega n}\} \tag{1.44}$$

where B is defined in terms of A and ϕ by Eq. (1.39). Figure 1.32 shows the two-dimensional representation of the complex exponential $e^{j\Omega n}$ for $\Omega = \pi/4$ and $n = 0, 1, \ldots, 7$. The projection of each value on the real axis is $\cos(\Omega n)$, while the projection on the imaginary axis is $\sin(\Omega n)$.

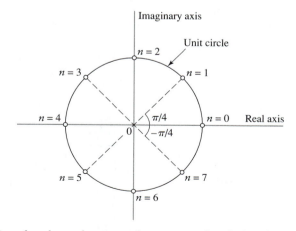

FIGURE 1.32 Complex plane, showing eight points uniformly distributed on the unit circle.

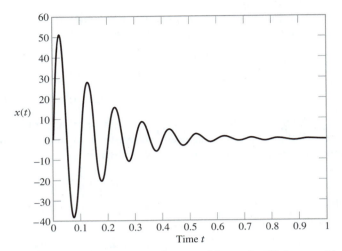

FIGURE 1.33　Exponentially damped sinusoidal signal $e^{-\alpha t} \sin \omega t$, with $\alpha > 0$.

■ **EXPONENTIALLY DAMPED SINUSOIDAL SIGNALS**

The multiplication of a sinusoidal signal by a real-valued decaying exponential signal results in a new signal referred to as an exponentially damped sinusoidal signal. Specifically, multiplying the continuous-time sinusoidal signal $A \sin(\omega t + \phi)$ by the exponential $e^{-\alpha t}$ results in the exponentially damped sinusoidal signal

$$x(t) = A e^{-\alpha t} \sin(\omega t + \phi), \qquad \alpha > 0 \tag{1.45}$$

Figure 1.33 shows the waveform of this signal for $A = 60$, $\alpha = 6$, and $\phi = 0$. For increasing time t, the amplitude of the sinusoidal oscillations decreases in an exponential fashion, approaching zero for infinite time.

To illustrate the generation of an exponentially damped sinusoidal signal, consider the parallel circuit of Fig. 1.34, consisting of a capacitor of capacitance C, an inductor of inductance L, and a resistor of resistance R. The resistance R represents the combined effect of losses associated with the inductor and the capacitor. Let V_0 denote the voltage developed across the capacitor at time $t = 0$. The operation of the circuit in Fig. 1.34 is described by

$$C \frac{d}{dt} v(t) + \frac{1}{R} v(t) + \frac{1}{L} \int_{-\infty}^{t} v(\tau) \, d\tau = 0 \tag{1.46}$$

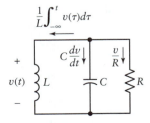

FIGURE 1.34　Parallel LCR circuit, with inductor L, capacitor C, and resistor R all assumed to be ideal.

where $v(t)$ is the voltage across the capacitor at time $t \geq 0$. Equation (1.46) is an *integro-differential equation*. Its solution is given by

$$v(t) = V_0 e^{-t/2CR} \cos(\omega_0 t) \tag{1.47}$$

where

$$\omega_0 = \sqrt{\frac{1}{LC} - \frac{1}{4C^2 R^2}} \tag{1.48}$$

In Eq. (1.48) it is assumed that $4CR^2 > L$. Comparing Eq. (1.47) with (1.45), we have $A = V_0$, $\alpha = 1/2CR$, $\omega = \omega_0$, and $\phi = \pi/2$.

The circuits of Figs. 1.27, 1.30, and 1.34 served as examples in which an exponential signal, a sinusoidal signal, and an exponentially damped sinusoidal signal, respectively, arose naturally as solutions to physical problems. The operations of these circuits are described by the differential equations (1.29), (1.33), and (1.46), whose solutions were simply stated. Methods for solving these differential equations are presented in subsequent chapters.

Returning to the subject matter at hand, the discrete-time version of the exponentially damped sinusoidal signal of Eq. (1.45) is described by

$$x[n] = Br^n \sin[\Omega n + \phi] \tag{1.49}$$

For the signal of Eq. (1.49) to decay exponentially with time, the parameter r must lie in the range $0 < |r| < 1$.

▶ **Drill Problem 1.15**   Is it possible for an exponentially damped sinusoidal signal of whatever kind to be periodic?

Answer: No.   ◀

■ STEP FUNCTION

The discrete-time version of the step function, commonly denoted by $u[n]$, is defined by

$$u[n] = \begin{cases} 1, & n \geq 0 \\ 0, & n < 0 \end{cases} \tag{1.50}$$

which is illustrated in Fig. 1.35.

The continuous-time version of the step function, commonly denoted by $u(t)$, is defined by

$$u(t) = \begin{cases} 1, & t \geq 0 \\ 0, & t < 0 \end{cases} \tag{1.51}$$

Figure 1.36 presents a portrayal of the step function $u(t)$. It is said to exhibit a discontinuity at $t = 0$, since the value of $u(t)$ changes instantaneously from 0 to 1 when $t = 0$.

The step function $u(t)$ is a particularly simple signal to apply. Electrically, a battery or dc source is applied at $t = 0$ by closing a switch, for example. As a test signal, it is

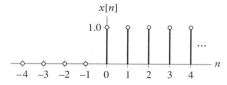

FIGURE 1.35   Discrete-time version of step function of unit amplitude.

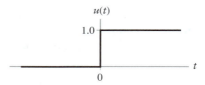

FIGURE 1.36   Continuous-time version of step function of unit amplitude.

useful because the output of a system due to a step input reveals a great deal about how quickly the system responds to an abrupt change in the input signal. A similar remark applies to $u[n]$ in the context of a discrete-time system.

The step function $u(t)$ may also be used to construct other discontinuous waveforms, as illustrated in the following example.

> **EXAMPLE 1.7**   Consider the rectangular pulse $x(t)$ shown in Fig. 1.37(a). This pulse has an amplitude A and duration T. Express $x(t)$ as a weighted sum of two step functions.
>
> **Solution:**   The rectangular pulse $x(t)$ may be written in mathematical terms as follows:
>
> $$x(t) = \begin{cases} A, & 0 \le |t| < T/2 \\ 0, & |t| > T/2 \end{cases} \qquad (1.52)$$
>
> where $|t|$ denotes the magnitude of time t. The rectangular pulse $x(t)$ is represented as the difference between two time-shifted step functions, as illustrated in Fig. 1.37(b). On the basis of this figure, we may express $x(t)$ as
>
> $$x(t) = Au\left(t + \frac{T}{2}\right) - Au\left(t - \frac{T}{2}\right) \qquad (1.53)$$
>
> where $u(t)$ is the step function. For the purpose of illustration, we have set $T = 1$s in Fig. 1.37.

▶ **Drill Problem 1.16**   A discrete-time signal $x[n]$ is defined by

$$x[n] = \begin{cases} 1, & 0 \le n \le 9 \\ 0, & \text{otherwise} \end{cases}$$

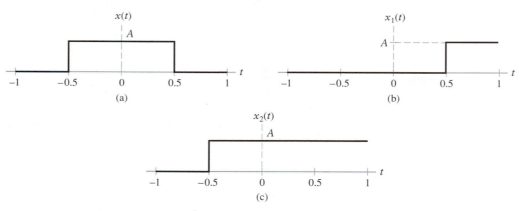

FIGURE 1.37   (a) Rectangular pulse $x(t)$ of amplitude A and duration $T = 1$s symmetric about the origin. (b) Representation of $x(t)$ as the superposition of two step functions of amplitude A, with one step function shifted to the left by $T/2$ and the other shifted to the right by $T/2$; these two shifted signals are denoted by $x_1(t)$ and $x_2(t)$, respectively.

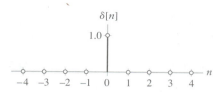

FIGURE 1.38   Discrete-time form of impulse.

Using $u[n]$, describe $x[n]$ as the superposition of two step functions.

Answer: $x[n] = u[n] - u[n - 10]$.   ◀

■ IMPULSE FUNCTION

The discrete-time version of the impulse, commonly denoted by $\delta[n]$, is defined by

$$\delta[n] = \begin{cases} 1, & n = 0 \\ 0, & n \neq 0 \end{cases} \tag{1.54}$$

which is illustrated in Fig. 1.38.

The continuous-time version of the unit impulse, commonly denoted by $\delta(t)$, is defined by the following pair of relations:

$$\delta(t) = 0 \quad \text{for } t \neq 0 \tag{1.55}$$

and

$$\int_{-\infty}^{\infty} \delta(t) \, dt = 1 \tag{1.56}$$

Equation (1.55) says that the impulse $\delta(t)$ is zero everywhere except at the origin. Equation (1.56) says that the total area under the unit impulse is unity. The impulse $\delta(t)$ is also referred to as the *Dirac delta function*. Note that the impulse $\delta(t)$ is the derivative of the step function $u(t)$ with respect to time t. Conversely, the step function $u(t)$ is the integral of the impulse $\delta(t)$ with respect to time t.

A graphical description of the impulse $\delta[n]$ for discrete time is straightforward, as shown in Fig. 1.38. In contrast, visualization of the unit impulse $\delta(t)$ for continuous time requires more detailed attention. One way to visualize $\delta(t)$ is to view it as the limiting form of a rectangular pulse of unit area, as illustrated in Fig. 1.39(a). Specifically, the duration of the pulse is decreased and its amplitude is increased such that the area under the pulse

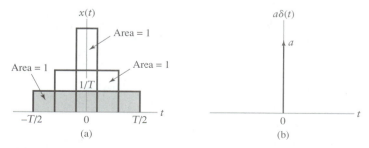

FIGURE 1.39   (a) Evolution of a rectangular pulse of unit area into an impulse of unit strength. (b) Graphical symbol for an impulse of strength a.

is maintained constant at unity. As the duration decreases, the rectangular pulse better approximates the impulse. Indeed, we may generalize this result by stating that

$$\delta(t) = \lim_{T \to 0} g_T(t) \tag{1.57}$$

where $g_T(t)$ is any pulse that is an even function of time t, with duration T, and unit area. The area under the pulse defines the *strength* of the impulse. Thus when we speak of the impulse function $\delta(t)$, in effect we are saying that its strength is unity. The graphical symbol for an impulse is depicted in Fig. 1.39(b). The strength of the impulse is denoted by the label next to the arrow.

From the defining equation (1.55), it immediately follows that the unit impulse $\delta(t)$ is an even function of time t, as shown by

$$\delta(-t) = \delta(t) \tag{1.58}$$

For the unit impulse $\delta(t)$ to have mathematical meaning, however, it has to appear as a factor in the integrand of an integral with respect to time and then, strictly speaking, only when the other factor in the integrand is a continuous function of time at which the impulse occurs. Let $x(t)$ be such a function, and consider the product of $x(t)$ and the time-shifted delta function $\delta(t - t_0)$. In light of the two defining equations (1.55) and (1.56), we may express the integral of this product as follows:

$$\int_{-\infty}^{\infty} x(t)\delta(t - t_0)\, dt = x(t_0) \tag{1.59}$$

The operation indicated on the left-hand side of Eq. (1.59) sifts out the value $x(t_0)$ of the function $x(t)$ at time $t = t_0$. Accordingly, Eq. (1.59) is referred to as the *sifting property* of the unit impulse. This property is sometimes used as the definition of a unit impulse; in effect, it incorporates Eqs. (1.55) and (1.56) into a single relation.

Another useful property of the unit impulse $\delta(t)$ is the *time-scaling property*, described by

$$\delta(at) = \frac{1}{a}\delta(t), \qquad a > 0 \tag{1.60}$$

To prove this property, we replace t in Eq. (1.57) with at and so write

$$\delta(at) = \lim_{T \to 0} g_T(at) \tag{1.61}$$

To represent the function $g_T(t)$, we use the rectangular pulse shown in Fig. 1.40(a), which has duration T, amplitude $1/T$, and therefore unit area. Correspondingly, the time-scaled

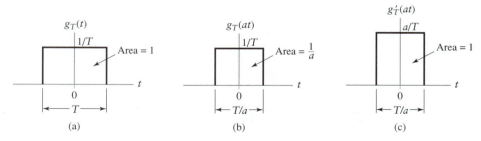

FIGURE 1.40 Steps involved in proving the time-scaling property of the unit impulse. (a) Rectangular pulse $g_T(t)$ of amplitude $1/T$ and duration T, symmetric about the origin. (b) Pulse $g_T(t)$ compressed by factor a. (c) Amplitude scaling of the compressed pulse, restoring it to unit area.

function $g_T(at)$ is shown in Fig. 1.40(b) for $a > 1$. The amplitude of $g_T(at)$ is left unchanged by the time-scaling operation. Therefore, in order to restore the area under this pulse to unity, the amplitude of $g_T(at)$ is scaled by the same factor a, as indicated in Fig. 1.40(c). The time function in Fig. 1.40(c) is denoted by $g_T'(at)$; it is related to $g_T(at)$ by

$$g_T'(at) = a g_T(at) \tag{1.62}$$

Substituting Eq. (1.62) in (1.61), we get

$$\delta(at) = \frac{1}{a} \lim_{t \to 0} g_T'(at) \tag{1.63}$$

Since, by design, the area under the function $g_T'(at)$ is unity, it follows that

$$\delta(t) = \lim_{T \to 0} g_T'(at) \tag{1.64}$$

Accordingly, the use of Eq. (1.64) in (1.63) results in the time-scaling property described in Eq. (1.60).

Having defined what a unit impulse is and described its properties, there is one more question that needs to be addressed: What is the practical use of a unit impulse? We cannot generate a physical impulse function, since that would correspond to a signal of infinite amplitude at $t = 0$ and that is zero elsewhere. However, the impulse function serves a mathematical purpose by providing an approximation to a physical signal of extremely short duration and high amplitude. The response of a system to such an input reveals much about the character of the system. For example, consider the parallel LCR circuit of Fig. 1.34, assumed to be initially at rest. Suppose now a voltage signal approximating an impulse function is applied to the circuit at time $t = 0$. The current through an inductor cannot change instantaneously, but the voltage across a capacitor can. It follows therefore that the voltage across the capacitor suddenly rises to a value equal to V_0, say, at time $t = 0^+$. Here $t = 0^+$ refers to the instant of time when energy in the input signal is expired. Thereafter, the circuit operates without additional input. The resulting value of the voltage $v(t)$ across the capacitor is defined by Eq. (1.47). The response $v(t)$ is called the *transient* or *natural response* of the circuit, the evaluation of which is facilitated by the application of an impulse function as the test signal.

■ RAMP FUNCTION

The impulse function $\delta(t)$ is the derivative of the step function $u(t)$ with respect to time. By the same token, the integral of the step function $u(t)$ is a ramp function of unit slope. This latter test signal is commonly denoted by $r(t)$, which is formally defined as follows:

$$r(t) = \begin{cases} t, & t \geq 0 \\ 0, & t < 0 \end{cases} \tag{1.65}$$

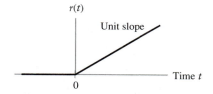

FIGURE 1.41   Ramp function of unit slope.

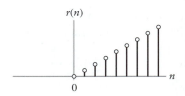

FIGURE 1.42   Discrete-time version of the ramp function.

Equivalently, we may write

$$r(t) = tu(t) \tag{1.66}$$

The ramp function $r(t)$ is shown graphically in Fig. 1.41.

In mechanical terms, a ramp function may be visualized as follows. If the input variable is represented as the angular displacement of a shaft, then the constant-speed rotation of the shaft provides a representation of the ramp function. As a test signal, the ramp function enables us to evaluate how a continuous-time system would respond to a signal that increases linearly with time.

The discrete-time version of the ramp function is defined by

$$r[n] = \begin{cases} n, & n \geq 0 \\ 0, & n < 0 \end{cases} \tag{1.67}$$

or, equivalently,

$$r[n] = nu[n] \tag{1.68}$$

It is illustrated in Fig. 1.42.

1.7 Systems Viewed as Interconnections of Operations

In mathematical terms, a system may be viewed as an *interconnection of operations* that transforms an input signal into an output signal with properties different from those of the input signal. The signals may be of the continuous-time or discrete-time variety, or a mixture of both. Let the overall *operator H* denote the action of a system. Then the application of a continuous-time signal $x(t)$ to the input of the system yields the output signal described by

$$y(t) = H\{x(t)\} \tag{1.69}$$

Figure 1.43(a) shows a block diagram representation of Eq. (1.69). Correspondingly, for the discrete-time case, we may write

$$y[n] = H\{x[n]\} \tag{1.70}$$

$$x(t) \longrightarrow \boxed{H} \longrightarrow y(t) \qquad\qquad x[n] \longrightarrow \boxed{H} \longrightarrow y[n]$$

$$\text{(a)} \qquad\qquad\qquad\qquad \text{(b)}$$

FIGURE 1.43   Block diagram representation of operator H for (a) continuous time and (b) discrete time.

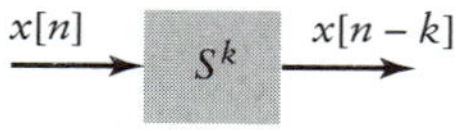

$$\xrightarrow{\;x[n]\;}\boxed{S^k}\xrightarrow{\;x[n-k]\;}$$

FIGURE 1.44 Discrete-time shift operator S^k, operating on the discrete-time signal $x[n]$ to produce $x[n-k]$.

where the discrete-time signals $x[n]$ and $y[n]$ denote the input and output signals, respectively, as depicted in Fig. 1.43(b).

EXAMPLE 1.8 Consider a discrete-time system whose output signal $y[n]$ is the average of the three most recent values of the input signal $x[n]$, as shown by

$$y[n] = \tfrac{1}{3}(x[n] + x[n-1] + x[n-2])$$

Such a system is referred to as a *moving-average system* for two reasons. First, $y[n]$ is the average of the sample values $x[n]$, $x[n-1]$, and $x[n-2]$. Second, the value of $y[n]$ changes as n moves along the discrete-time axis. Formulate the operator H for this system; hence, develop a block diagram representation for it.

Solution: Let the operator S^k denote a system that time shifts the input $x[n]$ by k time units to produce an output equal to $x[n-k]$, as depicted in Fig. 1.44. Accordingly, we may define the overall operator H for the moving-average system as

$$H = \tfrac{1}{3}(1 + S + S^2)$$

Two different implementations of the operator H (i.e., the moving-average system) that suggest themselves are presented in Fig. 1.45. The implementation shown in part (a) of the figure uses the *cascade* connection of two identical unity time shifters, namely, $S^1 = S$. On the other hand, the implementation shown in part (b) of the figure uses two different time shifters, S and S^2, connected in *parallel*. In both cases, the moving-average system is made up of an interconnection of three functional blocks, namely, two time shifters, an adder, and a scalar multiplication.

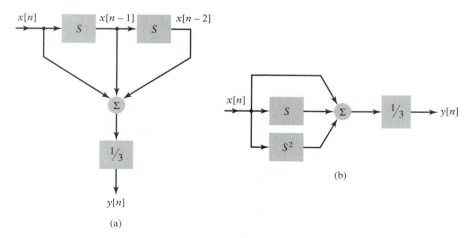

FIGURE 1.45 Two different (but equivalent) implementations of the moving-average system: (a) cascade form of implementation, and (b) parallel form of implementation.

▶ **Drill Problem 1.17**   Express the operator that describes the input–output relation

$$y[n] = \tfrac{1}{3}(x[n + 1] + x[n] + x[n - 1])$$

in terms of the time-shift operator S.

Answer: $H = \tfrac{1}{3}(S^{-1} + 1 + S^1).$ ◀

In the interconnected systems shown in Figs. 1.45(a) and (b), the signal flows through each one of them in the forward direction only. Another possible way of combining systems is through the use of *feedback* connections. Figure 1.4 shows an example of a *feedback system*, which is characterized by two paths. The forward path involves the cascade connection of the controller and plant. The feedback path is made possible through the use of a sensor connected to the output of the system at one end and the input at the other end. The use of feedback has many desirable benefits and gives rise to problems of its own that require special attention; the subject of feedback is discussed in Chapter 9.

1.8 *Properties of Systems*

The properties of a system describe the characteristics of the operator H representing the system. In what follows, we study some of the most basic properties of systems.

■ STABILITY

A system is said to be *bounded input–bounded output (BIBO) stable* if and only if every bounded input results in a bounded output. The output of such a system does not diverge if the input does not diverge.

To put the condition for BIBO stability on a formal basis, consider a continuous-time system whose input–output relation is as described in Eq. (1.69). The operator H is BIBO stable if the output signal $y(t)$ satisfies the condition

$$|y(t)| \le M_y < \infty \quad \text{for all } t$$

whenever the input signals $x(t)$ satisfy the condition

$$|x(t)| \le M_x < \infty \quad \text{for all } t$$

Both M_x and M_y represent some finite positive numbers. We may describe the condition for the BIBO stability of a discrete-time system in a similar manner.

From an engineering perspective, it is important that a system of interest remains stable under all possible operating conditions. It is only then that the system is guaranteed to produce a bounded output for a bounded input. Unstable systems are usually to be avoided, unless some mechanism can be found to stabilize them.

One famous example of an unstable system is the first Tacoma Narrows suspension bridge that collapsed on November 7, 1940, at approximately 11:00 a.m., due to wind-induced vibrations. Situated on the Tacoma Narrows in Puget Sound, near the city of Tacoma, Washington, the bridge had only been open for traffic a few months before it collapsed; see Fig. 1.46 for photographs taken just prior to failure of the bridge and soon thereafter.

(a)

(b)

FIGURE 1.46 Dramatic photographs showing the collapse of the Tacoma Narrows suspension bridge on November 7, 1940. (a) Photograph showing the twisting motion of the bridge's center span just before failure. (b) A few minutes after the first piece of concrete fell, this second photograph shows a 600-ft section of the bridge breaking out of the suspension span and turning upside down as it crashed in Puget Sound, Washington. Note the car in the top right-hand corner of the photograph. (Courtesy of the Smithsonian Institution.)

EXAMPLE 1.9 Show that the moving-average system described in Example 1.8 is BIBO stable.

Solution: Assume that

$$|x[n]| < M_x < \infty \quad \text{for all } n$$

Using the given input–output relation

$$y[n] = \tfrac{1}{3}(x[n] + x[n-1] + x[n-2])$$

we may write

$$
\begin{aligned}
|y[n]| &= \tfrac{1}{3}|x[n] + x[n-1] + x[n-2]| \\
&\leq \tfrac{1}{3}(|x[n]| + |x[n-1]| + |x[n-2]|) \\
&\leq \tfrac{1}{3}(M_x + M_x + M_x) \\
&= M_x
\end{aligned}
$$

Hence the absolute value of the output signal $y[n]$ is always less than the maximum absolute value of the input signal $x[n]$ for all n, which shows that the moving-average system is stable.

▶ **Drill Problem 1.18** Show that the moving-average system described by the input–output relation

$$y[n] = \tfrac{1}{3}(x[n+1] + x[n] + x[n-1])$$

is BIBO stable.   ◀

EXAMPLE 1.10 Consider a discrete-time system whose input–output relation is defined by

$$y[n] = r^n x[n]$$

where $r > 1$. Show that this system is unstable.

Solution: Assume that the input signal $x[n]$ satisfies the condition

$$|x[n]| \leq M_x < \infty \quad \text{for all } n$$

We then find that

$$
\begin{aligned}
|y[n]| &= |r^n x[n]| \\
&= |r^n| \cdot |x[n]|
\end{aligned}
$$

With $r > 1$, the multiplying factor r^n diverges for increasing n. Accordingly, the condition that the input signal is bounded is not sufficient to guarantee a bounded output signal, and so the system is unstable. To prove stability, we need to establish that all bounded inputs produce a bounded output.

■ MEMORY

A system is said to possess *memory* if its output signal depends on past values of the input signal. The temporal extent of past values on which the output depends defines how far the memory of the system extends into the past. In contrast, a system is said to be *memoryless* if its output signal depends only on the present value of the input signal.

For example, a resistor is memoryless since the current $i(t)$ flowing through it in response to the applied voltage $v(t)$ is defined by

$$i(t) = \frac{1}{R} v(t)$$

where R is the resistance of the resistor. On the other hand, an inductor has memory, since the current $i(t)$ flowing through it is related to the applied voltage $v(t)$ as follows:

$$i(t) = \frac{1}{L} \int_{-\infty}^{t} v(\tau) \, d\tau$$

where L is the inductance of the inductor. That is, unlike a resistor, the current through an inductor at time t depends on all past values of the voltage $v(t)$; the memory of an inductor extends into the infinite past.

The moving-average system of Example 1.8 described by the input–output relation

$$y[n] = \tfrac{1}{3}(x[n] + x[n-1] + x[n-2])$$

has memory, since the value of the output signal $y[n]$ at time n depends on the present and two past values of the input signal $x[n]$. On the other hand, a system described by the input–output relation

$$y[n] = x^2[n]$$

is memoryless, since the value of the output signal $y[n]$ at time n depends only on the present value of the input signal $x[n]$.

▶ **Drill Problem 1.19**   How far does the memory of the moving-average system described by the input–output relation

$$y[n] = \tfrac{1}{3}(x[n] + x[n-1] + x[n-2])$$

extend into the past?

Answer: Two time units.   ◀

▶ **Drill Problem 1.20**   The input–output relation of a semiconductor diode is represented by

$$i(t) = a_0 + a_1 v(t) + a_2 v^2(t) + a_3 v^3(t) + \cdots$$

where $v(t)$ is the applied voltage, $i(t)$ is the current flowing through the diode, and $a_0, a_1, a_3, \ldots$ are constants. Does this diode have memory?

Answer: No.   ◀

▶ **Drill Problem 1.21**   The input–output relation of a capacitor is described by

$$v(t) = \frac{1}{C} \int_{-\infty}^{t} i(\tau) \, d\tau$$

What is its memory?

Answer: Memory extends from time t to the infinite past.   ◀

■ CAUSALITY

A system is said to be *causal* if the present value of the output signal depends only on the present and/or past values of the input signal. In contrast, the output signal of a *noncausal* system depends on future values of the input signal.

For example, the moving-average system described by

$$y[n] = \tfrac{1}{3}(x[n] + x[n-1] + x[n-2])$$

is causal. On the other hand, the moving-average system described by

$$y[n] = \tfrac{1}{3}(x[n+1] + x[n] + x[n-1])$$

is noncausal, since the output signal $y[n]$ depends on a future value of the input signal, namely $x[n+1]$.

▶ **Drill Problem 1.22**   Consider the *RC* circuit shown in Fig. 1.47. Is it causal or noncausal?

Answer: Causal.   ◀

▶ **Drill Problem 1.23**   Suppose k in the operator of Fig. 1.44 is replaced by $-k$. Is the resulting system causal or noncausal for positive k?

Answer: Noncausal.   ◀

■ INVERTIBILITY

A system is said to be *invertible* if the input of the system can be recovered from the system output. We may view the set of operations needed to recover the input as a second system connected in cascade with the given system, such that the output signal of the second system is equal to the input signal applied to the given system. To put the notion of invertibility on a formal basis, let the operator H represent a continuous-time system, with the input signal $x(t)$ producing the output signal $y(t)$. Let the output signal $y(t)$ be applied to a second continuous-time system represented by the operator H^{-1}, as illustrated in Fig. 1.48. The output signal of the second system is defined by

$$H^{-1}\{y(t)\} = H^{-1}\{H\{x(t)\}\}$$
$$= H^{-1}H\{x(t)\}$$

where we have made use of the fact that two operators H and H^{-1} connected in cascade are equivalent to a single operator $H^{-1}H$. For this output signal to equal the original input signal $x(t)$, we require that

$$H^{-1}H = I \tag{1.71}$$

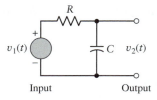

FIGURE 1.47   Series *RC* circuit driven from an ideal voltage source $v_1(t)$, producing output voltage $v_2(t)$.

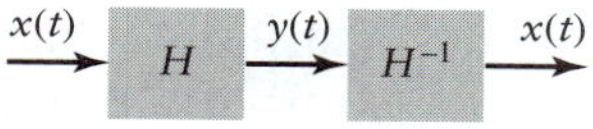

$$x(t) \xrightarrow{\quad} \boxed{H} \xrightarrow{y(t)} \boxed{H^{-1}} \xrightarrow{x(t)}$$

FIGURE 1.48   The notion of system invertibility. The second operator H^{-1} is the inverse of the first operator H. Hence the input $x(t)$ is passed through the cascade correction of H and H^{-1} completely unchanged.

where I denotes the *identity operator*. The output of a system described by the identity operator is exactly equal to the input. Equation (1.71) is the condition that the new operator H^{-1} must satisfy in relation to the given operator H for the original input signal $x(t)$ to be recovered from $y(t)$. The operator H^{-1} is called the *inverse operator*, and the associated system is called the *inverse system*. Note that H^{-1} is not the reciprocal of the operator H; rather, the use of the superscript -1 is intended to be merely a flag indicating "inverse." In general, the problem of finding the inverse of a given system is a difficult one. In any event, a system is not invertible unless distinct inputs applied to the system produce distinct outputs. That is, there must be a one-to-one mapping between input and output signals for a system to be invertible. Identical conditions must hold for a discrete-time system to be invertible.

The property of invertibility is of particular importance in the design of communication systems. As remarked in Section 1.3, when a transmitted signal propagates through a communication channel, it becomes distorted due to the physical characteristics of the channel. A widely used method of compensating for this distortion is to include in the receiver a network called an *equalizer*, which is connected in cascade with the channel in a manner similar to that described in Fig. 1.48. By designing the equalizer to be the inverse of the channel, the transmitted signal is restored to its original form, assuming ideal conditions.

EXAMPLE 1.11   Consider the time-shift system described by the input–output relation

$$y(t) = x(t - t_0) = S^{t_0}\{x(t)\}$$

where the operator S^{t_0} represents a time shift of t_0 seconds. Find the inverse of this system.

Solution: For this example, the inverse of a time shift of t_0 seconds is a time shift of $-t_0$ seconds. We may represent the time shift of $-t_0$ by the operator S^{-t_0}. Thus applying S^{-t_0} to the output signal of the given time-shift system, we get

$$S^{-t_0}\{y(t)\} = S^{-t_0}\{S^{t_0}\{x(t)\}\}$$
$$= S^{-t_0}S^{t_0}\{x(t)\}$$

For this output signal to equal the original input signal $x(t)$, we require that

$$S^{-t_0}S^{t_0} = I$$

which is in perfect accord with the condition for invertibility described in Eq. (1.71).

▶ **Drill Problem 1.24**   An inductor is described by the input–output relation

$$y(t) = \frac{1}{L} \int_{-\infty}^{t} x(\tau) \, d\tau$$

Find the operation representing the inverse system.

Answer: $L \dfrac{d}{dt}$ ◀

EXAMPLE 1.12 Show that a square-law system described by the input–output relation

$$y(t) = x^2(t)$$

is not invertible.

Solution: We note that the square-law system violates a necessary condition for invertibility, which postulates that distinct inputs must produce distinct outputs. Specifically, the distinct inputs $x(t)$ and $-x(t)$ produce the same output $y(t)$. Accordingly, the square-law system is not invertible.

■ **TIME INVARIANCE**

A system is said to be *time invariant* if a time delay or time advance of the input signal leads to an identical time shift in the output signal. This implies that a time-invariant system responds identically no matter when the input signal is applied. Stated in another way, the characteristics of a time-invariant system do not change with time. Otherwise, the system is said to be *time variant*.

Consider a continuous-time system whose input–output relation is described by Eq. (1.69), reproduced here for convenience of presentation:

$$y(t) = H\{x(t)\}$$

Suppose the input signal $x(t)$ is shifted in time by t_0 seconds, resulting in the new input $x(t - t_0)$. This operation may be described by writing

$$x(t - t_0) = S^{t_0}\{x(t)\}$$

where the operator S^{t_0} represents a time shift equal to t_0 seconds. Let $y_i(t)$ denote the output signal of the system produced in response to the time-shifted input $x(t - t_0)$. We may then write

$$
\begin{aligned}
y_i(t) &= H\{x(t - t_0)\} \\
&= H\{S^{t_0}\{x(t)\}\} \\
&= HS^{t_0}\{x(t)\}
\end{aligned}
\tag{1.72}
$$

which is represented by the block diagram shown in Fig. 1.49(a). Now suppose $y_o(t)$ represents the output of the original system shifted in time by t_0 seconds, as shown by

$$
\begin{aligned}
y_o(t) &= S^{t_0}\{y(t)\} \\
&= S^{t_0}\{H\{x(t)\}\} \\
&= S^{t_0}H\{x(t)\}
\end{aligned}
\tag{1.73}
$$

which is represented by the block diagram shown in Fig. 1.49(b). The system is time invariant if the outputs $y_i(t)$ and $y_o(t)$ defined in Eqs. (1.72) and (1.73) are equal for any identical input signal $x(t)$. Hence we require

$$HS^{t_0} = S^{t_0}H \tag{1.74}$$

That is, for a system described by the operator H to be time invariant, the system operator H and the time-shift operator S^{t_0} must *commute* with each other for all t_0. A similar relation must hold for a discrete-time system to be time invariant.

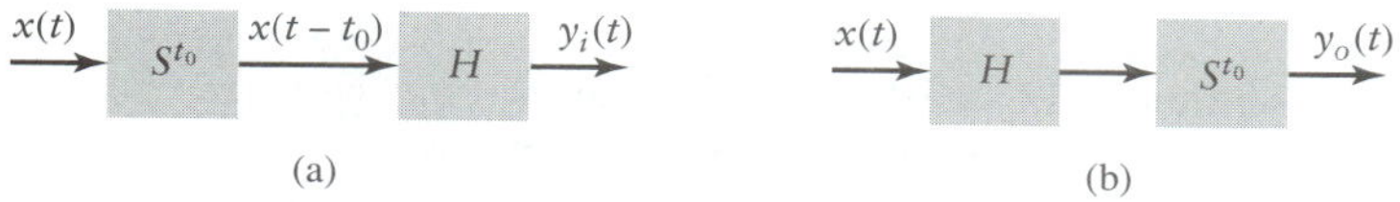

FIGURE 1.49   The notion of time invariance. (a) Time-shift operator S^{t_0} preceding operator H. (b) Time-shift operator S^{t_0} following operator H. These two situations are equivalent, provided that H is time invariant.

EXAMPLE 1.13   Use the voltage $v(t)$ across an inductor to represent the input signal $x(t)$, and the current $i(t)$ flowing through it to represent the output signal $y(t)$. Thus the inductor is described by the input–output relation

$$y(t) = \frac{1}{L} \int_{-\infty}^{t} x(\tau)\, d\tau$$

where L is the inductance. Show that the inductor so described is time invariant.

Solution: Let the input $x(t)$ be shifted by t_0 seconds, yielding $x(t - t_0)$. The response $y_i(t)$ of the inductor to $x(t - t_0)$ is

$$y_i(t) = \frac{1}{L} \int_{-\infty}^{t} x(\tau - t_0)\, d\tau$$

Next, let $y_o(t)$ denote the original output of the inductor shifted by t_0 seconds, as shown by

$$y_o(t) = y(t - t_0)$$
$$= \frac{1}{L} \int_{-\infty}^{t - t_0} x(\tau)\, d\tau$$

Though at first examination $y_i(t)$ and $y_o(t)$ look different, they are in fact equal, as shown by a simple change in the variable for integration. Let

$$\tau' = \tau - t_0$$

For a constant t_0, we have $d\tau' = d\tau$. Hence changing the limits of integration, the expression for $y_i(t)$ may be rewritten as

$$y_i(t) = \frac{1}{L} \int_{-\infty}^{t - t_0} x(\tau')\, d\tau'$$

which, in mathematical terms, is identical to $y_o(t)$. It follows therefore that an ordinary inductor is time invariant.

EXAMPLE 1.14   A thermistor has a resistance that varies with time due to temperature changes. Let $R(t)$ denote the resistance of the thermistor, expressed as a function of time. Associating the input signal $x(t)$ with the voltage applied across the thermistor, and the output signal $y(t)$ with the current flowing through it, we may express the input–output relation of the thermistor as

$$y(t) = \frac{x(t)}{R(t)}$$

Show that the thermistor so described is time variant.

Solution: Let $y_i(t)$ denote the response of the thermistor produced by a time-shifted version $x(t - t_0)$ of the original input signal. We may then write

$$y_i(t) = \frac{x(t - t_0)}{R(t)}$$

Next, let $y_o(t)$ denote the original output of the thermistor shifted in time by t_0, as shown by

$$y_o(t) = y(t - t_0)$$
$$= \frac{x(t - t_0)}{R(t - t_0)}$$

We now see that since, in general, $R(t) \neq R(t - t_0)$ for $t_0 \neq 0$, then

$$y_o(t) \neq y_i(t) \quad \text{for } t_0 \neq 0$$

Hence a thermistor is time variant, which is intuitively satisfying.

▶ **Drill Problem 1.25**   Is a discrete-time system described by the input–output relation

$$y(n) = r^n x(n)$$

time invariant?

Answer: No.   ◀

■ **LINEARITY**

A system is said to be *linear* if it satisfies the *principle of superposition*. That is, the response of a linear system to a weighted sum of input signals is equal to the same weighted sum of output signals, each output signal being associated with a particular input signal acting on the system independently of all the other input signals. A system that violates the principle of superposition is said to be *nonlinear*.

Let the operator H represent a continuous-time system. Let the signal applied to the system input be defined by the weighted sum

$$x(t) = \sum_{i=1}^{N} a_i x_i(t) \tag{1.75}$$

where $x_1(t), x_2(t), \ldots, x_N(t)$ denote a set of input signals, and $a_1, a_2, \ldots, a_N$ denote the corresponding weighting factors. The resulting output signal is written as

$$y(t) = H\{x(t)\}$$
$$= H\left\{ \sum_{i=1}^{N} a_i x_i(t) \right\} \tag{1.76}$$

If the system is linear, we may (in accordance with the principle of superposition) express the output signal $y(t)$ of the system as

$$y(t) = \sum_{i=1}^{N} a_i y_i(t) \tag{1.77}$$

where $y_i(t)$ is the output of the system in response to the input $x_i(t)$ acting alone; that is,

$$y_i(t) = H\{x_i(t)\} \tag{1.78}$$

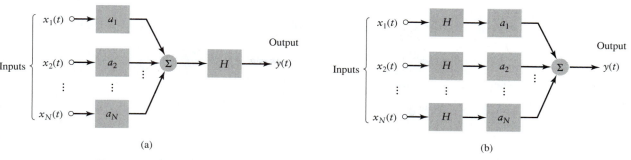

(a)   (b)

FIGURE 1.50   The linearity property of a system. (a) The combined operation of amplitude scaling and summation precedes the operator H for multiple inputs. (b) The operator H precedes amplitude scaling for each input; the resulting outputs are summed to produce the overall output $y(t)$. If these two configurations produce the same output $y(t)$, the operator H is linear.

The weighted sum of Eq. (1.77) describing the output signal $y(t)$ is of the same mathematical form as that of Eq. (1.75), describing the input signal $x(t)$. Substituting Eq. (1.78) into (1.77), we get

$$y(t) = \sum_{i=1}^{N} a_i H\{x_i(t)\} \tag{1.79}$$

In order to write Eq. (1.79) in the same form as Eq. (1.76), the system operation described by H must *commute* with the summation and amplitude scaling in Eq. (1.79), as illustrated in Fig. 1.50. Indeed, Eqs. (1.78) and (1.79), viewed together, represent a mathematical statement of the principle of superposition. For a linear discrete-time system, the principle of superposition is described in a similar manner.

EXAMPLE 1.15   Consider a discrete-time system described by the input–output relation

$$y[n] = nx[n]$$

Show that this system is linear.

Solution: Let the input signal $x[n]$ be expressed as the weighted sum

$$x[n] = \sum_{i=1}^{N} a_i x_i[n]$$

We may then express the resulting output signal of the system as

$$y[n] = n \sum_{i=1}^{N} a_i x_i[n]$$

$$= \sum_{i=1}^{N} a_i n x_i[n]$$

$$= \sum_{i=1}^{N} a_i y_i[n]$$

where

$$y_i[n] = nx_i[n]$$

is the output due to each input acting independently. We thus see that the given system satisfies the principle of superposition and is therefore linear.

EXAMPLE 1.16   Consider next the continuous-time system described by the input–output relation

$$y(t) = x(t)x(t - 1)$$

Show that this system is nonlinear.

Solution: Let the input signal $x(t)$ be expressed as the weighted sum

$$x(t) = \sum_{i=1}^{N} a_i x_i(t)$$

Correspondingly, the output signal of the system is given by the double summation

$$y(t) = \sum_{i=1}^{N} a_i x_i(t) \sum_{j=1}^{N} a_j x_j(t - 1)$$

$$= \sum_{i=1}^{N} \sum_{j=1}^{N} a_i a_j x_i(t) x_j(t - 1)$$

The form of this equation is radically different from that describing the input signal $x(t)$. That is, here we cannot write $y(t) = \sum_{i=1}^{N} a_i y_i(t)$. Thus the system violates the principle of superposition and is therefore nonlinear.

▶ **Drill Problem 1.26**   Show that the moving-average system described by

$$y[n] = \tfrac{1}{3}(x[n] + x[n - 1] + x[n - 2])$$

is a linear system.   ◀

▶ **Drill Problem 1.27**   Is it possible for a linear system to be noncausal?

Answer: Yes.   ◀

▶ **Drill Problem 1.28**   The hard limiter is a memoryless device whose output y is related to the input x by

$$y = \begin{cases} 1, & x \geq 0 \\ 0, & x < 0 \end{cases}$$

Is the hard limiter linear?

Answer: No.   ◀

1.9 *Exploring Concepts with MATLAB*

The basic *object* used in MATLAB is a rectangular numerical matrix with possibly complex elements. The kinds of data objects encountered in the study of signals and systems are all well suited to matrix representations. In this section we use MATLAB to explore the generation of elementary signals described in previous sections. The exploration of systems and more advanced signals is deferred to subsequent chapters.

The MATLAB Signal Processing Toolbox has a large variety of functions for generating signals, most of which require that we begin with the vector representation of time t or n. To generate a vector t of time values with a *sampling interval* $\mathcal{T}$ of 1 ms on the interval from 0 to 1s, for example, we use the command:

```
t = 0:.001:1;
```

This corresponds to 1000 time samples for each second or a *sampling rate* of 1000 Hz. To generate a vector n of time values for discrete-time signals, say, from $n = 0$ to $n = 1000$, we use the command:

```
n = 0:1000;
```

Given t or n, we may then proceed to generate the signal of interest.

In MATLAB, a discrete-time signal is represented *exactly*, because the values of the signal are described as the elements of a vector. On the other hand, MATLAB provides only an *approximation* to a continuous-time signal. The approximation consists of a vector whose individual elements are samples of the underlying continuous-time signal. When using this approximate approach, it is important that we choose the sampling interval $\mathcal{T}$ sufficiently small so as to ensure that the samples capture all the details of the signal.

In this section, we consider the generation of both continuous-time and discrete-time signals of various kinds.

■ PERIODIC SIGNALS

It is an easy matter to generate periodic signals such as square waves and triangular waves using MATLAB. Consider first the generation of a square wave of amplitude A, fundamental frequency w0 (measured in radians per second), and duty cycle rho. That is, rho is the fraction of each period for which the signal is positive. To generate such a signal, we use the basic command:

```
A*square(w0*t + rho);
```

The square wave shown in Fig. 1.13(a) was thus generated using the following complete set of commands:

```
≫ A = 1;
≫ w0 = 10*pi;
≫ rho = 0.5;
≫ t = 0:.001:1;
≫ sq = A*square(w0*t + rho);
≫ plot(t, sq)
```

In the second command, pi is a built-in MATLAB function that returns the floating-point number closest to π. The last command is used to view the square wave. The command plot draws lines connecting the successive values of the signal and thus gives the appearance of a continuous-time signal.

Consider next the generation of a triangular wave of amplitude A, fundamental frequency w0 (measured in radians per second), and width W. Let the period of the triangular wave be T, with the first maximum value occurring at t = WT. The basic command for generating this second periodic signal is

```
A*sawtooth(w0*t + W);
```

Thus to generate the symmetric triangular wave shown in Fig. 1.14, we used the following commands:

```
≫ A = 1;
≫ w0 = 10*pi;
≫ W = 0.5;
≫ t = 0:0.001:1;
≫ tri = A*sawtooth(w0*t + W);
≫ plot(t, tri)
```

As mentioned previously, a signal generated on MATLAB is inherently of a discrete-time nature. To visualize a discrete-time signal, we may use the **stem** command. Specifically, **stem(n, x)** depicts the data contained in vector **x** as a discrete-time signal at the time values defined by **n**. The vectors **n** and **x** must, of course, have compatible dimensions.

Consider, for example, the discrete-time square wave shown in Fig. 1.15. This signal is generated using the following commands:

```
≫ A = 1;
≫ omega = pi/4;
≫ rho = 0.5;
≫ n = -10:10;
≫ x = A*square(omega*n + rho);
≫ stem(n, x)
```

▶ **Drill Problem 1.29** Use MATLAB to generate the triangular wave depicted in Fig. 1.14.   ◀

■ EXPONENTIAL SIGNALS

Moving on to exponential signals, we have decaying exponentials and growing exponentials. The MATLAB command for generating a decaying exponential is

```
B*exp(-a*t);
```

To generate a growing exponential, we use the command

```
B*exp(a*t);
```

In both cases, the exponential parameter **a** is positive. The following commands were used to generate the decaying exponential signal shown in Fig. 1.26(a):

```
≫ B = 5;
≫ a = 6;
≫ t = 0:.001:1;
≫ x = B*exp(-a*t);   % decaying exponential
≫ plot(t, x)
```

The growing exponential signal shown in Figure 1.26(b) was generated using the commands

```
≫ B = 1;
≫ a = 5;
≫ t = 0:0.001:1;
≫ x = B*exp(a*t);   % growing exponential
≫ plot(t, x)
```

Consider next the exponential sequence defined in Eq. (1.31). The growing form of this exponential is shown in Fig. 1.28(b). This figure was generated using the following commands:

```
≫ B = 1;
≫ r = 0.85
≫ n = -10:10;
≫ x = B*r.^n;   % decaying exponential
≫ stem(n, x)
```

Note that, in this example, the base `r` is a scalar but the exponent is a vector, hence the use of the symbol `.^` to denote *element-by-element powers*.

▶ **Drill Problem 1.30** Use MATLAB to generate the decaying exponential sequence depicted in Fig. 1.28(a). ◀

■ SINUSOIDAL SIGNALS

MATLAB also contains trigonometric functions that can be used to generate sinusoidal signals. A cosine signal of amplitude `A`, frequency `w0` (measured in radians per second), and phase angle `phi` (in radians) is obtained by using the command

```
A*cos(w0*t + phi);
```

Alternatively, we may use the sine function to generate a sinusoidal signal by using the command

```
A*sin(w0*t + phi);
```

These two commands were used as the basis of generating the sinusoidal signals shown in Fig. 1.29. Specifically, for the cosine signal shown in Fig. 1.29(a), we used the following commands:

```
≫ A = 4;
≫ w0 = 20*pi;
≫ phi = pi/6;
≫ t = 0:.001:1;
≫ cosine = A*cos(w0*t + phi);
≫ plot(t, cosine)
```

▶ **Drill Problem 1.31** Use MATLAB to generate the sine signal shown in Fig. 1.29(b). ◀

Consider next the discrete-time sinusoidal signal defined in Eq. (1.36). This periodic signal is plotted in Fig. 1.31. The figure was generated using the following commands:

```
≫ A = 1;
≫ omega = 2*pi/12;  % angular frequency
≫ phi = 0;
≫ n = -10:10;
≫ y = A*cos(omega*n);
≫ stem(n, y)
```

■ EXPONENTIALLY DAMPED SINUSOIDAL SIGNALS

In all of the signal-generation commands described above, we have generated the desired amplitude by multiplying a scalar, `A`, into a vector representing a unit-amplitude signal (e.g., `sin(w0*t + phi)`). This operation is described by using an asterisk. We next consider the generation of a signal that requires *element-by-element multiplication* of two vectors.

Suppose we multiply a sinusoidal signal by an exponential signal to produce an exponentially damped sinusoidal signal. With each signal component being represented by a vector, the generation of such a product signal requires the multiplication of one vector by another vector on an element-by-element basis. MATLAB represents element-by-element multiplication by using a dot followed by an asterisk. Thus the command for generating the exponentially damped sinusoidal signal

$$x(t) = A \sin(\omega_0 t + \phi) \exp(-at)$$

is as follows:

```
A*sin(w0*t + phi).*exp(-a*t);
```

For a decaying exponential, a is positive. This command was used in the generation of the waveform shown in Fig. 1.33. The complete set of commands is as follows:

```
≫ A = 60;
≫ w0 = 20*pi;
≫ phi = 0;
≫ a = 6;
≫ t = 0:.001:1;
≫ expsin = A*sin(w0*t + phi).*exp(-a*t);
≫ plot(t, expsin)
```

Consider next the exponentially damped sinusoidal sequence depicted in Fig. 1.51. This sequence is obtained by multiplying the sinusoidal sequence $x[n]$ of Fig. 1.31 by the decaying exponential sequence $y[n]$ of Fig. 1.28(a). Both of these sequences are defined for $n = -10:10$. Thus using $z[n]$ to denote this product sequence, we may use the following commands to generate and visualize it:

```
≫ z = x.*y;  % elementwise multiplication
≫ stem(n, z)
```

Note that there is no need to include the definition of n in the generation of z as it is already included in the commands for both x and y.

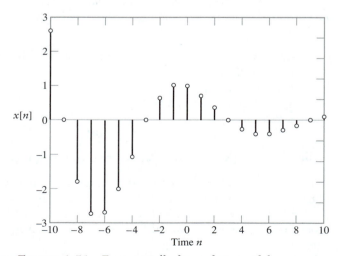

FIGURE 1.51   Exponentially damped sinusoidal sequence.

▶ **Drill Problem 1.32**   Use MATLAB to generate a signal defined as the product of the growing exponential of Fig. 1.28(b) and the sinusoidal signal of Fig. 1.31.   ◀

■ STEP, IMPULSE, AND RAMP FUNCTIONS

In MATLAB, `ones (M, N)` is an *M*-by-*N* matrix of ones, and `zeros (M, N)` is an *M*-by-*N* matrix of zeros. We may use these two matrices to generate two commonly used signals, as follows:

▶ *Step function.* A unit-amplitude step function is generated by writing

```
u = [zeros(1, 50), ones(1, 50)];
```

▶ *Discrete-time impulse.* A unit-amplitude discrete-time impulse is generated by writing

```
delta = [zeros(1, 49), 1, zeros(1, 49)];
```

To generate a ramp sequence, we simply write

```
ramp = n;
```

In Fig. 1.37, we illustrated how a pair of step functions shifted in time relative to each other may be used to produce a rectangular pulse. In light of the procedure illustrated therein, we may formulate the following set of commands for generating a rectangular pulse centered on the origin:

```
t = -1:1/500:1;
u1 = [zeros(1, 250), ones(1, 751)];
u2 = [zeros(1, 751), ones(1, 250)];
u = u1 - u2;
```

The first command defines time running from -1 second to 1 second in increments of 2 milliseconds. The second command generates a step function `u1` of unit amplitude, onset at time $t = -0.5$ second. The third command generates a second step function `u2`, onset at time $t = 0.5$ second. The fourth command subtracts `u2` from `u1` to produce a rectangular pulse of unit amplitude and unit duration centered on the origin.

■ USER-DEFINED FUNCTION

An important feature of the MATLAB environment is that it permits us to create our own *M-files* or subroutines. Two types of M-files exist, namely, scripts and functions. *Scripts*, or *script files*, automate long sequences of commands. On the other hand, *functions*, or *function files*, provide extensibility to MATLAB by allowing us to add new functions. Any variables used in function files do not remain in memory. For this reason, input and output variables must be declared explicitly.

We may thus say that a function M-file is a separate entity characterized as follows:

1. It begins with a statement defining the function name, its input arguments, and its output arguments.
2. It also includes additional statements that compute the values to be returned.
3. The inputs may be scalars, vectors, or matrices.

Consider, for example, the generation of the rectangular pulse depicted in Fig. 1.37 using an M-file. This pulse has unit amplitude and unit duration. To generate it, we create a file called `rect.m` containing the following statements:

```
function g = rect(x)
g = zeros(size(x));
set1 = find(abs(x)<= 0.5);
g(set1) = ones(size(set1));
```

In the last three statements of this M-file, we have introduced two useful functions:

▶ The function `size` returns a two-element vector containing the row and column dimensions of a matrix.
▶ The function `find` returns the indices of a vector or matrix that satisfy a prescribed relational condition. For the example at hand, `find(abs(x)<= T)` returns the indices of the vector `x` where the absolute value of `x` is less than or equal to `T`.

The new function `rect.m` can be used like any other MATLAB function. In particular, we may use it to generate a rectangular pulse, as follows:

```
t = -1:1/500:1;
plot(t, rect(0.5))
```

1.10 *Summary*

In this chapter we presented an overview of signals and systems, setting the stage for the rest of the book. A particular theme that stands out in the discussion presented herein is that signals may be of the continuous-time or discrete-time variety, and likewise for systems, as summarized here:

▶ A continuous-time signal is defined for all values of time. In contrast, a discrete-time signal is defined only for discrete instants of time.
▶ A continuous-time system is described by an operator that changes a continuous-time input signal into a continuous-time output signal. In contrast, a discrete-time system is described by an operator that changes a discrete-time input signal into a discrete-time output signal.

In practice, many systems mix continuous-time and discrete-time components. Analysis of *mixed* systems is an important part of the material presented in Chapters 4, 5, 8, and 9.

In discussing the various properties of signals and systems, we took special care in treating these two classes of signals and systems side by side. In so doing, much is gained by emphasizing the similarities and differences between continuous-time signals/systems and their discrete-time counterparts. This practice is followed in later chapters too, as appropriate.

Another noteworthy point is that, in the study of systems, particular attention is given to the analysis of *linear time-invariant systems*. Linearity means that the system obeys the principle of superposition. Time invariance means that the characteristics of the system do not change with time. By invoking these two properties, the analysis of systems becomes mathematically tractable. Indeed, a rich set of tools has been developed for the analysis of linear time-invariant systems, which provides direct motivation for much of the material on system analysis presented in this book.

In this chapter, we also explored the use of MATLAB for the generation of elementary waveforms, representing the continuous-time and discrete-time variety. MATLAB provides a powerful environment for exploring concepts and testing system designs, as will be illustrated in subsequent chapters.

FURTHER READING

1. For a readable account of signals, their representations, and use in communication systems, see the book:
   ▶ Pierce, J. R., and A. M. Noll, *Signals: The Science of Telecommunications* (Scientific American Library, 1990)

2. For examples of control systems, see Chapter 1 of the book:
   ▶ Kuo, B. C., *Automatic Control Systems*, Seventh Edition (Prentice-Hall, 1995)
   and Chapters 1 and 2 of the book:
   ▶ Phillips, C. L., and R. D. Harbor, *Feedback Control Systems*, Third Edition (Prentice-Hall, 1996)

3. For a general discussion of remote sensing, see the book:
   ▶ Hord, R. M., *Remote Sensing: Methods and Applications* (Wiley, 1986)
   For material on the use of spaceborne radar for remote sensing, see the book:
   ▶ Elachi, C., *Introduction to the Physics and Techniques of Remote Sensing* (Wiley, 1987)
   For detailed description of synthetic aperture radar and the role of signal processing in its implementation, see the book:
   ▶ Curlander, J. C., and R. N. McDonough, *Synthetic Aperture Radar: Systems and Signal Processing* (Wiley, 1991)

4. For a collection of essays on biological signal processing, see the book:
   ▶ Weitkunat, R., editor, *Digital Biosignal Processing* (Elsevier, 1991)

5. For detailed discussion of the auditory system, see the following:
   ▶ Dallos, P., A. N. Popper, and R. R. Fay, editors, *The Cochlea* (Springer-Verlag, 1996)
   ▶ Hawkins, H. L., and T. McMullen, editors, *Auditory Computation* (Springer-Verlag, 1996)
   ▶ Kelly, J. P., "Hearing." In E. R. Kandel, J. H. Schwartz, and T. M. Jessell, *Principles of Neural Science*, Third Edition (Elsevier, 1991)
   The cochlea has provided a source of motivation for building an electronic version of it, using silicon integrated circuits. Such an artificial implementation is sometimes referred to as a "silicon cochlea." For a discussion of the silicon cochlea, see:
   ▶ Lyon, R. F., and C. Mead, "Electronic Cochlea." In C. Mead, *Analog VLSI and Neural Systems* (Addison-Wesley, 1989)

6. For an account of the legendary story of the first Tacoma Narrows suspension bridge, see the report:
   ▶ Smith, D., "A Case Study and Analysis of the Tacoma Narrows Bridge Failure," 99.497 Engineering Project, Department of Mechanical Engineering, Carleton University, March 29, 1974 (supervised by Professor G. Kardos)

7. For a textbook treatment of MATLAB, see:
   ▶ Etter, D. M., *Engineering Problem Solving with MATLAB* (Prentice-Hall, 1993)

PROBLEMS

1.1 Find the even and odd components of each of the following signals:
(a) $x(t) = \cos(t) + \sin(t) + \sin(t)\cos(t)$
(b) $x(t) = 1 + t + 3t^2 + 5t^3 + 9t^4$
(c) $x(t) = 1 + t\cos(t) + t^2\sin(t) + t^3\sin(t)\cos(t)$
(d) $x(t) = (1 + t^3)\cos^3(10t)$

1.2 Determine whether the following signals are periodic. If they are periodic, find the fundamental period.
(a) $x(t) = (\cos(2\pi t))^2$
(b) $x(t) = \sum_{k=-5}^{5} w(t - 2k)$ for $w(t)$ depicted in Fig. P1.2b.
(c) $x(t) = \sum_{k=-\infty}^{\infty} w(t - 3k)$ for $w(t)$ depicted in Fig. P1.2b.
(d) $x[n] = (-1)^n$
(e) $x[n] = (-1)^{n^2}$
(f) $x[n]$ depicted in Fig. P1.2f.
(g) $x(t)$ depicted in Fig. P1.2g.
(h) $x[n] = \cos(2n)$
(i) $x[n] = \cos(2\pi n)$

1.3 The sinusoidal signal

$$x(t) = 3\cos(200t + \pi/6)$$

is passed through a square-law device defined by the input–output relation

$$y(t) = x^2(t)$$

Using the trigonometric identity

$$\cos^2\theta = \tfrac{1}{2}(\cos 2\theta + 1)$$

show that the output $y(t)$ consists of a dc component and a sinusoidal component.
(a) Specify the dc component.
(b) Specify the amplitude and fundamental frequency of the sinusoidal component in the output $y(t)$.

1.4 Categorize each of the following signals as an energy or power signal, and find the energy or power of the signal.

(a) $x(t) = \begin{cases} t, & 0 \le t \le 1 \\ 2 - t, & 1 \le t \le 2 \\ 0, & \text{otherwise} \end{cases}$

(b) $x[n] = \begin{cases} n, & 0 \le n \le 5 \\ 10 - n, & 5 \le n \le 10 \\ 0, & \text{otherwise} \end{cases}$

(c) $x(t) = 5\cos(\pi t) + \sin(5\pi t)$, $-\infty < t < \infty$

(d) $x(t) = \begin{cases} 5\cos(\pi t), & -1 \le t \le 1 \\ 0, & \text{otherwise} \end{cases}$

(e) $x(t) = \begin{cases} 5\cos(\pi t), & -0.5 \le t \le 0.5 \\ 0, & \text{otherwise} \end{cases}$

(f) $x[n] = \begin{cases} \sin(\pi/2\, n), & -4 \le n \le 4 \\ 0, & \text{otherwise} \end{cases}$

(g) $x[n] = \begin{cases} \cos(\pi n), & -4 \le n \le 4 \\ 0, & \text{otherwise} \end{cases}$

(h) $x[n] = \begin{cases} \cos(\pi n), & n \ge 0 \\ 0, & \text{otherwise} \end{cases}$

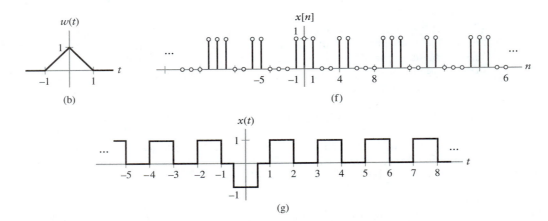

FIGURE P1.2

1.5 Consider the sinusoidal signal

$$x(t) = A \cos(\omega t + \phi)$$

Determine the average power of $x(t)$.

1.6 The angular frequency Ω of the sinusoidal signal

$$x[n] = A \cos(\Omega n + \phi)$$

satisfies the condition for $x[n]$ to be periodic. Determine the average power of $x[n]$.

1.7 The raised-cosine pulse $x(t)$ shown in Fig. P1.7 is defined as

$$x(t) = \begin{cases} \frac{1}{2}[\cos(\omega t) + 1], & -\pi/\omega \le t \le \pi/\omega \\ 0, & \text{otherwise} \end{cases}$$

Determine the total energy of $x(t)$.

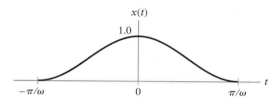

FIGURE P1.7

1.8 The trapezoidal pulse $x(t)$ shown in Fig. P1.8 is defined by

$$x(t) = \begin{cases} 5 - t, & 4 \le t \le 5 \\ 1, & -4 \le t \le 4 \\ t + 5 & -5 \le t \le -4 \\ 0, & \text{otherwise} \end{cases}$$

Determine the total energy of $x(t)$.

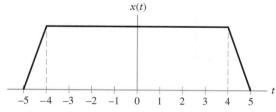

FIGURE P1.8

1.9 The trapezoidal pulse $x(t)$ of Fig. P1.8 is applied to a differentiator, defined by

$$y(t) = \frac{d}{dt} x(t)$$

(a) Determine the resulting output $y(t)$ of the differentiator.

(b) Determine the total energy of $y(t)$.

1.10 A rectangular pulse $x(t)$ is defined by

$$x(t) = \begin{cases} A, & 0 \le t \le T \\ 0, & \text{otherwise} \end{cases}$$

The pulse $x(t)$ is applied to an integrator defined by

$$y(t) = \int_0^t x(\tau) \, d\tau$$

Find the total energy of the output $y(t)$.

1.11 The trapezoidal pulse $x(t)$ of Fig. P1.8 is time scaled, producing

$$y(t) = x(at)$$

Sketch $y(t)$ for (a) $a = 5$ and (b) $a = 0.2$.

1.12 A triangular pulse signal $x(t)$ is depicted in Fig. P1.12. Sketch each of the following signals derived from $x(t)$:

(a) $x(3t)$

(b) $x(3t + 2)$

(c) $x(-2t - 1)$

(d) $x(2(t + 2))$

(e) $x(2(t - 2))$

(f) $x(3t) + x(3t + 2)$

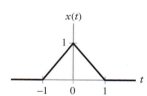

FIGURE P1.12

1.13 Sketch the trapezoidal pulse $y(t)$ that is related to that of Fig. P1.8 as follows:

$$y(t) = x(10t - 5)$$

1.14 Let $x(t)$ and $y(t)$ be given in Figs. P1.14(a) and (b), respectively. Carefully sketch the following signals:

(a) $x(t)y(t - 1)$

(b) $x(t - 1)y(-t)$

(c) $x(t + 1)y(t - 2)$

(d) $x(t)y(-1 - t)$

(e) $x(t)y(2 - t)$

(f) $x(2t)y(\frac{1}{2}t + 1)$

(g) $x(4 - t)y(t)$

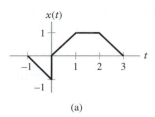

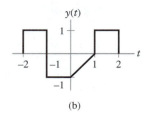

(a)   (b)

FIGURE P1.14

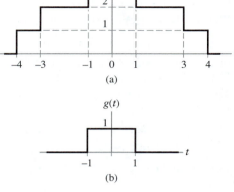

(a)

(b)

FIGURE P1.17

1.15 Figure P1.15(a) shows a staircase-like signal $x(t)$ that may be viewed as the superposition of four rectangular pulses. Starting with the rectangular pulse $g(t)$ shown in Fig. P1.15(b), construct this waveform, and express $x(t)$ in terms of $g(t)$.

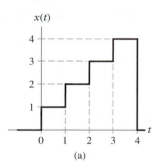

(a)   (b)

FIGURE P1.15

(e) $x[n - 2] + y[n + 2]$
(f) $x[2n] + y[n - 4]$
(g) $x[n + 2]y[n - 2]$
(h) $x[3 - n]y[n]$
(i) $x[-n]y[-n]$
(j) $x[n]y[-2 - n]$
(k) $x[n + 2]y[6 - n]$

1.16 Sketch the waveforms of the following signals:
  (a) $x(t) = u(t) - u(t - 2)$
  (b) $x(t) = u(t + 1) - 2u(t) + u(t - 1)$
  (c) $x(t) = -u(t + 3) + 2u(t + 1) - 2u(t - 1) + u(t - 3)$
  (d) $y(t) = r(t + 1) - r(t) + r(t - 2)$
  (e) $y(t) = r(t + 2) - r(t + 1) - r(t - 1) + r(t - 2)$

1.17 Figure P1.17(a) shows a pulse $x(t)$ that may be viewed as the superposition of three rectangular pulses. Starting with the rectangular pulse $g(t)$ of Fig. P1.17(b), construct this waveform, and express $x(t)$ in terms of $g(t)$.

1.18 Let $x[n]$ and $y[n]$ be given in Figs. P1.18(a) and (b), respectively. Carefully sketch the following signals:
  (a) $x[2n]$
  (b) $x[3n - 1]$
  (c) $y[1 - n]$
  (d) $y[2 - 2n]$

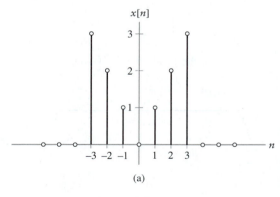

(a)

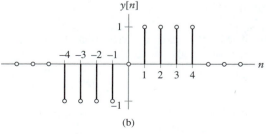

(b)

FIGURE P1.18

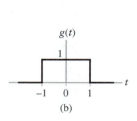

1.19 Consider the sinusoidal signal

$$x[n] = 10 \cos\left(\frac{4\pi}{31} n + \frac{\pi}{5}\right)$$

Determine the fundamental period of $x(n)$.

1.20 The sinusoidal signal $x[n]$ has fundamental period $N = 10$ samples. Determine the smallest angular frequency Ω for which $x[n]$ is periodic.

1.21 Determine whether the following signals are periodic. If they are periodic, find the fundamental period.

(a) $x[n] = \cos(\frac{8}{15}\pi n)$
(b) $x[n] = \cos(\frac{7}{15}\pi n)$
(c) $x(t) = \cos(2t) + \sin(3t)$
(d) $x(t) = \Sigma_{k=-\infty}^{\infty}(-1)^k \delta(t - 2k)$
(e) $x[n] = \Sigma_{k=-\infty}^{\infty}\{\delta[n - 3k] + \delta[n - k^2]\}$
(f) $x(t) = \cos(t)u(t)$
(g) $x(t) = v(t) + v(-t)$, where $v(t) = \cos(t)u(t)$
(h) $x(t) = v(t) + v(-t)$, where $v(t) = \sin(t)u(t)$
(i) $x[n] = \cos(\frac{1}{5}\pi n) \sin(\frac{1}{3}\pi n)$

1.22 A complex sinusoidal signal $x(t)$ has the following components:

$$\text{Re}\{x(t)\} = x_R(t) = A \cos(\omega t + \phi)$$
$$\text{Im}\{x(t)\} = x_I(t) = A \sin(\omega t + \phi)$$

The amplitude of $x(t)$ is defined by the square root of $x_R^2(t) + x_I^2(t)$. Show that this amplitude equals A, independent of the phase angle ϕ.

1.23 Consider the complex-valued exponential signal

$$x(t) = Ae^{\alpha t + j\omega t}, \qquad \alpha > 0$$

Evaluate the real and imaginary components of $x(t)$.

1.24 Consider the continuous-time signal

$$x(t) = \begin{cases} t/T + 0.5, & -T/2 \le t \le T/2 \\ 1, & t \ge T/2 \\ 0, & t < -T/2 \end{cases}$$

which is applied to a differentiator. Show that the output of the differentiator approaches the unit impulse $\delta(t)$ as T approaches zero.

1.25 In this problem, we explore what happens when a unit impulse is applied to a differentiator. Consider a triangular pulse $x(t)$ of duration T and amplitude $1/2T$, as depicted in Fig. P1.25. The area under the pulse is unity. Hence as the duration T approaches zero, the triangular pulse approaches a unit impulse.

(a) Suppose the triangular pulse $x(t)$ is applied to a differentiator. Determine the output $y(t)$ of the differentiator.

(b) What happens to the differentiator output $y(t)$ as T approaches zero? Use the definition of a unit impulse $\delta(t)$ to express your answer.

(c) What is the total area under the differentiator output $y(t)$ for all T? Justify your answer.

Based on your findings in parts (a) to (c), describe in succinct terms the result of differentiating a unit impulse.

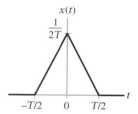

FIGURE P1.25

1.26 The derivative of impulse function $\delta(t)$ is referred to as a *doublet*. It is denoted by $\delta'(t)$. Show that $\delta'(t)$ satisfies the sifting property

$$\int_{-\infty}^{\infty} \delta'(t - t_0) f(t) \, dt = f'(t_0)$$

where

$$f'(t_0) = \frac{d}{dt} f(t) \bigg|_{t=t_0}$$

Assume that the function $f(t)$ has a continuous derivative at time $t = t_0$.

1.27 A system consists of several subsystems connected as shown in Fig. P1.27. Find the operator H relating $x(t)$ to $y(t)$ for the subsystem operators given by:
$H_1: y_1(t) = x_1(t)x_1(t - 1)$
$H_2: y_2(t) = |x_2(t)|$
$H_3: y_3(t) = 1 + 2x_3(t)$
$H_4: y_4(t) = \cos(x_4(t))$

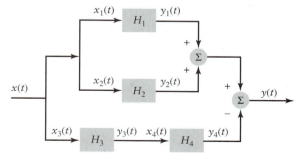

FIGURE P1.27

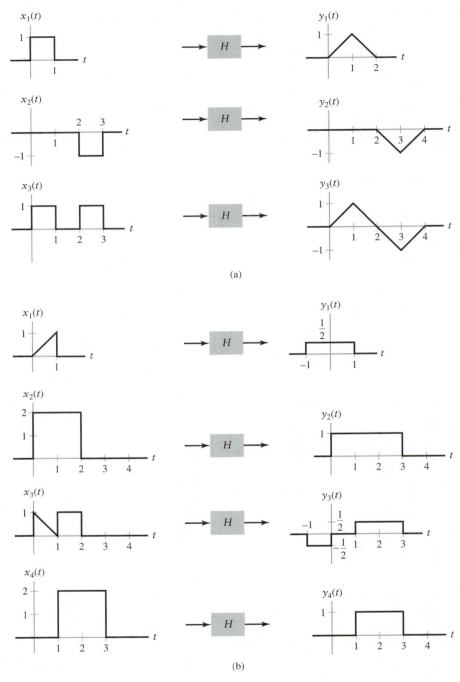

(a)

(b)

FIGURE P1.39

1.28 The systems given below have input $x(t)$ or $x[n]$ and output $y(t)$ or $y[n]$, respectively. Determine whether each of them is (i) memoryless, (ii) stable, (iii) causal, (iv) linear, and (v) time invariant.

(a) $y(t) = \cos(x(t))$

(b) $y[n] = 2x[n]u[n]$

(c) $y[n] = \log_{10}(|x[n]|)$

(d) $y(t) = \int_{-\infty}^{t/2} x(\tau)\, d\tau$

(e) $y[n] = \sum_{k=-\infty}^{n} x[k+2]$

(f) $y(t) = \dfrac{d}{dt} x(t)$

(g) $y[n] = \cos(2\pi x[n+1]) + x[n]$

(h) $y(t) = \dfrac{d}{dt}\{e^{-t}x(t)\}$

(i) $y(t) = x(2-t)$

(j) $y[n] = x[n] \sum_{k=-\infty}^{\infty} \delta[n-2k]$

(k) $y(t) = x(t/2)$

(l) $y[n] = 2x[2^n]$

1.29 The output of a discrete-time system is related to its input $x[n]$ as follows:

$$y[n] = a_0 x[n] + a_1 x[n-1] + a_2 x[n-2] + a_3 x[n-3]$$

Let the operator S^k denote a system that shifts the input $x[n]$ by k time units to produce $x[n-k]$. Formulate the operator H for the system relating $y[n]$ to $x[n]$. Hence develop a block diagram representation for H, using (a) cascade implementation and (b) parallel implementation.

1.30 Show that the system described in Problem 1.29 is BIBO stable for all a_0, a_1, a_2, and a_3.

1.31 How far does the memory of the discrete-time system described in Problem 1.29 extend into the past?

1.32 Is it possible for a noncausal system to possess memory? Justify your answer.

1.33 The output signal $y[n]$ of a discrete-time system is related to its input signal $x[n]$ as follows:

$$y[n] = x[n] + x[n-1] + x[n-2]$$

Let the operator S denote a system that shifts its input by one time unit.

(a) Formulate the operator H for the system relating $y[n]$ to $x[n]$.

(b) The operator H^{-1} denotes a discrete-time system that is the inverse of this system. How is H^{-1} defined?

1.34 Show that the discrete-time system described in Problem 1.29 is time invariant, independent of the coefficients a_0, a_1, a_2, and a_3.

1.35 Is it possible for a time-variant system to be linear? Justify your answer.

1.36 Show that an Nth power-law device defined by the input–output relation

$$y(t) = x^N(t), \quad N \text{ integer and } N \neq 0, 1$$

is nonlinear.

1.37 A linear time-invariant system may be causal or noncausal. Give an example for each one of these two possibilities.

1.38 Figure 1.50 shows two equivalent system configurations on condition that the system operator H is linear. Which of these two configurations is simpler to implement? Justify your answer.

1.39 A system H has its input–output pairs given. Determine whether the system could be memoryless, causal, linear, and time invariant for (a) signals depicted in Fig. P1.39(a) and (b) signals depicted in Fig. P1.39(b). For all cases, justify your answers.

1.40 A linear system H has the input–output pairs depicted in Fig. P1.40(a). Determine the following and explain your answers:

(a) Is this system causal?

(b) Is this system time invariant?

(c) Is this system memoryless?

(d) Find the output for the input depicted in Fig. P1.40(b).

1.41 A discrete-time system is both linear and time invariant. Suppose the output due to an input $x[n] = \delta[n]$ is given in Fig. P1.41(a).

(a) Find the output due to an input $x[n] = \delta[n-1]$.

(b) Find the output due to an input $x[n] = 2\delta[n] - \delta[n-2]$.

(c) Find the output due to the input depicted in Fig. P1.41(b).

▶ *Computer Experiments*

1.42 Write a set of MATLAB commands for approximating the following continuous-time periodic waveforms:

(a) Square wave of amplitude 5 volts, fundamental frequency 20 Hz, and duty cycle 0.6.

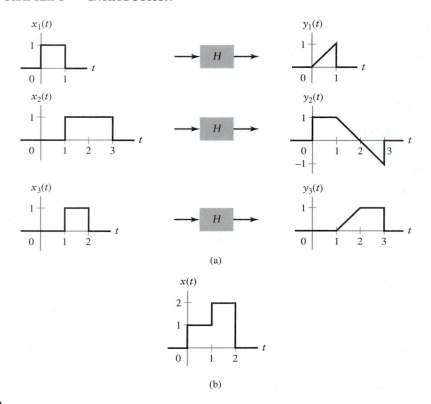

(a)

(b)

FIGURE P1.40

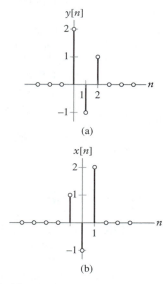

(a)

(b)

FIGURE P1.41

(b) Sawtooth wave of amplitude 5 volts, and fundamental frequency 20 Hz.

Hence plot five cycles of each of these two waveforms.

1.43 (a) The solution to a linear differential equation is given by

$$x(t) = 10e^{-t} - 5e^{-0.5t}$$

Using MATLAB, plot $x(t)$ versus t for t = 0:0.01:5.

(b) Repeat the problem for

$$x(t) = 10e^{-t} + 5e^{-0.5t}$$

1.44 An exponentially damped sinusoidal signal is defined by

$$x(t) = 20 \sin(2\pi \times 1000t - \pi/3) \exp(-at)$$

where the exponential parameter a is variable; it takes on the following set of values: $a = 500$,

750, 1000. Using MATLAB, investigate the effect of varying a on the signal $x(t)$ for $-2 \leq t \leq 2$ milliseconds.

1.45 A raised-cosine sequence is defined by

$$w[n] = \begin{cases} \cos(2\pi Fn), & -1/2F \leq n \leq 1/2F \\ 0, & \text{otherwise} \end{cases}$$

Use MATLAB to plot $w[n]$ versus n for $F = 0.1$.

1.46 A rectangular pulse $x(t)$ is defined by

$$x(t) = \begin{cases} 10, & 0 \leq t \leq 5 \\ 0, & \text{otherwise} \end{cases}$$

Generate $x(t)$ using:

(a) A pair of time-shifted step functions.

(b) An M-file.

2 | Time-Domain Representations for Linear Time-Invariant Systems

![decorative wave pattern]

2.1 Introduction

In this chapter we consider several methods for describing the relationship between the input and output of linear time-invariant (LTI) systems. The focus here is on system descriptions that relate the output signal to the input signal when both signals are represented as functions of time, hence the terminology "time domain" in the chapter title. Methods for relating system output and input in domains other than time are presented in later chapters. The descriptions developed in this chapter are useful for analyzing and predicting the behavior of LTI systems and for implementing discrete-time systems on a computer.

We begin by characterizing a LTI system in terms of its impulse response. The impulse response is the system output associated with an impulse input. Given the impulse response, we determine the output due to an arbitrary input by expressing the input as a weighted superposition of time-shifted impulses. By linearity and time invariance, the output must be a weighted superposition of time-shifted impulse responses. The term "convolution" is used to describe the procedure for determining the output from the input and the impulse response.

The second method considered for characterizing the input–output behavior of LTI systems is the linear constant-coefficient differential or difference equation. Differential equations are used to represent continuous-time systems, while difference equations represent discrete-time systems. We focus on characterizing differential and difference equation solutions with the goal of developing insight into system behavior.

The third system representation we discuss is the block diagram. A block diagram represents the system as an interconnection of three elementary operations: scalar multiplication, addition, and either a time shift for discrete-time systems or integration for continuous-time systems.

The final time-domain representation discussed in this chapter is the state-variable description. The state-variable description is a series of coupled first-order differential or difference equations that represent the behavior of the system's "state" and an equation that relates the state to the output. The state is a set of variables associated with energy storage or memory devices in the system.

All four of these time-domain system representations are equivalent in the sense that identical outputs result from a given input. However, each relates the input and output in a different manner. Different representations offer different views of the system, with each offering different insights into system behavior. Each representation has advantages and disadvantages for analyzing and implementing systems. Understanding how different representations are related and determining which offers the most insight and straightforward solution in a particular problem are important skills to develop.

2.2 Convolution: Impulse Response Representation for LTI Systems

The *impulse response* is the output of a LTI system due to an impulse input applied at time $t = 0$ or $n = 0$. The impulse response completely characterizes the behavior of any LTI system. This may seem surprising, but it is a basic property of all LTI systems. The impulse response is often determined from knowledge of the system configuration and dynamics or, in the case of an unknown system, can be measured by applying an approximate impulse to the system input. Generation of a discrete-time impulse sequence for testing an unknown system is straightforward. In the continuous-time case, a true impulse of zero width and infinite amplitude cannot actually be generated and usually is physically approximated as a pulse of large amplitude and narrow width. Thus the impulse response may be interpreted as the system behavior in response to a high-amplitude, extremely short-duration input.

If the input to a linear system is expressed as a weighted superposition of time-shifted impulses, then the output is a weighted superposition of the system response to each time-shifted impulse. If the system is also time invariant, then the system response to a time-shifted impulse is a time-shifted version of the system response to an impulse. Hence the output of a LTI system is given by a weighted superposition of time-shifted impulse responses. This weighted superposition is termed the *convolution sum* for discrete-time systems and the *convolution integral* for continuous-time systems.

We begin by considering the discrete-time case. First an arbitrary signal is expressed as a weighted superposition of time-shifted impulses. The convolution sum is then obtained by applying a signal represented in this manner to a LTI system. A similar procedure is used to obtain the convolution integral for continuous-time systems later in this section.

■ THE CONVOLUTION SUM

Consider the product of a signal $x[n]$ and the impulse sequence $\delta[n]$, written as

$$x[n]\delta[n] = x[0]\delta[n]$$

Generalize this relationship to the product of $x[n]$ and a time-shifted impulse sequence to obtain

$$x[n]\delta[n - k] = x[k]\delta[n - k]$$

In this expression n represents the time index; hence $x[n]$ denotes a signal, while $x[k]$ represents the value of the signal $x[n]$ at time k. We see that multiplication of a signal by a time-shifted impulse results in a time-shifted impulse with amplitude given by the value of the signal at the time the impulse occurs. This property allows us to express $x[n]$ as the following weighted sum of time-shifted impulses:

$$x[n] = \cdots + x[-2]\delta[n + 2] + x[-1]\delta[n + 1] + x[0]\delta[n]$$
$$+ x[1]\delta[n - 1] + x[2]\delta[n - 2] + \cdots$$

We may rewrite this representation for $x[n]$ in concise form as

$$x[n] = \sum_{k=-\infty}^{\infty} x[k]\delta[n - k] \tag{2.1}$$

A graphical illustration of Eq. (2.1) is given in Fig. 2.1.

Let the operator H denote the system to which the input $x[n]$ is applied. Then using Eq. (2.1) to represent the input $x[n]$ to the system results in the output

$$y[n] = H\left\{ \sum_{k=-\infty}^{\infty} x[k]\delta[n - k] \right\}$$

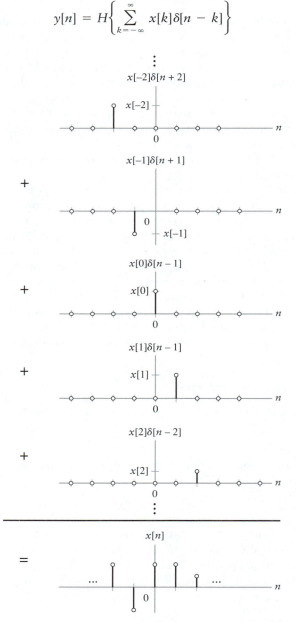

FIGURE 2.1   Graphical example illustrating the representation of a signal $x[n]$ as a weighted sum of time-shifted impulses.

Now use the linearity property to interchange the system operator H with the summation and signal values $x[k]$ to obtain

$$y[n] = \sum_{k=-\infty}^{\infty} x[k]H\{\delta[n - k]\}$$

$$= \sum_{k=-\infty}^{\infty} x[k]h_k[n] \tag{2.2}$$

where $h_k[n] = H\{\delta[n - k]\}$ is the response of the system to a time-shifted impulse. If we further assume the system is time invariant, then a time shift in the input results in a time shift in the output. This implies that the output due to a time-shifted impulse is a time-shifted version of the output due to an impulse; that is, $h_k[n] = h_0[n - k]$. Letting $h[n] = h_0[n]$ be the impulse response of the LTI system H, Eq. (2.2) is rewritten as

$$y[n] = \sum_{k=-\infty}^{\infty} x[k]h[n - k] \tag{2.3}$$

Thus the output of a LTI system is given by a weighted sum of time-shifted impulse responses. This is a direct consequence of expressing the input as a weighted sum of time-shifted impulses. The sum in Eq. (2.3) is termed the *convolution sum* and is denoted by the symbol $*$; that is,

$$x[n] * h[n] = \sum_{k=-\infty}^{\infty} x[k]h[n - k]$$

The convolution process is illustrated in Fig. 2.2. Figure 2.2(a) depicts the impulse response of an arbitrary LTI system. In Fig. 2.2(b) the input is represented as a sum of weighted and time-shifted impulses, $p_k[n] = x[k]\delta[n - k]$. The output of the system associated with each input $p_k[n]$ is

$$v_k[n] = x[k]h[n - k]$$

Here $v_k[n]$ is obtained by time-shifting the impulse response k units and multiplying by $x[k]$. The output $y[n]$ in response to the input $x[n]$ is obtained by summing all the sequences $v_k[n]$:

$$y[n] = \sum_{k=-\infty}^{\infty} v_k[n]$$

That is, for each value of n, we sum the values along the k axis indicated on the right side of Fig. 2.2(b). The following example illustrates this process.

EXAMPLE 2.1   Assume a LTI system H has impulse response

$$h[n] = \begin{cases} 1, & n = \pm 1 \\ 2, & n = 0 \\ 0, & \text{otherwise} \end{cases}$$

Determine the output of this system in response to the input

$$x[n] = \begin{cases} 2, & n = 0 \\ 3, & n = 1 \\ -2, & n = 2 \\ 0, & \text{otherwise} \end{cases}$$

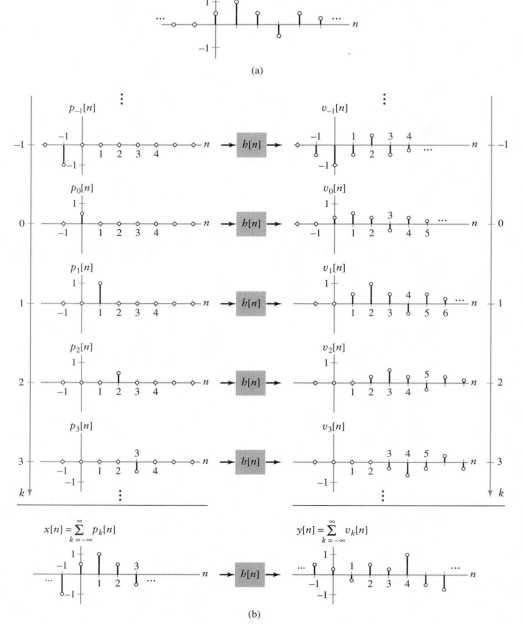

FIGURE 2.2   Illustration of the convolution sum. (a) Impulse response of a system. (b) Decomposition of the input $x[n]$ into a weighted sum of time-shifted impulses results in an output $y[n]$ given by a weighted sum of time-shifted impulse responses. Here $p_k[n]$ is the weighted (by $x[k]$) and time-shifted (by k) impulse input, and $v_k[n]$ is the weighted and time-shifted impulse response output. The dependence of both $p_k[n]$ and $v_k[n]$ on k is depicted by the k axis shown on the left- and right-hand sides of the figure. The output is obtained by summing $v_k[n]$ over all values of k.

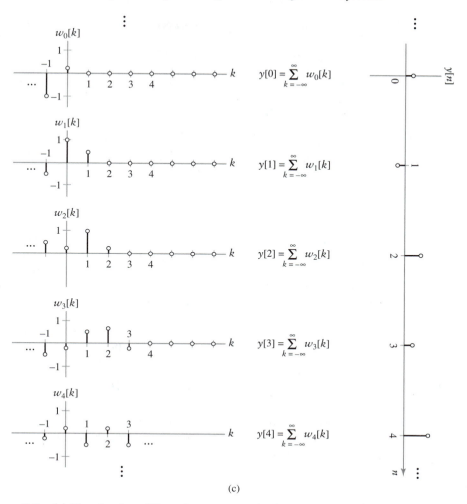

(c)

FIGURE 2.2   (c) The signals $w_n[k]$ used to compute the output at time n for several values of n. Here we have redrawn the right-hand side of Fig. 2.2(b) so that the k axis is horizontal. The output is obtained for $n = n_0$ by summing $w_{n0}[k]$ over all values of k.

Solution: First write $x[n]$ as the weighted sum of time-shifted impulses

$$x[n] = 2\delta[n] + 3\delta[n - 1] - 2\delta[n - 2]$$

Here $p_0[n] = 2\delta[n]$, $p_1[n] = 3\delta[n - 1]$, and $p_2[n] = -2\delta[n - 2]$. All other time-shifted $p_k[n]$ are zero because the input is zero for $n < 0$ and $n > 2$. Since a weighted, time-shifted, impulse input, $a\delta[n - k]$, results in a weighted, time-shifted, impulse response output, $ah[n - k]$, the system output may be written as

$$y[n] = 2h[n] + 3h[n - 1] - 2h[n - 2]$$

Here $v_0[n] = 2h[n]$, $v_1[n] = 3h[n-1]$, $v_2[n] = -2h[n-2]$, and all other $v_k[n] = 0$. Summation of the weighted and time-shifted impulse responses over k gives

$$y[n] = \begin{cases} 0, & n \leq -2 \\ 2, & n = -1 \\ 7, & n = 0 \\ 6, & n = 1 \\ -1, & n = 2 \\ -2, & n = 3 \\ 0, & n \geq 4 \end{cases}$$

In Example 2.1, we found all the $v_k[n]$ and then summed over k to determine $y[n]$. This approach illustrates the principles that underlie convolution and is very effective when the input is of short duration so that only a small number of signals $v_k[n]$ need to be determined. When the input has a long duration, then a very large, possibly infinite, number of signals $v_k[n]$ must be evaluated before $y[n]$ can be found and this procedure can be cumbersome.

An alternative approach for evaluating the convolution sum is obtained by a slight change in perspective. Consider evaluating the output at a fixed time n_o

$$y[n_o] = \sum_{k=-\infty}^{\infty} v_k[n_o]$$

That is, we sum along the k or vertical axis on the right-hand side of Fig. 2.2(b) at a fixed time $n = n_o$. Suppose we define a signal representing the values at $n = n_o$ as a function of the independent variable k, $w_{n_o}[k] = v_k[n_o]$. The output is now obtained by summing over the independent variable k:

$$y[n_o] = \sum_{k=-\infty}^{\infty} w_{n_o}[k]$$

Note that here we need only determine one signal, $w_{n_o}[k]$, to evaluate the output at $n = n_o$. Figure 2.2(c) depicts $w_{n_o}[k]$ for several different values of n_o and the corresponding output. Here the horizontal axis corresponds to k and the vertical axis corresponds to n. We may view $v_k[n]$ as representing the kth row on the right-hand side of Fig. 2.2(b), while $w_n[k]$ represents the nth column. In Fig. 2.2(c), $w_n[k]$ is the nth row, while $v_k[n]$ is the kth column.

We have defined the intermediate sequence $w_n[k] = x[k]h[n-k]$ as the product of $x[k]$ and $h[n-k]$. Here k is the independent variable and n is treated as a constant. Hence $h[n-k] = h[-(k-n)]$ is a reflected and time-shifted (by $-n$) version of $h[k]$. The time shift n determines the time at which we evaluate the output of the system, since

$$y[n] = \sum_{k=-\infty}^{\infty} w_n[k] \tag{2.4}$$

Note that now we need only determine one signal, $w_n[k]$, for each time at which we desire to evaluate the output.

EXAMPLE 2.2   A LTI system has the impulse response

$$h[n] = (\tfrac{3}{4})^n u[n]$$

Use Eq. (2.4) to determine the output of the system at times $n = -5$, $n = 5$, and $n = 10$ when the input is $x[n] = u[n]$.

Solution: Here the impulse response and input are of infinite duration so the procedure followed in Example 2.1 would require determining an infinite number of signals $v_k[n]$. By using Eq. (2.4) we only form one signal, $w_n[k]$, for each n of interest. Figure 2.3(a) depicts $x[k]$, while Fig. 2.3(b) depicts the reflected and time-shifted impulse response $h[n - k]$. We see that

$$h[n - k] = \begin{cases} (\tfrac{3}{4})^{n-k}, & k \leq n \\ 0, & \text{otherwise} \end{cases}$$

Figures 2.3(c), (d), and (e) depict the product $w_n[k]$ for $n = -5$, $n = 5$, and $n = 10$, respectively. We have

$$w_{-5}[k] = 0$$

and thus Eq. (2.4) gives $y[-5] = 0$. For $n = 5$, we have

$$w_5[k] = \begin{cases} (\tfrac{3}{4})^{5-k}, & 0 \leq k \leq 5 \\ 0, & \text{otherwise} \end{cases}$$

and so Eq. (2.4) gives

$$y[5] = \sum_{k=0}^{5} \left(\frac{3}{4}\right)^{5-k}$$

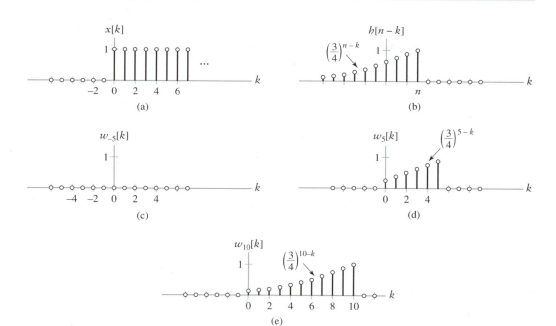

FIGURE 2.3   Evaluation of Eq. (2.4) in Example 2.2. (a) The input signal $x[k]$ depicted as a function of k. (b) The reflected and time-shifted impulse response, $h[n - k]$, as a function of k. (c) The product signal $w_{-5}[k]$ used to evaluate $y[-5]$. (d) The product signal $w_5[k]$ used to evaluate $y[5]$. (e) The product signal $w_{10}[k]$ used to evaluate $y[10]$.

Factor $(\frac{3}{4})^5$ from the sum and apply the formula for the sum of a finite geometric series to obtain

$$y[5] = \left(\frac{3}{4}\right)^5 \sum_{k=0}^{5} \left(\frac{4}{3}\right)^k$$

$$= \left(\frac{3}{4}\right)^5 \frac{1 - (\frac{4}{3})^6}{1 - (\frac{4}{3})}$$

Lastly, for $n = 10$ we see that

$$w_{10}[k] = \begin{cases} (\frac{3}{4})^{10-k}, & 0 \le k \le 10 \\ 0, & \text{otherwise} \end{cases}$$

and Eq. (2.4) gives

$$y[10] = \sum_{k=0}^{10} \left(\frac{3}{4}\right)^{10-k}$$

$$= \left(\frac{3}{4}\right)^{10} \sum_{k=0}^{10} \left(\frac{4}{3}\right)^k$$

$$= \left(\frac{3}{4}\right)^{10} \frac{1 - (\frac{4}{3})^{11}}{1 - (\frac{4}{3})}$$

Note that in this example $w_n[k]$ has only two different functional forms. For $n < 0$, we have $w_n[k] = 0$ since there is no overlap between the nonzero portions of $x[k]$ and $h[n - k]$. When $n \ge 0$ the nonzero portions of $x[k]$ and $h[n - k]$ overlap on the interval $0 \le k \le n$ and we may write

$$w_n[k] = \begin{cases} (\frac{3}{4})^{n-k}, & 0 \le k \le n \\ 0, & \text{otherwise} \end{cases}$$

Hence we may determine the output for an arbitrary n by using the appropriate functional form for $w_n[k]$ in Eq. (2.4).

This example suggests that in general we may determine $y[n]$ for all n without evaluating Eq. (2.4) at an infinite number of distinct time shifts n. This is accomplished by identifying intervals of n on which $w_n[k]$ has the same functional form. We then only need to evaluate Eq. (2.4) using the $w_n[k]$ associated with each interval. Often it is very helpful to graph both $x[k]$ and $h[n - k]$ when determining $w_n[k]$ and identifying the appropriate intervals of time shifts. This procedure is now summarized:

1. Graph both $x[k]$ and $h[n - k]$ as a function of the independent variable k. To determine $h[n - k]$, first reflect $h[k]$ about $k = 0$ to obtain $h[-k]$ and then time shift $h[-k]$ by $-n$.

2. Begin with the time shift n large and negative.

3. Write the functional form for $w_n[k]$.

4. Increase the time shift n until the functional form for $w_n[k]$ changes. The value of n at which the change occurs defines the end of the current interval and the beginning of a new interval.

5. Let n be in the new interval. Repeat steps 3 and 4 until all intervals of time shifts n and the corresponding functional forms for $w_n[k]$ are identified. This usually implies increasing n to a very large positive number.

6. For each interval of time shifts n, sum all the values of the corresponding $w_n[k]$ to obtain $y[n]$ on that interval.

The effect of varying n from $-\infty$ to ∞ is to slide $h[-k]$ past $x[k]$ from left to right. Transitions in the intervals of n identified in step 4 generally occur when a change point in the representation for $h[-k]$ slides through a change point in the representation for $x[k]$. Alternatively, we can sum all the values in $w_n[k]$ as each interval of time shifts is identified, that is, after step 4, rather than waiting until all intervals are identified. The following examples illustrate this procedure for evaluating the convolution sum.

EXAMPLE 2.3   A LTI system has impulse response given by

$$h[n] = u[n] - u[n - 10]$$

and depicted in Fig. 2.4(a). Determine the output of this system when the input is the rectangular pulse defined as

$$x[n] = u[n - 2] - u[n - 7]$$

and shown in Fig. 2.4(b).

Solution: First we graph $x[k]$ and $h[n - k]$, treating n as a constant and k as the independent variable as depicted in Figs. 2.4(c) and (d). Now identify intervals of time shifts n on which the product signal $w_n[k]$ has the same functional form. Begin with n large and negative, in which case $w_n[k] = 0$ because there is no overlap in the nonzero portions of $x[k]$ and $h[n - k]$. By increasing n, we see that $w_n[k] = 0$ provided $n < 2$. Hence the first interval of time shifts is $n < 2$.

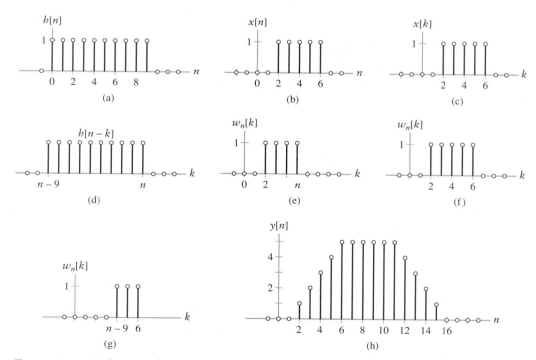

FIGURE 2.4   Evaluation of the convolution sum for Example 2.3. (a) The system impulse response $h[n]$. (b) The input signal $x[n]$. (c) The input depicted as a function of k. (d) The reflected and time-shifted impulse response $h[n - k]$ depicted as a function of k. (e) The product signal $w_n[k]$ for the interval of time shifts $2 \leq n \leq 6$. (f) The product signal $w_n[k]$ for the interval of time shifts $6 < n \leq 11$. (g) The product signal $w_n[k]$ for the interval of time shifts $12 < n \leq 15$. (h) The output $y[n]$.

When $n = 2$ the right edge of $h[n - k]$ slides past the left edge of $x[k]$ and a transition occurs in the functional form for $w_n[k]$. For $n \geq 2$,

$$w_n[k] = \begin{cases} 1, & 2 \leq k \leq n \\ 0, & \text{otherwise} \end{cases}$$

This functional form is correct until $n > 6$ and is depicted in Fig. 2.4(e). When $n > 6$ the right edge of $h[n - k]$ slides past the right edge of $x[k]$ so the form of $w_n[k]$ changes. Hence our second interval of time shifts is $2 \leq n \leq 6$.

For $n > 6$, the functional form of $w_n[k]$ is given by

$$w_n[k] = \begin{cases} 1, & 2 \leq k \leq 6 \\ 0, & \text{otherwise} \end{cases}$$

as depicted in Fig. 2.4(f). This form holds until $n - 9 = 2$, or $n = 11$, since at that value of n the left edge of $h[n - k]$ slides past the left edge of $x[k]$. Hence our third interval of time shifts is $6 < n \leq 11$.

Next, for $n > 11$, the functional form for $w_n[k]$ is given by

$$w_n[k] = \begin{cases} 1, & n - 9 \leq k \leq 6 \\ 0, & \text{otherwise} \end{cases}$$

as depicted in Fig. 2.4(g). This form holds until $n - 9 = 6$, or $n = 15$, since for $n > 15$ the left edge of $h[n - k]$ lies to the right of $x[k]$ and the functional form for $w_n[k]$ again changes. Hence the fourth interval of time shifts is $11 < n \leq 15$.

For all values of $n > 15$, we see that $w_n[k] = 0$. Thus the last interval of time shifts in this problem is $n > 15$.

The output of the system on each interval of n is obtained by summing the values of the corresponding $w_n[k]$ according to Eq. (2.4). Beginning with $n < 2$ we have $y[n] = 0$. Next, for $2 \leq n \leq 6$, we have

$$y[n] = \sum_{k=2}^{n} 1$$
$$= n - 1$$

On the third interval, $6 < n \leq 11$, Eq. (2.4) gives

$$y[n] = \sum_{k=2}^{6} 1$$
$$= 5$$

For $11 < n \leq 15$, Eq. (2.4) gives

$$y[n] = \sum_{k=n-9}^{6} 1$$
$$= 16 - n$$

Lastly, for $n > 15$, we see that $y[n] = 0$. Figure 2.4(h) depicts the output $y[n]$ obtained by combining the results on each interval.

EXAMPLE 2.4   Let the input, $x[n]$, to a LTI system H be given by

$$x[n] = \alpha^n\{u[n] - u[n - 10]\}$$

and the impulse response of the system be given by

$$h[n] = \beta^n u[n]$$

where $0 < \beta < 1$. Find the output of this system.

Solution: First we graph $x[k]$ and $h[n - k]$, treating n as a constant and k as the independent variable as depicted in Figs. 2.5(a) and (b). We see that

$$x[k] = \begin{cases} \alpha^k, & 0 \le k \le 9 \\ 0, & \text{otherwise} \end{cases}$$

$$h[n - k] = \begin{cases} \beta^{n-k}, & k \le n \\ 0, & \text{otherwise} \end{cases}$$

Now identify intervals of time shifts n on which the functional form of $w_n[k]$ is the same. Begin by considering n large and negative. We see that for $n < 0$, $w_n[k] = 0$ since there are no values k such that $x[k]$ and $h[n - k]$ are both nonzero. Hence the first interval is $n < 0$.

When $n = 0$ the right edge of $h[n - k]$ slides past the left edge of $x[k]$ so a transition occurs in the form of $w_n[k]$. For $n \ge 0$,

$$w_n[k] = \begin{cases} \alpha^k \beta^{n-k}, & 0 \le k \le n \\ 0, & \text{otherwise} \end{cases}$$

This form is correct provided $0 \le n \le 9$ and is depicted in Fig. 2.5(c). When $n = 9$ the right edge of $h[n - k]$ slides past the right edge of $x[k]$ so the form of $w_n[k]$ again changes.

Now for $n > 9$ we have a third form for $w_n[k]$,

$$w_n[k] = \begin{cases} \alpha^k \beta^{n-k}, & 0 \le k \le 9 \\ 0, & \text{otherwise} \end{cases}$$

Figure 2.5(d) depicts this $w_n[k]$ for the third and last interval in this problem, $n > 9$.

We now determine the output $y[n]$ for each of these three sets of time shifts by summing

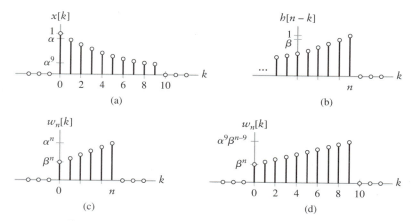

FIGURE 2.5   Evaluation of the convolution sum for Example 2.4. (a) The input signal $x[k]$ depicted as a function of k. (b) Reflected and time-shifted impulse response, $h[n - k]$. (c) The product signal $w_n[k]$ for $0 \le n \le 9$. (d) The product signal $w_n[k]$ for $n > 9$.

$w_n[k]$ over all k. Starting with the first interval, $n < 0$, we have $w_n[k] = 0$, and thus $y[n] = 0$. For the second interval, $0 \leq n \leq 9$, we have

$$y[n] = \sum_{k=0}^{n} \alpha^k \beta^{n-k}$$

Here the index of summation is limited from $k = 0$ to n because these are the only times k for which $w_n[k]$ is nonzero. Combining terms raised to the kth power, we have

$$y[n] = \beta^n \sum_{k=0}^{n} \left(\frac{\alpha}{\beta}\right)^k$$

Next, apply the formula for summing a geometric series of $(n + 1)$ terms to obtain

$$y[n] = \beta^n \frac{1 - (\alpha/\beta)^{n+1}}{1 - \alpha/\beta}$$

Now considering the third interval, $n \geq 10$, we have

$$y[n] = \sum_{k=0}^{9} \alpha^k \beta^{n-k}$$

$$= \beta^n \sum_{k=0}^{9} \left(\frac{\alpha}{\beta}\right)^k$$

$$= \beta^n \frac{1 - (\alpha/\beta)^{10}}{1 - \alpha/\beta}$$

where, again, the index of summation is limited from $k = 0$ to 9 because these are the only times for which $w_n[k]$ is nonzero. The last equality also follows from the formula for a finite geometric series. Combining the solutions for each interval of shifts gives the system output as

$$y[n] = \begin{cases} 0, & n < 0, \\ \beta^n \dfrac{1 - (\alpha/\beta)^{n+1}}{1 - \alpha/\beta}, & 0 \leq n \leq 9 \\ \beta^n \dfrac{1 - (\alpha/\beta)^{10}}{1 - \alpha/\beta}, & n > 9 \end{cases}$$

▶ **Drill Problem 2.1**   Repeat the convolution in Example 2.1 by directly evaluating the convolution sum.

Answer: See Example 2.1.   ◀

▶ **Drill Problem 2.2**   Let the input to a LTI system with impulse response $h[n] = \alpha^n\{u[n - 2] - u[n - 13]\}$ be $x[n] = 2\{u[n + 2] - u[n - 12]\}$. Find the output $y[n]$.

Answer:

$$y[n] = \begin{cases} 0, & n < 0 \\ 2\alpha^{n+2} \dfrac{1 - (\alpha)^{-1-n}}{1 - \alpha^{-1}}, & 0 \leq n \leq 10 \\ 2\alpha^{12} \dfrac{1 - (\alpha)^{-11}}{1 - \alpha^{-1}}, & 11 \leq n \leq 13 \\ 2\alpha^{12} \dfrac{1 - (\alpha)^{n-24}}{1 - \alpha^{-1}}, & 14 \leq n \leq 23 \\ 0, & n \geq 24 \end{cases}$$ ◀

▶ **Drill Problem 2.3** Suppose the input $x[n]$ and impulse response $h[n]$ of a LTI system H are given by

$$x[n] = -u[n] + 2u[n - 3] - u[n - 6]$$
$$h[n] = u[n + 1] - u[n - 10]$$

Find the output of this system, $y[n]$.

Answer:

$$y[n] = \begin{cases} 0, & n < -1 \\ -(n + 2), & -1 \le n \le 1 \\ n - 4, & 2 \le n \le 4 \\ 0, & 5 \le n \le 9 \\ n - 9, & 10 \le n \le 11 \\ 15 - n, & 12 \le n \le 14 \\ 0, & n > 14 \end{cases}$$

◀

The next example in this subsection uses the convolution sum to obtain an equation directly relating the input and output of a system with a finite-duration impulse response.

EXAMPLE 2.5 Consider a LTI system with impulse response

$$h[n] = \begin{cases} \frac{1}{4}, & 0 \le n \le 3 \\ 0, & \text{otherwise} \end{cases}$$

Find an expression that directly relates an arbitrary input $x[n]$, to the output of this system, $y[n]$.

Solution: Figures 2.6 (a) and (b) depict an arbitrary input $x[k]$ and the reflected, time shifted impulse response $h[n - k]$. For any time shift n we have

$$w_n[k] = \begin{cases} \frac{1}{4}x[k], & n - 3 \le k \le n \\ 0, & \text{otherwise} \end{cases}$$

Summing $w_n[k]$ over all k gives the output

$$y[n] = \tfrac{1}{4}(x[n] + x[n - 1] + x[n - 2] + x[n - 3])$$

The output of the system in Example 2.5 is the arithmetic average of the four most recent inputs. In Chapter 1 such a system was termed a moving-average system. The

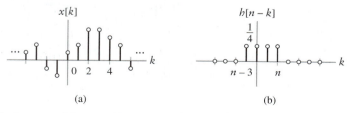

(a)   (b)

FIGURE 2.6 Evaluation of the convolution sum for Example 2.5. (a) An arbitrary input signal depicted as a function of k. (b) Reflected and time-shifted impulse response, $h[n - k]$.

effect of the averaging in this system is to smooth out short-term fluctuations in the input data. Such systems are often used to identify trends in data.

> **EXAMPLE 2.6**   Apply the average January temperature data depicted in Fig. 2.7 to the following moving-average systems:
>
> **(a)** $h[n] = \begin{cases} \frac{1}{2}, & 0 \le n \le 1 \\ 0, & \text{otherwise} \end{cases}$
>
> **(b)** $h[n] = \begin{cases} \frac{1}{4}, & 0 \le n \le 3 \\ 0, & \text{otherwise} \end{cases}$
>
> **(c)** $h[n] = \begin{cases} \frac{1}{8}, & 0 \le n \le 7 \\ 0, & \text{otherwise} \end{cases}$
>
> ***Solution:***   In case (a) the output is the average of the two most recent inputs, in case (b) the four most recent inputs, and in case (c) the eight most recent inputs. The system output for cases (a), (b), and (c) is depicted in Figs. 2.8(a), (b), and (c), respectively. As the impulse response duration increases, the degree of smoothing introduced by the system increases because the output is computed as an average of a larger number of inputs. The input to the system prior to 1900 is assumed to be zero, so the output near 1900 involves an average with some of the values zero. This leads to low values of the output, a phenomenon most evident in case (c).

In general, the output of any discrete-time system with a finite-duration impulse response is given by a weighted sum of the input signal values. Such weighted sums can easily be implemented in a computer to process discrete-time signals. The effect of the system on the signal depends on the weights or values of the system impulse response. The weights are usually chosen to enhance some feature of the data, such as an underlying trend, or to impart a particular characteristic. These issues are discussed throughout later chapters of the text.

■ THE CONVOLUTION INTEGRAL

The output of a continuous-time LTI system may also be determined solely from knowledge of the input and the system's impulse response. The approach and result are analogous to the discrete-time case. We first express an arbitrary input signal as a weighted superposition of time-shifted impulses. Here the superposition is an integral instead of a sum

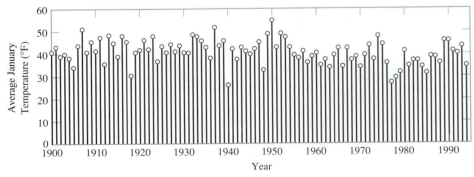

FIGURE 2.7   Average January temperature from 1900 to 1994.

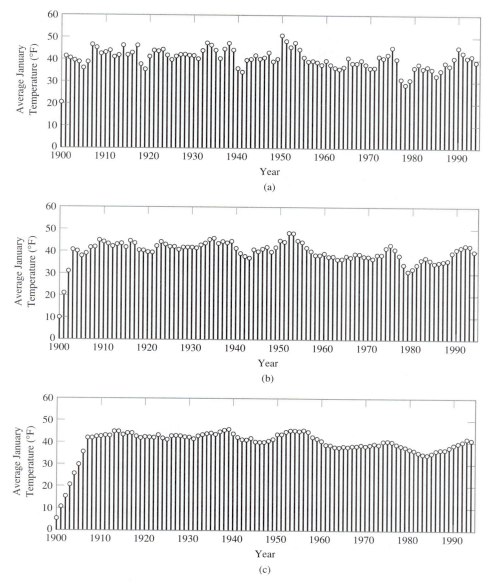

FIGURE 2.8 Result of passing average January temperature data through several moving-average systems. (a) Output of two-point moving-average system. (b) Output of four-point moving-average system. (c) Output of eight-point moving-average system.

due to the continuous nature of the input. We then apply this input to a LTI system to write the output as a weighted superposition of time-shifted impulse responses, an expression termed the convolution integral.

The convolution sum was derived by expressing the input signal $x[n]$ as a weighted sum of time-shifted impulses as shown by

$$x[n] = \sum_{k=-\infty}^{\infty} x[k]\delta[n-k]$$

Similarly, we may express a continuous-time signal as the weighted superposition of time-shifted impulses:

$$x(t) = \int_{-\infty}^{\infty} x(\tau)\delta(t - \tau) \, d\tau \tag{2.5}$$

Here the superposition is an integral and the time shifts are given by the continuous variable τ. The weights $x(\tau) \, d\tau$ are derived from the value of the signal $x(t)$ at the time at which each impulse occurs, τ. Equation (2.5) is a statement of the sifting property of the impulse function.

Define the impulse response $h(t) = H\{\delta(t)\}$ as the output of the system in response to an impulse input. If the system is time invariant, then $H\{\delta(t - \tau)\} = h(t - \tau)$. That is, a time-shifted impulse input generates a time-shifted impulse response output. Now consider the system output in response to a general input expressed as the weighted superposition in Eq. (2.5), as shown by

$$y(t) = H\left\{ \int_{-\infty}^{\infty} x(\tau)\delta(t - \tau) \, d\tau \right\}$$

Using the linearity property of the system we obtain

$$y(t) = \int_{-\infty}^{\infty} x(\tau) H\{\delta(t - \tau)\} \, d\tau$$

and, since the system is time invariant, we have

$$y(t) = \int_{-\infty}^{\infty} x(\tau) h(t - \tau) \, d\tau$$

Hence the output of a LTI system in response to an input of the form of Eq. (2.5) may be expressed as

$$y(t) = \int_{-\infty}^{\infty} x(\tau) h(t - \tau) \, d\tau \tag{2.6}$$

The output $y(t)$ is given as a weighted superposition of impulse responses time shifted by τ. The weights are $x(\tau) \, d\tau$. Equation (2.6) is termed the *convolution integral* and, as before, is denoted by the symbol $*$; that is,

$$x(t) * h(t) = \int_{-\infty}^{\infty} x(\tau) h(t - \tau) \, d\tau$$

The convolution process is illustrated in Fig. 2.9. Figure 2.9(a) depicts the impulse response of a system. In Fig. 2.9(b) the input to this system is represented as an integral of weighted and time-shifted impulses, $p_\tau(t) = x(\tau)\delta(t - \tau)$. These weighted and time-shifted impulses are depicted for several values of τ on the left-hand side of Fig. 2.9. The output associated with each input $p_\tau(t)$ is the weighted and time-shifted impulse response:

$$v_\tau(t) = x(\tau) h(t - \tau)$$

The right-hand side of Fig. 2.9(b) depicts $v_\tau(t)$ for several values of τ. Note that $v_\tau(t)$ is a function of two independent variables, τ and t. On the right-hand side of Fig. 2.9(b), the variation with t is shown on the horizontal axis, while the variation with τ occurs vertically,

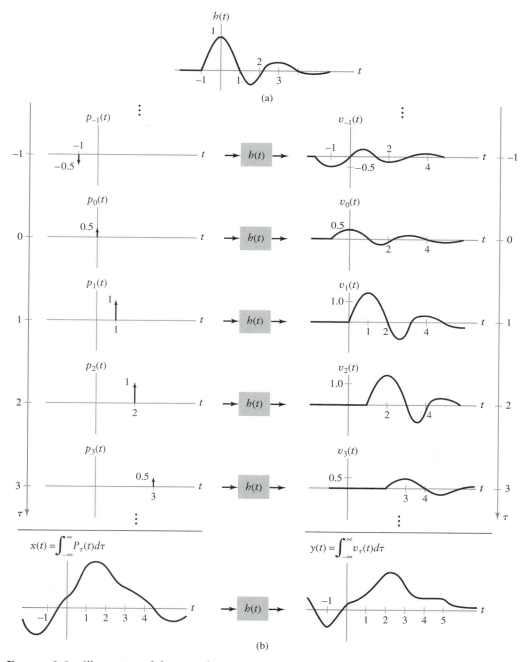

FIGURE 2.9 Illustration of the convolution integral. (a) Impulse response of a continuous-time system. (b) Decomposition of $x(t)$ into a weighted integral of time-shifted impulses results in an output $y(t)$ given by a weighted integral of time-shifted impulse responses. Here $p_\tau(t)$ is the weighted (by $x(\tau)$) and time-shifted (by τ) impulse input, and $v_\tau(t)$ is the weighted and time-shifted impulse response output. Both $p_\tau(t)$ and $v_\tau(t)$ are depicted only at integer values of τ. The dependence of both $p_\tau(t)$ and $v_\tau(t)$ on τ is depicted by the τ axis shown on the left- and right-hand sides of the figure. The output is obtained by integrating $v_\tau(t)$ over τ.

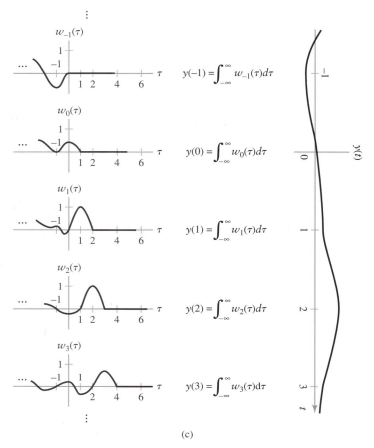

FIGURE 2.9   (c) The signals $w_t(\tau)$ used to compute the output at time t correspond to vertical slices of $v_\tau(t)$. Here we have redrawn the right-hand side of Fig. 2.9(b) so that the τ axis is horizontal. The output is obtained for $t = t_o$ by integrating $w_t(\tau)$ over τ.

as shown by the vertical axis on the right-hand side. The system output at time $t = t_o$ is obtained by integrating over τ, as shown by

$$y(t_o) = \int_{-\infty}^{\infty} v_\tau(t_o)\ d\tau$$

That is, we integrate along the vertical or τ axis on the right-hand side of Fig. 2.9(b) at a fixed time, $t = t_o$.

   Define a signal $w_{t_o}(\tau)$ to represent the variation of $v_\tau(t)$ along the τ axis for a fixed time $t = t_o$. This implies $w_{t_o}(\tau) = v_\tau(t_o)$. Examples of this signal for several values of t_o are depicted in Fig. 2.9(c). The corresponding system output is now obtained by integrating $w_{t_o}(\tau)$ over τ from $-\infty$ to ∞. Note that the horizontal axis in Fig. 2.9(b) is t and the vertical axis is τ. In Fig. 2.9(c) we have in effect redrawn the right-hand side of Fig. 2.9(b) with τ as the horizontal axis and t as the vertical axis.

   We have defined the intermediate signal $w_t(\tau) = x(\tau)h(t - \tau)$ as the product of $x(\tau)$ and $h(t - \tau)$. In this definition τ is the independent variable and t is treated as a constant. This is explicitly indicated by writing t as a subscript and τ within the parentheses of $w_t(\tau)$. Hence $h(t - \tau) = h(-(\tau - t))$ is a reflected and time-shifted (by $-t$) version of $h(\tau)$. The time shift t determines the time at which we evaluate the output of the system since Eq. (2.6) becomes

$$y(t) = \int_{-\infty}^{\infty} w_t(\tau)\, d\tau \qquad (2.7)$$

The system output at any time t is the area under the signal $w_t(\tau)$.

In general, the functional form for $w_t(\tau)$ will depend on the value of t. As in the discrete-time case, we may avoid evaluating Eq. (2.7) at an infinite number of values of t by identifying intervals of t on which $w_t(\tau)$ has the same functional form. We then only need to evaluate Eq. (2.7) using the $w_t(\tau)$ associated with each interval. Often it is very helpful to graph both $x(\tau)$ and $h(t - \tau)$ when determining $w_t(\tau)$ and identifying the appropriate interval of time shifts. This procedure is summarized as follows:

1. Graph $x(\tau)$ and $h(t - \tau)$ as a function of the independent variable τ. To obtain $h(t - \tau)$, reflect $h(\tau)$ about $\tau = 0$ to obtain $h(-\tau)$ and then time $h(-t)$ shift by $-t$.
2. Begin with the time shift t large and negative.
3. Write the functional form for $w_t(\tau)$.
4. Increase the time shift t until the functional form for $w_t(\tau)$ changes. The value t at which the change occurs defines the end of the current interval and the beginning of a new interval.
5. Let t be in the new interval. Repeat steps 3 and 4 until all intervals of time shifts t and the corresponding functional forms for $w_t(\tau)$ are identified. This usually implies increasing t to a large and positive value.
6. For each interval of time shifts t, integrate $w_t(\tau)$ from $\tau = -\infty$ to $\tau = \infty$ to obtain $y(t)$ on that interval.

The effect of increasing t from a large negative value to a large positive value is to slide $h(-\tau)$ past $x(\tau)$ from left to right. Transitions in the intervals of t associated with the same form of $w_t(\tau)$ generally occur when a transition in $h(-\tau)$ slides through a transition in $x(\tau)$. Alternatively, we can integrate $w_t(\tau)$ as each interval of time shifts is identified, that is, after step 4, rather than waiting until all intervals are identified. The following examples illustrate this procedure for evaluating the convolution integral.

EXAMPLE 2.7 Consider the RC circuit depicted in Fig. 2.10 and assume the circuit's time constant is $RC = 1$ s. Determine the voltage across the capacitor, $y(t)$, resulting from an input voltage $x(t) = e^{-3t}\{u(t) - u(t - 2)\}$.

Solution: The circuit is linear and time invariant, so the output is the convolution of the input and the impulse response. That is, $y(t) = x(t) * h(t)$. The impulse response of this circuit is

$$h(t) = e^{-t}u(t)$$

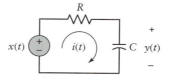

FIGURE 2.10 RC circuit system with the voltage source $x(t)$ as input and the voltage measured across the capacitor, $y(t)$, as output.

To evaluate the convolution integral, first graph $x(\tau)$ and $h(t - \tau)$ as a function of the independent variable τ while treating t as a constant. We see from Figs. 2.11(a) and (b) that

$$x(\tau) = \begin{cases} e^{-3\tau}, & 0 < \tau < 2 \\ 0, & \text{otherwise} \end{cases}$$

and

$$h(t - \tau) = \begin{cases} e^{-(t-\tau)}, & \tau < t \\ 0, & \text{otherwise} \end{cases}$$

Now identify the intervals of time shifts t for which the functional form of $w_t(\tau)$ does not change. Begin with t large and negative. Provided $t < 0$, we have $w_t(\tau) = 0$ since there are no values τ for which both $x(\tau)$ and $h(t - \tau)$ are both nonzero. Hence the first interval of time shifts is $t < 0$.

Note that at $t = 0$ the right edge of $h(t - \tau)$ intersects the left edge of $x(\tau)$. For $t > 0$

$$w_t(\tau) = \begin{cases} e^{-t-2\tau}, & 0 < \tau < t \\ 0, & \text{otherwise} \end{cases}$$

This form for $w_t(\tau)$ is depicted in Fig. 2.11(c). It does not change until $t > 2$, at which point the right edge of $h(t - \tau)$ passes through the right edge of $x(\tau)$. The second interval of time shifts t is thus $0 \le t < 2$.

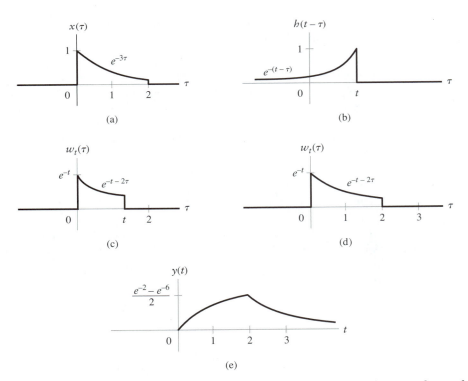

FIGURE 2.11 Evaluation of the convolution integral for Example 2.7. (a) The input depicted as a function of τ. (b) Reflected and time-shifted impulse response, $h(t - \tau)$. (c) The product signal $w_t(\tau)$ for $0 \le t < 2$. (d) The product signal $w_t(\tau)$ for $t \ge 2$. (e) System output $y(t)$.

For $t > 2$ we have a third form for $w_t(\tau)$, which is written as

$$w_t(\tau) = \begin{cases} e^{-t-2\tau}, & 0 < \tau < 2 \\ 0, & \text{otherwise} \end{cases}$$

Figure 2.11(d) depicts $w_t(\tau)$ for this third interval of time shifts, $t \geq 2$.

We now determine the output $y(t)$ for each of these three intervals of time shifts by integrating $w_t(\tau)$ from $\tau = -\infty$ to $\tau = \infty$. Starting with the first interval, $t \leq 0$, we have $w_t(\tau) = 0$ and thus $y(t) = 0$. For the second interval, $0 \leq t < 2$, we have

$$\begin{aligned} y(t) &= \int_0^t e^{-t-2\tau}\, d\tau \\ &= e^{-t}\{-\tfrac{1}{2}e^{-2\tau}|_0^t \\ &= \tfrac{1}{2}(e^{-t} - e^{-3t}) \end{aligned}$$

For the third interval, $t \geq 2$, we have

$$\begin{aligned} y(t) &= \int_0^2 e^{-t-2\tau}\, d\tau \\ &= e^{-t}\{-\tfrac{1}{2}e^{-2\tau}|_0^2 \\ &= \tfrac{1}{2}(1 - e^{-4})\, e^{-t} \end{aligned}$$

Combining the solutions for each interval of time shifts gives the output

$$y(t) = \begin{cases} 0, & t < 0 \\ \tfrac{1}{2}(1 - e^{-2t})\, e^{-t}, & 0 \leq t < 2 \\ \tfrac{1}{2}(1 - e^{-4})\, e^{-t}, & t \geq 2 \end{cases}$$

as depicted in Fig. 2.11(e).

EXAMPLE 2.8   Suppose the input $x(t)$ and impulse response $h(t)$ of a LTI system are given by

$$\begin{aligned} x(t) &= 2u(t - 1) - 2u(t - 3) \\ h(t) &= u(t + 1) - 2u(t - 1) + u(t - 3) \end{aligned}$$

Find the output of this system.

Solution: Graphical representations for $x(\tau)$ and $h(t - \tau)$ are given in Figs. 2.12(a) and (b). From these we can determine the intervals of time shifts t on which the functional form of $w_t(\tau)$ is the same. Begin with t large and negative. For $t + 1 < 1$ or $t < 0$ the right edge of $h(t - \tau)$ is to the left of the nonzero portion of $x(\tau)$ and consequently $w_t(\tau) = 0$.

For $t > 0$ the right edge of $h(t - \tau)$ overlaps with the nonzero portion of $x(\tau)$ and we have

$$w_t(\tau) = \begin{cases} 2, & 1 < \tau < t + 1 \\ 0, & \text{otherwise} \end{cases}$$

This form for $w_t(\tau)$ holds provided $t + 1 < 3$, or $t < 2$, and is depicted in Fig. 2.12(c).

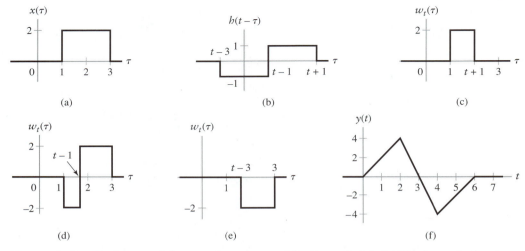

FIGURE 2.12   Evaluation of the convolution integral for Example 2.8. (a) The input depicted as a function of τ. (b) Reflected and time-shifted impulse response, $h(t - \tau)$. (c) The product signal $w_t(\tau)$ for $0 \leq t < 2$. (d) The product signal $w_t(\tau)$ for $2 \leq t < 4$. (e) The product signal $w_t(\tau)$ for $4 \leq t < 6$. (f) System output $y(t)$.

For $t > 2$ the right edge of $h(t - \tau)$ is to the right of the nonzero portion of $x(\tau)$. In this case we have

$$w_t(\tau) = \begin{cases} -2, & 1 < \tau < t - 1 \\ 2, & t - 1 < \tau < 3 \\ 0, & \text{otherwise} \end{cases}$$

This form for $w_t(\tau)$ holds provided $t - 1 < 3$, or $t < 4$, and is depicted in Fig. 2.12(d).

For $t > 4$ the leftmost edge of $h(t - \tau)$ is within the nonzero portion of $x(\tau)$ and we have

$$w_t(\tau) = \begin{cases} -2, & t - 3 < \tau < 3 \\ 0, & \text{otherwise} \end{cases}$$

This form for $w_t(\tau)$ is depicted in Fig. 2.12(e) and holds provided $t - 3 < 3$, or $t < 6$.

For $t > 6$, no nonzero portions of $x(\tau)$ and $h(t - \tau)$ overlap and consequently $w_t(\tau) = 0$.

The system output $y(t)$ is obtained by integrating $w_t(\tau)$ from $\tau = -\infty$ to $\tau = \infty$ for each interval of time shifts identified above. Beginning with $t < 0$, we have $y(t) = 0$ since $w_t(\tau) = 0$. For $0 \leq t < 2$ we have

$$y(t) = \int_1^{t+1} 2 \, d\tau$$

$$= 2t$$

On the next interval, $2 \leq t < 4$, we have

$$y(t) = \int_1^{t-1} -2 \, d\tau + \int_{t-1}^3 2 \, d\tau$$

$$= -4t + 12$$

Now considering $4 \leq t < 6$ the output is

$$y(t) = \int_{t-3}^{3} -2 \, d\tau$$
$$= 2t - 12$$

Lastly, for $t > 6$, we have $y(t) = 0$ since $w_t(\tau) = 0$. Combining the outputs for each interval of time shifts gives the result

$$y(t) = \begin{cases} 0, & t < 0 \\ 2t, & 0 \leq t < 2 \\ -4t + 12, & 2 \leq t < 4 \\ 2t - 12, & 4 \leq t < 6 \\ 0, & t \geq 6 \end{cases}$$

as depicted in Fig. 2.12(f).

▶ **Drill Problem 2.4**   Let the impulse response of a LTI system be $h(t) = e^{-2(t+1)}u(t + 1)$. Find the output $y(t)$ if the input is $x(t) = e^{-|t|}$.

Answer: For $t < -1$,

$$w_t(\tau) = \begin{cases} e^{-2(t+1)}e^{3\tau}, & -\infty < \tau < t + 1 \\ 0, & \text{otherwise} \end{cases}$$
$$y(t) = \tfrac{1}{3}e^{t+1}$$

For $t > -1$,

$$w_t(\tau) = \begin{cases} e^{-2(t+1)}e^{3\tau}, & -\infty < \tau < 0 \\ e^{-2(t+1)}e^{\tau}, & 0 < \tau < t + 1 \\ 0, & \text{otherwise} \end{cases}$$
$$y(t) = e^{-(t+1)} - \tfrac{2}{3}e^{-2(t+1)}$$ ◀

▶ **Drill Problem 2.5**   Let the impulse response of a LTI system be given by $y(t) = u(t - 1) - u(t - 4)$. Find the output of this system in response to an input $x(t) = u(t) + u(t - 1) - 2u(t - 2)$.

Answer:

$$y(t) = \begin{cases} 0, & t < 1 \\ t - 1, & 1 \leq t < 2 \\ 2t - 3, & 2 \leq t < 3 \\ 3, & 3 \leq t < 4 \\ 7 - t, & 4 \leq t < 5 \\ 12 - 2t, & 5 \leq t < 6 \\ 0, & t \geq 6 \end{cases}$$ ◀

The convolution integral describes the behavior of a continuous-time system. The system impulse response can provide insight into the nature of the system. We will develop this insight in the next section and subsequent chapters. To glimpse some of the insight offered by the impulse response, consider the following example.

EXAMPLE 2.9   Let the impulse response of a LTI system be $h(t) = \delta(t - a)$. Determine the output of this system in response to an input $x(t)$.

Solution: Consider first obtaining $h(t - \tau)$. Reflecting $h(\tau) = \delta(\tau - a)$ about $\tau = 0$ gives $h(-\tau) = \delta(\tau + a)$ since the impulse function has even symmetry. Now shift the independent variable τ by $-t$ to obtain $h(t - \tau) = \delta(\tau - (t - a))$. Substitute this expression for $h(t - \tau)$ in the convolution integral of Eq. (2.6) and use the sifting property of the impulse function to obtain

$$y(t) = \int_{-\infty}^{\infty} x(\tau)\delta(\tau - (t - a)) \, d\tau$$
$$= x(t - a)$$

Note that the identity system is represented for $a = 0$ since in this case the output is equal to the input. When $a \neq 0$, the system time shifts the input. If a is positive the input is delayed, and if a is negative the input is advanced. Hence the location of the impulse response relative to the time origin determines the amount of delay introduced by the system.

2.3 *Properties of the Impulse Response Representation for LTI Systems*

The impulse response completely characterizes the input–output behavior of a LTI system. Hence properties of a system, such as memory, causality, and stability, are related to its impulse response. Also, the impulse response of an interconnection of LTI systems is related to the impulse response of the constituent systems. In this section we examine the impulse response of interconnected systems and relate the impulse response to system properties. These relationships tell us how the impulse response characterizes system behavior. The results for continuous- and discrete-time systems are obtained using nearly identical approaches, so we derive one and simply state the results for the other.

■ PARALLEL CONNECTION OF SYSTEMS

Consider two LTI systems with impulse responses $h_1(t)$ and $h_2(t)$ connected in parallel as illustrated in Fig. 2.13(a). The output of this connection of systems, $y(t)$, is the sum of the outputs of each system

$$y(t) = y_1(t) + y_2(t)$$
$$= x(t) * h_1(t) + x(t) * h_2(t)$$

Substitute the integral representation for each convolution

$$y(t) = \int_{-\infty}^{\infty} x(\tau)h_1(t - \tau) \, d\tau + \int_{-\infty}^{\infty} x(\tau)h_2(t - \tau) \, d\tau$$

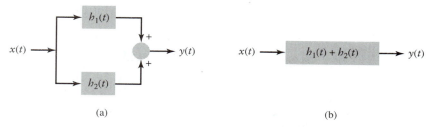

FIGURE 2.13   Interconnection of two systems. (a) Parallel connection of two systems. (b) Equivalent system.

and combine the integrals to obtain

$$y(t) = \int_{-\infty}^{\infty} x(\tau)\{h_1(t - \tau) + h_2(t - \tau)\}\, d\tau$$

$$= \int_{-\infty}^{\infty} x(\tau)h(t - \tau)\, d\tau$$

$$= x(t) * h(t)$$

where $h(t) = h_1(t) + h_2(t)$. We identify $h(t)$ as the impulse response of the parallel connection of two systems. This equivalent system is depicted in Fig. 2.13(b). The impulse response of two systems connected in parallel is the sum of the individual impulse responses.

Mathematically, this implies that convolution possesses the *distributive property*:

$$x(t) * h_1(t) + x(t) * h_2(t) = x(t) * \{h_1(t) + h_2(t)\} \tag{2.8}$$

Identical results hold for the discrete-time case:

$$x[n] * h_1[n] + x[n] * h_2[n] = x[n] * \{h_1[n] + h_2[n]\} \tag{2.9}$$

■ CASCADE CONNECTION OF SYSTEMS

Now consider the cascade connection of two LTI systems illustrated in Fig. 2.14(a). Let $z(t)$ be the output of the first system and the input to the second system in the cascade. The output $y(t)$ is expressed in terms of $z(t)$ as

$$y(t) = z(t) * h_2(t) \tag{2.10}$$

$$= \int_{-\infty}^{\infty} z(\tau)h_2(t - \tau)\, d\tau \tag{2.11}$$

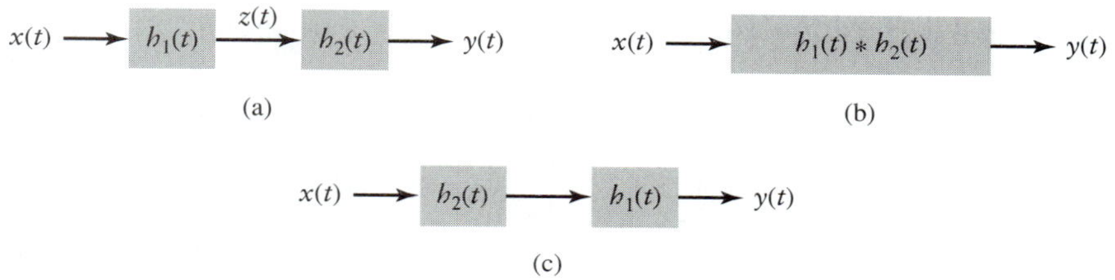

$$x(t) \longrightarrow \boxed{h_1(t)} \xrightarrow{z(t)} \boxed{h_2(t)} \longrightarrow y(t) \qquad\qquad x(t) \longrightarrow \boxed{h_1(t) * h_2(t)} \longrightarrow y(t)$$

$$\text{(a)} \qquad\qquad\qquad\qquad\qquad\qquad \text{(b)}$$

$$x(t) \longrightarrow \boxed{h_2(t)} \longrightarrow \boxed{h_1(t)} \longrightarrow y(t)$$

$$\text{(c)}$$

FIGURE 2.14   Interconnection of two systems. (a) Cascade connection of two systems. (b) Equivalent system. (c) Equivalent system: interchange system order.

However, $z(\tau)$ is the output of the first system and is expressed in terms of the input $x(\tau)$ as

$$
\begin{aligned}
z(\tau) &= x(\tau) * h_1(\tau) \\
&= \int_{-\infty}^{\infty} x(\nu)h_1(\tau - \nu)\, d\nu
\end{aligned}
\tag{2.12}
$$

Here ν is used as the variable of integration in the convolution integral. Substituting Eq. (2.12) for $z(\tau)$ in Eq. (2.11) gives

$$
y(t) = \int_{-\infty}^{\infty} \int_{-\infty}^{\infty} x(\nu)h_1(\tau - \nu)h_2(t - \tau)\, d\nu\, d\tau
$$

Now perform the change of variable $\eta = \tau - \nu$ and interchange integrals to obtain

$$
y(t) = \int_{-\infty}^{\infty} x(\nu) \left[\int_{-\infty}^{\infty} h_1(\eta)h_2(t - \nu - \eta)\, d\eta \right] d\nu
\tag{2.13}
$$

The inner integral is identified as the convolution of $h_1(t)$ with $h_2(t)$ evaluated at $t - \nu$. That is, if we define $h(t) = h_1(t) * h_2(t)$, then

$$
\int_{-\infty}^{\infty} h_1(\eta)h_2(t - \nu - \eta)\, d\eta = h(t - \nu)
$$

Substituting this relationship into Eq. (2.13) yields

$$
\begin{aligned}
y(t) &= \int_{-\infty}^{\infty} x(\nu)h(t - \nu)\, d\nu \\
&= x(t) * h(t)
\end{aligned}
\tag{2.14}
$$

Hence the impulse response of two LTI systems connected in cascade is the convolution of the individual impulse responses. The cascade connection is input–output equivalent to the single system represented by the impulse response $h(t)$ as shown in Fig. 2.14(b).

Substituting $z(t) = x(t) * h_1(t)$ into the expression for $y(t)$ given in Eq. (2.10) and $h(t) = h_1(t) * h_2(t)$ into the alternative expression for $y(t)$ given in Eq. (2.14) establishes that convolution possesses the *associative property*

$$
\{x(t) * h_1(t)\} * h_2(t) = x(t) * \{h_1(t) * h_2(t)\}
\tag{2.15}
$$

A second important property for the cascade connection of systems concerns the ordering of the systems. Write $h(t) = h_1(t) * h_2(t)$ as the integral

$$
h(t) = \int_{-\infty}^{\infty} h_1(\tau)h_2(t - \tau)\, d\tau
$$

and perform the change of variable $\nu = t - \tau$ to obtain

$$
\begin{aligned}
h(t) &= \int_{-\infty}^{\infty} h_1(t - \nu)h_2(\nu)\, d\nu \\
&= h_2(t) * h_1(t)
\end{aligned}
\tag{2.16}
$$

Hence the convolution of $h_1(t)$ and $h_2(t)$ can be performed in either order. This corresponds to interchanging the order of the systems in the cascade as shown in Fig. 2.14(c). Since

$$
x(t) * \{h_1(t) * h_2(t)\} = x(t) * \{h_2(t) * h_1(t)\}
$$

we conclude that the output of a cascade combination of LTI systems is independent of the order in which the systems are connected. Mathematically, we say that the convolution operation possesses the *commutative property*

$$h_1(t) * h_2(t) = h_2(t) * h_1(t) \qquad (2.17)$$

The commutative property is often used to simplify evaluation or interpretation of the convolution integral.

Discrete-time systems and convolution have identical properties to their continuous-time counterparts. The impulse response of a cascade connection of LTI systems is given by the convolution of the individual impulse responses and the output of a cascade combination of LTI systems is independent of the order in which the systems are connected. Discrete-time convolution is associative

$$\{x[n] * h_1[n]\} * h_2[n] = x[n] * \{h_1[n] * h_2[n]\} \qquad (2.18)$$

and commutative

$$h_1[n] * h_2[n] = h_2[n] * h_1[n] \qquad (2.19)$$

The following example demonstrates the use of convolution properties for finding a single system that is input–output equivalent to an interconnected system.

EXAMPLE 2.10 Consider the interconnection of LTI systems depicted in Fig. 2.15. The impulse response of each system is given by

$$h_1[n] = u[n]$$
$$h_2[n] = u[n + 2] - u[n]$$
$$h_3[n] = \delta[n - 2]$$
$$h_4[n] = \alpha^n u[n]$$

Find the impulse response of the overall system, $h[n]$.

Solution: We first derive an expression for the overall impulse response in terms of the impulse response of each system. Begin with the parallel combination of $h_1[n]$ and $h_2[n]$. The equivalent system has impulse response $h_{12}[n] = h_1[n] + h_2[n]$. This system is in series with $h_3[n]$, so the equivalent system for the upper branch has impulse response $h_{123}[n] = h_{12}[n] * h_3[n]$. Substituting for $h_{12}[n]$, we have $h_{123}[n] = (h_1[n] + h_2[n]) * h_3[n]$. The upper branch is in parallel with the lower branch, characterized by $h_4[n]$; hence the overall system impulse response is $h[n] = h_{123}[n] - h_4[n]$. Substituting for $h_{123}[n]$ yields

$$h[n] = (h_1[n] + h_2[n]) * h_3[n] - h_4[n]$$

Now substitute the specific forms of $h_1[n]$ and $h_2[n]$ to obtain

$$h_{12}[n] = u[n] + u[n + 2] - u[n]$$
$$= u[n + 2]$$

Convolving $h_{12}[n]$ with $h_3[n]$ gives

$$h_{123}[n] = u[n + 2] * \delta[n - 2]$$
$$= u[n]$$

Lastly, we obtain the overall impulse response by summing $h_{123}[n]$ and $h_4[n]$

$$h[n] = \{1 - \alpha^n\} u[n]$$

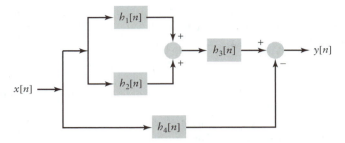

FIGURE 2.15   Interconnection of systems for Example 2.10.

Interconnections of systems occur naturally in analysis. Often it is easier to break a complex system into simpler subsystems, analyze each subsystem, and then study the entire system as an interconnection of subsystems than it is to analyze the overall system directly. This is an example of the "divide-and-conquer" approach to problem solving and is possible due to the assumptions of linearity and time invariance. Interconnections of systems are also useful in system implementation, since systems that are equivalent in the input–output sense are not necessarily equivalent in other senses. For example, the computational complexity of two input–output equivalent systems for processing data in a computer may differ significantly. The fact that many different interconnections of LTI systems are input–output equivalent can be exploited to optimize some other implementation criterion such as computation.

■ MEMORYLESS SYSTEMS

Recall that the output of a memoryless system depends only on the present input. Exploiting the commutative property of convolution, the output of a LTI discrete-time system may be expressed as

$$y[n] = h[n] * x[n]$$

$$= \sum_{k=-\infty}^{\infty} h[k]x[n-k]$$

For this system to be memoryless, $y[n]$ must depend only on $x[n]$ and cannot depend on $x[n-k]$ for $k \neq 0$. This condition implies that $h[k] = 0$ for $k \neq 0$. Hence a LTI discrete-time system is memoryless if and only if $h[k] = c\delta[k]$, where c is an arbitrary constant.

Writing the output of a continuous-time system as

$$y(t) = \int_{-\infty}^{\infty} h(\tau)x(t-\tau)\,d\tau$$

we see that, analogous to the discrete-time case, a continuous-time system is memoryless if and only if $h(\tau) = c\delta(\tau)$ for c an arbitrary constant.

The memoryless condition places severe restrictions on the form of the impulse response. All memoryless LTI systems perform scalar multiplication on the input.

■ CAUSAL SYSTEMS

The output of a causal system depends only on past or present values of the input. Again write the convolution sum as

$$y[n] = \sum_{k=-\infty}^{\infty} h[k]x[n-k]$$

Past and present values of the input, $x[n]$, $x[n - 1]$, $x[n - 2]$, ..., are associated with indices $k \geq 0$ in the convolution sum, while future values of the input are associated with indices $k < 0$. In order for $y[n]$ to depend only on past or present values of the input, we require $h[k] = 0$ for $k < 0$. Hence, for a causal system, $h[k] = 0$ for $k < 0$, and the convolution sum is rewritten

$$y[n] = \sum_{k=0}^{\infty} h[k]x[n - k]$$

The causality condition for a continuous-time system follows in an analogous manner from the convolution integral

$$y(t) = \int_{-\infty}^{\infty} h(\tau)x(t - \tau)\, d\tau$$

A causal continuous-time system has impulse response that satisfies $h(\tau) = 0$ for $\tau < 0$. The output of a causal system is thus expressed as the convolution integral

$$y(t) = \int_{0}^{\infty} h(\tau)x(t - \tau)\, d\tau$$

The causality condition is intuitively satisfying. Recall that the impulse response is the output of a system in response to an impulse input applied at time $t = 0$. Causal systems are nonanticipative: that is, they cannot generate an output before the input is applied. Requiring the impulse response to be zero for negative time is equivalent to saying the system cannot respond prior to application of the impulse.

■ STABLE SYSTEMS

Recall from Chapter 1 that a system is bounded input–bounded output (BIBO) stable if the output is guaranteed to be bounded for every bounded input. Formally, if the input to a stable discrete-time system satisfies $|x[n]| \leq M_x < \infty$, then the output must satisfy $|y[n]| \leq M_y < \infty$. We shall derive conditions on $h[n]$ that guarantee stability of the system by bounding the convolution sum. The magnitude of the output is given by

$$|y[n]| = |h[n] * x[n]|$$

$$= \left| \sum_{k=-\infty}^{\infty} h[k]x[n - k] \right|$$

We seek an upper bound on $|y[n]|$ that is a function of the upper bound on $|x[n]|$ and the impulse response. Since the magnitude of a sum of numbers is less than or equal to the sum of the magnitudes, that is, $|a + b| \leq |a| + |b|$, we may write

$$|y[n]| \leq \sum_{k=-\infty}^{\infty} |h[k]x[n - k]|$$

Furthermore, the magnitude of a product is equal to the product of the magnitudes, that is, $|ab| = |a||b|$, and so we have

$$|y[n]| \leq \sum_{k=-\infty}^{\infty} |h[k]||x[n - k]|$$

If we assume that the input is bounded, $|x[n]| \leq M_x < \infty$, then $|x[n - k]| < M_x$ and

$$|y[n]| \leq M_x \sum_{k=\infty}^{\infty} |h[k]| \tag{2.20}$$

Hence the output is bounded, $|y[n]| < \infty$, provided that the impulse response of the system is absolutely summable. We conclude that the impulse response of a stable system satisfies the bound

$$\sum_{k=-\infty}^{\infty} |h[k]| < \infty$$

Our derivation so far has established absolute summability of the impulse response as a sufficient condition for BIBO stability. The reader is asked to show that this is also a necessary condition for BIBO stability in Problem 2.13.

A similar set of steps may be used to establish that a continuous-time system is BIBO stable if and only if the impulse response is absolutely integrable, that is,

$$\int_{-\infty}^{\infty} |h(\tau)| \, d\tau < \infty$$

EXAMPLE 2.11   A discrete-time system has impulse response

$$h[n] = a^n u[n + 2]$$

Is this system BIBO stable, causal, and memoryless?

Solution: Stability is determined by checking whether the impulse response is absolutely summable, as shown by

$$\sum_{k=-\infty}^{\infty} |h[n]| = \sum_{k=-2}^{\infty} |a^n|$$

$$= |a^{-2}| + |a^{-1}| + \sum_{k=0}^{\infty} |a|^n$$

The infinite geometric sum in the second line converges only if $|a| < 1$. Hence the system is stable provided $0 < |a| < 1$. The system is not causal, since the impulse response $h[n]$ is nonzero for $n = -1, -2$. The system is not memoryless because $h[n]$ is nonzero for some values $n \neq 0$.

▶ **Drill Problem 2.6**   Determine the conditions on a such that the continuous-time system with impulse response $h(t) = e^{at}u(t)$ is stable, causal, and memoryless.

Answer: The system is stable provided $a < 0$, causal for all a, and there is no a for which the system is memoryless.   ◀

We emphasize that a system can be unstable even though the impulse response is finite valued. For example, the impulse response $h[n] = u[n]$ is never greater than one, but is not absolutely summable and thus the system is unstable. To demonstrate this, use the convolution sum to express the output of this system in terms of the input as

$$y[n] = \sum_{k=-\infty}^{n} x[k]$$

Although the output is bounded for some bounded inputs $x[n]$, it is not bounded for every bounded $x[n]$. In particular, the constant input $x[n] = c$ clearly results in an unbounded output.

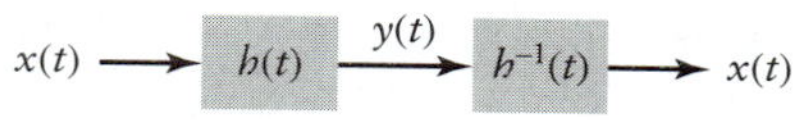

$$x(t) \longrightarrow \boxed{h(t)} \xrightarrow{y(t)} \boxed{h^{-1}(t)} \longrightarrow x(t)$$

FIGURE 2.16 Cascade of LTI system with impulse response $h(t)$ and inverse system with impulse response $h^{-1}(t)$.

■ INVERTIBLE SYSTEMS AND DECONVOLUTION

A system is *invertible* if the input to the system can be recovered from the output. This implies existence of an inverse system that takes the output of the original system as its input and produces the input of the original system. We shall limit ourselves here to consideration of inverse systems that are LTI. Figure 2.16 depicts the cascade of a LTI system having impulse response $h(t)$ with a LTI inverse system whose impulse response is denoted as $h^{-1}(t)$.

The process of recovering $x(t)$ from $h(t) * x(t)$ is termed *deconvolution*, since it corresponds to reversing or undoing the convolution operation. An inverse system has output $x(t)$ in response to input $y(t) = h(t) * x(t)$ and thus solves the deconvolution problem. Deconvolution and inverse systems play an important role in many signal-processing and systems problems. A common problem is that of reversing or "equalizing" the distortion introduced by a nonideal system. For example, consider using a high-speed modem to communicate over telephone lines. Distortion introduced by the telephone network places severe restrictions on the rate at which information can be transmitted, so an equalizer is incorporated into the modem. The equalizer reverses the telephone network distortion and permits much higher data rates to be achieved. In this case the equalizer represents an inverse system for the telephone network. We will discuss equalization in more detail in Chapters 5 and 8.

The relationship between the impulse response of a system, $h(t)$, and the corresponding inverse system, $h^{-1}(t)$, is easily derived. The impulse response of the cascade connection in Fig. 2.16 is the convolution of $h(t)$ and $h^{-1}(t)$. We require the output of the cascade to equal the input, or

$$x(t) * (h(t) * h^{-1}(t)) = x(t)$$

This implies that

$$h(t) * h^{-1}(t) = \delta(t) \tag{2.21}$$

Similarly, the impulse response of a discrete-time LTI inverse system, $h^{-1}[n]$, must satisfy

$$h[n] * h^{-1}[n] = \delta[n] \tag{2.22}$$

In many equalization applications an exact inverse system may be difficult to find or implement. Determination of an approximate solution to Eq. (2.21) or Eq. (2.22) is often sufficient in such cases. The following example illustrates a case where an exact inverse system is obtained by directly solving Eq. (2.22).

EXAMPLE 2.12 Consider designing a discrete-time inverse system to eliminate the distortion associated with an undesired echo in a data transmission problem. Assume the echo is represented as attenuation by a constant a and a delay corresponding to one time unit of the input sequence. Hence the distorted received signal, $y[n]$, is expressed in terms of the transmitted signal $x[n]$ as

$$y[n] = x[n] + ax[n - 1]$$

Find a causal inverse system that recovers $x[n]$ from $y[n]$. Check if this inverse system is stable.

Solution: First we identify the impulse response of the system relating $y[n]$ and $x[n]$. Writing the convolution sum as

$$y[n] = \sum_{k=-\infty}^{\infty} h[k]x[n-k]$$

we identify

$$h[k] = \begin{cases} 1, & k = 0 \\ a, & k = 1 \\ 0, & \text{otherwise} \end{cases}$$

as the impulse response of the system that models direct transmission plus the echo. The inverse system $h^{-1}[n]$ must satisfy $h[n] * h^{-1}[n] = \delta[n]$. Substituting for $h[n]$, we desire to find $h^{-1}[n]$ that satisfies the equation

$$h^{-1}[n] + ah^{-1}[n-1] = \delta[n] \tag{2.23}$$

Consider solving this equation for several different values of n. For $n < 0$, we must have $h^{-1}[n] = 0$ in order to obtain a causal inverse system. For $n = 0$, $\delta[n] = 1$ and Eq. (2.23) implies

$$h^{-1}[0] + ah^{-1}[-1] = 1$$

so $h^{-1}[0] = 1$. For $n > 0$, $\delta[n] = 0$ and Eq. (2.23) implies

$$h^{-1}[n] + ah^{-1}[n-1] = 0$$

or $h^{-1}[n] = -ah^{-1}[n-1]$. Since $h^{-1}[0] = 1$, we have $h^{-1}[1] = -a$, $h^{-1}[2] = a^2$, $h^{-1}[3] = -a^3$, and so on. Hence the inverse system has the impulse response

$$h^{-1}[n] = (-a)^n u[n]$$

To check for stability, we determine whether $h^{-1}[n]$ is absolutely summable, as shown by

$$\sum_{k=-\infty}^{\infty} |h^{-1}[k]| = \sum_{k=0}^{\infty} |a|^k$$

This geometric series converges and hence the system is stable provided $|a| < 1$. This implies that the inverse system is stable if the echo attenuates the transmitted signal $x[n]$, but unstable if the echo amplifies $x[n]$.

Obtaining an inverse system by directly solving Eq. (2.21) or Eq. (2.22) is difficult in general. Furthermore, not every LTI system has a stable and causal inverse. Methods developed in later chapters provide additional insight into the existence and determination of inverse systems.

■ **STEP RESPONSE**

The response of a LTI system to a step characterizes how the system responds to sudden changes in the input. The *step response* is easily expressed in terms of the impulse response using convolution by assuming that the input is a step function. Let a discrete-time system have impulse response $h[n]$ and denote the step response as $s[n]$. We have

$$s[n] = h[n] * u[n]$$

$$= \sum_{k=-\infty}^{\infty} h[k]u[n-k]$$

Now, since $u[n - k] = 0$ for $k > n$ and $u[n - k] = 1$ for $k \leq n$, we have

$$s[n] = \sum_{k=-\infty}^{n} h[k]$$

That is, the step response is the running sum of the impulse response. Similarly, the step response, $s(t)$, for a continuous-time system is expressed as the running integral of the impulse response

$$s(t) = \int_{-\infty}^{t} h(\tau) \, d\tau \tag{2.24}$$

Note that we may invert these relationships to express the impulse response in terms of the step response as

$$h[n] = s[n] - s[n - 1]$$

and

$$h(t) = \frac{d}{dt} s(t)$$

EXAMPLE 2.13   Find the step response of the RC circuit depicted in Fig. 2.10 having impulse response

$$h(t) = \frac{1}{RC} e^{-t/RC} u(t)$$

Solution: Apply Eq. (2.24) to obtain

$$s(t) = \int_{-\infty}^{t} \frac{1}{RC} e^{-\tau/RC} u(\tau) \, d\tau$$

Now simplify the integral as

$$y(t) = \begin{cases} 0, & t \leq 0 \\ \dfrac{1}{RC} \displaystyle\int_{0}^{t} e^{-\tau/RC} \, d\tau, & t > 0 \end{cases}$$

$$= \begin{cases} 0, & t \leq 0 \\ 1 - e^{-t/RC}, & t > 0 \end{cases}$$

▶ **Drill Problem 2.7**   Find the step response of a discrete-time system with impulse response

$$h[n] = (-a)^n u[n]$$

assuming $|a| < 1$.

Answer:

$$s[n] = \frac{1 - (-a)^{n+1}}{1 + a} u[n]$$

◀

■ **SINUSOIDAL STEADY-STATE RESPONSE**

Sinusoidal input signals are often used to characterize the response of a system. Here we examine the relationship between the impulse response and the steady-state response of a

LTI system to a complex sinusoidal input. This relationship is easily established using convolution and a complex sinusoid input signal. Consider the output of a discrete-time system with impulse response $h[n]$ and unit-amplitude complex sinusoidal input $x[n] = e^{j\Omega n}$, given by

$$y[n] = \sum_{k=-\infty}^{\infty} h[k]x[n-k]$$

$$= \sum_{k=-\infty}^{\infty} h[k]e^{j\Omega(n-k)}$$

Factor $e^{j\Omega n}$ from the sum to obtain

$$y[n] = e^{j\Omega n} \sum_{k=-\infty}^{\infty} h[k]e^{-j\Omega k}$$

$$= H(e^{j\Omega})e^{j\Omega n}$$

where we have defined

$$H(e^{j\Omega}) = \sum_{k=-\infty}^{\infty} h[k]e^{-j\Omega k} \tag{2.25}$$

Hence the output of the system is a complex sinusoid of the same frequency as the input multiplied by the complex number $H(e^{j\Omega})$. This relationship is depicted in Fig. 2.17. The quantity $H(e^{j\Omega})$ is not a function of time, n, but is only a function of frequency, Ω, and is termed the *frequency response* of the discrete-time system.

Similar results are obtained for continuous-time systems. Let the impulse response of a system be $h(t)$ and the input be $x(t) = e^{j\omega t}$. The convolution integral gives the output as

$$y(t) = \int_{-\infty}^{\infty} h(\tau)e^{j\omega(t-\tau)} \, d\tau$$

$$= e^{j\omega t} \int_{-\infty}^{\infty} h(\tau)e^{-j\omega\tau} \, d\tau \tag{2.26}$$

$$= H(j\omega)e^{j\omega t}$$

where we define

$$H(j\omega) = \int_{-\infty}^{\infty} h(\tau)e^{-j\omega\tau} \, d\tau \tag{2.27}$$

The output of the system is a complex sinusoid of the same frequency as the input multiplied by the complex constant $H(j\omega)$. $H(j\omega)$ is a function of only frequency, ω, and not time, t. It is termed the *frequency response* of the continuous-time system.

An intuitive interpretation of the sinusoidal steady-state response is obtained by writing the complex number $H(j\omega)$ in polar form. Recall that if $c = a + jb$ is a complex number, then we may write c in polar form as $c = |c|e^{j \, \arg\{c\}}$, where $|c| = \sqrt{a^2 + b^2}$ and $\arg\{c\} = \arctan(b/a)$. Hence we have $H(j\omega) = |H(j\omega)|e^{j \, \arg\{H(j\omega)\}}$. Here $|H(j\omega)|$ is termed

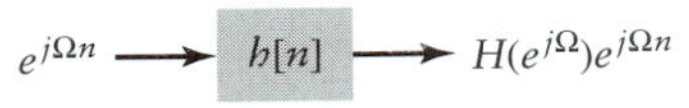

$$e^{j\Omega n} \longrightarrow \boxed{h[n]} \longrightarrow H(e^{j\Omega})e^{j\Omega n}$$

FIGURE 2.17 A complex sinusoidal input to a LTI system results in a complex sinusoidal output of the same frequency multiplied by the frequency response of the system.

the *magnitude response* and arg$\{H(j\omega)\}$ is termed the *phase response* of the system. Substituting this polar form in Eq. (2.26), the output $y(t)$ is expressed as

$$y(t) = |H(j\omega)| e^{j(\omega t + \arg\{H(j\omega)\})}$$

The system modifies the amplitude of the input by $|H(j\omega)|$ and the phase by arg$\{H(j\omega)\}$. The sinusoidal steady-state response has a similar interpretation for real-valued sinusoids. Write

$$x(t) = A\cos(\omega t + \phi)$$
$$= \frac{A}{2} e^{j(\omega t + \phi)} + \frac{A}{2} e^{-j(\omega t + \phi)}$$

and use linearity to obtain the output as

$$y(t) = |H(j\omega)| \frac{A}{2} e^{j(\omega t + \phi + \arg\{H(j\omega)\})} + |H(-j\omega)| \frac{A}{2} e^{-j(\omega t + \phi - \arg\{H(-j\omega)\})}$$

Assuming that $h(t)$ is real valued, Eq. (2.27) implies that $H(j\omega)$ possesses conjugate symmetry, that is, $H^*(j\omega) = H(-j\omega)$. This implies that $|H(j\omega)|$ is an even function of ω while arg$\{H(j\omega)\}$ is odd. Exploiting these symmetry conditions and simplifying yields

$$y(t) = |H(j\omega)| A\cos(\omega t + \phi + \arg\{H(j\omega)\})$$

As with a complex sinusoidal input, the system modifies the input sinusoid's amplitude by $|H(j\omega)|$ and the phase by arg$\{H(j\omega)\}$. This modification is illustrated in Fig. 2.18.

Similar results are obtained for discrete-time systems using the polar form for $H(e^{j\Omega})$. Specifically, if $x[n] = e^{j\Omega n}$ is the input, then

$$y[n] = |H(e^{j\Omega})| e^{j(\Omega n + \arg\{H(e^{j\Omega})\})}$$

Furthermore, if $x[n] = A\cos(\Omega n + \phi)$ is the input to a discrete-time system with real-valued impulse response, then

$$y[n] = |H(e^{j\Omega})| A\cos(\Omega n + \phi + \arg\{H(e^{j\Omega})\})$$

Once again, the system changes the amplitude of the sinusoidal input by $|H(e^{j\Omega})|$ and its phase by arg$\{H(e^{j\Omega})\}$.

The frequency response characterizes the steady-state response of the system to sinusoidal inputs as a function of the sinusoid's frequency. We say this is a steady-state response because the input sinusoid is assumed to exist for all time and thus the system is in an equilibrium or steady-state condition. The frequency response provides a great deal of information about the system and is useful for both understanding and analyzing systems, topics that are explored in depth in later chapters. It is easily measured with a

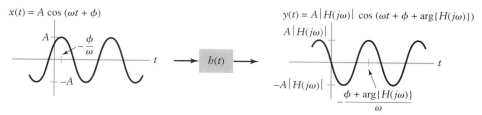

FIGURE 2.18   A sinusoidal input to a LTI system results in a sinusoidal output of the same frequency with the amplitude and phase modified by the system's frequency response.

sinusoidal oscillator and oscilloscope by using the oscilloscope to measure the amplitude and phase change between the input and output sinusoids for different oscillator frequencies.

It is standard practice to represent the frequency response graphically by separately displaying the magnitude and phase response as functions of frequency, as illustrated in the following examples.

EXAMPLE 2.14   The impulse responses of two discrete-time systems are given by

$$h_1[n] = \tfrac{1}{2}(\delta[n] + \delta[n-1])$$
$$h_2[n] = \tfrac{1}{2}(\delta[n] - \delta[n-1])$$

Find the frequency response of each system and plot the magnitude responses.

Solution: Substitute $h_1[n]$ into Eq. (2.25) to obtain

$$H_1(e^{j\Omega}) = \frac{1 + e^{-j\Omega}}{2}$$

which may be rewritten as

$$H_1(e^{j\Omega}) = e^{-j\Omega/2}\,\frac{e^{j\Omega/2} + e^{-j\Omega/2}}{2}$$
$$= e^{-j\Omega/2}\cos(\Omega/2)$$

Hence the magnitude response is expressed as

$$|H_1(e^{j\Omega})| = |\cos(\Omega/2)|$$

and the phase response is expressed as

$$\arg\{H_1(e^{j\Omega})\} = \begin{cases} -\Omega/2 & \text{for } \cos(\Omega/2) > 0 \\ -\Omega/2 - \pi & \text{for } \cos(\Omega/2) < 0 \end{cases}$$

Similarly, the frequency response of the second system is given by

$$H_2(e^{j\Omega}) = \frac{1 - e^{-j\Omega}}{2}$$
$$= je^{-j\Omega/2}\,\frac{e^{j\Omega/2} - e^{-j\Omega/2}}{2j}$$
$$= je^{-j\Omega/2}\sin(\Omega/2)$$

In this case the magnitude response is expressed as

$$|H_2(e^{j\Omega})| = |\sin(\Omega/2)|$$

and the phase response is expressed as

$$\arg\{H_2(e^{j\Omega})\} = \begin{cases} -\Omega/2 + \pi/2 & \text{for } \sin(\Omega/2) > 0 \\ -\Omega/2 - \pi/2 & \text{for } \sin(\Omega/2) < 0 \end{cases}$$

Figures 2.19(a) and (b) depict the magnitude response of each system on the interval $-\pi < \Omega \le \pi$. This interval is chosen because it corresponds to the range of frequencies for which the complex sinusoid $e^{j\Omega n}$ is a unique function of frequency. The convolution sum indicates that $h_1[n]$ averages successive inputs, while $h_2[n]$ takes the difference of successive inputs. Thus we expect $h_1[n]$ to pass low-frequency signals while attenuating high frequencies. This characteristic is reflected by the magnitude response. In contrast, the differencing operation implemented by $h_2[n]$ has the effect of attenuating low frequencies and passing high frequencies, as indicated by its magnitude response.

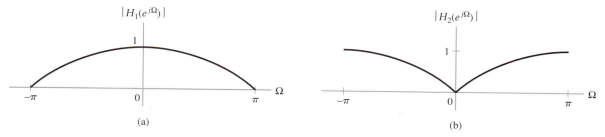

FIGURE 2.19   The magnitude responses of two simple discrete-time systems. (a) A system that averages successive inputs tends to attenuate high frequencies. (b) A system that forms the difference of successive inputs tends to attenuate low frequencies.

EXAMPLE 2.15   The impulse response of the system relating the input voltage to the voltage across the capacitor in Fig. 2.10 is given by

$$h(t) = \frac{1}{RC} e^{-t/RC} u(t)$$

Find an expression for the frequency response and plot the magnitude and phase response.

Solution: Substituting $h(t)$ into Eq. (2.27) gives

$$H(j\omega) = \frac{1}{RC} \int_{-\infty}^{\infty} e^{-\tau/RC} u(\tau) e^{-j\omega\tau} \, d\tau$$

$$= \frac{1}{RC} \int_{0}^{\infty} e^{-(j\omega + 1/RC)\tau} \, d\tau$$

$$= \frac{1}{RC} \frac{-1}{(j\omega + 1/RC)} e^{-(j\omega + 1/RC)\tau} \Big|_{0}^{\infty}$$

$$= \frac{1}{RC} \frac{-1}{(j\omega + 1/RC)} (0 - 1)$$

$$= \frac{1/RC}{j\omega + 1/RC}$$

The magnitude response is

$$|H(j\omega)| = \frac{\dfrac{1}{RC}}{\sqrt{\omega^2 + \left(\dfrac{1}{RC}\right)^2}}$$

while the phase response is

$$\arg\{H(j\omega)\} = -\arctan(\omega RC)$$

The magnitude response and phase response are presented in Figs. 2.20(a) and (b), respectively. The magnitude response indicates that the RC circuit tends to attenuate high-frequency sinusoids. This agrees with our intuition from circuit analysis. The circuit cannot respond to rapid changes in the input voltage. High-frequency sinusoids also experience a phase shift of $\pi/2$ radians. Low-frequency sinusoids are passed by the circuit with much higher gain and experience relatively little phase shift.

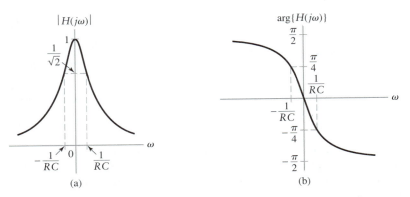

FIGURE 2.20   Frequency response of the RC circuit in Fig. 2.10. (a) Magnitude response. (b) Phase response.

▶ **Drill Problem 2.8**   Find an expression for the frequency response of the discrete-time system with impulse response

$$h[n] = (-a)^n u[n]$$

assuming $|a| < 1$.

Answer:

$$H(e^{j\Omega}) = \frac{1}{1 + ae^{-j\Omega}}$$

◀

2.4 Differential and Difference Equation Representations for LTI Systems

Linear constant-coefficient difference and differential equations provide another representation for the input–output characteristics of LTI systems. Difference equations are used to represent discrete-time systems, while differential equations represent continuous-time systems. The general form of a linear constant-coefficient differential equation is

$$\sum_{k=0}^{N} a_k \frac{d^k}{dt^k} y(t) = \sum_{k=0}^{M} b_k \frac{d^k}{dt^k} x(t) \tag{2.28}$$

Here $x(t)$ is the input to the system and $y(t)$ is the output. A linear constant-coefficient difference equation has a similar form, with the derivatives replaced by delayed values of the input $x[n]$ and output $y[n]$, as shown by

$$\sum_{k=0}^{N} a_k y[n - k] = \sum_{k=0}^{M} b_k x[n - k] \tag{2.29}$$

The integer N is termed the *order* of the differential or difference equation and corresponds to the highest derivative or maximum memory involving the system output, respectively. The order represents the number of energy storage devices in the system.

As an example of a differential equation that describes the behavior of a physical system, consider the RLC circuit depicted in Fig. 2.21(a). Assume the input is the voltage

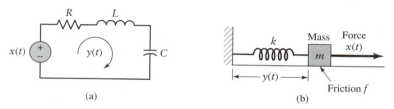

FIGURE 2.21 Examples of systems described by differential equations. (a) *RLC* circuit. (b) Spring-mass-damper system.

source $x(t)$ and the output is the current around the loop, $y(t)$. Summing the voltage drops around the loop gives

$$Ry(t) + L\frac{dy(t)}{dt} + \frac{1}{C}\int_{-\infty}^{t} y(\tau)\,d\tau = x(t)$$

Differentiating both sides of this equation with respect to t gives

$$\frac{1}{C}y(t) + R\frac{dy(t)}{dt} + L\frac{d^2y(t)}{dt^2} = \frac{dx(t)}{dt}$$

This differential equation describes the relationship between the current $y(t)$ and voltage $x(t)$ in the circuit. In this example, the order is $N = 2$ and we note that the circuit contains two energy storage devices, a capacitor and an inductor.

Mechanical systems may also be described in terms of differential equations using Newton's laws. In the system depicted in Fig. 2.21(b), the applied force, $x(t)$, is the input and the position of the mass, $y(t)$, is the output. The force associated with the spring is directly proportional to position, the force due to friction is directly proportional to velocity, and the force due to mass is proportional to acceleration. Equating the forces on the mass gives

$$m\frac{d^2}{dt^2}y(t) + f\frac{d}{dt}y(t) + ky(t) = x(t)$$

This differential equation relates position to the applied force. The system contains two energy storage mechanisms, a spring and a mass, and the order is $N = 2$.

An example of a second-order difference equation is

$$y[n] + y[n-1] + \tfrac{1}{4}y[n-2] = x[n] + 2x[n-1] \qquad (2.30)$$

This difference equation might represent the relationship between the input and output signals for a system that processes data in a computer. In this example the order is $N = 2$ because the difference equation involves $y[n-2]$, implying a maximum memory in the system output of 2. Memory in a discrete-time system is analogous to energy storage in a continuous-time system.

Difference equations are easily rearranged to obtain recursive formulas for computing the current output of the system from the input signal and past outputs. Rewrite Eq. (2.29) so that $y[n]$ is alone on the left-hand side, as shown by

$$y[n] = \frac{1}{a_0}\sum_{k=0}^{M} b_k x[n-k] - \frac{1}{a_0}\sum_{k=1}^{N} a_k y[n-k]$$

This equation indicates how to obtain $y[n]$ from the input and past values of the output. Such equations are often used to implement discrete-time systems in a computer. Consider computing $y[n]$ for $n \geq 0$ from $x[n]$ for the example second-order difference equation given in Eq. (2.30). We have

$$y[n] = x[n] + 2x[n-1] - y[n-1] - \tfrac{1}{4}y[n-2]$$

Beginning with $n = 0$, we may determine the output by evaluating the sequence of equations

$$y[0] = x[0] + 2x[-1] - y[-1] - \tfrac{1}{4}y[-2]$$
$$y[1] = x[1] + 2x[0] - y[0] - \tfrac{1}{4}y[-1]$$
$$y[2] = x[2] + 2x[1] - y[1] - \tfrac{1}{4}y[0]$$
$$y[3] = x[3] + 2x[2] - y[2] - \tfrac{1}{4}y[1]$$
$$\vdots$$

In each equation the current output is computed from the input and past values of the output. In order to begin this process at time $n = 0$, we must know the two most recent past values of the output, namely, $y[-1]$ and $y[-2]$. These values are known as *initial conditions*. This technique for finding the output of a system is very useful for computation but does not provide much insight into the relationship between the difference equation description and system characteristics.

The initial conditions summarize all the information about the system's past that is needed to determine future outputs. No additional information about the past output or input is necessary. Note that in general the number of initial conditions required to determine the output is equal to the order of the system. Initial conditions are also required to solve differential equations. In this case, the initial conditions are the values of the first N derivatives of the output

$$y(t), \frac{dy(t)}{dt}, \frac{d^2y(t)}{dt^2}, \ldots, \frac{d^{N-1}y(t)}{dt^{N-1}}$$

evaluated at the time t_0 after which we desire to determine $y(t)$. The initial conditions in a differential-equation description for a LTI system are directly related to the initial values of the energy storage devices in the system, such as initial voltages on capacitors and initial currents through inductors. As in the discrete-time case, the initial conditions summarize all information about the past of the system that can impact future outputs. Hence initial conditions also represent the "memory" of continuous-time systems.

EXAMPLE 2.16   A system is described by the difference equation

$$y[n] - 1.143y[n-1] + 0.4128y[n-2] = 0.0675x[n] + 0.1349x[n-1] + 0.0675x[n-2]$$

Write a recursive formula to compute the present output from the past outputs and current inputs. Determine the step response of the system, the system output when the input is zero and the initial conditions are $y[-1] = 1$, $y[-2] = 2$, and the output in response to the sinusoidal inputs $x_1[n] = \cos(\tfrac{1}{10}\pi n)$, $x_2[n] = \cos(\tfrac{1}{5}\pi n)$, and $x_3[n] = \cos(\tfrac{7}{10}\pi n)$ assuming zero initial conditions. Lastly, find the output of the system if the input is the average January temperature data depicted in Fig. 2.22(f).

Solution: We rewrite the difference equation as shown by

$$y[n] = 1.143y[n-1] - 0.4128y[n-2] + 0.0675x[n] + 0.1349x[n-1] + 0.0675x[n-2]$$

This equation is evaluated in a recursive manner to determine the system output from the input and initial conditions $y[-1]$ and $y[-2]$.

The step response of the system is evaluated by assuming the input is a step, $x[n] = u[n]$, and that the system is initially at rest, so the initial conditions are zero. Figure 2.22(a)

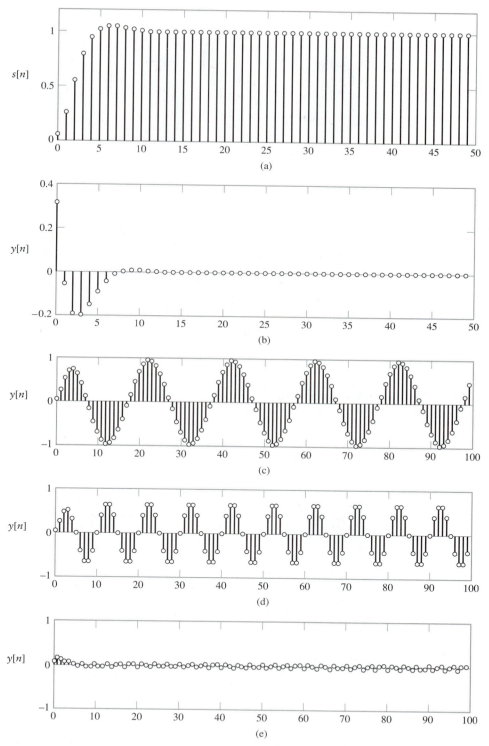

FIGURE 2.22 Illustration of the solution to Example 2.16. (a) Step response of system. (b) Output due to nonzero initial conditions with zero input. (c) Output due to $x_1[n] = \cos(\frac{1}{10}\pi n)$. (d) Output due to $x_2[n] = \cos(\frac{1}{5}\pi n)$. (e) Output due to $x_3[n] = \cos(\frac{7}{10}\pi n)$.

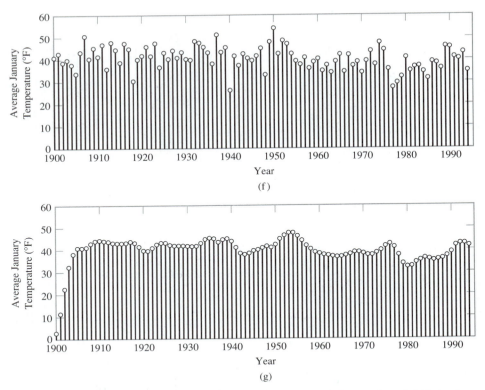

FIGURE 2.22   (f) Input signal consisting of average January temperature data. (g) Output associated with average January temperature data.

depicts the first 50 values of the step response. This system responds to a step by initially rising to a value slightly greater than the input amplitude and then decreasing to the value of the input at about $n = 13$. For n sufficiently large, we may consider the step to be a dc or constant input. Since the output amplitude is equal to the input amplitude, we see that this system has unit gain to constant inputs.

The response of the system to the initial conditions $y[-1] = 1$, $y[-2] = 2$ and zero input is shown in Fig. 2.22(b). Although the recursive nature of the difference equation suggests that the initial conditions affect all future values of the output, we see that the significant portion of the output due to the initial conditions lasts until about $n = 13$.

The outputs due to the sinusoidal inputs $x_1[n]$, $x_2[n]$, and $x_3[n]$ are depicted in Figs. 2.22(c), (d), and (e), respectively. Once we are distant from the initial conditions and enter a steady-state condition, we see that the system output is a sinusoid of the same frequency as the input. Recall that the ratio of the steady-state output to input sinusoid amplitude is the magnitude response of the system. The magnitude response at frequency $\frac{1}{10}\pi$ is unity, is about 0.7 at frequency $\frac{1}{5}\pi$, and is near zero at frequency $\frac{7}{10}\pi$. These results suggest that the magnitude response of this system decreases as frequency increases: that is, the system attenuates the components of the input that vary rapidly, while passing with unit gain those that vary slowly. This characteristic is evident in the output of the system in response to the average January temperature input shown in Fig. 2.22(g). We see that the output initially increases gradually in the same manner as the step response. This is a consequence of assuming the input is zero prior to 1900. After about 1906, the system has a smoothing effect since it attenuates rapid fluctuations in the input and passes constant terms with zero gain.

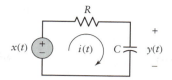

FIGURE 2.23   *RC* circuit.

▶ **Drill Problem 2.9**   Write a differential equation describing the relationship between the input voltage $x(t)$ and voltage $y(t)$ across the capacitor in Fig. 2.23.

Answer:

$$RC\,\frac{dy(t)}{dt} + y(t) = x(t)$$

◀

■ SOLVING DIFFERENTIAL AND DIFFERENCE EQUATIONS

We now briefly review a method for solving differential and difference equations. This offers a general characterization of solutions that provides insight into system behavior.

It is convenient to express the output of a system described by a differential or difference equation as a sum of two components: one associated only with initial conditions, and a second due only to the input. We shall term the component of the output associated with the initial conditions the *natural response* of the system and denote it as $y^{(n)}$. The component of the output due only to the input is termed the *forced response* of the system and denoted as $y^{(f)}$. The natural response is the system output for zero input, while the forced response is the system output assuming zero initial conditions. A system with zero initial conditions is said to be at rest, since there is no stored energy or memory in the system. The natural response describes the manner in which the system dissipates any energy or memory of the past represented by nonzero initial conditions. The forced response describes the system behavior that is "forced" by the input when the system is at rest.

The Natural Response

The natural response is the system output when the input is zero. Hence for a continuous-time system the natural response, $y^{(n)}(t)$, is the solution to the homogeneous equation

$$\sum_{k=0}^{N} a_k \frac{d^k}{dt^k} y^{(n)}(t) = 0$$

The natural response for a continuous-time system is of the form

$$y^{(n)}(t) = \sum_{i=1}^{N} c_i e^{r_i t} \tag{2.31}$$

where the r_i are the N roots of the system's characteristic equation

$$\sum_{k=0}^{N} a_k r^k = 0 \tag{2.32}$$

Substitution of Eq. (2.31) into the homogeneous equation establishes that $y^{(n)}(t)$ is a solution for any set of constants c_i.

In discrete time the natural response, $y^{(n)}[n]$, is the solution to the homogeneous equation

$$\sum_{k=0}^{N} a_k y^{(n)}[n-k] = 0$$

It is of the form

$$y^{(n)}[n] = \sum_{i=1}^{N} c_i r_i^n \tag{2.33}$$

where the r_i are the N roots of the discrete-time system's characteristic equation

$$\sum_{k=0}^{N} a_k r^{N-k} = 0 \tag{2.34}$$

Again, substitution of Eq. (2.33) into the homogeneous equation establishes that $y^{(n)}[n]$ is a solution. In both cases, the c_i are determined so that the solution $y^{(n)}$ satisfies the initial conditions. Note that the continuous-time and discrete-time characteristic equations differ.

The form of the natural response changes slightly when the characteristic equation described by Eq. (2.32) or Eq. (2.34) has repeated roots. If a root r_j is repeated p times, then we include p distinct terms in the solutions Eqs. (2.31) and (2.33) associated with r_j. They involve the p functions

$$e^{r_j t}, \; t e^{r_j t}, \; \ldots, \; t^{p-1} e^{r_j t}$$

and

$$r_j^n, \; n r_j^n, \; \ldots, \; n^{p-1} r_j^n$$

respectively.

The nature of each term in the natural response depends on whether the roots r_i are real, imaginary, or complex. Real roots lead to real exponentials, imaginary roots to sinusoids, and complex roots to exponentially damped sinusoids.

EXAMPLE 2.17 Consider the RL circuit depicted in Fig. 2.24 as a system whose input is the applied voltage $x(t)$ and output is the current $y(t)$. Find a differential equation that describes this system and determine the natural response of the system for $t > 0$ assuming the current through the inductor at $t = 0$ is $y(0) = 2$ A.

Solution: Summing the voltages around the loop gives the differential equation

$$Ry(t) + L\frac{dy(t)}{dt} = x(t)$$

The natural response is the solution of the homogeneous equation

$$Ry(t) + L\frac{dy(t)}{dt} = 0$$

The solution is given by Eq. (2.31) for $N = 1$,

$$y^{(n)}(t) = c_1 e^{r_1 t} \text{ A}$$

where r_1 is the root of the equation

$$R + Lr = 0$$

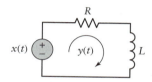

FIGURE 2.24   *RL* circuit.

Hence $r_1 = -R/L$. The coefficient c_1 is determined so that the response satisfies the initial condition $y(0) = 2$. This implies $c_1 = 2$ and the natural response of this system is

$$y^{(n)}(t) = 2e^{-(R/L)t} \text{ A}, \; t \geq 0$$

▶ **Drill Problem 2.10**   Determine the form of the natural response for the system described by the difference equation

$$y[n] + \tfrac{1}{4}y[n-2] = x[n] + 2x[n-2]$$

Answer:

$$y^{(n)}[n] = c_1(\tfrac{1}{2}e^{j\pi/2})^n + c_2(\tfrac{1}{2}e^{-j\pi/2})^n \qquad \blacktriangleleft$$

▶ **Drill Problem 2.11**   Determine the form of the natural response for the *RLC* circuit depicted in Fig. 2.21(a) as a function of R, L, and C. Indicate the conditions on R, L, and C so that the natural response consists of real exponentials, complex sinusoids, and exponentially damped sinusoids.

Answer: For $R^2 \neq 4L/C$,

$$y^{(n)}(t) = c_1 e^{r_1 t} + c_2 e^{r_2 t}$$

where

$$r_1 = \frac{-R + \sqrt{R^2 - 4L/C}}{2L}, \qquad r_2 = \frac{-R - \sqrt{R^2 - 4L/C}}{2L}$$

For $R^2 = 4L/C$,

$$y^{(n)}(t) = c_1 e^{-(R/2L)t} + c_2 t e^{-(R/2L)t}$$

For real exponentials $R^2 > 4L/C$, for complex sinusoids $R = 0$, and for exponentially damped sinusoids $R^2 < 4L/C$. $\qquad \blacktriangleleft$

The Forced Response

The forced response is the solution to the differential or difference equation for the given input assuming the initial conditions are zero. It consists of the sum of two components: a term of the same form as the natural response, and a particular solution.

The particular solution is denoted as $y^{(p)}$ and represents any solution to the differential or difference equation for the given input. It is usually obtained by assuming the system output has the same general form as the input. For example, if the input to a discrete-time system is $x[n] = \alpha^n$, then we assume the output is of the form $y^{(p)}[n] = c\alpha^n$ and find the constant c so that $y^{(p)}[n]$ is a solution to the system's difference equation. If the input is $x[n] = A\cos(\Omega n + \phi)$, then we assume a general sinusoidal response of the

TABLE 2.1   *Form of a Particular Solution Corresponding to Several Common Inputs*

Continuous Time		Discrete Time	
Input	Particular Solution	Input	Particular Solution
1	c	1	c
e^{-at}	ce^{-at}	α^n	$c\alpha^n$
$\cos(\omega t + \phi)$	$c_1 \cos(\omega t) + c_2 \sin(\omega t)$	$\cos(\Omega n + \phi)$	$c_1 \cos(\Omega n) + c_2 \sin(\Omega n)$

form $y^{(p)}[n] = c_1 \cos(\Omega n) + c_2 \sin(\Omega n)$, where c_1 and c_2 are determined so that $y^{(p)}[n]$ satisfies the system's difference equation. Assuming an output of the same form as the input is consistent with our expectation that the output of the system be directly related to the input.

The form of the particular solution associated with common input signals is given in Table 2.1. More extensive tables are given in books devoted to solving difference and differential equations, such as those listed at the end of this chapter. The procedure for identifying a particular solution is illustrated in the following example.

EXAMPLE 2.18   Consider the *RL* circuit of Example 2.17 and depicted in Fig. 2.24. Find a particular solution for this system with an input $x(t) = \cos(\omega_0 t)$ V.

Solution: The differential equation describing this system was obtained in Example 2.17 as

$$Ry(t) + L\frac{dy(t)}{dt} = x(t)$$

We assume a particular solution of the form $y^{(p)}(t) = c_1 \cos(\omega_0 t) + c_2 \sin(\omega_0 t)$. Replacing $y(t)$ in the differential equation by $y^{(p)}(t)$ and $x(t)$ by $\cos(\omega_0 t)$ gives

$$Rc_1 \cos(\omega_0 t) + Rc_2 \sin(\omega_0 t) - L\omega_0 c_1 \sin(\omega_0 t) + L\omega_0 c_2 \cos(\omega_0 t) = \cos(\omega_0 t)$$

The coefficients c_1 and c_2 are obtained by separately equating the coefficients of $\cos(\omega_0 t)$ and $\sin(\omega_0 t)$. This gives a system of two equations in two unknowns, as shown by

$$Rc_1 + L\omega_0 c_2 = 1$$
$$-L\omega_0 c_1 + Rc_2 = 0$$

Solving these for c_1 and c_2 gives

$$c_1 = \frac{R}{R^2 + L^2\omega_0^2}$$

$$c_2 = \frac{L\omega_0}{R^2 + L^2\omega_0^2}$$

Hence the particular solution is

$$y^{(p)}(t) = \frac{R}{R^2 + L^2\omega_0^2} \cos(\omega_0 t) + \frac{L\omega_0}{R^2 + L^2\omega_0^2} \sin(\omega_0 t)   \text{A}$$

This approach for finding a particular solution is modified when the input is of the same form as one of the components of the natural response. In this case we must assume a particular solution that is independent of all terms in the natural response in order to obtain the forced response of the system. This is accomplished analogously to the procedure for generating independent natural response components when there are repeated roots in the characteristic equation. Specifically, we multiply the form of the particular solution by the lowest power of t or n that will give a response component not included in the natural response. For example, if the natural response contains the terms e^{-at} and te^{-at} due to a second-order root at $-a$, and the input is $x(t) = e^{-at}$, then we assume a particular solution of the form $y^{(p)}(t) = ct^2 e^{-at}$.

The forced response of the system is obtained by summing the particular solution with the form of the natural response and finding the unspecified coefficients in the natural response so that the combined response satisfies zero initial conditions. Assuming the input is applied at time $t = 0$ or $n = 0$, this procedure is as follows:

1. Find the form of the natural response $y^{(n)}$ from the roots of the characteristic equation.

2. Find a particular solution $y^{(p)}$ by assuming it is of the same form as the input yet independent of all terms in the natural response.

3. Determine the coefficients in the natural response so that the forced response $y^{(f)} = y^{(p)} + y^{(n)}$ has zero initial conditions at $t = 0$ or $n = 0$. The forced response is valid for $t \geq 0$ or $n \geq 0$.

In the discrete-time case, the zero initial conditions, $y^{(f)}[-N], \ldots, y^{(f)}[-1]$, must be translated to times $n \geq 0$, since the forced response is valid only for times $n \geq 0$. This is accomplished by using the recursive form of the difference equation, the input, and the at-rest conditions $y^{(f)}[-N] = 0, \ldots, y^{(f)}[-1] = 0$ to obtain translated initial conditions $y^{(f)}[0], y^{(f)}[1], \ldots, y^{(f)}[N - 1]$. These are then used to determine the unknown coefficients in the natural response component of $y^{(f)}[n]$.

EXAMPLE 2.19 Find the forced response of the RL circuit depicted in Fig. 2.24 to an input $x(t) = \cos(t)$ V assuming normalized values $R = 1 \ \Omega$ and $L = 1$ H.

Solution: The form of the natural response was obtained in Example 2.17 as

$$y^{(n)}(t) = ce^{-(R/L)t} \quad \text{A}$$

A particular solution was obtained in Example 2.18 for this input as

$$y^{(p)}(t) = \frac{R}{R^2 + L^2} \cos(t) + \frac{L}{R^2 + L^2} \sin(t) \quad \text{A}$$

where we have used $\omega_0 = 1$. Substituting $R = 1 \ \Omega$ and $L = 1$ H, the forced response for $t \geq 0$ is

$$y^{(f)}(t) = ce^{-t} + \tfrac{1}{2} \cos t + \tfrac{1}{2} \sin t \quad \text{A}$$

The coefficient c is now determined from the initial condition $y(0) = 0$

$$0 = ce^{-0} + \tfrac{1}{2} \cos 0 + \tfrac{1}{2} \sin 0$$
$$= c + \tfrac{1}{2}$$

and so we find that $c = -\tfrac{1}{2}$.

▶ **Drill Problem 2.12** A system described by the difference equation

$$y[n] - \tfrac{1}{4}y[n - 2] = 2x[n] + x[n - 1]$$

has input signal $x[n] = u[n]$. Find the forced response of the system. *Hint:* Use $y[n] = \frac{1}{4}y[n-2] + 2x[n] + x[n-1]$ with $x[n] = u[n]$ and $y^{(f)}[-2] = 0$, $y^{(f)}[-1] = 0$ to determine $y^{(f)}[0]$ and $y^{(f)}[1]$.

Answer:

$$y[n] = (-2(\tfrac{1}{2})^n + 4)\, u[n]$$ ◄

The Complete Response

The complete response of the system is the sum of the natural response and the forced response. If there is no need to separately obtain the natural and the forced response, then the complete response of the system may be obtained directly by repeating the three-step procedure for determining the forced response using the actual initial conditions instead of zero initial conditions. This is illustrated in the following example.

EXAMPLE 2.20   Find the current through the *RL* circuit depicted in Fig. 2.24 for an applied voltage $x(t) = \cos(t)$ V assuming normalized values $R = 1\ \Omega$, $L = 1$ H and that the initial condition is $y(0) = 2$ A.

Solution: The form of the forced response was obtained in Example 2.19 as

$$y(t) = ce^{-t} + \tfrac{1}{2}\cos t + \tfrac{1}{2}\sin t \quad \text{A}$$

We obtain the complete response of the system by solving for c so that the initial condition $y(0) = 2$ is satisfied. This implies

$$2 = c + \tfrac{1}{2}(1) + \tfrac{1}{2}(0)$$

or $c = \tfrac{3}{2}$. Hence

$$y(t) = \tfrac{3}{2}e^{-t} + \tfrac{1}{2}\cos t + \tfrac{1}{2}\sin t \quad \text{A}, \; t \geq 0$$

Note that this corresponds to the sum of the natural and forced responses. In Example 2.17 we obtained

$$y^{(n)}(t) = 2e^{-t} \quad \text{A}, \; t \geq 0$$

while in Example 2.19 we obtained

$$y^{(f)}(t) = -\tfrac{1}{2}e^{-t} + \tfrac{1}{2}\cos t + \tfrac{1}{2}\sin t \quad \text{A}, \; t \geq 0$$

The sum, $y(t) = y^{(n)}(t) + y^{(f)}(t)$, is given by

$$y(t) = \tfrac{3}{2}e^{-t} + \tfrac{1}{2}\cos t + \tfrac{1}{2}\sin t \quad \text{A}, \; t \geq 0$$

and is exactly equal to the response we obtained by directly solving for the complete response. Figure 2.25 depicts the natural, forced, and complete responses of the system.

▶ **Drill Problem 2.13**   Find the response of the *RC* circuit depicted in Fig. 2.23 to $x(t) = u(t)$, assuming the initial voltage across the capacitor is $y(0) = -1$ V.

Answer:

$$y(t) = (1 - 2e^{-t/RC}) \quad \text{V}, \; t \geq 0$$ ◄

The Impulse Response

The method described thus far for solving differential and difference equations cannot be used to find the impulse response directly. However, the impulse response is easily

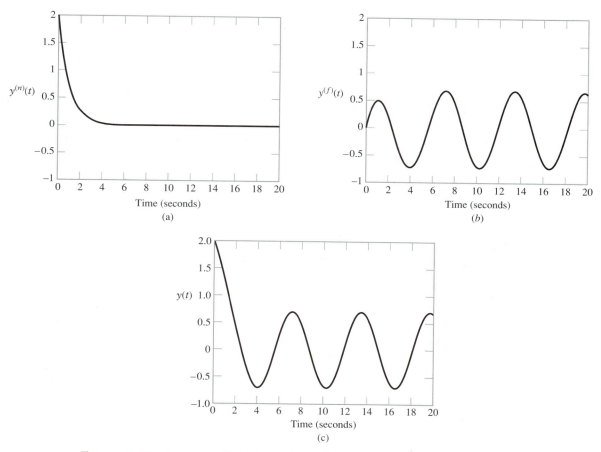

FIGURE 2.25 Response of *RL* circuit depicted in Fig. 2.24 to input $x(t) = \cos(t)$ V when $y(0) = 2$ A. (See Example 2.20.) (a) Natural response. (b) Forced response. (c) Complete response.

determined by first finding the step response and then exploiting the relationship between the impulse and step response. The definition of the step response assumes the system is at rest, so it represents the forced response of the system to a step input. For a continuous-time system, the impulse response, $h(t)$, is related to the step response, $s(t)$, as $h(t) = \frac{d}{dt} s(t)$. For a discrete-time system we have $h[n] = s[n] - s[n-1]$. Thus the impulse response is obtained by differentiating or differencing the step response. The differentiation and differencing operations eliminate the constant term associated with the particular solution in the step response and change only the constants associated with the exponential terms in the natural response component. This implies that the impulse response is only a function of the terms in the natural response.

■ CHARACTERISTICS OF SYSTEMS DESCRIBED BY DIFFERENTIAL AND DIFFERENCE EQUATIONS

The forced response of a LTI system described by a differential or difference equation is linear with respect to the input. If $y_1^{(f)}$ is the forced response associated with an input x_1 and $y_2^{(f)}$ is the forced response associated with an input x_2, then the input $\alpha x_1 + \beta x_2$

generates a forced response given by $\alpha y_1^{(f)} + \beta y_2^{(f)}$. Similarly, the natural response is linear with respect to the initial conditions. If $y_1^{(n)}$ is the natural response associated with initial conditions I_1 and $y_2^{(n)}$ is the natural response associated with initial conditions I_2, then the initial condition $\alpha I_1 + \beta I_2$ results in a natural response $\alpha y_1^{(n)} + \beta y_2^{(n)}$. The forced response is also time invariant. A time shift in the input results in a time shift in the output since the system is initially at rest. In general, the complete response of a system described by a differential or difference equation is not time invariant, since the initial conditions will result in an output term that does not shift with a time shift of the input. Lastly, we observe that the forced response is also causal. Since the system is initially at rest, the output does not begin prior to the time at which the input is applied to the system.

The forced response depends on both the input and the roots of the characteristic equation since it involves both the basic form of the natural response and a particular solution to the differential or difference equation. The basic form of the natural response is dependent entirely on the roots of the characteristic equation. The impulse response of the system also depends on the roots of the characteristic equation since it contains the identical terms as the natural response. Thus the roots of the characteristic equation provide considerable information about the system behavior.

For example, the stability characteristics of a system are directly related to the roots of the system's characteristic equation. To see this, note that the output of a stable system in response to zero input must be bounded for any set of initial conditions. This follows from the definition of BIBO stability and implies that the natural response of the system must be bounded. Thus each term in the natural response must be bounded. In the discrete-time case we must have $|r_i^n|$ bounded, or $|r_i| < 1$. When $|r_i| = 1$, the natural response does not decay and the system is said to be on the verge of instability. For continuous-time systems we require that $|e^{r_i t}|$ be bounded, which implies $\text{Re}\{r_i\} < 0$. Here again, when $\text{Re}\{r_i\} = 0$, the system is said to be on the verge of instability. These results imply that a discrete-time system is unstable if any root of the characteristic equation has magnitude greater than unity, and a continuous-time system is unstable if the real part of any root of the characteristic equation is positive.

This discussion establishes that the roots of the characteristic equation indicate when a system is unstable. In later chapters we establish that a discrete-time causal system is stable if and only if all roots of the characteristic equation have magnitude less than unity, and a continuous-time causal system is stable if and only if the real parts of all roots of the characteristic equation are negative. These stability conditions imply that the natural response of a system goes to zero as time approaches infinity since each term in the natural response is a decaying exponential. This "decay to zero" is consistent with our intuitive concept of a system's zero input behavior. We expect a zero output when the input is zero. The initial conditions represent any energy present in the system; in a stable system with zero input this energy eventually dissipates and the output approaches zero.

The response time of a system is also determined by the roots of the characteristic equation. Once the natural response has decayed to zero, the system behavior is governed only by the particular solution—which is of the same form as the input. Thus the natural response component describes the transient behavior of the system: that is, it describes the transition of the system from its initial condition to an equilibrium condition determined by the input. Hence the transient response time of a system is determined by the time it takes the natural response to decay to zero. Recall that natural response contains terms of the form r_i^n for a discrete-time system and $e^{r_i t}$ for a continuous-time system. The transient response time of a discrete-time system is therefore proportional to the magnitude of the largest root of the characteristic equation, while that of a continuous-time system is determined by the root whose real component is closest to zero. In order to have a continuous-time system with a fast response time, all the roots of the characteristic equation must have large and negative real parts.

The impulse response of the system can be determined directly from the differential- or difference-equation description of a system, although it is generally much easier to obtain the impulse response indirectly using methods described in later chapters. Note that there is no provision for initial conditions when using the impulse response; it applies only to systems that are initially at rest or when the input is known for all time. Differential and difference equation system descriptions are more flexible in this respect, since they apply to systems either at rest or with nonzero initial conditions.

2.5 Block Diagram Representations

In this section we examine block diagram representations for LTI systems described by differential and difference equations. A *block diagram* is an interconnection of elementary operations that act on the input signal. The block diagram is a more detailed representation of the system than the impulse response or difference- and differential-equation descriptions since it describes how the system's internal computations or operations are ordered. The impulse response and difference- or differential-equation descriptions represent only the input–output behavior of a system. We shall show that a system with a given input–output characteristic can be represented with different block diagrams. Each block diagram representation describes a different set of internal computations used to determine the system output.

Block diagram representations consist of an interconnection of three elementary operations on signals:

1. Scalar multiplication: $y(t) = cx(t)$ or $y[n] = cx[n]$, where c is a scalar.
2. Addition: $y(t) = x(t) + w(t)$ or $y[n] = x[n] + w[n]$.
3. Integration for continuous-time systems: $y(t) = \int_{-\infty}^{t} x(\tau)\, d\tau$; or a time shift for discrete-time systems: $y[n] = x[n-1]$.

Figure 2.26 depicts the block diagram symbols used to represent each of these operations. In order to express a continuous-time system in terms of integration, we shall convert the differential equation into an integral equation. The operation of integration is usually used in block diagrams for continuous-time systems instead of differentiation because integrators are more easily built from analog components than are differentiators. Also, integrators smooth out noise in the system, while differentiators accentuate noise.

The integral or difference equation corresponding to the system behavior is obtained by expressing the sequence of operations represented by the block diagram in equation

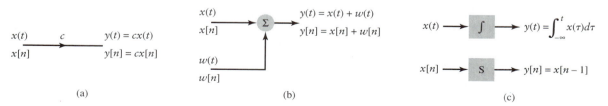

(a) (b) (c)

FIGURE 2.26 Symbols for elementary operations in block diagram descriptions for systems. (a) Scalar multiplication. (b) Addition. (c) Integration for continuous-time systems and time shift for discrete-time systems.

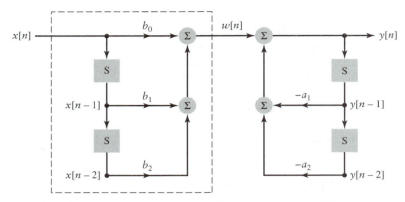

FIGURE 2.27   Block diagram representation for a discrete-time system described by a second-order difference equation.

form. Begin with the discrete-time case. A discrete-time system is depicted in Fig. 2.27. Consider writing an equation corresponding to the portion of the system within the dashed box. The output of the first time shift is $x[n - 1]$. The second time shift has output $x[n - 2]$. The scalar multiplications and summations imply

$$w[n] = b_0 x[n] + b_1 x[n - 1] + b_2 x[n - 2] \qquad (2.35)$$

Now we may write an expression for $y[n]$ in terms of $w[n]$. The block diagram indicates that

$$y[n] = w[n] - a_1 y[n - 1] - a_2 y[n - 2] \qquad (2.36)$$

The output of this system may be expressed as a function of the input $x[n]$ by substituting Eq. (2.35) for $w[n]$ in Eq. (2.36). We have

$$y[n] = -a_1 y[n - 1] - a_2 y[n - 2] + b_0 x[n] + b_1 x[n - 1] + b_2 x[n - 2]$$

or

$$y[n] + a_1 y[n - 1] + a_2 y[n - 2] = b_0 x[n] + b_1 x[n - 1] + b_2 x[n - 2] \quad (2.37)$$

Thus the block diagram in Fig. 2.27 describes a system whose input–output characteristic is represented by a second-order difference equation.

Note that the block diagram explicitly represents the operations involved in computing the output from the input and tells us how to simulate the system on a computer. The operations of scalar multiplication and addition are easily evaluated using a computer. The outputs of the time-shift operations correspond to memory locations in a computer. In order to compute the current output from the current input, we must have saved the past values of the input and output in memory. To begin a computer simulation at a specified time we must know the input and the past two values of the output. The past values of the output are the initial conditions required to solve the difference equation directly.

▶ **Drill Problem 2.14**   Determine the difference equation corresponding to the block diagram description of the system depicted in Fig. 2.28.

Answer:

$$y[n] + \tfrac{1}{2}y[n - 1] - \tfrac{1}{3}y[n - 3] = x[n] + 2x[n - 2]$$

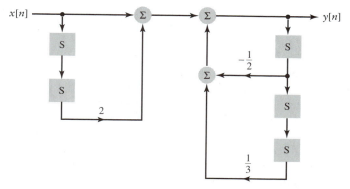

FIGURE 2.28 Block diagram representation for Drill Problem 2.14. ◀

The block diagram description for a system is not unique. We illustrate this by developing a second block diagram description for the system described by the second-order difference equation given by Eq. (2.37). We may view the system in Fig. 2.27 as a cascade of two systems: one with input $x[n]$ and output $w[n]$ described by Eq. (2.35) and a second with input $w[n]$ and output $y[n]$ described by Eq. (2.36). Since these are LTI systems, we may interchange their order without changing the input–output behavior of the cascade. Interchange their order and denote the output of the new first system as $f[n]$. This output is obtained from Eq. (2.36) and the input $x[n]$ as shown by

$$f[n] = -a_1 f[n - 1] - a_2 f[n - 2] + x[n] \tag{2.38}$$

The signal $f[n]$ is also the input to the second system. The output of the second system, $y[n]$, is obtained from Eq. (2.35) as

$$y[n] = b_0 f[n] + b_1 f[n - 1] + b_2 f[n - 2] \tag{2.39}$$

Both systems involve time-shifted versions of $f[n]$. Hence only one set of time shifts is needed in the block diagram for this system. We may represent the system described by Eqs. (2.38) and (2.39) as the block diagram illustrated in Fig. 2.29.

The block diagrams in Figs. 2.27 and 2.29 represent different implementations for a system with input–output behavior described by Eq. (2.37). The system in Fig. 2.27 is termed a "direct form I" implementation. The system in Fig. 2.29 is termed a "direct form

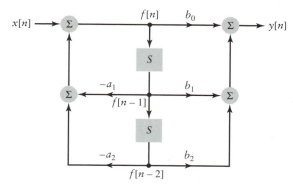

FIGURE 2.29 Alternative block diagram representation for a system described by a second-order difference equation.

II" implementation. The direct form II implementation uses memory more efficiently, since for this example it requires only two memory locations compared to the four required for the direct form I.

There are many different implementations for a system whose input–output behavior is described by a difference equation. They are obtained by manipulating either the difference equation or the elements in a block diagram representation. While these different systems are equivalent from an input–output perspective, they will generally differ with respect to other criteria such as memory requirements, the number of computations required per output value, or numerical accuracy.

Analogous results hold for continuous-time systems. We may simply replace the time-shift operations in Figs. 2.27 and 2.29 with time differentiation to obtain block diagram representations for systems described by differential equations. However, in order to depict the continuous-time system in terms of the more easily implemented integration operation, we must first rewrite the differential equation description

$$\sum_{k=0}^{N} a_k \frac{d^k}{dt^k} y(t) = \sum_{k=0}^{M} b_k \frac{d^k}{dt^k} x(t) \tag{2.40}$$

as an integral equation.

We define the integration operation in a recursive manner to simplify the notation. Let $v^{(0)}(t) = v(t)$ be an arbitrary signal and set

$$v^{(n)}(t) = \int_{-\infty}^{t} v^{(n-1)}(\tau) \, d\tau, \qquad n = 1, 2, 3, \ldots$$

Hence $v^{(n)}(t)$ is the n-fold integral of $v(t)$ with respect to time. This definition integrates over all past values of time. We may rewrite this in terms of an initial condition on the integrator as

$$v^{(n)}(t) = \int_{0}^{t} v^{(n-1)}(\tau) \, d\tau + v^{(n)}(0), \qquad n = 1, 2, 3, \ldots$$

If we assume zero initial conditions, then integration and differentiation are inverse operations; that is,

$$\frac{d}{dt} v^{(n)}(t) = v^{(n-1)}(t), \qquad t > 0 \text{ and } n = 1, 2, 3, \ldots$$

Hence if $N \geq M$ and we integrate Eq. (2.40) N times, we obtain the integral equation description for the system:

$$\sum_{k=0}^{N} a_k y^{(N-k)}(t) = \sum_{k=0}^{M} b_k x^{(N-k)}(t) \tag{2.41}$$

For a second-order system with $a_2 = 1$, Eq. (2.41) may be written

$$y(t) = -a_1 y^{(1)}(t) - a_0 y^{(2)}(t) + b_2 x(t) + b_1 x^{(1)}(t) + b_0 x^{(2)}(t) \tag{2.42}$$

Direct form I and direct form II implementations of this system are depicted in Figs. 2.30(a) and (b). The reader is asked to show that these block diagrams implement the integral equation in Problem 2.25. Note that the direct form II implementation uses fewer integrators than the direct form I implementation.

Block diagram representations for continuous-time systems may be used to specify analog computer simulations of systems. In such a simulation, signals are represented as

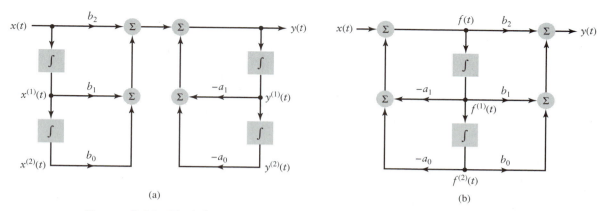

FIGURE 2.30   Block diagram representations for continuous-time system described by a second-order integral equation. (a) Direct form I. (b) Direct form II.

voltages, resistors are used to implement scalar multiplication, and the integrators are constructed using operational amplifiers, resistors, and capacitors. Initial conditions are specified as initial voltages on integrators. Analog computer simulations are much more cumbersome than digital computer simulations and suffer from drift, however, so it is common to simulate continuous-time systems on digital computers by using numerical approximations to either integration or differentiation operations.

2.6 *State-Variable Descriptions for LTI Systems*

The state-variable description for a LTI system consists of a series of coupled first-order differential or difference equations that describe how the state of the system evolves and an equation that relates the output of the system to the current state variables and input. These equations are written in matrix form. The *state* of a system may be defined as a minimal set of signals that represent the system's entire memory of the past. That is, given only the value of the state at a point in time n_o (or t_o) and the input for times $n \geq n_o$ (or $t \geq t_o$), we can determine the output for all times $n \geq n_o$ (or $t \geq t_o$). We shall see that the selection of signals comprising the state of a system is not unique and that there are many possible state-variable descriptions corresponding to a system with a given input–output characteristic. The ability to represent a system with different state-variable descriptions is a powerful attribute that finds application in advanced methods for control system analysis and discrete-time system implementation.

■ THE STATE-VARIABLE DESCRIPTION

We shall develop the general state-variable description by starting with the direct form II implementation for a second-order LTI system depicted in Fig. 2.31. In order to determine the output of the system for $n \geq n_o$, we must know the input for $n \geq n_o$ and the outputs of the time-shift operations labeled $q_1[n]$ and $q_2[n]$ at time $n = n_o$. This suggests that we may choose $q_1[n]$ and $q_2[n]$ as the state of the system. Note that since $q_1[n]$ and $q_2[n]$ are the outputs of the time-shift operations, the next value of the state, $q_1[n + 1]$ and $q_2[n + 1]$, must correspond to the variables at the input to the time-shift operations.

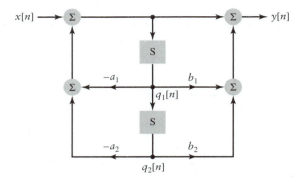

FIGURE 2.31   Direct form II representation for a second-order discrete-time system depicting state variables $q_1[n]$ and $q_2[n]$.

The block diagram indicates that the next value of the state is obtained from the current state and the input via the equations

$$q_1[n + 1] = -a_1 q_1[n] - a_2 q_2[n] + x[n] \tag{2.43}$$

$$q_2[n + 1] = q_1[n] \tag{2.44}$$

The block diagram also indicates that the system output is expressed in terms of the input and state as

$$y[n] = x[n] - a_1 q_1[n] - a_2 q_2[n] + b_1 q_1[n] + b_2 q_2[n]$$

or

$$y[n] = (b_1 - a_1)q_1[n] + (b_2 - a_2)q_2[n] + x[n] \tag{2.45}$$

We write Eqs. (2.43) and (2.44) in matrix form as

$$\begin{bmatrix} q_1[n + 1] \\ q_2[n + 1] \end{bmatrix} = \begin{bmatrix} -a_1 & -a_2 \\ 1 & 0 \end{bmatrix} \begin{bmatrix} q_1[n] \\ q_2[n] \end{bmatrix} + \begin{bmatrix} 1 \\ 0 \end{bmatrix} x[n] \tag{2.46}$$

while Eq. (2.45) is expressed as

$$y[n] = \begin{bmatrix} b_1 - a_1 & b_2 - a_2 \end{bmatrix} \begin{bmatrix} q_1[n] \\ q_2[n] \end{bmatrix} + [1]x[n] \tag{2.47}$$

If we define the state vector as the column vector

$$\mathbf{q}[n] = \begin{bmatrix} q_1[n] \\ q_2[n] \end{bmatrix}$$

then we can rewrite Eqs. (2.46) and (2.47) as

$$\mathbf{q}[n + 1] = \mathbf{A}\mathbf{q}[n] + \mathbf{b}x[n] \tag{2.48}$$

$$y[n] = \mathbf{c}\mathbf{q}[n] + Dx[n] \tag{2.49}$$

where the matrix $\mathbf{A}$, vectors $\mathbf{b}$ and $\mathbf{c}$, and scalar D are given by

$$\mathbf{A} = \begin{bmatrix} -a_1 & -a_2 \\ 1 & 0 \end{bmatrix}, \qquad \mathbf{b} = \begin{bmatrix} 1 \\ 0 \end{bmatrix}$$

$$\mathbf{c} = \begin{bmatrix} b_1 - a_1 & b_2 - a_2 \end{bmatrix}, \qquad D = [1]$$

Equations (2.48) and (2.49) are the general form for a state-variable description corresponding to a discrete-time system. The matrix $\mathbf{A}$, vectors $\mathbf{b}$ and $\mathbf{c}$, and scalar D represent another description for the system. Systems having different internal structures will be represented by different **A**, **b**, **c**, and D. The state-variable description is the only analytic system representation capable of specifying the internal structure of the system. Thus the state-variable description is used in any problem in which the internal system structure needs to be considered.

If the input–output characteristics of the system are described by an Nth order difference equation, then the state vector $\mathbf{q}[n]$ is N-by-1, **A** is N-by-N, **b** is N-by-1, and **c** is 1-by-N. Recall that solution of the difference equation requires N initial conditions. The N initial conditions represent the system's memory of the past, as does the N-dimensional state vector. Also, an Nth order system contains at least N time-shift operations in its block diagram representation. If the block diagram for a system has a minimal number of time shifts, then a natural choice for the states are the outputs of the unit delays, since the unit delays embody the memory of the system. This choice is illustrated in the following example.

EXAMPLE 2.21   Find the state-variable description corresponding to the second-order system depicted in Fig. 2.32 by choosing the state variables to be the outputs of the unit delays.

Solution: The block diagram indicates that the states are updated according to the equations

$$q_1[n + 1] = \alpha q_1[n] + \delta_1 x[n]$$
$$q_2[n + 1] = \gamma q_1[n] + \beta q_2[n] + \delta_2 x[n]$$

and the output is given by

$$y[n] = \eta_1 q_1[n] + \eta_2 q_2[n]$$

These equations are expressed in the state-variable forms of Eqs. (2.48) and (2.49) if we define

$$\mathbf{q}[n] = \begin{bmatrix} q_1[n] \\ q_2[n] \end{bmatrix}$$

and

$$\mathbf{A} = \begin{bmatrix} \alpha & 0 \\ \gamma & \beta \end{bmatrix}, \qquad \mathbf{b} = \begin{bmatrix} \delta_1 \\ \delta_2 \end{bmatrix}$$
$$\mathbf{c} = \begin{bmatrix} \eta_1 & \eta_2 \end{bmatrix}, \qquad D = [0]$$

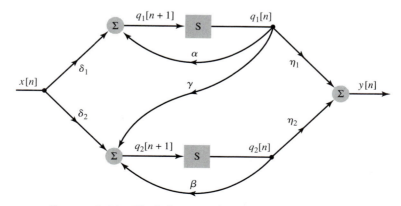

FIGURE 2.32   Block diagram of system for Example 2.21.

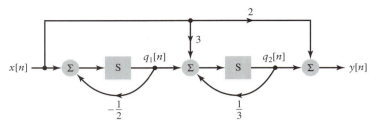

FIGURE 2.33 Block diagram of system for Drill Problem 2.15.

▶ **Drill Problem 2.15** Find the state-variable description corresponding to the block diagram in Fig. 2.33. Choose the state variables to be the outputs of the unit delays, $q_1[n]$ and $q_2[n]$, as indicated in the figure.

Answer:

$$\mathbf{A} = \begin{bmatrix} -\frac{1}{2} & 0 \\ 1 & \frac{1}{3} \end{bmatrix}, \qquad \mathbf{b} = \begin{bmatrix} 1 \\ 3 \end{bmatrix}$$
$$\mathbf{c} = \begin{bmatrix} 0 & 1 \end{bmatrix}, \qquad D = \begin{bmatrix} 2 \end{bmatrix}$$ ◀

The state-variable description for continuous-time systems is analogous to that for discrete-time systems, with the exception that the state equation given by Eq. (2.48) is expressed in terms of a derivative. We thus write

$$\frac{d}{dt}\,\mathbf{q}(t) = \mathbf{A}\mathbf{q}(t) + \mathbf{b}x(t) \tag{2.50}$$

$$y(t) = \mathbf{c}\mathbf{q}(t) + Dx(t) \tag{2.51}$$

Once again, the matrix **A**, vectors **b** and **c**, and scalar D describe the internal structure of the system.

The memory of a continuous-time system is contained within the system's energy storage devices. Hence state variables are usually chosen as the physical quantities associated with the energy storage devices. For example, in electrical systems the energy storage devices are capacitors and inductors. We may choose state variables to correspond to the voltage across capacitors or the current through inductors. In a mechanical system the energy-storing devices are springs and masses. State variables may be chosen as spring displacement or mass velocity. In a block diagram representation energy storage devices are integrators. The state-variable equations represented by Eqs. (2.50) and (2.51) are obtained from the equations that relate the behavior of the energy storage devices to the input and output. This procedure is demonstrated in the following examples.

EXAMPLE 2.22 Consider the electrical circuit depicted in Fig. 2.34. Derive a state-variable description for this system if the input is the applied voltage $x(t)$ and the output is the current through the resistor labeled $y(t)$.

Solution: Choose the state variables as the voltage across each capacitor. Summing the voltage drops around the loop involving $x(t)$, R_1, and C_1 gives

$$x(t) = y(t)R_1 + q_1(t)$$

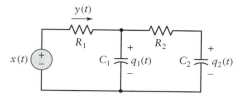

FIGURE 2.34   Circuit diagram of system for Example 2.22.

or

$$y(t) = -\frac{1}{R_1} q_1(t) + \frac{1}{R_1} x(t) \tag{2.52}$$

This equation expresses the output as a function of the state variables and input. Let $i_2(t)$ be the current through R_2. Summing the voltage drops around the loop involving C_1, R_2, and C_2 we obtain

$$q_1(t) = R_2 i_2(t) + q_2(t)$$

or

$$i_2(t) = \frac{1}{R_2} q_1(t) - \frac{1}{R_2} q_2(t) \tag{2.53}$$

However, we also know that

$$i_2(t) = C_2 \frac{d}{dt} q_2(t)$$

Substitute Eq. (2.53) for $i_2(t)$ to obtain

$$\frac{d}{dt} q_2(t) = \frac{1}{C_2 R_2} q_1(t) - \frac{1}{C_2 R_2} q_2(t) \tag{2.54}$$

Lastly, we need a state equation for $q_1(t)$. This is obtained by applying Kirchhoff's current law to the node between R_1 and R_2. Letting $i_1(t)$ be the current through C_1, we have

$$y(t) = i_1(t) + i_2(t)$$

Now substitute Eq. (2.52) for $y(t)$, Eq. (2.53) for $i_2(t)$, and

$$i_1(t) = C_1 \frac{d}{dt} q_1(t)$$

for $i_1(t)$, and rearrange to obtain

$$\frac{d}{dt} q_1(t) = -\left(\frac{1}{C_1 R_1} + \frac{1}{C_1 R_2}\right) q_1(t) + \frac{1}{C_1 R_2} q_2(t) + \frac{1}{C_1 R_1} x(t) \tag{2.55}$$

The state-variable description is now obtained from Eqs. (2.52), (2.54), and (2.55) as

$$\mathbf{A} = \begin{bmatrix} -\left(\dfrac{1}{C_1 R_1} + \dfrac{1}{C_1 R_2}\right) & \dfrac{1}{C_1 R_2} \\ \dfrac{1}{C_2 R_2} & -\dfrac{1}{C_2 R_2} \end{bmatrix}, \quad \mathbf{b} = \begin{bmatrix} \dfrac{1}{C_1 R_1} \\ 0 \end{bmatrix}$$

$$\mathbf{c} = \begin{bmatrix} -\dfrac{1}{R_1} & 0 \end{bmatrix}, \quad D = \begin{bmatrix} \dfrac{1}{R_1} \end{bmatrix}$$

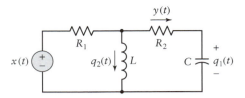

FIGURE 2.35   Circuit diagram of system for Drill Problem 2.16.

▶ **Drill Problem 2.16**   Find the state-variable description for the circuit depicted in Fig. 2.35. Choose state variables $q_1(t)$ and $q_2(t)$ as the voltage across the capacitor and the current through the inductor, respectively.

Answer:

$$
\mathbf{A} = \begin{bmatrix} \dfrac{-1}{(R_1 + R_2)C} & \dfrac{-R_1}{(R_1 + R_2)C} \\[2mm] \dfrac{R_1}{(R_1 + R_2)L} & \dfrac{-R_1 R_2}{(R_1 + R_2)L} \end{bmatrix}, \qquad \mathbf{b} = \begin{bmatrix} \dfrac{1}{(R_1 + R_2)C} \\[2mm] \dfrac{R_2}{(R_1 + R_2)L} \end{bmatrix}
$$

$$
\mathbf{c} = \begin{bmatrix} \dfrac{-1}{R_1 + R_2} & \dfrac{-R_1}{R_1 + R_2} \end{bmatrix} \qquad D = \begin{bmatrix} \dfrac{1}{R_1 + R_2} \end{bmatrix}
$$
◀

In a block diagram representation for a continuous-time system the state variables correspond to the outputs of the integrators. Thus the input to the integrator is the derivative of the corresponding state variable. The state-variable description is obtained by writing equations that correspond to the operations in the block diagram. This procedure is illustrated in the following example.

EXAMPLE 2.23   Determine the state-variable description corresponding to the block diagram in Fig. 2.36. The choice of state variables is indicated on the diagram.

Solution: The block diagram indicates that

$$
\frac{d}{dt} q_1(t) = 2q_1(t) - q_2(t) + x(t)
$$

$$
\frac{d}{dt} q_2(t) = q_1(t)
$$

$$
y(t) = 3q_1(t) + q_2(t)
$$

Hence the state-variable description is

$$
\mathbf{A} = \begin{bmatrix} 2 & -1 \\ 1 & 0 \end{bmatrix}, \qquad \mathbf{b} = \begin{bmatrix} 1 \\ 0 \end{bmatrix}
$$

$$
\mathbf{c} = [3 \quad 1], \qquad D = [0]
$$

■ **TRANSFORMATIONS OF THE STATE**

We have claimed that there is no unique state-variable description for a system with a given input–output characteristic. Different state-variable descriptions may be obtained

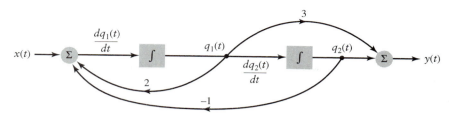

FIGURE 2.36   Block diagram of system for Example 2.23.

by transforming the state variables. This transformation is accomplished by defining a new set of state variables that are a weighted sum of the original state variables. This changes the form of **A**, **b**, **c**, and D but does not change the input–output characteristics of the system. To illustrate this, reconsider Example 2.23. Define new states $q_2'(t) = q_1(t)$ and $q_1'(t) = q_2(t)$. Here we simply have interchanged the state variables: $q_2'(t)$ is the output of the first integrator and $q_1'(t)$ is the output of the second integrator. We have not changed the structure of the block diagram, so clearly the input–output characteristic of the system remains the same. The state-variable description is different, however, since now we have

$$\mathbf{A}' = \begin{bmatrix} 0 & 1 \\ -1 & 2 \end{bmatrix}, \qquad \mathbf{b}' = \begin{bmatrix} 0 \\ 1 \end{bmatrix}$$
$$\mathbf{c}' = \begin{bmatrix} 1 & 3 \end{bmatrix}, \qquad D' = \begin{bmatrix} 0 \end{bmatrix}$$

The example in the previous paragraph employs a particularly simple transformation of the original state. In general, we may define a new state vector as a transformation of the original state vector, or $\mathbf{q}' = \mathbf{Tq}$. We define $\mathbf{T}$ as the state transformation matrix. Here we have dropped the time index (t) or $[n]$ in order to treat both continuous- and discrete-time cases simultaneously. In order for the new state to represent the entire system's memory, the relationship between $\mathbf{q}'$ and $\mathbf{q}$ must be one to one. This implies that $\mathbf{T}$ must be a nonsingular matrix, or that the inverse matrix $\mathbf{T}^{-1}$ exists. Hence $\mathbf{q} = \mathbf{T}^{-1}\mathbf{q}'$. The original state-variable description is

$$\dot{\mathbf{q}} = \mathbf{Aq} + \mathbf{b}x$$
$$y = \mathbf{cq} + Dx$$

where the dot over **q** denotes differentiation in continuous time or time advance in discrete time. The new state-variable description **A**′, **b**′, **c**′, and D' is derived by noting

$$\dot{\mathbf{q}}' = \mathbf{T}\dot{\mathbf{q}}$$
$$= \mathbf{TAq} + \mathbf{Tb}x$$
$$= \mathbf{TAT}^{-1}\mathbf{q}' + \mathbf{Tb}x$$

and

$$y = \mathbf{cq} + Dx$$
$$= \mathbf{cT}^{-1}\mathbf{q}' + Dx$$

Hence if we set

$$\mathbf{A}' = \mathbf{TAT}^{-1} \qquad \mathbf{b}' = \mathbf{Tb}$$
$$\mathbf{c}' = \mathbf{cT}^{-1} \qquad D' = D \tag{2.56}$$

then

$$\dot{\mathbf{q}}' = \mathbf{A}'\mathbf{q}' + \mathbf{b}'x$$
$$y = \mathbf{c}'\mathbf{q}' + D'x$$

is the new state-variable description.

EXAMPLE 2.24   A discrete-time system has the state-variable description

$$\mathbf{A} = \frac{1}{10}\begin{bmatrix} -1 & 4 \\ 4 & -1 \end{bmatrix}, \qquad \mathbf{b} = \begin{bmatrix} 2 \\ 4 \end{bmatrix}$$
$$\mathbf{c} = \tfrac{1}{2}\begin{bmatrix} 1 & 1 \end{bmatrix}, \qquad D = \begin{bmatrix} 2 \end{bmatrix}$$

Find the state-variable description $\mathbf{A}'$, $\mathbf{b}'$, $\mathbf{c}'$, and D' corresponding to the new states $q_1'[n] = -\frac{1}{2}q_1[n] + \frac{1}{2}q_2[n]$ and $q_2'[n] = \frac{1}{2}q_1[n] + \frac{1}{2}q_2[n]$.

Solution: Write the new state vector as $\mathbf{q}' = \mathbf{T}\mathbf{q}$, where

$$\mathbf{T} = \frac{1}{2}\begin{bmatrix} -1 & 1 \\ 1 & 1 \end{bmatrix}$$

This matrix is nonsingular, and its inverse is

$$\mathbf{T}^{-1} = \begin{bmatrix} -1 & 1 \\ 1 & 1 \end{bmatrix}$$

Hence substituting for $\mathbf{T}$ and $\mathbf{T}^{-1}$ in Eq. (2.56) gives

$$\mathbf{A}' = \begin{bmatrix} -\frac{1}{2} & 0 \\ 0 & \frac{3}{10} \end{bmatrix}, \qquad \mathbf{b}' = \begin{bmatrix} 1 \\ 3 \end{bmatrix}$$
$$\mathbf{c}' = \begin{bmatrix} 0 & 1 \end{bmatrix}, \qquad D' = \begin{bmatrix} 2 \end{bmatrix}$$

Note that this choice for $\mathbf{T}$ results in $\mathbf{A}'$ being a diagonal matrix and thus separates the state update into the two decoupled first-order difference equations as shown by

$$q_1[n + 1] = -\tfrac{1}{2}q_1[n] + x[n]$$
$$q_2[n + 1] = \tfrac{3}{10}q_2[n] + 3x[n]$$

The decoupled form of the state-variable description is particularly useful for analyzing systems because of its simple structure.

▶ **Drill Problem 2.17**   A continuous-time system has the state-variable description

$$\mathbf{A} = \begin{bmatrix} -2 & 0 \\ 1 & -1 \end{bmatrix}, \qquad \mathbf{b} = \begin{bmatrix} 1 \\ 1 \end{bmatrix}$$
$$\mathbf{c} = \begin{bmatrix} 0 & 2 \end{bmatrix}, \qquad D = \begin{bmatrix} 1 \end{bmatrix}$$

Find the state-variable description $\mathbf{A}'$, $\mathbf{b}'$, $\mathbf{c}'$, and D' corresponding to the new states $q_1'(t) = 2q_1(t) + q_2(t)$ and $q_2'(t) = q_1(t) - q_2(t)$.

Answer:

$$A' = \frac{1}{3}\begin{bmatrix} -4 & -1 \\ -2 & -5 \end{bmatrix}, \qquad b' = \begin{bmatrix} 3 \\ 0 \end{bmatrix}$$
$$c' = \tfrac{1}{3}[2 \quad -4], \qquad D' = [1] \qquad\qquad \blacktriangleleft$$

Note that each nonsingular transformation **T** generates a different state-variable description for a system with a given input–output behavior. The ability to transform the state-variable description without changing the input–output characteristics of the system is a powerful tool. It is used to analyze systems and identify implementations of systems that optimize some performance criteria not directly related to input–output behavior, such as the numerical effects of roundoff in a computer-based system implementation.

2.7 *Exploring Concepts with MATLAB*

Digital computers are ideally suited to implementing time-domain descriptions of discrete-time systems, because computers naturally store and manipulate sequences of numbers. For example, the convolution sum describes the relationship between the input and output of a discrete-time system and is easily evaluated with a computer as the sum of products of numbers. In contrast, continuous-time systems are described in terms of continuous functions, which are not easily represented or manipulated in a digital computer. For example, the output of a continuous-time system is described by the convolution integral. Evaluation of the convolution integral with a computer requires use of either numerical integration or symbolic manipulation techniques, both of which are beyond the scope of this book. Hence our exploration with MATLAB focuses on discrete-time systems.

A second limitation on exploring signals and systems is imposed by the finite memory or storage capacity and nonzero computation times inherent to all digital computers. Consequently, we can only manipulate finite-duration signals. For example, if the impulse response of a system has infinite duration and the input is of infinite duration, then the convolution sum involves summing an infinite number of products. Of course, even if we could store the infinite-length signals in the computer, the infinite sum could not be computed in a finite amount of time. In spite of this limitation, the behavior of a system in response to an infinite-length signal may often be inferred from its response to a carefully chosen finite-length signal.

Both the MATLAB Signal Processing Toolbox and Control System Toolbox are used in this section.

■ CONVOLUTION

Recall that the convolution sum expresses the output of a discrete-time system in terms of the input and impulse response of the system. MATLAB has a function named `conv` that evaluates the convolution of finite-duration discrete-time signals. If x and h are vectors representing signals, then the MATLAB command y = conv(x, h) generates a vector y representing the convolution of the signals represented by x and h. The number of elements in y is given by the sum of the number of elements in x and h minus one. Note that we must know the time origin of the signals represented by x and h in order to determine the time origin of their convolution. In general, if the first element of x corresponds to time $n = k_x$ and the first element of h corresponds to time $n = k_h$, then the first element of y corresponds to time $n = k_x + k_h$.

To illustrate this, consider repeating Example 2.1 using MATLAB. Here the first nonzero value in the impulse response occurs at time $n = -1$ and the first element of the input x occurs at time $n = 0$. We evaluate this convolution in MATLAB as follows:

```
>> h = [1, 2, 1];
>> x = [2, 3, -2];
>> y = conv(x,h)
y =
    2 7 6 -1 -2
```

The first element in the vector y corresponds to time $n = 0 + (-1) = -1$.

In Example 2.3 we used hand calculation to determine the output of a system with impulse response given by

$$h[n] = u[n] - u[n - 10]$$

and input

$$x[n] = u[n - 2] - u[n - 7]$$

We may use the MATLAB command c o n v to perform the convolution as follows. In this case, the impulse response consists of ten consecutive ones beginning at time $n = 0$, and the input consists of five consecutive ones beginning at time $n = 2$. These signals may be defined in MATLAB using the commands

```
>> h = ones(1,10);
>> x = ones(1,5);
```

The output is obtained and graphed using the commands

```
>> n = 2:15;
>> y = conv(x,h);
>> stem(n,y); xlabel('Time'); ylabel('Amplitude')
```

Here the first element of the vector y corresponds to time $n = 2 + 0 = 2$ as depicted in Fig. 2.37.

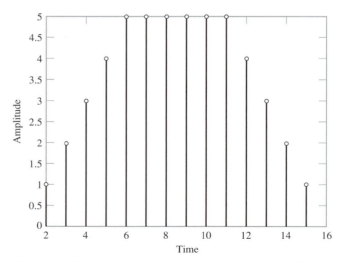

FIGURE 2.37   Convolution sum computed using MATLAB.

▶ **Drill Problem 2.18** Use MATLAB to solve Drill Problem 2.2 for $\alpha = 0.9$. That is, find the output of the system with input $x[n] = 2\{u[n + 2] - u[n - 12]\}$ and impulse response $h[n] = 0.9^n\{u[n - 2] - u[n - 13]\}$.

Answer: See Fig. 2.38. ◀

■ **STEP AND SINUSOIDAL STEADY-STATE RESPONSES**

The step response is the output of a system in response to a step input and is infinite in duration in general. However, we can evaluate the first p values of the step response using the `conv` function if the system impulse response is zero for times $n < k_h$, by convolving the first p values of $h[n]$ with a finite-duration step of length p. That is, we construct a vector `h` from the first p nonzero values of the impulse response, define the step `u = ones(1, p)`, and evaluate `s = conv(u, h)`. The first element of `s` corresponds to time k_h, and the first p values of `s` represent the first p values of the step response. The remaining values of `s` do not correspond to the step response, but are an artifact of convolving finite-duration signals.

For example, we may determine the first 50 values of the step response of the system with impulse response given in Drill Problem 2.7:

$$h[n] = (-a)^n u[n]$$

with $a = 0.9$ by using the MATLAB commands

```
>> h = (-0.9).^ [0:49];
>> u = ones(1,50);
>> s = conv(u,h);
```

The vector `s` has 99 values, the first 50 of which represent the step response and are depicted in Fig. 2.39. This figure is obtained using the MATLAB command `stem([0:49], s(1:50))`.

The sinusoidal steady-state response of a discrete-time system is given by the amplitude and phase change experienced by the infinite-duration complex sinusoidal input signal

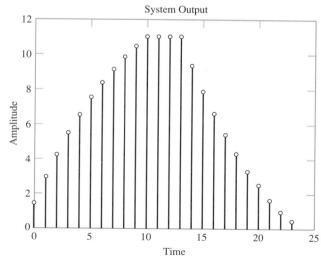

FIGURE 2.38 Solution to Drill Problem 2.18.

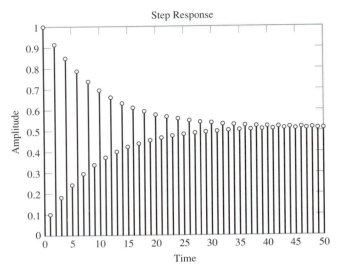

FIGURE 2.39   Step response computed using MATLAB.

$x[n] = e^{j\Omega n}$. The sinusoidal steady-state response of a system with finite-duration impulse response may be determined using a finite-duration sinusoid provided the sinusoid is sufficiently long to drive the system to a steady-state condition. To show this, suppose $h[n] = 0$ for $n < n_1$ and $n > n_2$, and let the system input be the finite-duration sinusoid $v[n] = e^{j\Omega n}(u[n] - u[n - n_v])$. We may write the system output as

$$y[n] = h[n] * v[n]$$

$$= \sum_{k=n_1}^{n_2} h[k]e^{j\Omega(n-k)}, \qquad n_2 \leq n < n_1 + n_v$$

$$= h[n] * e^{j\Omega n}, \qquad n_2 \leq n < n_1 + n_v$$

Hence the system output in response to a finite-duration sinusoidal input corresponds to the sinusoidal steady-state response on the interval $n_2 \leq n < n_1 + n_v$. The magnitude and phase response of the system may be determined from $y[n]$, $n_2 \leq n < n_1 + n_v$, by noting that

$$y[n] = |H(e^{j\Omega})|e^{j(\Omega n + \arg\{H(e^{j\Omega})\})}, \qquad n_2 \leq n < n_1 + n_v$$

Take the magnitude and phase of $y[n]$ to obtain

$$|y[n]| = |H(e^{j\Omega})|, \qquad n_2 \leq n < n_1 + n_v$$

and

$$\arg\{y[n]\} - \Omega n = \arg\{H(e^{j\Omega})\}, \qquad n_2 \leq n < n_1 + n_v$$

We may use this approach to evaluate the sinusoidal steady-state response of one of the systems given in Example 2.14. Consider the system with impulse response

$$h[n] = \begin{cases} \frac{1}{2}, & n = 0 \\ -\frac{1}{2}, & n = 1 \\ 0, & \text{otherwise} \end{cases}$$

We shall determine the frequency response and 50 values of the sinusoidal steady-state response of this system for input frequencies $\Omega = \frac{1}{4}\pi$ and $\frac{3}{4}\pi$.

Here $n_1 = 0$ and $n_2 = 1$, so to obtain 50 values of the sinusoidal steady-state response we require $n_v \geq 51$. The sinusoidal steady-state responses are obtained by MATLAB commands

```
>> Omega1 = pi/4;  Omega2 = 3*pi/4;
>> v1 = exp(j*Omega1*[0:50]);
>> v2 = exp(j*Omega2*[0:50]);
>> h = [0.5, -0.5];
>> y1 = conv(v1,h);  y2 = conv(v2,h);
```

Figures 2.40(a) and (b) depict the real and imaginary components of **y1**, respectively, and may be obtained with the commands

```
>> subplot(2,1,1)
>> stem([0:51],real(y1))
>> xlabel('Time'); ylabel('Amplitude');
>> title('Real(y1)')
>> subplot(2,1,2)
>> stem([0:51],imag(y1))
>> xlabel('Time'); ylabel('Amplitude');
>> title('Imag(y1)')
```

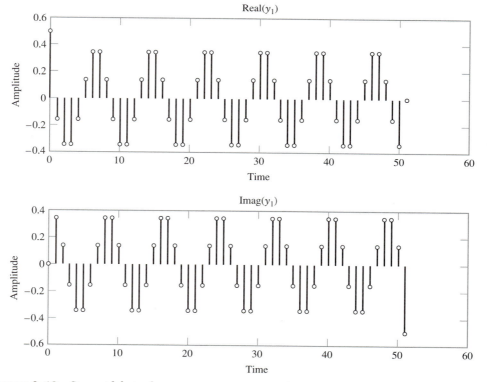

FIGURE 2.40   Sinusoidal steady-state response computed using MATLAB. The values at times 1 through 50 represent the sinusoidal steady-state response.

The sinusoidal steady-state response is represented by the values at time indices 1 through 50.

We may now obtain the magnitude and phase responses from any element of the vectors y1 and y2 except for the first one or the last one. Using the fifth element, we use the commands

```
≫ H1mag = abs(y1(5))
H1mag =
     0.3287
≫ H2mag = abs(y2(5))
H2mag =
     0.9239
≫ H1phs = angle(y1(5)) - Omega1*5
H1phs =
     -5.8905
≫ H2phs = angle(y2(5)) - Omega2*5
H2phs =
     -14.5299
```

The phase response is measured in radians. Note that the **angle** command always returns a value between $-\pi$ and π radians. Hence measuring phase with the command **angle(y1(n)) - Omega1*n** may result in answers that differ by integer multiples of 2π when different values of **n** are used.

▶ **Drill Problem 2.19**   Evaluate the frequency response and 50 values of the sinusoidal steady-state response of the system with impulse response

$$h[n] = \begin{cases} \frac{1}{4}, & 0 \le n \le 3 \\ 0, & \text{otherwise} \end{cases}$$

at frequency $\Omega = \frac{1}{3}\pi$.

Answer: The steady-state response is given by the values at time indices 3 through 52 in Fig. 2.41. Using the fourth element of the steady-state response gives $|H(e^{j\pi/3})| = 0.4330$ and $\arg\{H(e^{j\pi/3})\} = -1.5708$ radians.   ◀

■ SIMULATING DIFFERENCE EQUATIONS

In Section 2.4, we expressed the difference-equation description for a system in a recursive form that allowed the system output to be computed from the input signal and past outputs. The **filter** command performs a similar function. Define vectors $\mathbf{a} = [a_0, a_1, \ldots, a_N]$ and $\mathbf{b} = [b_0, b_1, \ldots, b_M]$ representing the coefficients of the difference equation given by Eq. (2.29). If **x** is a vector representing the input signal, then the command **y = filter(b,a,x)** results in a vector **y** representing the output of the system for zero initial conditions. The number of output values in **y** corresponds to the number of input values in **x**. Nonzero initial conditions are incorporated by using the alternative command syntax **y = filter(b,a,x,zi)**, where **zi** represents the initial conditions required by **filter**. The initial conditions used by **filter** are not the past values of the output since **filter** uses a modified form of the difference equation to determine the output. These initial conditions are obtained from knowledge of the past outputs using

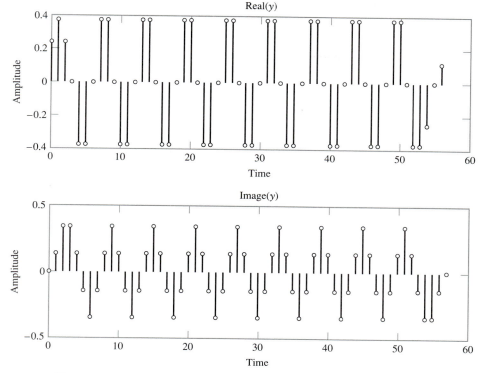

FIGURE 2.41   Sinusoidal steady-state response for Drill Problem 2.19.

the command $zi = filtic(b,a,yi)$, where yi is a vector containing the initial conditions in the order $[y(-1), y(-2), \ldots, y(-N)]$.

We illustrate use of the **filter** command by revisiting Example 2.16. The system of interest is described by the difference equation

$$y[n] - 1.143y[n-1] + 0.4128y[n-2] = 0.0675x[n]$$
$$+ 0.1349x[n-1] + 0.0675x[n-2] \qquad (2.57)$$

We determine the output in response to zero input and initial conditions $y[-1] = 1$, $y[-2] = 2$ using the commands

```
>> a = [1, -1.143, 0.4128];
>> b = [0.0675, 0.1349, 0.0675];
>> x = zeros(1, 50);
>> zi = filtic(b,a,[1, 2]);
>> y = filter(b,a,x,zi);
```

The result is depicted in Fig. 2.22(b). We may determine the system response to an input consisting of the average January temperature data with the commands

```
>> load Jantemp;
>> filttemp = filter(b,a,Jantemp);
```

Here we have assumed the average January temperature data are in the file **Jantemp.mat**. The result is depicted in Fig. 2.22(g).

▶ **Drill Problem 2.20**   Use `filter` to determine the first 50 values of the step response of the system described by Eq. (2.57) and the first 100 values of the response to the input $x[n] = \cos(\frac{1}{5}\pi n)$ assuming zero initial conditions.

Answer: See Figs. 2.22(a) and (d).   ◀

The command `[h,t] = impz(b,a,n)` evaluates n values of the impulse response of a system described by a difference equation. The difference-equation coefficients are contained in the vectors `b` and `a` as for `filter`. The vector `h` contains the values of the impulse response and `t` contains the corresponding time indices.

■ STATE-VARIABLE DESCRIPTIONS

The MATLAB Control System Toolbox contains numerous routines for manipulating state-variable descriptions. A key feature of the Control System Toolbox is the use of LTI objects, which are customized data structures that enable manipulation of LTI system descriptions as single MATLAB variables. If `a`, `b`, `c`, and `d` are MATLAB arrays representing the **A**, **b**, **c**, and D matrices in the state-variable description, then the command `sys = ss(a,b,c,d,-1)` produces a LTI object `sys` that represents the discrete-time system in state-variable form. Note that a continuous-time system is obtained by omitting the `-1`, that is, using `sys = ss(a,b,c,d)`. LTI objects corresponding to other system representations are discussed in Sections 6.9 and 7.10.

Systems are manipulated in MATLAB by operations on their LTI objects. For example, if `sys1` and `sys2` are objects representing two systems in state-variable form, then `sys = sys1 + sys2` produces the state-variable description for the parallel combination of `sys1` and `sys2`, while `sys = sys1*sys2` represents the cascade combination.

The function `lsim` simulates the output of a system in response to a specified input. For a discrete-time system, the command has the form `y = lsim(sys,x)`, where `x` is a vector containing the input and `y` represents the output. The command `h = impulse(sys,N)` places the first N values of the impulse response in `h`. Both of these may also be used for continuous-time systems, although the command syntax changes slightly. In the continuous-time case, numerical methods are used to approximate the continuous-time system response.

Recall that there is no unique state-variable description for a given system. Different state-variable descriptions for the same system are obtained by transforming the state. Transformations of the state may be computed in MATLAB using the routine `ss2ss`. The state transformation is identical for both continuous- and discrete-time systems, so the same command is used for transforming either type of system. The command is of the form `sysT = ss2ss(sys,T)`, where `sys` represents the original state-variable description, `T` is the state transformation matrix, and `sysT` represents the transformed state-variable description.

Consider using `ss2ss` to transform the state-variable description of Example 2.24

$$\mathbf{A} = \frac{1}{10}\begin{bmatrix} -1 & 4 \\ 4 & -1 \end{bmatrix}, \qquad \mathbf{b} = \begin{bmatrix} 2 \\ 4 \end{bmatrix}$$

$$\mathbf{c} = \tfrac{1}{2}[1 \quad 1], \qquad D = [2]$$

using the state transformation matrix

$$T = \frac{1}{2}\begin{bmatrix} -1 & 1 \\ 1 & 1 \end{bmatrix}$$

The following commands produce the desired result:

```
>> a = [-0.1, 0.4; 0.4, -0.1];  b = [2; 4];
>> c = [0.5, 0.5];  d = 2;
>> sys = ss(a,b,c,d,-1); % define the state-space object sys
>> T = 0.5*[-1, 1; 1, 1];
>> sysT = ss2ss(sys,T)
a =

                 x1          x2
     x1    -0.50000           0
     x2           0     0.30000
b =

                 u1
     x1     1.00000
     x2     3.00000
c =

                 x1          x2
     y1           0     1.00000
d =

                 u1
     y1     2.00000
Sampling time: unspecified
Discrete-time system.
```

This result agrees with Example 2.24. We may verify that the two systems represented by sys and sysT have identical input–output characteristics by comparing their impulse responses via the following commands:

```
>> h = impulse(sys,10);  hT = impulse(sysT,10);
>> subplot(2,1,1)
>> stem([0:9],h)
>> title('Original System Impulse Response');
>> xlabel('Time');  ylabel('Amplitude')
>> subplot(2,1,2)
>> stem([0:9],hT)
>> title('Transformed System Impulse Response');
>> xlabel('Time');  ylabel('Amplitude')
```

Figure 2.42 depicts the first 10 values of the impulse responses of the original and transformed systems produced by this sequence of commands. We may verify that the original and transformed systems have the (numerically) identical impulse response by computing the error err = h-hT.

▶ **Drill Problem 2.21**   Solve Drill Problem 2.17 using MATLAB.   ◀

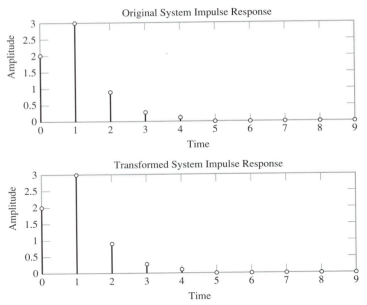

FIGURE 2.42   Impulse responses associated with the original and transformed state-variable descriptions computed using MATLAB.

2.8 Summary

There are many different methods for describing the action of a LTI system on an input signal. In this chapter we have examined four different descriptions for LTI systems: the impulse response, difference- and differential-equation, block diagram, and state-variable descriptions. All four are equivalent in the input–output sense; for a given input, each description will produce the identical output. However, different descriptions offer different insights into system characteristics and use different techniques for obtaining the output from the input. Thus each description has its own advantages and disadvantages for solving a particular system problem.

The impulse response is the output of a system when the input is an impulse. The output of a linear time-invariant system in response to an arbitrary input is expressed in terms of the impulse response as a convolution operation. System properties, such as causality and stability, are directly related to the impulse response. The impulse response also offers a convenient framework for analyzing interconnections of systems. The input must be known for all time in order to determine the output of a system using the impulse response and convolution.

The input and output of a LTI system may also be related using either a differential or difference equation. Differential equations often follow directly from the physical principles that define the behavior and interaction of continuous-time system components. The order of a differential equation reflects the maximum number of energy storage devices in the system, while the order of a difference equation represents the system's maximum memory of past outputs. In contrast to impulse response descriptions, the output of a system from a given point in time forward can be determined without knowledge of all past inputs provided initial conditions are known. Initial conditions are the initial values of energy storage or system memory and summarize the effect of all past inputs up to the starting time of interest. The solution to a differential or difference equation can be separated into a natural and forced response. The natural response describes the behavior of the system due to the initial conditions. The forced response describes the behavior of the system in response to the input alone.

The block diagram represents the system as an interconnection of elementary operations on signals. The manner in which these operations are interconnected defines the internal structure of the system. Different block diagrams can represent systems with identical input–output characteristics.

The state-variable description is a series of coupled first-order differential or difference equations representing the system behavior, which are written in matrix form. It consists of two equations: one equation describes how the state of the system evolves and a second equation relates the state to the output. The state represents the system's entire memory of the past. The number of states corresponds to the number of energy storage devices or maximum memory of past outputs present in the system. The choice of state is not unique; an infinite number of different state-variable descriptions can be used to represent systems with the same input–output characteristic. The state-variable description can be used to represent the internal structure of a physical system and thus provides a more detailed characterization of systems than the impulse response or differential (difference) equations.

FURTHER READING

1. A concise summary and many worked problems for much of the material presented in this and later chapters is found in:
   ▶ Hsu, H. P., *Signals and Systems*, Schaum's Outline Series (McGraw-Hill, 1995)

2. The notation $H(e^{j\Omega})$ and $H(j\omega)$ for the sinusoidal steady-state response of a discrete- and a continuous-time system, respectively, may seem unnatural at first glance. Indeed, the alternative notations $H(\Omega)$ and $H(\omega)$ are sometimes used in engineering practice. However, our notation is more commonly used as it allows the sinusoidal steady-state response to be defined naturally in terms of the z-transform (Chapter 7) and the Laplace transform (Chapter 6).

3. A general treatment of differential equations is given in:
   ▶ Boyce, W. E., and R. C. DiPrima, *Elementary Differential Equations*, Sixth Edition (Wiley, 1997)

4. The role of difference equations and block diagram descriptions for discrete-time systems in signal processing are described in:
   ▶ Proakis, J. G., and D. G. Manolakis, *Introduction to Digital Signal Processing* (Macmillan, 1988)
   ▶ Oppenheim, A. V., and R. W. Schafer, *Discrete-Time Signal Processing* (Prentice Hall, 1989)

5. The role of differential equations, block diagram descriptions, and state-variable descriptions in control systems is described in:
   ▶ Dorf, R. C., and R. H. Bishop, *Modern Control Systems*, Seventh Edition (Addison-Wesley, 1995)
   ▶ Phillips, C. L., and R. D. Harbor, *Feedback Control Systems*, Third Edition (Prentice Hall, 1996)

6. State variable descriptions in control systems are discussed in:
   ▶ Chen, C. T., *Linear System Theory and Design* (Holt, Rinehart, and Winston, 1984)

▶ B. Friedland, *Control System Design: An Introduction to State-Space Methods* (McGraw-Hill, 1986)

A thorough, yet advanced, treatment of state-variable descriptions in the context of signal processing is given in:

▶ Roberts, R. A., and C. T. Mullis, *Digital Signal Processing* (Addison-Wesley, 1987)

PROBLEMS

2.1 A discrete-time LTI system has the impulse response $h[n]$ depicted in Fig. P2.1(a). Use linearity and time invariance to determine the system output $y[n]$ if the input $x[n]$ is:

(a) $x[n] = 2\delta[n] - \delta[n-1]$

(b) $x[n] = u[n] - u[n-3]$

(c) $x[n]$ as given in Fig. P2.1(b)

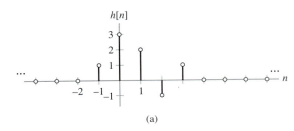

(a)

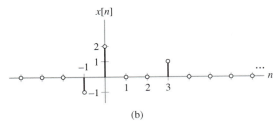

(b)

FIGURE P2.1

2.2 Evaluate the discrete-time convolution sums given below.

(a) $y[n] = u[n] * u[n-3]$

(b) $y[n] = 2^n u[-n+2] * u[n-3]$

(c) $y[n] = (\frac{1}{2})^n u[n-2] * u[n]$

(d) $y[n] = \cos(\frac{1}{2}\pi n)u[n] * u[n-1]$

(e) $y[n] = \cos(\frac{1}{2}\pi n) * 2^n u[-n+2]$

(f) $y[n] = \cos(\frac{1}{2}\pi n) * (\frac{1}{2})^n u[n-2]$

(g) $y[n] = \beta^n u[n] * u[n-3], \quad |\beta| < 1$

(h) $y[n] = \beta^n u[n] * \alpha^n u[n], \quad |\beta| < 1, |\alpha| < 1$

(i) $y[n] = (u[n+10] - 2u[n+5] + u[n-6]) * u[n-2]$

(j) $y[n] = (u[n+10] - 2u[n+5] + u[n-6]) * \beta^n u[n], \quad |\beta| < 1$

(k) $y[n] = (u[n+10] - 2u[n+5] + u[n-6]) * \cos(\frac{1}{2}\pi n)$

(l) $y[n] = u[n] * \sum_{p=0}^{\infty}\delta[n-2p]$

(m) $y[n] = \beta^n u[n] * \sum_{p=0}^{\infty}\delta[n-2p], \quad |\beta| < 1$

(n) $y[n] = u[n-2] * h[n]$, where $h[n] = \begin{cases} \gamma^n, & n < 0, \quad |\gamma| > 1 \\ \eta^n, & n \geq 0, \quad |\eta| < 1 \end{cases}$

(o) $y[n] = (\frac{1}{2})^n u[n+2] * h[n]$, where $h[n]$ is defined in part (n)

2.3 Consider the discrete-time signals depicted in Fig. P2.3. Evaluate the convolution sums indicated below.

(a) $m[n] = x[n] * z[n]$

(b) $m[n] = x[n] * y[n]$

(c) $m[n] = x[n] * f[n]$

(d) $m[n] = x[n] * g[n]$

(e) $m[n] = y[n] * z[n]$

(f) $m[n] = y[n] * g[n]$

(g) $m[n] = y[n] * w[n]$

(h) $m[n] = y[n] * f[n]$

(i) $m[n] = z[n] * g[n]$

(j) $m[n] = w[n] * g[n]$

(k) $m[n] = f[n] * g[n]$

2.4 A LTI system has impulse response $h(t)$ depicted in Fig. P2.4. Use linearity and time invariance to determine the system output $y(t)$ if the input $x(t)$ is:

(a) $x(t) = 2\delta(t+1) - \delta(t-1)$

(b) $x(t) = \delta(t-1) + \delta(t-2) + \delta(t-3)$

(c) $x(t) = \sum_{p=-\infty}^{\infty}(-1)^p \delta(t-2p)$

2.5 Evaluate the continuous-time convolution integrals given below.

(a) $y(t) = u(t+1) * u(t-2)$

(b) $y(t) = e^{-2t}u(t) * u(t+2)$

(c) $y(t) = \cos(\pi t)(u(t+1) - u(t-3)) * u(t)$

(d) $y(t) = (u(t+2) - u(t-1)) * u(-t+2)$

(e) $y(t) = (tu(t) + (10 - 2t)u(t-5) - (10-t)u(t-10)) * u(t)$

(f) $y(t) = (t + 2t^2)(u(t+1) - u(t-1)) * 2u(t+2)$

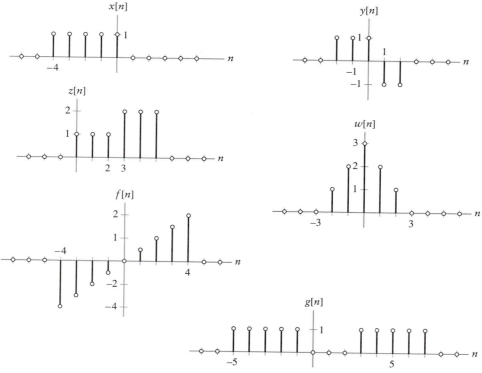

FIGURE P2.3

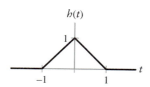

FIGURE P2.4

(g) $y(t) = \cos(\pi t)(u(t + 1) - u(t - 3)) *$
$(u(t + 2) - u(t - 1))$

(h) $y(t) = \cos(\pi t)(u(t + 1) - u(t - 3)) *$
$e^{-2t}u(t)$

(i) $y(t) = (2\delta(t) + \delta(t - 5)) * u(t + 1)$

(j) $y(t) = (\delta(t + 2) + \delta(t - 5)) * (tu(t) + (10 - 2t)u(t - 5) - (10 - t)u(t - 10))$

(k) $y(t) = e^{-\gamma t}u(t) * (u(t + 2) - u(t - 2))$

(l) $y(t) = e^{-\gamma t}u(t) * \sum_{p=0}^{\infty}(\frac{1}{2})^p\delta(t - p)$

(m) $y(t) = (2\delta(t) + \delta(t - 5)) * \sum_{p=0}^{\infty}(\frac{1}{2})^p\delta(t - p)$

(n) $y(t) = e^{-\gamma t}u(t) * e^{\beta t}u(-t) \; \gamma > 0, \beta > 0$

(o) $y(t) = u(t - 1) * b(t)$, where $b(t) = \begin{cases} e^{2t}, & t < 0 \\ e^{-3t}, & t \geq 0 \end{cases}$

2.6 Consider the continuous-time signals depicted in Fig. P2.6. Evaluate the convolution integrals indicated below.

(a) $m(t) = x(t) * y(t)$

(b) $m(t) = x(t) * z(t)$

(c) $m(t) = x(t) * f(t)$

(d) $m(t) = x(t) * b(t)$

(e) $m(t) = x(t) * a(t)$

(f) $m(t) = y(t) * z(t)$

(g) $m(t) = y(t) * w(t)$

(h) $m(t) = y(t) * g(t)$

(i) $m(t) = y(t) * c(t)$

(j) $m(t) = z(t) * f(t)$

(k) $m(t) = z(t) * g(t)$

(l) $m(t) = z(t) * b(t)$

(m) $m(t) = w(t) * g(t)$

(n) $m(t) = w(t) * a(t)$

(o) $m(t) = f(t) * g(t)$

(p) $m(t) = f(t) * c(t)$

(q) $m(t) = f(t) * d(t)$

(r) $m(t) = x(t) * d(t)$

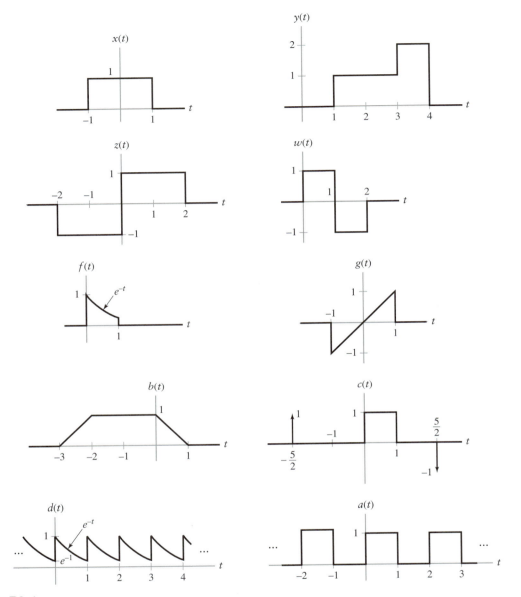

FIGURE P2.6

2.7 Use the definition of the convolution sum to prove the following properties:

(a) Distributive: $x[n] * (h[n] + g[n]) = x[n] * h[n] + x[n] * g[n]$

(b) Associative: $x[n] * (h[n] * g[n]) = (x[n] * h[n]) * g[n]$

(c) Commutative: $x[n] * h[n] = h[n] * x[n]$

2.8 A LTI system has the impulse response depicted in Fig. P2.8.

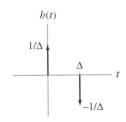

FIGURE P2.8

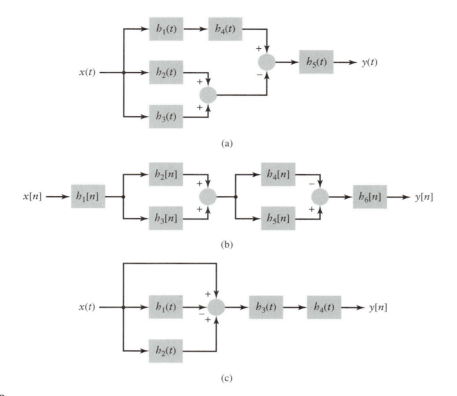

(a)

(b)

(c)

FIGURE P2.9

(a) Express the system output $y(t)$ as a function of the input $x(t)$.

(b) Identify the mathematical operation performed by this system in the limit as $\Delta \to 0$.

(c) Let $g(t) = \lim_{\Delta \to 0} h(t)$. Use the results of (b) to express the output of a LTI system with impulse response

$$h^n(t) = \underbrace{g(t) * g(t) * \cdots * g(t)}_{n \text{ times}}$$

as a function of the input $x(t)$.

2.9 Find the expression for the impulse response relating the input $x[n]$ or $x(t)$ to the output $y[n]$ or $y(t)$ in terms of the impulse response of each subsystem for the LTI systems depicted in:

(a) Fig. P2.9(a)

(b) Fig. P2.9(b)

(c) Fig. P2.9(c)

2.10 Let $h_1(t)$, $h_2(t)$, $h_3(t)$, and $h_4(t)$ be impulse responses of LTI systems. Construct a system with impulse response $h(t)$ using $h_1(t)$, $h_2(t)$, $h_3(t)$, and $h_4(t)$ as subsystems. Draw the interconnection of systems required to obtain:

(a) $h(t) = h_1(t) + \{h_2(t) + h_3(t)\} * h_4(t)$

(b) $h(t) = h_1(t) * h_2(t) + h_3(t) * h_4(t)$

(c) $h(t) = h_1(t) * \{h_2(t) + h_3(t) + h_4(t)\}$

2.11 An interconnection of LTI systems is depicted in Fig. P2.11. The impulse responses are $h_1[n] = (\frac{1}{2})^n(u[n + 2] - u[n - 3])$, $h_2[n] = \delta[n]$, and $h_3[n] = u[n - 1]$. Let the impulse response of the overall system from $x[n]$ to $y[n]$ be denoted as $h[n]$.

(a) Express $h[n]$ in terms of $h_1[n]$, $h_2[n]$, and $h_3[n]$.

(b) Evaluate $h[n]$ using the results of (a).

In parts (c)–(e) determine whether the system corresponding to each impulse response is (i) stable, (ii) causal, and (iii) memoryless.

(c) $h_1[n]$

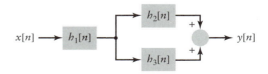

FIGURE P2.11

(d) $h_2[n]$

(e) $h_3[n]$

2.12 For each impulse response listed below, determine whether the corresponding system is (i) memoryless, (ii) causal, and (iii) stable.

(a) $h(t) = e^{-2|t|}$

(b) $h(t) = e^{2t}u(t - 1)$

(c) $h(t) = u(t + 1) - 2u(t - 1)$

(d) $h(t) = 3\delta(t)$

(e) $h(t) = \cos(\pi t)u(t)$

(f) $h[n] = 2^n u[-n]$

(g) $h[n] = e^{2n}u[n - 1]$

(h) $h[n] = \cos(\frac{1}{8}\pi n)\{u[n] - u[n - 10]\}$

(i) $h[n] = 2u[n] - 2u[n - 1]$

(j) $h[n] = \sin(\frac{1}{2}\pi n)$

(k) $h[n] = \delta[n] + \sin(\pi n)$

*2.13 Prove that absolute summability of the impulse response is a necessary condition for stability of a discrete-time system. *Hint:* Find a bounded input $x[n]$ such that the output at some time n_o satisfies $|y[n_o]| = \sum_{k=-\infty}^{\infty}|h[k]|$.

2.14 Evaluate the step response for the LTI systems represented by the following impulse responses:

(a) $h[n] = (\frac{1}{2})^n u[n]$

(b) $h[n] = \delta[n] - \delta[n - 1]$

(c) $h[n] = (-1)^n\{u[n + 2] - u[n - 3]\}$

(d) $h[n] = u[n]$

(e) $h(t) = e^{-|t|}$

(f) $h(t) = \delta(t) - \delta(t - 1)$

(g) $h(t) = u(t + 1) - u(t - 1)$

(h) $h(t) = tu(t)$

2.15 Evaluate the frequency response for the LTI systems represented by the following impulse responses:

(a) $h[n] = (\frac{1}{2})^n u[n]$

(b) $h[n] = \delta[n] - \delta[n - 1]$

(c) $h[n] = (-1)^n\{u[n + 2] - u[n - 3]\}$

(d) $h[n] = (.9)^n e^{j(\pi/2)n}u[n]$

(e) $h(t) = e^{-|t|}$

(f) $h(t) = -\delta(t + 1) + \delta(t) - \delta(t - 1)$

(g) $h(t) = \cos(\pi t)\{u(t + 3) - u(t - 3)\}$

(h) $h(t) = e^{2t}u(-t)$

2.16 Write a differential equation description relating the output to the input of the following electrical circuits:

(a) Fig. P2.16(a)

(b) Fig. P2.16(b)

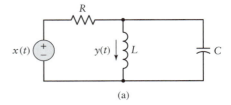

(a)

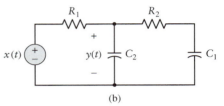

(b)

FIGURE P2.16

2.17 Determine the natural response for the systems described by the following differential equations:

(a) $5\frac{d}{dt}y(t) + 10y(t) = 2x(t), \quad y(0) = 3$

(b) $\frac{d^2}{dt^2}y(t) + 5\frac{d}{dt}y(t) + 6y(t) = 2x(t) + \frac{d}{dt}x(t), \quad y(0) = 2, \frac{d}{dt}y(t)\Big|_{t=0} = 1$

(c) $\frac{d^2}{dt^2}y(t) + 3\frac{d}{dt}y(t) + 2y(t) = x(t) + \frac{d}{dt}x(t), \quad y(0) = 0, \frac{d}{dt}y(t)\Big|_{t=0} = 1$

(d) $\frac{d^2}{dt^2}y(t) + 2\frac{d}{dt}y(t) + y(t) = \frac{d}{dt}x(t), \quad y(0) = 1, \frac{d}{dt}y(t)\Big|_{t=0} = 1$

(e) $\frac{d^2}{dt^2}y(t) + 4y(t) = 3\frac{d}{dt}x(t), \quad y(0) = -1, \frac{d}{dt}y(t)\Big|_{t=0} = 1$

(f) $\frac{d^2}{dt^2}y(t) + 2\frac{d}{dt}y(t) + 2y(t) = \frac{d}{dt}x(t), \quad y(0) = 1, \frac{d}{dt}y(t)\Big|_{t=0} = 0$

2.18 Determine the natural response for the systems described by the following difference equations:

(a) $y[n] - \alpha y[n - 1] = 2x[n], \quad y[-1] = 3$

(b) $y[n] - \frac{9}{16}y[n - 2] = x[n - 1], \quad y[-1] = 1, y[-2] = -1$

(c) $y[n] = -\frac{1}{4}y[n-1] - \frac{1}{8}y[n-2] = x[n] + x[n-1]$, $\quad y[-1] = 0$, $y[-2] = 1$

(d) $y[n] + \frac{9}{16}y[n-2] = x[n-1]$, $\quad y[-1] = 1$, $y[-2] = -1$

(e) $y[n] + y[n-1] + \frac{1}{2}y[n-2] = x[n] + 2x[n-1]$, $\quad y[-1] = -1$, $y[-2] = 1$

2.19 Determine the forced response for the systems described by the following differential equations for the given inputs:

(a) $5\frac{d}{dt}y(t) + 10y(t) = 2x(t)$

   (i) $x(t) = 2u(t)$

   (ii) $x(t) = e^{-t}u(t)$

   (iii) $x(t) = \cos(3t)u(t)$

(b) $\frac{d^2}{dt^2}y(t) + 5\frac{d}{dt}y(t) + 6y(t) = 2x(t) + \frac{d}{dt}x(t)$

   (i) $x(t) = -2u(t)$

   (ii) $x(t) = 2e^{-t}u(t)$

   (iii) $x(t) = \sin(3t)u(t)$

   (iv) $x(t) = 5e^{-2t}u(t)$

(c) $\frac{d^2}{dt^2}y(t) + 3\frac{d}{dt}y(t) + 2y(t) = x(t) + \frac{d}{dt}x(t)$

   (i) $x(t) = 5u(t)$

   (ii) $x(t) = e^{2t}u(t)$

   (iii) $x(t) = (\cos(t) + \sin(t))u(t)$

   (iv) $x(t) = e^{-t}u(t)$

(d) $\frac{d^2}{dt^2}y(t) + 2\frac{d}{dt}y(t) + y(t) = \frac{d}{dt}x(t)$

   (i) $x(t) = e^{-3t}u(t)$

   (ii) $x(t) = 2e^{-t}u(t)$

   (iii) $x(t) = 2\sin(t)u(t)$

2.20 Determine the forced response for the systems described by the following difference equations for the given inputs:

(a) $y[n] - \frac{2}{5}y[n-1] = 2x[n]$

   (i) $x[n] = 2u[n]$

   (ii) $x[n] = -(\frac{1}{2})^n u[n]$

   (iii) $x[n] = \cos(\frac{1}{5}\pi n)u[n]$

(b) $y[n] - \frac{9}{16}y[n-2] = x[n-1]$

   (i) $x[n] = u[n]$

   (ii) $x[n] = -(\frac{1}{2})^n u[n]$

   (iii) $x[n] = (\frac{3}{4})^n u[n]$

(c) $y[n] - \frac{1}{4}y[n-1] - \frac{1}{8}y[n-2] = x[n] + x[n-1]$

   (i) $x[n] = -2u[n]$

   (ii) $x[n] = (\frac{1}{8})^n u[n]$

   (iii) $x[n] = e^{j(\pi/4)n}u[n]$

   (iv) $x[n] = (\frac{1}{2})^n u[n]$

(d) $y[n] + y[n-1] + \frac{1}{2}y[n-2] = x[n] + 2x[n-1]$

   (i) $x[n] = u[n]$

   (ii) $x[n] = (-\frac{1}{2})^n u[n]$

2.21 Determine the output of the systems described by the following differential equations with input and initial conditions as specified:

(a) $\frac{d}{dt}y(t) + 10y(t) = 2x(t)$, $\quad y(0) = 1$, $x(t) = u(t)$

(b) $\frac{d^2}{dt^2}y(t) + 5\frac{d}{dt}y(t) + 4y(t) = \frac{d}{dt}x(t)$, $y(0) = 0$, $\frac{d}{dt}y(t)\Big|_{t=0} = 1$, $x(t) = e^{-2t}u(t)$

(c) $\frac{d^2}{dt^2}y(t) + 3\frac{d}{dt}y(t) + 2y(t) = 2x(t)$, $y(0) = -1$, $\frac{d}{dt}y(t)\Big|_{t=0} = 1$, $x(t) = \cos(t)u(t)$

(d) $\frac{d^2}{dt^2}y(t) + y(t) = 3\frac{d}{dt}x(t)$, $\quad y(0) = -1$, $\frac{d}{dt}y(t)\Big|_{t=0} = 1$, $x(t) = 2e^{-t}u(t)$

2.22 Determine the output of the systems described by the following difference equations with input and initial conditions as specified:

(a) $y[n] - \frac{1}{2}y[n-1] = 2x[n]$, $\quad y[-1] = 3$, $x[n] = 2(-\frac{1}{2})^n u[n]$

(b) $y[n] - \frac{1}{9}y[n-2] = x[n-1]$, $\quad y[-1] = 1$, $y[-2] = 0$, $x[n] = u[n]$

(c) $y[n] - \frac{1}{4}y[n-1] - \frac{1}{8}y[n-2] = x[n] + x[n-1]$, $\quad y[-1] = 2$, $y[-2] = -1$, $x[n] = 2^n u[n]$

(d) $y[n] - \frac{3}{4}y[n-1] + \frac{1}{8}y[n-2] = 2x[n]$, $\quad y[-1] = 1$, $y[-2] = -1$, $x[n] = 2u[n]$

2.23 Find difference-equation descriptions for the four systems depicted in Fig. P2.23.

2.24 Draw direct form I and direct form II implementations for the following difference equations:

(a) $y[n] - \frac{1}{2}y[n-1] = 2x[n]$

(b) $y[n] + \frac{1}{4}y[n-1] - \frac{1}{8}y[n-2] = x[n] + x[n-1]$

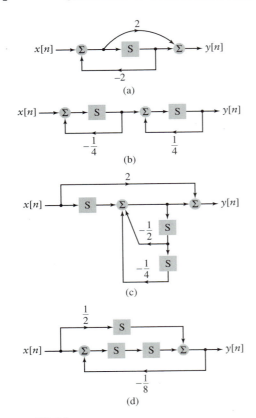

FIGURE P2.23

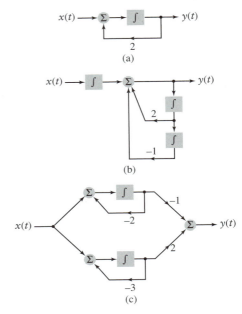

FIGURE P2.27

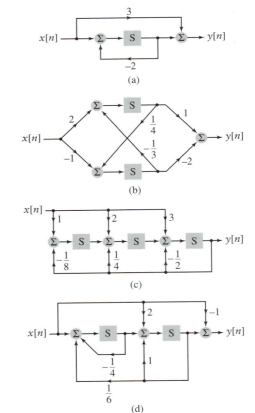

(c) $y[n] - \frac{1}{9}y[n-2] = 2x[n] + x[n-1]$

(d) $y[n] + \frac{1}{2}y[n-1] - y[n-3] = 3x[n-1] + 2x[n-2]$

2.25 Show that the direct form I and II implementations depicted in Fig. 2.27 implement the second-order integral equation given by Eq. (2.42).

2.26 Convert the following differential equations to integral equations and draw direct form I and direct form II implementations of the corresponding systems:

(a) $\frac{d}{dt}y(t) + 10y(t) = 2x(t)$

(b) $\frac{d^2}{dt^2}y(t) + 5\frac{d}{dt}y(t) + 4y(t) = \frac{d}{dt}x(t)$

(c) $\frac{d^2}{dt^2}y(t) + y(t) = 3\frac{d}{dt}x(t)$

(d) $\frac{d^3}{dt^3}y(t) + 2\frac{d}{dt}y(t) + 3y(t) = x(t) + 3\frac{d}{dt}x(t)$

2.27 Find differential-equation descriptions for the three systems depicted in Fig. P2.27.

FIGURE P2.28

2.28 Determine a state-variable description for the four discrete-time systems depicted in Fig. P2.28.

2.29 Draw block diagram system representations corresponding to the following discrete-time state-variable descriptions.

(a) $\mathbf{A} = \begin{bmatrix} 0 & -\frac{1}{2} \\ \frac{1}{3} & 0 \end{bmatrix}$, $\mathbf{b} = \begin{bmatrix} 2 \\ 0 \end{bmatrix}$, $\mathbf{c} = [1 \ -1]$, $D = [0]$

(b) $\mathbf{A} = \begin{bmatrix} 1 & -\frac{1}{2} \\ \frac{1}{3} & 0 \end{bmatrix}$, $\mathbf{b} = \begin{bmatrix} 1 \\ 2 \end{bmatrix}$, $\mathbf{c} = [1 \ -1]$, $D = [0]$

(c) $\mathbf{A} = \begin{bmatrix} 0 & -\frac{1}{2} \\ \frac{1}{3} & -1 \end{bmatrix}$, $\mathbf{b} = \begin{bmatrix} 0 \\ 1 \end{bmatrix}$, $\mathbf{c} = [1 \ 0]$, $D = [1]$

(d) $\mathbf{A} = \begin{bmatrix} 0 & 0 \\ 0 & 1 \end{bmatrix}$, $\mathbf{b} = \begin{bmatrix} 2 \\ 3 \end{bmatrix}$, $\mathbf{c} = [1 \ -1]$, $D = [0]$

2.30 Determine a state-variable description for the five continuous-time systems depicted in Fig. P2.30.

2.31 Draw block diagram system representations corresponding to the following continuous-time state-variable descriptions:

(a) $\mathbf{A} = \begin{bmatrix} \frac{1}{3} & 0 \\ 0 & -\frac{1}{2} \end{bmatrix}$, $\mathbf{b} = \begin{bmatrix} -1 \\ 2 \end{bmatrix}$, $\mathbf{c} = [1 \ 1]$, $D = [0]$

(b) $\mathbf{A} = \begin{bmatrix} 1 & 1 \\ 1 & 0 \end{bmatrix}$, $\mathbf{b} = \begin{bmatrix} -1 \\ 2 \end{bmatrix}$, $\mathbf{c} = [0 \ -1]$, $D = [0]$

(c) $\mathbf{A} = \begin{bmatrix} 1 & -1 \\ 0 & -1 \end{bmatrix}$, $\mathbf{b} = \begin{bmatrix} 0 \\ 5 \end{bmatrix}$, $\mathbf{c} = [1 \ 0]$, $D = [0]$

(d) $\mathbf{A} = \begin{bmatrix} 1 & -2 \\ 1 & 1 \end{bmatrix}$, $\mathbf{b} = \begin{bmatrix} 2 \\ 3 \end{bmatrix}$, $\mathbf{c} = [1 \ 1]$, $D = [0]$

2.32 Let a discrete-time system have the state-variable description

$$\mathbf{A} = \begin{bmatrix} 1 & -\frac{1}{2} \\ \frac{1}{3} & 0 \end{bmatrix}, \quad \mathbf{b} = \begin{bmatrix} 1 \\ 2 \end{bmatrix},$$
$$\mathbf{c} = [1 \ -1], \quad D = [0]$$

(a) Define new states $q_1'[n] = 2q_1[n]$, $q_2'[n] = 3q_2[n]$. Find the new state-variable description $\mathbf{A}', \mathbf{b}', \mathbf{c}', D'$.

(b) Define new states $q_1'[n] = 3q_2[n]$, $q_2'[n] = 2q_1[n]$. Find the new state-variable description $\mathbf{A}', \mathbf{b}', \mathbf{c}', D'$.

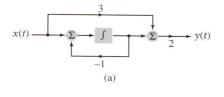

(a)

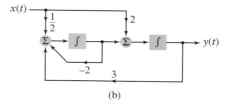

(b)

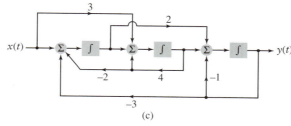

(c)

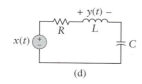

(d)

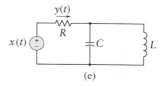

(e)

Figure P2.30

(c) Define new states $q_1'[n] = q_1[n] + q_2[n]$, $q_2'[n] = q_1[n] - q_2[n]$. Find the new state-variable description $\mathbf{A}'$, $\mathbf{b}'$, $\mathbf{c}'$, D'.

2.33 Consider the continuous-time system depicted in Fig. P2.33.

(a) Find the state variable description for this system assuming the states $q_1(t)$ and $q_2(t)$ are as labeled.

(b) Define new states $q_1'(t) = q_1(t) - q_2(t)$, $q_2'(t) = 2q_1(t)$. Find the new state-variable description $\mathbf{A}'$, $\mathbf{b}'$, $\mathbf{c}'$, D'.

(c) Draw a block diagram corresponding to the new state-variable description in (b).

(d) Define new states $q_1'(t) = (1/b_1)q_1(t)$, $q_2'(t) = b_2q_1(t) - b_1q_2(t)$. Find the new state-variable description $\mathbf{A}'$, $\mathbf{b}'$, $\mathbf{c}'$, D'.

(e) Draw a block diagram corresponding to the new state-variable description in (d).

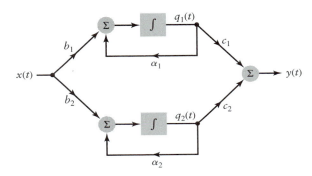

FIGURE P2.33

*2.34 We may develop the convolution integral using linearity, time invariance, and the limiting form of a stairstep approximation to the input signal. Define $g_\Delta(t)$ as the unit area rectangular pulse depicted in Fig. P2.34(a).

(a) A stairstep approximation to a signal $x(t)$ is depicted in Fig. P2.34(b). Express $\tilde{x}(t)$ as a weighted sum of shifted pulses $g_\Delta(t)$. Does the approximation quality improve as Δ decreases?

(b) Let the response of a LTI system to an input $g_\Delta(t)$ be $h_\Delta(t)$. If the input to this system is $\tilde{x}(t)$, find an expression for the output of this system in terms of $h_\Delta(t)$.

(c) In the limit as Δ goes to zero, $g_\Delta(t)$ satisfies the properties of an impulse and we may interpret $h(t) = \lim_{\Delta \to 0} h_\Delta(t)$ as the impulse response of the system. Show that the ex-

pression for the system output derived in (b) reduces to $x(t) * h(t)$ in the limit as Δ goes to zero.

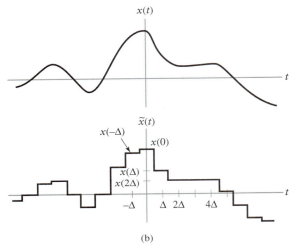

(a)

(b)

FIGURE P2.34

*2.35 In this problem we use linearity, time invariance, and representation of an impulse as the limiting form of a pulse to obtain the impulse response of a simple RC circuit. The voltage across the capacitor, $y(t)$, in the RC circuit of Fig. P2.35(a) in response to an applied voltage $x(t) = u(t)$ is given by

$$s(t) = \{1 - e^{-t/RC}\}u(t)$$

(See Drill Problems 2.8 and 2.12.) We wish to find the impulse response of the system relating the input voltage $x(t)$ to the voltage across the capacitor $y(t)$.

(a) Write the pulse input $x(t) = g_\Delta(t)$ depicted in Fig. P2.35(b) as a weighted sum of step functions.

(b) Use linearity, time invariance, and knowledge of the step response of this circuit to express the output of the circuit in response to the input $x(t) = g_\Delta(t)$ in terms of $s(t)$.

(c) In the limit as $\Delta \to 0$ the pulse input $g_\Delta(t)$ approaches an impulse. Obtain the impulse response of the circuit by taking the limit as $\Delta \to 0$ of the output obtained in (b). *Hint:* Use the definition of the derivative

$$\frac{d}{dt} z(t) = \lim_{\Delta \to 0} \frac{z\left(t + \frac{\Delta}{2}\right) - z\left(t - \frac{\Delta}{2}\right)}{\Delta}$$

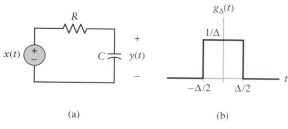

(a)   (b)

FIGURE P2.35

*2.36 The cross-correlation between two real signals $x(t)$ and $y(t)$ is defined as

$$r_{xy}(t) = \int_{-\infty}^{\infty} x(\tau) y(\tau - t) \, d\tau$$

This is the area under the product of $x(t)$ and a shifted version of $y(t)$. Note that the independent variable $\tau - t$ is the negative of that found in the definition of convolution. The autocorrelation, $r_{xx}(t)$, of a signal $x(t)$ is obtained by replacing $y(t)$ with $x(t)$.

(a) Show that $r_{xy}(t) = x(t) * y(-t)$.

(b) Derive a step-by-step procedure for evaluating the cross-correlation analogous to the one for evaluating convolution integral given in Section 2.2.

(c) Evaluate the cross-correlation between the following signals:
   (i) $x(t) = e^{-t} u(t)$, $y(t) = e^{-3t} u(t)$
   (ii) $x(t) = \cos(\pi t)[u(t + 2) - u(t - 2)]$, $y(t) = \cos(2\pi t)[u(t + 2) - u(t - 2)]$
   (iii) $x(t) = u(t) - 2u(t - 1) + u(t - 2)$, $y(t) = u(t + 1) - u(t)$
   (iv) $x(t) = u(t - a) - u(t - a - 1)$, $y(t) = u(t) - u(t - 1)$

(d) Evaluate the autocorrelation of the following signals:
   (i) $x(t) = e^{-t} u(t)$
   (ii) $x(t) = \cos(\pi t)[u(t + 2) - u(t - 2)]$

   (iii) $x(t) = u(t) - 2u(t - 1) + u(t - 2)$
   (iv) $x(t) = u(t - a) - u(t - a - 1)$

(e) Show that $r_{xy}(t) = r_{yx}(-t)$.

(f) Show that $r_{xx}(t) = r_{xx}(-t)$.

▶ **Computer Experiments**

2.37 Repeat Problem 2.3 using the MATLAB command `conv`.

2.38 Use MATLAB to repeat Example 2.6.

2.39 Use MATLAB to evaluate the first 20 values of the step response for the systems in Problem 2.14(a)–(d).

2.40 Consider the three moving-average systems defined in Example 2.6.

(a) Use MATLAB to evaluate and plot 50 values of the sinusoidal steady-state response at frequencies of $\Omega = \pi/3$ and $\Omega = 2\pi/3$ for each system.

(b) Use the results of (a) to determine the magnitude and phase response of each system at frequencies $\Omega = \pi/3$ and $\Omega = 2\pi/3$.

(c) Obtain a closed-form expression for the magnitude response of each system and plot it on $-\pi < \Omega \le \pi$ using MATLAB.

2.41 Consider the two systems having impulse responses

$$h_1[n] = \begin{cases} \frac{1}{4}, & 0 \le n \le 3 \\ 0, & \text{otherwise} \end{cases}$$

$$h_2[n] = \begin{cases} \frac{1}{4}, & n = 0, 2 \\ -\frac{1}{4}, & n = 1, 3 \\ 0, & \text{otherwise} \end{cases}$$

(a) Use the MATLAB command `conv` to plot the first 20 values of the step response.

(b) Obtain a closed-form expression for the magnitude response and plot it on $-\pi < \Omega \le \pi$ using MATLAB.

2.42 Use the MATLAB commands `filter` and `filtic` to repeat Example 2.16.

2.43 Use the MATLAB commands `filter` and `filtic` to determine the first 50 output values in Problem 2.22.

2.44 The magnitude response of a system described by a difference equation may be obtained from the output $y[n]$ by applying an input $x[n] = e^{j\Omega n} u[n]$ to a system that is initially at rest. Once the natural response of the system has decayed to a negligible value, $y[n]$ is due only to the input and we have $y[n] \approx H(e^{j\Omega})e^{j\Omega n}$.

(a) Determine the value n_o for which each term in the natural response of the system in Example 2.16 is a factor of 1000 smaller than its value at time $n = 0$.

(b) Show that $|H(e^{j\Omega})| = |y[n_o]|$.

(c) Use the results in (a) and (b) to experimentally determine the magnitude response of this system with the MATLAB command `filter`. Plot the magnitude response for input frequencies in the range $-\pi < \Omega \leq \pi$.

2.45 Use the MATLAB command `impz` to determine the first 30 values of the impulse response for the systems described in Problem 2.22.

2.46 Use the MATLAB command `ss2ss` to solve Problem 2.32.

2.47 A system has the state-variable description

$$\mathbf{A} = \begin{bmatrix} \frac{1}{2} & -\frac{1}{2} \\ \frac{1}{3} & 0 \end{bmatrix}, \quad \mathbf{b} = \begin{bmatrix} 1 \\ 2 \end{bmatrix},$$
$$\mathbf{c} = \begin{bmatrix} 1 & -1 \end{bmatrix}, \quad D = [0]$$

(a) Use the MATLAB commands `lsim` and `impulse` to determine the first 30 values of the step and impulse responses of this system.

(b) Define new states $q_1[n] = q_1[n] + q_2[n]$ and $q_2[n] = 2q_1[n] - q_2[n]$. Repeat part (a) for the transformed system.

3 Fourier Representations for Signals

3.1 Introduction

In this chapter we consider representing a signal as a weighted superposition of complex sinusoids. If such a signal is applied to a linear system, then the system output is a weighted superposition of the system response to each complex sinusoid. A similar application of the linearity property was exploited in the previous chapter to develop the convolution integral and convolution sum. In Chapter 2, the input signal was expressed as a weighted superposition of time-shifted impulses; the output was then given by a weighted superposition of time-shifted versions of the system's impulse response. The expression for the output that resulted from expressing signals in terms of impulses was termed "convolution." By representing signals in terms of sinusoids, we will obtain an alternative expression for the input–output behavior of a LTI system.

Representation of signals as superpositions of complex sinusoids not only leads to a useful expression for the system output but also provides a very insightful characterization of signals and systems. The focus of this chapter is representation of signals using complex sinusoids and the properties of such representations. Applications of these representations to system and signal analysis are emphasized in the following chapter.

The study of signals and systems using sinusoidal representations is termed Fourier analysis after Joseph Fourier (1768–1830) for his contributions to the theory of representing functions as weighted superpositions of sinusoids. Fourier methods have widespread application beyond signals and systems; they are used in every branch of engineering and science.

■ COMPLEX SINUSOIDS AND LTI SYSTEMS

The sinusoidal steady-state response of a LTI system was introduced in Section 2.3. We showed that a complex sinusoid input to a LTI system generates an output equal to the sinusoidal input multiplied by the system frequency response. That is, in discrete time, the input $x[n] = e^{j\Omega n}$ results in the output

$$y[n] = H(e^{j\Omega})e^{j\Omega n}$$

where the *frequency response* $H(e^{j\Omega})$ is defined in terms of the impulse response $h[n]$ as

$$H(e^{j\Omega}) = \sum_{k=-\infty}^{\infty} h[k]e^{-j\Omega k}$$

In continuous time, the input $x(t) = e^{j\omega t}$ results in the output

$$y(t) = H(j\omega)e^{j\omega t}$$

where the frequency response $H(j\omega)$ is defined in terms of the impulse response $h(t)$ as

$$H(j\omega) = \int_{-\infty}^{\infty} h(\tau)e^{-j\omega\tau} \, d\tau$$

We say that the complex sinusoid $\psi(t) = e^{j\omega t}$ is an *eigenfunction* of the system H associated with the *eigenvalue* $\lambda = H(j\omega)$ because it satisfies an eigenvalue problem described by

$$H\{\psi(t)\} = \lambda\psi(t)$$

This eigenrelation is illustrated in Fig. 3.1. The effect of the system on an eigenfunction input signal is one of scalar multiplication—the output is given by the product of the input and a complex number. This eigenrelation is analogous to the more familiar matrix eigenvalue problem. If $\mathbf{e}_k$ is an eigenvector of a matrix $\mathbf{A}$ with eigenvalue λ_k, then we have

$$\mathbf{A}\mathbf{e}_k = \lambda_k\mathbf{e}_k$$

Multiplying $\mathbf{e}_k$ by the matrix $\mathbf{A}$ is equivalent to multiplying $\mathbf{e}_k$ by the scalar λ_k.

Signals that are eigenfunctions of systems play an important role in systems theory. By representing arbitrary signals as weighted superpositions of eigenfunctions, we transform the operation of convolution to one of multiplication. To see this, consider expressing the input to a LTI system as the weighted sum of M complex sinusoids

$$x(t) = \sum_{k=1}^{M} a_k e^{j\omega_k t}$$

If $e^{j\omega_k t}$ is an eigenfunction of the system with eigenvalue $H(j\omega_k)$, then each term in the input, $a_k e^{j\omega_k t}$, produces an output term, $a_k H(j\omega_k)e^{j\omega_k t}$. Hence we express the output of the system as

$$y(t) = \sum_{k=1}^{M} a_k H(j\omega_k)e^{j\omega_k t}$$

The output is a weighted sum of M complex sinusoids, with the weights, a_k, modified by the system frequency response, $H(j\omega_k)$. The operation of convolution, $h(t) * x(t)$, becomes multiplication, $a_k H(j\omega_k)$, because $x(t)$ is expressed as a sum of eigenfunctions. The analogous relationship holds in the discrete-time case.

This property is a powerful motivation for representing signals as weighted superpositions of complex sinusoids. In addition, the weights provide an alternative interpretation of the signal. Rather than describing the signal behavior as a function of time, the

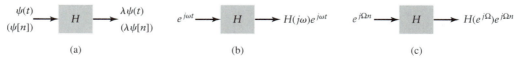

(a)   (b)   (c)

FIGURE 3.1   Illustration of the eigenfunction property of linear systems. The action of the system on an eigenfunction input is one of multiplication by the corresponding eigenvalue. (a) General eigenfunction $\psi(t)$ or $\psi[n]$ and eigenvalue λ. (b) Complex sinusoid eigenfunction $e^{j\omega t}$ and eigenvalue $H(j\omega)$. (c) Complex sinusoid eigenfunction $e^{j\Omega n}$ and eigenvalue $H(e^{j\Omega})$.

weights describe the signal as a function of frequency. The general notion of describing complicated signals as a function of frequency is commonly encountered in music. For example, the musical score for an orchestra contains parts for instruments having different frequency ranges, such as a string bass, which produces very low frequency sound, and a piccolo, which produces very high frequency sound. The sound that we hear when listening to an orchestra represents the superposition of sounds generated by each instrument. Similarly, the score for a choir contains bass, tenor, alto, and soprano parts, each of which contributes to a different frequency range in the overall sound. The signal representations developed in this chapter can be viewed analogously: the weight associated with a sinusoid of a given frequency represents the contribution of that sinusoid to the overall signal. A frequency-domain view of signals is very informative, as we shall see in what follows.

■ FOURIER REPRESENTATIONS FOR FOUR SIGNAL CLASSES

There are four distinct Fourier representations, each applicable to a different class of signals. These four classes are defined by the periodicity properties of a signal and whether it is continuous or discrete time. Periodic signals have Fourier series representations. The Fourier series (FS) applies to continuous-time periodic signals and the discrete-time Fourier series (DTFS) applies to discrete-time periodic signals. Nonperiodic signals have Fourier transform representations. If the signal is continuous time and nonperiodic, the representation is termed the Fourier transform (FT). If the signal is discrete time and nonperiodic, then the discrete-time Fourier transform (DTFT) is used. Table 3.1 illustrates the relationship between the time properties of a signal and the appropriate Fourier representation. The DTFS is often referred to as the discrete Fourier transform or DFT; however, this terminology does not correctly reflect the series nature of the DTFS and often leads to confusion with the DTFT so we adopt the more descriptive DTFS terminology.

TABLE 3.1 *Relationship Between Time Properties of a Signal and the Appropriate Fourier Representation*

Time Property	Periodic	Nonperiodic
Continuous	Fourier Series (FS)	Fourier Transform (FT)
Discrete	Discrete-Time Fourier Series (DTFS)	Discrete-Time Fourier Transform (DTFT)

Periodic Signals: Fourier Series Representations

Consider representing a periodic signal as a weighted superposition of complex sinusoids. Since the weighted superposition must have the same period as the signal, each sinusoid in the superposition must have the same period as the signal. This implies that the frequency of each sinusoid must be an integer multiple of the signal's fundamental frequency. If $x[n]$ is a discrete-time signal of fundamental period N, then we seek to represent $x[n]$ by the DTFS

$$\hat{x}[n] = \sum_k A[k]e^{jk\Omega_o n} \tag{3.1}$$

where $\Omega_o = 2\pi/N$ is the fundamental frequency of $x[n]$. The frequency of the kth sinusoid in the superposition is $k\Omega_o$. Similarly, if $x(t)$ is a continuous-time signal of fundamental period T, we represent $x(t)$ by the FS

$$\hat{x}(t) = \sum_k A[k]e^{jk\omega_o t} \tag{3.2}$$

where $\omega_o = 2\pi/T$ is the fundamental frequency of $x(t)$. Here the frequency of the kth sinusoid is $k\omega_o$. In both Eqs. (3.1) and (3.2), $A[k]$ is the weight applied to the kth complex sinusoid and the hat ˆ denotes approximate value, since we do not yet assume that either $x[n]$ or $x(t)$ can be represented exactly by a series of this form.

How many terms and weights should we use in each sum? Beginning with the DTFS described in Eq. (3.1), the answer to this question becomes apparent if we recall that complex sinusoids with distinct frequencies are not always distinct. In particular, the complex sinusoids $e^{jk\Omega_o n}$ are N periodic in the frequency index k. We have

$$e^{j(N+k)\Omega_o n} = e^{jN\Omega_o n}e^{jk\Omega_o n}$$
$$= e^{j2\pi n}e^{jk\Omega_o n}$$
$$= e^{jk\Omega_o n}$$

Thus there are only N distinct complex sinusoids of the form $e^{jk\Omega_o n}$. A unique set of N distinct complex sinusoids is obtained by letting the frequency index k take on any N consecutive values. Hence we may rewrite Eq. (3.1) as

$$\hat{x}[n] = \sum_{k=\langle N \rangle} A[k]e^{jk\Omega_o n} \tag{3.3}$$

where the notation $k = \langle N \rangle$ implies letting k range over any N consecutive values. The set of N consecutive values over which k varies is arbitrary and is usually chosen to simplify the problem by exploiting symmetries in the signal $x[n]$. Common choices are $k = 0$ to $N - 1$ and, for N even, $k = -N/2$ to $N/2 - 1$.

In order to determine the weights or coefficients $A[k]$, we shall minimize the mean-squared error (MSE) between the signal and its series representation. The construction of the series representation ensures that both the signal and the representation are periodic with the same period. Hence the MSE is the average of the squared difference between the signal and its representation over any one period. In the discrete-time case only N consecutive values of $\hat{x}[n]$ and $x[n]$ are required since both are N periodic. We have

$$MSE = \frac{1}{N} \sum_{n=\langle N \rangle} |x[n] - \hat{x}[n]|^2 \tag{3.4}$$

where we again use the notation $n = \langle N \rangle$ to indicate summation over any N consecutive values. We leave the interval for evaluating the MSE unspecified since it will later prove convenient to choose different intervals in different problems.

In contrast to the discrete-time case, continuous-time complex sinusoids $e^{jk\omega_o t}$ with distinct frequencies $k\omega_o$ are always distinct. Hence there are potentially an infinite number of distinct terms in the series of Eq. (3.2) and we approximate $x(t)$ as

$$\hat{x}(t) = \sum_{k=-\infty}^{\infty} A[k]e^{jk\omega_o t} \tag{3.5}$$

We seek coefficients $A[k]$ so that $\hat{x}(t)$ is a good approximation to $x(t)$.

Nonperiodic Signals: Fourier Transform Representations

In contrast to the periodic signal case, there are no restrictions on the period of the sinusoids used to represent nonperiodic signals. Hence the Fourier transform representations employ complex sinusoids having a continuum of frequencies. The signal is represented as a weighted integral of complex sinusoids where the variable of integration is the sinusoid's frequency. Discrete-time sinusoids are used to represent discrete-time signals in the DTFT, while continuous-time sinusoids are used to represent continuous-time signals in the FT. Continuous-time sinusoids with distinct frequencies are distinct, so the FT involves sinusoidal frequencies from $-\infty$ to ∞. Discrete-time sinusoids are only unique over a 2π interval of frequency, since discrete-time sinusoids with frequencies separated by an integer multiple of 2π are identical. Hence the DTFT involves sinusoidal frequencies within a 2π interval.

The next four sections of this chapter develop, in sequence, the DTFS, FS, DTFT, and FT. The remainder of the chapter explores the properties of these four representations. All four representations are based on complex sinusoidal basis functions and thus have analogous properties.

■ ORTHOGONALITY OF COMPLEX SINUSOIDS

The orthogonality of complex sinusoids plays a key role in Fourier representations. We say that two signals are *orthogonal* if their *inner product* is zero. For discrete-time periodic signals, the inner product is defined as the sum of values in their product. If $\phi_k[n]$ and $\phi_m[n]$ are two N periodic signals, their inner product is

$$I_{k,m} = \sum_{n=\langle N \rangle} \phi_k[n]\phi_m^*[n]$$

Note that the inner product is defined using complex conjugation when the signals are complex valued. If $I_{k,m} = 0$ for $k \neq m$, then $\phi_k[n]$ and $\phi_m[n]$ are orthogonal. Correspondingly, for continuous-time signals with period T, the inner product is defined in terms of an integral, as shown by

$$I_{k,m} = \int_{\langle T \rangle} \phi_k(t)\phi_m^*(t) \, dt$$

where the notation $\langle T \rangle$ implies integration over any interval of length T. As in discrete time, if $I_{k,m} = 0$ for $k \neq m$, then we say $\phi_k(t)$ and $\phi_m(t)$ are orthogonal.

Beginning with the discrete-time case, let $\phi_k[n] = e^{jk\Omega_o n}$ be a complex sinusoid with frequency $k\Omega_o$. Choosing the interval $n = 0$ to $n = N - 1$, the inner product is given by

$$I_{k,m} = \sum_{n=0}^{N-1} e^{j(k-m)\Omega_o n}$$

Assuming k and m are restricted to the same interval of N consecutive values, this is a finite geometric series whose sum depends on whether $k = m$ or $k \neq m$, as shown by

$$\sum_{n=0}^{N-1} e^{j(k-m)\Omega_o n} = \begin{cases} N, & k = m \\ \dfrac{1 - e^{jk2\pi}}{1 - e^{jk\Omega_o}}, & k \neq m \end{cases}$$

Now use $e^{-jk2\pi} = 1$ to obtain

$$\sum_{n=0}^{N-1} e^{j(k-m)\Omega_o n} = \begin{cases} N, & k = m \\ 0, & k \neq m \end{cases} \tag{3.6}$$

This result indicates that complex sinusoids with frequencies separated by an integer multiple of the fundamental frequency are orthogonal. We shall use this result in deriving the DTFS representation.

Continuous-time complex sinusoids with frequencies separated by an integer multiple of the fundamental frequency are also orthogonal. Letting $\phi_k(t) = e^{jk\omega_o t}$, the inner product between $e^{jk\omega_o t}$ and $e^{jm\omega_o t}$ is expressed as

$$I_{k,m} = \int_0^T e^{j(k-m)\omega_o t}\, dt$$

This integral takes on two values, depending on the value $k - m$, as shown by

$$\int_0^T e^{j(k-m)\omega_o t}\, dt = \begin{cases} T, & k = m \\ \dfrac{1}{j(k-m)\omega_o} e^{j(k-m)\omega_o t}\Big|_0^T, & k \neq m \end{cases}$$

Using the fact $e^{j(k-m)\omega_o T} = e^{j(k-m)2\pi} = 1$, we obtain

$$\int_0^T e^{j(k-m)\omega_o t}\, dt = \begin{cases} T, & k = m \\ 0 & k \neq m \end{cases} \tag{3.7}$$

This property is central to determining the FS coefficients.

3.2 Discrete-Time Periodic Signals: The Discrete-Time Fourier Series

∗ ■ DERIVATION

The DTFS represents an N periodic discrete-time signal $x[n]$ as the series of Eq. (3.3):

$$\hat{x}[n] = \sum_{k=\langle N \rangle} A[k] e^{jk\Omega_o n}$$

where $\Omega_o = 2\pi/N$.

In order to choose the DTFS coefficients $A[k]$, we now minimize the MSE defined in Eq. (3.4), rewritten as

$$MSE = \frac{1}{N} \sum_{n=\langle N \rangle} |x[n] - \hat{x}[n]|^2$$

$$= \frac{1}{N} \sum_{n=\langle N \rangle} \left| x[n] - \sum_{k=\langle N \rangle} A[k] e^{jk\Omega_o n} \right|^2$$

Minimization of the MSE is instructive, although it involves tedious algebraic manipulation. The end result is an expression for $A[k]$ in terms of $x[n]$. Also, by examining the minimum value of the MSE, we are able to establish the accuracy with which $\hat{x}[n]$ approximates $x[n]$.

The magnitude squared of a complex number c is given by $|c|^2 = cc^*$. Expanding the magnitude squared in the sum using $|a + b|^2 = (a + b)(a + b)^*$ yields

$$MSE = \frac{1}{N} \sum_{n=\langle N \rangle} \left\{ \left(x[n] - \sum_{k=\langle N \rangle} A[k]e^{jk\Omega_o n} \right) \left(x[n] - \sum_{m=\langle N \rangle} A[m]e^{jm\Omega_o n} \right)^* \right\}$$

Now multiply each term to obtain

$$MSE = \frac{1}{N} \sum_{n=\langle N \rangle} |x[n]|^2 - \sum_{m=\langle N \rangle} A^*[m] \left(\frac{1}{N} \sum_{n=\langle N \rangle} x[n]e^{-jm\Omega_o n} \right)$$

$$- \sum_{k=\langle N \rangle} A[k] \left(\frac{1}{N} \sum_{n=\langle N \rangle} x^*[n]e^{jk\Omega_o n} \right)$$

$$+ \sum_{k=\langle N \rangle} \sum_{m=\langle N \rangle} A^*[m]A[k] \left(\frac{1}{N} \sum_{n=\langle N \rangle} e^{j(k-m)\Omega_o n} \right)$$

Define

$$X[k] = \frac{1}{N} \sum_{n=\langle N \rangle} x[n]e^{-jk\Omega_o n} \tag{3.8}$$

and apply the orthogonality property of discrete-time complex sinusoids, Eq. (3.6), to the last term in the MSE. Hence we may write the MSE as

$$MSE = \frac{1}{N} \sum_{n=\langle N \rangle} |x[n]|^2 - \sum_{k=\langle N \rangle} A^*[k]X[k] - \sum_{k=\langle N \rangle} A[k]X^*[k] + \sum_{k=\langle N \rangle} |A[k]|^2$$

Now use the technique of "completing the square" to write the MSE as a perfect square in the DTFS coefficients $A[k]$. Add and subtract $\sum_{k=\langle N \rangle} |X[k]|^2$ to the right-hand side of the MSE, so that it may be written as

$$MSE = \frac{1}{N} \sum_{n=\langle N \rangle} |x[n]|^2 + \sum_{k=\langle N \rangle} (|A[k]|^2 - A^*[k]X[k] - A[k]X^*[k] + |X[k]|^2)$$

$$- \sum_{k=\langle N \rangle} |X[k]|^2$$

Rewrite the middle sum as a square to obtain

$$MSE = \frac{1}{N} \sum_{n=\langle N \rangle} |x[n]|^2 + \sum_{k=\langle N \rangle} |A[k] - X[k]|^2 - \sum_{k=\langle N \rangle} |X[k]|^2 \tag{3.9}$$

The dependence of the MSE on the unknown DTFS coefficients $A[k]$ is confined to the middle term of Eq. (3.9), and this term is always nonnegative. Hence the MSE is minimized by forcing the middle term to zero with the choice

$$A[k] = X[k]$$

These coefficients minimize the MSE between $\hat{x}[n]$ and $x[n]$.

Note that $X[k]$ is N periodic in k, since

$$X[k + N] = \frac{1}{N} \sum_{n=\langle N \rangle} x[n]e^{-j(k+N)\Omega_o n}$$

$$= \frac{1}{N} \sum_{n=\langle N \rangle} x[n]e^{-jk\Omega_o n}e^{-jN\Omega_o n}$$

Using the fact that $e^{-jN\Omega_o n} = e^{-j2\pi n} = 1$ we obtain

$$X[k + N] = \frac{1}{N} \sum_{n=\langle N \rangle} x[n]e^{-jk\Omega_o n}$$
$$= X[k]$$

which establishes that $X[k]$ is N periodic.

The value of the minimum MSE determines how well $\hat{x}[n]$ approximates $x[n]$. We determine the minimum MSE by substituting $A[k] = X[k]$ into Eq. (3.9) to obtain

$$MSE = \frac{1}{N} \sum_{n=\langle N \rangle} |x[n]|^2 - \sum_{k=\langle N \rangle} |X[k]|^2 \tag{3.10}$$

We next substitute Eq. (3.8) into the second term of Eq. (3.10) to obtain

$$\sum_{k=\langle N \rangle} |X[k]|^2 = \sum_{k=\langle N \rangle} \frac{1}{N^2} \sum_{n=\langle N \rangle} \sum_{m=\langle N \rangle} x[n]x^*[m]e^{j(m-n)\Omega_o k}$$

Interchange the order of summation to write

$$\sum_{k=\langle N \rangle} |X[k]|^2 = \frac{1}{N} \sum_{n=\langle N \rangle} \sum_{m=\langle N \rangle} x[n]x^*[m] \frac{1}{N} \sum_{k=\langle N \rangle} e^{j(m-n)\Omega_o k} \tag{3.11}$$

Equation (3.11) is simplified by recalling that $e^{jm\Omega_o k}$ and $e^{jn\Omega_o k}$ are orthogonal. Referring to Eq. (3.6), we have

$$\frac{1}{N} \sum_{k=\langle N \rangle} e^{j(m-n)\Omega_o k} = \begin{cases} 1, & n = m \\ 0, & n \neq m \end{cases}$$

This reduces the double sum over m and n on the right-hand side of Eq. (3.11) to the single sum

$$\sum_{k=\langle N \rangle} |X[k]|^2 = \frac{1}{N} \sum_{n=\langle N \rangle} |x[n]|^2$$

Substituting this result into Eq. (3.10) gives MSE = 0. That is, if the DTFS coefficients are given by Eq. (3.8), then the MSE between $\hat{x}[n]$ and $x[n]$ is zero. Since the MSE is zero, the error is zero for each value of n and thus $\hat{x}[n] = x[n]$.

■ THE **DTFS** REPRESENTATION

The DTFS representation for $x[n]$ is given by

$$x[n] = \sum_{k=\langle N \rangle} X[k]e^{jk\Omega_o n} \tag{3.12}$$

$$X[k] = \frac{1}{N} \sum_{n=\langle N \rangle} x[n]e^{-jk\Omega_o n} \tag{3.13}$$

where $x[n]$ has fundamental period N and $\Omega_o = 2\pi/N$. We say that $x[n]$ and $X[k]$ are a DTFS pair and denote this relationship as

$$x[n] \xleftrightarrow{\;DTFS;\,\Omega_o\;} X[k]$$

From N values of $X[k]$ we may determine $x[n]$ using Eq. (3.12), and from N values of $x[n]$ we may determine $X[k]$ using Eq. (3.13). Either $X[k]$ or $x[n]$ provides a complete description of the signal. We shall see that in some problems it is advantageous to represent the signal using its time values $x[n]$, while in others the DTFS coefficients $X[k]$ offer a more convenient description of the signal. The DTFS coefficient representation is also known as a frequency-domain representation because each DTFS coefficient is associated with a complex sinusoid of a different frequency.

Before presenting several examples illustrating the DTFS, we remind the reader that the starting values of the indices k and n in Eqs. (3.12) and (3.13) are arbitrary because both $x[n]$ and $X[k]$ are N periodic. The range for the indices may thus be chosen to simplify the problem at hand.

EXAMPLE 3.1 Find the DTFS representation for

$$x[n] = \cos(\tfrac{\pi}{8}n + \phi)$$

Solution: The fundamental period of $x[n]$ is $N = 16$. Hence $\Omega_o = 2\pi/16$. We could determine the DTFS coefficients using Eq. (3.13); however, in this case it is easier to find them by inspection. Write

$$x[n] = \frac{e^{j[(\pi/8)n+\phi]} + e^{-j[(\pi/8)n+\phi]}}{2}$$

$$= \tfrac{1}{2}e^{-j\phi}e^{-j(\pi/8)n} + \tfrac{1}{2}e^{j\phi}e^{j(\pi/8)n} \tag{3.14}$$

and compare this to the DTFS of Eq. (3.12) written using a starting index $k = -7$

$$x[n] = \sum_{k=-7}^{8} X[k]e^{jk(\pi/8)n} \tag{3.15}$$

Equating the terms in Eq. (3.14) and Eq. (3.15) having equal frequencies, $k\pi/8$, gives

$$x[n] \xleftrightarrow{\ DTFS;\ 2\pi/16\ } X[k] = \begin{cases} \tfrac{1}{2}e^{-j\phi}, & k = -1 \\ \tfrac{1}{2}e^{j\phi}, & k = 1 \\ 0, & -7 \le k \le 8 \text{ and } k \ne \pm 1 \end{cases}$$

Since $X[k]$ has period $N = 16$, we have $X[15] = X[31] = \cdots = \tfrac{1}{2}e^{-j\phi}$ and similarly $X[17] = X[33] = \cdots = \tfrac{1}{2}e^{j\phi}$ with all other values of $X[k]$ equal to zero. Plots of the magnitude and phase of $X[k]$ are depicted in Fig. 3.2.

In general it is easiest to determine the DTFS coefficients by inspection when the signal consists of a sum of sinusoids.

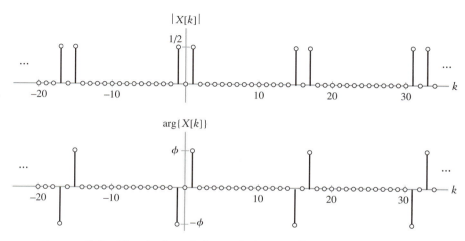

FIGURE 3.2 Magnitude and phase of DTFS coefficients for Example 3.1.

The magnitude of $X[k]$, $|X[k]|$, is known as the *magnitude spectrum* of $x[n]$. Similarly, the phase of $X[k]$, $\arg\{X[k]\}$, is known as the *phase spectrum* of $x[n]$. In the previous example all the components of $x[n]$ are concentrated at two frequencies, Ω_o ($k = 1$) and $-\Omega_o$ ($k = -1$).

▶ **Drill Problem 3.1**   Determine the DTFS coefficients by inspection for the signal

$$x[n] = 1 + \sin\left(\frac{1}{12}\pi n + \frac{3\pi}{8}\right)$$

Answer:

$$x[n] \xleftrightarrow{\ DTFS;\ 2\pi/24\ } X[k] = \begin{cases} -\dfrac{e^{-j(3\pi/8)}}{2j}, & k = -1 \\[2mm] 1, & k = 0 \\[2mm] \dfrac{e^{j(3\pi/8)}}{2j}, & k = 1 \\[2mm] 0, & \text{otherwise on } -11 \le k \le 12 \end{cases}$$ ◀

The next example directly evaluates Eq. (3.13) to determine the DTFS coefficients.

EXAMPLE 3.2   Find the DTFS coefficients for the N periodic square wave depicted in Fig. 3.3.

Solution: The period is N, so $\Omega_o = 2\pi/N$. It is convenient in this case to evaluate Eq. (3.13) over indices $n = -M$ to $n = N - M - 1$. We thus have

$$X[k] = \frac{1}{N}\sum_{n=-M}^{N-M-1} x[n]e^{-jk\Omega_o n}$$

$$= \frac{1}{N}\sum_{n=-M}^{M} e^{-jk\Omega_o n}$$

Perform the change of variable on the index of summation, $m = n + M$, to obtain

$$X[k] = \frac{1}{N}e^{jk\Omega_o M}\sum_{m=0}^{2M} e^{-jk\Omega_o m}$$

Summing the geometric series yields

$$X[k] = \frac{e^{jk\Omega_o M}}{N}\left(\frac{1 - e^{-jk\Omega_o(2M+1)}}{1 - e^{-jk\Omega_o}}\right), \qquad k \ne 0, \pm N, \pm 2N, \ldots$$

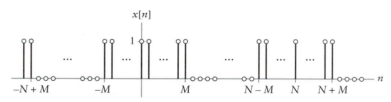

FIGURE 3.3   Square wave for Example 3.2.

which may be rewritten as

$$X[k] = \frac{1}{N}\left(\frac{e^{jk\Omega_o(2M+1)/2}}{e^{jk\Omega_o/2}}\right)\left(\frac{1 - e^{-jk\Omega_o(2M+1)}}{1 - e^{-jk\Omega_o}}\right),$$

$$= \frac{1}{N}\left(\frac{e^{jk\Omega_o(2M+1)/2} - e^{-jk\Omega_o(2M+1)/2}}{e^{jk\Omega_o/2} - e^{-jk\Omega_o/2}}\right), \qquad k \neq 0, \pm N, \pm 2N, \ldots$$

At this point we may divide the numerator and denominator by $2j$ to express $X[k]$ as a ratio of two sine functions, as shown by

$$X[k] = \frac{1}{N}\frac{\sin\left(k\frac{\Omega_o}{2}(2M+1)\right)}{\sin\left(k\frac{\Omega_o}{2}\right)}, \qquad k \neq 0, \pm N, \pm 2N, \ldots$$

An alternative expression for $X[k]$ is obtained by substituting $\Omega_o = 2\pi/N$, yielding

$$X[k] = \frac{1}{N}\frac{\sin\left(k\frac{\pi}{N}(2M+1)\right)}{\sin\left(k\frac{\pi}{N}\right)}, \qquad k \neq 0, \pm N, \pm 2N, \ldots$$

The technique used here to write the finite geometric sum expression for $X[k]$ as a ratio of sine functions involves symmetrizing both the numerator, $1 - e^{-jk\Omega_o(2M+1)}$, and denominator, $1 - e^{-jk\Omega_o}$, with the appropriate power of $e^{jk\Omega_o}$. Now, for $k = 0, \pm N, \pm 2N, \ldots$, we have

$$X[k] = \frac{1}{N}\sum_{m=-M}^{M} 1$$

$$= \frac{2M+1}{N}$$

and the expression for $X[k]$ is

$$X[k] = \begin{cases} \dfrac{1}{N}\dfrac{\sin\left(k\dfrac{\pi}{N}(2M+1)\right)}{\sin\left(k\dfrac{\pi}{N}\right)}, & k \neq 0, \pm N, \pm 2N, \ldots \\[4mm] \dfrac{2M+1}{N}, & k = 0, \pm N, \pm 2N, \ldots \end{cases}$$

Using L'Hopital's rule, it is easy to show that

$$\lim_{k \to 0, \pm N, \pm 2N, \ldots}\left(\frac{1}{N}\frac{\sin\left(k\frac{\pi}{N}(2M+1)\right)}{\sin\left(k\frac{\pi}{N}\right)}\right) = \frac{2M+1}{N}$$

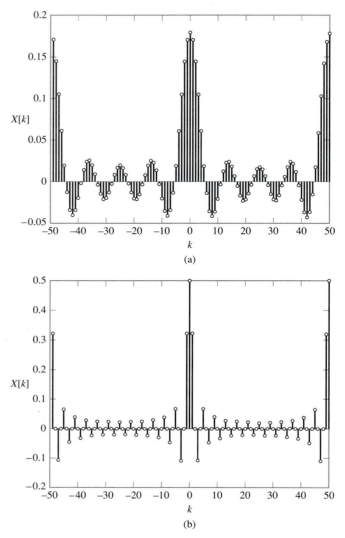

FIGURE 3.4   The DTFS coefficients for a square wave: (a) $M = 4$ and (b) $M = 12$.

For this reason, it is common to write the expression for $X[k]$ as

$$X[k] = \frac{1}{N} \frac{\sin\left(k \frac{\pi}{N} (2M + 1)\right)}{\sin\left(k \frac{\pi}{N}\right)}$$

In this form it is understood that the value $X[k]$ for $k = 0, \pm N, \pm 2N, \ldots$ is obtained from the limit as $k \to 0$. A plot of two periods of $X[k]$ as a function of k is depicted in Fig. 3.4 for $M = 4$ and $M = 12$ assuming $N = 50$. Note that in this example $X[k]$ is real; hence the magnitude spectrum is the absolute value of $X[k]$ and the phase spectrum is 0 when $X[k]$ is positive and π when $X[k]$ is negative.

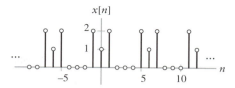

FIGURE 3.5 Signal $x[n]$ for Drill Problem 3.2.

▶ **Drill Problem 3.2** Determine the DTFS coefficients for the periodic signal depicted in Fig. 3.5.

Answer:

$$x[n] \xleftrightarrow{\text{DTFS; } 2\pi/6} X[k] = \frac{1}{6} + \frac{2}{3} \cos k \frac{\pi}{3}$$

◀

Each term in the DTFS of Eq. (3.12) associated with a nonzero coefficient $X[k]$ contributes to the representation of the signal. We now examine this representation by considering the contribution of each term for the square wave in Example 3.2. In this example the DTFS coefficients have even symmetry, $X[k] = X[-k]$, and we may rewrite the DTFS of Eq. (3.12) as a series involving harmonically related cosines. General conditions under which the DTFS coefficients have even or odd symmetry are discussed in Section 3.6. Assume for convenience that N is even so that $N/2$ is integer and let k range from $-N/2 + 1$ to $N/2$, and thus write

$$x[n] = \sum_{k=-N/2+1}^{N/2} X[k]e^{jk\Omega_o n}$$

$$= X[0] + \sum_{m=1}^{N/2-1} (X[m]e^{jm\Omega_o n} + X[-m]e^{-jm\Omega_o n}) + X[N/2]e^{j(N/2)\Omega_o n}$$

Now exploit $X[m] = X[-m]$ and $N\Omega_o = 2\pi$ to obtain

$$x[n] = X[0] + \sum_{m=1}^{N/2-1} 2X[m]\left(\frac{e^{jm\Omega_o n} + e^{-jm\Omega_o n}}{2}\right) + X[N/2]e^{j\pi n}$$

$$= X[0] + \sum_{m=1}^{N/2-1} 2X[m] \cos(m\Omega_o n) + X[N/2] \cos(\pi n)$$

where we have also used $e^{j\pi n} = \cos(\pi n)$. If we define the new set of coefficients

$$B[k] = \begin{cases} X[k], & k = 0, N/2 \\ 2X[k], & k = 1, 2, \ldots, N/2 - 1 \end{cases}$$

then we may write the DTFS in terms of a series of harmonically related cosines as

$$x[n] = \sum_{k=0}^{N/2} B[k] \cos(k\Omega_o n)$$

EXAMPLE 3.3 Define a partial sum approximation to $x[n]$ as

$$\hat{x}_J[n] = \sum_{k=0}^{J} B[k] \cos(k\Omega_o n)$$

where $J \leq N/2$. This approximation contains the first $2J + 1$ terms centered on $k = 0$ in Eq. (3.12). Evaluate one period of the Jth term in the sum and $\hat{x}_J[n]$ for $J = 1, 3, 5, 23$, and 25, assuming $N = 50$ and $M = 12$ for the square wave in Example 3.2.

Solution: Figure 3.6 depicts the *J*th term in the sum, $B[J] \cos(J\Omega_o n)$, and one period of $\hat{x}_J[n]$ for the specified values of *J*. Only odd values for *J* are considered because the even indexed coefficients $B[k]$ are zero. Note that the approximation improves as *J* increases, with exact representation of $x[n]$ when $J = N/2 = 25$. In general, the coefficients $B[k]$ associated with values of *k* near zero represent the low-frequency or slowly varying features in the signal, while the coefficients associated with the values of *k* near $\pm N/2$ represent the high-frequency or rapidly varying features in the signal.

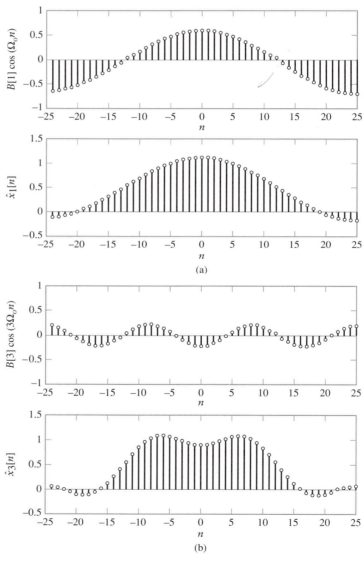

FIGURE 3.6 Individual terms in the DTFS expansion for a square wave (top panel) and the corresponding partial sum approximations $\hat{x}_J[n]$ (bottom panel). The $J = 0$ term is $\hat{x}_0[n] = \frac{1}{2}$ and is not shown. (a) $J = 1$. (b) $J = 3$.

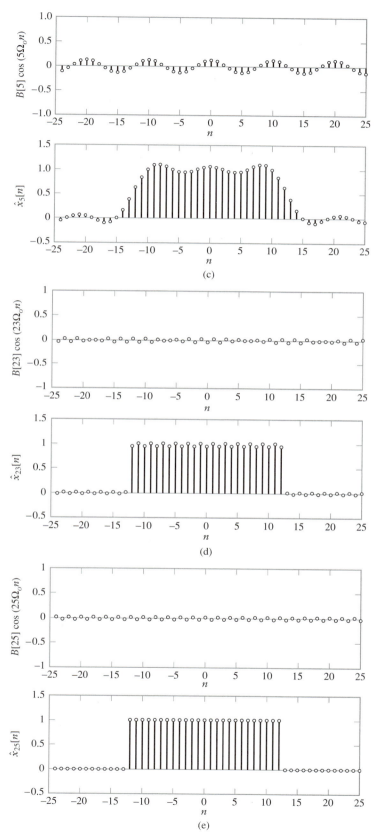

FIGURE 3.6 (continued) (c) $J = 5$. (d) $J = 23$. (e) $J = 25$.

The DTFS is the only Fourier representation that can be numerically evaluated and manipulated in a computer. This is because both the time-domain, $x[n]$, and frequency-domain, $X[k]$, representations of the signal are exactly characterized by a finite set of N numbers. The computational tractability of the DTFS is of great significance. The DTFS finds extensive use in numerical signal analysis and system implementation and is often

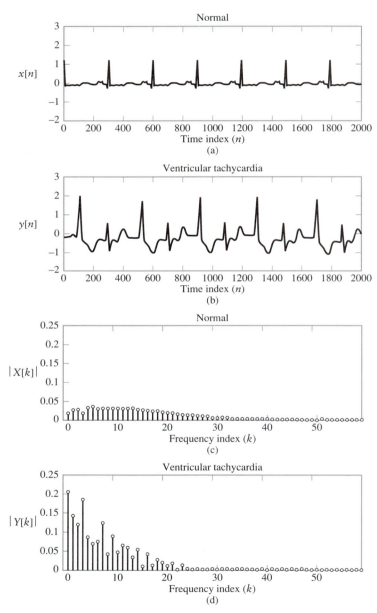

FIGURE 3.7 Electrocardiograms for two different heartbeats and the first 60 coefficients of their magnitude spectra. (a) Normal heartbeat. (b) Ventricular tachycardia. (c) Magnitude spectrum for the normal heartbeat. (d) Magnitude spectrum for ventricular tachycardia.

used to numerically approximate the other three Fourier representations. These issues are explored in the next chapter.

EXAMPLE 3.4   In this example we evaluate the DTFS representations of two different electrocardiogram (ECG) waveforms. Figures 3.7(a) and (b) depict the ECG of a normal heart and one experiencing ventricular tachycardia, respectively. These sequences are drawn as continuous functions due to the difficulty of depicting all 2000 values in each case. Both of these appear nearly periodic, with very slight variations in the amplitude and length of each period. The DTFS of one period of each ECG may be computed numerically. The period of the normal ECG is $N = 305$, while the period of the ventricular tachycardia ECG is $N = 421$. One period of each waveform is available. Evaluate the DTFS coefficients for each and plot their magnitude spectrum.

Solution: The magnitude spectrum of the first 60 DTFS coefficients is depicted in Figs. 3.7(c) and (d). The higher indexed coefficients are very small and thus not shown.

   The time waveforms differ, as do the DTFS coefficients. The normal ECG is dominated by a sharp spike or impulsive feature. Recall that the DTFS coefficients for a unit impulse have constant magnitude. The DTFS coefficients of the normal ECG are approximately constant, showing a gradual decrease in amplitude as the frequency increases. They also have a fairly small magnitude, since there is relatively little power in the impulsive signal. In contrast, the ventricular tachycardia ECG is not as impulsive but has smoother features. Consequently, the DTFS coefficients have greater dynamic range with the low-frequency coefficients dominating. The ventricular tachycardia ECG has greater power than the normal ECG and thus the DTFS coefficients have larger amplitude.

3.3 *Continuous-Time Periodic Signals: The Fourier Series*

* ■ DERIVATION

We begin our derivation of the FS by approximating a signal $x(t)$ having fundamental period T using the series of Eq. (3.5):

$$\hat{x}(t) = \sum_{k=-\infty}^{\infty} A[k]e^{jk\omega_o t} \tag{3.16}$$

where $\omega_o = 2\pi/T$.

   We shall now use the orthogonality property, Eq. (3.7), to find the FS coefficients. We begin by assuming we can find coefficients $A[k]$ so that $x(t) = \hat{x}(t)$. If $x(t) = \hat{x}(t)$, then

$$\int_{\langle T \rangle} x(t)e^{-jm\omega_o t}\, dt = \int_{\langle T \rangle} \hat{x}(t)e^{-jm\omega_o t}\, dt$$

Substitute the series expression for $\hat{x}(t)$ in this equality to obtain the expression

$$\int_{\langle T \rangle} x(t)e^{-jm\omega_o t}\, dt = \int_{\langle T \rangle} \sum_{k=-\infty}^{\infty} A[k]e^{jk\omega_o t}e^{-jm\omega_o t}\, dt$$

$$= \sum_{k=-\infty}^{\infty} A[k] \int_{\langle T \rangle} e^{jk\omega_o t}e^{-jm\omega_o t}\, dt$$

The orthogonality property of Eq. (3.7) implies that the integral on the right-hand side is zero except for $k = m$, and so we have

$$\int_{\langle T \rangle} x(t)e^{-jm\omega_o t} \, dt = A[m]T$$

We conclude that if $x(t) = \hat{x}(t)$, then the mth coefficient is given by

$$A[m] = \frac{1}{T}\int_{\langle T \rangle} x(t)e^{-jm\omega_o t} \, dt \qquad (3.17)$$

Problem 3.32 establishes that this value also minimizes the MSE between $x(t)$ and the $2J + 1$ term, truncated approximation

$$\hat{x}_J(t) = \sum_{k=-J}^{J} A[k]e^{jk\omega_o t}$$

Suppose we choose the coefficients according to Eq. (3.17). Under what conditions does the infinite series of Eq. (3.16) actually converge to $x(t)$? A detailed analysis of this question is beyond the scope of this book. However, we can state several results. First, if $x(t)$ is square integrable, that is,

$$\frac{1}{T}\int_{\langle T \rangle} |x(t)|^2 \, dt < \infty$$

then the MSE between $x(t)$ and $\hat{x}(t)$ is zero. This is a useful result that applies to a very broad class of signals encountered in engineering practice. Note that in contrast to the discrete-time case, zero MSE does not imply that $x(t)$ and $\hat{x}(t)$ are equal pointwise (at each value of t); it simply implies that there is zero energy in their difference.

Pointwise convergence is guaranteed at all values of t except those corresponding to discontinuities if the Dirichlet conditions are satisfied:

▶ $x(t)$ is bounded.
▶ $x(t)$ has a finite number of local maxima and minima in one period.
▶ $x(t)$ has a finite number of discontinuities in one period.

If a signal $x(t)$ satisfies the Dirichlet conditions and is not continuous, then the FS representation of Eq. (3.16) converges to the midpoint of $x(t)$ at each discontinuity.

■ THE FS REPRESENTATION

We may write the FS as

$$x(t) = \sum_{k=-\infty}^{\infty} X[k]e^{jk\omega_o t} \qquad (3.18)$$

$$X[k] = \frac{1}{T}\int_{\langle T \rangle} x(t)e^{-jk\omega_o t} \, dt \qquad (3.19)$$

where $x(t)$ has fundamental period T and $\omega_o = 2\pi/T$. We say that $x(t)$ and $X[k]$ are a FS pair and denote this relationship as

$$x(t) \xleftrightarrow{\ FS;\ \omega_o\ } X[k]$$

From the FS coefficients $X[k]$ we may determine $x(t)$ using Eq. (3.18) and from $x(t)$ we may determine $X[k]$ using Eq. (3.19). We shall see later that in some problems it is advantageous to represent the signal in the time domain as $x(t)$, while in others the FS coefficients $X[k]$ offer a more convenient description. The FS coefficient representation is also known as a frequency-domain representation because each FS coefficient is associated with a complex sinusoid of a different frequency. The following examples illustrate determination of the FS representation.

EXAMPLE 3.5 Determine the FS representation for the signal

$$x(t) = 3\,\cos\!\left(\frac{\pi}{2}\,t + \frac{\pi}{4}\right)$$

Solution: The fundamental period of $x(t)$ is $T = 4$. Hence $\omega_o = 2\pi/4 = \pi/2$ and we seek to express $x(t)$ as

$$x(t) = \sum_{k=-\infty}^{\infty} X[k]e^{jk(\pi/2)t}$$

One approach to finding $X[k]$ is to use Eq. (3.19). However, in this case $x(t)$ is expressed in terms of sinusoids, so it is easier to obtain $X[k]$ by inspection. Write

$$x(t) = 3\,\cos\!\left(\frac{\pi}{2}\,t + \frac{\pi}{4}\right)$$

$$= 3\,\frac{e^{j(\pi/2)t + \pi/4} + e^{-[j(\pi/2)t + \pi/4]}}{2}$$

$$= \frac{3}{2}\,e^{-j\pi/4}e^{-j(\pi/2)t} + \frac{3}{2}\,e^{j\pi/4}e^{j(\pi/2)t}$$

This last expression is in the form of the Fourier series. We may thus identify

$$X[k] = \begin{cases} \frac{3}{2}e^{-j\pi/4}, & k = -1 \\ \frac{3}{2}e^{j\pi/4}, & k = 1 \\ 0, & \text{otherwise} \end{cases}$$

The magnitude and phase of $X[k]$ are depicted in Fig. 3.8.

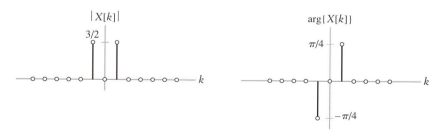

FIGURE 3.8 Magnitude and phase spectra for Example 3.5.

▶ **Drill Problem 3.3**   Determine the FS representation for

$$x(t) = 2\sin(2\pi t - 3) + \sin(6\pi t)$$

Answer:

$$x(t) \xleftrightarrow{\text{FS; } 2\pi} X[k] = \begin{cases} j/2, & k = -3 \\ je^{j3}, & k = -1 \\ -je^{-j3}, & k = 1 \\ -j/2, & k = 3 \\ 0, & \text{otherwise} \end{cases}$$ ◀

As in the DTFS, the magnitude of $X[k]$ is known as the magnitude spectrum of $x(t)$, while the phase of $X[k]$ is known as the phase spectrum of $x(t)$. In the previous example all the power in $x(t)$ is concentrated at two frequencies, ω_o and $-\omega_o$. In the next example the power in $x(t)$ is distributed across many frequencies.

EXAMPLE 3.6   Determine the FS representation for the square wave depicted in Fig. 3.9.

Solution: The period is T, so $\omega_o = 2\pi/T$. It is convenient in this problem to use the integral formula Eq. (3.19) to determine the FS coefficients. We integrate over the period $t = -T/2$ to $t = T/2$ to exploit the even symmetry of $x(t)$ and obtain for $k \neq 0$

$$X[k] = \frac{1}{T} \int_{-T/2}^{T/2} x(t)e^{-jk\omega_o t}\, dt$$

$$= \frac{1}{T} \int_{-T_s}^{T_s} e^{-jk\omega_o t}\, dt$$

$$= \frac{-1}{Tjk\omega_o} e^{-jk\omega_o t} \Big|_{-T_s}^{T_s},$$

$$= \frac{2}{Tk\omega_o} \left(\frac{e^{jk\omega_o T_s} - e^{-jk\omega_o T_s}}{2j} \right),$$

$$= \frac{2\sin(k\omega_o T_s)}{Tk\omega_o}, \qquad k \neq 0$$

For $k = 0$, we have

$$X[0] = \frac{1}{T} \int_{-T_s}^{T_s} dt$$

$$= \frac{2T_s}{T}$$

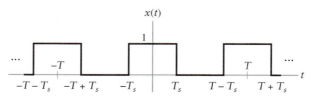

FIGURE 3.9   Square wave for Example 3.6.

Using L'Hopital's rule it is straightforward to show that

$$\lim_{k \to 0} \frac{2 \sin(k\omega_o T_s)}{Tk\omega_o} = \frac{2T_s}{T}$$

and thus we write

$$X[k] = \frac{2 \sin(k\omega_o T_s)}{Tk\omega_o}$$

with the understanding that $X[0]$ is obtained as a limit. In this problem $X[k]$ is real valued. Substituting $\omega_o = 2\pi/T$ gives $X[k]$ as a function of the ratio T_s/T, as shown by

$$X[k] = \frac{2 \sin\left(k \dfrac{2\pi T_s}{T}\right)}{k2\pi} \qquad (3.20)$$

Figure 3.10 depicts $X[k]$, $- 50 \leq k \leq 50$, for $T_s/T = \frac{1}{4}$ and $T_s/T = \frac{1}{16}$. Note that as T_s/T decreases, the signal becomes more concentrated in time within each period while the FS representation becomes less concentrated in frequency. We shall explore the inverse relationship between time- and frequency-domain concentrations of signals more fully in the sections that follow.

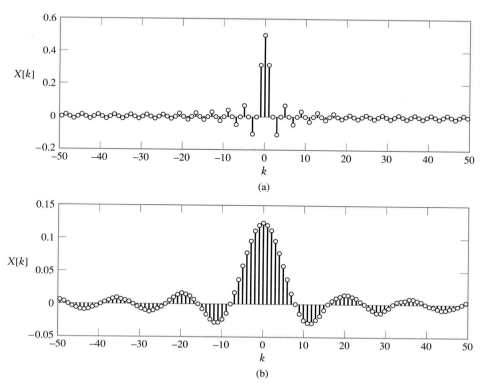

FIGURE 3.10   The FS coefficients, $X[k]$, $-50 \leq k \leq 50$, for two square waves: (a) $T_s/T = \frac{1}{4}$ and (b) $T_s/T = \frac{1}{16}$.

The functional form $\sin(\pi u)/\pi u$ occurs sufficiently often in Fourier analysis that we give it a special name:

$$\text{sinc}(u) = \frac{\sin(\pi u)}{\pi u} \qquad (3.21)$$

A graph of $\text{sinc}(u)$ is depicted in Fig. 3.11. The maximum of the sinc function is unity at $u = 0$, the zero crossings occur at integer values of u, and the magnitude dies off as $1/u$. The portion of the sinc function between the zero crossings at $u = \pm 1$ is known as the *mainlobe* of the sinc function. The smaller ripples outside the mainlobe are termed *sidelobes*. The FS coefficients in Eq. (3.20) are expressed using the sinc function notation as

$$X[k] = \frac{2T_s}{T} \text{sinc}\left(k\,\frac{2T_s}{T}\right)$$

Each term in the FS of Eq. (3.18) associated with a nonzero coefficient $X[k]$ contributes to the representation of the signal. The square wave of the previous example provides a convenient illustration of how the individual terms in the FS contribute to the representation of $x(t)$. As with the DTFS square wave representation, we exploit the even symmetry of $X[k]$ to write the FS as a sum of harmonically related cosines. Since $X[k] = X[-k]$, we have

$$\begin{aligned}
x(t) &= \sum_{k=-\infty}^{\infty} X[k]e^{jk\omega_o t} \\
&= X[0] + \sum_{m=1}^{\infty} (X[m]e^{jm\omega_o t} + X[-m]e^{-jm\omega_o t}) \\
&= X[0] + \sum_{m=1}^{\infty} 2X[m]\,\frac{e^{jm\omega_o t} + e^{-jm\omega_o t}}{2} \\
&= X[0] + \sum_{m=1}^{\infty} 2X[m]\cos(m\omega_o t)
\end{aligned}$$

If we define $B[0] = X[0]$ and $B[k] = 2X[k]$, $k \neq 0$, then

$$x(t) = \sum_{k=0}^{\infty} B[k]\cos(k\omega_o t)$$

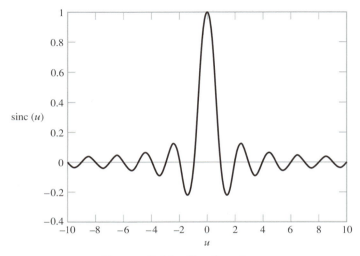

FIGURE 3.11 Sinc function.

EXAMPLE 3.7   We define the partial sum approximation to the FS representation for the square wave, as shown by

$$\hat{x}_J(t) = \sum_{k=0}^{J} B[k] \cos(k\omega_o t)$$

Assume $T = 1$ and $T_s/T = \frac{1}{4}$. Note that in this case we have

$$B[k] = \begin{cases} \dfrac{1}{2}, & k = 0 \\ \dfrac{2(-1)^{(k-1)/2}}{k\pi}, & k \text{ odd} \\ 0, & k \text{ even} \end{cases}$$

so the even indexed coefficients are zero. Depict one period of the Jth term in this sum and $\hat{x}_J(t)$ for $J = 1, 3, 7, 29,$ and 99.

Solution: The individual terms and partial sum approximations are depicted in Fig. 3.12. The behavior of the partial sum approximation in the vicinity of the square wave discontinuities at $t = \pm\frac{1}{4}$ is of particular interest. We note that each partial sum approximation passes through the average value ($\frac{1}{2}$) of the discontinuity, as stated in our convergence discussion. On each side of the discontinuity the approximation exhibits ripple. As J increases, the maximum height of the ripples does not appear to change. In fact, it can be shown for any finite J that the maximum ripple is 9% of the discontinuity. This ripple near discontinuities in partial sum FS approximations is termed the *Gibbs phenomenon* in honor of the mathematical physicist J. Willard Gibbs for his explanation of this phenomenon in 1899. The square wave satisfies the Dirichlet conditions and so we know that the FS approximation ultimately converges to the square wave for all values of t except at the discontinuities. However, for finite J the ripple is always present. As J increases, the ripple in the partial sum approximations becomes more and more concentrated near the discontinuities. Hence, for any given J, the accuracy of the partial sum approximation is best at times distant from discontinuities and worst near the discontinuities.

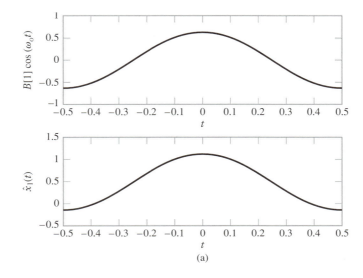

FIGURE 3.12   Individual terms in FS expansion for a square wave (top panel) and the corresponding partial sum approximations $\hat{x}_J(t)$ (bottom panel). The $J = 0$ term is $\hat{x}_0(t) = \frac{1}{2}$ and is not shown. (a) $J = 1$.

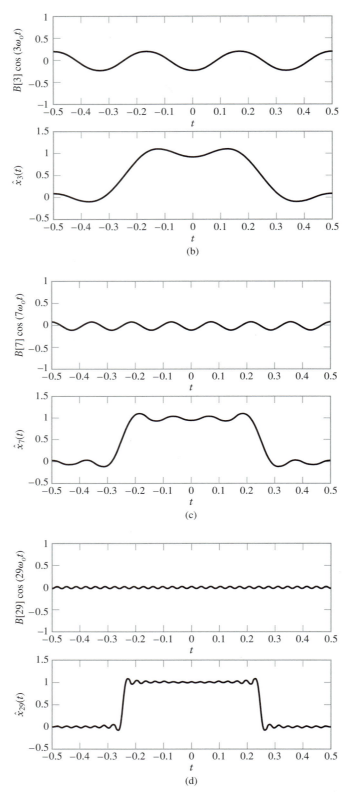

FIGURE 3.12 (continued) (b) $J = 3$. (c) $J = 7$. (d) $J = 29$.

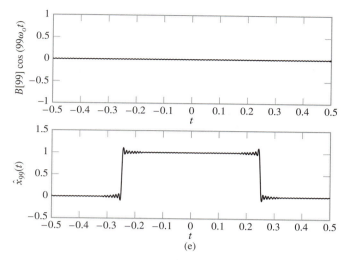

FIGURE 3.12 (continued) (e) *J* = 99.

▶ **Drill Problem 3.4** Find the FS representation for the sawtooth wave depicted in Fig. 3.13. *Hint:* Use integration by parts.

Answer: Integrate *t* from $-\frac{1}{2}$ to 1 in Eq. (3.19) to obtain

$$x[n] \xleftrightarrow{\text{FS; } 4\pi/3} X[k] = \begin{cases} \frac{1}{4}, & k = 0 \\ \frac{-2}{3jk\omega_o}\left(e^{-jk\omega_o} + \frac{1}{2}e^{j\frac{k\omega_o}{2}}\right) + \frac{2}{3k^2\omega_o^2}\left(e^{-jk\omega_o} + e^{j\frac{k\omega_o}{2}}\right), & \text{otherwise} \end{cases}$$

◀

The following example exploits linearity and the FS representation for the square wave to determine the output of a LTI system.

> **EXAMPLE 3.8** Here we wish to find the FS representation for the output, *y*(*t*), of the *RC* circuit depicted in Fig. 3.14 in response to the square wave input depicted in Fig. 3.9 assuming $T_s/T = \frac{1}{4}$, *T* = 1 s, and *RC* = 0.1 s.
>
> *Solution:* If the input to a LTI system is expressed as a weighted sum of sinusoids, then the output is also a weighted sum of sinusoids. The *k*th weight in the output sum is given by

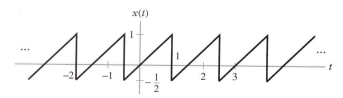

FIGURE 3.13 Periodic signal for Drill Problem 3.4.

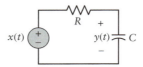

FIGURE 3.14   *RC* circuit for Example 3.8.

the product of the *k*th weight in the input sum and system frequency response evaluated at the *k*th sinusoid's frequency. Hence if

$$x(t) = \sum_{k=-\infty}^{\infty} X[k]e^{jk\omega_o t}$$

then the output *y(t)* is

$$y(t) = \sum_{k=-\infty}^{\infty} H(jk\omega_o)X[k]e^{jk\omega_o t}$$

where *H(jω)* is the frequency response of the system. Thus

$$y(t) \xleftrightarrow{\;FS;\; \omega_o\;} Y[k] = H(jk\omega_o)X[k]$$

The frequency response of the *RC* circuit was computed in Example 2.15 as

$$H(j\omega) = \frac{1/RC}{j\omega + 1/RC}$$

and the FS coefficients for the square wave are given in Eq. (3.20). Substituting for $H(jk\omega_o)$ with $RC = 0.1$ s, $\omega_o = 2\pi$, and using $T_s/T = \frac{1}{4}$ gives

$$Y[k] = \frac{10}{j2\pi k + 10}\frac{\sin(k\pi/2)}{k\pi}$$

The magnitude spectrum $|Y[k]|$ goes to zero in proportion to $1/k^2$ as *k* increases, so a reasonably accurate representation for *y(t)* may be determined using a modest number of terms in the FS. Determine *y(t)* using

$$y(t) \approx \sum_{k=-100}^{100} Y[k]e^{jk\omega_o t}$$

The magnitude and phase of *Y[k]* for $-25 \le k \le 25$ are depicted in Figs. 3.15(a) and (b), respectively. Comparing *Y[k]* to *X[k]* as depicted in Fig. 3.10(a), we see that the circuit attenuates the amplitude of *X[k]* when $|k| \ge 1$. The degree of attenuation increases as frequency, $k\omega_o$, increases. The circuit also introduces a frequency-dependent phase shift. One period of the time waveform *y(t)* is shown in Fig. 3.15(c). This result is consistent with our intuition from circuit analysis. When the input switches from 0 to 1, the charge on the capacitor increases and the voltage exhibits an exponential rise. When the input switches from 1 to 0, the capacitor discharges and the voltage exhibits an exponential decay.

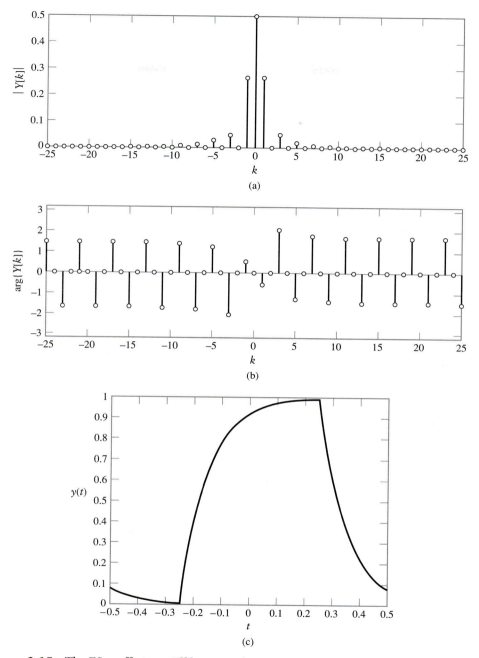

FIGURE 3.15   The FS coefficients, $Y[k]$, $-25 \leq k \leq 25$, for the RC circuit output in response to a square wave input. (a) Magnitude spectrum. (b) Phase spectrum. (c) One period of the output, $y(t)$.

3.4 Discrete-Time Nonperiodic Signals: The Discrete-Time Fourier Transform

✳ ▪ DERIVATION

A rigorous derivation of the DTFT is complex, so we employ an intuitive approach. We develop the DTFT from the DTFS by describing a nonperiodic signal as the limit of a periodic signal whose period, N, approaches infinity. For this approach to be meaningful, we assume that the nonperiodic signal is represented by a single period of the periodic signal that is centered on the origin, and that the limit as N approaches infinity is taken in a symmetric manner. Let $\tilde{x}[n]$ be a periodic signal with period $N = 2M + 1$. Define the finite-duration nonperiodic signal $x[n]$ as one period of $\tilde{x}[n]$, as shown by

$$x[n] = \begin{cases} \tilde{x}[n], & -M \leq n \leq M \\ 0, & |n| > M \end{cases}$$

This relationship is illustrated in Fig. 3.16. Note that as M increases, the periodic replicates of $x[n]$ that are present in $\tilde{x}[n]$ move farther and farther away from the origin. Eventually, as $M \to \infty$, these replicates are removed to infinity. Thus we may write

$$x[n] = \lim_{M \to \infty} \tilde{x}[n] \tag{3.22}$$

Begin with the DTFS representation for the periodic signal $\tilde{x}[n]$. We have the DTFS pair

$$\tilde{x}[n] = \sum_{k=-M}^{M} X[k]e^{jk\Omega_o n} \tag{3.23}$$

$$X[k] = \frac{1}{2M + 1} \sum_{n=-M}^{M} \tilde{x}[n]e^{-jk\Omega_o n} \tag{3.24}$$

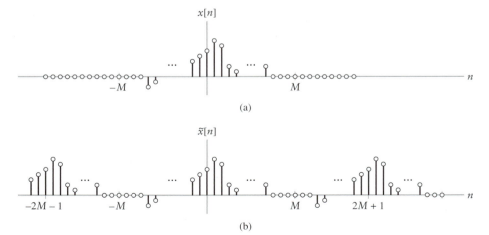

FIGURE 3.16 Approximation of a nonperiodic signal with a periodic signal. (a) Nonperiodic signal $x[n]$. (b) Periodic approximation $\tilde{x}[n]$.

Since $\tilde{x}[n] = x[n]$ for $-M \le n \le M$, we may rewrite Eq. (3.24) in terms of $x[n]$ as

$$X[k] = \frac{1}{2M+1} \sum_{n=-M}^{M} x[n] e^{-jk\Omega_o n}$$

$$= \frac{1}{2M+1} \sum_{n=-\infty}^{\infty} x[n] e^{-jk\Omega_o n}$$

where the second line follows from the fact that $x[n] = 0$ for $|n| > M$. We now define a continuous function of frequency, $X(e^{j\Omega})$, whose samples at $k\Omega_o$ are equal to the DTFS coefficients normalized by $2M + 1$. That is,

$$X(e^{j\Omega}) = \sum_{n=-\infty}^{\infty} x[n] e^{-j\Omega n} \tag{3.25}$$

so that $X[k] = X(e^{jk\Omega_o})/(2M+1)$. Substitute this definition for $X[k]$ into Eq. (3.23) to obtain

$$\tilde{x}[n] = \frac{1}{2M+1} \sum_{k=-M}^{M} X(e^{jk\Omega_o}) e^{jk\Omega_o n}$$

Using the relationship $\Omega_o = 2\pi/(2M+1)$, we write

$$\tilde{x}[n] = \frac{1}{2\pi} \sum_{k=-M}^{M} X(e^{jk\Omega_o}) e^{jk\Omega_o n} \Omega_o \tag{3.26}$$

At this point we invoke the fact that $x[n]$ is the limiting value of $\tilde{x}[n]$ as $M \to \infty$. However, let us first consider the effect of $M \to \infty$ on the fundamental frequency, Ω_o. As M increases, Ω_o decreases and the spacing between harmonics in the DTFS decreases. This decrease in harmonic spacing is illustrated in Fig. 3.17 by depicting $X(e^{jk\Omega_o})$ for increasing values of M. Note that $X(e^{j\Omega})$ is 2π periodic in Ω. This follows from Eq. (3.25) and the 2π periodicity of $e^{-j\Omega n}$. Combining Eq. (3.22) with Eq. (3.26), we have

$$x[n] = \lim_{M \to \infty} \frac{1}{2\pi} \sum_{k=-M}^{M} X(e^{jk\Omega_o}) e^{jk\Omega_o n} \Omega_o \tag{3.27}$$

In Eq. (3.27) we are summing values of a function $X(e^{j\Omega}) e^{j\Omega n}$ evaluated at $k\Omega_o$ multiplied by the width between samples, Ω_o. This is the rectangular rule approximation to an integral. Taking the limit and identifying $\Omega = k\Omega_o$ so that $d\Omega = \Omega_o$, the sum in Eq. (3.27) passes to the integral

$$x[n] = \frac{1}{2\pi} \int_{-\pi}^{\pi} X(e^{j\Omega}) e^{j\Omega n} \, d\Omega$$

The limits on the integral are obtained by noting that $\lim_{M \to \infty} M\Omega_o = \pi$. We have thus expressed $x[n]$ as a weighted superposition of discrete-time sinusoids. In this case the superposition is an integral and the weighting on each sinusoid is $(1/2\pi)X(e^{j\Omega}) \, d\Omega$.

■ THE DTFT REPRESENTATION

The DTFT representation is expressed as

$$x[n] = \frac{1}{2\pi} \int_{-\pi}^{\pi} X(e^{j\Omega}) e^{j\Omega n} \, d\Omega \tag{3.28}$$

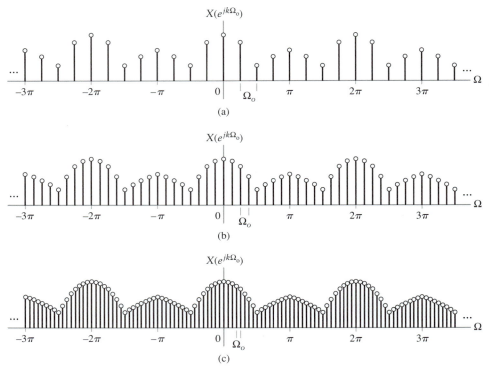

FIGURE 3.17   Example spectra $X(e^{jk\Omega_o})$ for increasing values of M. M increases from (a) to (c), so Ω_o decreases.

where

$$X(e^{j\Omega}) = \sum_{n=-\infty}^{\infty} x[n]e^{-j\Omega n} \qquad (3.29)$$

We say that $X(e^{j\Omega})$ and $x[n]$ are a DTFT pair and write

$$x[n] \xleftrightarrow{\ \ DTFT\ \ } X(e^{j\Omega})$$

The transform $X(e^{j\Omega})$ describes the signal $x[n]$ as a function of sinusoidal frequency Ω and is termed the frequency-domain representation of $x[n]$. We say that Eq. (3.29) is the DTFT of $x[n]$ since it converts the time-domain signal into its frequency-domain representation. Equation (3.28) is termed the inverse DTFT since it converts the frequency-domain representation back into the time domain.

In deriving the DTFT we assumed that $x[n]$ has finite duration. We may apply these results to infinite-duration signals but must then address the conditions under which the infinite sum in Eq. (3.29) converges. If $x[n]$ is absolutely summable, that is,

$$\sum_{n=-\infty}^{\infty} |x[n]| < \infty$$

then the sum in Eq. (3.29) converges uniformly to a continuous function of Ω. If $x[n]$ is not absolutely summable, but does have finite energy, that is,

$$\sum_{n=-\infty}^{\infty} |x[n]|^2 < \infty$$

then it can be shown that the sum in Eq. (3.29) converges in a mean-squared error sense but does not converge pointwise.

Many physical signals encountered in engineering practice satisfy these conditions. However, several common nonperiodic signals, such as the unit step, $u[n]$, do not. In some of these cases we can define a transform pair that behaves like the DTFT by including impulses in the transform. This enables us to use the DTFT as a problem-solving tool even though strictly speaking it does not converge. One example of this is given later in the section; others are presented in Chapter 4.

We now consider several examples illustrating determination of the DTFT for common signals.

EXAMPLE 3.9   *Exponential Sequence.* Find the DTFT of the sequence $x[n] = \alpha^n u[n]$.

Solution: Using Eq. (3.29), we have

$$X(e^{j\Omega}) = \sum_{n=-\infty}^{\infty} \alpha^n u[n] e^{-j\Omega n}$$

$$= \sum_{n=0}^{\infty} \alpha^n e^{-j\Omega n}$$

This sum diverges for $|\alpha| \geq 1$. For $|\alpha| < 1$ we have the convergent geometric series:

$$X(e^{j\Omega}) = \sum_{n=0}^{\infty} (\alpha e^{-j\Omega})^n$$

$$= \frac{1}{1 - \alpha e^{-j\Omega}}, \qquad |\alpha| < 1$$

If α is real valued, we may write the magnitude and phase of $X(e^{j\Omega})$ as

$$|X(e^{j\Omega})| = \frac{1}{((1 - \alpha \cos \Omega)^2 + \alpha^2 \sin^2 \Omega)^{1/2}}$$

$$= \frac{1}{(\alpha^2 + 1 - 2\alpha \cos \Omega)^{1/2}}$$

$$\arg\{X(e^{j\Omega})\} = -\arctan \left(\frac{\alpha \sin \Omega}{1 - \alpha \cos \Omega} \right)$$

The magnitude and phase of $X(e^{j\Omega})$ are depicted graphically in Fig. 3.18 for $\alpha = 0.5$. Note that both are 2π periodic.

As in the other Fourier representations, the magnitude spectrum of a signal is the magnitude of $X(e^{j\Omega})$ depicted as a function of Ω. The phase spectrum is the phase of $X(e^{j\Omega})$.

▶ **Drill Problem 3.5**   Find the DTFT of $x[n] = 2(3)^n u[-n]$.

Answer:

$$X(e^{j\Omega}) = \frac{2}{1 - e^{j\Omega}/3}$$

◀

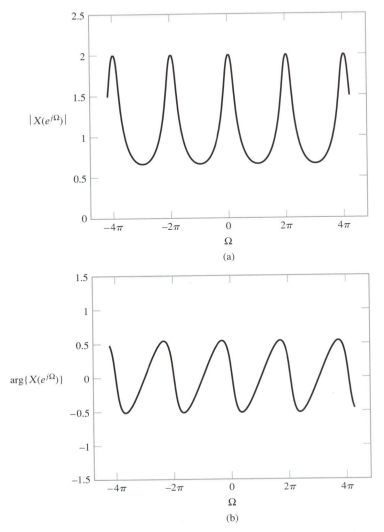

FIGURE 3.18 The DTFT of $x[n] = (\frac{1}{2})^n u[n]$. (a) Magnitude spectrum. (b) Phase spectrum.

EXAMPLE 3.10 *Rectangular Pulse.* Let

$$x[n] = \begin{cases} 1, & |n| \leq M \\ 0, & |n| > M \end{cases}$$

as depicted in Fig. 3.19(a). Find the DTFT of $x[n]$.

Solution: Substitute for $x[n]$ in Eq. (3.29) to obtain

$$X(e^{j\Omega}) = \sum_{n=-M}^{M} 1 e^{-j\Omega n}$$

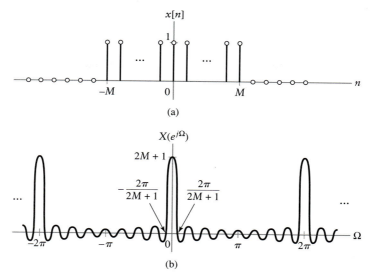

FIGURE 3.19   (a) Rectangular pulse in time. (b) DTFT.

Now perform the change of variable, $m = n + M$, obtaining

$$X(e^{j\Omega}) = \sum_{m=0}^{2M} e^{-j\Omega(m-M)}$$

$$= e^{j\Omega M} \sum_{m=0}^{2M} e^{-j\Omega m}$$

$$= \begin{cases} e^{j\Omega M} \dfrac{1 - e^{-j\Omega(2M+1)}}{1 - e^{-j\Omega}}, & \Omega \neq 0, \pm 2\pi, \pm 4\pi, \ldots \\ 2M + 1, & \Omega = 0, \pm 2\pi, \pm 4\pi, \ldots \end{cases}$$

The expression for $X(e^{j\Omega})$, when $\Omega \neq 0, \pm 2\pi, \pm 4\pi, \ldots$, may be simplified by symmetrizing the powers of the exponential in the numerator and denominator, as shown by

$$X(e^{j\Omega}) = e^{j\Omega M} \frac{e^{-j\Omega(2M+1)/2}(e^{j\Omega(2M+1)/2} - e^{-j\Omega(2M+1)/2})}{e^{-j\Omega/2}(e^{j\Omega/2} - e^{-j\Omega/2})}$$

$$= \frac{\sin\left(\dfrac{\Omega}{2}(2M+1)\right)}{\sin\left(\dfrac{\Omega}{2}\right)}$$

Note that L'Hopital's rule gives

$$\lim_{\Omega \to 0, \pm 2\pi, \pm 4\pi, \ldots} \frac{\sin\left(\Omega \dfrac{2M+1}{2}\right)}{\sin\left(\dfrac{\Omega}{2}\right)} = 2M + 1$$

Hence rather than writing $X(e^{j\Omega})$ as two forms dependent on the value of Ω, we simply write

$$X(e^{j\Omega}) = \frac{\sin\left(\Omega\,\dfrac{2M+1}{2}\right)}{\sin\left(\dfrac{\Omega}{2}\right)}$$

with the understanding that $X(e^{j\Omega})$, for $\Omega = 0, \pm 2\pi, \pm 4\pi, \ldots$, is obtained as a limit. In this example $X(e^{j\Omega})$ is purely real. A graph of $X(e^{j\Omega})$ as a function of Ω is given in Fig. 3.19(b). We see that as M increases, the time extent of $x[n]$ increases while the energy in $X(e^{j\Omega})$ becomes more concentrated near $\Omega = 0$.

EXAMPLE 3.11   *Discrete-Time sinc Function.* Find the inverse DTFT of

$$X(e^{j\Omega}) = \begin{cases} 1, & |\Omega| \le W \\ 0, & W < |\Omega| < \pi \end{cases}$$

as depicted in Fig. 3.20(a).

Solution: First note that $X(e^{j\Omega})$ is specified only for $-\pi < \Omega < \pi$. This is all that is needed, since $X(e^{j\Omega})$ is always 2π periodic and the inverse DTFT depends only on the values in the interval $-\pi < \Omega < \pi$. Substituting for $X(e^{j\Omega})$ in Eq. (3.28) gives

$$x[n] = \frac{1}{2\pi} \int_{-W}^{W} e^{j\Omega n}\, d\Omega$$

$$= \frac{1}{2\pi nj} e^{j\Omega n} \Big|_{-W}^{W}, \qquad n \ne 0$$

$$= \frac{1}{\pi n} \sin(Wn), \qquad n \ne 0$$

For $n = 0$, the integrand is unity and we have $x[0] = W/\pi$. It is easy to show using L'Hopital's rule that

$$\lim_{n \to 0} \frac{1}{\pi n} \sin(Wn) = \frac{W}{\pi}$$

and thus we usually write

$$x[n] = \frac{1}{\pi n} \sin(Wn)$$

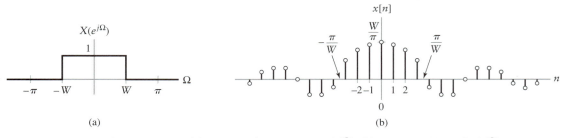

FIGURE 3.20   (a) Rectangular spectrum $X(e^{j\Omega})$. (b) Inverse DTFT of $X(e^{j\Omega})$.

as the inverse DTFT of $X(e^{j\Omega})$ with the understanding that the value at $n = 0$ is obtained as the limit. We may also write

$$x[n] = \frac{W}{\pi} \operatorname{sinc}\left(\frac{Wn}{\pi}\right)$$

using the sinc function notation defined in Eq. (3.21). A graph depicting $x[n]$ is given in Fig. 3.20(b).

EXAMPLE 3.12 *The Impulse.* Find the DTFT of $x[n] = \delta[n]$.

Solution: For $x[n] = \delta[n]$, we have

$$X(e^{j\Omega}) = \sum_{n=-\infty}^{\infty} \delta[n]e^{-j\Omega n}$$

$$= 1$$

Hence

$$\delta[n] \xleftrightarrow{\quad DTFT \quad} 1$$

This DTFT pair is depicted in Fig. 3.21.

EXAMPLE 3.13 Find the inverse DTFT of $X(e^{j\Omega}) = \delta(\Omega), -\pi < \Omega \leq \pi$.

Solution: By definition,

$$x[n] = \frac{1}{2\pi} \int_{-\pi}^{\pi} \delta(\Omega)e^{j\Omega n}\, d\Omega$$

Use the sifting property of the impulse function to obtain $x[n] = 1/2\pi$, and thus write

$$\frac{1}{2\pi} \xleftrightarrow{\quad DTFT \quad} \delta(\Omega), \qquad -\pi < \Omega \leq \pi$$

In this example we have again defined only one period of $X(e^{j\Omega})$. We can alternatively define $X(e^{j\Omega})$ over all Ω by writing it as an infinite sum of delta functions shifted by integer multiples of 2π

$$X(e^{j\Omega}) = \sum_{k=-\infty}^{\infty} \delta(\Omega - k2\pi)$$

Both definitions are common. This DTFT pair is depicted in Fig. 3.22.

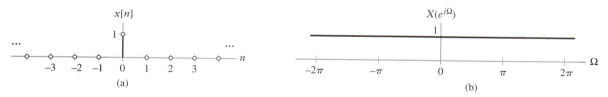

FIGURE 3.21 (a) Impulse in time. (b) DTFT.

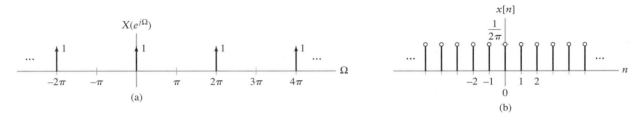

FIGURE 3.22   (a) Impulse in frequency. (b) Inverse DTFT.

This last example presents an interesting dilemma. The DTFT of $x[n] = 1/2\pi$ does not converge, since it is not a square summable signal, yet $x[n]$ is a valid inverse DTFT. This is a direct consequence of allowing impulses in $X(e^{j\Omega})$. We shall treat $x[n]$ and $X(e^{j\Omega})$ as a DTFT pair despite this apparent quandary, because they do satisfy all the properties of a DTFT pair. Indeed, we can greatly expand the class of signals that can be represented by the DTFT if we allow impulses in the transform. Strictly speaking, the DTFTs of these signals do not exist since the sum in Eq. (3.29) does not converge. However, as in this example, we can identify transform pairs using the inverse transform of Eq. (3.28) and thus use the DTFT as a problem-solving tool. Additional examples illustrating the use of impulses in the DTFT are presented in Chapter 4.

▶ **Drill Problem 3.6**   Find the inverse DTFT of

$$X(e^{j\Omega}) = 2 \cos(2\Omega)$$

Answer:

$$x[n] = \begin{cases} 1, & n = \pm 2 \\ 0, & \text{otherwise} \end{cases}$$ ◀

▶ **Drill Problem 3.7**   Find the DTFT of

$$x[n] = \begin{cases} 2^n, & 0 \le n \le 9 \\ 0, & \text{otherwise} \end{cases}$$

Answer:

$$X(e^{j\Omega}) = \frac{1 - 2^{10}e^{-j10\Omega}}{1 - 2e^{-j\Omega}}$$ ◀

3.5 Continuous-Time Nonperiodic Signals: The Fourier Transform

The Fourier transform (FT) is used to represent a continuous-time nonperiodic signal as a superposition of complex sinusoids. We shall simply present the FT in this section. An interpretation of the FT as the limiting form of the FS is developed as Problems 3.34. Recall from Section 3.1 that the continuous nonperiodic nature of a time signal implies that the superposition of complex sinusoids involves a continuum of frequencies ranging from $-\infty$ to ∞. Thus the FT representation for a time signal involves an integral over frequency, as shown by

$$x(t) = \frac{1}{2\pi} \int_{-\infty}^{\infty} X(j\omega)e^{j\omega t}\, d\omega \tag{3.30}$$

where

$$X(j\omega) = \int_{-\infty}^{\infty} x(t)e^{-j\omega t}\, dt \qquad (3.31)$$

In Eq. (3.30) we have expressed $x(t)$ as a weighted superposition of sinusoids. The superposition is an integral and the weight on each sinusoid is $(1/2\pi)X(j\omega)\, d\omega$. We say that $x(t)$ and $X(j\omega)$ are a FT pair and write

$$x(t) \xleftrightarrow{\ FT\ } X(j\omega)$$

The transform $X(j\omega)$ describes the signal $x(t)$ as a function of sinusoidal frequency ω and is termed the frequency-domain representation for $x(t)$. Equation (3.31) is termed the FT of $x(t)$ since it converts the time-domain signal into its frequency-domain representation. Equation (3.30) is termed the inverse FT since it converts the frequency-domain representation $X(j\omega)$ back into the time domain.

The integrals in Eqs. (3.30) and (3.31) may not converge for all functions $x(t)$ and $X(j\omega)$. An analysis of convergence is beyond the scope of this book, so we simply state several convergence conditions on the time-domain signal $x(t)$. Define

$$\hat{x}(t) = \frac{1}{2\pi} \int_{-\infty}^{\infty} X(j\omega)e^{j\omega t}\, d\omega$$

where $X(j\omega)$ is given in terms of $x(t)$ by Eq. (3.31). It can be shown that the MSE between $x(t)$ and $\hat{x}(t)$, given by

$$\int_{-\infty}^{\infty} |x(t) - \hat{x}(t)|^2\, dt$$

is zero if $x(t)$ is square integrable, that is, if

$$\int_{-\infty}^{\infty} |x(t)|^2\, dt < \infty$$

Zero MSE does not imply pointwise convergence, or $x(t) = \hat{x}(t)$ at all values of t, but rather that there is zero energy in their difference.

Pointwise convergence is guaranteed at all values of t except those corresponding to discontinuities if $x(t)$ satisfies the Dirichlet conditions for nonperiodic signals:

▶ $x(t)$ is absolutely integrable:

$$\int_{-\infty}^{\infty} |x(t)|\, dt < \infty$$

▶ $x(t)$ has a finite number of local maxima, minima, and discontinuities in any finite interval.

▶ The size of each discontinuity is finite.

Almost all physical signals encountered in engineering practice satisfy the second and third conditions. However, many common signals, such as the unit step, are not absolutely or square integrable. In some of these cases we can define a transform pair that satisfies FT properties through the use of impulses. In this way we can still use the FT as a problem-solving tool, even though the FT does not converge for such signals in a strict sense.

The following examples illustrate determination of the FT and inverse FT for several common signals.

EXAMPLE 3.14   *Real Exponential.* Find the FT of $x(t) = e^{-at}u(t)$.

Solution: The FT does not converge for $a \leq 0$ since $x(t)$ is not absolutely integrable, as shown by

$$\int_0^\infty e^{-at}\, dt = \infty, \qquad a \leq 0$$

For $a > 0$, we have

$$X(j\omega) = \int_{-\infty}^\infty e^{-at}u(t)e^{-j\omega t}\, dt$$

$$= \int_0^\infty e^{-(a+j\omega)t}\, dt$$

$$= -\frac{1}{a + j\omega} e^{-(a+j\omega)t}\Big|_0^\infty$$

$$= \frac{1}{a + j\omega}$$

Converting to polar form, the magnitude and phase of $X(j\omega)$ are given by

$$|X(j\omega)| = \frac{1}{(a^2 + \omega^2)^{1/2}}$$

$$\arg\{X(j\omega)\} = -\arctan\left(\frac{\omega}{a}\right)$$

and are depicted in Fig. 3.23.

As before, the magnitude of $X(j\omega)$ is termed the magnitude spectrum and the phase of $X(j\omega)$ is termed the phase spectrum of the signal.

▶ **Drill Problem 3.8**   Find the FT of $x(t) = e^{at}u(-t)$ assuming $a > 0$.

Answer: $X(j\omega) = -1/(j\omega - a)$.   ◀

EXAMPLE 3.15   *Rectangular Pulse.* Consider the rectangular pulse depicted in Fig. 3.24(a) and defined as

$$x(t) = \begin{cases} 1, & -T \leq t \leq T \\ 0, & |t| > T \end{cases}$$

Find the FT of $x(t)$.

Solution: The rectangular pulse $x(t)$ is absolutely integrable provided $T < \infty$. For $\omega \neq 0$ we have

$$X(j\omega) = \int_{-\infty}^\infty x(t)e^{-j\omega t}\, dt$$

$$= \int_{-T}^T e^{-j\omega t}\, dt$$

$$= -\frac{1}{j\omega} e^{-j\omega t}\Big|_{-T}^T$$

$$= \frac{2}{\omega} \sin(\omega T), \qquad \omega \neq 0$$

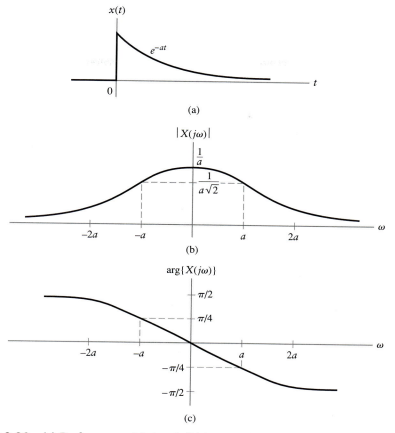

(a)

(b)

(c)

FIGURE 3.23   (a) Real exponential signal. (b) Magnitude spectrum. (c) Phase spectrum.

For $\omega = 0$, the integral simplifies to $2T$. It is straightforward to show using L'Hopital's rule that

$$\lim_{\omega \to 0} \frac{2}{\omega} \sin(\omega T) = 2T$$

Thus we write for all ω

$$X(j\omega) = \frac{2}{\omega} \sin(\omega T)$$

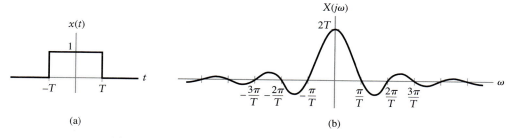

(a)

(b)

FIGURE 3.24   (a) Rectangular pulse in time. (b) FT.

with the understanding that the value at $\omega = 0$ is obtained by evaluating a limit. In this case $X(j\omega)$ is real. It is depicted in Fig. 3.24(b). The magnitude spectrum is

$$|X(j\omega)| = 2\left|\frac{\sin(\omega T)}{\omega}\right|$$

and the phase spectrum is

$$\arg\{X(j\omega)\} = \begin{cases} 0, & \dfrac{\sin(\omega T)}{\omega} > 0 \\ \pi, & \dfrac{\sin(\omega T)}{\omega} < 0 \end{cases}$$

We may write $X(j\omega)$ using the sinc function notation as

$$X(j\omega) = 2T \operatorname{sinc}\left(\frac{\omega T}{\pi}\right)$$

This example illustrates a very important property of the Fourier transform. Consider the effect of changing T. As T increases, $x(t)$ becomes less concentrated about the time origin, while $X(j\omega)$ becomes more concentrated about the frequency origin. Conversely, as T decreases, $x(t)$ becomes more concentrated about the time origin, while $X(j\omega)$ becomes less concentrated about the frequency origin. In a certain sense, the "width" of $x(t)$ is inversely related to the "width" of $X(j\omega)$. As a general principle, we shall see that signals concentrated in one domain are spread out in the other domain.

EXAMPLE 3.16 *The sinc Function.* Find the inverse FT of the rectangular spectrum depicted in Fig. 3.25(a) and given by

$$X(j\omega) = \begin{cases} 1, & -W \le \omega \le W \\ 0, & |\omega| > W \end{cases}$$

Solution: Using Eq. (3.30) for the inverse FT gives for $t \ne 0$

$$\begin{aligned} x(t) &= \frac{1}{2\pi} \int_{-W}^{W} e^{j\omega t}\, d\omega \\ &= \frac{1}{2j\pi t} e^{j\omega t}\Big|_{-W}^{W} \\ &= \frac{1}{\pi t} \sin(Wt) \\ &= \frac{W}{\pi} \operatorname{sinc}\left(\frac{Wt}{\pi}\right), \qquad t \ne 0 \end{aligned}$$

When $t = 0$, the integral simplifies to W/π. Since

$$\lim_{t \to 0} \frac{1}{\pi t} \sin(Wt) = \frac{W}{\pi}$$

we write for all t

$$x(t) = \frac{1}{\pi t} \sin(Wt)$$

with the understanding that the value at $t = 0$ is obtained as a limit. Figure 3.25(b) depicts $x(t)$.

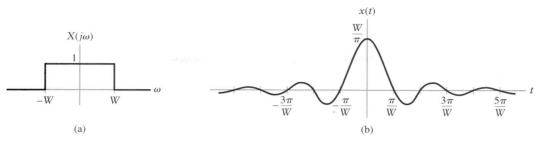

FIGURE 3.25   (a) Rectangular pulse in frequency. (b) Inverse FT.

Note again the inverse relationship between the concentration of the signal about the origin in the time and frequency domains. As W increases, the frequency-domain representation becomes less concentrated about $\omega = 0$, while the time-domain representation becomes more concentrated about $t = 0$. Another interesting observation can be made by considering this and the previous example. In the previous example a rectangular time-domain pulse is transformed to a sinc function in frequency. In this example, a sinc function in time is transformed to a rectangular function in frequency. This "duality" is a consequence of the similarity between the forward transform in Eq. (3.31) and inverse transform in Eq. (3.30) and is studied further in Section 3.6. The next two examples also exhibit this property.

EXAMPLE 3.17   *The Impulse.* Find the FT of $x(t) = \delta(t)$.

Solution: This $x(t)$ does not satisfy the Dirichlet conditions, since the impulse $\delta(t)$ is only defined within an integral. We attempt to proceed in spite of this potential problem using Eq. (3.31) to write

$$X(j\omega) = \int_{-\infty}^{\infty} \delta(t)e^{-j\omega t}\, dt$$
$$= 1$$

The second line follows from the sifting property of the impulse function. Hence

$$\delta(t) \xleftrightarrow{\;\;FT\;\;} 1$$

and the impulse contains unity contributions from complex sinusoids of all frequencies, from $\omega = -\infty$ to $\omega = \infty$.

EXAMPLE 3.18   *DC Signal.* Find the inverse FT of $X(j\omega) = 2\pi\delta(\omega)$.

Solution: Here again we may expect convergence irregularities since $X(j\omega)$ has an infinite discontinuity at the origin. Using Eq. (3.30), we find

$$x(t) = \frac{1}{2\pi} \int_{-\infty}^{\infty} 2\pi\delta(\omega)e^{j\omega t}\, d\omega$$
$$= 1$$

Hence we identify

$$1 \xleftarrow{\;\;FT\;\;} 2\pi\delta(\omega)$$

as a FT pair. This implies that the frequency content of a dc signal is concentrated entirely at $\omega = 0$, which is intuitively satisfying.

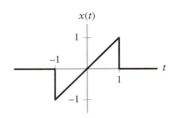

FIGURE 3.26 Time-domain signal for Drill Problem 3.10.

While convergence of the FT cannot be guaranteed in the previous two examples, the transform pairs do satisfy the properties of a FT pair and are thus useful for analysis. In both cases the transform pairs are consequences of the properties of the impulse function. By permitting the use of impulse functions we greatly expand the class of signals that are representable by the FT and thus enhance the power of the FT as a problem-solving tool. In Chapter 4 we shall use impulses to obtain FT representations for both periodic and discrete-time signals.

▶ **Drill Problem 3.9**   Find the inverse FT of

$$X(j\omega) = \begin{cases} 2\cos\omega, & |\omega| \le \pi \\ 0, & |\omega| > \pi \end{cases}$$

Answer:

$$x(t) = \frac{\sin(\pi(t+1))}{\pi(t+1)} + \frac{\sin(\pi(t-1))}{\pi(t-1)}$$ ◀

▶ **Drill Problem 3.10**   Find the FT of the time signal depicted in Fig. 3.26 and given by

$$x(t) = \begin{cases} t, & |t| \le 1 \\ 0, & |t| > 1 \end{cases}$$

Hint: Use integration by parts.

Answer:

$$X(j\omega) = j\frac{2}{\omega}\cos\omega - j\frac{2}{\omega^2}\sin\omega$$ ◀

3.6 *Properties of Fourier Representations*

The four Fourier representations discussed in this chapter are summarized in Table 3.2. This table provides a convenient reference for both the definition of each transform and identification of the class of signals to which each applies. All four Fourier representations are based on the complex sinusoid. Consequently, they all share a set of common properties that follow from the characteristics of complex sinusoids. This section examines the properties of Fourier representations. In many cases, we derive a property for one representation and simply state it for the other representations. The reader is asked to prove some of these properties in the problem section of this chapter. A comprehensive table of all properties is given in Appendix C.

▌TABLE 3.2   *The Four Fourier Representations*

Time Domain	Periodic	Nonperiodic	
C o n t i n u o u s	**Fourier Series** $$x(t) = \sum_{k=-\infty}^{\infty} X[k]e^{jk\omega_o t}$$ $$X[k] = \frac{1}{T} \int_{\langle T \rangle} x(t)e^{-jk\omega_o t} \, dt$$ $x(t)$ has period T $$\omega_o = \frac{2\pi}{T}$$	**Fourier Transform** $$x(t) = \frac{1}{2\pi} \int_{-\infty}^{\infty} X(j\omega)e^{j\omega t} \, d\omega$$ $$X(j\omega) = \int_{-\infty}^{\infty} x(t)e^{-j\omega t} \, dt$$	N o n p e r i o d i c
D i s c r e t e	**Discrete-Time Fourier Series** $$x[n] = \sum_{k=\langle N \rangle} X[k]e^{jk\Omega_o n}$$ $$X[k] = \frac{1}{N} \sum_{n=\langle N \rangle} x[n]e^{-jk\Omega_o n}$$ $x[n]$ and $X[k]$ have period N $$\Omega_o = \frac{2\pi}{N}$$	**Discrete-Time Fourier Transform** $$x[n] = \frac{1}{2\pi} \int_{-\pi}^{\pi} X(e^{j\Omega}) \, e^{j\Omega n} \, d\Omega$$ $$X(e^{j\Omega}) = \sum_{n=-\infty}^{\infty} x[n]e^{-j\Omega n}$$ $X(e^{j\Omega})$ has period 2π	P e r i o d i c
	Discrete	*Continuous*	*Frequency Domain*

■ PERIODICITY PROPERTIES

The borders of Table 3.2 summarize the periodicity properties of the four representations by denoting time-domain characteristics on the top and left sides with the corresponding frequency-domain characteristics on the bottom and right sides.

Continuous- or discrete-time periodic signals have a Fourier series representation. In a Fourier series the signal is represented as a weighted sum of complex sinusoids having the same period as the signal. A discrete set of frequencies is involved in the series; hence the frequency-domain representation involves a discrete set of weights or coefficients. In contrast, for nonperiodic signals both continuous- and discrete-time Fourier transform representations involve weighted integrals of complex sinusoids over a continuum of frequencies. Hence the frequency-domain representation for nonperiodic signals is a continuous function of frequency. Signals that are periodic in time have discrete frequency-domain representations, while nonperiodic time signals have continuous frequency-domain representations. This is the correspondence indicated on the top and bottom of Table 3.2.

We also observe that the Fourier representations for discrete-time signals, either the DTFS or DTFT, are periodic functions of frequency. This is because the discrete-time complex sinusoids used to represent discrete-time signals are 2π periodic functions of frequency: that is, discrete-time sinusoids whose frequencies differ by integer multiples of 2π are identical. In contrast, Fourier representations for continuous-time signals involve superpositions of continuous-time sinusoids. Continuous-time sinusoids with distinct frequencies are always distinct; thus the frequency-domain representations for continuous-

TABLE 3.3   *Fourier Representation Periodicity Properties*

Time-Domain Property	Frequency-Domain Property
Continuous	Nonperiodic
Discrete	Periodic
Periodic	Discrete
Nonperiodic	Continuous

time signals are nonperiodic. Summarizing, discrete-time signals have periodic frequency-domain representations, while continuous-time signals have nonperiodic frequency-domain representations. This is the correspondence indicated on the left and right sides of Table 3.2.

In general, representations that are continuous in one domain, either time or frequency, are nonperiodic in the other domain. Conversely, representations that are discrete in one domain, either time or frequency, are periodic in the other domain. These relationships are indicated in Table 3.3.

■ LINEARITY

It is a straightforward exercise to show that all four Fourier representations involve linear operations. Specifically, they satisfy the linearity property

$$z(t) = ax(t) + by(t) \xleftarrow{\ FT\ } Z(j\omega) = aX(j\omega) + bY(j\omega)$$

$$z(t) = ax(t) + by(t) \xleftarrow{\ FS;\ \omega_o\ } Z[k] = aX[k] + bY[k]$$

$$z[n] = ax[n] + by[n] \xleftarrow{\ DTFT\ } Z(e^{j\Omega}) = aX(e^{j\Omega}) + bY(e^{j\Omega})$$

$$z[n] = ax[n] + by[n] \xleftarrow{\ DTFS;\ \Omega_o\ } Z[k] = aX[k] + bY[k]$$

In the above relationships we assume that the uppercase symbols denote the Fourier representation of the corresponding lowercase symbol. Furthermore, in the FS and DTFS cases the signals being summed are assumed to have the same fundamental period.

EXAMPLE 3.19   Suppose $z(t)$ is the periodic signal depicted in Fig. 3.27(a). Use the linearity property and the results of Example 3.6 to determine the FS coefficients $Z[k]$.

Solution: Write $z(t)$ as the sum of signals

$$z(t) = \tfrac{3}{2}x(t) + \tfrac{1}{2}y(t)$$

where $x(t)$ and $y(t)$ are depicted in Figs. 3.27(b) and (c), respectively. From Example 3.6 we have

$$x(t) \xleftarrow{\ FS;\ 2\pi\ } X[k] = \frac{1}{k\pi} \sin\left(k\,\frac{\pi}{4} \right)$$

$$y(t) \xleftarrow{\ FS;\ 2\pi\ } Y[k] = \frac{1}{k\pi} \sin\left(k\,\frac{\pi}{2} \right)$$

The linearity property implies that

$$z(t) \xleftarrow{\ FS;\ 2\pi\ } Z[k] = \frac{3}{2k\pi} \sin\left(k\,\frac{\pi}{4} \right) + \frac{1}{2k\pi} \sin\left(k\,\frac{\pi}{2} \right)$$

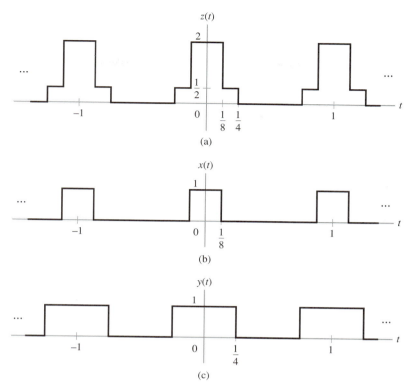

FIGURE 3.27 Representation of the periodic signal $z(t)$ as a weighted sum of periodic square waves: $z(t) = \frac{3}{2}x(t) + \frac{1}{2}y(t)$. (a) $z(t)$. (b) $x(t)$. (c) $y(t)$.

The linearity property is also the basis of the partial fraction method for determining the inverse FT of frequency-domain representations $X(j\omega)$ given by a ratio of polynomials in $j\omega$. Let

$$X(j\omega) = \frac{b_M(j\omega)^M + \cdots + b_1(j\omega) + b_0}{(j\omega)^N + a_{N-1}(j\omega)^{N-1} + \cdots + a_1(j\omega) + a_0}$$

$$= \frac{B(j\omega)}{A(j\omega)}$$

Frequency-domain representations of this form occur frequently in analysis of systems that are described by linear constant-coefficient differential equations. The idea is to write $X(j\omega)$ as a sum of terms for which the inverse FT is known. We may accomplish this using a partial fraction expansion of $X(j\omega)$.

We assume that $M < N$. If $M \geq N$, then we may use long division to express $X(j\omega)$ in the form

$$X(j\omega) = \sum_{k=0}^{M-N} f_k(j\omega)^k + \frac{\tilde{B}(j\omega)}{A(j\omega)}$$

The numerator polynomial $\tilde{B}(j\omega)$ now has order one less than that of the denominator and the partial fraction expansion is applied to determine the inverse Fourier transform of $\dfrac{\tilde{B}(j\omega)}{A(j\omega)}$. The inverse Fourier transform of the terms in the sum are obtained from the pair $\delta(t) \xleftrightarrow{\;FT\;} 1$ and the differentiation property, which is introduced later in this section.

Let the roots of the denominator polynomial in $(j\omega)$ be d_k, $k = 1, 2, \ldots, N$. These roots are found by replacing $j\omega$ with a generic variable u and determining the roots of the polynomial.

$$u^N + a_{N-1}u^{N-1} + \cdots + a_1 u + a_0 = 0$$

We may then write

$$X(j\omega) = \frac{\sum_{k=0}^{M} b_k (j\omega)^k}{\prod_{k=1}^{N} (j\omega - d_k)}$$

Assuming all the roots d_k, $k = 1, 2, \ldots, N$ are distinct, we may express $X(j\omega)$ as the sum

$$X(j\omega) = \sum_{k=1}^{N} \frac{C_k}{j\omega - d_k}$$

where the coefficients C_k, $k = 1, 2, \ldots, N$ are determined by either solving a system of linear equations or by the method of residues. These methods and the expansion for repeated roots are reviewed in Appendix B. In Example 3.14 we derived the FT pair

$$e^{dt}u(t) \xleftrightarrow{\ FT\ } \frac{1}{j\omega - d} \quad \text{for } d < 0$$

The reader may verify that this pair is valid even if d is complex, provided $\text{Re}\{d\} < 0$. Assuming the real part of each d_k, $k = 1, 2, \ldots, N$, is negative, we use linearity to write

$$x(t) = \sum_{k=1}^{N} C_k e^{d_k t} u(t) \xleftrightarrow{\ FT\ } X(j\omega) = \sum_{k=1}^{N} \frac{C_k}{j\omega - d_k}$$

The following example illustrates this technique.

EXAMPLE 3.20   Find the inverse FT of

$$X(j\omega) = \frac{j\omega + 1}{(j\omega)^2 + 5j\omega + 6}$$

Solution: First find the partial fraction expansion of $X(j\omega)$. The roots of the denominator polynomial are $d_1 = -2$ and $d_2 = -3$. Hence we write $X(j\omega)$ as the sum

$$\frac{j\omega + 1}{(j\omega)^2 + 5j\omega + 6} = \frac{C_1}{j\omega + 2} + \frac{C_2}{j\omega + 3}$$

We may solve for C_1 and C_2 using the method of residues, as shown by

$$C_1 = (j\omega + 2) \frac{j\omega + 1}{(j\omega)^2 + 5j\omega + 6}\bigg|_{j\omega = -2}$$

$$= \frac{j\omega + 1}{j\omega + 3}\bigg|_{j\omega = -2}$$

$$= -1$$

$$C_2 = (j\omega + 3) \frac{j\omega + 1}{(j\omega)^2 + 5j\omega + 6}\bigg|_{j\omega = -3}$$

$$= \frac{j\omega + 1}{j\omega + 2}\bigg|_{j\omega = -3}$$

$$= 2$$

Thus the partial fraction expansion of $X(j\omega)$ is

$$X(j\omega) = \frac{-1}{j\omega + 2} + \frac{2}{j\omega + 3}$$

Now using linearity, we obtain

$$X(j\omega) \xleftrightarrow{\text{FT}} x(t) = 2e^{-3t}u(t) - e^{-2t}u(t)$$

▶ **Drill Problem 3.11** Use partial fraction expansion and linearity to determine the inverse FT of

$$X(j\omega) = \frac{-j\omega}{(j\omega)^2 + 3j\omega + 2}$$

Answer:

$$x(t) = e^{-t}u(t) - 2e^{-2t}u(t)$$

◀

Partial fraction expansions and linearity are also used to determine the inverse DTFT of frequency-domain representations given as a ratio of polynomials in $e^{-j\Omega}$. Let

$$X(e^{j\Omega}) = \frac{b_M e^{-j\Omega M} + \cdots + b_1 e^{-j\Omega} + b_0}{a_N e^{-j\Omega N} + a_{N-1}e^{-j\Omega(N-1)} + \cdots + a_1 e^{-j\Omega} + 1}$$

Representations of this form occur frequently in the study of systems described by linear constant-coefficient difference equations. Note that the constant term in the denominator polynomial has been normalized to unity. As before, we rewrite $X(e^{j\Omega})$ as a sum of terms whose inverse DTFT is known using a partial fraction expansion. We factor the denominator polynomial as

$$a_N e^{-j\Omega N} + a_{N-1}e^{-j\Omega(N-1)} + \cdots + a_1 e^{-j\Omega} + 1 = \prod_{k=1}^{N} (1 - d_k e^{-j\Omega})$$

Partial fraction expansions based on this factorization are reviewed in Appendix B. In this case, the d_k are roots of the polynomial

$$u^N + a_1 u^{N-1} + a_2 u^{N-2} + \cdots + a_{N-1}u + a_N = 0$$

Assuming $M < N$, we may express $X(e^{j\Omega})$ as the sum

$$X(e^{j\Omega}) = \sum_{k=1}^{N} \frac{C_k}{1 - d_k e^{-j\Omega}}$$

This form again assumes all of the d_k are distinct. Expansions for repeated roots are treated in Appendix B. Since

$$(d_k)^n u[n] \xleftrightarrow{\text{DTFT}} \frac{1}{1 - d_k e^{-j\Omega}}$$

the linearity property implies

$$x[n] = \sum_{k=1}^{N} C_k (d_k)^n u[n]$$

EXAMPLE 3.21   Find the inverse DTFT of

$$X(e^{j\Omega}) = \frac{-\frac{5}{6}e^{-j\Omega} + 5}{1 + \frac{1}{6}e^{-j\Omega} - \frac{1}{6}e^{-j\Omega 2}}$$

Solution: The roots of the polynomial

$$u^2 + \frac{1}{6}u - \frac{1}{6} = 0$$

are $d_1 = -\frac{1}{2}$ and $d_2 = \frac{1}{3}$. We seek coefficients C_1 and C_2 so that

$$\frac{-\frac{5}{6}e^{-j\Omega} + 5}{1 + \frac{1}{6}e^{-j\Omega} - \frac{1}{6}e^{-j\Omega 2}} = \frac{C_1}{1 + \frac{1}{2}e^{-j\Omega}} + \frac{C_2}{1 - \frac{1}{3}e^{-j\Omega}}$$

Using the method of residues, we obtain

$$C_1 = (1 + \tfrac{1}{2}e^{-j\Omega}) \left. \frac{-\frac{5}{6}e^{-j\Omega} + 5}{1 + \frac{1}{6}e^{-j\Omega} - \frac{1}{6}e^{-j\Omega 2}} \right|_{e^{-j\Omega} = -2}$$

$$= \left. \frac{-\frac{5}{6}e^{-j\Omega} + 5}{1 - \frac{1}{3}e^{-j\Omega}} \right|_{e^{-j\Omega} = -2}$$

$$= 4$$

$$C_2 = (1 - \tfrac{1}{3}e^{-j\Omega}) \left. \frac{-\frac{5}{6}e^{-j\Omega} + 5}{1 + \frac{1}{6}e^{-j\Omega} - \frac{1}{6}e^{-j\Omega 2}} \right|_{e^{-j\Omega} = 3}$$

$$= \left. \frac{-\frac{5}{6}e^{-j\Omega} + 5}{1 + \frac{1}{2}e^{-j\Omega}} \right|_{e^{-j\Omega} = 3}$$

$$= 1$$

Hence

$$x[n] = 4(-\tfrac{1}{2})^n u[n] + (\tfrac{1}{3})^n u[n]$$

■ SYMMETRY PROPERTIES—REAL AND IMAGINARY SIGNALS

We develop the symmetry properties using the FT. Results for the other three Fourier representations may be obtained in an analogous manner and are simply stated.

First, suppose $x(t)$ is real. This implies that $x(t) = x^*(t)$. Consider $X^*(j\omega)$, defined by

$$X^*(j\omega) = \left[\int_{-\infty}^{\infty} x(t)e^{-j\omega t}\, dt \right]^*$$

$$= \int_{-\infty}^{\infty} x^*(t)e^{j\omega t}\, dt$$

We now may substitute $x(t)$ for $x^*(t)$ since $x(t)$ is real, obtaining

$$X^*(j\omega) = \int_{-\infty}^{\infty} x(t)e^{-j(-\omega)t}\, dt$$

$$= X(-j\omega)$$

This shows that $X(j\omega)$ is complex-conjugate symmetric or $X^*(j\omega) = X(-j\omega)$. Taking the real and imaginary parts of this expression gives $\text{Re}\{X(j\omega)\} = \text{Re}\{X(-j\omega)\}$ and $\text{Im}\{X(j\omega)\} = -\text{Im}\{X(-j\omega)\}$. In words, if $x(t)$ is real valued, then the real part of the transform is an even function of frequency, while the imaginary part is an odd function of frequency. This also implies that the magnitude spectrum is an even function while the phase spectrum is an odd function. The symmetry conditions in all four Fourier representations of real-valued

TABLE 3.4 Fourier Representation Symmetry Properties for Real-Valued Time Signals

	Complex Form	Rectangular Form	Polar Form
FT	$X^*(j\omega) = X(-j\omega)$	$\mathrm{Re}\{X(j\omega)\} = \mathrm{Re}\{X(-j\omega)\}$ $\mathrm{Im}\{X(j\omega)\} = -\mathrm{Im}\{X(-j\omega)\}$	$\lvert X(j\omega)\rvert = \lvert X(-j\omega)\rvert$ $\arg\{X(j\omega)\} = -\arg\{X(-j\omega)\}$
FS	$X^*[k] = X[-k]$	$\mathrm{Re}\{X[k]\} = \mathrm{Re}\{X[-k]\}$ $\mathrm{Im}\{X[k]\} = -\mathrm{Im}\{X[-k]\}$	$\lvert X[k]\rvert = \lvert X[-k]\rvert$ $\arg\{X[k]\} = -\arg\{X[-k]\}$
DTFT	$X^*(e^{j\Omega}) = X(e^{-j\Omega})$	$\mathrm{Re}\{X(e^{j\Omega})\} = \mathrm{Re}\{X(e^{-j\Omega})\}$ $\mathrm{Im}\{X(e^{j\Omega})\} = -\mathrm{Im}\{X(e^{-j\Omega})\}$	$\lvert X(e^{j\Omega})\rvert = \lvert X(e^{-j\Omega})\rvert$ $\arg\{X(e^{j\Omega})\} = -\arg\{X(e^{-j\Omega})\}$
DTFS	$X^*[k] = X[-k]$	$\mathrm{Re}\{X[k]\} = \mathrm{Re}\{X[-k]\}$ $\mathrm{Im}\{X[k]\} = -\mathrm{Im}\{X[-k]\}$	$\lvert X[k]\rvert = \lvert X[-k]\rvert$ $\arg\{X[k]\} = -\arg\{X[-k]\}$

signals are indicated in Table 3.4. In each case the real part of the Fourier representation has even symmetry and the imaginary part has odd symmetry. Hence the magnitude spectrum has even symmetry and the phase spectrum has odd symmetry. Note that the conjugate symmetry property for the DTFS may also be written as $X^*[k] = X[N - k]$, because the DTFS coefficients are N periodic, satisfying $X[k] = X[N + k]$.

Now suppose $x(t)$ is purely imaginary so that $x(t) = -x^*(t)$. In this case, we may write

$$X^*(j\omega) = \left[\int_{-\infty}^{\infty} x(t)e^{-j\omega t}\, dt\right]^*$$

$$= \int_{-\infty}^{\infty} x^*(t)e^{j\omega t}\, dt$$

$$= -\int_{-\infty}^{\infty} x(t)e^{-j(-\omega)t}\, dt$$

$$= -X(-j\omega)$$

Examining the real and imaginary parts of this relationship gives $\mathrm{Re}\{X(j\omega)\} = -\mathrm{Re}\{X(-j\omega)\}$ and $\mathrm{Im}\{X(j\omega)\} = \mathrm{Im}\{X(-j\omega)\}$. That is, if $x(t)$ is purely imaginary, then the real part of the FT has odd symmetry and the imaginary part has even symmetry. The corresponding symmetry relationships for all four Fourier representations are given in Table 3.5. Note that the magnitude and phase spectra have the same symmetry as given in Table 3.4.

TABLE 3.5 Fourier Representation Symmetry Properties for Imaginary-Valued Time Signals

	Complex Form	Rectangular Form
FT	$X^*(j\omega) = -X(-j\omega)$	$\mathrm{Re}\{X(j\omega)\} = -\mathrm{Re}\{X(-j\omega)\}$ $\mathrm{Im}\{X(j\omega)\} = \mathrm{Im}\{X(-j\omega)\}$
FS	$X^*[k] = -X[-k]$	$\mathrm{Re}\{X[k]\} = -\mathrm{Re}\{X[-k]\}$ $\mathrm{Im}\{X[k]\} = \mathrm{Im}\{X[-k]\}$
DTFT	$X^*(e^{j\Omega}) = -X(e^{-j\Omega})$	$\mathrm{Re}\{X(e^{j\Omega})\} = -\mathrm{Re}\{X(e^{-j\Omega})\}$ $\mathrm{Im}\{X(e^{j\Omega})\} = \mathrm{Im}\{X(e^{-j\Omega})\}$
DTFS	$X^*[k] = -X[-k]$	$\mathrm{Re}\{X[k]\} = -\mathrm{Re}\{X[-k]\}$ $\mathrm{Im}\{X[k]\} = \mathrm{Im}\{X[-k]\}$

■ SYMMETRY PROPERTIES—EVEN AND ODD SIGNALS

Assume $x(t)$ is real valued and has even symmetry. These conditions imply $x^*(t) = x(t)$ and $x(-t) = x(t)$, respectively. Using these relationships we may write

$$X^*(j\omega) = \int_{-\infty}^{\infty} x^*(t)e^{j\omega t} \, dt$$

$$= \int_{-\infty}^{\infty} x(t)e^{j\omega t} \, dt$$

$$= \int_{-\infty}^{\infty} x(-t)e^{-j\omega(-t)} \, dt$$

Now perform the change of variable $\tau = -t$ to obtain

$$X^*(j\omega) = \int_{-\infty}^{\infty} x(\tau)e^{-j\omega\tau} \, d\tau$$
$$= X(j\omega)$$

The only way that the condition $X^*(j\omega) = X(j\omega)$ holds is for the imaginary part of $X(j\omega)$ to be zero. Hence if $x(t)$ is real and even, then $X(j\omega)$ is real. Similarly, we may show that if $x(t)$ is real and odd, then $X^*(j\omega) = -X(j\omega)$ and $X(j\omega)$ is imaginary.

The identical symmetry relationships hold for all four Fourier representations. If the time signal is real and even, then the frequency-domain representation is also real. If the time signal is real and odd, then the frequency-domain representation is imaginary. Note that since we have assumed real-valued time signals in deriving these symmetry properties, we may combine the results of this subsection with those of the previous subsection. That is, real and even time signals have real and even frequency-domain representations, and real and odd time signals have imaginary and odd frequency-domain representations.

■ TIME-SHIFT PROPERTIES

In this section we consider the effect of a time shift on the Fourier representation. As before, we derive the result for the FT and state the results for the other three representations.

Let $z(t) = x(t - t_o)$ be a time-shifted version of $x(t)$. The goal is to relate the FT of $z(t)$ to the FT of $x(t)$. We have

$$Z(j\omega) = \int_{-\infty}^{\infty} z(t)e^{-j\omega t} \, dt$$

$$= \int_{-\infty}^{\infty} x(t - t_o)e^{-j\omega t} \, dt$$

Now perform the change of variable $\tau = t - t_o$, obtaining

$$Z(j\omega) = \int_{-\infty}^{\infty} x(\tau)e^{-j\omega(\tau + t_o)} \, d\tau$$

$$= e^{-j\omega t_o} \int_{-\infty}^{\infty} x(\tau)e^{-j\omega\tau} \, d\tau$$

The result of time shifting by t_o is to multiply the transform by $e^{-j\omega t_o}$. Note that $|Z(j\omega)| = |X(j\omega)|$ and $\arg\{Z(j\omega)\} = \arg\{X(j\omega)\} - \omega t_o$. Hence a shift in time leaves the magnitude spectrum unchanged and introduces a phase shift that is a linear function of frequency. The slope of this linear phase term is equal to the time shift. A similar property

> **TABLE 3.6** *Time-Shift Properties of Fourier Representations*

$$x(t - t_o) \xleftrightarrow{\text{FT}} e^{-j\omega t_o}X(j\omega)$$

$$x(t - t_o) \xleftrightarrow{\text{FS; } \omega_o} e^{-jk\omega_o t_o}X[k]$$

$$x[n - n_o] \xleftrightarrow{\text{DTFT}} e^{-j\Omega n_o}X(e^{j\Omega})$$

$$x[n - n_o] \xleftrightarrow{\text{DTFS; } \Omega_o} e^{-jk\Omega_o n_o}X[k]$$

holds for the other three Fourier representations, as indicated in Table 3.6. These properties are a direct consequence of the time-shift properties of complex sinusoids used in Fourier representations. Time shifting a complex sinusoid results in a complex sinusoid of the same frequency whose phase is shifted by the product of the time shift and the sinusoid's frequency.

EXAMPLE 3.22   Use the FT of the rectangular pulse $x(t)$ depicted in Fig. 3.28(a) to determine the FT of the time-shifted rectangular pulse $z(t)$ depicted in Fig. 3.28(b).

Solution: First we note that $z(t) = x(t - T)$, so the time-shift property implies that $Z(j\omega) = e^{-j\omega T}X(j\omega)$. In Example 3.15 we obtained

$$X(j\omega) = \frac{2}{\omega} \sin(\omega T)$$

Thus we have

$$Z(j\omega) = e^{-j\omega T} \frac{2}{\omega} \sin(\omega T)$$

▶ **Drill Problem 3.12**   Use the DTFS of the periodic square wave depicted in Fig. 3.29(a) as derived in Example 3.2 to determine the DTFS of the periodic square wave depicted in Fig. 3.29(b).

Answer:

$$Z[k] = e^{-jk(6\pi/7)} \frac{\sin\left(k\,\frac{5\pi}{7}\right)}{7 \sin\left(k\,\frac{\pi}{7}\right)}$$

◀

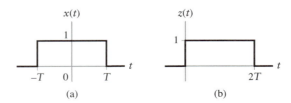

FIGURE 3.28   Application of the time-shift property for Example 3.22.

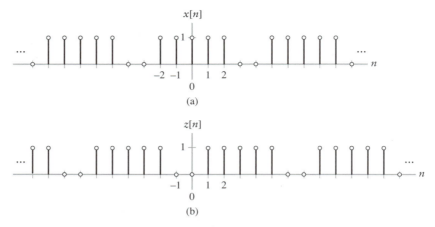

FIGURE 3.29   Original and time-shifted square waves for Drill Problem 3.12.

■ FREQUENCY-SHIFT PROPERTIES

In the previous subsection we considered the effect of a time shift on the frequency-domain representation. In this section we consider the effect of a frequency shift on the time-domain signal. Suppose $x(t) \xleftrightarrow{\ FT\ } X(j\omega)$. The problem is to express the inverse FT of $Z(j\omega) = X(j(\omega - \gamma))$ in terms of $x(t)$. Let $z(t) \xleftrightarrow{\ FT\ } Z(j\omega)$. By the FT definition, we have

$$z(t) = \frac{1}{2\pi} \int_{-\infty}^{\infty} Z(j\omega)e^{j\omega t}\, d\omega$$

$$= \frac{1}{2\pi} \int_{-\infty}^{\infty} X(j(\omega - \gamma))e^{j\omega t}\, d\omega$$

Perform the substitution of variables $\eta = \omega - \gamma$, obtaining

$$z(t) = \frac{1}{2\pi} \int_{-\infty}^{\infty} X(j\eta)e^{j(\eta + \gamma)t}\, d\eta$$

$$= e^{j\gamma t} \frac{1}{2\pi} \int_{-\infty}^{\infty} X(j\eta)e^{j\eta t}\, d\eta$$

$$= e^{j\gamma t} x(t)$$

Hence a frequency shift corresponds to multiplication in the time domain by a complex sinusoid whose frequency is equal to the shift.

This property is a consequence of the frequency-shift properties of the complex sinusoid. A shift in the frequency of a complex sinusoid is equivalent to multiplication of the original complex sinusoid by another complex sinusoid whose frequency is equal to the shift. Since all the Fourier representations are based on complex sinusoids, they all share this property as summarized in Table 3.7. Note that the frequency shift must be integer valued in both Fourier series cases. This leads to multiplication by a complex sinusoid whose frequency is an integer multiple of the fundamental frequency. The other observation is that the frequency-shift property is the "dual" of the time-shift property. We may summarize both properties by stating that a shift in one domain, either frequency or time, leads to multiplication by a complex sinusoid in the other domain.

> **TABLE 3.7** *Frequency-Shift Properties of Fourier Representations*

$$e^{j\gamma t}x(t) \xleftrightarrow{\text{FT}} X(j(\omega - \gamma))$$

$$e^{jk_o\omega_o t}x(t) \xleftrightarrow{\text{FS}; \omega_o} X[k - k_o]$$

$$e^{j\Gamma n}x[n] \xleftrightarrow{\text{DTFT}} X(e^{j(\Omega-\Gamma)})$$

$$e^{jk_o\Omega_o n}x[n] \xleftrightarrow{\text{DTFS}; \Omega_o} X[k - k_o]$$

EXAMPLE 3.23   Use the frequency-shift property to determine the FT of the complex sinusoidal pulse

$$z(t) = \begin{cases} e^{j10t}, & |t| \le \pi \\ 0, & \text{otherwise} \end{cases}$$

Solution: We may express $z(t)$ as the product of a complex sinusoid, e^{j10t}, and a rectangular pulse

$$x(t) = \begin{cases} 1, & |t| \le \pi \\ 0, & \text{otherwise} \end{cases}$$

Using the results of Example 3.15, we write

$$x(t) \xleftrightarrow{\text{FT}} X(j\omega) = \frac{2}{\omega}\sin(\omega\pi)$$

and using the frequency-shift property

$$e^{j10t}x(t) \xleftrightarrow{\text{FT}} X(j(\omega - 10))$$

we obtain

$$z(t) \xleftrightarrow{\text{FT}} \frac{2}{\omega - 10}\sin((\omega - 10)\pi)$$

▶ **Drill Problem 3.13**   Use the frequency-shift property to find the inverse DTFT of

$$Z(e^{j\Omega}) = \frac{1}{1 - \alpha e^{j(\Omega + \pi/4)}}$$

Assume $|\alpha| < 1$.

Answer:

$$z[n] = e^{-j(\pi/4)n}\alpha^n u[n]$$ ◀

■ **SCALING PROPERTIES**

Now consider the effect of scaling the time variable on the frequency-domain representation of a signal. Beginning with the FT, let $z(t) = x(at)$. By definition, we have

$$Z(j\omega) = \int_{-\infty}^{\infty} z(t)e^{-j\omega t}\,dt$$

$$= \int_{-\infty}^{\infty} x(at)e^{-j\omega t}\,dt$$

Perform the substitution $\tau = at$ to obtain

$$Z(j\omega) = \begin{cases} \dfrac{1}{a} \displaystyle\int_{-\infty}^{\infty} x(\tau)e^{-j(\omega/a)\tau}\, d\tau, & a > 0 \\[2mm] \dfrac{1}{a} \displaystyle\int_{\infty}^{-\infty} x(\tau)e^{-j(\omega/a)\tau}\, d\tau, & a < 0 \end{cases}$$

These two integrals may be combined into the single integral

$$\begin{aligned} Z(j\omega) &= \frac{1}{|a|} \int_{-\infty}^{\infty} x(\tau)e^{-j(\omega/a)\tau}\, d\tau \\[2mm] &= \frac{1}{|a|} X\!\left(j\,\frac{\omega}{a}\right) \end{aligned} \tag{3.32}$$

Hence scaling the signal in time introduces inverse scaling in the frequency-domain representation and an amplitude scaling.

This effect may be experienced by playing a recorded sound at a speed different from that at which it was recorded. If we play the sound back at a higher speed, corresponding to $a > 1$, we compress the time signal. The inverse scaling in the frequency domain expands the Fourier representation over a broader frequency band and explains the increase in the perceived pitch of the sound. Conversely, playing the sound back at a slower speed corresponds to expanding the time signal, since $0 < a < 1$. The inverse scaling in the frequency domain compresses the Fourier representation and explains the decrease in the perceived pitch of the sound.

EXAMPLE 3.24   Let $x(t)$ be the rectangular pulse

$$x(t) = \begin{cases} 1, & |t| \leq 1 \\ 0, & |t| > 1 \end{cases}$$

Use the FT of $x(t)$ and the scaling property to find the FT of the scaled rectangular pulse

$$y(t) = \begin{cases} 1, & |t| \leq 2 \\ 0, & |t| > 2 \end{cases}$$

Solution: Substituting $T = 1$ into the result of Example 3.15 gives

$$X(j\omega) = \frac{2}{\omega}\sin(\omega)$$

Note that $y(t) = x(\tfrac{1}{2}t)$. Hence application of the scaling property of Eq. (3.32) with $a = \tfrac{1}{2}$ gives

$$\begin{aligned} Y(j\omega) &= 2X(j2\omega) \\[1mm] &= \frac{2}{\omega}\sin(2\omega) \end{aligned}$$

This answer may also be obtained by substituting $T = 2$ into the result of Example 3.15. Figure 3.30 illustrates the scaling between time and frequency that occurs in this example.

If $x(t)$ is a periodic signal, then $z(t) = x(at)$ is also periodic and the FS is the appropriate Fourier representation. For convenience we assume that a is positive. In this case, scaling changes the fundamental period of the signal. If $x(t)$ has fundamental period T,

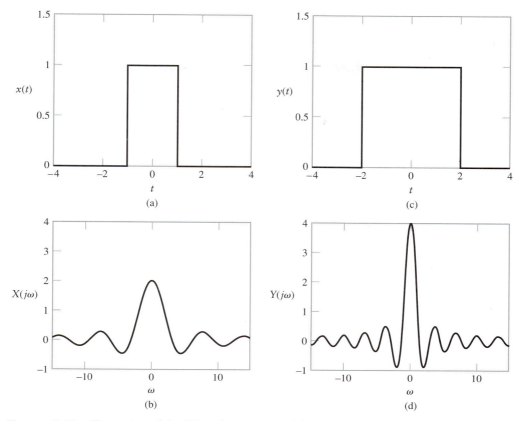

FIGURE 3.30 Illustration of the FT scaling property. (a) Original time signal. (b) Original FT. (c) Scaled time signal $y(t) = x(\frac{1}{2}t)$. (d) Scaled FT $Y(j\omega) = 2X(j2\omega)$.

then $z(t)$ has fundamental period T/a. Hence if the fundamental frequency of $x(t)$ is ω_o, then the fundamental frequency of $z(t)$ is $a\omega_o$. By definition, the FS coefficients for $z(t)$ are given by

$$Z[k] = \frac{a}{T} \int_{\langle T/a \rangle} z(t)e^{-jka\omega_o t}\, dt$$

Substituting $x(at)$ for $z(t)$ and performing the change of variable as in the FT case, we obtain

$$x(at) = z(t) \xleftrightarrow{\;FS;\; a\omega_o\;} Z[k] = X[k], \qquad a > 0$$

That is, the FS coefficients of $x(t)$ and $x(at)$ are identical; the scaling operation simply changes the harmonic spacing from ω_o to $a\omega_o$.

The scaling operation has a slightly different character in discrete time than in continuous time. First, $z[n] = x[pn]$ is defined only for integer values of p. Second, if $|p| > 1$, then the scaling operation discards information since it retains only every pth value of $x[n]$. This loss of information prevents us from expressing the DTFT or DTFS of $z[n]$ in terms of the DTFT or DTFS of $x[n]$ in a manner similar to the continuous-time results

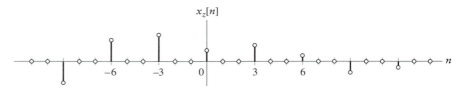

FIGURE 3.31   Example signal that is zero except at multiples of the scaling parameter $p = 3$.

derived above. To proceed, we define a sequence that is zero except at integer multiples of the scaling parameter p. That is, let

$$x_z[n] = 0, \quad \text{unless } \frac{n}{p} \text{ is integer}$$

An example of such a sequence is given in Fig. 3.31 for $p = 3$.

If $x_z[n]$ is nonperiodic, then the scaled sequence $z[n] = x_z[pn]$ is also nonperiodic and the DTFT is the appropriate representation. Proceeding from the definition of the DTFT, we may show that

$$Z(e^{j\Omega}) = X_z(e^{j\Omega/p})$$

In the special case of a reflection, $p = -1$, the assumption of zeros at intermediate values in the time signal is unnecessary and we have

$$x[-n] \overset{DTFT}{\longleftrightarrow} X(e^{-j\Omega})$$

▶ **Drill Problem 3.14**   Use the DTFT of the signal $w[n]$ depicted in Fig. 3.32(a) and the scaling property to determine the DTFT of the signal $f[n]$ depicted in Fig. 3.32(b).

Answer:

$$F(e^{j\Omega}) = \frac{\sin(7\Omega)}{\sin(\Omega)} \qquad \blacktriangleleft$$

If $x_z[n]$ is periodic with fundamental period N, then $z[n] = x_z[pn]$ has fundamental period N/p. For convenience we assume p is positive. Note that N is always an integer multiple of p as a consequence of the definition of $x_z[n]$ and thus the fundamental period

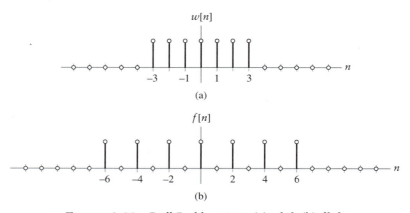

FIGURE 3.32   Drill Problem 3.14. (a) $w[n]$. (b) $f[n]$.

of $z[n]$ will always be an integer. Let Ω_o be the fundamental frequency of $x_z[n]$. In this case we have

$$x_z[pn] = z[n] \xleftrightarrow{\ DTFS;\ p\Omega_o\ } Z[k] = pX_z[k], \qquad p > 0$$

The scaling operation changes the harmonic spacing from Ω_o to $p\Omega_o$ and amplifies the DTFS coefficients by p.

■ DIFFERENTIATION AND INTEGRATION

Differentiation and integration are operations that apply to continuous functions. Hence we may consider the effect of differentiation and integration with respect to time for a continuous-time signal, or with respect to frequency in the FT and DTFT since these are continuous functions of frequency. We derive integration and differentiation properties for several of these cases. The analogous discrete-valued operations of differencing and summation are much less commonly used and only briefly discussed.

Differentiation in Time

Consider the effect of differentiating a nonperiodic signal $x(t)$. First, recall that $x(t)$ and its FT, $X(j\omega)$, are related by

$$x(t) = \frac{1}{2\pi} \int_{-\infty}^{\infty} X(j\omega)e^{j\omega t}\,d\omega$$

Differentiating both sides of this equation with respect to t yields

$$\frac{d}{dt}x(t) = \frac{1}{2\pi} \int_{-\infty}^{\infty} X(j\omega)j\omega e^{j\omega t}\,d\omega$$

which implies that

$$\frac{d}{dt}x(t) \xleftrightarrow{\ FT\ } j\omega X(j\omega)$$

That is, differentiating in time corresponds to multiplying by $j\omega$ in the frequency domain. This operation accentuates the high-frequency components of the signal. Note that differentiation destroys any dc component of $x(t)$ and, consequently, the FT of the differentiated signal at $\omega = 0$ is zero.

EXAMPLE 3.25 The differentiation property implies that

$$\frac{d}{dt}(e^{-at}u(t)) \xleftrightarrow{\ FT\ } \frac{j\omega}{a + j\omega}$$

Verify this result by differentiating and finding the FT directly.

Solution: Using the product rule for differentiation we have

$$\frac{d}{dt}(e^{-at}u(t)) = -ae^{-at}u(t) + e^{-at}\delta(t)$$

$$= -ae^{-at}u(t) + \delta(t)$$

Taking the FT of each term and using linearity, we may write

$$\frac{d}{dt}(e^{-at}u(t)) \overset{FT}{\longleftrightarrow} \frac{-a}{a+j\omega} + 1$$

$$= \frac{j\omega}{a+j\omega}$$

If $x(t)$ is a periodic signal, then we have the FS representation

$$x(t) = \sum_{k=-\infty}^{\infty} X[k]e^{jk\omega_o t}$$

Differentiating both sides of this equation gives

$$\frac{d}{dt}x(t) = \sum_{k=-\infty}^{\infty} X[k]jk\omega_o e^{jk\omega_o t}$$

and thus we conclude

$$\frac{d}{dt}x(t) \overset{FS;\ \omega_o}{\longleftrightarrow} jk\omega_o X[k]$$

Once again, differentiation forces the average value of the differentiated signal to be zero; hence the FS coefficient for $k = 0$ is zero.

EXAMPLE 3.26   Use the differentiation property to find the FS representation for the triangular wave depicted in Fig. 3.33(a).

Solution: Define a waveform

$$z(t) = \frac{d}{dt}f(t)$$

Figure 3.33(b) illustrates $z(t)$. The FS coefficients for a periodic square wave were derived in Example 3.6. The signal $z(t)$ corresponds to the square wave $x(t)$ of Example 3.6 provided we subtract a constant term of two units, scale the amplitude of $z(t)$ by a factor of 4, and set $T_s/T = \frac{1}{4}$. That is, $z(t) = 4x(t) - 2$. Thus $Z[k] = 4X[k] - 2\delta[k]$ and we may write

$$z(t) \overset{FS;\ \omega_o}{\longleftrightarrow} Z[k] = \begin{cases} 0, & k = 0 \\ \dfrac{4\sin(k\pi/2)}{k\pi}, & k \neq 0 \end{cases}$$

The differentiation property implies that $Z[k] = jk\omega_o F[k]$. Hence we may determine $F[k]$ from $Z[k]$ as $F[k] = (1/jk\omega_o)Z[k]$, except for $k = 0$. The quantity $F[0]$ is the average value of $x(t)$ and is determined by inspection of Fig. 3.33(a) to be $T/2 = \pi/\omega_o$. Therefore

$$f(t) \overset{FS;\ \omega_o}{\longleftrightarrow} F[k] = \begin{cases} \dfrac{\pi}{\omega_o}, & k = 0 \\ \dfrac{4\sin(k\pi/2)}{jk^2\pi\omega_o}, & k \neq 0 \end{cases}$$

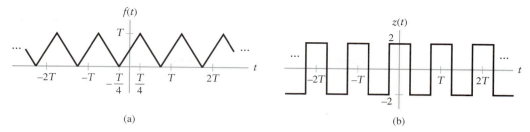

(a)   (b)

FIGURE 3.33 Signals for Example 3.26. (a) Triangular wave $f(t)$. (b) Derivative of $f(t)$ is the square wave $z(t)$.

Differentiation in Frequency

Next consider the effect of differentiating the frequency-domain representation of a signal. Beginning with the FT,

$$X(j\omega) = \int_{-\infty}^{\infty} x(t)e^{-j\omega t}\,dt$$

we differentiate both sides of this equation with respect to ω, and obtain

$$\frac{d}{d\omega}X(j\omega) = \int_{-\infty}^{\infty} -jtx(t)e^{-j\omega t}\,dt$$

which implies

$$-jtx(t) \xleftrightarrow{\ FT\ } \frac{d}{d\omega}X(j\omega)$$

Differentiation in frequency corresponds to multiplication in time by $-jt$.

EXAMPLE 3.27 Use the differentiation in time and differentiation in frequency properties to determine the FT of the *Gaussian pulse*, defined by $g(t) = (1/\sqrt{2\pi})e^{-t^2/2}$ and depicted in Fig. 3.34.

Solution: We note that the derivative of $g(t)$ with respect to time is given by

$$\frac{d}{dt}g(t) = \frac{-t}{\sqrt{2\pi}}e^{-t^2/2}$$

$$= -tg(t)$$

(3.33)

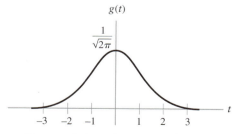

FIGURE 3.34 Gaussian pulse $g(t)$.

The differentiation in time property says that

$$\frac{d}{dt}g(t) \overset{FT}{\longleftrightarrow} j\omega G(j\omega)$$

and thus Eq. (3.33) implies that

$$-tg(t) \overset{FT}{\longleftrightarrow} j\omega G(j\omega) \tag{3.34}$$

The differentiation in frequency property, as shown by

$$-jtg(t) \overset{FT}{\longleftrightarrow} \frac{d}{d\omega}G(j\omega)$$

indicates that

$$-tg(t) \overset{FT}{\longleftrightarrow} \frac{1}{j}\frac{d}{d\omega}G(j\omega) \tag{3.35}$$

Since the left-hand sides of Eqs. (3.34) and (3.35) are equal, the right-hand sides must also be equal and thus

$$\frac{d}{d\omega}G(j\omega) = -\omega G(j\omega)$$

This is a differential-equation description for $G(j\omega)$ that has the same form as the differential-equation description for $g(t)$, given in Eq. (3.33). This implies that the functional form of $G(j\omega)$ is the same as that of $g(t)$, and thus

$$G(j\omega) = ce^{-\omega^2/2}$$

The constant c is determined by noting (see Appendix A.4) that

$$G(j0) = \int_{-\infty}^{\infty} \frac{1}{\sqrt{2\pi}}e^{-t^2/2}\,dt$$

$$= 1$$

This indicates that $c = 1$ and we conclude that a Gaussian pulse is its own Fourier transform, as shown by

$$\frac{1}{\sqrt{2\pi}}e^{-t^2/2} \overset{FT}{\longleftrightarrow} e^{-\omega^2/2}$$

▶ **Drill Problem 3.15** Use the frequency-differentiation property to find the FT of

$$x(t) = te^{-at}u(t)$$

Answer:

$$X(j\omega) = \frac{1}{(a + j\omega)^2} \qquad \blacktriangleleft$$

The operation of differentiation does not apply to discrete-valued quantities, and thus a frequency-domain differentiation property for the FS or DTFS does not exist. However, a frequency-domain differentiation property does exist for the DTFT. By definition,

$$X(e^{j\Omega}) = \sum_{n=-\infty}^{\infty} x[n]e^{-j\Omega n}$$

Differentiation of both sides of this expression with respect to frequency leads to the property

$$-jnx[n] \overset{DTFT}{\longleftrightarrow} \frac{d}{d\Omega} X(e^{j\Omega})$$

Integration

The operation of integration applies only to continuous independent variables. Hence we may integrate with respect to time in both the FT and FS and with respect to frequency in the FT and DTFT. We limit our consideration here to integration of nonperiodic signals with respect to time. Define

$$y(t) = \int_{-\infty}^{t} x(\tau) \, d\tau$$

That is, the value of y at time t is the integral of x over all time prior to t. Note that

$$\frac{d}{dt} y(t) = x(t)$$

so the differentiation property would suggest

$$Y(j\omega) = \frac{1}{j\omega} X(j\omega) \tag{3.36}$$

This relationship is indeterminate at $\omega = 0$, a consequence of the differentiation operation destroying any dc component of $y(t)$ and implying $X(j0)$ must be zero. Hence Eq. (3.36) applies only to signals with zero average value, that is, $X(j0) = 0$.

In general, we desire to apply the integration property to signals that do not have zero average value. However, if the average value of $x(\tau)$ is not zero, then it is possible that $y(t)$ is not square integrable and consequently the FT of $y(t)$ does not converge. We may get around this problem by including impulses in the transform. We know Eq. (3.36) holds for all ω except possibly $\omega = 0$. The value at $\omega = 0$ is modified by adding a term $c\delta(\omega)$, where the constant c depends on the average value of $x(\tau)$. The correct result is obtained by setting $c = \pi X(j0)$. This gives the integration property:

$$\int_{-\infty}^{t} x(\tau) \, d\tau \overset{FT}{\longleftrightarrow} \frac{1}{j\omega} X(j\omega) + \pi X(j0)\delta(\omega) \tag{3.37}$$

where it is understood that the first term on the right-hand side is zero at $\omega = 0$. Integration may be viewed as an averaging operation and thus it tends to smooth signals in time. This smoothing in time corresponds to deemphasizing the high-frequency components of the signal, as indicated in Eq. (3.37) by the ω term in the denominator.

We may demonstrate this property by deriving the FT of the unit step. The unit step may be expressed as the integral of the impulse function

$$u(t) = \int_{-\infty}^{t} \delta(\tau) \, d\tau$$

Since $\delta(t) \overset{FT}{\longleftrightarrow} 1$, Eq. (3.37) suggests

$$u(t) \overset{FT}{\longleftrightarrow} U(j\omega) = \frac{1}{j\omega} + \pi\delta(\omega)$$

Let us check this result by independently deriving $U(j\omega)$. First, express the unit step as the sum of two functions

$$u(t) = \tfrac{1}{2} + \tfrac{1}{2}\text{sgn}(t)$$

where the *signum* function, sgn(t), is defined as

$$\text{sgn}(t) = \begin{cases} -1, & t < 0 \\ 0, & t = 0 \\ 1, & t > 0 \end{cases}$$

This representation is illustrated in Fig. 3.35. Using the results of Example 3.18, we have $\tfrac{1}{2} \xleftrightarrow{\text{FT}} \pi\delta(\omega)$. The transform of sgn(t) is derived using the differentiation property. Let $\text{sgn}(t) \xleftrightarrow{\text{FT}} S(j\omega)$. We have

$$\frac{d}{dt}\text{sgn}(t) = 2\delta(t)$$

Hence

$$j\omega S(j\omega) = 2$$

We know that $S(j0) = 0$ because sgn(t) is an odd function and thus has zero average value. This knowledge removes the indeterminacy at $\omega = 0$ associated with the differentiation property, and we conclude

$$S(j\omega) = \begin{cases} 2/j\omega, & \omega \neq 0 \\ 0, & \omega = 0 \end{cases}$$

It is common to write this as $S(j\omega) = 2/j\omega$ with the understanding that $S(j0) = 0$. Now use linearity to obtain the FT of $u(t)$ as

$$u(t) \xleftrightarrow{\text{FT}} \frac{1}{j\omega} + \pi\delta(\omega)$$

This agrees exactly with the transform of the step function obtained using the integration property.

Summation and Differencing

The discrete-time analog of integration is summation: that is, define

$$y[n] = \sum_{k=-\infty}^{n} x[k]$$

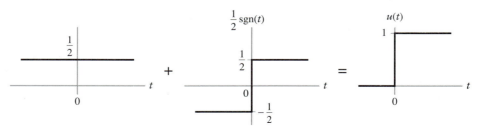

FIGURE 3.35 Representation of step function as the sum of a constant and signum function.

The discrete-time analog of differentiation is differencing. The differencing operation recovers $x[n]$ from $y[n]$, as shown by

$$x[n] = y[n] - y[n - 1]$$

This gives the differencing property, assuming that $x[k]$ is nonperiodic, as shown by

$$x[n] \xleftrightarrow{\text{DTFT}} X(e^{j\Omega}) = (1 - e^{-j\Omega})Y(e^{j\Omega})$$

We may invert this equation to express $Y(e^{j\Omega})$ as a function of $X(e^{j\Omega})$; however, we cannot determine $Y(e^{j0})$. As with the integration property, we may add an impulse to account for a nonzero average value in $x[k]$ and obtain the summation property as

$$y[n] \xleftrightarrow{\text{DTFT}} Y(e^{j\Omega}) = \frac{X(e^{j\Omega})}{1 - e^{-j\Omega}} + \pi X(e^{j0})\delta(\Omega), \qquad -\pi < \Omega \leq \pi$$

where the first term in $Y(e^{j\Omega})$ is assumed zero for $\Omega = 0$. The quantity $Y(e^{j\Omega})$ is 2π periodic, and so we may alternatively express this property for all values of Ω as

$$y[n] \xleftrightarrow{\text{DTFT}} Y(e^{j\Omega}) = \frac{X(e^{j\Omega})}{1 - e^{-j\Omega}} + \pi X(e^{j0}) \sum_{k=-\infty}^{\infty} \delta(\Omega - k2\pi)$$

Table 3.8 summarizes the differentiation, integration, and summation properties of Fourier representations.

EXAMPLE 3.28 This example illustrates the use of multiple Fourier representation properties. Find $x(t)$ if

$$X(j\omega) = j\frac{d}{d\omega}\left\{\frac{e^{j2\omega}}{1 + j\omega/3}\right\}$$

Solution: We identify three different properties that may be of use in finding $x(t)$: differentiation in frequency, time shifting, and scaling. These must be applied in the proper order to obtain the correct result. Use the transform pair

$$s(t) = e^{-t}u(t) \xleftrightarrow{\text{FT}} S(j\omega) = \frac{1}{1 + j\omega}$$

TABLE 3.8 *Commonly Used Differentiation, Integration, and Summation Properties*

$$\frac{d}{dt}x(t) \xleftrightarrow{\text{FT}} j\omega X(j\omega)$$

$$\frac{d}{dt}x(t) \xleftrightarrow{\text{FS; }\omega_o} jk\omega_o X[k]$$

$$-jtx(t) \xleftrightarrow{\text{FT}} \frac{d}{d\omega}X(j\omega)$$

$$-jnx[n] \xleftrightarrow{\text{DTFT}} \frac{d}{d\Omega}X(e^{j\Omega})$$

$$\int_{-\infty}^{t} x(\tau)\,d\tau \xleftrightarrow{\text{FT}} \frac{1}{j\omega}X(j\omega) + \pi X(j0)\delta(\omega)$$

$$\sum_{k=-\infty}^{n} x[k] \xleftrightarrow{\text{DTFT}} \frac{X(e^{j\Omega})}{1 - e^{-j\Omega}} + \pi X(e^{j0})\sum_{k=-\infty}^{\infty}\delta(\Omega - k2\pi)$$

to express $X(j\omega)$ as

$$X(j\omega) = j\frac{d}{d\omega}\left\{e^{j2\omega}S\left(j\frac{\omega}{3}\right)\right\}$$

Performing the innermost property first, we scale, then time shift, and lastly apply the differentiation property. If we define $Y(j\omega) = S(j\omega/3)$, then the scaling property gives

$$y(t) = 3s(3t)$$
$$= 3e^{-3t}u(3t)$$
$$= 3e^{-3t}u(t)$$

Now define $W(j\omega) = e^{j2\omega}Y(j\omega)$ and apply the time-shift property to obtain

$$w(t) = y(t + 2)$$
$$= 3e^{-3(t+2)}u(t + 2)$$

Lastly, since

$$X(j\omega) = j\frac{d}{d\omega}W(j\omega)$$

the differentiation property yields

$$x(t) = tw(t)$$
$$= 3te^{-3(t+2)}u(t + 2)$$

▶ **Drill Problem 3.16**   Show that the DTFT of

$$x[n] = ne^{j(\pi/8)n}\alpha^{n-3}u[n - 3]$$

is

$$X(e^{j\Omega}) = j\frac{d}{d\Omega}\left\{\frac{e^{-j3(\Omega-\pi/8)}}{1 - \alpha e^{-j(\Omega-\pi/8)}}\right\}$$ ◀

■ **CONVOLUTION AND MODULATION PROPERTIES**

Two of the most important properties of Fourier representations are the convolution and modulation properties. An important form of *modulation* refers to multiplication of two signals; one of the signals changes or "modulates" the amplitude of the other. We shall show that convolution in the time domain is transformed to modulation in the frequency domain, and that modulation in the time domain is transformed to convolution in the frequency domain. Hence we may analyze the input–output behavior of a linear system in the frequency domain using multiplication of transforms instead of convolving time signals. This can significantly simplify system analysis and offers considerable insight into system behavior. Both the convolution and modulation properties are a consequence of complex sinusoids being eigenfunctions of LTI systems. We now present these properties for each of the four Fourier representations, beginning with nonperiodic signals.

Nonperiodic Convolution

Consider the convolution of two nonperiodic continuous-time signals $x(t)$ and $h(t)$. Define

$$y(t) = h(t) * x(t)$$

$$= \int_{-\infty}^{\infty} h(\tau)x(t - \tau) \, d\tau$$

Now express $x(t - \tau)$ in terms of its FT, as shown by

$$x(t - \tau) = \frac{1}{2\pi} \int_{-\infty}^{\infty} X(j\omega)e^{j\omega(t-\tau)} \, d\omega$$

Substitute this expression into the convolution integral to obtain

$$y(t) = \int_{-\infty}^{\infty} h(\tau) \frac{1}{2\pi} \int_{-\infty}^{\infty} X(j\omega)e^{j\omega t}e^{-j\omega\tau} \, d\omega \, d\tau$$

$$= \frac{1}{2\pi} \int_{-\infty}^{\infty} \int_{-\infty}^{\infty} h(\tau)e^{-j\omega\tau} \, d\tau \, X(j\omega)e^{j\omega t} \, d\omega$$

We recognize the inner integral over τ as the FT of $h(\tau)$, or $H(j\omega)$. Hence $y(t)$ may be rewritten as

$$y(t) = \frac{1}{2\pi} \int_{-\infty}^{\infty} H(j\omega)X(j\omega)e^{j\omega t} \, d\omega$$

so $y(t)$ is the inverse FT of $H(j\omega)X(j\omega)$. We conclude that convolution of signals in time corresponds to multiplication of transforms in the frequency domain, as described by

$$y(t) = h(t) * x(t) \xleftarrow{\quad FT \quad} Y(j\omega) = X(j\omega)H(j\omega) \tag{3.38}$$

The following examples illustrate applications of this property.

EXAMPLE 3.29   Let $x(t) = (1/\pi t) \sin(\pi t)$ and $h(t) = (1/\pi t) \sin(2\pi t)$. Find $y(t) = x(t) * h(t)$.

Solution: This problem is extremely difficult to solve in the time domain. However, it is simple to solve in the frequency domain using the convolution property. We have

$$x(t) \xleftarrow{\quad FT \quad} X(j\omega) = \begin{cases} 1, & |\omega| \leq \pi \\ 0, & \text{otherwise} \end{cases}$$

$$h(t) \xleftarrow{\quad FT \quad} H(j\omega) = \begin{cases} 1, & |\omega| \leq 2\pi \\ 0, & \text{otherwise} \end{cases}$$

Since $y(t) = x(t) * h(t) \xleftarrow{\quad FT \quad} Y(j\omega) = X(j\omega)H(j\omega)$, we have

$$Y(j\omega) = \begin{cases} 1, & |\omega| \leq \pi \\ 0, & \text{otherwise} \end{cases}$$

and conclude that $y(t) = (1/\pi t) \sin(\pi t)$.

EXAMPLE 3.30 Use the convolution property to find $x(t)$, where

$$x(t) \xrightarrow{\ FT\ } X(j\omega) = \frac{4}{\omega^2} \sin^2(\omega)$$

Solution: We may write $X(j\omega)$ as the product $Z(j\omega)Z(j\omega)$, where

$$Z(j\omega) = \frac{2}{\omega} \sin(\omega)$$

The convolution property states that $z(t) * z(t) \xrightarrow{\ FT\ } Z(j\omega)Z(j\omega)$ so $x(t) = z(t) * z(t)$. We have

$$z(t) = \begin{cases} 1, & |t| \le 1 \\ 0, & \text{otherwise} \end{cases} \xrightarrow{\ FT\ } Z(j\omega)$$

as depicted in Fig. 3.36(a). Performing the convolution of $z(t)$ with itself gives the triangular waveform depicted in Fig. 3.36(b) as the solution for $x(t)$.

▶ **Drill Problem 3.17** Let the input to a system with impulse response $h(t) = 2e^{-2t}u(t)$ be $x(t) = 3e^{-t}u(t)$. Use the convolution property to find the output of the system, $y(t)$.

Answer:
$$y(t) = 6e^{-t}u(t) - 6e^{-2t}u(t)$$
◀

A similar property holds for convolution of discrete-time nonperiodic signals. If $x[n] \xrightarrow{\ DTFT\ } X(e^{j\Omega})$ and $h[n] \xrightarrow{\ DTFT\ } H(e^{j\Omega})$, then

$$y[n] = x[n] * h[n] \xrightarrow{\ DTFT\ } Y(e^{j\Omega}) = X(e^{j\Omega})H(e^{j\Omega}) \tag{3.39}$$

The proof of this result closely parallels that of the continuous-time case.

EXAMPLE 3.31 Reconsider the problem addressed in Example 2.12. In this problem a distorted, received signal $y[n]$ is expressed in terms of a transmitted signal $x[n]$ as

$$y[n] = x[n] + ax[n - 1], \qquad |a| < 1$$

Find the impulse response of an inverse system that will recover $x[n]$ from $y[n]$.

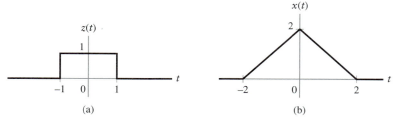

(a)　　　　　(b)

FIGURE 3.36 Signals for Example 3.30. (a) Rectangular pulse $z(t)$. (b) Convolution of $z(t)$ with itself gives $x(t)$.

Solution: In Example 2.12 we solved this problem using convolution. We now solve it in the frequency domain using the convolution property. Write the output as the convolution of the input with the system impulse response as $y[n] = x[n] * h[n]$, where the impulse response $h[n]$ is given by

$$h[n] = \begin{cases} 1, & n = 0 \\ a, & n = 1 \\ 0, & \text{otherwise} \end{cases}$$

The impulse response of an inverse system, $h^{-1}[n]$, must satisfy

$$h^{-1}[n] * h[n] = \delta[n]$$

Taking the DTFT of both sides of this equation and using the convolution property gives

$$H^{-1}(e^{j\Omega})H(e^{j\Omega}) = 1$$

which implies the frequency response of the inverse system is given by

$$H^{-1}(e^{j\Omega}) = \frac{1}{H(e^{j\Omega})}$$

Substitution of $h[n]$ into the definition of the DTFT yields

$$h[n] \xleftrightarrow{\text{DTFT}} H(e^{j\Omega}) = 1 + ae^{-j\Omega}$$

Hence

$$H^{-1}(e^{j\Omega}) = \frac{1}{1 + ae^{-j\Omega}}$$

Taking the inverse DTFT of $H^{-1}(e^{j\Omega})$ gives the impulse response of the inverse system

$$h^{-1}[n] = (-a)^n u[n]$$

▶ **Drill Problem 3.18**   Let the impulse response of a discrete-time system be given by $h[n] = (1/\pi n)\sin((\pi/4)n)$. Find the output $y[n]$ in response to the input (a) $x[n] = (1/\pi n)\sin((\pi/8)n)$, and (b) $x[n] = (1/\pi n)\sin((\pi/2)n)$.

Answer:

(a)
$$y[n] = \frac{1}{\pi n}\sin\left(\frac{\pi}{8}n\right)$$

(b)
$$y[n] = \frac{1}{\pi n}\sin\left(\frac{\pi}{4}n\right)$$

◀

Modulation

If $x(t)$ and $z(t)$ are nonperiodic signals, then we wish to express the FT of the product $y(t) = x(t)z(t)$ in terms of the FT of $x(t)$ and $z(t)$. Represent $x(t)$ and $z(t)$ in terms of their FTs as

$$x(t) = \frac{1}{2\pi}\int_{-\infty}^{\infty} X(j\nu)e^{j\nu t}\, d\nu$$

$$z(t) = \frac{1}{2\pi}\int_{-\infty}^{\infty} Z(j\eta)e^{j\eta t}\, d\eta$$

The product term, $y(t)$, may thus be written in the form

$$y(t) = \frac{1}{(2\pi)^2} \int_{-\infty}^{\infty} \int_{-\infty}^{\infty} X(j\nu)Z(j\eta)e^{j(\eta+\nu)t} \, d\eta \, d\nu$$

Now perform a change of variable on η, substituting $\eta = \omega - \nu$, to obtain

$$y(t) = \frac{1}{2\pi} \int_{-\infty}^{\infty} \frac{1}{2\pi} \int_{-\infty}^{\infty} X(j\nu)Z(j(\omega - \nu)) \, d\nu \, e^{j\omega t} \, d\omega$$

The inner integral over ν represents the convolution of $Z(j\omega)$ and $X(j\omega)$, while the outer integral over ω is of the form of the Fourier representation for $y(t)$. Hence we identify this convolution, scaled by $1/2\pi$, as $Y(j\omega)$, shown by

$$y(t) = x(t)z(t) \xleftrightarrow{\;FT\;} Y(j\omega) = \frac{1}{2\pi} X(j\omega) * Z(j\omega) \tag{3.40}$$

where

$$X(j\omega) * Z(j\omega) = \int_{-\infty}^{\infty} X(j\nu)Z(j(\omega - \nu)) \, d\nu$$

Multiplication in the time domain leads to convolution in the frequency domain.

Similarly, if $x[n]$ and $z[n]$ are discrete-time nonperiodic signals, then the DTFT of the product $y[n] = x[n]z[n]$ is given by the convolution of $X(e^{j\Omega})$ and $Z(e^{j\Omega})$ although the definition of convolution changes slightly because $X(e^{j\Omega})$ and $Z(e^{j\Omega})$ are periodic. Specifically,

$$y[n] = x[n]z[n] \xleftrightarrow{\;DTFT\;} Y(e^{j\Omega}) = \frac{1}{2\pi} X(e^{j\Omega}) \circledast Z(e^{j\Omega}) \tag{3.41}$$

where the symbol $\circledast$ denotes periodic convolution. The difference between periodic and nonperiodic convolution is that the integration in a periodic convolution is performed over a single period of the signals involved, whereas in the nonperiodic case it is performed over the entire interval. Here $X(e^{j\Omega})$ and $Z(e^{j\Omega})$ are 2π periodic, so we evaluate the convolution over a 2π interval, as shown by

$$X(e^{j\Omega}) \circledast Z(e^{j\Omega}) = \int_{\langle 2\pi \rangle} X(e^{j\theta})Z(e^{j(\Omega-\theta)}) \, d\theta$$

An important application of the modulation property is understanding the effects of truncating a signal on its frequency-domain representation. The process of truncating a signal is also known as *windowing*, since it corresponds to viewing the signal through a window. The portion of the signal that is not visible through the window is truncated. The windowing operation is represented mathematically by multiplying the signal, say, $x(t)$, by a window function $w(t)$ that is zero outside the time range of interest. Denoting the windowed signal by $y(t)$, we have $y(t) = x(t)w(t)$. This operation is illustrated in Fig. 3.37 for a window function that truncates $x(t)$ to the time interval $-T < t < T$. The FT of $y(t)$ is related to the FTs of $x(t)$ and $w(t)$ through the modulation property:

$$y(t) \xleftrightarrow{\;FT\;} Y(j\omega) = \frac{1}{2\pi} X(j\omega) * W(j\omega)$$

If $w(t)$ is the rectangular window depicted in Fig. 3.37, then

$$W(j\omega) = \frac{2}{\omega} \sin(\omega T)$$

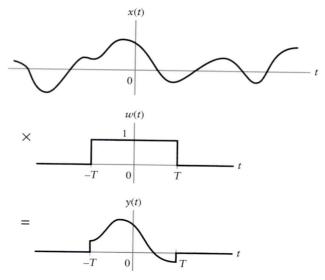

FIGURE 3.37   Truncating a signal using a window function $w(t)$.

Figure 3.38 illustrates the effect of windowing with a rectangular window in the frequency domain. The general effect of the window is to smooth detail in $X(j\omega)$ and introduce oscillations near discontinuities in $X(j\omega)$. The smoothing is a consequence of the $2\pi/T$ width of the mainlobe of $W(j\omega)$ while the oscillations are due to the oscillations in the sidelobes of $W(j\omega)$. The following example illustrates the effect of windowing the impulse response of an ideal discrete-time system.

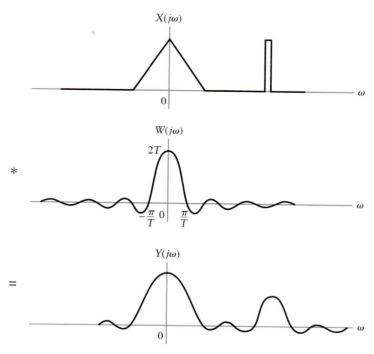

FIGURE 3.38   Convolution of signal and window FTs resulting from truncation in time.

EXAMPLE 3.32   The frequency response $H(e^{j\Omega})$ of an ideal discrete-time system is depicted in Fig. 3.39(a). Describe the frequency response of a system whose impulse response is the ideal impulse response truncated to the interval $-M \leq n \leq M$.

Solution: The ideal impulse response is the inverse DTFT of $H(e^{j\Omega})$, as shown by

$$h[n] = \frac{1}{\pi n} \sin\left(\frac{\pi n}{2}\right)$$

This response is infinite in extent. Let $h_t[n]$ be the truncated impulse response,

$$h_t[n] = \begin{cases} h[n], & |n| \leq M \\ 0, & \text{otherwise} \end{cases}$$

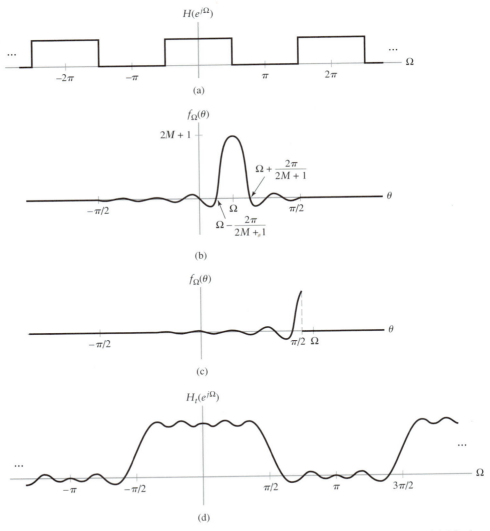

FIGURE 3.39   Effect of truncating the impulse response of a discrete-time system. (a) Ideal system frequency response. (b) $f_\Omega(\theta)$ for Ω near zero. (c) $f_\Omega(\theta)$ for Ω slightly greater than $\pi/2$. (d) Frequency response of system with truncated impulse response.

We may express $h_t[n]$ as the product of $h[n]$ and a window function $w[n]$, $h_t[n] = h[n]w[n]$, where

$$w[n] = \begin{cases} 1, & |n| \le M \\ 0, & \text{otherwise} \end{cases}$$

Let $h_t[n] \xleftrightarrow{\text{DTFT}} H_t(e^{j\Omega})$, and use the modulation property to obtain

$$H_t(e^{j\Omega}) = \frac{1}{2\pi} \int_{(2\pi)} H(e^{j\theta}) W(e^{j(\Omega-\theta)}) \, d\theta$$

Choose the 2π interval of integration to be $-\pi < \theta < \pi$. Now use

$$H(e^{j\theta}) = \begin{cases} 1, & |\theta| \le \pi/2 \\ 0, & \pi/2 < |\theta| < \pi \end{cases}$$

and

$$W(e^{j(\Omega-\theta)}) = \frac{\sin\left((\Omega - \theta)\dfrac{2M + 1}{2}\right)}{\sin\left(\dfrac{\Omega - \theta}{2}\right)}$$

to obtain

$$H_t(e^{j\Omega}) = \frac{1}{2\pi} \int_{-\pi/2}^{\pi/2} f_\Omega(\theta) \, d\theta$$

where we have defined

$$f_\Omega(\theta) = \begin{cases} W(e^{j(\Omega-\theta)}), & |\theta| < \pi/2 \\ 0, & \text{otherwise} \end{cases}$$

Figure 3.39(b) depicts $f_\Omega(\theta)$ for $\Omega < \pi/2$. $H_t(e^{j\Omega})$ is the area under $f_\Omega(\theta)$ between $\theta = -\pi/2$ and $\theta = \pi/2$. To visualize the behavior of $H_t(e^{j\Omega})$, consider the area under $f_\Omega(\theta)$ as Ω increases starting from $\Omega = 0$. As Ω increases, the small oscillations in $f_\Omega(\theta)$ move through the boundary at $\theta = \pi/2$. When a positive oscillation moves through the boundary, the net area under $f_\Omega(\theta)$ decreases; the net area increases when a negative oscillation moves through this boundary. Oscillations also move through the boundary at $\theta = -\pi/2$. However, these are smaller than those on the right because they are further away from Ω and thus have much less of an effect. The effect of the oscillations in $f_\Omega(\theta)$ moving through the boundary at $\theta = \pi/2$ is to introduce oscillations in $H_t(e^{j\Omega})$. These increase in size as Ω increases. As Ω approaches $\pi/2$, the area under $f_\Omega(\theta)$ decreases rapidly because the mainlobe moves through $\theta = \pi/2$. Figure 3.39(c) depicts $f_\Omega(\theta)$ for Ω slightly larger than $\pi/2$. As Ω continues to increase, the oscillations to the left of the mainlobe move through the boundary at $\theta = \pi/2$, causing additional oscillations in the area under $f_\Omega(\theta)$. However, now the net area oscillates about zero because the mainlobe of $f_\Omega(\theta)$ is no longer included.

Thus $H_t(e^{j\Omega})$ takes on the form depicted in Fig. 3.39(d). Truncation of the ideal impulse response introduces ripple into the frequency response and widens the transitions at $\Omega = \pm\pi/2$. These effects decrease as M increases, since then the mainlobe of $W(e^{j\Omega})$ becomes narrower and the oscillations decay more quickly.

▶ **Drill Problem 3.19** Use the modulation property to find the FT of

$$x(t) = \frac{4}{\pi^2 t^2} \sin^2(2t)$$

Answer: See Fig. 3.40. ◀

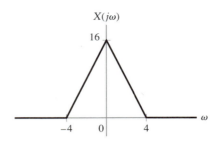

FIGURE 3.40   Solution to Drill Problem 3.19.

Convolution and Modulation for Signals That Are Periodic in Time

This subsection addresses the convolution and modulation properties for signals that are periodic functions of time. Convolution of periodic signals does not occur naturally in the context of evaluating the input–output relationships for systems, since any system with a periodic impulse response is unstable. However, convolution of periodic signals is a useful signal analysis and manipulation tool. We have already encountered an example of periodic convolution in the modulation property for the DTFT.

Define the periodic convolution of two continuous-time signals $x(t)$ and $z(t)$ each having period T as

$$y(t) = x(t) \circledast z(t)$$

$$= \int_{\langle T \rangle} x(\tau)z(t - \tau) \, d\tau$$

Here again the $\circledast$ symbol denotes that integration is performed over a single period of the signals involved. $y(t)$ is also periodic with period T; hence the FS is the appropriate representation for all three signals, $x(t)$, $z(t)$, and $y(t)$.

Substitution of the FS representation for $z(t)$ into the convolution integral leads to the property

$$y(t) = x(t) \circledast z(t) \xleftrightarrow{\; FS;\ 2\pi/T \;} Y[k] = TX[k]Z[k] \tag{3.42}$$

Again we see that convolution in time transforms to multiplication of the frequency-domain representations. This property explains the origin of the Gibbs phenomenon that was observed in Example 3.7. A partial sum approximation to the FS representation for $x(t)$ may be obtained using FS coefficients $Y[k]$ given by the product of $X[k]$ and a function $Z[k]$ that is 1 for $-J \leq k \leq J$ and zero otherwise. In the time domain, $y(t)$ is the periodic convolution of $x(t)$ and $z(t)$, where

$$z(t) = \frac{\sin\left(t\,\dfrac{2J + 1}{2}\right)}{\sin\left(\dfrac{t}{2}\right)}$$

The signal $z(t)$ corresponds exactly to $X(e^{j\Omega})$ depicted in Fig. 3.19(b) if we replace Ω by t and M by J. The periodic convolution of $x(t)$ and $z(t)$ is the area under shifted versions of $z(t)$ on $|t| < \frac{1}{2}$. The ripples in the partial sum approximation of $x(t)$ are a consequence of the variations in this area associated with shifting sidelobes of $z(t)$ into and out of the interval $|t| < \frac{1}{2}$.

The discrete-time convolution of two N periodic sequences $x[n]$ and $z[n]$ is defined as

$$y[n] = x[n] \circledast z[n]$$
$$= \sum_{k=\langle N \rangle} x[k]z[n-k]$$

This is the periodic convolution of $x[n]$ and $z[n]$. The signal $y[n]$ is N periodic so the DTFS is the appropriate representation for all three signals, $x[n]$, $z[n]$, and $y[n]$. Substitution of the DTFS representation for $z[n]$ results in the property

$$y[n] = x[n] \circledast z[n] \xleftrightarrow{\;DTFS;\; 2\pi/N\;} Y[k] = NX[k]Z[k] \qquad (3.43)$$

Convolution of time signals is transformed to multiplication of DTFS coefficients.

The modulation property for periodic signals is also analogous to that of nonperiodic signals. Multiplication of periodic time signals corresponds to convolution of the Fourier representations. Specifically, in continuous time we have

$$y(t) = x(t)z(t) \xleftrightarrow{\;FS;\; 2\pi/T\;} Y[k] = X[k] * Z[k] \qquad (3.44)$$

where

$$X[k] * Z[k] = \sum_{m=-\infty}^{\infty} X[m]Z[k-m]$$

is the nonperiodic convolution of the FS coefficients. All three time-domain signals have common fundamental period, T. In discrete time

$$y[n] = x[n]z[n] \xleftrightarrow{\;DTFS;\; 2\pi/N\;} Y[k] = X[k] \circledast Z[k] \qquad (3.45)$$

where

$$X[k] \circledast Z[k] = \sum_{m=\langle N \rangle} X[m]Z[k-m]$$

is the periodic convolution of DTFS coefficients. Again, all three time-domain signals have a common fundamental period, N.

EXAMPLE 3.33 Evaluate the periodic convolution of the sinusoidal signal

$$x(t) = 2\cos(2\pi t) + \sin(4\pi t)$$

with the T periodic square wave $z(t)$ depicted in Fig. 3.41.

Solution: Both $x(t)$ and $z(t)$ have fundamental period $T = 1$. Let $y(t) = x(t) \circledast z(t)$. The convolution property indicates that $y(t) \xleftrightarrow{\;FS;\; 2\pi\;} Y[k] = X[k]Z[k]$. The FS representation for $x(t)$ has coefficients

$$X[k] = \begin{cases} 1, & k = \pm 1 \\ 1/2j, & k = 2 \\ -1/2j, & k = -2 \\ 0, & \text{otherwise} \end{cases}$$

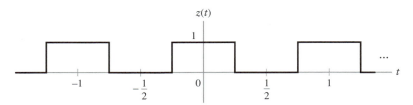

FIGURE 3.41   Square wave for Example 3.33.

The FS coefficients for $z(t)$ may be obtained from Example 3.6 as

$$Z[k] = \frac{2\,\sin(k\pi/2)}{k2\pi}$$

Hence the FS coefficients for $y(t)$ are

$$Y[k] = \begin{cases} 1/\pi, & k = \pm 1 \\ 0, & \text{otherwise} \end{cases}$$

which implies

$$y(t) = \frac{2}{\pi}\cos(2\pi t)$$

The convolution and modulation properties are summarized for all four Fourier representations in Table 3.9. In general, convolution in one domain corresponds to multiplication or modulation in the other domain. Periodic convolution is used for periodic signals and nonperiodic convolution is used for nonperiodic signals.

We have not yet considered several important cases of convolution and modulation that occur when there is a mixing of signal classes. Examples include the modulation of a nonperiodic signal with a periodic signal and the convolution of a periodic and a nonperiodic signal such as occurs when applying a periodic signal to a linear system. The properties derived here can be applied to these cases if we use a Fourier transform representation for periodic signals. This representation is developed in Chapter 4.

TABLE 3.9   *Convolution and Modulation Properties*

Convolution		Modulation	
$x(t) * z(t) \xleftrightarrow{\;FT\;} X(j\omega)Z(j\omega)$		$x(t)z(t) \xleftrightarrow{\;FT\;} \dfrac{1}{2\pi} X(j\omega) * Z(j\omega)$	
$x(t) \circledast z(t) \xleftrightarrow{\;FS;\,\omega_o\;} TX[k]Z[k]$		$x(t)z(t) \xleftrightarrow{\;FS;\,\omega_o\;} X[k] * Z[k]$	
$x[n] * z[n] \xleftrightarrow{\;DTFT\;} X(e^{j\Omega})Z(e^{j\Omega})$		$x[n]z[n] \xleftrightarrow{\;DTFT\;} \dfrac{1}{2\pi} X(e^{j\Omega}) \circledast Z(e^{j\Omega})$	
$x[n] \circledast z[n] \xleftrightarrow{\;DTFS;\,\Omega_o\;} NX[k]Z[k]$		$x[n]z[n] \xleftrightarrow{\;DTFS;\,\Omega_o\;} X[k] \circledast Z[k]$	

EXAMPLE 3.34 Find the FT of the signal

$$x(t) = \frac{d}{dt}\{(e^{-3t}u(t)) * (e^{-2t}u(t-2))\}$$

Solution: First we break the problem up into a series of simpler problems. Let $w(t) = e^{-3t}u(t)$ and $v(t) = e^{-2t}u(t-2)$ so that we may write

$$x(t) = \frac{d}{dt}\{w(t) * v(t)\}$$

Hence applying the differentiation and convolution properties, we obtain

$$X(j\omega) = j\omega W(j\omega)V(j\omega)$$

The transform pair

$$e^{-at}u(t) \xleftrightarrow{\text{FT}} \frac{1}{a + j\omega}$$

implies

$$W(j\omega) = \frac{1}{3 + j\omega}$$

We use the same transform pair and the time-shift property to find $V(j\omega)$ by first writing

$$v(t) = e^{-4}e^{-2(t-2)}u(t-2)$$

Thus

$$V(j\omega) = \frac{e^{-4}e^{-j2\omega}}{2 + j\omega}$$

and

$$X(j\omega) = e^{-4}\frac{j\omega e^{-j2\omega}}{(2 + j\omega)(3 + j\omega)}$$

▶ **Drill Problem 3.20** Find $x[n]$ if

$$X(e^{j\Omega}) = \left(\frac{e^{-j3\Omega}}{1 + \frac{1}{2}e^{-j\Omega}}\right) \circledast \left(\frac{\sin(21\Omega/2)}{\sin(\Omega/2)}\right)$$

Answer:

$$x[n] = 2\pi \left(-\tfrac{1}{2}\right)^{n-3}(u[n-3] - u[n-11])$$ ◀

■ **PARSEVAL RELATIONSHIPS**

The Parseval relationships state that the energy or power in the time-domain representation of a signal is equal to the energy or power in the frequency-domain representation. Hence energy or power is conserved in the Fourier representation. We shall derive this result for the FT and simply state it for the other three cases.

The energy in a continuous-time nonperiodic signal is

$$E_x = \int_{-\infty}^{\infty} |x(t)|^2 \, dt$$

where it is assumed that $x(t)$ may be complex valued in general. Note that $|x(t)|^2 = x(t)x^*(t)$ and that $x^*(t)$ is expressed in terms of its FT $X(j\omega)$ as

$$x^*(t) = \frac{1}{2\pi} \int_{-\infty}^{\infty} X^*(j\omega)e^{-j\omega t} \, d\omega$$

Substitute this into the expression for E_x to obtain

$$E_x = \int_{-\infty}^{\infty} x(t) \frac{1}{2\pi} \int_{-\infty}^{\infty} X^*(j\omega)e^{-j\omega t} \, d\omega \, dt$$

Now interchange the order of integration

$$E_x = \frac{1}{2\pi} \int_{-\infty}^{\infty} X^*(j\omega) \left\{ \int_{-\infty}^{\infty} x(t)e^{-j\omega t} \, dt \right\} d\omega$$

Observing that the term in braces is the FT of $x(t)$, we obtain

$$E_x = \frac{1}{2\pi} \int_{-\infty}^{\infty} X^*(j\omega)X(j\omega) \, d\omega$$

and so conclude

$$\int_{-\infty}^{\infty} |x(t)|^2 \, dt = \frac{1}{2\pi} \int_{-\infty}^{\infty} |X(j\omega)|^2 \, d\omega \tag{3.46}$$

Hence the energy in the time-domain representation of the signal is equal to the energy in the frequency-domain representation normalized by 2π. The quantity $|X(j\omega)|^2$ is termed the *energy spectrum* of the signal. Equation (3.46) is also referred to as Rayleigh's energy theorem.

Analogous results hold for the other three Fourier representations as summarized in Table 3.10. The energy or power in the time-domain representation is equal to the energy or power in the frequency-domain representation. Energy is used for nonperiodic time-domain signals, while power applies to periodic time-domain signals. Recall that power is defined as the integral or sum of the magnitude squared over one period normalized by the length of the period. The power or energy spectrum of a signal is defined as the square of the magnitude spectrum. They indicate how the power or energy in the signal is distributed as a function of frequency.

TABLE 3.10   *Parseval Relationships for the Four Fourier Representations*

Representation	Parseval Relation
FT	$\int_{-\infty}^{\infty} |x(t)|^2 \, dt = \frac{1}{2\pi} \int_{-\infty}^{\infty} |X(j\omega)|^2 \, d\omega$
FS	$\frac{1}{T} \int_{\langle T \rangle} |x(t)|^2 \, dt = \sum_{k=-\infty}^{\infty} |X[k]|^2$
DTFT	$\sum_{n=-\infty}^{\infty} |x[n]|^2 = \frac{1}{2\pi} \int_{\langle 2\pi \rangle} |X(e^{j\Omega})|^2 \, d\Omega$
DTFS	$\frac{1}{N} \sum_{n=\langle N \rangle} |x[n]|^2 = \sum_{k=\langle N \rangle} |X[k]|^2$

EXAMPLE 3.35 Use Parseval's theorem to evaluate

$$\chi = \sum_{n=-\infty}^{\infty} \frac{\sin^2(Wn)}{\pi^2 n^2}$$

Solution: Let

$$x[n] = \frac{\sin(Wn)}{\pi n}$$

so that $\chi = \sum_{n=-\infty}^{\infty} |x[n]|^2$. By Parseval's theorem, we have

$$\chi = \frac{1}{2\pi} \int_{\langle 2\pi \rangle} |X(e^{j\Omega})|^2 \, d\Omega$$

Since

$$x[n] \xleftrightarrow{\text{DTFT}} X(e^{j\Omega}) = \begin{cases} 1, & |\Omega| \leq W \\ 0, & W < |\Omega| < \pi \end{cases}$$

we have

$$\chi = \frac{1}{2\pi} \int_{-W}^{W} 1 \, d\Omega$$

$$= \frac{W}{\pi}$$

Note that direct calculation of χ is very difficult.

▶ **Drill Problem 3.21** Use Parseval's theorem to evaluate

$$\chi = \int_{-\infty}^{\infty} \frac{2}{|j\omega + 2|^2} \, d\omega$$

Answer:

$$\chi = \pi$$ ◀

■ DUALITY

Throughout this chapter, we have observed a consistent symmetry between the time- and frequency-domain representations of signals. For example, a continuous rectangular pulse in either time or frequency corresponds to a sinc function in either frequency or time, as illustrated in Fig. 3.42. An impulse in time transforms to a constant in frequency, while a constant in time transforms to an impulse in frequency. We have also observed symmetries in Fourier representation properties: convolution in one domain corresponds to modulation in the other domain, differentiation in one domain corresponds to multiplication by the independent variable in the other domain, and so on. These symmetries are a consequence of the symmetry in the definitions of time- and frequency-domain representations. If we are careful, we may interchange time and frequency. This interchangeability property is termed *duality*.

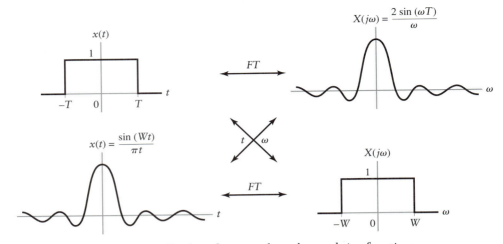

FIGURE 3.42   Duality of rectangular pulses and sinc functions.

Begin with the FT, and recall

$$x(t) = \frac{1}{2\pi} \int_{-\infty}^{\infty} X(j\omega)e^{j\omega t} \, d\omega$$

$$X(j\omega) = \int_{-\infty}^{\infty} x(t)e^{-j\omega t} \, dt$$

The difference between the expression for $x(t)$ and that for $X(j\omega)$ is the factor 2π and the sign change in the complex sinusoid. Both can be expressed in terms of the general equation

$$y(\nu) = \frac{1}{2\pi} \int_{-\infty}^{\infty} z(\eta)e^{j\nu\eta} \, d\eta \tag{3.47}$$

If we choose $\nu = t$ and $\eta = \omega$, then Eq. (3.47) implies that

$$y(t) = \frac{1}{2\pi} \int_{-\infty}^{\infty} z(\omega)e^{j\omega t} \, d\omega$$

Therefore we conclude that

$$y(t) \xleftrightarrow{\ FT\ } z(\omega) \tag{3.48}$$

Conversely, if we interchange the roles of time and frequency by setting $\nu = -\omega$ and $\eta = t$, then Eq. (3.47) implies that

$$y(-\omega) = \frac{1}{2\pi} \int_{-\infty}^{\infty} z(t)e^{-j\omega t} \, dt$$

and we have

$$z(t) \xleftrightarrow{\ FT\ } 2\pi y(-\omega) \tag{3.49}$$

The relationships of Eqs. (3.48) and (3.49) imply a certain symmetry between the roles of time and frequency. Specifically, if we are given a FT pair,

$$f(t) \xleftrightarrow{\ FT\ } F(j\omega) \tag{3.50}$$

we may interchange the roles of time and frequency to obtain the new FT pair,

$$F(jt) \xleftrightarrow{\text{FT}} 2\pi f(-\omega) \tag{3.51}$$

The notation $F(jt)$ implies evaluation of $F(j\omega)$ in Eq. (3.50) with frequency ω replaced by time t, while $f(-\omega)$ means we evaluate $f(t)$ as a function of reflected frequency $-\omega$. The duality relationship described by Eqs. (3.50) and (3.51) is illustrated in Fig. 3.43.

EXAMPLE 3.36   Use duality to evaluate the FT of

$$x(t) = \frac{1}{1 + jt}$$

Solution: First recognize that

$$f(t) = e^{-t}u(t) \xleftrightarrow{\text{FT}} F(j\omega) = \frac{1}{1 + j\omega}$$

Replacing ω by t we obtain

$$F(jt) = \frac{1}{1 + jt}$$

Hence we have expressed $x(t)$ as $F(jt)$ and using the duality property

$$F(jt) \xleftrightarrow{\text{FT}} 2\pi f(-\omega)$$

which indicates that

$$X(j\omega) = 2\pi f(-\omega)$$
$$= 2\pi e^{\omega}u(-\omega)$$

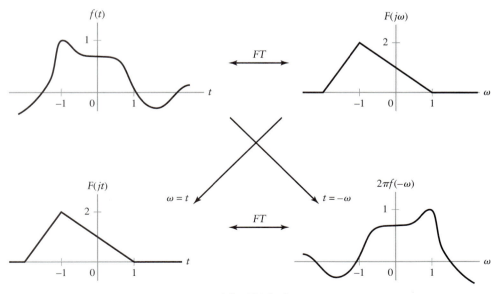

FIGURE 3.43   The FT duality property.

▶ **Drill Problem 3.22**   Use duality to evaluate the inverse FT of the step function in frequency, $X(j\omega) = u(\omega)$.

Answer:

$$x(t) = \frac{-1}{2\pi jt} + \frac{\delta(-t)}{2}$$ ◀

The FT stays entirely within its signal class; it maps a continuous-time nonperiodic function into a continuous-frequency nonperiodic function. The DTFS also stays entirely within its signal class, since discrete periodic functions are mapped into discrete periodic functions. The DTFS possesses a duality property analogous to the FT. Recall that

$$x[n] = \sum_{k=\langle N \rangle} X[k]e^{jk\Omega_o n}$$

and

$$X[k] = \frac{1}{N} \sum_{n=\langle N \rangle} x[n]e^{-jk\Omega_o n}$$

Here the difference between the forms of the forward and inverse transforms is the factor N and the change in sign of the complex sinusoidal frequencies. The DTFS duality property is stated as follows. If

$$x[n] \xleftrightarrow{\ DTFS;\ 2\pi/N\ } X[k] \tag{3.52}$$

then

$$X[n] \xleftrightarrow{\ DTFS;\ 2\pi/N\ } \frac{1}{N}x[-k] \tag{3.53}$$

Here n is the time index and k is the frequency index. The notation $X[n]$ implies evaluation of $X[k]$ in Eq. (3.52) as a function of the time index n, while $x[-k]$ means we evaluate $x[n]$ in Eq. (3.52) as a function of frequency index $-k$.

The DTFT and FS do not stay within their signal class. The FS maps a continuous periodic function into a discrete nonperiodic function. Conversely, the DTFT maps a discrete nonperiodic function into a continuous periodic function. Compare the FS expansion of a periodic continuous-time signal $z(t)$, as shown by

$$z(t) = \sum_{k=-\infty}^{\infty} Z[k]e^{jk\omega_o t}$$

and the DTFT of a nonperiodic discrete-time signal $x[n]$, as shown by

$$X(e^{j\Omega}) = \sum_{n=-\infty}^{\infty} x[n]e^{-j\Omega n}$$

If we force $z(t)$ to have the same period as $X(e^{j\Omega})$, 2π, then $\omega_o = 1$ and we see that Ω in the DTFT corresponds to t in the FS, while n in the DTFT corresponds to $-k$ in the FS. Similarly, the expression for the FS coefficients parallels the expression for the DTFT representation of $x[n]$, as we now show:

$$Z[k] = \frac{1}{2\pi} \int_{\langle 2\pi \rangle} z(t)e^{-jkt}\,dt$$

$$x[n] = \frac{1}{2\pi} \int_{\langle 2\pi \rangle} X(e^{j\Omega})e^{jn\Omega}\,d\Omega$$

The roles of Ω and n in the DTFT again correspond to those of t and $-k$ in the FS. The duality property here is between the FS and the DTFT. If

$$x[n] \xleftrightarrow{\ DTFT\ } X(e^{j\Omega}) \tag{3.54}$$

then

$$X(e^{jt}) \xleftrightarrow{\ FS;\ 1\ } x[-k] \tag{3.55}$$

EXAMPLE 3.37   Use the FS–DTFT duality property and the results of Example 3.26 to determine the inverse DTFT of the triangular spectrum $Y(e^{j\Omega})$ depicted in Fig. 3.44(a).

Solution: Define a time function $z(t) = Y(e^{jt})$. The duality property of Eq. (3.55) implies that if $z(t) \xleftrightarrow{\ FS;\ 1\ } Z[k]$, then $y[n] = Z[-n]$. Hence we seek the FS coefficients, $Z[k]$, associated with $z(t)$. $z(t)$ is a time-shifted version of the triangular wave $f(t)$ considered in Example 3.26 assuming $T = 2\pi$. Specifically, $z(t) = f(t + \pi/2)$. Using the time-shift property we have

$$Z[k] = e^{jk\pi/2}F[k]$$

$$= \begin{cases} \pi, & k = 0 \\ e^{j(k-1)\pi/2}\dfrac{4\sin(k\pi/2)}{\pi k^2}, & k \neq 0 \end{cases}$$

Consequently, using $y[n] = Z[-n]$, we have

$$y[n] = \begin{cases} \pi, & n = 0 \\ e^{-j(n+1)\pi/2}\dfrac{-4\sin(n\pi/2)}{\pi n^2}, & n \neq 0 \end{cases}$$

Figure 3.44(b) depicts $y[n]$.

■ TIME–BANDWIDTH PRODUCT

We have observed an inverse relationship between the time and frequency extent of a signal. For example, recall that

$$x(t) = \begin{cases} 1, & |t| \leq T \\ 0, & |t| > T \end{cases} \xleftrightarrow{\ FT\ } X(j\omega) = 2T\text{sinc}\left(\frac{\omega T}{\pi}\right)$$

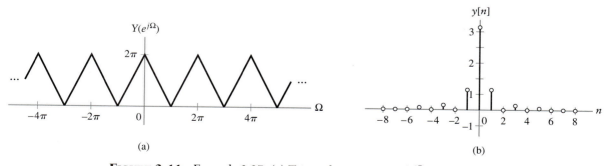

FIGURE 3.44   Example 3.37. (a) Triangular spectrum $Y(e^{j\Omega})$. (b) Inverse DTFT of $Y(e^{j\Omega})$.

As depicted in Fig. 3.45, the signal $x(t)$ has time extent $2T$. Its FT, $X(j\omega)$, is actually of infinite extent in frequency but has the majority of its energy contained in the interval associated with the mainlobe of the sinc function, $|\omega| < \pi/T$. As T decreases, the signal's time extent decreases, while the frequency extent increases. In fact, the product of the time extent and mainlobe width is a constant.

The general nature of the inverse relationship between time and frequency extent is demonstrated by the scaling property. Compressing a signal in time leads to expansion in the frequency domain and vice versa. This inverse relationship may formally be stated in terms of the signal's time–bandwidth product.

The *bandwidth* of a signal refers to the significant frequency content of the signal for positive frequencies. It is difficult to define bandwidth, especially for signals having infinite frequency extent, because the meaning of the term "significant" is not mathematically precise. In spite of this difficulty, there are several definitions for bandwidth in common use. One such definition applies to signals that have a frequency-domain representation characterized by a mainlobe bounded by nulls. If the signal is *lowpass*, that is, the mainlobe is centered on the origin, then the bandwidth is defined as one-half the mainlobe width. Using this definition, the signal depicted in Fig. 3.45 has bandwidth π/T. If the signal is *bandpass*, meaning the mainlobe is centered on $\pm\omega_c$, then the bandwidth is equal to the mainlobe width. Another commonly used definition of bandwidth is based on the frequency at which the magnitude spectrum is $1/\sqrt{2}$ times its peak value. At this frequency the energy spectrum has a value of one-half its peak value. Note that similar difficulty is encountered in precisely defining the time extent or duration of a signal.

The above definitions of bandwidth and duration are not well suited for analytic evaluation. We may analytically describe the inverse relationship between the duration and bandwidth of arbitrary signals by defining root mean-square measures of effective duration and bandwidth. We shall formally define the duration of a signal $x(t)$ as

$$T_d = \left[\frac{\int_{-\infty}^{\infty} t^2 |x(t)|^2 \, dt}{\int_{-\infty}^{\infty} |x(t)|^2 \, dt} \right]^{1/2} \tag{3.56}$$

and the bandwidth as

$$B_w = \left[\frac{\int_{-\infty}^{\infty} \omega^2 |X(j\omega)|^2 \, d\omega}{\int_{-\infty}^{\infty} |X(j\omega)|^2 \, d\omega} \right]^{1/2} \tag{3.57}$$

These definitions assume $x(t)$ is centered about the origin and is lowpass. The interpretation of T_d as an effective duration follows from examination of Eq. (3.56). The integral in the numerator is the second moment of the signal about the origin. The integrand weights the square of the value of $x(t)$ at each time instant by the square of its distance from $t = 0$. Hence if $x(t)$ is large for large values of t, then the duration will be larger than if $x(t)$ is large for small values of t. This integral is normalized by the total energy in $x(t)$. A similar interpretation applies to B_w. Note that while the root mean-square definitions offer certain

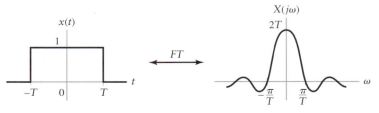

FIGURE 3.45   Rectangular pulse illustrating the inverse relationship between the time and frequency extent of a signal.

analytic tractability, they are not easily measured from a given signal and its magnitude spectrum.

The time–bandwidth product for any signal is lower bounded, as shown by

$$T_d B_w \geq \tfrac{1}{2} \tag{3.58}$$

This bound indicates that we cannot simultaneously decrease the duration and bandwidth of a signal. Gaussian pulses are the only signals that satisfy this relationship with equality. Equation (3.58) is also known as the *uncertainty principle* for its application in modern physics, which states that the exact position and exact momentum of an electron cannot be determined simultaneously. This result generalizes to alternative definitions of bandwidth and duration. The product of bandwidth and duration is always lower bounded by a constant, with the value of this constant dependent on the definitions of bandwidth and duration.

Analogous results can be derived for the other Fourier representations.

3.7 *Exploring Concepts with MATLAB*

■ THE DTFS

The DTFS is the only Fourier representation that is discrete valued in both time and frequency and hence suited for direct MATLAB implementation. While Eqs. (3.12) and (3.13) are easily implemented as M-files, the built-in MATLAB commands f f t and i f f t may also be used to evaluate the DTFS. Given a length N vector x representing one period of an N periodic signal $x[n]$, the command

```
≫ X = fft(x)/N
```

produces a length N vector X containing the DTFS coefficients, $X[k]$. MATLAB assumes the summations in Eqs. (3.12) and (3.13) run from 0 to $N - 1$, so the first elements of x and X correspond to $x[0]$ and $X[0]$, respectively, while the last elements correspond to $x[N - 1]$ and $X[N - 1]$. Note that division by N is necessary because f f t evaluates the sum in Eq. (3.13) without dividing by N. Similarly, given DTFS coefficients in a vector X, the command

```
≫ x = ifft(X)*N
```

produces a vector x that represents one period for the time waveform. Note that i f f t must be multiplied by N to evaluate Eq. (3.12). Both f f t and i f f t are computed using a numerically efficient or *fast* algorithm termed the *fast Fourier transform*. Development of this algorithm is discussed in Section 4.11.

Consider using MATLAB to solve Drill Problem 3.1 for the DTFS coefficients. The signal is

$$x[n] = 1 + \sin\left(\frac{\pi}{12} n + \frac{3\pi}{8}\right)$$

This signal has period 24, so we define one period and evaluate the DTFS coefficients using the commands

```
≫ x = ones(1,24) + sin([0:23]*pi/12 + 3*pi/8);
≫ X = fft(x)/24
X =
Columns 1 through 4
1.0000   0.4619 - 0.1913i   0.0000 + 0.0000i
-0.0000 + 0.0000i
```

```
Columns 5 through 8
0.0000 = 0.0000i  -0.0000 - 0.0000i  0.0000 - 0.0000i
-0.0000 - 0.0000i
Columns 9 through 12
-0.0000 - 0.0000i  -0.0000 - 0.0000i  -0.0000 - 0.0000i
0.0000 - 0.0000i
Columns 13 through 16
0.0000 + 0.0000i  0.0000 + 0.0000i  -0.0000 + 0.0000i
0.0000 - 0.0000i
Columns 17 through 20
-0.0000 - 0.0000i  -0.0000 - 0.0000i  0.0000 + 0.0000i
-0.0000 + 0.0000i
Columns 21 through 24
-0.0000 + 0.0000i  -0.0000 - 0.0000i  0.0000 - 0.0000i
0.4619 + 0.1913i
```

Note that MATLAB uses i to denote the square root of -1. Hence we conclude that

$$X[k] = \begin{cases} 1, & k = 0 \\ 0.4619 - j0.1913, & k = 1 \\ 0.4619 + j0.1913, & k = 23 \\ 0 & \text{otherwise on } 0 \le k \le 23 \end{cases}$$

which corresponds to the answer to Drill Problem 3.1 expressed in rectangular form. We may reconstruct the time-domain signal using **ifft** and evaluate the first four values of the reconstructed signal using the commands

```
≫ xrecon = ifft(X)*24;
≫ xrecon(1:4);
ans =
1.9239 - 0.0000i  1.9914 + 0.0000i  1.9914 + 0.0000i
1.9239 - 0.0000i
```

Note that the reconstructed signal has an imaginary component, albeit a very small one, even though the original signal was purely real. The imaginary component is an artifact of numerical rounding errors in the computations performed by **fft** and **ifft** and may be ignored.

▶ **Drill Problem 3.23**   Repeat Drill Problem 3.2 using MATLAB.   ◀

The partial sum approximation used in Example 3.3 is easily evaluated using MATLAB as follows:

```
≫ k = 1:24;
≫ n = -24:25;
≫ B(1) = 25/50;   %coeff for k = 0
≫ B(2:25) = 2*sin(k*pi*25/50)./(50*sin(k*pi/50));
≫ B(26) = sin(25*pi*25/50)/(50*sin(25*pi/50));
≫ xJhat(1,:) = B(1)*cos(n*0*pi/25); %term in sum for k = 0
%accumulate partial sums
≫ for k = 2:26
xJhat(k,:) = xJhat(k-1,:) + B(k)*cos(n*(k-1)*pi/25);
end
```

This set of commands produces a matrix **xJhat** whose $(J + 1)$st row corresponds to $\hat{x}_J[n]$.

■ THE FS

The partial sum approximation to the FS in Example 3.7 is evaluated analogously to that of the DTFS with one important additional consideration. The signal $\hat{x}_J(t)$ and the cosines in the partial sum approximation are continuous functions of time. Since MATLAB represents these functions as vectors consisting of discrete points, we must use sufficiently closely spaced samples to capture the details in $\hat{x}_J(t)$. This is assured by sampling the functions closely enough so that the highest frequency term in the sum, $\cos(J_{\max}\omega_o t)$, is well approximated by the sampled signal, $\cos(J_{\max}\omega_o n\Delta)$. The sampled cosine provides a visually pleasing approximation to the continuous cosine using MATLAB's `plot` command if there are on the order of 20 samples per period. Using 20 samples per period, we obtain $\Delta = T/20J_{\max}$. Note that the total number of samples in one period is then $20J_{\max}$. Assuming $J_{\max} = 99$ and $T = 1$, we may compute the partial sums given $B[k]$ by using the commands:

```
≫ t = [-(10*Jmax-1):10*Jmax]*Delta;
≫ xJhat(1,:) = B(1)*cos(t*0*2*pi/T);
≫ for k = 2:100
xJhat(k,:) = xJhat(k-1,:) + B(k)*cos(t*(k-1)*2*pi/T);
end
```

Since the rows of `xJhat` represent samples of a continuous-valued function, we display them using `plot` instead of `stem`. For example, the partial sum for $J = 5$ is displayed with the command `plot(t,xJhat(6,:))`.

■ TIME–BANDWIDTH PRODUCT

The `fft` command may be used to evaluate the DTFS and explore the time–bandwidth product property for discrete-time periodic signals. Since the DTFS applies to signals that are periodic in both time and frequency, we define both duration and bandwidth based on the extent of the signal within one period. For example, consider the period N square wave studied in Example 3.2. One period of the time-domain signal is defined as

$$x[n] = \begin{cases} 1, & |n| \le M \\ 0, & \text{otherwise on } -M < n < N - M - 1 \end{cases}$$

and the DTFS coefficients are given by

$$X[k] = \frac{1}{N} \frac{\sin\left(k\frac{\pi}{N}(2M+1)\right)}{\sin\left(k\frac{\pi}{N}\right)}$$

If we define the time duration T_d as the nonzero portion of one period, then $T_d = 2M + 1$. If we further define the bandwidth B_w as the width of the mainlobe of $X[k]$ over one period, then we have $B_w \approx 2N/(2M + 1)$ and we see that the time–bandwidth product for the square wave is independent of M: $T_d B_w \approx 2N$.

The following set of MATLAB commands may be used to verify this result:

```
≫ x = [ones(1,M+1), zeros(1,N-2*M-1), ones(1,M)];
≫ X = fft(x)/N;
≫ k = [-N/2+1:N/2];  %frequency index for N even
≫ stem(k,real(fftshift(X)))
```

Here we define one period of an even square wave on the interval $0 \leq n \leq N - 1$, find the DTFS coefficients using the `fft` command, and display them using `stem`. The command `fftshift` swaps the left and right halves of the vector `X` so that the zero frequency index is in the center of the vector. That is, the frequency indices are changed from `0` to `N − 1` to `−N/2 + 1` to `N/2` (assuming `N` is even). The `real` command is used to suppress any small imaginary components resulting from numerical rounding. We then determine the effective bandwidth by counting the number of DTFS coefficients in the mainlobe. One of the computer experiments evaluates the time–bandwidth product in this fashion.

The formal definitions of duration and bandwidth given in Eqs. (3.56) and (3.57) may be generalized to discrete-time periodic signals by replacing the integrals with sums over one period as follows:

$$
T_d = \left[\frac{\displaystyle\sum_{n=-(N-1)/2}^{(N-1)/2} n^2 |x[n]|^2}{\displaystyle\sum_{n=-(N-1)/2}^{(N-1)/2} |x[n]|^2} \right]^{1/2}
\tag{3.59}
$$

$$
B_w = \left[\frac{\displaystyle\sum_{k=-(N-1)/2}^{(N-1)/2} k^2 |X[k]|^2}{\displaystyle\sum_{k=-(N-1)/2}^{(N-1)/2} |X[k]|^2} \right]^{1/2}
\tag{3.60}
$$

Here we assume that N is odd and the majority of the energy in $x[n]$ and $X[k]$ is centered on the origin.

The following MATLAB function evaluates the product $T_d B_w$ based on Eqs. (3.59) and (3.60):

```
function TBP = TdBw(x)
% Compute the Time-Bandwidth product using the DTFS
% One period must be less than 1025 points
N=1025;
M = (N - max(size(x)))/2;
xc = [zeros(1,M),x,zeros(1,M)]; %center pulse within a period
n = [-(N-1)/2:(N-1)/2];
n2 = n.*n;
Td = sqrt((xc.*xc)*n2'/(xc*xc'));
X = fftshift(fft(xc)/N);   %evaluates DTFS and centers
Bw = sqrt(real((X.*conj(X))*n2'/(X*X')));
TBP = Td*Bw;
```

This function assumes that the length of the input signal `x` is odd and centers `x` within a 1025-point period before computing `Td` and `Bw`. Note that `.*` is used to perform the element-by-element product. The `*` operation computes the inner product when placed between a row vector and a column vector and the apostrophe `'` indicates complex conjugate transpose. Hence the command `X*X'` performs the inner product of `X` and the complex conjugate of `X`, that is, the sum of the magnitude squared of each element of `X`.

We may use the function `TdBw` to evaluate the time–bandwidth product for two rectangular, raised cosine, and Gaussian pulse trains as follows:

```
>> x = ones(1,101);   % 101 point rectangular pulse
>> TdBw(x)
ans =
788.0303
>> x = ones(1,301);   % 301 point rectangular pulse
>> TdBw(x)
ans =
1.3604e+03
>> x = 0.5*ones(1,101) + cos(2*pi*[-50:50]/101);
% 101 point raised cosine
>> TdBw(x)
ans =
277.7327
>> x = 0.5*ones(1,301) + cos(2*pi*[-150:150]/301);
% 301 point raised cosine
>> TdBw(x)
ans =
443.0992
>> n = [-500:500];
>> x = exp(-0.001*(n.*n));   % narrow Gaussian pulse
>> TdBw(x)
ans =
81.5669
>> x = exp(-0.0001*(n.*n));   % broad Gaussian pulse
>> TdBw(x)
ans =
81.5669
```

Note that the Gaussian pulse trains have the smallest time–bandwidth product. Furthermore, the time–bandwidth product is identical for both the narrow and broad Gaussian pulse trains. These observations offer evidence that the time–bandwidth product for periodic discrete-time signals is lower bounded by that of a Gaussian pulse train. Such a result would not be too surprising given that the Gaussian pulses attain the lower bound for continuous-time nonperiodic signals. This issue is revisited as a computer experiment in Chapter 4.

3.8 *Summary*

In this chapter we have developed techniques for representing signals as weighted superpositions of complex sinusoids. The weights are a function of the complex sinusoidal frequencies and provide a frequency-domain description of the signal. There are four distinct representations applicable to four different signal classes.

▶ The DTFS applies to discrete-time N periodic signals and represents the signal as a weighted sum of N discrete-time complex sinusoids whose frequencies are integer multiples of the fundamental frequency of the signal. This frequency-domain representation is a discrete and N periodic function of frequency. The DTFS is the only Fourier representation that can be computed numerically.

▶ The FS applies to continuous-time periodic signals. It represents the signal as a weighted sum of an infinite number of continuous-time complex sinusoids whose frequencies are integer multiples of the signal's fundamental frequency. Here the frequency-domain representation is a discrete and nonperiodic function of frequency.

▶ The DTFT represents nonperiodic discrete-time signals as a weighted integral of discrete-time complex sinusoids whose frequencies vary continuously over a 2π interval. This frequency-domain representation is a continuous and 2π periodic function of frequency.

▶ The FT represents nonperiodic continuous-time signals as a weighted integral of continuous-time complex sinusoids whose frequencies vary continuously from $-\infty$ to ∞. Here the frequency-domain representation is a continuous and nonperiodic function of frequency.

Fourier representation properties relate the effect of an action on the time-domain signal to the corresponding change in the frequency-domain representation. They are a consequence of the properties of complex sinusoids. Since all four representations employ complex sinusoids, all four share similar properties. The properties provide insight into the nature of both time- and frequency-domain signal representations. They also provide a powerful set of tools for manipulating signals in both the time and frequency domain. Often it is much simpler to use the properties to determine a time- or frequency-domain signal representation than it is to use the defining equation.

The frequency domain offers an alternative perspective of signals and the systems with which they interact. Certain signal characteristics are more easily identified in the frequency domain than in the time domain, and vice versa. Also, some systems problems are more easily solved in the frequency domain than in the time domain, and vice versa. Both the time- and frequency-domain representations have their own advantages and disadvantages. Where one may excel, the other may be very cumbersome. Determining which domain is the most advantageous for solving a particular problem is an important skill to develop and can only be done through experience. The next chapter offers a start on this journey by examining applications of Fourier representations in signal and systems problems.

FURTHER READING

1. Joseph Fourier studied the flow of heat in the early nineteenth century. Understanding heat flow was a problem of both practical and scientific significance at that time and required solving a partial differential equation called the heat equation. Fourier developed a technique for solving partial differential equations based on assuming the solution was a weighted sum of harmonically related sinusoids with unknown coefficients, which we now term the Fourier series. Fourier's initial work on heat conduction was submitted as a paper to the Academy of Sciences of Paris in 1807 and rejected after review by Lagrange, Laplace, and Legendre. Fourier persisted in developing his ideas in spite of being criticized for a lack of rigor by his contemporaries. Eventually, in 1822, he published a book containing much of his work, *Theorie analytique de la chaleur*, which is now regarded as one of the classics of mathematics.

2. The DTFS is most often referred to in the literature as the discrete Fourier transform or DFT. We have adopted the DTFS terminology in this text because it is more descriptive and less likely to lead to confusion with the DTFT. The reader should be aware that they will likely encounter the DFT terminology in other texts and references.

3. A general treatment of Fourier transforms is offered in:
   ▸ Bracewell, R. N., *The Fourier Transform and Its Applications*, Second Edition (McGraw-Hill, 1978)
   ▸ Papoulis, A., *The Fourier Integral and Its Applications* (McGraw-Hill, 1962)

4. An introductory treatment of the FS, FT, and DTFS is offered in:
   ▸ Morrison, N., *Introduction to Fourier Analysis* (Wiley, 1994)

5. The role of the FS and FT in solving the heat equation, wave equation, and potential equation is described in:
   ▸ Powers, D. L., *Boundary Value Problems*, Second Edition (Academic Press, 1979)

6. The uncertainty principle, Eq. (3.58), is proved in:
   ▸ Bracewell, R. N., *The Fourier Transform and Its Applications*, Second Edition (McGraw-Hill, 1978)

▌ PROBLEMS

3.1 Use the defining equation for the DTFS coefficients to evaluate the DTFS coefficients for the following signals. Sketch the magnitude and phase spectra.

(a) $x[n] = \cos\left(\dfrac{6\pi}{13} n + \dfrac{\pi}{6}\right)$

(b) $x[n] = \sin\left(\dfrac{4\pi}{21} n\right) + \cos\left(\dfrac{10\pi}{21} n\right) + 1$

(c) $x[n] = \sum_{m=-\infty}^{\infty} \delta[n - 2m] + \delta[n + 3m]$

(d) $x[n]$ as depicted in Fig. P3.1(a)

(e) $x[n]$ as depicted in Fig. P3.1(b)

(f) $x[n]$ as depicted in Fig. P3.1(c)

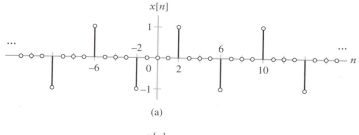

(a)

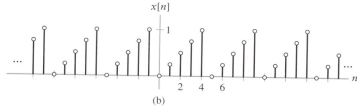

(b)

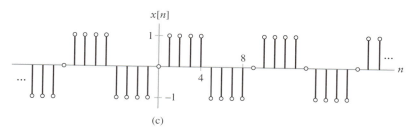

(c)

FIGURE P3.1

3.2 Use the definition of the DTFS to determine the time signals represented by the following DTFS coefficients.

(a) $X[k] = \cos\left(\frac{6\pi}{17} k\right)$

(b) $X[k] = \cos\left(\frac{10\pi}{21} k\right) + j \sin\left(\frac{4\pi}{21} k\right)$

(c) $X[k] = \sum_{m=-\infty}^{\infty} \delta[k - 2m] - 2\delta[k + 3m]$

(d) $X[k]$ as depicted in Fig. P3.2(a)

(e) $X[k]$ as depicted in Fig. P3.2(b)

(f) $X[k]$ as depicted in Fig. P3.2(c)

3.3 Use the defining equation for the FS coefficients to evaluate the FS coefficients for the following signals. Sketch the magnitude and phase spectra.

(a) $x(t) = \sin(2\pi t) + \cos(3\pi t)$

(b) $x(t) = \sum_{m=-\infty}^{\infty} \delta(t - \frac{1}{2}m) + \delta(t - \frac{3}{2}m)$

(c) $x(t) = \sum_{m=-\infty}^{\infty} e^{j(2\pi/3)m}\delta(t - 2m)$

(d) $x(t)$ as depicted in Fig. P3.3(a)

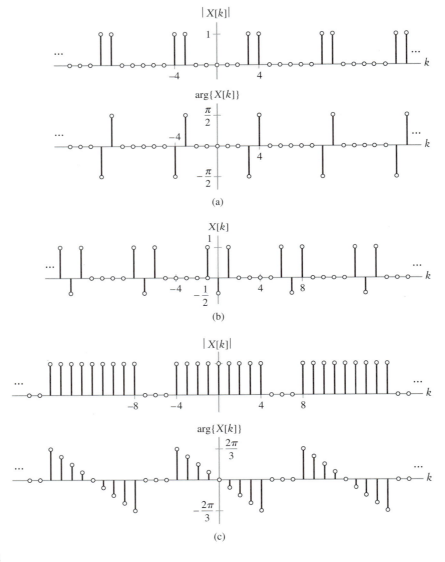

(a)

(b)

(c)

FIGURE P3.2

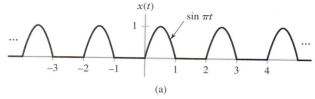

(a)

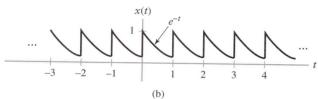

(b)

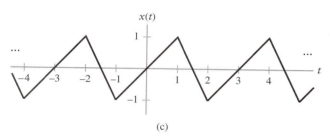

(c)

FIGURE P3.3

(e) $x(t)$ as depicted in Fig. P3.3(b)

(f) $x(t)$ as depicted in Fig. P3.3(c)

3.4 Use the definition of the FS to determine the time signals represented by the following FS coefficients.

(a) $X[k] = j\delta[k - 1] - j\delta[k + 1] + \delta[k - 3] + \delta[k + 3], \quad \omega_o = \pi$

(b) $X[k] = j\delta[k - 1] - j\delta[k + 1] + \delta[k - 3] + \delta[k + 3], \quad \omega_o = 3\pi$

(c) $X[k] = (-\frac{1}{2})^{|k|}, \quad \omega_o = 1$

(d) $X[k]$ as depicted in Fig. P3.4(a), $\omega_o = \pi$

(e) $X[k]$ as depicted in Fig. P3.4(b), $\omega_o = 2\pi$

(f) $X[k]$ as depicted in Fig. 3.4(c), $\omega_o = \pi$

3.5 Use the defining equation for the DTFT to evaluate the frequency-domain representations for the following signals. Sketch the magnitude and phase spectra.

(a) $x[n] = (\frac{1}{2})^n u[n - 4]$

(b) $x[n] = a^{|n|}, \quad |a| < 1$

(c) $x[n] = \begin{cases} \frac{1}{2} + \frac{1}{2} \cos\left(\frac{\pi}{N} n\right), & |n| \leq N \\ 0, & \text{otherwise} \end{cases}$

(d) $x[n] = \delta[6 - 3n]$

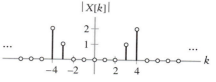

(a)

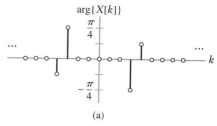

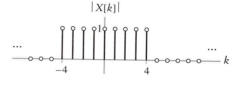

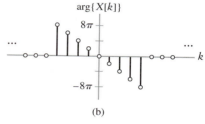

(b)

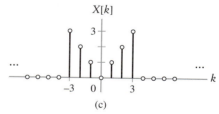

(c)

FIGURE P3.4

(e) $x[n]$ as depicted in Fig. P3.5(a)

(f) $x[n]$ as depicted in Fig. P3.5(b)

3.6 Use the equation describing the DTFT representation to determine the time-domain signals corresponding to the following DTFTs.

(a) $X(e^{j\Omega}) = \cos(\Omega) + j \sin(\Omega)$

(b) $X(e^{j\Omega}) = \sin(\Omega) + \cos(\Omega/2)$

(c) $|X(e^{j\Omega})| = \begin{cases} 1, & \pi/2 < |\Omega| < \pi \\ 0, & \text{otherwise} \end{cases}$
$\arg\{X(e^{j\Omega})\} = -4\Omega$

(d) $X(e^{j\Omega})$ as depicted in Fig. P3.6(a)

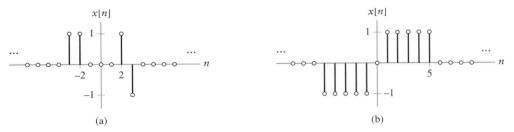

FIGURE P3.5

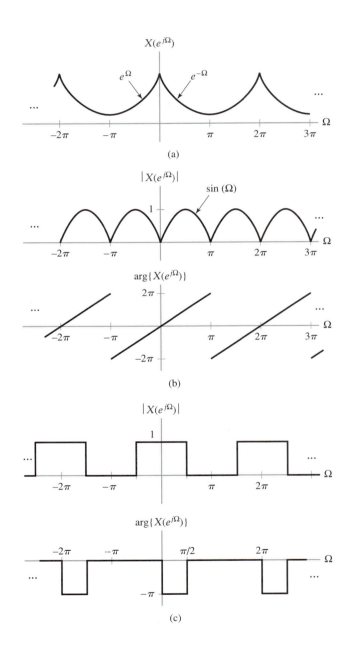

FIGURE P3.6

(e) $X(e^{j\Omega})$ as depicted in Fig. P3.6(b)

(f) $X(e^{j\Omega})$ as depicted in Fig. P3.6(c)

3.7 Use the defining equation for the FT to evaluate the frequency-domain representations for the following signals. Sketch the magnitude and phase spectra.

(a) $x(t) = e^{-3t}u(t-1)$

(b) $x(t) = e^{-|t|}$

(c) $x(t) = te^{-2t}u(t)$

(d) $x(t) = \sum_{m=0}^{\infty} a^m \delta(t-m)$, $|a| < 1$

(e) $x(t)$ as depicted in Fig. P3.7(a)

(f) $x(t)$ as depicted in Fig. P3.7(b)

3.8 Use the equation describing the FT representation to determine the time-domain signals corresponding to the following FTs.

(a) $X(j\omega) = \begin{cases} \cos(\omega), & |\omega| < \pi/2 \\ 0, & \text{otherwise} \end{cases}$

(b) $X(j\omega) = e^{-2\omega}u(\omega)$

(c) $X(j\omega) = e^{-|\omega|}$

(d) $X(j\omega)$ as depicted in Fig. P3.8(a)

(e) $X(j\omega)$ as depicted in Fig. P3.8(b)

(f) $X(j\omega)$ as depicted in Fig. P3.8(c)

3.9 Determine the appropriate Fourier representation for the following time-domain signals using the defining equations.

(a) $x(t) = e^{-3t}\cos(\pi t)u(t)$

(b) $x[n] = \begin{cases} \cos\left(\dfrac{\pi}{5}n\right) + j\sin\left(\dfrac{\pi}{5}n\right), & |n| < 10 \\ 0, & \text{otherwise} \end{cases}$

(c) $x[n]$ as depicted in Fig. P3.9(a)

(d) $x(t) = e^{1+t}u(-t+2)$

(e) $x(t) = |\sin(2\pi t)|$

(f) $x[n]$ as depicted in Fig. P3.9(b)

(g) $x(t)$ as depicted in Fig. P3.9(c)

3.10 The following are frequency-domain representations for signals. Determine the time-domain signal corresponding to each.

(a) $X[k] = \begin{cases} e^{-jk\pi}, & |k| < 10 \\ 0, & \text{otherwise} \end{cases}$

Fundamental period of time domain signal is $T = 1$

(b) $X[k]$ as depicted in Fig. P3.10(a)

(c) $X(j\omega) = \begin{cases} \cos(\omega/2) + j\sin(\omega/2), & |\omega| < \pi \\ 0, & \text{otherwise} \end{cases}$

(d) $X(j\omega)$ as depicted in Fig. P3.10(b)

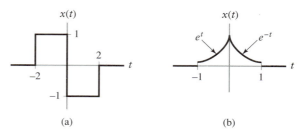

FIGURE P3.7

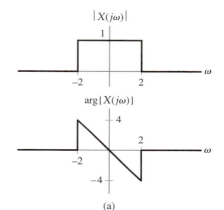

(a)

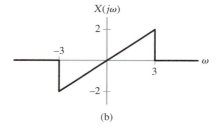

(b)

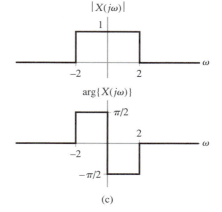

(c)

FIGURE P3.8

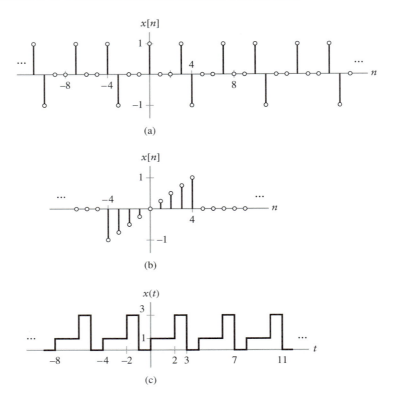

FIGURE P3.9

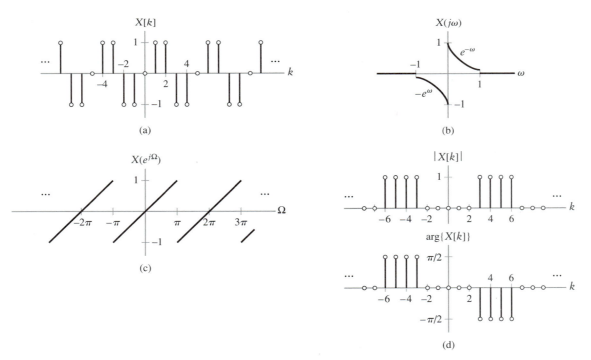

FIGURE P3.10

(e) $X(e^{j\Omega})$ as depicted in Fig. P3.10(c)

(f) $X[k]$ as depicted in Fig. P3.10(d)

(g) $X(e^{j\Omega}) = |\sin(\Omega)|$

3.11 Use partial fraction expansions to determine the inverse FT for the following signals:

(a) $X(j\omega) = \dfrac{5j\omega + 12}{(j\omega)^2 + 5j\omega + 6}$

(b) $X(j\omega) = \dfrac{4}{-\omega^2 + 4j\omega + 3}$

(c) $X(j\omega) = \dfrac{-j\omega}{(j\omega)^2 + 3j\omega + 2}$

(d) $X(j\omega) = \dfrac{-(j\omega)^2 - 4j\omega - 6}{[(j\omega)^2 + 3j\omega + 2](j\omega + 4)}$

(e) $X(j\omega) = \dfrac{2(j\omega)^2 + 12j\omega + 14}{(j\omega)^2 + 6j\omega + 5}$

(f) $X(j\omega) = \dfrac{2j\omega + 1}{(j\omega + 2)^2}$

3.12 Use partial fraction expansions to determine the inverse DTFT for the following signals:

(a) $X(e^{j\Omega}) = \dfrac{3 - \frac{1}{4}e^{-j\Omega}}{-\frac{1}{16}e^{-j2\Omega} + 1}$

(b) $X(e^{j\Omega}) = \dfrac{3 - \frac{5}{4}e^{-j\Omega}}{\frac{1}{8}e^{-j2\Omega} - \frac{3}{4}e^{-j\Omega} + 1}$

(c) $X(e^{j\Omega}) = \dfrac{6}{e^{-j2\Omega} - 5e^{-j\Omega} + 6}$

(d) $X(e^{j\Omega}) = \dfrac{6 - 2e^{-j\Omega} + \frac{1}{2}e^{-j2\Omega}}{(-\frac{1}{4}e^{-j2\Omega} + 1)(1 - \frac{1}{4}e^{-j\Omega})}$

(e) $X(e^{j\Omega}) = \dfrac{6 - \frac{2}{3}e^{-j\Omega} - \frac{1}{6}e^{-j2\Omega}}{-\frac{1}{6}e^{-j2\Omega} + \frac{1}{6}e^{-j\Omega} + 1}$

3.13 Use the tables of transforms and properties to find the FTs of the following signals:

(a) $x(t) = \sin(\pi t)e^{-2t}u(t)$

(b) $x(t) = e^{-3|t-2|}$

(c) $x(t) = \left[\dfrac{2\sin(\pi t)}{\pi t}\right]\left[\dfrac{\sin(2\pi t)}{\pi t}\right]$

(d) $x(t) = \dfrac{d}{dt}(te^{-2t}\sin(t)u(t))$

(e) $x(t) = \int_{-\infty}^{t} \dfrac{\sin(\pi\tau)}{\pi\tau}\,d\tau$

(f) $x(t) = e^{-2t+1}u\left(\dfrac{t-4}{2}\right)$

(g) $x(t) = \dfrac{d}{dt}\left[\left(\dfrac{\sin(t)}{\pi t}\right) * \left(\dfrac{\sin(2t)}{\pi t}\right)\right]$

3.14 Use the tables of transforms and properties to find the inverse FTs of the following signals:

(a) $X(j\omega) = \dfrac{j\omega}{(2 + j\omega)^2}$

(b) $X(j\omega) = \dfrac{4\sin(2\omega - 2)}{2\omega - 2} + \dfrac{4\sin(2\omega + 2)}{2\omega + 2}$

(c) $X(j\omega) = \dfrac{1}{j\omega(j\omega + 1)} + 2\pi\delta(\omega)$

(d) $X(j\omega) = \dfrac{d}{d\omega}\left[4\cos(3\omega)\dfrac{\sin(2\omega)}{\omega}\right]$

(e) $X(j\omega) = \dfrac{2\sin(\omega)}{\omega(j\omega + 1)}$

(f) $X(j\omega) = \text{Im}\left\{e^{-j3\omega}\dfrac{1}{j\omega + 2}\right\}$

(g) $X(j\omega) = \dfrac{4\sin^2(\omega)}{\omega^2}$

3.15 Use the tables of transforms and properties to find the DTFTs of the following signals:

(a) $x[n] = (\frac{1}{2})^n u[n - 2]$

(b) $x[n] = (n - 2)(u[n - 5] - u[n - 6])$

(c) $x[n] = \sin\left(\dfrac{\pi}{4}n\right)\left(\dfrac{1}{4}\right)^n u[n - 1]$

(d) $x[n] = \left[\dfrac{\sin\left(\frac{\pi}{4}n\right)}{\pi n}\right] * \left[\dfrac{\sin\left(\frac{\pi}{4}(n - 2)\right)}{\pi(n - 2)}\right]$

(e) $x[n] = \left[\dfrac{\sin\left(\frac{\pi}{2}n\right)}{\pi n}\right]^2$

3.16 Use the tables of transforms and properties to find the inverse DTFTs of the following signals:

(a) $X(e^{j\Omega}) = \cos(2\Omega) + 1$

(b) $X(e^{j\Omega}) = \left[\dfrac{\sin(\frac{15}{2}\Omega)}{\sin(\Omega/2)}\right] \circledast \left[e^{-j3\Omega}\dfrac{\sin(\frac{7}{2}\Omega)}{\sin(\Omega/2)}\right]$

(c) $X(e^{j\Omega}) = \cos(\Omega)\left[\dfrac{\sin(\frac{3}{2}\Omega)}{\sin(\Omega/2)}\right]$

(d) $X(e^{j\Omega}) = \begin{cases} 1, & \pi/4 < |\Omega| < 3\pi/4 \\ 0, & \text{otherwise for } |\Omega| < \pi \end{cases}$

(e) $X(e^{j\Omega}) = e^{-j(4\Omega + \pi/2)}\dfrac{d}{d\Omega}\left[\dfrac{2}{1 + \frac{1}{4}e^{-j(\Omega - \pi/8)}} + \dfrac{2}{1 + \frac{1}{4}e^{-j(\Omega + \pi/8)}}\right]$

3.17 Use the FT pair

$$x(t) = \begin{cases} 1, & |t| \le 1 \\ 0, & \text{otherwise} \end{cases} \xleftrightarrow{\ FT\ } X(j\omega) = \frac{2\sin(\omega)}{\omega}$$

and the FT properties to evaluate the frequency-domain representations for the signals depicted in Fig. P3.17(a)–(i).

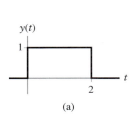

(a)

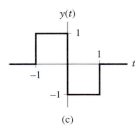

(c)

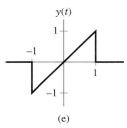

(e)

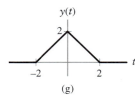

(g)

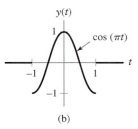

(b)

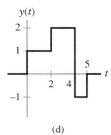

(d)

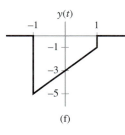

(f)

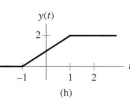

(h)

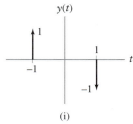

(i)

FIGURE P3.17

3.18 We have $x[n] = n(-\frac{1}{2})^n u[n] \xleftrightarrow{\ DTFT\ } X(e^{j\Omega})$. Without evaluating $X(e^{j\Omega})$, find $y[n]$ if $Y(e^{j\Omega})$ is given by:

(a) $Y(e^{j\Omega}) = e^{j3\Omega} X(e^{j\Omega})$

(b) $Y(e^{j\Omega}) = \text{Re}\{X(e^{j\Omega})\}$

(c) $Y(e^{j\Omega}) = \dfrac{d}{d\Omega} X(e^{j\Omega})$

(d) $Y(e^{j\Omega}) = X(e^{j\Omega}) \circledast X(e^{j(\Omega - \pi)})$

(e) $Y(e^{j\Omega}) = \dfrac{d}{d\Omega}[e^{j2\Omega} X(e^{j\Omega})]$

(f) $Y(e^{j\Omega}) = X(e^{j\Omega}) + X(e^{-j\Omega})$

(g) $Y(e^{j\Omega}) = \dfrac{d}{d\Omega}\{e^{-j2\Omega}[X(e^{j(\Omega + \pi/4)}) - X(e^{j(\Omega - \pi/4)})]\}$

3.19 A periodic signal has FS representation $x(t) \xleftrightarrow{\ FS;\ \pi\ } X[k] = -k2^{-|k|}$. Without determining $x(t)$, find the FS representation ($Y[k]$ and ω_o) if $y(t)$ is given by:

(a) $y(t) = x(2t)$

(b) $y(t) = \dfrac{d}{dt} x(t)$

(c) $y(t) = x(t - \frac{1}{4})$

(d) $y(t) = \text{Re}\{x(t)\}$

(e) $y(t) = \cos(2\pi t)x(t)$

(f) $y(t) = x(t) \circledast x(t - \frac{1}{2})$

3.20 Given

$$x[n] = \frac{\sin\left(\dfrac{11\pi}{20} n\right)}{\sin\left(\dfrac{\pi}{20} n\right)} \xleftrightarrow{\ DTFS;\ \pi/10\ } X[k]$$

evaluate the time-domain signal $y[n]$ with the following DTFS coefficients using only DTFS properties.

(a) $Y[k] = X[k - 5] + X[k + 5]$

(b) $Y[k] = \cos(k\pi/5) X[k]$

(c) $Y[k] = X[k] \circledast X[k]$

(d) $Y[k] = \text{Re}\{X[k]\}$

3.21 Evaluate the following quantities:

(a) $\displaystyle\int_{-\pi}^{\pi} \frac{1}{|1 - \frac{1}{4}e^{-j\Omega}|^2}\, d\Omega$

(b) $\displaystyle\sum_{k=-\infty}^{\infty} \frac{\sin^2(k\pi/4)}{k^2}$

(c) $\displaystyle\int_{-\infty}^{\infty} \frac{4}{(\omega^2 + 1)^2}\, d\omega$

(d) $\Sigma_{k=0}^{19} \dfrac{\sin^2\left(\dfrac{11\pi}{20} k\right)}{\sin^2\left(\dfrac{\pi}{20} k\right)}$

(e) $\int_{-\infty}^{\infty} \dfrac{\sin^2(\pi t)}{\pi t^2} dt$

3.22 Use the duality property of Fourier representations to evaluate the following:

(a) $x(t) \xleftrightarrow{\ FT\ } e^{-2\omega} u(\omega)$

(b) $\dfrac{1}{1+t^2} \xleftrightarrow{\ FT\ } X(j\omega)$

(c) $\dfrac{\sin\left(\dfrac{11\pi}{20} n\right)}{\sin\left(\dfrac{\pi}{20} n\right)} \xleftrightarrow{\ DTFS;\ \pi/10\ } X[k]$

3.23 For the signal $x(t)$ shown in Fig. P3.23, evaluate the following quantities without explicitly computing $X(j\omega)$:

(a) $\int_{-\infty}^{\infty} X(j\omega)\, d\omega$

(b) $\int_{-\infty}^{\infty} |X(j\omega)|^2\, d\omega$

(c) $\int_{-\infty}^{\infty} X(j\omega)e^{j2\omega}\, d\omega$

(d) $\arg\{X(j\omega)\}$

(e) $X(j0)$

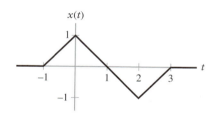

FIGURE P3.23

3.24 Let $x[n] \xleftrightarrow{\ DTFT\ } X(e^{j\Omega})$, where $X(e^{j\Omega})$ is depicted in Fig. P3.24. Evaluate the following without explicitly computing $x[n]$:

(a) $x[0]$

(b) $\arg\{x[n]\}$

(c) $\sum_{n=-\infty}^{\infty} x[n]$

(d) $\sum_{n=-\infty}^{\infty} |x[n]|^2$

(e) $\sum_{n=-\infty}^{\infty} x[n]e^{j(\pi/4)n}$

3.25 Prove the following properties:

(a) The FS symmetry properties for:
   (i) Real-valued time signals.
   (ii) Real and even time signals.

(b) The DTFT time-shift property.

(c) The DTFS frequency-shift property.

(d) Linearity for the FT.

(e) The DTFT convolution property.

(f) The DTFT modulation property.

(g) The DTFS convolution property.

(h) The FS modulation property.

(i) The Parseval relationship for the FS.

3.26 In this problem we show that Gaussian pulses achieve the lower bound in the time–bandwidth product. *Hint:* Use the definite integrals in Appendix A.4.

(a) Let $x(t) = e^{-t^2/2}$. Find the effective duration, T_d, and bandwidth, B_w, and evaluate the time–bandwidth product.

(b) Let $x(t) = e^{-t^2/2a^2}$. Find the effective duration, T_d, and bandwidth, B_w, and evaluate the time–bandwidth product. What happens to T_d, B_w, and $T_d B_w$ as a increases?

3.27 Let

$$x(t) = \begin{cases} 1, & |t| \le T \\ 0, & \text{otherwise} \end{cases}$$

Use the uncertainty principle to place a bound on the effective bandwidth of the following signals:

(a) $x(t)$

(b) $x(t) * x(t)$

3.28 Show that the time–bandwidth product, $T_d B_w$, of a signal $x(t)$ is invariant to scaling. That is, use the definitions of T_d and B_w to show that

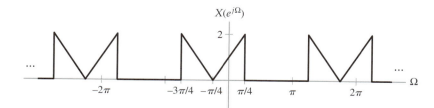

FIGURE P3.24

$x(t)$ and $x(at)$ have the same time–bandwidth product.

3.29 The form of the FS representation presented in this chapter,

$$x(t) = \sum_{k=-\infty}^{\infty} X[k]e^{jk\omega_o t}$$

is termed the exponential FS. In this problem we explore several alternative, yet equivalent, ways of expressing the FS representation for real-valued periodic signals.

(a) Trigonometric form.

(i) Show that the FS for a real-valued signal $x(t)$ can be written as

$$x(t) = a_0 +$$

$$\sum_{k=1}^{\infty} a_k \cos(k\omega_o t) + b_k \sin(k\omega_o t)$$

where a_k and b_k are real-valued coefficients.

(ii) Express a_k and b_k in terms of $X[k]$. Also express $X[k]$ in terms of a_k and b_k.

(iii) Show that

$$a_0 = \frac{1}{T} \int_{\langle T \rangle} x(t)\, dt$$

$$a_k = \frac{2}{T} \int_{\langle T \rangle} x(t) \cos k\omega_o t\, dt$$

$$b_k = \frac{2}{T} \int_{\langle T \rangle} x(t) \sin k\omega_o t\, dt$$

(iv) Show that $b_k = 0$ if $x(t)$ is even and $a_k = 0$ if $x(t)$ is odd.

(b) Polar form.

(i) Show that the FS for a real-valued signal $x(t)$ can be written as

$$x(t) = c_0 + \sum_{k=1}^{\infty} c_k \cos(k\omega_o t + \theta_k)$$

where c_k is the magnitude (positive) and θ_k is the phase of the kth harmonic.

(ii) Express c_k and θ_k as functions of $X[k]$.

(iii) Express c_k and θ_k as functions of a_k and b_k from (a).

*3.30 A signal with fundamental period T is said to possess half-wave symmetry if it satisfies $x(t) = -x(t - T/2)$. That is, half of one period of the signal is the negative of the other half. Show that the FS coefficients associated with even harmonics, $X[2k]$, are zero for all signals with half-wave symmetry.

*3.31 In this problem we explore a matrix representation for the DTFS. The DTFS expresses the N time-domain values of an N periodic signal $x[n]$ as a function of N frequency-domain values, $X[k]$. Define vectors

$$\mathbf{x} = \begin{bmatrix} x[0] \\ x[1] \\ \vdots \\ x[N-1] \end{bmatrix}, \quad \mathbf{X} = \begin{bmatrix} X[0] \\ X[1] \\ \vdots \\ X[N-1] \end{bmatrix}$$

(a) Show that the DTFS representation

$$x[n] = \sum_{k=0}^{N-1} X[k]e^{jk\Omega_o n},$$

$$n = 0, 1, \ldots, N - 1$$

can be written in matrix form as $\mathbf{x} = \mathbf{VX}$, where $\mathbf{V}$ is an N by N matrix. Find the elements of $\mathbf{V}$.

(b) Show that the expression for the DTFS coefficients

$$X[k] = \frac{1}{N} \sum_{n=0}^{N-1} x[n]e^{-jk\Omega_o n},$$

$$k = 0, 1, \ldots, N - 1$$

can be written in matrix vector form as $\mathbf{X} = \mathbf{Wx}$, where $\mathbf{W}$ is an N by N matrix. Find the elements of $\mathbf{W}$.

(c) The expression $\mathbf{x} = \mathbf{VX}$ implies that $\mathbf{X} = \mathbf{V}^{-1}\mathbf{x}$ provided $\mathbf{V}$ is a nonsingular matrix. Comparing this equation to the results of (b) we conclude that $\mathbf{W} = \mathbf{V}^{-1}$. Show that this is true by establishing $\mathbf{WV} = \mathbf{I}$ where $\mathbf{I}$ is the identity matrix. *Hint:* Use the definitions of $\mathbf{V}$ and $\mathbf{W}$ determined in (a) and (b) to obtain an expression for the element in the lth row and mth column of $\mathbf{WV}$.

*3.32 In this problem we find the FS coefficients $X[k]$ by minimizing the mean squared error (MSE) between the signal $x(t)$ and its FS approximation. Define the J term FS

$$\hat{x}_J(t) = \sum_{k=-J}^{J} A[k]e^{jk\omega_o t}$$

and the J term MSE as the average squared difference over one period

$$MSE_J = \frac{1}{T} \int_{\langle T \rangle} |x(t) - \hat{x}_J(t)|^2\, dt$$

(a) Substitute the series representation for $\hat{x}_J(t)$ and expand the magnitude squared using the identity $|a + b|^2 = (a + b)(a^* + b^*)$ to obtain

$$MSE_J = \frac{1}{T}\int_{\langle T \rangle} |x(t)|^2 \, dt$$

$$- \sum_{k=-J}^{J} A^*[k]\left(\frac{1}{T}\int_{\langle T \rangle} x(t)e^{-jk\omega_o t} \, dt\right)$$

$$- \sum_{k=-J}^{J} A[k]\left(\frac{1}{T}\int_{\langle T \rangle} x^*(t)e^{jk\omega_o t} \, dt\right)$$

$$+ \sum_{m=-J}^{J} \sum_{k=-J}^{J} A^*[k]A[m]\left(\frac{1}{T}\int_{\langle T \rangle} e^{-jk\omega_o t}e^{jm\omega_o t} \, dt\right)$$

(b) Define

$$X[k] = \frac{1}{T}\int_{\langle T \rangle} x(t)e^{-jk\omega_o t} \, dt$$

and use the orthogonality of $e^{jk\omega_o t}$ and $e^{jm\omega_o t}$ to show that

$$MSE_J = \frac{1}{T}\int_{\langle T \rangle} |x(t)|^2 \, dt - \sum_{k=-J}^{J} A^*[k]X[k]$$

$$- \sum_{k=-J}^{J} A[k]X^*[k] + \sum_{k=-J}^{J} |A[k]|^2$$

(c) Use the technique of completing the square to show that

$$MSE_J = \frac{1}{T}\int_{\langle T \rangle} |x(t)|^2 \, dt$$

$$+ \sum_{k=-J}^{J} |A[k] - X[k]|^2 - \sum_{k=-J}^{J} |X[k]|^2$$

(d) Find the value for $A[k]$ that minimizes MSE_J.

(e) Express the minimum MSE_J as a function of $x(t)$ and $X[k]$. What happens to MSE_J as J increases?

*3.33 *Generalized Fourier Series.* The concept of the Fourier series may be generalized to sums of signals other than complex sinusoids. That is, given a signal $x(t)$ we may approximate $x(t)$ on an interval $[t_1, t_2]$ as a weighted sum of N functions $\phi_0(t), \phi_2(t), \ldots, \phi_{N-1}(t)$:

$$x(t) \approx \sum_{k=0}^{N-1} c_k \phi_k(t)$$

We shall assume that these N functions are mutually orthogonal on $[t_1, t_2]$. This means that

$$\int_{t_1}^{t_2} \phi_k(t)\phi_l^*(t) \, dt = \begin{cases} 0, & k \neq l \\ f_k, & k = l \end{cases}$$

where f_k is some constant. The MSE using this approximation is

$$MSE = \frac{1}{t_2 - t_1}\int_{t_1}^{t_2} \left| x(t) - \sum_{k=0}^{N-1} c_k \phi_k(t) \right|^2 \, dt$$

(a) Show that the MSE is minimized by choosing $c_k = (1/f_k) \int_{t_1}^{t_2} x(t)\phi_k^*(t) \, dt$. *Hint:* Generalize steps outlined in Problem 3.32(a)–(d) to this problem.

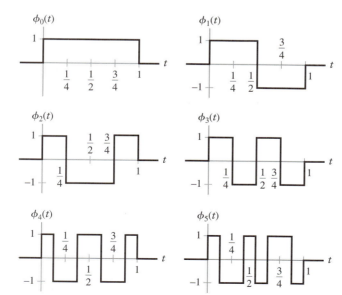

FIGURE P3.33

(b) Show that the MSE is zero if

$$\int_{t_1}^{t_2} |x(t)|^2 \, dt = \sum_{k=0}^{N-1} f_k |c_k|^2$$

If this relationship holds for all $x(t)$ in a given class of functions, then the basis functions $\phi_0(t)$, $\phi_2(t)$, ..., $\phi_{N-1}(t)$ are said to be "complete" for that class.

(c) The Walsh functions are one set of orthogonal functions that are used for signal representation on [0, 1]. Determine the c_k and MSE obtained by approximating the following signals with the first six Walsh functions depicted in Fig. P3.33. Sketch the signal and the Walsh function approximation.

(i) $x(t) = \begin{cases} 2, & \frac{1}{2} \le t \le \frac{3}{4} \\ 0, & 0 < t < \frac{1}{2}, \frac{3}{4} < t < 1 \end{cases}$

(ii) $x(t) = \sin(2\pi t)$

(d) The Legendre polynomials are another set of orthogonal functions on the interval $[-1, 1]$. They are obtained from the difference equation

$$\phi_k(t) = \frac{2k - 1}{k} t\phi_{k-1}(t) - \frac{k - 1}{k} \phi_{k-2}(t)$$

using the initial functions $\phi_0(t) = 1$ and $\phi_1(t) = t$. Determine the c_k and MSE obtained by approximating the following signals with the first six Legendre polynomials.

(i) $x(t) = \begin{cases} 2, & 0 < t < \frac{1}{2} \\ 0, & -1 < t < 0, \frac{1}{2} < t < 1 \end{cases}$

(ii) $x(t) = \sin(\pi t)$

*3.34 We may derive the FT from the FS by describing a nonperiodic signal as the limiting form of a periodic signal whose period, T, approaches infinity. In order to take this approach, we assume that the FS of the periodic version of the signal exists, that the nonperiodic signal is zero for $|t| > T/2$, and that the limit as T approaches infinity is taken in a symmetric manner. Define the finite-duration nonperiodic signal $x(t)$ as one period of the T periodic signal $\tilde{x}(t)$

$$x(t) = \begin{cases} \tilde{x}(t), & -T/2 \le t \le T/2 \\ 0, & |t| > T/2 \end{cases}$$

(a) Graph an example of $x(t)$ and $\tilde{x}(t)$ to demonstrate that as T increases, the periodic replicates of $x(t)$ in $\tilde{x}(t)$ are moved farther and farther away from the origin. Eventu-

ally, as T approaches infinity, these replicates are removed to infinity. Thus we write

$$x(t) = \lim_{T \to \infty} \tilde{x}(t)$$

(b) The FS representation for the periodic signal $\tilde{x}(t)$ is

$$\tilde{x}(t) = \sum_{k=-\infty}^{\infty} X[k]e^{jk\omega_o t}$$

$$X[k] = \frac{1}{T} \int_{-T/2}^{T/2} \tilde{x}(t)e^{-jk\omega_o t} \, dt$$

Show that $X[k] = \frac{1}{T} X(jk\omega_o)$, where

$$X(j\omega) = \int_{-\infty}^{\infty} x(t)e^{-j\omega t} \, dt$$

(c) Substitute this definition for $X[k]$ into the expression for $\tilde{x}(t)$ in (b) and show that

$$\tilde{x}(t) = \frac{1}{2\pi} \sum_{k=-\infty}^{\infty} X(jk\omega_o)e^{jk\omega_o t}\omega_o$$

(d) Use the limiting expression for $x(t)$ in (a) and define $\omega \approx k\omega_o$ to express the limiting form of the sum in (c) as the integral

$$x(t) = \frac{1}{2\pi} \int_{-\infty}^{\infty} X(j\omega)e^{j\omega t} \, d\omega$$

▶ *Computer Experiments*

3.35 Use MATLAB to repeat Example 3.3 for $N = 50$ and (a) $M = 12$, (b) $M = 5$, and (c) $M = 20$.

3.36 Use the MATLAB command `fft` to repeat Problem 3.1.

3.37 Use the MATLAB command `ifft` to repeat Problem 3.2.

3.38 Use the MATLAB command `fft` to repeat Example 3.4.

3.39 Use MATLAB to repeat Example 3.7. Evaluate the peak overshoot for $J = 29, 59$, and 99.

3.40 Let $x(t)$ be the triangular wave depicted in Fig. P3.40.

(a) Find the FS coefficients, $X[k]$.

(b) Show that the FS representation for $x(t)$ can be expressed in the form

$$x(t) = \sum_{k=0}^{\infty} B[k] \cos(k\omega_o t)$$

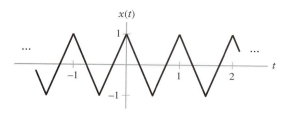

FIGURE P3.40

(c) Define the *J*-term partial sum approximation to $x(t)$ as

$$\hat{x}_J(t) = \sum_{k=0}^{J} B[k] \cos(k\omega_o t)$$

Use MATLAB to evaluate and plot one pe-
riod of the *J*th term in this sum and $\hat{x}_J(t)$ for
$J = 1, 3, 7, 29,$ and 99.

3.41 Repeat Problem 3.40 for the impulse train given
by

$$x(t) = \sum_{n=-\infty}^{\infty} \delta(t - n)$$

3.42 Use MATLAB to repeat Example 3.8 using the
following values for the time constant:
(a) $RC = 0.01$ s.
(b) $RC = 0.1$ s.
(c) $RC = 1$ s.

3.43 Evaluate the frequency response of the truncated filter in Example 3.32. You may do this in
MATLAB by writing an M-file to evaluate

$$H_t(e^{j\Omega}) = \sum_{n=-M}^{M} h[n]e^{-j\Omega n}$$

for a large number (>1000) of values of Ω in
the interval $-\pi < \Omega \le \pi$. Plot the frequency
response magnitude in dB ($20 \log_{10}|H_t(e^{j\Omega})|$)
for the following values of M:
(a) $M = 4$

(b) $M = 10$
(c) $M = 25$
(d) $M = 50$

Discuss the effect of increasing M on the accuracy with which $H_t(e^{j\Omega})$ approximates $H(e^{j\Omega})$.

3.44 Use MATLAB to verify that the time–bandwidth product for a discrete-time square wave
is approximately independent of the number of
nonzero values in each period when duration is
defined as the number of nonzero values in the
square wave and bandwidth is defined as the
mainlobe width. Define one period of the square
wave as

$$x[n] = \begin{cases} 1, & 0 \le n < M \\ 0, & M \le n \le 999 \end{cases}$$

Evaluate the bandwidth by first using the commands **fft** and **abs** to obtain the magnitude
spectrum and then count the number of DTFS
coefficients in the mainlobe for $M = 10, 20,$
40, 50, 100, and 200.

3.45 Use the MATLAB function **TdBw** introduced
in Section 3.7 to evaluate and plot the time–
bandwidth product as a function of duration
for the following classes of signals:

(a) Rectangular pulse trains. Let the pulse in a
single period be of length M and vary M
from 51 to 701 in steps of 50.

(b) Raised cosine pulse trains. Let the pulse in
a single period be of length M and vary M
from 51 to 701 in steps of 50.

(c) Gaussian pulse trains. Let $x[n] = e^{-an^2}$,
$-500 \le n \le 500$, represent the Gaussian
pulse in a single period. Vary the pulse duration by letting a take the following values:
0.00005, 0.0001, 0.0002, 0.0005, 0.001,
0.002, and 0.005.

4

Applications of Fourier Representations

4.1 Introduction

In the previous chapter we developed Fourier representations for four distinct signal classes: the discrete-time Fourier series (DTFS) for periodic discrete-time signals, the Fourier series (FS) for periodic continuous-time signals, the discrete-time Fourier transform (DTFT) for nonperiodic discrete-time signals, and the Fourier transform (FT) for nonperiodic continuous-time signals. We now focus on applications of these Fourier representations. The two most common applications are (1) analysis of the interaction between signals and systems, and (2) numerical evaluation of signal properties or system behavior. The FT and DTFT are most commonly used for analysis applications, while the DTFS is the primary representation used for computational applications. The first and major portion of this chapter is devoted to the presentation of analysis applications; computational applications are discussed briefly at the end of the chapter.

An important aspect of applying Fourier representations is dealing with situations in which there is a mixing of signal classes. For example, if we apply a periodic signal to a stable system, the convolution representation for the system output involves a mixing of nonperiodic (impulse response) and periodic (input) signal classes. A system that samples continuous-time signals involves both continuous- and discrete-time signals. In order to use Fourier methods to analyze such interactions, we must build bridges between the Fourier representations for different signal classes. We establish these relationships in this chapter. Specifically, we develop FT and DTFT representations for continuous- and discrete-time periodic signals, respectively. We may then use the FT to analyze continuous-time applications that involve a mixture of periodic and nonperiodic signals. Similarly, the DTFT may be used to analyze mixtures of discrete-time periodic and nonperiodic signals. Lastly, we develop a FT representation for discrete-time signals to address problems involving mixtures of continuous- and discrete-time signals.

We begin this chapter by relating the frequency response description of a LTI system to the time-domain descriptions presented in Chapter 2. We then revisit convolution and modulation, considering applications in which periodic and nonperiodic signals interact using FT and DTFT representations for periodic signals. Next, we analyze the process of sampling signals and reconstruction of continuous-time signals from samples using the FT representation for discrete signals. These issues are of fundamental importance whenever a computer is used to manipulate continuous-time signals. Computers are used to manipulate signals in communication systems (Chapter 5), for the purpose of filtering (Chapter 8), and control (Chapter 9). Our analysis reveals the limitations associated with discrete-time processing of continuous-time signals and suggests a practical system that minimizes them.

Recall that the DTFS is the only Fourier representation that can be evaluated exactly on a computer. Consequently, it finds extensive use in numerical algorithms for signal processing. We conclude the chapter by examining two common uses of the DTFS: numerical approximation of the FT and efficient implementation of discrete-time convolution. In both of these, a clear understanding of the relationship between the Fourier representations for different signal classes is essential for correct interpretation of the results.

In the course of applying Fourier representations, we discover the relationships between all four. A thorough understanding of these relationships is a critical first step in using Fourier methods to solve problems involving signals and systems.

4.2 *Frequency Response of LTI Systems*

In this section we use the FT and DTFT to explore the relationships between the time-domain system descriptions introduced in Chapter 2 and the corresponding frequency response of the system. The frequency response offers a useful and intuitive characterization of the input–output behavior of the system. This is because convolution in the time domain transforms to multiplication in the frequency domain: the output of a system is obtained simply by multiplying the Fourier representation of the input with the system frequency response. It is easy to visualize and interpret the operation of multiplying two frequency-domain functions. We begin the discussion by examining the relationship between the frequency response and the impulse response, and then proceed to derive the relationships between the frequency response and differential/difference-equation and state-variable description representations.

■ IMPULSE RESPONSE

We established in the previous chapter that the impulse response and frequency response of a continuous- or discrete-time system constitute a FT or DTFT pair, respectively. Recall that the impulse response of a bounded input, bounded output stable system $h(t)$ is absolutely integrable, as defined by

$$\int_{-\infty}^{\infty} |h(t)| \, dt < \infty$$

Thus the Dirichlet conditions are satisfied and the FT of $h(t)$ exists if $h(t)$ has a finite number of local maxima, minima, and discontinuities with each discontinuity of finite size. These conditions are met by many physical systems. Similarly, a stable discrete-time system has an absolutely summable impulse response, as defined by

$$\sum_{n=-\infty}^{\infty} |h[n]| < \infty$$

This condition is sufficient to guarantee the existence of the DTFT. We conclude that the frequency response exists for stable systems.

The convolution property relates the input and output of a system as follows:

$$y(t) = x(t) * h(t) \xleftarrow{\quad FT \quad} Y(j\omega) = H(j\omega)X(j\omega)$$

$$y[n] = x[n] * h[n] \xleftarrow{\quad DTFT \quad} Y(e^{j\Omega}) = H(e^{j\Omega})X(e^{j\Omega})$$

The multiplication that occurs in the frequency-domain representation gives rise to the notion of *filtering*. The system filters the input signal by presenting a different response to components of the input at different frequencies. We often describe systems in terms of the type of filtering that they perform on the input signal. A *lowpass filter* attenuates high-frequency components of the input and passes the lower-frequency components. In contrast, a *highpass filter* attenuates low frequencies and passes the high frequencies. A *bandpass filter* passes signals within a certain frequency band and attenuates signals outside this band. Figures 4.1(a)–(c) illustrate *ideal* lowpass, highpass, and bandpass filters, respectively, corresponding to both continuous- and discrete-time systems. Note that characterization of the discrete-time filter is based on its behavior in the frequency range $-\pi < \Omega \leq \pi$ because its frequency response is 2π periodic. Hence a highpass discrete-time filter passes frequencies near π and attenuates frequencies near zero.

The *passband* of a filter is the band of frequencies that are passed by the system, while the *stopband* refers to the range of frequencies that are attenuated by the system. It is impossible to build a practical system that has the discontinuous frequency response characteristics of the ideal systems depicted in Fig. 4.1. Realistic filters always have a gradual transition from the passband to the stopband. The range of frequencies over which this occurs is known as the *transition band*. Furthermore, realistic filters do not have zero gain over the entire stopband, but instead have a very small gain relative to that of the passband. In general, filters with sharp transitions from passband to stopband are more difficult to implement. Detailed treatment of filters is deferred to Chapter 8.

The magnitude response of a filter is commonly described in units of decibels or dB, defined as

$$20 \log_{10} |H(j\omega)| \quad \text{or} \quad 20 \log_{10} |H(e^{j\Omega})|$$

The magnitude response in the stopband is normally much smaller than that in the passband and the details of the stopband response are difficult to visualize on a linear scale.

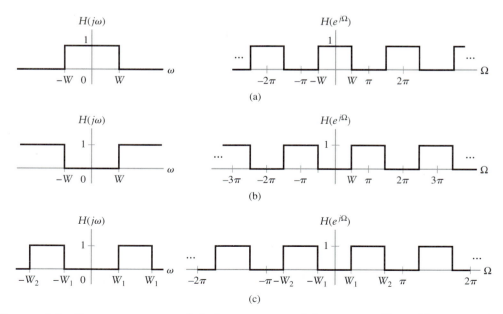

FIGURE 4.1 Frequency response of ideal continuous- and discrete-time filters. (a) Lowpass characteristic. (b) Highpass characteristic. (c) Bandpass characteristic.

By using units of dB, we display the magnitude response on a logarithmic scale and are able to examine the details of the response in both the passband and the stopband. Note that unit gain corresponds to 0 dB. Hence the magnitude response in the filter passband is normally close to 0 dB. The edge of the passband is usually defined by the frequencies for which the response is -3 dB. At these frequencies the magnitude response is $1/\sqrt{2}$. Since the energy spectrum of the filter output is given by

$$|Y(j\omega)|^2 = |H(j\omega)|^2|X(j\omega)|^2$$

we see that the -3-dB point corresponds to frequencies at which the filter passes only $\frac{1}{2}$ of the input power. These are often termed the *cutoff* frequencies of the filter.

EXAMPLE 4.1   The impulse response of the RC circuit in Fig. 4.2(a) was derived as

$$h(t) = \frac{1}{RC}\,e^{-t/RC}u(t)$$

Plot the magnitude response of this system on a linear scale and in dB, and characterize this system as a filter.

Solution: The frequency response of this system is

$$H(j\omega) = \frac{1}{j\omega RC + 1}$$

Figure 4.2(b) depicts the magnitude response, $|H(j\omega)|$. Figure 4.2(c) illustrates the magnitude response in dB. This system has unit gain at low frequencies and tends to attenuate high frequencies. Hence it has a lowpass filtering characteristic. We see that the cutoff frequency is $\omega_c = 1/RC$ since the magnitude response is -3 dB at ω_c. Therefore the filter passband is from 0 to $1/RC$.

The convolution property implies that the frequency response of a system may be expressed as the ratio of the FT or DTFT of the output to that of the input. Specifically, for a continuous-time system we may write

$$H(j\omega) = \frac{Y(j\omega)}{X(j\omega)} \tag{4.1}$$

and for a discrete-time system

$$H(e^{j\Omega}) = \frac{Y(e^{j\Omega})}{X(e^{j\Omega})} \tag{4.2}$$

Both of these expressions are of the indeterminate form 0/0 at frequencies where $X(j\omega)$ or $X(e^{j\Omega})$ is zero. Hence if the input spectra are nonzero at all frequencies, we can determine the frequency response of a system from knowledge of the input and output spectra.

Note that if $H(j\omega)$ and $H(e^{j\Omega})$ are nonzero, then Eqs. (4.1) and (4.2) also imply that we can recover the input of the system from the output as

$$X(j\omega) = H^{-1}(j\omega)Y(j\omega)$$

and

$$X(e^{j\Omega}) = H^{-1}(e^{j\Omega})Y(e^{j\Omega})$$

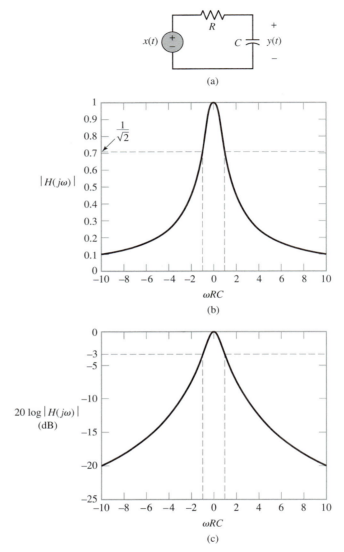

FIGURE 4.2 (a) RC circuit with input $x(t)$ and output $y(t)$. (b) Magnitude response of RC circuit. (c) Magnitude response of RC circuit in dB.

where $H^{-1}(j\omega) = 1/H(j\omega)$ and $H^{-1}(e^{j\Omega}) = 1/H(e^{j\Omega})$ are the frequency responses of the respective inverse systems. An inverse system is also known as an *equalizer* and the process of recovering the input from the output is known as *equalization*. In practice it is often difficult or impossible to build an exact inverse system, so an approximate inverse is used. For example, a communication channel may introduce a time delay in addition to distorting the signal's magnitude and phase spectra. An exact equalizer would have to introduce a time advance, which implies it is noncausal and cannot be implemented. However, we may choose to build an approximate equalizer, one that compensates for all the distortion except for the time delay. An introduction to equalizer design is given in Chapter 8.

EXAMPLE 4.2   The output of a system in response to an input $x(t) = e^{-2t}u(t)$ is $y(t) = e^{-t}u(t)$. Find the frequency response and the impulse response of this system.

Solution: Take the FT of $x(t)$ and $y(t)$, obtaining

$$X(j\omega) = \frac{1}{j\omega + 2}$$

and

$$Y(j\omega) = \frac{1}{j\omega + 1}$$

Now use the definition

$$H(j\omega) = \frac{Y(j\omega)}{X(j\omega)}$$

to obtain the system frequency response

$$H(j\omega) = \frac{j\omega + 2}{j\omega + 1}$$

which may be rewritten as

$$H(j\omega) = \left(\frac{j\omega + 1}{j\omega + 1}\right) + \frac{1}{j\omega + 1}$$

$$= 1 + \frac{1}{j\omega + 1}$$

Take the inverse FT of each term to obtain the impulse response of the system:

$$h(t) = \delta(t) + e^{-t}u(t)$$

■ **DIFFERENTIAL- AND DIFFERENCE-EQUATION DESCRIPTIONS**

By definition, the frequency response is the amplitude and phase change the system imparts to a complex sinusoid. The sinusoid is assumed to exist for all time; it does not have a starting or ending time. This implies that the frequency response is the system's steady-state response to a sinusoid. In contrast to differential- and difference-equation descriptions for a system, the frequency response description cannot represent initial conditions; it can only describe a system in a steady-state condition.

The differential-equation representation for a continuous-time system is

$$\sum_{k=0}^{N} a_k \frac{d^k}{dt^k} y(t) = \sum_{k=0}^{M} b_k \frac{d^k}{dt^k} x(t)$$

Take the FT of both sides of this equation and repeatedly apply the differentiation property

$$\frac{d}{dt} g(t) \xleftrightarrow{\ FT\ } j\omega G(j\omega)$$

to obtain

$$\sum_{k=0}^{N} a_k (j\omega)^k Y(j\omega) = \sum_{k=0}^{M} b_k (j\omega)^k X(j\omega)$$

Rearrange this equation as the ratio of the FT of the output to the input, obtaining

$$\frac{Y(j\omega)}{X(j\omega)} = \frac{\sum_{k=0}^{M} b_k (j\omega)^k}{\sum_{k=0}^{N} a_k (j\omega)^k}$$

Hence Eq. (4.1) implies that the frequency response of the system is

$$H(j\omega) = \frac{\sum_{k=0}^{M} b_k (j\omega)^k}{\sum_{k=0}^{N} a_k (j\omega)^k} \qquad (4.3)$$

The frequency response of a system described by a linear constant-coefficient differential equation is a ratio of two polynomials in $j\omega$. Note that we can reverse this process and determine a differential-equation description for the system from the frequency response provided the frequency response is expressed as a ratio of polynomials in $j\omega$.

The difference-equation representation for a discrete-time system is of the form

$$\sum_{k=0}^{N} a_k y[n - k] = \sum_{k=0}^{M} b_k x[n - k]$$

Take the DTFT of both sides of this equation, using the time-shift property

$$g[n - k] \xleftarrow{\quad DTFT \quad} e^{-jk\Omega} G(e^{j\Omega})$$

to obtain

$$\sum_{k=0}^{N} a_k (e^{-j\Omega})^k Y(e^{j\Omega}) = \sum_{k=0}^{M} b_k (e^{-j\Omega})^k X(e^{j\Omega})$$

Rewrite this equation as the ratio

$$\frac{Y(e^{j\Omega})}{X(e^{j\Omega})} = \frac{\sum_{k=0}^{M} b_k (e^{-j\Omega})^k}{\sum_{k=0}^{N} a_k (e^{-j\Omega})^k}$$

Identifying this ratio with Eq. (4.2), we have

$$H(e^{j\Omega}) = \frac{\sum_{k=0}^{M} b_k (e^{-j\Omega})^k}{\sum_{k=0}^{N} a_k (e^{-j\Omega})^k} \qquad (4.4)$$

In the discrete-time case, the frequency response is a ratio of polynomials in $e^{-j\Omega}$. Given a frequency response of the form described in Eq. (4.4), we may reverse our derivation to determine a difference-equation description for the system, if so desired.

EXAMPLE 4.3   Find the frequency response and impulse response of the system described by the differential equation

$$\frac{d^2}{dt^2} y(t) + 3 \frac{d}{dt} y(t) + 2y(t) = 2 \frac{d}{dt} x(t) + x(t)$$

Solution: Here we have $N = 2$, $M = 1$. Substituting the coefficients of this differential equation into Eq. (4.3), we obtain the frequency response

$$H(j\omega) = \frac{2j\omega + 1}{(j\omega)^2 + 3j\omega + 2}$$

The impulse response is given by the inverse FT of $H(j\omega)$. Rewrite $H(j\omega)$ using the partial fraction expansion:

$$\frac{2j\omega + 1}{(j\omega)^2 + 3j\omega + 2} = \frac{A}{j\omega + 1} + \frac{B}{j\omega + 2}$$

Solving for A and B we obtain $A = -1$, and $B = 3$. Hence

$$H(j\omega) = \frac{-1}{j\omega + 1} + \frac{3}{j\omega + 2}$$

The inverse FT gives the impulse response

$$h(t) = 3e^{-2t}u(t) - e^{-t}u(t)$$

EXAMPLE 4.4   The mechanical system depicted in Fig. 4.3(a) has the applied force $x(t)$ as its input and position $y(t)$ as its output. The relationship between $x(t)$ and $y(t)$ is governed by the differential equation

$$m\frac{d^2}{dt^2}y(t) + f\frac{d}{dt}y(t) + ky(t) = x(t)$$

Find the frequency response of this system and plot the magnitude response in dB for $m = 0.5\ kg$, $f = 0.1\ N \cdot s/m$, and $k = 50\ N/m$.

Solution: Application of Eq. (4.3) gives

$$H(j\omega) = \frac{1}{m(j\omega)^2 + fj\omega + k}$$

Figure 4.3(b) depicts the magnitude response in dB for the specified values of m, f, and k. This system tends to attenuate frequencies both below and above 10 rad/s. Hence it has a very narrow bandpass filtering characteristic and effectively only responds to input frequency components near 10 rad/s.

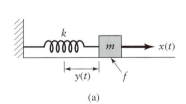

(a)

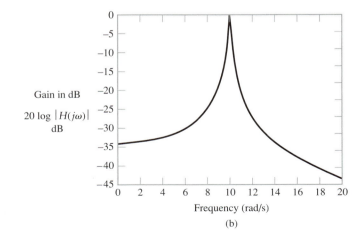

(b)

FIGURE 4.3   (a) Mechanical system with input $x(t)$ and output $y(t)$. (b) System magnitude response.

▶ **Drill Problem 4.1**   Find the frequency response and the impulse response of a discrete-time system described by the difference equation

$$y[n - 2] + 5y[n - 1] + 6y[n] = 8x[n - 1] + 18x[n]$$

Answer:

$$H(e^{j\Omega}) = \frac{8e^{-j\Omega} + 18}{(e^{-j\Omega})^2 + 5e^{-j\Omega} + 6}$$

$$h[n] = 2(-\tfrac{1}{3})^n u[n] + (-\tfrac{1}{2})^n u[n]$$

◀

▶ **Drill Problem 4.2**   Write the differential equation relating the input and output of the *RC* circuit depicted in Fig. 4.2(a). Use the results of this subsection to identify the frequency response and impulse response from the differential equation.

Answer: See Example 4.1.   ◀

■ STATE-VARIABLE DESCRIPTION

The state-variable description for a continuous-time system is

$$\frac{d}{dt}\mathbf{q}(t) = \mathbf{A}\mathbf{q}(t) + \mathbf{b}x(t) \tag{4.5}$$

$$y(t) = \mathbf{c}\mathbf{q}(t) + Dx(t) \tag{4.6}$$

We shall determine the frequency response in terms of (**A**, **b**, **c**, *D*) by taking the FT of both sides of these equations using the differentiation property. Beginning with Eq. (4.5), we have

$$j\omega\mathbf{q}(j\omega) = \mathbf{A}\mathbf{q}(j\omega) + \mathbf{b}X(j\omega) \tag{4.7}$$

where we have defined the FT of the state vector as the vector containing the FT of each state variable. That is,

$$\mathbf{q}(j\omega) = \begin{bmatrix} Q_1(j\omega) \\ Q_2(j\omega) \\ \vdots \\ Q_N(j\omega) \end{bmatrix}$$

where the *i*th entry in $\mathbf{q}(j\omega)$ is the FT of the *i*th state variable, $q_i(t) \xleftrightarrow{\ FT\ } Q_i(j\omega)$. We may rewrite Eq. (4.7) as

$$j\omega\mathbf{q}(j\omega) - \mathbf{A}\mathbf{q}(j\omega) = \mathbf{b}X(j\omega)$$

$$(j\omega\mathbf{I} - \mathbf{A})\mathbf{q}(j\omega) = \mathbf{b}X(j\omega)$$

and thus write

$$\mathbf{q}(j\omega) = (j\omega\mathbf{I} - \mathbf{A})^{-1}\mathbf{b}X(j\omega) \tag{4.8}$$

Here **I** is the *N* by *N* identity matrix. Now take the FT of Eq. (4.6) to obtain

$$Y(j\omega) = \mathbf{c}\mathbf{q}(j\omega) + DX(j\omega)$$

and substitute Eq. (4.8) to obtain

$$Y(j\omega) = (\mathbf{c}(j\omega\mathbf{I} - \mathbf{A})^{-1}\mathbf{b} + D)X(j\omega) \tag{4.9}$$

Since by definition

$$H(j\omega) = \frac{Y(j\omega)}{X(j\omega)}$$

we therefore conclude that

$$H(j\omega) = \mathbf{c}(j\omega\mathbf{I} - \mathbf{A})^{-1}\mathbf{b} + D \qquad (4.10)$$

is the expression for the frequency response defined in terms of the state-variable description $(\mathbf{A}, \mathbf{b}, \mathbf{c}, D)$.

We may derive the frequency response of a discrete-time system in terms of $(\mathbf{A}, \mathbf{b}, \mathbf{c}, D)$ by following an analogous set of steps and using the time-shift property in place of the differentiation property. The result is

$$H(e^{j\Omega}) = \mathbf{c}(e^{j\Omega}\mathbf{I} - \mathbf{A})^{-1}\mathbf{b} + D \qquad (4.11)$$

EXAMPLE 4.5 Determine the frequency response of the continuous-time system with state-variable description

$$\mathbf{A} = \begin{bmatrix} 2 & -1 \\ 1 & 0 \end{bmatrix}, \qquad \mathbf{b} = \begin{bmatrix} 1 \\ 0 \end{bmatrix}$$
$$\mathbf{c} = [3 \quad 1], \qquad D = [0]$$

Solution: The frequency response is determined by substituting $(\mathbf{A}, \mathbf{b}, \mathbf{c}, D)$ into Eq. (4.10). Begin by evaluating $(j\omega\mathbf{I} - \mathbf{A})^{-1}$. For this example we have

$$(j\omega\mathbf{I} - \mathbf{A})^{-1} = \begin{bmatrix} j\omega - 2 & 1 \\ -1 & j\omega \end{bmatrix}^{-1}$$

$$= \frac{1}{(j\omega)^2 - 2j\omega + 1} \begin{bmatrix} j\omega & -1 \\ 1 & j\omega - 2 \end{bmatrix}$$

Now substitute $\mathbf{c}$, $(j\omega\mathbf{I} - \mathbf{A})^{-1}$, $\mathbf{b}$, and D into Eq. (4.10) to obtain

$$H(j\omega) = [3 \quad 1] \frac{1}{(j\omega)^2 - 2j\omega + 1} \begin{bmatrix} j\omega & -1 \\ 1 & j\omega - 2 \end{bmatrix} \begin{bmatrix} 1 \\ 0 \end{bmatrix} + 0$$

$$= \frac{1}{(j\omega)^2 - 2j\omega + 1} [3j\omega + 1 \quad j\omega - 5] \begin{bmatrix} 1 \\ 0 \end{bmatrix}$$

$$= \frac{3j\omega + 1}{(j\omega)^2 - 2j\omega + 1}$$

▶ **Drill Problem 4.3** Find the frequency response of a discrete-time system with state-variable description

$$\mathbf{A} = \begin{bmatrix} -2 & 0 \\ 1 & -1 \end{bmatrix}, \qquad \mathbf{b} = \begin{bmatrix} 1 \\ 1 \end{bmatrix}$$
$$\mathbf{c} = [0 \quad 2], \qquad D = [1]$$

Answer:

$$H(e^{j\Omega}) = \frac{(e^{j\Omega})^2 + 5e^{j\Omega} + 8}{(e^{j\Omega})^2 + 3e^{j\Omega} + 2}$$

◀

In Chapter 2 we noted that there are many different state-variable descriptions for a system with a given input–output characteristic. These different state-variable descriptions are obtained by transforming the system's state vector with a nonsingular matrix **T**. Since the frequency response of a system is an input–output description, it should also be invariant to transformations of the state vector. We now prove this important property of a LTI system.

Let

$$H(j\omega) = \mathbf{c}(j\omega\mathbf{I} - \mathbf{A})^{-1}\mathbf{b} + D$$

be the frequency response of a continuous-time system with state-variable description (**A**, **b**, **c**, D). Now transform the state vector with a nonsingular matrix **T** to obtain a new state-variable description (**A**′, **b**′, **c**′, D′). The frequency response of the transformed system is

$$H'(j\omega) = \mathbf{c}'(j\omega\mathbf{I} - \mathbf{A}')^{-1}\mathbf{b}' + D'$$

In Chapter 2 we established that $\mathbf{A}' = \mathbf{TAT}^{-1}$, $\mathbf{b}' = \mathbf{Tb}$, $\mathbf{c}' = \mathbf{cT}^{-1}$, and $D' = D$. Substitute these relationships into the expression for $H'(j\omega)$ to obtain

$$H'(j\omega) = \mathbf{cT}^{-1}(j\omega\mathbf{I} - \mathbf{TAT}^{-1})^{-1}\mathbf{Tb} + D$$

Write $\mathbf{I} = \mathbf{TT}^{-1}$ and substitute it in the above equation to obtain

$$
\begin{aligned}
H'(j\omega) &= \mathbf{cT}^{-1}(j\omega\mathbf{TT}^{-1} - \mathbf{TAT}^{-1})^{-1}\mathbf{Tb} + D \\
&= \mathbf{cT}^{-1}(\mathbf{T}(j\omega\mathbf{I} - \mathbf{A})\mathbf{T}^{-1})^{-1}\mathbf{Tb} + D
\end{aligned}
$$

Now use the identity for the inverse of a product of invertible matrices $(\mathbf{FGH})^{-1} = \mathbf{H}^{-1}\mathbf{G}^{-1}\mathbf{F}^{-1}$ to write

$$
\begin{aligned}
H'(j\omega) &= \mathbf{cT}^{-1}\mathbf{T}(j\omega\mathbf{I} - \mathbf{A})^{-1}\mathbf{T}^{-1}\mathbf{Tb} + D \\
&= \mathbf{c}(j\omega\mathbf{I} - \mathbf{A})^{-1}\mathbf{b} + D
\end{aligned}
$$

Hence we have shown that $H'(j\omega) = H(j\omega)$ and conclude that the frequency response of a LTI system is invariant to transformations of its state vector. An analogous result holds for discrete-time systems.

4.3 Fourier Transform Representations for Periodic Signals

Recall that the FS and DTFS have been derived as the Fourier representations for periodic signals. Strictly speaking, neither the FT nor DTFT converges for periodic signals. However, by incorporating impulses into the FT and DTFT in the appropriate manner, we may develop FT and DTFT representations for periodic signals. These representations satisfy the properties expected of the FT and DTFT. Hence we may use these representations and the properties of the FT or DTFT to analyze problems involving mixtures of periodic and nonperiodic signals. Our derivation also indicates the relationship between Fourier series representations and Fourier transform representations. We begin with the continuous-time case.

■ RELATING THE FT TO THE FS

The FS representation for a periodic signal $x(t)$ is

$$x(t) = \sum_{k=-\infty}^{\infty} X[k]e^{jk\omega_o t} \tag{4.12}$$

where ω_o is the fundamental frequency of $x(t)$. Now note that the inverse FT of a frequency-shifted impulse, $\delta(\omega - k\omega_o)$, is a complex sinusoid with frequency $k\omega_o$, as shown by

$$\frac{1}{2\pi} e^{jk\omega_o t} \xleftrightarrow{\ FT\ } \delta(\omega - k\omega_o) \tag{4.13}$$

Although $e^{jk\omega_o t}$ is a periodic function and thus does not have a convergent FT, we obtain this FT pair as a consequence of the sifting property of the impulse function.

Substitute the FT pair Eq. (4.13) into the FS representation Eq. (4.12) and use the linearity property of the FT to obtain

$$x(t) = \sum_{k=-\infty}^{\infty} X[k]e^{jk\omega_o t} \xleftrightarrow{\ FT\ } X(j\omega) = 2\pi \sum_{k=-\infty}^{\infty} X[k]\delta(\omega - k\omega_o) \tag{4.14}$$

Hence the FT of a periodic signal is a series of impulses spaced by the fundamental frequency ω_o. The kth impulse has strength $2\pi X[k]$, where $X[k]$ is the kth FS coefficient. Figure 4.4 illustrates this relationship. Throughout this chapter we denote the strength of impulses in the figures by their height as indicated by the labels on the vertical axis. This is done solely for convenience in presenting the large number of impulses that occur in this material. Using this convention we see that the shape of $X(j\omega)$ is identical to that of $X[k]$.

Equation (4.14) also indicates how to convert between FT and FS representations for periodic signals. The FT is obtained from the FS by placing impulses at integer multiples of ω_o and weighting them by 2π times the corresponding FS coefficient. Given a FT consisting of impulses that are uniformly spaced in ω, we obtain FS coefficients by dividing the impulse strengths by 2π. The fundamental frequency corresponds to the spacing between impulses.

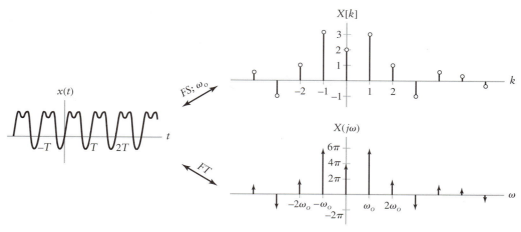

FIGURE 4.4   FS and FT representations for a periodic continuous-time signal.

EXAMPLE 4.6   Find the FT representation for $x(t) = \cos(\omega_o t)$.

Solution: The FS representation for $x(t)$ is

$$\cos(\omega_o t) \xleftrightarrow{\;FS;\;\omega_o\;} X[k] = \begin{cases} \frac{1}{2}, & k = \pm 1 \\ 0, & k \neq \pm 1 \end{cases}$$

Substituting these coefficients into Eq. (4.14) gives

$$\cos(\omega_o t) \xleftrightarrow{\;FT\;} X(j\omega) = \pi\delta(\omega - \omega_o) + \pi\delta(\omega + \omega_o)$$

This pair is depicted graphically in Fig. 4.5.

EXAMPLE 4.7   Find the FT of the *impulse train*

$$p(t) = \sum_{n=-\infty}^{\infty} \delta(t - n\mathcal{T})$$

Solution: We note that $p(t)$ is periodic with fundamental period $\mathcal{T}$, so $\omega_o = 2\pi/\mathcal{T}$ and the FS coefficients are given by

$$P[k] = \frac{1}{\mathcal{T}} \int_{-\mathcal{T}/2}^{\mathcal{T}/2} \delta(t)e^{-jk\omega_o t}\,dt$$

$$= \frac{1}{\mathcal{T}}$$

Substitute these values into Eq. (4.14) to obtain

$$P(j\omega) = \frac{2\pi}{\mathcal{T}} \sum_{k=-\infty}^{\infty} \delta(\omega - k\omega_o)$$

Hence the FT of an impulse train is another impulse train. The spacing between the impulses in frequency is inversely related to the spacing between impulses in time. This FT pair is depicted in Fig. 4.6.

▶ **Drill Problem 4.4**   Determine the FT representation for the periodic square wave depicted in Fig. 4.7.

Answer:

$$X(j\omega) = \sum_{k=-\infty}^{\infty} \frac{2\,\sin(k\pi/2)}{k} \delta\left(\omega - k\frac{\pi}{2}\right)$$

◀

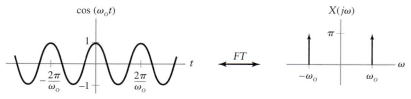

FIGURE 4.5   FT of a cosine function.

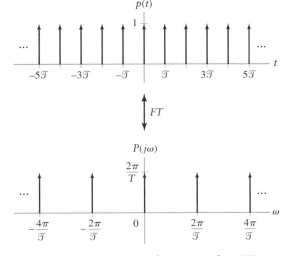

FIGURE 4.6 An impulse train and its FT.

■ RELATING THE **DTFT** TO THE **DTFS**

The method for deriving the DTFT of a discrete-time periodic signal parallels that of the continuous-time case. The DTFS expression for an N periodic signal $x[n]$ is

$$x[n] = \sum_{k=\langle N\rangle} X[k]e^{jk\Omega_o n} \tag{4.15}$$

As in the FS case, the key observation is that the inverse DTFT of a frequency-shifted impulse is a discrete-time complex sinusoid. The DTFT is a 2π periodic function of frequency, and so we may express a frequency-shifted impulse either by expressing one period

$$\delta(\Omega - k\Omega_o), \quad -\pi < \Omega \leq \pi, \quad -\pi < k\Omega_o \leq \pi$$

or by using an infinite series of shifted impulses separated by an interval of 2π to obtain the 2π periodic function

$$\sum_{m=-\infty}^{\infty} \delta(\Omega - k\Omega_o - m2\pi) \tag{4.16}$$

which is depicted in Fig. 4.8. The inverse DTFT of Eq. (4.16) is evaluated using the sifting property of the impulse function. We have

$$\frac{1}{2\pi} e^{jk\Omega_o n} \xleftarrow{\quad DTFT \quad} \sum_{m=-\infty}^{\infty} \delta(\Omega - k\Omega_o - m2\pi) \tag{4.17}$$

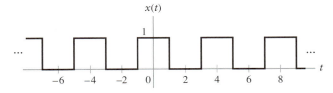

FIGURE 4.7 Square wave for Drill Problem 4.4.

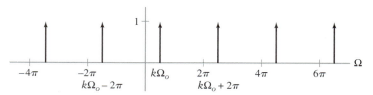

FIGURE 4.8   Infinite series of frequency-shifted impulses that is 2π periodic in frequency Ω.

Hence we identify the complex sinusoid and the frequency-shifted impulse as a DTFT pair. This relationship is a direct consequence of the properties of impulse functions.

Use linearity and substitute Eq. (4.17) into Eq. (4.15) to obtain the DTFT of the periodic signal $x[n]$ as

$$x[n] = \sum_{k=\langle N \rangle} X[k]e^{jk\Omega_o n} \xleftrightarrow{\ DTFT\ } X(e^{j\Omega}) = 2\pi \sum_{k=\langle N \rangle} X[k] \sum_{m=-\infty}^{\infty} \delta(\Omega - k\Omega_o - m2\pi)$$

Since $X[k]$ is N periodic and $N\Omega_o = 2\pi$, we may rewrite the DTFT of $x[n]$ as

$$x[n] = \sum_{k=\langle N \rangle} X[k]e^{jk\Omega_o n} \xleftrightarrow{\ DTFT\ } X(e^{j\Omega}) = 2\pi \sum_{k=-\infty}^{\infty} X[k]\delta(\Omega - k\Omega_o) \quad (4.18)$$

Thus the DTFT representation for a periodic signal is a series of impulses spaced by the fundamental frequency Ω_o. The kth impulse has strength $2\pi X[k]$, where $X[k]$ is the kth DTFS coefficient for $x[n]$. Figure 4.9 depicts both DTFS and DTFT representations for a periodic discrete-time signal. Here again we see that the DTFS $X[k]$ and the corresponding DTFT $X(e^{j\Omega})$ are similar.

Equation (4.18) establishes the relationship between DTFS and DTFT. Given the DTFS coefficients and the fundamental frequency Ω_o, we obtain the DTFT representation by placing impulses at integer multiples of Ω_o and weighting them by 2π times the corresponding DTFS coefficient. We reverse this process to obtain the DTFS coefficients from the DTFT representation. If the DTFT consists of impulses that are uniformly spaced in Ω, then we obtain DTFS coefficients by dividing the impulse strengths by 2π. The fundamental frequency is the spacing between the impulses.

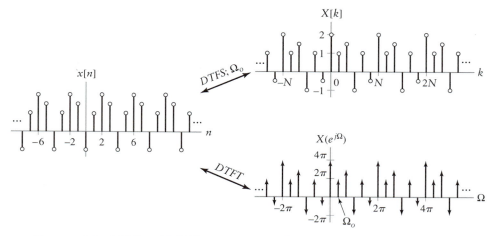

FIGURE 4.9   DTFS and DTFT representations for a periodic discrete-time signal.

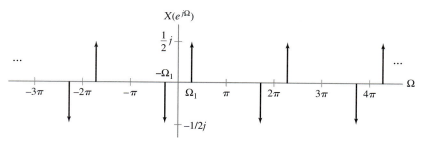

FIGURE 4.10 DTFT of periodic signal for Example 4.8.

EXAMPLE 4.8 Determine the inverse DTFT of the frequency-domain representation depicted in Fig. 4.10.

Solution: Express one period of $X(e^{j\Omega})$ as

$$X(e^{j\Omega}) = \frac{1}{2j}\,\delta(\Omega - \Omega_1) - \frac{1}{2j}\,\delta(\Omega + \Omega_1), \qquad -\pi < \Omega \le \pi$$

Now take the inverse DTFT of each frequency-shifted impulse to obtain

$$x[n] = \frac{1}{2\pi}\left[\frac{1}{2j}\,(e^{j\Omega_1 n} - e^{-j\Omega_1 n})\right]$$

$$= \frac{1}{2\pi}\,\sin(\Omega_1 n)$$

▶ **Drill Problem 4.5** Find both the DTFS and DTFT representations for the periodic signal

$$x[n] = 2\,\cos\!\left(\frac{3\pi}{8}\,n + \frac{\pi}{3}\right) + 4\,\sin\!\left(\frac{\pi}{2}\,n\right)$$

Answer: DTFS: $\Omega_o = 2\pi/16$

$$X[k] = \begin{cases} -2/j, & k = -4 \\ e^{-j\pi/3}, & k = -3 \\ e^{j\pi/3}, & k = 3 \\ 2/j, & k = 4 \\ 0, & \text{otherwise for } -7 \le k \le 8 \end{cases}$$

DTFT: one period

$$X(e^{j\Omega}) = -\frac{4\pi}{j}\,\delta\!\left(\Omega + \frac{\pi}{2}\right) + 2\pi\,e^{-j\pi/3}\delta\!\left(\Omega + \frac{3\pi}{8}\right)$$

$$+ 2\pi\,e^{j\pi/3}\delta\!\left(\Omega - \frac{3\pi}{8}\right) + \frac{4\pi}{j}\,\delta\!\left(\Omega - \frac{\pi}{2}\right), \qquad \pi < \Omega \le \pi$$

◀

4.4 *Convolution and Modulation with Mixed Signal Classes*

In this section we use the FT and DTFT representations of periodic signals to analyze problems involving mixtures of periodic and nonperiodic signals. It is common to have mixing of periodic and nonperiodic signals in convolution and modulation problems. For example, if a periodic signal is applied to a stable filter, the output is expressed as the convolution of the periodic input signal and the nonperiodic impulse response. The tool we use to analyze problems involving mixtures of periodic and nonperiodic continuous-time signal classes is the FT. The DTFT applies to mixtures of periodic and nonperiodic discrete-time signal classes. This analysis is possible since we now have FT and DTFT representations for both periodic and nonperiodic signals. We begin examining convolution of periodic and nonperiodic signals and then focus on modulation applications.

■ CONVOLUTION OF PERIODIC AND NONPERIODIC SIGNALS

In Section 3.6 we established that convolution in the time domain corresponds to multiplication in the frequency domain: that is,

$$y(t) = x(t) * h(t) \overset{FT}{\longleftrightarrow} Y(j\omega) = X(j\omega)H(j\omega)$$

This property may be applied to problems in which one of the time-domain signals, say, $x(t)$, is periodic by using its FT representation. Recall that the FT of a periodic signal $x(t)$ is

$$x(t) \overset{FT}{\longleftrightarrow} X(j\omega) = 2\pi \sum_{k=-\infty}^{\infty} X[k]\delta(\omega - k\omega_o)$$

where $X[k]$ are the FS coefficients. Substitute this representation into the convolution property to obtain

$$y(t) = x(t) * h(t) \overset{FT}{\longleftrightarrow} Y(j\omega) = 2\pi \sum_{k=-\infty}^{\infty} X[k]\delta(\omega - k\omega_o)H(j\omega) \qquad (4.19)$$

$$\overset{FT}{\longleftrightarrow} Y(j\omega) = 2\pi \sum_{k=-\infty}^{\infty} H(jk\omega_o)X[k]\delta(\omega - k\omega_o) \qquad (4.20)$$

where in the last line we have used the sifting property of the impulse function. Figure 4.11 illustrates the multiplication of $X(j\omega)$ and $H(j\omega)$ that occurs in Eq. (4.20). The strength of the kth impulse is adjusted by the value of $H(j\omega)$ evaluated at the frequency at which it is located, that is, $H(jk\omega_o)$. The form of $Y(j\omega)$ corresponds to a periodic signal. Hence $y(t)$ is periodic with the same period as $x(t)$. The most common application of this property is in determining the output of a filter with impulse response $h(t)$ and periodic input $x(t)$.

EXAMPLE 4.9 Let the input to a system with impulse response $h(t) = (1/\pi t)\sin(\pi t)$ be the periodic square wave depicted in Fig. 4.7. Use the convolution property to find the output of this system.

Solution: The frequency response of the system is

$$h(t) \overset{FT}{\longleftrightarrow} H(j\omega) = \begin{cases} 1, & |\omega| \leq \pi \\ 0, & \text{otherwise} \end{cases}$$

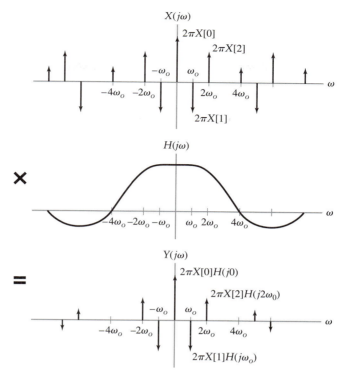

FIGURE 4.11   Convolution property for mixture of periodic and nonperiodic signals.

The FT of the periodic input signal was obtained in Drill Problem 4.4 and is given by

$$X(j\omega) = \sum_{k=-\infty}^{\infty} \frac{2\,\sin(k\pi/2)}{k}\,\delta\!\left(\omega - k\frac{\pi}{2}\right)$$

The FT of the system output is $Y(j\omega) = H(j\omega)X(j\omega)$. This product is depicted in Fig. 4.12, where we see that

$$Y(j\omega) = 2\delta\!\left(\omega + \frac{\pi}{2}\right) + \pi\delta(\omega) + 2\delta\!\left(\omega - \frac{\pi}{2}\right)$$

In effect, the system described by $H(j\omega)$ acts as a lowpass filter, passing the discrete-frequency components at $-\pi/2$, 0, and $\pi/2$, while suppressing all others. Taking the inverse FT of $Y(j\omega)$ gives the output

$$y(t) = \frac{1}{2} + \frac{2}{\pi}\cos\!\left(t\,\frac{\pi}{2}\right)$$

An analogous result is obtained in the discrete-time case. The convolution property is

$$y[n] = x[n] * h[n] \xleftrightarrow{\;DTFT\;} Y(e^{j\Omega}) = X(e^{j\Omega})H(e^{j\Omega})$$

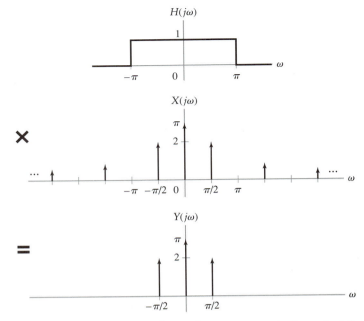

FIGURE 4.12   Application of convolution property in Example 4.9.

We may use this property when $x[n]$ is periodic by substituting the DTFT representation given in Eq. (4.18) for periodic signals,

$$X(e^{j\Omega}) = 2\pi \sum_{k=-\infty}^{\infty} X[k]\delta(\Omega - k\Omega_o)$$

and so obtain

$$y[n] = x[n] * h[n] \xleftrightarrow{\ DTFT\ } Y(e^{j\Omega}) = 2\pi \sum_{k=-\infty}^{\infty} H(e^{jk\Omega_o})X[k]\delta(\Omega - k\Omega_o) \quad (4.21)$$

The form of $Y(e^{j\Omega})$ indicates that $y[n]$ is also periodic with the same period as $x[n]$. This property finds application in evaluating the input–output behavior of LTI systems.

▶ **Drill Problem 4.6**   Let the impulse response of a discrete-time system be

$$h[n] = (\tfrac{1}{2})^n u[n]$$

Determine the output of this system in response to the input

$$x[n] = 3 + \cos(\pi n + \pi/3)$$

Answer:

$$y[n] = 6 + \tfrac{2}{3}\cos(\pi n + \pi/3) \qquad\qquad ◀$$

■ MODULATION OF PERIODIC AND NONPERIODIC SIGNALS

Now consider the modulation property of the FT described by

$$y(t) = g(t)x(t) \xleftrightarrow{\ FT\ } Y(j\omega) = \frac{1}{2\pi} G(j\omega) * X(j\omega)$$

We may use this property even if $x(t)$ is periodic by employing its FT representation. Substituting Eq. (4.14) for $X(j\omega)$ gives

$$y(t) = g(t)x(t) \xleftrightarrow{\;FT\;} Y(j\omega) = G(j\omega) * \sum_{k=-\infty}^{\infty} X[k]\delta(\omega - k\omega_o)$$

The sifting property of the impulse function implies that convolution of any function with a shifted impulse results in a shifted version of the original function. Hence we have

$$y(t) = g(t)x(t) \xleftrightarrow{\;FT\;} Y(j\omega) = \sum_{k=-\infty}^{\infty} X[k]G(j(\omega - k\omega_o)) \qquad (4.22)$$

Modulation of $g(t)$ with the periodic function $x(t)$ gives a FT consisting of a weighted sum of shifted versions of $G(j\omega)$. This result is illustrated in Fig. 4.13. Note that the form of $Y(j\omega)$ corresponds to the FT of a nonperiodic signal. The product of periodic and nonperiodic signals is nonperiodic.

EXAMPLE 4.10 Consider a system with output $y(t) = g(t)x(t)$. Assume $x(t)$ is the square wave depicted in Fig. 4.7. (a) Find $Y(j\omega)$ in terms of $G(j\omega)$. (b) Sketch $Y(j\omega)$ if $g(t) = \cos(t/2)$.

Solution: The square wave has the FS representation

$$x(t) \xleftrightarrow{\;FS;\; \pi/2\;} X[k] = \frac{\sin(k\pi/2)}{k\pi}$$

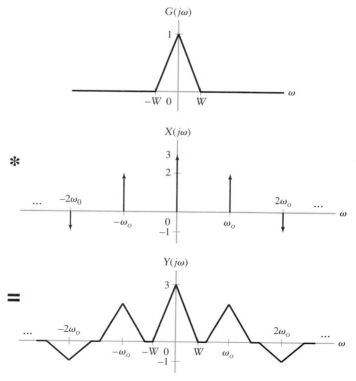

FIGURE 4.13 Modulation property for the combination of a periodic and nonperiodic signal.

(a) Substituting this result into Eq. (4.22) gives

$$Y(j\omega) = \sum_{k=-\infty}^{\infty} \frac{\sin(k\pi/2)}{k\pi} G\left(j\left(\omega - k\frac{\pi}{2}\right)\right)$$

(b) Here we have

$$G(j\omega) = \pi\delta(\omega - \tfrac{1}{2}) + \pi\delta(\omega + \tfrac{1}{2})$$

and thus $Y(j\omega)$ may be expressed as

$$Y(j\omega) = \sum_{k=-\infty}^{\infty} \frac{\sin(k\pi/2)}{k}\left[\delta\left(\omega - \frac{1}{2} - k\frac{\pi}{2}\right) + \delta\left(\omega + \frac{1}{2} - k\frac{\pi}{2}\right)\right]$$

Figure 4.14 depicts the terms constituting $Y(j\omega)$ in the sum near $k = 0$.

▶ **Drill Problem 4.7**   Use the modulation property to determine the frequency response of a system with impulse response

$$h(t) = \frac{\sin(\pi t)}{\pi t}\cos(3\pi t)$$

Answer:

$$H(j\omega) = \begin{cases} \frac{1}{2}, & 2\pi < |\omega| < 4\pi \\ 0, & \text{otherwise} \end{cases}$$ ◀

The discrete-time modulation property is

$$y[n] = x[n]g[n] \xrightarrow{DTFT} Y(e^{j\Omega}) = \frac{1}{2\pi} X(e^{j\Omega}) \circledast G(e^{j\Omega})$$

If $x[n]$ is periodic, then this property is still applicable provided that we use the DTFT representation for $x[n]$, given in Eq. (4.18), as shown by

$$X(e^{j\Omega}) = 2\pi \sum_{k=-\infty}^{\infty} X[k]\delta(\Omega - k\Omega_o)$$

where $X[k]$ are the DTFS coefficients. Substitute $X(e^{j\Omega})$ into the definition of periodic convolution to obtain

$$Y(e^{j\Omega}) = \int_{\langle 2\pi \rangle} \sum_{k=-\infty}^{\infty} X[k]\delta(\theta - k\Omega_o)G(e^{j(\Omega-\theta)}) \, d\theta$$

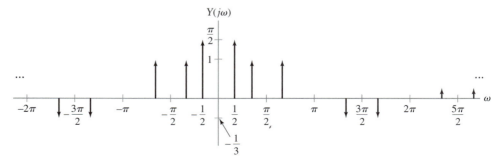

FIGURE 4.14   Solution for Example 4.10(b).

In any 2π interval of θ there are exactly N impulses of the form $\delta(\theta - k\Omega_o)$. This is because $\Omega_o = 2\pi/N$. Hence we can reduce the infinite sum to any N consecutive values of k. Interchanging the sum and integral gives

$$Y(e^{j\Omega}) = \sum_{k=\langle N \rangle} X[k] \int_{\langle 2\pi \rangle} \delta(\theta - k\Omega_o) G(e^{j(\Omega-\theta)}) \, d\theta$$

Now apply the sifting property of the impulse function to evaluate the integral and obtain

$$y[n] = x[n]g[n] \xleftrightarrow{\;DTFT\;} Y(e^{j\Omega}) = \sum_{k=\langle N \rangle} X[k]G(e^{j(\Omega-k\Omega_o)}) \tag{4.23}$$

Modulation of $g[n]$ with the periodic sequence $x[n]$ results in a DTFT consisting of a weighted sum of shifted versions of $G(e^{j\Omega})$. Note that $y[n]$ is nonperiodic since the product of a periodic signal and a nonperiodic signal is nonperiodic. Hence the form of $Y(e^{j\Omega})$ corresponds to a nonperiodic signal.

EXAMPLE 4.11   Consider the following signal:

$$x[n] = \cos\left(\frac{7\pi}{16}n\right) + \cos\left(\frac{9\pi}{16}n\right)$$

Use the modulation property to evaluate the effect of computing the DTFT using only the $2M + 1$ values $x[n]$, $|n| \le M$.

Solution: The DTFT of $x[n]$ is

$$X(e^{j\Omega}) = \pi\delta\left(\Omega + \frac{9\pi}{16}\right) + \pi\delta\left(\Omega + \frac{7\pi}{16}\right) + \pi\delta\left(\Omega - \frac{7\pi}{16}\right) + \pi\delta\left(\Omega - \frac{9\pi}{16}\right), \quad -\pi < \Omega \le \pi$$

$X(e^{j\Omega})$ consists of impulses at $\pm 7\pi/16$ and $\pm 9\pi/16$. Now define a signal $y[n] = x[n]w[n]$, where $w[n]$ is the window function

$$w[n] = \begin{cases} 1, & |n| \le M \\ 0, & |n| > M \end{cases}$$

This window function selects the $2M + 1$ values of $x[n]$ centered on $n = 0$. Comparison of the DTFTs of $y[n]$ and $x[n]$ establishes the effect of using only $2M + 1$ values. The modulation property implies

$$Y(e^{j\Omega}) = \tfrac{1}{2}\{W(e^{j(\Omega+9\pi/16)}) + W(e^{j(\Omega+7\pi/16)}) + W(e^{j(\Omega-7\pi/16)}) + W(e^{j(\Omega-9\pi/16)})\}$$

where

$$W(e^{j\Omega}) = \frac{\sin\left(\Omega \dfrac{(2M+1)}{2}\right)}{\sin\left(\dfrac{\Omega}{2}\right)}$$

We see that windowing introduces replicas of $W(e^{j\Omega})$ at the frequencies $7\pi/16$ and $9\pi/16$ instead of the impulses that are present in $X(e^{j\Omega})$. We may view this as a smearing or broadening of the original impulses. The energy in $Y(e^{j\Omega})$ is now smeared over a band centered on the frequencies of the cosines. The extent of the smearing depends on the width of the mainlobe of $W(e^{j\Omega})$, which is given by $4\pi/(2M+1)$ (see Fig. 3.19).

Figures 4.15(a)–(c) depict $Y(e^{j\Omega})$ for several decreasing values of M. When M is large enough so that the width of the mainlobe of $W(e^{j\Omega})$ is small relative to the separation between

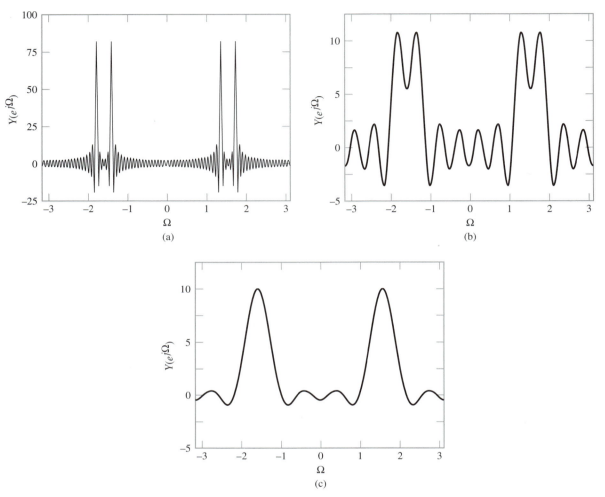

FIGURE 4.15   Effect of windowing a data record. $Y(e^{j\Omega})$ for different values of M assuming $\Omega_1 = 7\pi/16$ and $\Omega_2 = 9\pi/16$. (a) $M = 80$. (b) $M = 12$. (c) $M = 8$.

the frequencies $7\pi/16$ and $9\pi/16$, then $Y(e^{j\Omega})$ is a fairly good approximation to $X(e^{j\Omega})$. This case is depicted in Fig. 4.15(a) using $M = 80$. However, as M decreases and the mainlobe width becomes about the same as the separation between the frequencies $7\pi/16$ and $9\pi/16$, then the peaks associated with each individual shifted version of $W(e^{j\Omega})$ begin to overlap and merge into a single peak. This is illustrated in Figs. 4.15(b) and (c) by using values $M = 12$ and $M = 8$, respectively.

The problem of identifying sinusoidal signals of different frequencies in data is very important and occurs frequently in signal analysis. The preceding example illustrates that our ability to distinguish distinct sinusoids is limited by the length of the data record. If the number of available data points is small relative to the frequency separation, the DTFT is unable to distinguish the presence of two distinct sinusoids. In practice, we are always

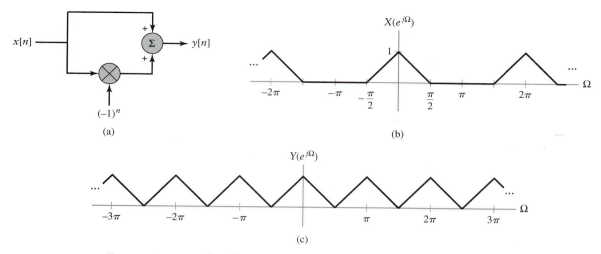

FIGURE 4.16 Drill Problem 4.8. (a) System. (b) Input spectrum. (c) Output spectrum.

restricted to finite-length data records in any signal analysis application. Thus it is important to recognize the effects of windowing.

▶ **Drill Problem 4.8** Consider the system depicted in Fig. 4.16(a). Determine an expression for $Y(e^{j\Omega})$, the DTFT of the output, $y[n]$, and sketch $Y(e^{j\Omega})$ assuming that $X(e^{j\Omega})$ is given in Fig. 4.16(b).

Answer:

$$Y(e^{j\Omega}) = X(e^{j\Omega}) + X(e^{j(\Omega - \pi)})$$

See Fig. 4.16(c) for the sketch. ◀

4.5 *Fourier Transform Representation for Discrete-Time Signals*

In this section we derive a FT representation for discrete-time signals by incorporating impulses into the signal description in the appropriate manner. This representation satisfies all the properties of the FT and thus converts the FT into a powerful tool for analyzing problems involving mixtures of discrete- and continuous-time signals. Our derivation also indicates the relationship between the FT and DTFT. Combining the results of this section with the Fourier transform representations for periodic signals derived in Section 4.3 enables the FT to be used as an analysis tool for any of the four signal classes.

We begin the discussion by establishing a correspondence between continuous-time frequency ω and discrete-time frequency Ω. Define complex sinusoids $x(t) = e^{j\omega t}$ and $g[n] = e^{j\Omega n}$. A connection between the frequencies of these sinusoids is established by requiring $g[n]$ to correspond to $x(t)$. Suppose we force $g[n]$ to be equal to the samples of $x(t)$ taken at intervals of $\mathcal{T}$, that is, $g[n] = x(n\mathcal{T})$. This implies

$$e^{j\Omega n} = e^{j\omega \mathcal{T} n}$$

and we may define $\Omega = \omega\mathcal{T}$. In words, discrete-time frequency Ω corresponds to continuous-time frequency ω multiplied by the sampling interval $\mathcal{T}$.

■ **RELATING THE FT TO THE DTFT**

Now consider the DTFT of an arbitrary discrete-time signal $x[n]$. We have

$$X(e^{j\Omega}) = \sum_{n=-\infty}^{\infty} x[n]e^{-j\Omega n} \tag{4.24}$$

We seek a FT pair $x_\delta(t) \overset{FT}{\longleftrightarrow} X_\delta(j\omega)$ that corresponds to the DTFT pair $x[n] \overset{DTFT}{\longleftrightarrow} X(e^{j\Omega})$. Substitute $\Omega = \omega\mathcal{T}$ into Eq. (4.24) to obtain the function of continuous-time frequency ω:

$$\begin{aligned} X_\delta(j\omega) &= X(e^{j\Omega})|_{\Omega=\omega\mathcal{T}} \\ &= \sum_{n=-\infty}^{\infty} x[n]e^{-j\omega\mathcal{T}n} \end{aligned} \tag{4.25}$$

Take the inverse FT of $X_\delta(j\omega)$ using linearity and the FT pair

$$\delta(t - n\mathcal{T}) \overset{FT}{\longleftrightarrow} e^{-j\omega\mathcal{T}n}$$

to obtain the continuous-time signal description

$$x_\delta(t) = \sum_{n=-\infty}^{\infty} x[n]\delta(t - n\mathcal{T}) \tag{4.26}$$

Hence

$$x_\delta(t) = \sum_{n=-\infty}^{\infty} x[n]\delta(t - n\mathcal{T}) \overset{FT}{\longleftrightarrow} X_\delta(j\omega) = \sum_{n=-\infty}^{\infty} x[n]e^{-j\omega\mathcal{T}n} \tag{4.27}$$

where $x_\delta(t)$ is a continuous-time signal that corresponds to $x[n]$, and the Fourier transform $X_\delta(j\omega)$ corresponds to the discrete-time Fourier transform $X(e^{j\Omega})$. We refer to Eq. (4.26) as the continuous-time representation of $x[n]$. This representation has an associated sampling interval $\mathcal{T}$ that determines the relationship between continuous- and discrete-time frequency: $\Omega = \omega\mathcal{T}$. Figure 4.17 illustrates the relationships between the signals $x[n]$ and

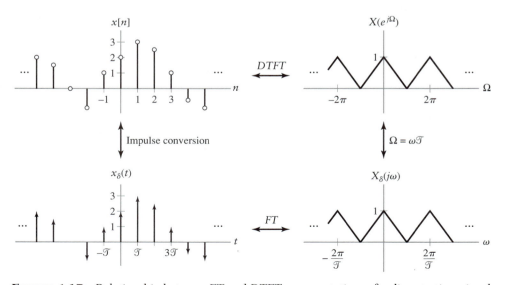

FIGURE 4.17   Relationship between FT and DTFT representations of a discrete-time signal.

$x_\delta(t)$ and the corresponding Fourier representations $X(e^{j\Omega})$ and $X_\delta(j\omega)$. The DTFT $X(e^{j\Omega})$ is 2π periodic in Ω, while the FT $X_\delta(j\omega)$ is $2\pi/\mathcal{T}$ periodic in ω. The discrete-time signal has values $x[n]$, while the corresponding continuous-time signal consists of a series of impulses separated by $\mathcal{T}$ with the nth impulse having strength $x[n]$.

EXAMPLE 4.12 Determine the FT pair associated with a signal whose DTFT is

$$X(e^{j\Omega}) = \frac{1}{1 - ae^{-j\Omega}}$$

Solution: Take the inverse DTFT to obtain

$$x[n] = a^n u[n]$$

Now use Eq. (4.26) to define the continuous-time signal

$$x_\delta(t) = \sum_{n=0}^{\infty} a^n \delta(t - n\mathcal{T})$$

Substituting $\Omega = \omega\mathcal{T}$ gives

$$x_\delta(t) \xleftrightarrow{\quad FT \quad} X_\delta(j\omega) = \frac{1}{1 - ae^{-j\omega\mathcal{T}}}$$

Note the many parallels between the continuous-time representation of a discrete-time signal given in Eq. (4.26) and the FT representation of a periodic signal given in Eq. (4.14). The FT representation is obtained from the FS coefficients by introducing impulses at integer multiples of the fundamental frequency ω_o with the strength of the kth impulse determined by the kth FS coefficient. The FS representation $X[k]$ is discrete valued while the corresponding FT representation $X(j\omega)$ is continuous in frequency. Here $x[n]$ is discrete valued, while $x_\delta(t)$ is continuous. The parameter $\mathcal{T}$ determines the separation between impulses in $x_\delta(t)$ just like ω_o does in $X(j\omega)$. These parallels between $x_\delta(t)$ and $X(j\omega)$ are a direct consequence of the FS–DTFT duality property discussed in Section 3.6. Duality states that the roles of time and frequency in Fourier analysis are interchangeable. Here $x_\delta(t)$ is a continuous-time signal whose FT is a $2\pi/\mathcal{T}$ periodic function of frequency, while $X(j\omega)$ is a continuous frequency signal whose inverse FT is a $2\pi/\omega_o$ periodic function of time.

■ RELATING THE FT TO THE DTFS

In Section 4.3 we derived the FT representation for a periodic continuous-time signal. Previously in this section we have shown how to represent a discrete-time nonperiodic signal with the FT. The remaining case, representation of a discrete-time periodic signal with the FT, is obtained by combining the DTFT representation for a discrete-time periodic signal derived in Section 4.3 with the results of the previous subsection. Once this is accomplished we may use the FT to represent any of the four signal classes.

Recall that the DTFT representation for an N periodic signal $x[n]$ is given in Eq. (4.18) as

$$X(e^{j\Omega}) = \sum_{k=-\infty}^{\infty} 2\pi X[k]\delta(\Omega - k\Omega_o)$$

where $X[k]$ are the DTFS coefficients. Perform the substitution $\Omega = \omega\mathcal{T}$ to obtain the FT representation

$$X_\delta(j\omega) = X(e^{j\omega\mathcal{T}})$$

$$= \sum_{k=-\infty}^{\infty} 2\pi X[k]\delta(\omega\mathcal{T} - k\Omega_o)$$

$$= \sum_{k=-\infty}^{\infty} 2\pi X[k]\delta\left(\mathcal{T}\left(\omega - k\frac{\Omega_o}{\mathcal{T}}\right)\right)$$

Now use the scaling property of the impulse function, $\delta(a\nu) = (1/a)\delta(\nu)$ derived in Chapter 1 to rewrite $X_\delta(j\omega)$ as

$$X_\delta(j\omega) = \frac{2\pi}{\mathcal{T}} \sum_{k=-\infty}^{\infty} X[k]\delta\left(\omega - k\frac{\Omega_o}{\mathcal{T}}\right) \tag{4.28}$$

Recall that $X[k]$ is an N periodic function. This implies that $X_\delta(j\omega)$ is periodic with period $N\Omega_o/\mathcal{T} = 2\pi/\mathcal{T}$. The signal $x_\delta(t)$ corresponding to this FT is most easily obtained by substituting the periodic signal $x[n]$ into Eq. (4.26), that is,

$$x_\delta(t) = \sum_{n=-\infty}^{\infty} x[n]\delta(t - n\mathcal{T}) \tag{4.29}$$

Note that the N periodic nature of $x[n]$ implies that $x_\delta(t)$ is also periodic with fundamental period $N\mathcal{T}$. Hence both $x_\delta(t)$ and $X_\delta(j\omega)$ are N periodic impulse trains as depicted in Fig. 4.18.

▶ **Drill Problem 4.9**   Determine the FT pair associated with the discrete-time periodic signal

$$x[n] = \cos\left(\frac{2\pi}{N}n\right)$$

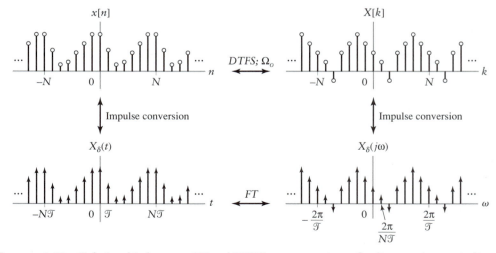

FIGURE 4.18   Relationship between FT and DTFS representations of a discrete-time periodic signal.

Answer:

$$x_\delta(t) = \sum_{n=-\infty}^{\infty} \cos\left(\frac{2\pi}{N}n\right)\delta(t-n\mathcal{T}) \xleftrightarrow{\;FT\;} X_\delta(j\omega) = \frac{\pi}{\mathcal{T}} \sum_{m=-\infty}^{\infty} \delta\left(\omega + \frac{2\pi}{N\mathcal{T}} - m\frac{2\pi}{\mathcal{T}}\right)$$

$$+ \delta\left(\omega - \frac{2\pi}{N\mathcal{T}} - m\frac{2\pi}{\mathcal{T}}\right) \quad\blacktriangleleft$$

4.6 Sampling

In this section we use the FT representation of discrete-time signals to analyze the effects of uniformly sampling a signal. The sampling operation generates a discrete-time signal from a continuous-time signal. Sampling of continuous-time signals is often performed in order to manipulate the signal with a computer or microprocessor. Such manipulations are common in communication, control, and signal-processing systems. We shall show how the DTFT of the sampled signal is related to the FT of the continuous-time signal. Sampling is also frequently performed on discrete-time signals to change the effective data rate, an operation termed *subsampling*. In this case the sampling process discards values of the signal. We examine the impact of subsampling by comparing the DTFT of the sampled signal to the DTFT of the original signal.

■ SAMPLING CONTINUOUS-TIME SIGNALS

Let $x(t)$ be a continuous-time signal. We define a discrete-time signal $x[n]$ that is equal to the "samples" of $x(t)$ at integer multiples of a sampling interval $\mathcal{T}$, that is, $x[n] = x(n\mathcal{T})$. The effect of sampling is evaluated by relating the DTFT of $x[n]$ to the FT of $x(t)$. Our tool for exploring this relationship is the FT representation of discrete-time signals.

Begin with the continuous-time representation for the discrete-time signal $x[n]$ given in Eq. (4.26),

$$x_\delta(t) = \sum_{n=-\infty}^{\infty} x[n]\delta(t-n\mathcal{T})$$

Now substitute $x(n\mathcal{T})$ for $x[n]$ to obtain

$$x_\delta(t) = \sum_{n=-\infty}^{\infty} x(n\mathcal{T})\delta(t-n\mathcal{T})$$

Since $x(t)\delta(t-n\mathcal{T}) = x(n\mathcal{T})\delta(t-n\mathcal{T})$, we may rewrite $x_\delta(t)$ as a product of time functions

$$x_\delta(t) = x(t)p(t) \tag{4.30}$$

where

$$p(t) = \sum_{n=-\infty}^{\infty} \delta(t-n\mathcal{T}) \tag{4.31}$$

Hence Eq. (4.30) implies that we may mathematically represent the sampled signal as the product of the original continuous-time signal and an impulse train, as depicted in Fig. 4.19. This representation is commonly termed *impulse sampling*.

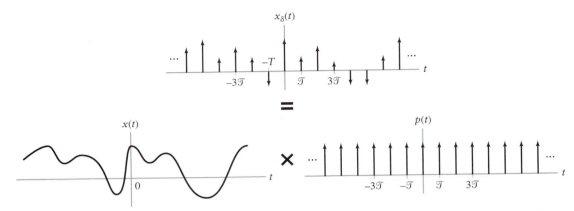

FIGURE 4.19 Mathematical representation of sampling as the product of a given time signal and an impulse train.

The effect of sampling is determined by relating the FT of $x_\delta(t)$ to the FT of $x(t)$. Since multiplication in the time domain corresponds to convolution in the frequency domain, we have

$$X_\delta(j\omega) = \frac{1}{2\pi} X(j\omega) * P(j\omega)$$

Substituting the value for $P(j\omega)$ determined in Example 4.7 into this relationship, we obtain

$$X_\delta(j\omega) = \frac{1}{2\pi} X(j\omega) * \frac{2\pi}{\mathcal{T}} \sum_{k=-\infty}^{\infty} \delta(\omega - k\omega_s)$$

where $\omega_s = 2\pi/\mathcal{T}$ is the sampling frequency. Now convolve $X(j\omega)$ with each of the shifted impulses to obtain

$$X_\delta(j\omega) = \frac{1}{\mathcal{T}} \sum_{k=-\infty}^{\infty} X(j(\omega - k\omega_s)) \tag{4.32}$$

The FT of the sampled signal is given by an infinite sum of shifted versions of the original signal's FT. The shifted versions are offset by integer multiples of ω_s. The shifted versions of $X(j\omega)$ may overlap with each other if ω_s is not large enough compared to the frequency extent of $X(j\omega)$. This effect is demonstrated in Fig. 4.20 by depicting Eq. (4.32) for several different values of $\mathcal{T}$. The frequency content of the signal $x(t)$ is assumed to lie within the frequency band $-W < \omega < W$ for purposes of illustration. In Figs. 4.20(b)–(d) we depict the cases $\omega_s = 3W$, $\omega_s = 2W$, and $\omega_s = \frac{3}{2}W$, respectively. The shifted replicates of $X(j\omega)$ associated with the kth term in Eq. (4.32) are labeled. Note that as $\mathcal{T}$ increases and ω_s decreases, the shifted replicates of $X(j\omega)$ move closer together. They overlap one another when $\omega_s < 2W$.

Overlap in the shifted replicas of the original spectrum is termed *aliasing*, which refers to the phenomenon of a high-frequency component taking on the identity of a low-frequency one. Aliasing distorts the spectrum of the original continuous-time signal. This effect is illustrated in Fig. 4.20(d). Overlap between the replicas of $X(j\omega)$ at $k = 0$ and $k = 1$ occurs for frequencies between $\omega_s - W$ and W. These replicas add, and thus the basic shape of the spectrum changes from portions of a triangle to a constant. The spectrum of the sampled signal no longer has a one-to-one correspondence to that of the original continuous-time signal. This means that we cannot use the spectrum of the sam-

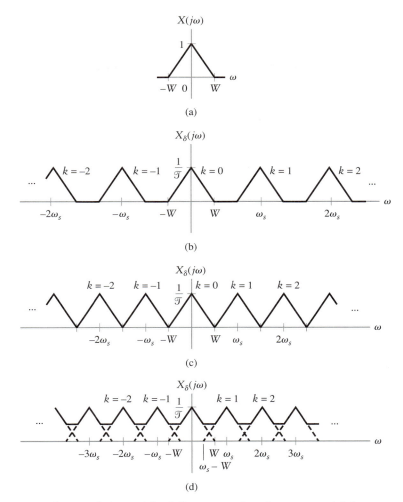

FIGURE 4.20   FT of a sampled signal for different sampling frequencies. (a) Spectrum of continuous-time signal. (b) Spectrum of sampled signal when $\omega_s = 3W$. (c) Spectrum of sampled signal when $\omega_s = 2W$. (d) Spectrum of sampled signal when $\omega_s = \frac{3}{2}W$.

pled signal to analyze the continuous-time signal, and we cannot uniquely reconstruct the continuous-time signal from its samples. The reconstruction problem is addressed in the following section. As Fig. 4.20 illustrates, aliasing is prevented by choosing the sampling interval $\mathcal{T}$ so that $\omega_s > 2W$, where W is the highest frequency component in the signal. This implies that we must satisfy the condition $\mathcal{T} < \pi/W$.

The DTFT of the sampled signal is obtained from $X_\delta(j\omega)$ using the relationship $\Omega = \omega\mathcal{T}$, that is,

$$x[n] \xleftrightarrow{\ DTFT\ } X(e^{j\Omega}) = X_\delta(j\omega)\big|_{\omega=\Omega/\mathcal{T}}$$

This scaling of the independent variable implies that $\omega = \omega_s$ corresponds to $\Omega = 2\pi$. Figures 4.21(a)–(c) depict the DTFTs of the sampled signals corresponding to the FTs in Figs. 4.20(b)–(d). Note that the shape is the same in each case; the only difference is a scaling of the frequency axis. The FTs have period ω_s, while the DTFTs have period 2π.

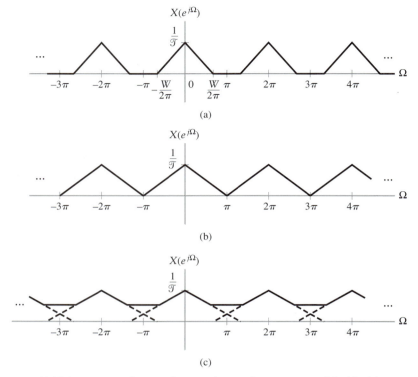

FIGURE 4.21 DTFTs corresponding to the FTs depicted in Figs. 4.20(b)–(d). (a) $\omega_s = 3W$. (b) $\omega_s = 2W$. (c) $\omega_s = \frac{3}{2}W$.

EXAMPLE 4.13 Consider the effect of sampling the sinusoidal signal

$$x(t) = \cos(\pi t)$$

Determine the FT of the sampled signal for the following sampling intervals: (a) $\mathcal{T} = \frac{1}{4}$, (b) $\mathcal{T} = 1$, and (c) $\mathcal{T} = \frac{3}{2}$.

Solution: Use Eq. (4.32) for each value of $\mathcal{T}$. In particular, note that

$$x(t) \xleftrightarrow{\;FT\;} X(j\omega) = \pi\delta(\omega + \pi) + \pi\delta(\omega - \pi)$$

Substitution of $X(j\omega)$ into Eq. (4.32) gives

$$X_\delta(j\omega) = \frac{\pi}{\mathcal{T}} \sum_{k=-\infty}^{\infty} \delta(\omega + \pi - k\omega_s) + \delta(\omega - \pi - k\omega_s)$$

Hence $X_\delta(j\omega)$ consists of pairs of impulses separated by 2π centered on integer multiples of the sampling frequency, ω_s. The sampling frequency is different in each case. Using $\omega_s = 2\pi/\mathcal{T}$ gives (a) $\omega_s = 8\pi$, (b) $\omega_s = 2\pi$, and (c) $\omega_s = 4\pi/3$, respectively. The sampled continuous-time signals and their FTs are depicted in Fig. 4.22.

In case (a) the impulses are clearly paired about multiples of 8π, as depicted in Fig. 4.22(b). As $\mathcal{T}$ increases and ω_s decreases, pairs of impulses associated with different values of k become closer together. In case (b), impulses associated with adjacent indices k superimpose on one another, as illustrated in Fig. 4.22(c). This corresponds to a sampling interval of one-half period. There is an ambiguity here since we cannot uniquely determine the original

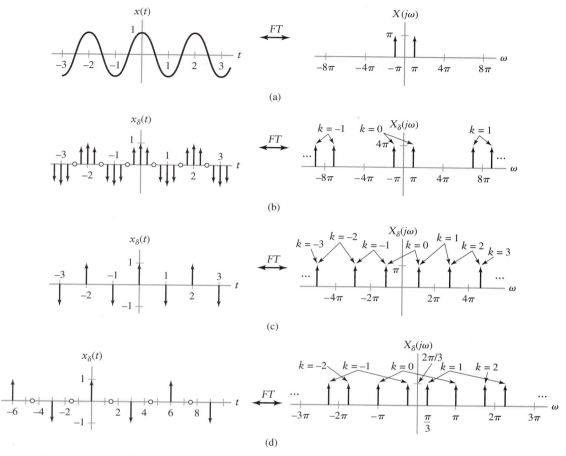

FIGURE 4.22 Effect of sampling a sinusoid at different rates. (a) Original signal and FT. (b) Sampled signal and FT for $\mathcal{T} = \frac{1}{4}$. (c) Sampled signal and FT for $\mathcal{T} = 1$. (d) Sampled signal and FT for $\mathcal{T} = \frac{3}{2}$.

from either $x_\delta(t)$ or $X_\delta(j\omega)$. For example, both $x_1(t) = \cos(\pi t)$ and $x_2(t) = e^{j\pi t}$ result in the same sequence $x[n] = (-1)^n$ for $\mathcal{T} = 1$. In case (c), shown in Fig. 4.22(d), the pairs of impulses associated with each index k are interspersed. In this case we also have an ambiguity. Both the original signal $x(t)$ and the signal $x_1(t) = \cos((\pi/3)t)$ are consistent with the sampled signal $x_\delta(t)$ and spectrum $X_\delta(j\omega)$. Sampling has caused the original sinusoid with frequency π to alias or appear as a new sinusoid of frequency $\pi/3$.

▶ **Drill Problem 4.10** Draw the FT of a sampled version of the continuous-time signal having FT depicted in Fig. 4.23(a) for (a) $\mathcal{T} = \frac{1}{2}$ and (b) $\mathcal{T} = 2$.

Answer: See Figs. 4.23(b) and (c). ◀

* ■ SUBSAMPLING: SAMPLING DISCRETE-TIME SIGNALS

The FT is also very helpful in analyzing the effect of sampling a discrete-time signal, or *subsampling*. Let $y[n] = x[qn]$ be a subsampled version of $x[n]$. We require q to be a

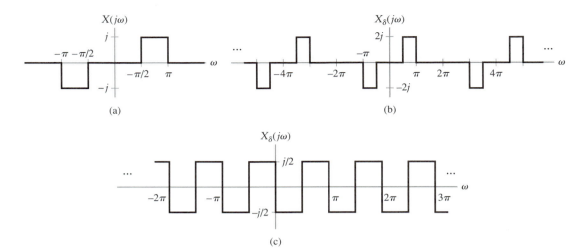

FIGURE 4.23 (a) Spectrum of original signal for Drill Problem 4.10. (b) Spectrum of sampled signal for $\mathcal{T} = \frac{1}{2}$. (c) Spectrum of sampled signal for $\mathcal{T} = 2$.

positive integer for this operation to be meaningful. Our goal is to relate the DTFT of $y[n]$ to the DTFT of $x[n]$. We accomplish this by using the FT to represent $x[n]$ as a sampled version of a continuous-time signal $x(t)$. We then express $y[n]$ as a sampled version of the same underlying continuous-time signal $x(t)$ obtained using a sampling interval q times that associated with $x[n]$.

Use Eq. (4.26) to represent $x[n]$ as the impulse sampled continuous-time signal with sampling interval $\mathcal{T}$, and thus write

$$x_\delta(t) = \sum_{n=-\infty}^{\infty} x[n]\delta(t - n\mathcal{T})$$

Suppose $x[n]$ are the samples of a continuous-time signal $x(t)$, obtained at integer multiples of $\mathcal{T}$. That is, $x[n] = x(n\mathcal{T})$. Let $x(t) \xleftrightarrow{\;FT\;} X(j\omega)$ and apply Eq. (4.32) to obtain

$$X_\delta(j\omega) = \frac{1}{\mathcal{T}} \sum_{k=-\infty}^{\infty} X(j(\omega - k\omega_s))$$

Since $y[n]$ is formed using every qth sample of $x[n]$, we may also express $y[n]$ as a sampled version of $x(t)$. We have

$$y[n] = x[qn]$$
$$= x(nq\mathcal{T})$$

Hence the effective sampling rate for $y[n]$ is $\mathcal{T}' = q\mathcal{T}$. Applying Eq. (4.32) to $y[n]$ gives

$$y_\delta(t) = x(t) \sum_{n=-\infty}^{\infty} \delta(t - n\mathcal{T}') \xleftrightarrow{\;FT\;} Y_\delta(j\omega) = \frac{1}{\mathcal{T}'} \sum_{k=-\infty}^{\infty} X(j(\omega - k\omega_s'))$$

Substituting $\mathcal{T}' = q\mathcal{T}$ and $\omega_s' = \omega_s/q$ in the right-hand member of this pair gives

$$Y_\delta(j\omega) = \frac{1}{q\mathcal{T}} \sum_{k=-\infty}^{\infty} X\left(j\left(\omega - \frac{k}{q}\omega_s\right)\right) \tag{4.33}$$

We have expressed both $X_\delta(j\omega)$ and $Y_\delta(j\omega)$ as functions of $X(j\omega)$. However, $X(j\omega)$ is unknown, since we only know $x[n]$ and do not know $x(t)$. Hence we shall further manipulate the expression for $Y_\delta(j\omega)$ in order to express $Y_\delta(j\omega)$ as a function of $X_\delta(j\omega)$, the FT of $x[n]$. We begin by writing k/q in Eq. (4.33) as a proper fraction, and thus set

$$\frac{k}{q} = l + \frac{m}{q}$$

where l is the integer portion of k/q and m is the remainder. Allowing k to range from $-\infty$ to ∞ corresponds to having l range from $-\infty$ to ∞ and m from 0 to $q - 1$. Hence we may rewrite Eq. (4.33) as the double sum

$$Y_\delta(j\omega) = \frac{1}{q} \sum_{m=0}^{q-1} \left\{ \frac{1}{\mathcal{T}} \sum_{l=-\infty}^{\infty} X\left(j\left(\omega - l\omega_s - \frac{m}{q}\omega_s\right)\right) \right\}$$

Recognizing that the term in braces corresponds to $X_\delta(j(\omega - (m/q)\omega_s))$, we rewrite $Y_\delta(j\omega)$ as

$$Y_\delta(j\omega) = \frac{1}{q} \sum_{m=0}^{q-1} X_\delta\left(j\left(\omega - \frac{m}{q}\omega_s\right)\right) \qquad (4.34)$$

which represents a sum of shifted versions of $X_\delta(j\omega)$ normalized by q.

At this point we convert from the FT representation back to the DTFT in order to express $Y(e^{j\Omega})$ as a function of $X(e^{j\Omega})$. The sampling interval associated with $Y_\delta(j\omega)$ is $\mathcal{T}'$. Using the relationship $\Omega = \omega\mathcal{T}'$ in Eq. (4.34), we have

$$Y(e^{j\Omega}) = Y_\delta(j\omega)\big|_{\omega=\Omega/\mathcal{T}'}$$

$$= \frac{1}{q} \sum_{m=0}^{q-1} X_\delta\left(j\left(\frac{\Omega}{\mathcal{T}'} - \frac{m}{q}\omega_s\right)\right)$$

Now substitute $\mathcal{T}' = q\mathcal{T}$ to obtain

$$Y(e^{j\Omega}) = \frac{1}{q} \sum_{m=0}^{q-1} X_\delta\left(j\left(\frac{\Omega}{q\mathcal{T}} - \frac{m}{q}\omega_s\right)\right)$$

$$= \frac{1}{q} \sum_{m=0}^{q-1} X_\delta\left(\frac{j}{\mathcal{T}}\left(\frac{\Omega}{q} - \frac{m}{q}2\pi\right)\right)$$

The sampling interval associated with $X_\delta(j\omega)$ is $\mathcal{T}$, and so $X(e^{j\Omega}) = X_\delta(j\Omega/\mathcal{T})$. Hence we may substitute $X(e^{j(\Omega/q - m2\pi/q)})$ for $X_\delta((j/\mathcal{T})(\Omega/q - m2\pi/q))$ and obtain

$$Y(e^{j\Omega}) = \frac{1}{q} \sum_{m=0}^{q-1} X(e^{j(\Omega/q - m2\pi/q)})$$

$$= \frac{1}{q} \sum_{m=0}^{q-1} X(e^{j(\Omega - m2\pi)/q}) \qquad (4.35)$$

This equation indicates that $Y(e^{j\Omega})$ is obtained by summing versions of the scaled DTFT $X_q(e^{j\Omega}) = X(e^{j\Omega/q})$ that are shifted by integer multiples of 2π. We may write this result explicitly as

$$Y(e^{j\Omega}) = \frac{1}{q} \sum_{m=0}^{q-1} X_q(e^{j(\Omega - m2\pi)})$$

Figure 4.24 illustrates the relationship between $Y(e^{j\Omega})$ and $X(e^{j\Omega})$ described in Eq. (4.35). Figure 4.24(a) depicts $X(e^{j\Omega})$. Figures 4.24(b)–(d) show the individual terms in the

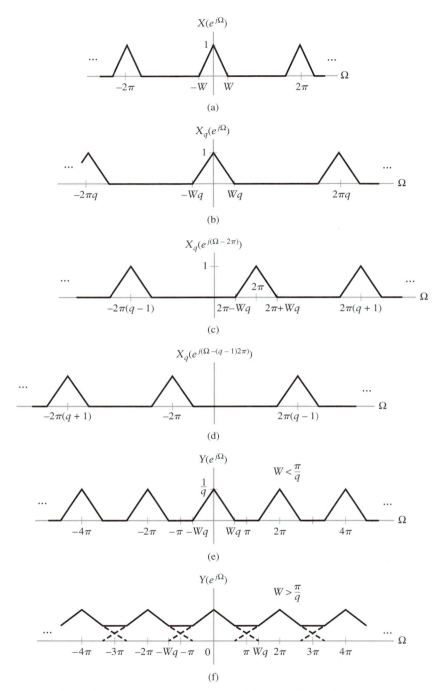

FIGURE 4.24   Effect of subsampling on the DTFT. (a) Original signal spectrum. (b) $m = 0$ term, $X_q(e^{j\Omega})$, in Eq. (4.35). (c) $m = 1$ term in Eq. (4.35). (d) $m = q - 1$ term in Eq. (4.35). (e) $Y(e^{j\Omega})$ assuming $W < \pi/q$. (f) $Y(e^{j\Omega})$ assuming $W > \pi/q$.

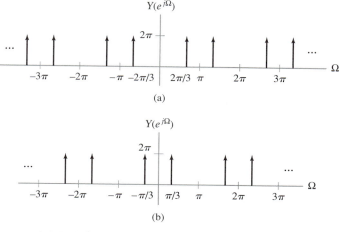

FIGURE 4.25 Solution to Drill Problem 4.11. (a) $q = 2$. (b) $q = 5$.

sum of Eq. (4.35) corresponding to $m = 0$, $m = 1$, and $m = q - 1$. In Figure 4.24(e) we depict $Y(e^{j\Omega})$ assuming that $W < \pi/q$, while Figure 4.24(f) shows $Y(e^{j\Omega})$ assuming that $W > \pi/q$. In this last case there is overlap between the scaled and shifted versions of $X(e^{j\Omega})$ involved in Eq. (4.35) and aliasing occurs. We conclude that aliasing can be prevented if W, the highest frequency component of $X(e^{j\Omega})$, is less than π/q.

▶ **Drill Problem 4.11** Depict the DTFT of the subsampled signal $y[n] = x[qn]$ for $q = 2$ and $q = 5$ assuming

$$x[n] = 2 \cos\left(\frac{\pi}{3} n\right)$$

Answer: See Figs. 4.25(a) and (b). ◀

4.7 *Reconstruction of Continuous-Time Signals from Samples*

The problem of reconstructing a continuous-time signal from samples also involves a mixture of continuous- and discrete-time signals. As illustrated in the block diagram of Fig. 4.26, a system that performs this operation has a discrete-time input signal and a continuous-time output signal. The FT is an ideal tool for analyzing this reconstruction problem, since it may be used to represent both continuous- and discrete-time signals. In this section we first consider the conditions that must be met in order to uniquely reconstruct a continuous-time signal from its samples. Assuming that these conditions are satisfied, we es-

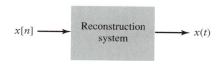

FIGURE 4.26 Block diagram illustrating conversion of a discrete-time signal to a continuous-time signal.

tablish a method for perfect reconstruction. Unfortunately, the ideal reconstruction approach cannot be implemented in any practical system. Hence the section concludes with an analysis of practical reconstruction techniques and their limitations.

▪ SAMPLING THEOREM

Our discussion of sampling indicated that the samples of a signal do not always uniquely determine the corresponding continuous-time signal. For example, if we sample a sinusoid at intervals of a period, then the sampled signal appears to be a constant and we cannot determine whether the original signal was a constant or the sinusoid. Figure 4.27 illustrates this problem by depicting two different continuous-time signals having the same set of samples. We have

$$x[n] = x_1(n\mathcal{T}) = x_2(n\mathcal{T})$$

The samples do not tell us anything about the behavior of the signal in between the sample times. In order to determine how the signal behaves in between the samples, we must specify additional constraints on the continuous-time signal. One such set of constraints that is very useful in practice involves requiring the signal to make smooth transitions from one sample to another. The smoothness, or rate at which the time-domain signal changes, is directly related to the maximum frequency present in the signal. Hence constraining smoothness in the time domain corresponds to limiting the signal bandwidth.

There is a one-to-one correspondence between the time-domain and frequency-domain representations for a signal. Thus we may also view the problem of reconstructing the continuous-time signal in the frequency domain. In order to uniquely reconstruct a continuous-time signal from its samples there must be a unique correspondence between the FTs of the continuous-time signal and the sampled signal. These FTs are uniquely related if the sampling process does not introduce aliasing. Aliasing distorts the spectrum of the original signal, as we discovered in the previous section, and destroys the one-to-one relationship between the continuous-time and sampled signal FTs. This suggests that a condition for unique correspondence between the continuous-time signal and its samples is equivalent to the condition for the prevention of aliasing. This requirement is formally stated as follows:

Sampling Theorem Let $x(t) \xleftrightarrow{\text{FT}} X(j\omega)$ represent a bandlimited signal so that $X(j\omega) = 0$ for $|\omega| > \omega_m$. If $\omega_s > 2\omega_m$, where $\omega_s = 2\pi/\mathcal{T}$ is the sampling frequency, then $x(t)$ is uniquely determined by its samples $x(n\mathcal{T})$, $n = 0, \pm 1, \pm 2, \ldots$.

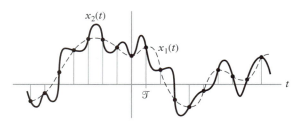

FIGURE 4.27 Two continuous-time signals, $x_1(t)$ (dashed line) and $x_2(t)$ (solid line), that have the same set of samples.

The minimum sampling frequency, $2\omega_m$, is often termed the *Nyquist sampling rate* or *Nyquist rate*. The actual sampling frequency, ω_s, is commonly referred to as the *Nyquist frequency* when discussing the FT of either the continuous-time or sampled signal. We note that in many problems it is more convenient to evaluate the sampling theorem using frequency expressed in units of hertz. If $f_m = \omega_m/2\pi$ is the highest frequency present in the signal and f_s denotes the sampling frequency, both expressed in units of hertz, then the sampling theorem states that $f_s > 2f_m$, where $f_s = 1/\mathcal{T}$. Equivalently, we must have $\mathcal{T} < 1/2f_m$ to satisfy the conditions of the sampling theorem.

EXAMPLE 4.14 Suppose $x(t) = \sin(10\pi t)/\pi t$. Determine the conditions on the sampling interval $\mathcal{T}$ so that $x(t)$ is uniquely represented by the discrete-time sequence $x[n] = x(n\mathcal{T})$.

Solution: In order to apply the sampling theorem, we must first determine the maximum frequency, ω_m, present in $x(t)$. Taking the FT, we have

$$X(j\omega) = \begin{cases} 1, & |\omega| \le 10\pi \\ 0, & |\omega| > 10\pi \end{cases}$$

as depicted in Fig. 4.28. We have $\omega_m = 10\pi$. Hence we require

$$\frac{2\pi}{\mathcal{T}} > 20\pi$$

or

$$\mathcal{T} < \tfrac{1}{10}$$

▶ **Drill Problem 4.12** Determine the conditions on the sampling interval $\mathcal{T}$ so that $x(t) = \cos(2\pi t)\sin(\pi t)/\pi t + 3\sin(6\pi t)\sin(2\pi t)/\pi t$ is uniquely represented by the discrete-time sequence $x[n] = x(n\mathcal{T})$.

Answer: $\mathcal{T} < \tfrac{1}{8}$. ◀

We are often interested in only the lower-frequency components of a signal and would like to sample the signal at a rate ω_s less than twice the highest frequency actually present. A reduced sampling rate can be used if the signal is passed through a continuous-time lowpass filter prior to sampling. Ideally, this filter passes frequency components below $\omega_s/2$ without distortion and suppresses any frequency components above $\omega_s/2$. Such a filter prevents aliasing and is thus termed an *anti-aliasing filter*. A practical anti-aliasing filter will change from passband to stopband gradually. To compensate for the filter's transition band, the passband is usually chosen to include the maximum signal frequency of interest, and the sampling frequency, ω_s, is chosen so that $\omega_s/2$ is in the anti-aliasing

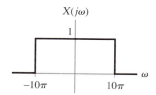

FIGURE 4.28 FT of continuous-time signal for Example 4.14.

filter stopband. This issue is discussed further in Section 4.8. An anti-aliasing filter is normally used even if the signal of interest is bandlimited to less than $\omega_s/2$ to avoid aliasing associated with the presence of measurement or electronic noise.

■ IDEAL RECONSTRUCTION

The sampling theorem indicates how fast we must sample a signal so that the samples uniquely represent the continuous-time signal. Now we consider the problem of reconstructing the continuous-time signal from these samples. This problem is solved most easily in the frequency domain using the FT. Recall that if $x(t) \overset{FT}{\longleftrightarrow} X(j\omega)$, then the FT representation for the sampled signal is given by Eq. (4.32), reproduced here as

$$X_\delta(j\omega) = \frac{1}{\mathcal{T}} \sum_{k=-\infty}^{\infty} X(j\omega - jk\omega_s)$$

Figures 4.29(a) and (b) depict $X(j\omega)$ and $X_\delta(j\omega)$, respectively, assuming that the conditions of the sampling theorem are satisfied.

The goal of reconstruction is to apply some operation to $X_\delta(j\omega)$ that converts it back to $X(j\omega)$. Any such operation must eliminate the replicates of $X(j\omega)$ that appear at $k\omega_s$. This is accomplished by multiplying $X_\delta(j\omega)$ by $H_r(j\omega)$, where

$$H_r(j\omega) = \begin{cases} \mathcal{T}, & |\omega| \le \omega_s/2 \\ 0, & |\omega| > \omega_s/2 \end{cases} \tag{4.36}$$

as depicted in Fig. 4.29(c). We now have

$$X(j\omega) = X_\delta(j\omega)H_r(j\omega) \tag{4.37}$$

Note that multiplication by $H_r(j\omega)$ will not recover $X(j\omega)$ from $X_\delta(j\omega)$ if the conditions of the sampling theorem are not met and aliasing occurs.

Multiplication in the frequency domain transforms to convolution in the time domain, and so Eq. (4.37) implies

$$x(t) = x_\delta(t) * h_r(t)$$

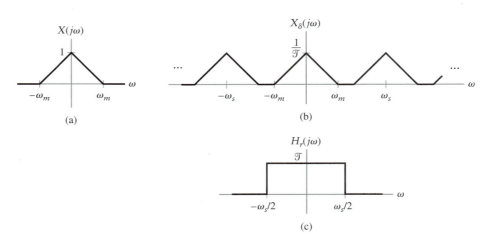

FIGURE 4.29 Ideal reconstruction. (a) Original signal spectrum. (b) Sampled signal spectrum. (c) Frequency response of reconstruction filter.

where $h_r(t) \overset{FT}{\longleftrightarrow} H_r(j\omega)$. Substituting for $x_\delta(t)$ in this relation gives

$$x(t) = h_r(t) * \sum_{n=-\infty}^{\infty} x[n]\delta(t - nT)$$

$$= \sum_{n=-\infty}^{\infty} x[n]h_r(t - nT)$$

Now use

$$h_r(t) = \frac{T \sin\left(\frac{\omega_s}{2} t\right)}{\pi t}$$

to obtain

$$x(t) = \sum_{n=-\infty}^{\infty} x[n] \; \text{sinc}\left(\frac{\omega_s}{2\pi}(t - nT)\right) \tag{4.38}$$

In the time domain we reconstruct $x(t)$ as a weighted sum of sinc functions shifted by the sampling interval. The weights correspond to the values of the discrete-time sequence. This reconstruction operation is illustrated in Fig. 4.30. The value of $x(t)$ at $t = nT$ is given by $x[n]$ because all of the shifted sinc functions go through zero at nT except the nth one, and its value is $x[n]$. The value of $x(t)$ in between integer multiples of T is determined by all of the values of the sequence $x[n]$.

The operation described in Eq. (4.38) is commonly referred to as *ideal bandlimited interpolation*, since it indicates how to interpolate in between the samples of a bandlimited signal. In practice this equation cannot be implemented. First, it represents a noncausal system. The output, $x(t)$, depends on past and future values of the input, $x[n]$. Second, the influence of each sample extends over an infinite amount of time because $h_r(t)$ has infinite time duration.

■ PRACTICAL RECONSTRUCTION—THE ZERO-ORDER HOLD

Practical reconstruction of continuous-time signals is often implemented with a device known as a *zero-order hold*, which simply maintains or holds the value $x[n]$ for T seconds, as depicted in Fig. 4.31. This causes sharp transitions in $x_o(t)$ at integer multiples of T and produces a stairstep approximation to the continuous-time signal. Once again, the FT offers a means for analyzing the quality of this approximation.

The zero-order hold is represented mathematically as a weighted sum of rectangular pulses shifted by integer multiples of the sampling interval. Let

$$h_o(t) = \begin{cases} 1, & 0 < t < T \\ 0, & \text{otherwise} \end{cases}$$

as depicted in Fig. 4.32. The output of the zero-order hold is expressed in terms of $h_o(t)$ as

$$x_o(t) = \sum_{n=-\infty}^{\infty} x[n]h_o(t - nT) \tag{4.39}$$

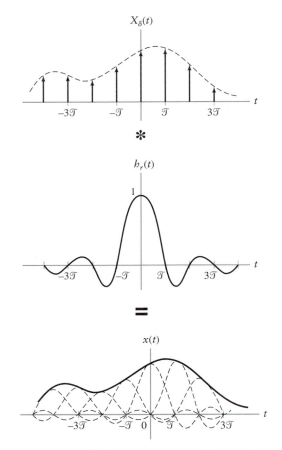

FIGURE 4.30   Ideal reconstruction in the time domain.

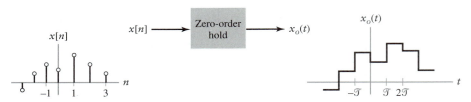

FIGURE 4.31   Zero-order hold.

FIGURE 4.32   Rectangular pulse used to analyze the zero-order hold.

We recognize Eq. (4.39) as the convolution of the impulse sampled signal $x_\delta(t)$ with $h_o(t)$, as shown by

$$x_o(t) = h_o(t) * \sum_{n=-\infty}^{\infty} x[n]\delta(t - n\mathcal{T})$$
$$= h_o(t) * x_\delta(t)$$

Now take the FT of $x_o(t)$ using the convolution–multiplication property of the FT to obtain

$$X_o(j\omega) = H_o(j\omega)X_\delta(j\omega)$$

where

$$h_o(t) \xleftrightarrow{\text{FT}} H_o(j\omega) = 2e^{-j\omega\mathcal{T}/2}\frac{\sin\left(\omega\dfrac{\mathcal{T}}{2}\right)}{\omega}$$

Figure 4.33 depicts the effect of the zero-order hold in the frequency domain, assuming that $\mathcal{T}$ is chosen to satisfy the sampling theorem. Comparing $X_o(j\omega)$ to $X(j\omega)$, we see that the zero-order hold introduces three forms of modification:

1. A linear phase shift corresponding to a time delay of $\mathcal{T}/2$ seconds.
2. The portion of $X_\delta(j\omega)$ between $-\omega_m$ and ω_m is distorted by the curvature of the mainlobe of $H_o(j\omega)$.
3. Distorted and attenuated versions of $X(j\omega)$ remain centered at nonzero multiples of ω_s.

By holding each value $x[n]$ for $\mathcal{T}$ seconds, we are introducing a time shift of $\mathcal{T}/2$ seconds into $x_o(t)$. This is the source of modification 1. The sharp transitions in $x_o(t)$ associated with the stairstep approximation suggest the presence of high-frequency components and are consistent with modification 3. Both modifications 1 and 2 are reduced by increasing ω_s or, equivalently, decreasing $\mathcal{T}$.

In some applications the modifications associated with the zero-order hold may be acceptable. In others, further processing of $x_o(t)$ may be desirable to reduce the distortion associated with modifications 2 and 3. Generally, a delay of $\mathcal{T}/2$ seconds is of no real consequence. The second and third modifications listed above are eliminated by passing $x_o(t)$ through a continuous-time compensation filter with frequency response

$$H_c(j\omega) = \begin{cases} \dfrac{\omega\mathcal{T}}{2\sin(\omega\mathcal{T}/2)}, & |\omega| \leq \omega_m \\ 0, & |\omega| > \omega_s - \omega_m \end{cases}$$

This frequency response is depicted in Fig. 4.34. On $|\omega| \leq \omega_m$ the compensation filter reverses the distortion introduced by the mainlobe curvature of $H_o(j\omega)$. On $|\omega| > \omega_s - \omega_m$, $H_c(j\omega)$ removes the energy in $X_o(j\omega)$ centered at nonzero multiples of ω_s. The value of $H_c(j\omega)$ does not matter on the frequency band $\omega_m < |\omega| < \omega_s - \omega_m$ since $X_o(j\omega)$ is zero. $H_c(j\omega)$ is often termed an *anti-imaging filter* because it eliminates the distorted "images" of $X(j\omega)$ present at nonzero multiples of ω_s. A block diagram representing the compensated zero-order hold reconstruction process is depicted in Fig. 4.35. The anti-imaging filter smooths out the step discontinuities in $x_o(t)$.

There are several practical issues that arise in designing and building an anti-imaging filter. We cannot obtain a causal anti-imaging filter that has zero phase. Hence a practical

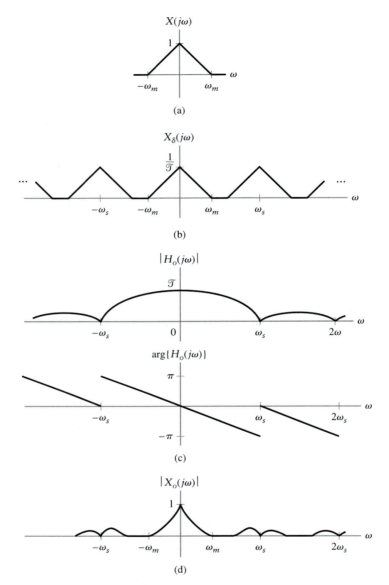

FIGURE 4.33   Effect of the zero-order hold in the frequency domain. (a) Original continuous-time signal spectrum. (b) FT of sampled signal. (c) Magnitude and phase of $H_o(j\omega)$. (d) Magnitude spectrum of signal reconstructed using zero-order hold.

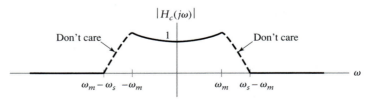

FIGURE 4.34   Frequency response of a compensation filter for eliminating some of the distortion introduced by the zero-order hold.

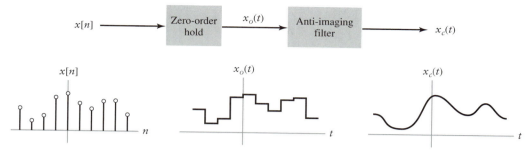

FIGURE 4.35 Block diagram of a practical reconstruction system.

filter will introduce some phase distortion. In many cases linear phase in the passband, $|\omega| \le \omega_m$, is acceptable, since linear phase distortion corresponds to an additional time delay. The difficulty of approximating $|H_c(j\omega)|$ depends on the separation between ω_m and $\omega_s - \omega_m$. If this distance, $\omega_s - 2\omega_m$, is large, then the mainlobe curvature of $H_o(j\omega)$ is very small, and a very good approximation is obtained by simply setting $|H_c(j\omega)| = 1$. Second, the region $\omega_m < \omega < \omega_s - \omega_m$ is used to make the transition from passband to stopband. If $\omega_s - 2\omega_m$ is large, then the transition band of the filter is large. Filters with large transition bands are much easier to design and build than those with small transition bands. Hence the requirements on an anti-imaging filter are greatly reduced by choosing $\mathcal{T}$ sufficiently small so that $\omega_s \gg 2\omega_m$. A more detailed discussion of filter design is given in Chapter 8.

It is common in practical reconstruction schemes to increase the effective sampling rate of the discrete-time signal prior to the zero-order hold. This technique is known as *oversampling*. It is done to relax the requirements on the anti-imaging filter, as illustrated in the following example. Although this increases the complexity of the discrete-time hardware, it usually produces a decrease in overall system cost for a given level of reconstruction quality.

EXAMPLE 4.15 In this example we explore the benefits of oversampling in an audio compact disc player. The maximum signal frequency is 20 kHz because we recall from Chapter 1 that the ear is sensitive to frequencies up to 20 kHz. Consider two cases: (a) basic sampling rate of $1/\mathcal{T}_1 = 44.1$ kHz; (b) 8 times oversampling for an effective sampling rate of $1/\mathcal{T}_2 = 352.8$ kHz. In each case, determine the constraints on the magnitude response of an anti-imaging filter so that the overall magnitude response of the zero-order hold reconstruction system is between 0.99 and 1.01 in the signal passband and less than 10^{-3} to the images of the signal spectrum that are located at multiples of the sampling frequency.

Solution: In this example, it is convenient to express frequency in units of hertz rather than radians per second. This is explicitly indicated by replacing ω with f and by representing the frequency responses $H_o(j\omega)$ and $H_c(j\omega)$ as $H_o'(jf)$ and $H_c'(jf)$. The overall magnitude response of the zero-order hold followed by an anti-imaging filter $H_c'(jf)$ is $|H_o'(jf)||H_c'(jf)|$. Our goal is to find the acceptable range of $|H_c'(jf)|$ so the product $|H_o'(jf)||H_c'(jf)|$ satisfies the response constraints. Figures 4.36(a) and (b) depict $|H_o'(jf)|$ assuming sampling rates of 44.1 kHz and 352.8 kHz, respectively. The dashed lines in each figure denote the signal passband and its images. At the lower sampling rate we see that the signal and its images occupy the majority of the spectrum; they are separated by 4.1 kHz. In the 8 times oversampling case the signal and its images occupy a very small portion of the spectrum; they are separated by 312.8 kHz.

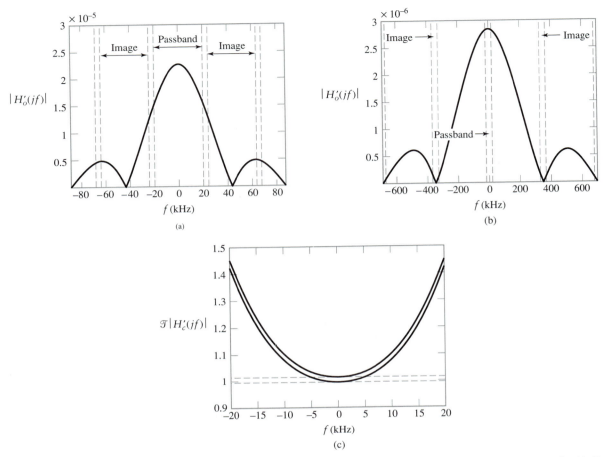

FIGURE 4.36   Anti-imaging filter design with and without oversampling. (a) Magnitude of $H_o'(jf)$ for 44.1-kHz sampling rate. Dashed lines denote signal passband and images. (b) Magnitude of $H_o'(jf)$ for 8 times oversampling (352.8-kHz sampling rate). Dashed lines denote signal passband and images. (c) Normalized constraints on passband response of anti-imaging filter. Solid lines assume 44.1-kHz sampling rate, dashed lines assume 8 times oversampling. The normalized filter response must lie between each pair of lines.

The passband constraint is $0.99 < |H_o'(jf)||H_c'(jf)| < 1.01$, which implies

$$\frac{0.99}{|H_o'(jf)|} < |H_c'(jf)| < \frac{1.01}{|H_o'(jf)|}, \qquad -20 \text{ kHz} < f < 20 \text{ kHz}$$

Figure 4.36(c) depicts these constraints for both cases. Here we have multiplied $|H_c'(jf)|$ by the sampling intervals $\mathcal{T}_1$ and $\mathcal{T}_2$ so that both cases may be displayed with the same scale. Note that case (a) requires substantial curvature in $|H_c'(jf)|$ to eliminate the passband distortion introduced by the mainlobe of $H_o'(jf)$. At the edge of the passband the bounds are

$$1.4257 < \mathcal{T}_1|H_c'(jf_m)| < 1.4545, \qquad f_m = 20 \text{ kHz}$$

for case (a) and

$$0.9953 < \mathcal{T}_2|H_c'(jf_m)| < 1.0154, \qquad f_m = 20 \text{ kHz}$$

for case (b).

The image rejection constraint implies $|H_o'(jf)||H_c'(jf)| < 0.001$ for all frequencies at which images are present. This condition is simplified somewhat by considering only the frequency at which $|H_o'(jf)|$ is largest. The maximum value of $|H_o'(jf)|$ in the image frequency bands occurs at the smallest frequency in the first image: 24.1 kHz in case (a) and 332.8 kHz in case (b). The values of $|H_o'(jf)|/\mathcal{T}_1$ and $|H_o'(jf)|/\mathcal{T}_2$ at these frequencies are 0.5763 and 0.0598, respectively. This implies the bounds

$$\mathcal{T}_1|H_c'(jf)| < 0.0017, \qquad f > 24.1 \text{ kHz}$$

and

$$\mathcal{T}_2|H_c'(jf)| < 0.0167, \qquad f > 332.8 \text{ kHz}$$

for cases (a) and (b), respectively.

Hence the anti-imaging filter for case (a) must show a transition from a value of $1.4257/\mathcal{T}_1$ to $0.0017/\mathcal{T}_1$ over an interval of 4.1 kHz. In contrast, with 8 times oversampling the filter must show a transition from $0.9953/\mathcal{T}_2$ to $0.0167/\mathcal{T}_2$ over a frequency interval of 312.8 kHz. Thus oversampling not only increases the transition width by a factor of almost 80, but also relaxes the stopband attenuation constraint by a factor of more than 10.

*4.8   *Discrete-Time Processing of Continuous-Time Signals*

In this section we use Fourier methods to discuss and analyze a typical system for discrete-time processing of continuous-time signals. There are several advantages to processing a continuous-time signal with a discrete-time system. These advantages result from the power and flexibility of discrete-time computing devices. First, a broad class of signal manipulations are more easily performed using the arithmetic operations of a computer than using analog components. Second, implementing a system in a computer only involves writing a set of instructions or program for the computer to execute. Third, the discrete-time system is easily changed by modifying the program. Often the system can be modified in real time to optimize some criterion associated with the processed signal. Yet another advantage of discrete-time processing is the direct dependence of the dynamic range and signal-to-noise ratio on the number of bits used to represent the discrete-time signal. These advantages have led to a proliferation of computing devices designed specifically for discrete-time signal processing.

A minimal system for discrete-time processing of continuous-time signals must contain a sampling device, and a computing device for implementing the discrete-time system. If the processed signal is to be converted back to continuous time, then reconstruction is also necessary. More sophisticated systems may also utilize oversampling, decimation, and interpolation. *Decimation* and *interpolation* are methods for changing the effective sampling rate of a discrete-time signal. Decimation reduces the effective sampling rate, while interpolation increases the effective sampling rate. Judicious use of these methods can reduce the cost of the overall system. We begin with an analysis of a basic system for processing continuous-time signals. We conclude by revisiting oversampling and examining the role of interpolation and decimation in systems that process continuous-time signals.

■ BASIC DISCRETE-TIME SIGNAL-PROCESSING SYSTEM

A typical system for processing continuous-time signals in discrete time is illustrated in Fig. 4.37(a). A continuous-time signal is first passed through a lowpass anti-aliasing filter

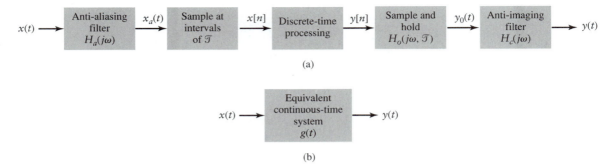

FIGURE 4.37   Block diagram for discrete-time processing of continuous-time signals. (a) A basic system. (b) Equivalent continuous-time system.

and then sampled at intervals of $\mathcal{T}$ to convert it to a discrete-time signal. The sampled signal is then processed by a discrete-time system to impart some desired effect to the signal. For example, the discrete-time system may represent a filter designed to have a specific frequency response, such as an equalizer. After processing, the signal is converted back to continuous-time format. A zero-order hold device converts the discrete-time signal back to continuous time, and an anti-imaging filter removes the distortion introduced by the zero-order hold.

This combination of operations may be reduced to an equivalent continuous-time filter using the FT as an analysis tool. The idea is to find a continuous-time system $g(t) \overset{FT}{\longleftrightarrow} G(j\omega)$ such that $Y(j\omega) = G(j\omega)X(j\omega)$ as depicted in Fig. 4.37(b). Hence $G(j\omega)$ has the same effect on the input as the system in Fig. 4.37(a). We assume for this analysis that the discrete-time processing operation is represented by a discrete-time system with frequency response $H(e^{j\Omega})$. Recall that $\Omega = \omega\mathcal{T}$, where $\mathcal{T}$ is the sampling interval, so the discrete-time system has a continuous-time frequency response $H(e^{j\omega\mathcal{T}})$. Also, the frequency response associated with the zero-order hold device is

$$H_o(j\omega, \mathcal{T}) = 2e^{-j\omega\mathcal{T}/2} \frac{\sin\left(\omega\dfrac{\mathcal{T}}{2}\right)}{\omega}$$

The first operation applied to $x(t)$ is the continuous-time anti-aliasing filter, whose output has FT given by

$$X_a(j\omega) = H_a(j\omega)X(j\omega)$$

After sampling, Eq. (4.32) indicates that the FT representation for $x[n]$ is

$$\begin{aligned}
X_\delta(j\omega) &= \frac{1}{\mathcal{T}} \sum_{k=-\infty}^{\infty} X_a(j(\omega - k\omega_s)) \\
&= \frac{1}{\mathcal{T}} \sum_{k=-\infty}^{\infty} H_a(j(\omega - k\omega_s))X(j(\omega - k\omega_s))
\end{aligned} \qquad (4.40)$$

where $\omega_s = 2\pi/\mathcal{T}$ is the sampling frequency. The discrete-time system modifies $X_\delta(j\omega)$ by $H(e^{j\omega\mathcal{T}})$, producing

$$Y_\delta(j\omega) = \frac{1}{\mathcal{T}} H(e^{j\omega\mathcal{T}}) \sum_{k=-\infty}^{\infty} H_a(j(\omega - k\omega_s))X(j(\omega - k\omega_s))$$

The reconstruction process modifies $Y_\delta(j\omega)$ by the product $H_o(j\omega, \mathcal{T})H_c(j\omega)$, and thus we may write

$$Y(j\omega) = \frac{1}{\mathcal{T}} H_o(j\omega, \mathcal{T})H_c(j\omega)H(e^{j\omega\mathcal{T}}) \sum_{k=-\infty}^{\infty} H_a(j(\omega - k\omega_s))X(j(\omega - k\omega_s))$$

Assuming that aliasing does not occur, the anti-imaging filter $H_c(j\omega)$ eliminates frequency components above $\omega_s/2$ and thus eliminates all the terms in the infinite sum except for the $k = 0$ term. We therefore have

$$Y(j\omega) = \frac{1}{\mathcal{T}} H_o(j\omega, \mathcal{T})H_c(j\omega)H(e^{j\omega\mathcal{T}})H_a(j\omega)X(j\omega)$$

This expression indicates that the overall system is equivalent to a continuous-time LTI system having the frequency response

$$G(j\omega) = \frac{1}{\mathcal{T}} H_o(j\omega, \mathcal{T})H_c(j\omega)H(e^{j\omega\mathcal{T}})H_a(j\omega) \qquad (4.41)$$

Note that this correspondence to a continuous-time LTI system assumes the absence of aliasing.

■ OVERSAMPLING

In Section 4.7 we noted that increasing the effective sampling rate associated with a discrete-time signal prior to the use of a zero-order hold for converting the discrete-time signal back to continuous time relaxes the requirements on the anti-imaging filter. Similarly, the requirements on the anti-aliasing filter are relaxed if the sampling rate is chosen significantly greater than the Nyquist rate. This allows a wide transition band in the anti-aliasing filter.

An anti-aliasing filter prevents aliasing by limiting the signal bandwidth prior to sampling. While the signal of interest may have maximum frequency W, in general the continuous-time signal will have energy at higher frequencies due to the presence of noise and other nonideal characteristics. Such a situation is illustrated in Fig. 4.38(a). The shaded area of the spectrum represents energy at frequencies above the maximum signal frequency; we shall refer to this component as noise. The anti-aliasing filter is chosen to prevent this noise from aliasing back down into the band of interest. The magnitude response of a practical anti-aliasing filter cannot go from unit gain to zero at frequency W, but instead goes from passband to stopband over a range of frequencies as depicted in Fig. 4.38(b). Here the stopband of the filter begins at W_s, and $W_t = W_s - W$ denotes the width of the transition band. The spectrum of the filtered signal $X_a(j\omega)$ now has maximum frequency W_s as depicted in Fig. 4.38(c). This signal is sampled at rate ω_s, resulting in the spectrum $X_\delta(j\omega)$ depicted in Fig. 4.38(d). We have drawn $X_\delta(j\omega)$ assuming that ω_s is large enough to prevent aliasing. As ω_s decreases, replicas of the original signal spectrum begin to overlap and aliasing occurs.

In order to prevent the noise from aliasing with itself, we require $\omega_s - W_s > W_s$ or $\omega_s > 2W_s$ as predicted by the sampling theorem. However, because of the subsequent discrete-time processing we often do not care if the noise aliases with itself, but rather wish to prevent the noise from aliasing back into the signal band $-W < \omega < W$. This implies that we must have

$$\omega_s - W_s > W$$

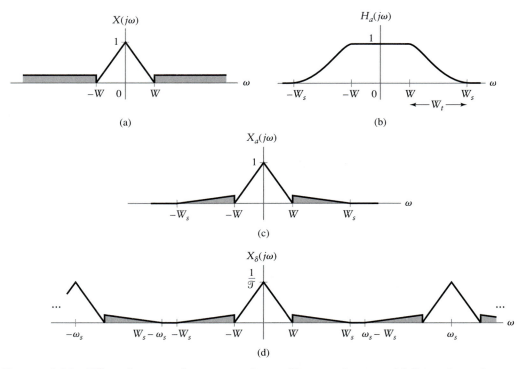

FIGURE 4.38   Effect of oversampling on anti-aliasing filter specifications. (a) Original signal spectrum. (b) Anti-aliasing filter frequency response magnitude. (c) Spectrum of signal at the anti-aliasing filter output. (d) Spectrum of the anti-aliasing filter output after sampling. The graph depicts the case $\omega_s > 2W_s$.

Substituting $W_s = W_t + W$ in this inequality, we may write

$$\omega_s - (W_t + W) > W$$

Rearranging terms to obtain the relationship between the transition band of the anti-aliasing filter and the sampling frequency, we have

$$W_t < \omega_s - 2W$$

Hence the transition band of the anti-aliasing filter must be less than the sampling frequency minus twice the highest signal frequency component of interest. Filters with small transition bands are difficult to design and expensive. By oversampling, or choosing $\omega_s \gg 2W$, we can greatly relax the requirements on the anti-aliasing filter transition band and consequently reduce its complexity and cost.

In both sampling and reconstruction, the difficulties of implementing practical analog filters suggest using the highest possible sampling rate. However, if this data set is processed with a discrete-time system, as depicted in Fig. 4.37(a), then high sampling rates lead to increased discrete-time system cost. The higher cost results because the discrete-time system must perform its computations at a faster rate. This conflict over the sampling rate is mitigated if we can somehow change the sampling rate such that a high rate is used for sampling and reconstruction, and a lower rate is used for discrete-time processing. Decimation and interpolation offer such a capability, as discussed next.

■ DECIMATION

We begin the discussion by considering the DTFTs obtained by sampling the identical continuous-time signal at different intervals, $\mathcal{T}_1$ and $\mathcal{T}_2$. Let the sampled signals be denoted as $x_1[n]$ and $x_2[n]$. We assume that $\mathcal{T}_1 = q\mathcal{T}_2$, where q is an integer, and that aliasing does not occur at either sampling rate. Figure 4.39 depicts the FT of a representative continuous-time signal and the DTFTs $X_1(e^{j\Omega})$ and $X_2(e^{j\Omega})$ associated with the sampling intervals $\mathcal{T}_1$ and $\mathcal{T}_2$. Decimation corresponds to changing $X_2(e^{j\Omega})$ to $X_1(e^{j\Omega})$. One way to do this is to convert the discrete-time sequence back to a continuous-time signal and then resample. Such an approach is subject to distortion introduced in the reconstruction operation. We can avoid this problem by using methods that operate directly on the discrete-time signals to change the sampling rate.

Subsampling is the key to reducing the sampling rate. If the sampling interval is $\mathcal{T}_2$ and we wish to increase it to $\mathcal{T}_1 = q\mathcal{T}_2$, we may accomplish this by selecting every qth sample of the sequence $x_2[n]$, that is, set $g[n] = x_2[qn]$. Equation (4.35) indicates that the relationship between $G(e^{j\Omega})$ and $X_2(e^{j\Omega})$ is

$$G(e^{j\Omega}) = \frac{1}{q} \sum_{m=0}^{q-1} X_2(e^{j(\Omega - m2\pi)/q})$$

That is, $G(e^{j\Omega})$ is a sum of shifted versions of $X_2(e^{j\Omega/q})$. The scaling spreads out $X_2(e^{j\Omega})$ by the factor q. Shifting these scaled versions of $X_2(e^{j\Omega})$ gives $G(e^{j\Omega})$ as depicted in

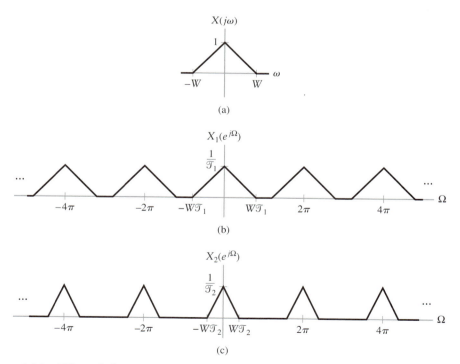

FIGURE 4.39   Effect of changing the sampling rate. (a) Underlying continuous-time signal FT. (b) DTFT of sampled data at sampling interval $\mathcal{T}_1$. (c) DTFT of sampled data at sampling interval $\mathcal{T}_2$.

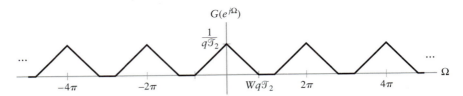

FIGURE 4.40   Spectrum that results from subsampling $X_2(e^{j\Omega})$ depicted in Fig. 4.39 by a factor of q.

Fig. 4.40. Identifying $\mathcal{T}_1 = q\mathcal{T}_2$, we see that $G(e^{j\Omega})$ corresponds to $X_1(e^{j\Omega})$ in Fig. 4.39(b). Hence subsampling by q changes the effective sampling rate by q.

This analysis assumes that the maximum frequency component of $X_2(e^{j\Omega})$ satisfies $W\mathcal{T}_2 < \pi/q$, so that aliasing does not occur as a consequence of the subsampling process. This assumption is rarely satisfied in practice; even if the signal of interest is bandlimited in this way, there will often be noise or other components present at higher frequencies. For example, if oversampling is used to obtain $x_2[n]$, then noise that passed through the transition band of the anti-aliasing will be present at frequencies above π/q. If we subsample $x_2[n]$ directly, then this noise will alias into frequencies $|\Omega| < W\mathcal{T}_1$ and distort the signal of interest. This aliasing problem is prevented by applying a lowpass discrete-time filter to $x_2[n]$ prior to subsampling.

Figure 4.41(a) depicts a decimation system including a lowpass discrete-time filter. The input signal $x[n]$ with DTFT shown in Fig. 4.41(b) corresponds to the oversampled signal, whose FT is depicted in Fig. 4.38(d). The shaded regions indicate noise energy. The lowpass filter shown in Fig. 4.41(c) removes most of the noise in producing the output signal depicted in Fig. 4.41(d). After subsampling, the noise does not alias into the signal band, as illustrated in Fig. 4.41(e). Note that this procedure is effective only if the discrete-time filter has a rapid transition from passband to stopband. Fortunately, a discrete-time filter with a narrow transition band is much easier to design and implement than a comparable continuous-time filter.

Decimation is also known as *downsampling*. It is often denoted by a downward arrow followed by the decimation factor as illustrated in the block diagram of Fig. 4.42.

■ INTERPOLATION

Interpolation is the process of increasing the sampling rate and requires that we somehow "interpolate" or assign values between the samples of the signal. In the frequency domain we seek to convert $X_1(e^{j\Omega})$ of Fig. 4.39(b) into $X_2(e^{j\Omega})$ of Fig. 4.39(c). We shall assume that we are increasing the sampling rate by an integer factor, that is, $\mathcal{T}_1 = q\mathcal{T}_2$.

The DTFT scaling property presented in Section 3.6 is the key to developing an interpolation procedure. Assume $x_1[n]$ is the sequence to be interpolated by the factor q. Define a new sequence $x_z[n]$ as follows:

$$x_z[n] = \begin{cases} x_1\left[\dfrac{n}{q}\right], & \dfrac{n}{q} \text{ integer} \\ 0, & \text{otherwise} \end{cases} \tag{4.42}$$

With this definition, we have $x_1[n] = x_z[qn]$ and the DTFT scaling property implies

$$X_z(e^{j\Omega}) = X_1(e^{jq\Omega})$$

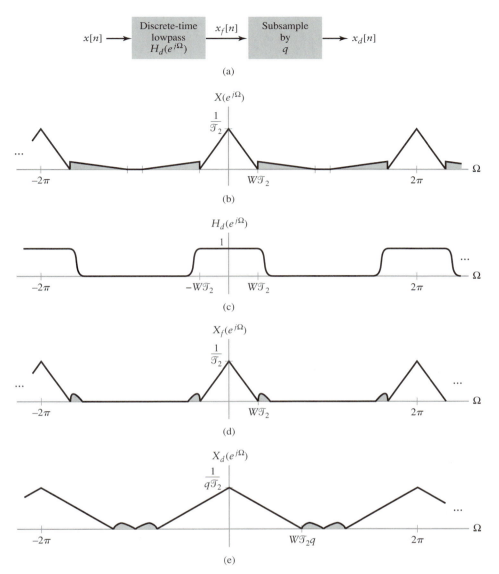

FIGURE 4.41   Frequency-domain interpretation of decimation. (a) Block diagram of decimation system. (b) Spectrum of oversampled input signal. Noise is depicted as the shaded portions of the spectrum. (c) Filter frequency response. (d) Spectrum of filter output. (e) Spectrum after subsampling.

FIGURE 4.42   Symbol for decimation by a factor of q.

That is, $X_z(e^{j\Omega})$ is a scaled version of $X_1(e^{j\Omega})$, as illustrated in Figs. 4.43(a) and (b). Identifying $\mathcal{T}_2 = \mathcal{T}_1/q$ we find that $X_z(e^{j\Omega})$ corresponds to $X_2(e^{j\Omega})$ in Fig. 4.39(c), except for the spectrum replicates at $\pm 2\pi/q, \pm 4\pi/q, \ldots, \pm(q-1)2\pi/q$. These can be removed by passing the signal $x_z[n]$ through a lowpass filter whose frequency response is depicted in Fig. 4.43(c). The passband of this filter is defined by $|\Omega| < W\mathcal{T}_2$ and the transition band must lie in the region $W\mathcal{T}_2 < |\Omega| < 2\pi/q - W\mathcal{T}_2$. The passband gain is chosen as q so that the interpolated signal has the correct amplitude. Figure 4.43(d) illustrates the spectrum of the filter output, $X_i(e^{j\Omega})$.

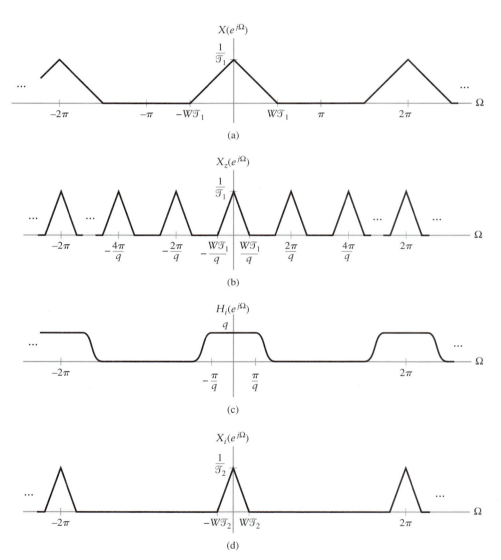

FIGURE 4.43 Frequency-domain interpretation of interpolation. (a) Spectrum of original sequence. (b) Spectrum after inserting $q - 1$ zeros in between every value of the original sequence. (c) Frequency response of a filter for removing undesired replicates located at $\pm 2\pi/q, \pm 4\pi/q, \ldots, \pm(q-1)2\pi/q$. (d) Spectrum of interpolated sequence.

Hence interpolation by the factor q is accomplished by inserting $q - 1$ zeros in between each sample of $x_1[n]$ and then lowpass filtering. A block diagram illustrating this procedure is depicted in Fig. 4.44(a). It is for this reason that interpolation is also known as *upsampling*; it is denoted by an upward arrow followed by the interpolation factor as depicted in the block diagram of Fig. 4.44(b).

This interpolation procedure has a time-domain interpretation analogous to that of the continuous-time signal reconstruction process. Let $h_i[n] \overset{DTFT}{\longleftrightarrow} H_i(e^{j\Omega})$ be the lowpass filter impulse response. We may then write

$$x_i[n] = x_z[n] * h_i[n]$$
$$= \sum_{k=-\infty}^{\infty} x_z[k]h_i[n - k] \tag{4.43}$$

Now suppose $H_i(e^{j\Omega})$ is an ideal lowpass filter with transition band of zero width. That is,

$$H_i(e^{j\Omega}) = \begin{cases} q, & |\Omega| < \pi/q \\ 0, & \pi/q < |\Omega| < \pi \end{cases}$$

Taking the inverse DTFT of $H_i(e^{j\Omega})$, we obtain

$$h_i[n] = q \, \frac{\sin\left(\dfrac{\pi}{q} n\right)}{\pi n}$$

Substituting $h_i[n]$ in Eq. (4.43) yields

$$x_i[n] = \sum_{k=-\infty}^{\infty} x_z[k] \, \frac{q \, \sin\left(\dfrac{\pi}{q}(n - k)\right)}{\pi(n - k)}$$

Now Eq. (4.42) indicates that $x_z[k] = 0$ unless $k = qm$, where m is an integer. Rewrite $x_i[n]$ by using only the nonzero terms in the sum

$$x_i[n] = \sum_{m=-\infty}^{\infty} x_z[qm] \, \frac{q \, \sin\left(\dfrac{\pi}{q}(n - qm)\right)}{\pi(n - qm)}$$

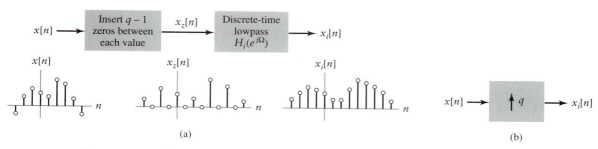

(a)   (b)

FIGURE 4.44   (a) Block diagram of an interpolation system with signals depicted assuming $q = 2$. (b) Symbol denoting interpolation by a factor of q.

Now substitute $x[m] = x_z[qm]$ and use sinc function notation to obtain

$$x_i[n] = \sum_{m=-\infty}^{\infty} x[m] \, \text{sinc}\left(\frac{1}{q}(n - qm)\right) \tag{4.44}$$

The interpolated sequence is given by a weighted sum of shifted discrete-time sinc functions. Equation (4.44) is the discrete-time version of Eq. (4.38).

A discrete-time signal-processing system that includes decimation and interpolation is depicted in Fig. 4.45. Consider the effect of the decimation and interpolation operations. Starting with decimation, we have

$$X_\delta^f(j\omega) = H_d(e^{j\omega \mathcal{T}_1}) X_\delta(j\omega)$$

Equation (4.34) implies the effect of the subsampling operation is

$$X_\delta^d(j\omega) = \frac{1}{q_1} \sum_{m=0}^{q_1-1} X_\delta^f\left(j\left(\omega - \frac{m}{q_1}\omega_s\right)\right)$$

Substitute for $X_\delta^f(j\omega)$ to obtain

$$\begin{aligned}
X_\delta^d(j\omega) &= \frac{1}{q_1} \sum_{m=0}^{q_1-1} H_d(e^{j(\omega - m\omega_s/q_1)\mathcal{T}_1}) X_\delta\left(j\left(\omega - \frac{m}{q_1}\omega_s\right)\right) \\
&= \frac{1}{q_1} \sum_{m=0}^{q_1-1} H_d(e^{j(\omega\mathcal{T}_1 - m2\pi/q_1)}) X_\delta\left(j\left(\omega - \frac{m}{q_1}\omega_s\right)\right)
\end{aligned} \tag{4.45}$$

The interpolation operation relates $Y_\delta(j\omega)$ and $Y_\delta^i(j\omega)$ as follows. First, the insertion of $q_2 - 1$ zeros in between each value of $y[n]$ corresponds to frequency-domain scaling, that is,

$$Y^f(j\omega) = Y_\delta(jq_2\omega)$$

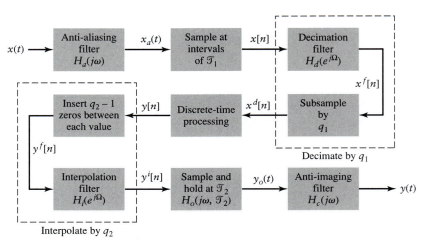

FIGURE 4.45 Block diagram of a system for discrete-time processing of continuous-time signals including decimation and interpolation.

The interpolation filter operates at the higher sampling rate $2\pi/\mathcal{T}_2$, so that the mapping between discrete- and continuous-time frequency is $\Omega = \omega\mathcal{T}_2$. We may therefore write

$$\begin{aligned} Y_i(j\omega) &= H_i(e^{j\omega\mathcal{T}_2})Y_f(j\omega) \\ &= H_i(e^{j\omega\mathcal{T}_2})Y_\delta(jq_2\omega) \end{aligned} \tag{4.46}$$

Although the expression for the equivalent continuous-time filter frequency response is too complex to provide much insight when interpolation and decimation are present, we may still use Eqs. (4.45) and (4.46) along with Eq. (4.40), $Y_\delta(j\omega) = H(e^{j\omega q_1 \mathcal{T}_1}) X_\delta^d(j\omega)$, and $Y(j\omega) = H_o(j\omega, \mathcal{T}_2)H_c(j\omega)Y_\delta^i(j\omega)$ to evaluate the equivalent continuous-time filtering operation represented by the system in Fig. 4.45.

4.9 *Fourier Series Representations for Finite-Duration Nonperiodic Signals*

The DTFS and FS are the Fourier representations for periodic signals. In this section we explore their use for representing finite-duration nonperiodic signals. The primary motivation for doing this is numerical computation of Fourier representations. Recall that the DTFS is the only Fourier representation that can be evaluated numerically. As a result, we often apply the DTFS to signals that are not periodic. It is important to understand the implications of applying a periodic representation to nonperiodic signals. A secondary benefit is further understanding of the relationship between the Fourier transform and corresponding Fourier series representations. We begin the discussion with the discrete-time case.

■ RELATING THE DTFS TO THE DTFT

Let $x[n]$ be a finite-duration signal of length M, that is,

$$x[n] = 0, \qquad n < 0 \text{ or } n \geq M$$

The DTFT of this signal is

$$X(e^{j\Omega}) = \sum_{n=0}^{M-1} x[n]e^{-j\Omega n}$$

Now suppose we evaluate $N \geq M$ DTFS coefficients using $x[n]$, $0 \leq n \leq N - 1$. We have $\Omega_o = 2\pi/N$ and

$$\tilde{X}[k] = \frac{1}{N} \sum_{n=0}^{N-1} x[n]e^{-jk\Omega_o n} \tag{4.47}$$

Since $x[n] = 0$ for $n \geq M$, we have

$$\tilde{X}[k] = \frac{1}{N} \sum_{n=0}^{M-1} x[n]e^{-jk\Omega_o n}$$

Comparison of $\tilde{X}[k]$ and $X(e^{j\Omega})$ reveals that

$$\tilde{X}[k] = \frac{1}{N} X(e^{j\Omega})\bigg|_{\Omega=k\Omega_o} \tag{4.48}$$

Hence the DTFS coefficients are samples of the DTFT divided by N and evaluated at intervals of $2\pi/N$. We may write the DTFS of $x[n]$ as $(1/N)X(e^{jk\Omega_o})$.

Now suppose we convert the DTFS coefficients in Eq. (4.48) back to a time-domain signal $\tilde{x}[n]$ using the DTFS expansion

$$\tilde{x}[n] = \sum_{k=\langle N \rangle} \tilde{X}[k] e^{jk\Omega_o n} \tag{4.49}$$

The complex sinusoids in Eq. (4.49) are all N periodic, so $\tilde{x}[n]$ is an N periodic signal. One period of $\tilde{x}[n]$ is obtained by recognizing that Eqs. (4.49) and (4.47) are inverses of each other. Hence

$$\tilde{x}[n] = x[n], \qquad 0 \leq n \leq N-1$$

This implies that the DTFS coefficients of $x[n]$ correspond to the DTFS coefficients of a periodically extended signal $\tilde{x}[n]$. In other words, *the effect of sampling the DTFT of a finite-duration nonperiodic signal is to periodically extend the signal in the time domain.* That is,

$$\tilde{x}[n] = \sum_{m=-\infty}^{\infty} x[n+mN] \xleftrightarrow{\;DTFS;\,\Omega_o\;} \tilde{X}[k] = \frac{1}{N} X(e^{jk\Omega_o})$$

Figure 4.46 illustrates these relationships in both time and frequency domains. They are the dual to sampling in time. Recall that sampling a signal in time generates shifted replicas

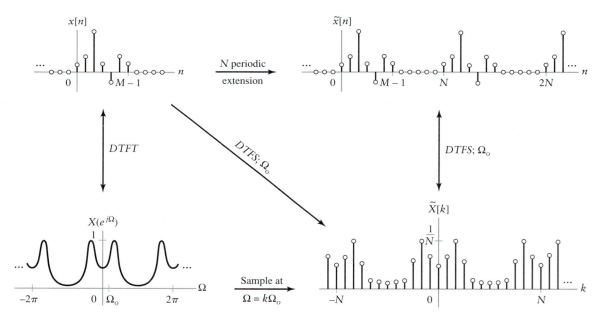

FIGURE 4.46   DTFS of a finite-duration nonperiodic signal.

of the original signal spectrum in the frequency domain. Here we sample in frequency. This generates shifted replicas of the original time signal in the time-domain representation. In order to prevent overlap or "aliasing" of these shifted replicas in time we require the frequency sampling interval Ω_o to be less than or equal to $2\pi/M$. Basically, this result corresponds to the sampling theorem applied in the frequency domain.

EXAMPLE 4.16   Consider the signal

$$x[n] = \begin{cases} \cos\left(\dfrac{3\pi}{8}\,n\right), & 0 \le n \le 31 \\ 0, & \text{otherwise} \end{cases}$$

Derive both the DTFT, $X(e^{j\Omega})$, and the DTFS, $X[k]$, assuming $N > 31$. Evaluate and plot $|X(e^{j\Omega})|$ and $N|X[k]|$ for $N = 32, 60,$ and 120.

Solution: First we evaluate the DTFT. Write $x[n] = g[n]w[n]$, where $g[n] = \cos\left(\frac{3}{8}\pi n\right)$ and $w[n]$ is the window function

$$w[n] = \begin{cases} 1, & 0 \le n \le 31 \\ 0, & \text{otherwise} \end{cases}$$

We have

$$G(e^{j\Omega}) = \pi\delta\left(\Omega + \frac{3\pi}{8}\right) + \pi\delta\left(\Omega - \frac{3\pi}{8}\right), \qquad -\pi < \Omega \le \pi$$

as one 2π period of $G(e^{j\Omega})$ and

$$W(e^{j\Omega}) = e^{-j(31/2)\Omega}\,\frac{\sin(16\Omega)}{\sin(\Omega/2)}$$

The modulation property implies that $X(e^{j\Omega}) = (1/2\pi)G(e^{j\Omega}) \circledast W(e^{j\Omega})$; for the problem at hand, this property yields

$$X(e^{j\Omega}) = \frac{e^{-j(31/2)(\Omega+3\pi/8)}}{2}\frac{\sin\left(16\left(\Omega + \frac{3\pi}{8}\right)\right)}{\sin\left(\frac{1}{2}\left(\Omega + \frac{3\pi}{8}\right)\right)} + \frac{e^{-j(31/2)(\Omega-3\pi/8)}}{2}\frac{\sin\left(16\left(\Omega - \frac{3\pi}{8}\right)\right)}{\sin\left(\frac{1}{2}\left(\Omega - \frac{3\pi}{8}\right)\right)}$$

Let $\Omega_o = 2\pi/N$, so the N DTFS coefficients are

$$X[k] = \frac{1}{N}\sum_{n=0}^{31}\cos\left(\frac{3\pi}{8}\,n\right)e^{-jk\Omega_o n}$$

$$= \frac{1}{2N}\sum_{n=0}^{31}e^{-j(k\Omega_o+3\pi/8)n} + \frac{1}{2N}\sum_{n=0}^{31}e^{-j(k\Omega_o-3\pi/8)n}$$

Summing each geometric series gives

$$X[k] = \frac{1}{2N}\frac{1 - e^{-j(k\Omega_o+3\pi/8)32}}{1 - e^{-j(k\Omega_o+3\pi/8)}} + \frac{1}{2N}\frac{1 - e^{-j(k\Omega_o-3\pi/8)32}}{1 - e^{-j(k\Omega_o-3\pi/8)}}$$

which we rewrite as

$$X[k] = \left(\frac{e^{-j(k\Omega_o + 3\pi/8)16}}{2Ne^{-j(k\Omega_o + 3\pi/8)/2}} \right) \frac{e^{j(k\Omega_o + 3\pi/8)16} - e^{-j(k\Omega_o + 3\pi/8)16}}{e^{j(k\Omega_o + 3\pi/8)/2} - e^{-j(k\Omega_o + 3\pi/8)/2}}$$

$$+ \left(\frac{e^{-j(k\Omega_o - 3\pi/8)16}}{2Ne^{-j(k\Omega_o - 3\pi/8)/2}} \right) \frac{e^{j(k\Omega_o - 3\pi/8)16} - e^{-j(k\Omega_o - 3\pi/8)16}}{e^{j(k\Omega_o - 3\pi/8)/2} - e^{-j(k\Omega_o - 3\pi/8)/2}}$$

$$= \left(\frac{e^{-j(31/2)(k\Omega_o + 3\pi/8)}}{2N} \right) \frac{\sin\left(16\left(k\Omega_o + \frac{3\pi}{8}\right)\right)}{\sin\left(\frac{1}{2}\left(k\Omega_o + \frac{3\pi}{8}\right)\right)}$$

$$+ \left(\frac{e^{-j(31/2)(k\Omega_o - 3\pi/8)}}{2N} \right) \frac{\sin\left(16\left(k\Omega_o - \frac{3\pi}{8}\right)\right)}{\sin\left(\frac{1}{2}\left(k\Omega_o - \frac{3\pi}{8}\right)\right)}$$

Comparison of $X[k]$ and $X(e^{j\Omega})$ indicates that Eq. (4.48) holds for this example.

Figures 4.47(a)–(c) depict $|X(e^{j\Omega})|$ as the dashed line and $N|X[k]|$ as the stems for $N = 32$, 60, and 120. As N increases, $X[k]$ samples $X(e^{j\Omega})$ more densely, and the shape of the DTFS coefficients resembles the underlying DTFT more closely.

In many applications, only M values of a signal $x[n]$ are available, and we have no knowledge of the signal's behavior outside this set of M values. The DTFS provides samples of the DTFT of this length M sequence. The practice of choosing $N > M$ when evaluating the DTFS is known as *zero-padding*, since it can be viewed as augmenting or padding the M available values of $x[n]$ with $N - M$ zeros. We emphasize that zero-padding does not overcome any of the limitations associated with knowing only M values of $x[n]$; it simply samples the underlying length M DTFT more densely, as illustrated in the previous example.

▶ **Drill Problem 4.13**   Use the DTFT of the finite-duration nonperiodic signal

$$x[n] = \begin{cases} 1, & 0 \le n \le 31 \\ 0, & \text{otherwise} \end{cases}$$

to find the DTFS coefficients of the period 64 signal $\tilde{x}[n]$ with one period given by

$$\tilde{x}[n] = \begin{cases} 1, & 0 \le n \le 31 \\ 0, & 32 \le n \le 63 \end{cases}$$

Answer:

$$\tilde{x}[n] \xleftrightarrow{\text{DTFS; } \pi/32} \tilde{X}[k]$$

where

$$\tilde{X}[k] = e^{-jk31\pi/64} \frac{\sin\left(k\frac{\pi}{2}\right)}{\sin\left(k\frac{\pi}{64}\right)}$$

◀

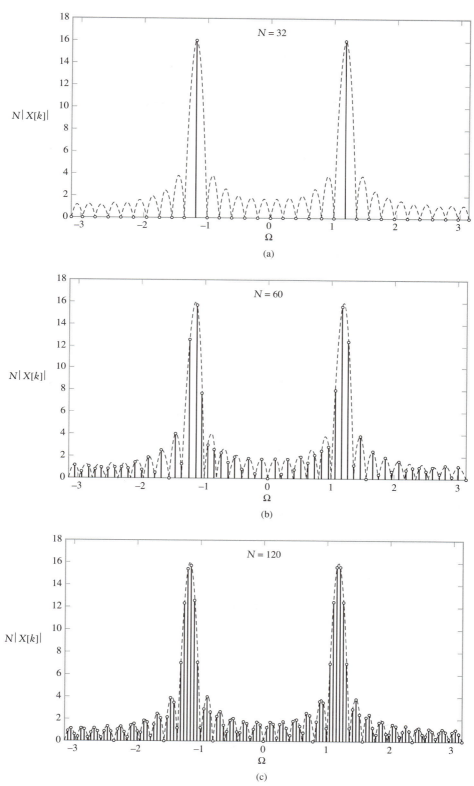

FIGURE 4.47 DTFT and length N DTFS of a 32-point cosine. The dashed line denotes $|X(e^{j\Omega})|$, the stems represent $N|X[k]|$. (a) $N = 32$, (b) $N = 60$, (c) $N = 120$.

■ RELATING THE FS TO THE FT

The relationship between the FS coefficients and the FT of a finite-duration nonperiodic signal is analogous to the discrete-time case discussed above. Let $x(t)$ have duration T_o so that

$$x(t) = 0, \qquad t < 0, \text{ or } t > T_o$$

Construct a periodic signal $\tilde{x}(t)$ with $T \geq T_o$ by periodically extending $x(t)$, as shown by

$$\tilde{x}(t) = \sum_{m=-\infty}^{\infty} x(t + mT)$$

The FS coefficients of $\tilde{x}(t)$ are

$$\tilde{X}[k] = \frac{1}{T} \int_0^T \tilde{x}(t) e^{-jk\omega_o t} \, dt$$

$$= \frac{1}{T} \int_0^{T_o} x(t) e^{-jk\omega_o t} \, dt$$

The FT of $x(t)$ is defined by

$$X(j\omega) = \int_{-\infty}^{\infty} x(t) e^{-j\omega t} \, dt$$

Hence comparing $\tilde{X}[k]$ with $X(j\omega)$, we conclude that

$$\tilde{X}[k] = \frac{1}{T} X(j\omega)\big|_{\omega = k\omega_o}$$

The FS coefficients are samples of the FT normalized by T.

*4.10 Computational Applications of the Discrete-Time Fourier Series

The DTFS involves a finite number of discrete-valued coefficients in both frequency and time domains. All the other Fourier representations are continuous in either the time or frequency domain. Hence the DTFS is the only Fourier representation that can be evaluated in a computer, and it is widely applied as a computational tool for manipulating signals. We begin this section by examining two common applications of the DTFS: approximating the FT and computing discrete-time convolution. In both of these the characteristics of the DTFS play a central role in correct interpretation of the result.

■ APPROXIMATING THE FT

The FT applies to continuous-time nonperiodic signals. The DTFS coefficients are computed using N values of a discrete-time signal. In order to use the DTFS to approximate the FT, we must sample the continuous-time signal and retain at most N samples. We assume that the sampling interval is $\mathcal{T}$ and that $M < N$ samples of the continuous-time signal are retained. Figure 4.48 depicts this sequence of steps. The problem at hand is to determine how well the DTFS coefficients $Y[k]$ approximate $X(j\omega)$, the FT of $x(t)$. Both the sampling and windowing operations are potential sources of error in the approximation.

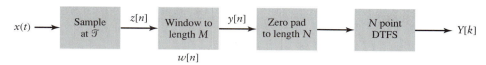

FIGURE 4.48   Block diagram depicting the sequence of operations involved in approximating the FT with the DTFS.

The potential error introduced by sampling is due to aliasing. Let $x[n] \xleftrightarrow{\text{FT}} X_\delta(j\omega)$ where $x[n] = x(nT)$ are the samples of $x(t)$. Equation (4.32) indicates

$$X_\delta(j\omega) = \frac{1}{T} \sum_{k=-\infty}^{\infty} X(j(\omega - k\omega_s)) \tag{4.50}$$

where $\omega_s = 2\pi/T$. Suppose we wish to approximate $X(j\omega)$ on the interval $-\omega_a < \omega < \omega_a$ and that $x(t)$ is bandlimited with maximum frequency ω_m where $\omega_m \geq \omega_a$. Aliasing in the band $-\omega_a < \omega < \omega_a$ is prevented by choosing T such that $\omega_s > \omega_m + \omega_a$, as illustrated in Fig. 4.49. That is, we require

$$T < \frac{2\pi}{\omega_m + \omega_a} \tag{4.51}$$

The windowing operation of length M corresponds to the periodic convolution

$$Y(e^{j\Omega}) = \frac{1}{2\pi} X(e^{j\Omega}) \circledast W(e^{j\Omega})$$

where $x[n] \xleftrightarrow{\text{DTFT}} X(e^{j\Omega})$ and $W(e^{j\Omega})$ is the window frequency response. We may rewrite this periodic convolution in terms of continuous-time frequency ω by performing the change of variable $\Omega = \omega T$ in the convolution integral. We thus have

$$Y_\delta(j\omega) = \frac{1}{\omega_s} X_\delta(j\omega) \circledast W_\delta(j\omega) \tag{4.52}$$

Here $X_\delta(j\omega)$ is given in Eq. (4.50), $y[n] \xleftrightarrow{\text{FT}} Y_\delta(j\omega)$ and $w[n] \xleftrightarrow{\text{FT}} W_\delta(j\omega)$. Both $X_\delta(j\omega)$ and $W_\delta(j\omega)$ have the same period, ω_s; hence the periodic convolution is performed over an interval of length ω_s. Since

$$w[n] = \begin{cases} 1, & 0 \leq n \leq M - 1 \\ 0, & \text{otherwise} \end{cases}$$

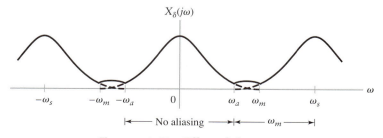

FIGURE 4.49   Effect of aliasing.

we have

$$W_\delta(j\omega) = e^{-j\omega\mathcal{T}(M-1)/2} \frac{\sin\left(M \frac{\omega\mathcal{T}}{2}\right)}{\sin\left(\frac{\omega\mathcal{T}}{2}\right)} \tag{4.53}$$

A plot of one period of $|W_\delta(j\omega)|$ is given in Fig. 4.50. The effect of the convolution is to smear or smooth the spectrum of $X_\delta(j\omega)$. This smearing limits our ability to resolve details in the spectrum. The degree of smearing depends on the mainlobe width of $W_\delta(j\omega)$. It is difficult to precisely quantify the loss in resolution resulting from windowing. Since we cannot resolve details in the spectrum that are closer than a mainlobe width apart, we define *resolution* as the mainlobe width ω_s/M. Hence to achieve a specified resolution ω_r we require

$$M \geq \frac{\omega_s}{\omega_r} \tag{4.54}$$

Substituting for ω_s, we may rewrite this explicitly as

$$M\mathcal{T} \geq \frac{2\pi}{\omega_r}$$

Recognizing that $M\mathcal{T}$ is the total time over which we sample $x(t)$, we see that this time interval must exceed $2\pi/\omega_r$.

The DTFS $y[n] \xleftrightarrow{\;DTFS;\; 2\pi/N\;} Y[k]$ samples the DTFT $Y(e^{j\Omega})$ at intervals of $2\pi/N$. That is, $Y[k] = (1/N)Y(e^{jk2\pi/N})$. In terms of continuous-time frequency ω the samples are spaced at intervals of $2\pi/N\mathcal{T} = \omega_s/N$, and so

$$Y[k] = \frac{1}{N} Y_\delta\left(jk \frac{\omega_s}{N}\right) \tag{4.55}$$

If the desired sampling interval is at least $\Delta\omega$, then we require

$$N \geq \frac{\omega_s}{\Delta\omega} \tag{4.56}$$

Hence if aliasing does not occur, and M is chosen large enough to prevent resolution loss due to windowing, then the DTFS approximation is related to the original signal spectrum according to

$$Y[k] \approx \frac{1}{N\mathcal{T}} X\left(jk \frac{\omega_s}{N}\right)$$

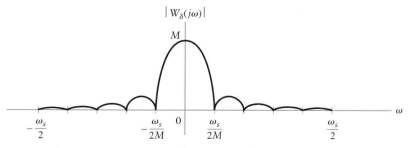

FIGURE 4.50   Magnitude response of M-point window.

The following example illustrates use of the guidelines given in Eqs. (4.51), (4.54), and (4.56) to approximate the FT with the DTFS.

EXAMPLE 4.17 Use the DTFS to approximate the FT of the signal

$$x(t) = e^{-t/10}u(t)(\cos(10t) + \cos(12t))$$

Assume the frequency band of interest is $-20 < \omega < 20$ and the desired sampling interval is $\Delta\omega = \pi/20$. Compare the DTFS approximation to the underlying FT for resolutions of (a) $\omega_r = 2\pi$, (b) $\omega_r = 2\pi/5$, and (c) $\omega_r = 2\pi/25$.

Solution: First evaluate the FT of $x(t)$. Let $f(t) = e^{-t/10}u(t)$ and $g(t) = \cos(10t) + \cos(12t)$ so that $x(t) = f(t)g(t)$. Use

$$F(j\omega) = \frac{1}{j\omega + \frac{1}{10}}$$

$$G(j\omega) = \pi\delta(\omega + 10) + \pi\delta(\omega - 10) + \pi\delta(\omega + 12) + \pi\delta(\omega - 12)$$

and the modulation property to obtain

$$X(j\omega) = \frac{1}{2}\left(\frac{1}{j(\omega + 10) + \frac{1}{10}} + \frac{1}{j(\omega - 10) + \frac{1}{10}} + \frac{1}{j(\omega + 12) + \frac{1}{10}} + \frac{1}{j(\omega - 12) + \frac{1}{10}}\right)$$

Now put the first and last two terms of $X(j\omega)$ over common denominators:

$$X(j\omega) = \frac{\frac{1}{10} + j\omega}{(\frac{1}{10} + j\omega)^2 + 10^2} + \frac{\frac{1}{10} + j\omega}{(\frac{1}{10} + j\omega)^2 + 12^2} \tag{4.57}$$

The maximum frequency of interest is given as 20, so $\omega_a = 20$. In order to use Eq. (4.51) to find the sampling interval we must also determine ω_m, the highest frequency present in $x(t)$. While $X(j\omega)$ in Eq. (4.57) is not strictly bandlimited, for $\omega \gg 12$ the magnitude spectrum $|X(j\omega)|$ decreases as $1/\omega$. We shall assume that $X(j\omega)$ is effectively bandlimited to $\omega_m = 500$ since $|X(j500)|$ is more than a factor of 10 less than $|X(j20)|$, the highest frequency of interest and nearest frequency at which aliasing occurs. This will not prevent aliasing in $-20 < \omega < 20$ but will ensure that the effect of aliasing in this region is small for all practical purposes. We require

$$\mathcal{T} < \frac{2\pi}{520}$$

$$\approx 0.0121$$

To satisfy this requirement, we choose $\mathcal{T} = 0.01$.

Given $\mathcal{T}$, the number of samples, M, is determined using Eq. (4.54):

$$M \geq \frac{200\pi}{\omega_r}$$

Hence for (a), $\omega_r = 2\pi$, we choose $M = 100$; for (b), $\omega_r = 2\pi/5$, we choose $M = 500$; and for (c), $\omega_r = 2\pi/25$, we choose $M = 2500$.

Finally, the length of the DTFS, N, must satisfy Eq. (4.56) as shown by

$$N \geq \frac{200\pi}{\Delta\omega}$$

Substitution of $\Delta\omega = \pi/20$ gives $N \geq 4000$, and so we choose $N = 4000$.

We compute the DTFS coefficients $Y[k]$ using these values of $\mathcal{T}$, M, and N. Figure 4.51 compares the FT to the DTFS approximation. The solid line in each plot is $|X(j\omega)|$ and the stems represent the DTFS approximation, $N\mathcal{T}|Y[k]|$. Both $|X(j\omega)|$ and $|Y[k]|$ have even symmetry because $x(t)$ is real, so we only need to depict the interval $0 < \omega < 20$. Figure 4.50(a)

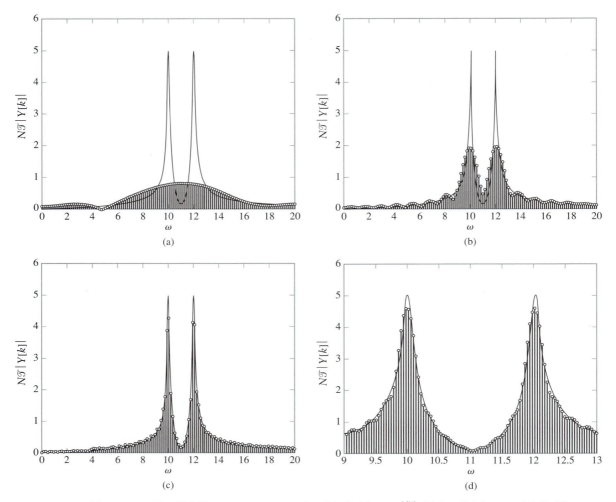

FIGURE 4.51   DTFS approximation to the FT of $x(t) = e^{-1/10}u(t)\,(\cos(10t) + \cos(12t))$. The solid line is the FT $|X(j\omega)|$, the stems denote the DTFS approximation $N\mathcal{T}|Y[k]|$. Both $|X(j\omega)|$ and $N\mathcal{T}|Y[k]|$ have even symmetry so only $0 < \omega < 20$ is displayed. (a) $M = 100$, $N = 4000$. (b) $M = 500$, $N = 4000$. (c) $M = 2500$, $N = 4000$. (d) $M = 2500$, $N = 16000$ for $9 < \omega < 13$.

depicts $M = 100$, (b) depicts $M = 500$, and (c) depicts $M = 2500$. As M increases and the resolution ω_r decreases, the quality of the approximation improves. In the $M = 100$ case, the resolution $(2\pi \approx 6)$ is larger than the separation between the two peaks and we cannot distinguish the presence of separate peaks. The only portions of the spectrum that are reasonably well approximated are the smooth sections away from the peaks. When $M = 500$, the resolution $(2\pi/5 \approx 1.25)$ is less than the separation between the peaks and distinct peaks are evident, although each peak is still blurred. As we move away from the peaks the quality of the approximation improves. In case (c), the resolution $(2\pi/25 \approx 0.25)$ is much less than the peak separation and a much better approximation is obtained over the entire frequency range.

It appears that the values at each peak are still not represented accurately in case (c). This could be due to the resolution limit imposed by M, or because we have not sampled the

DTFT at small enough intervals. In Fig. 4.51(d) we increase N to 16,000 while keeping $M = 2500$. The region of the spectrum near the peaks, $9 < \omega < 13$, is depicted. Increasing N by a factor of 4 reduces the frequency sampling interval by a factor of 4. We see that there is still some error in representing each peak value, although less than suggested by Fig. 4.51(c).

The quality of the DTFS approximation to the FT improves as $\mathcal{T}$ decreases, $M\mathcal{T}$ increases, and N increases. However, practical considerations such as memory limitations and hardware cost generally limit the range over which we can choose these parameters and force compromises. For example, if memory is limited, then we can increase $M\mathcal{T}$ to obtain better resolution only if we increase $\mathcal{T}$ and reduce the frequency range over which the approximation is valid.

Recall that the FT of periodic signals contains continuous-valued impulse functions whose areas are proportional to the value of the corresponding FS coefficients. The nature of the DTFS approximation to the FT of a periodic signal differs slightly from the non-periodic case because the DTFS coefficients are discrete valued and thus are not well suited to approximating continuous-valued impulses. In this case, the DTFS coefficients are proportional to the area under the impulses in the FT.

To illustrate this, consider using the DTFS to approximate the FT of a complex sinusoid with amplitude a, $x(t) = ae^{j\omega_o t}$. We have

$$x(t) \xleftrightarrow{\quad FT \quad} X(j\omega) = 2\pi a \delta(\omega - \omega_o)$$

Substitution for $X(j\omega)$ in Eq. (4.50) yields

$$X_\delta(j\omega) = \frac{2\pi}{\mathcal{T}} a \sum_{k=-\infty}^{\infty} \delta(\omega - \omega_o - k\omega_s)$$

Recognizing $\omega_s = 2\pi/\mathcal{T}$ and substituting for $X_\delta(j\omega)$ in Eq. (4.52) gives the FT of the sampled and windowed complex sinusoid as

$$Y_\delta(j\omega) = a W_\delta(j(\omega - \omega_o)) \tag{4.58}$$

where $W_\delta(j\omega)$ is given by Eq. (4.53).

Application of Eq. (4.55) indicates that the DTFS coefficients associated with the sampled and windowed complex sinusoid are given by

$$Y[k] = \frac{a}{N} W_\delta\left(j\left(k\frac{\omega_s}{N} - \omega_o\right)\right) \tag{4.59}$$

Hence the DTFS approximation to the FT of a complex sinusoid consists of samples of the FT of the window frequency response centered on ω_o, with amplitude proportional to a.

If we choose $N = M$ (no zero-padding) and the frequency of the complex sinusoid satisfies $\omega_o = m\omega_s/M$, then the DTFS samples $W_\delta(j(\omega - \omega_o))$ at the peak of its mainlobe and at its zero crossings. Consequently, we have

$$Y[k] = \begin{cases} a, & k = m \\ 0, & \text{otherwise for } 0 \le k \le M - 1 \end{cases}$$

In this special case, the continuous-valued impulse with strength $2\pi a$ in the FT is approximated by a discrete-valued impulse of amplitude a.

An arbitrary periodic signal is represented by the FS as a weighted sum of harmonically related complex sinusoids, so in general the DTFS approximation to the FT consists of samples of a weighted sum of shifted window frequency responses. The following example illustrates this effect.

EXAMPLE 4.18   Use the DTFS to approximate the FT of the periodic signal

$$x(t) = \cos(2\pi(0.4)t) + \tfrac{1}{2}\cos(2\pi(0.45)t)$$

Assume the frequency band of interest is $-10\pi < \omega < 10\pi$ and that the desired sampling interval is $\Delta\omega = 2\pi/M$. Evaluate the DTFS approximation for resolutions of (a) $\omega_r = \pi/2$ and (b) $\omega_r = \pi/100$.

Solution: First note that the FT of $x(t)$ is given by

$$X(j\omega) = \pi\delta(\omega + 0.8\pi) + \pi\delta(\omega - 0.8\pi) + \frac{\pi}{2}\delta(\omega + 0.9\pi) + \frac{\pi}{2}\delta(\omega - 0.9\pi)$$

The maximum frequency of interest is given as $\omega_a = 10\pi$ and this is much larger than the highest frequency in $X(j\omega)$, so aliasing is not a concern and we choose $\omega_s = 2\omega_a$. This gives $\mathcal{T} = 0.1$. The number of samples, M, is determined by substituting for ω_s into Eq. (4.54):

$$M \geq \frac{20\pi}{\omega_r}$$

To obtain the resolution specified in case (a) we require $M \geq 40$ samples, while in case (b) we choose $M \geq 2000$ samples. We shall choose $M = 40$ for case (a) and $M = 2000$ for case (b). The desired sampling interval in frequency indicates that $N = M$ and thus there is no zero-padding.

The signal is a weighted sum of complex sinusoids, so the underlying FT, $Y_\delta(j\omega)$, which will be sampled by the DTFS, is a weighted sum of shifted window frequency responses given by

$$Y_\delta(j\omega) = \tfrac{1}{2}W_\delta(j(\omega + 0.8\pi)) + \tfrac{1}{2}W_\delta(j(\omega - 0.8\pi)) + \tfrac{1}{4}W_\delta(j(\omega + 0.9\pi)) + \tfrac{1}{4}W_\delta(j(\omega - 0.9\pi))$$

In case (a), $W_\delta(j\omega)$ is given by

$$W_\delta(j\omega) = e^{-j\omega 39/20}\frac{\sin(2\omega)}{\sin\left(\dfrac{\omega}{20}\right)}$$

In case (b), it is given by

$$W_\delta(j\omega) = e^{-j\omega 1999/20}\frac{\sin(100\omega)}{\sin\left(\dfrac{\omega}{20}\right)}$$

The stems in Fig. 4.52(a) depict $|Y[k]|$ for $M = 40$ while the solid line depicts $(1/M)|Y_\delta(j\omega)|$ for positive frequencies. We have chosen to label the axis in units of Hz rather than rad/s for convenience. In this case, the minimum resolution of $\omega_r = \pi/2$ rad/s or 0.25 Hz is five times greater than the separation between the two sinusoidal components. Hence we cannot identify the presence of two sinusoids in either $|Y[k]|$ or $(1/M)|Y_\delta(j\omega)|$.

Figure 4.52(b) illustrates $|Y[k]|$ for $M = 2000$. We zoom in on the frequency band containing the sinusoids in Fig. 4.52(c), depicting $|Y[k]|$ with the stems and $(1/M)|Y_\delta(j\omega)|$ with the solid line. In this case, the minimum resolution is a factor of 10 times smaller than the separation between the two sinusoidal components, and we clearly see the presence of two

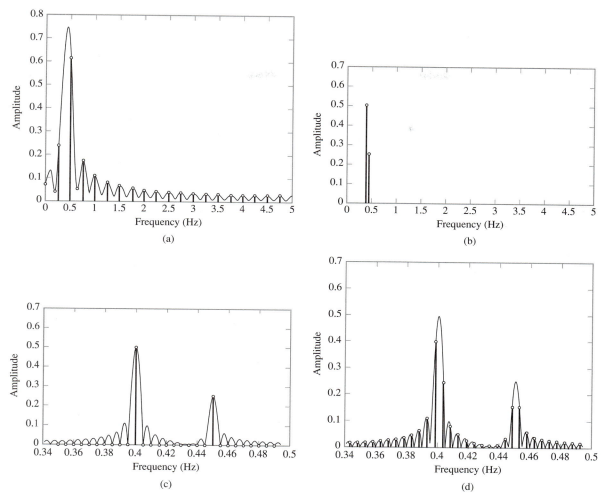

FIGURE 4.52 DTFS approximation to the FT of $x(t) = \cos(2\pi(0.4)t) + |\frac{1}{2}|\cos(2\pi(0.45)t)$. The stems denote $|Y[k]|$ while the solid line denotes $(1/M)|Y_\delta(j\omega)|$. The frequency axis is displayed in units of hertz for convenience and only positive frequencies are illustrated. (a) $M = 40$. (b) $M = 2000$. (c) Behavior in the vicinity of the sinusoidal frequencies for $M = 2000$. (d) Behavior in the vicinity of the sinusoidal frequencies for $M = 2010$.

sinusoids. The interval for which the DTFS samples $Y_\delta(j\omega)$ is $2\pi/200$ rad/s or 0.005 Hz. The frequency of each sinusoid is an integer multiple of the sampling interval, so $Y[k]$ samples $Y_\delta(j\omega)$ once at the peak of each mainlobe with the remainder of samples occurring at the zero crossings. Hence the amplitude of each component is correctly reflected in $|Y[k]|$.

Figure 4.52(d) depicts $|Y[k]|$ and $(1/M)|Y_\delta(j\omega)|$ assuming $M = 2010$. This results in slightly better resolution than $M = 2000$. However, in this case the frequency of each sinusoid is not an integer multiple of the interval at which the DTFS samples $Y_\delta(j\omega)$. Consequently, $Y_\delta(j\omega)$ is not sampled at the peak of each mainlobe and zero crossings. While the resolution is sufficient to reveal the presence of two components, we can no longer determine the amplitude of each component directly from $|Y[k]|$.

■ **COMPUTING DISCRETE-TIME CONVOLUTION WITH A FINITE-LENGTH SEQUENCE**

The DTFS may be used to efficiently compute the output of a system characterized by a finite-length impulse response. Discrete-time systems with finite-length impulse responses are often used to filter signals and are termed finite-duration impulse response (FIR) filters. The output of the filter is related to the input through the convolution sum. Instead of directly computing the convolution sum, the DTFS may be used to compute the output by performing multiplication in the frequency domain. If a fast algorithm is used to evaluate the DTFS (see the following section), then this approach offers significant savings in computation. While straightforward in concept, care must be exercised because we are using periodic convolution to compute ordinary or linear convolution: the filter output is given by the linear convolution of the impulse response with the input, but multiplication of DTFS coefficients corresponds to a periodic convolution.

Let $h[n]$ be an impulse response of length M; that is, assume $h[n] = 0$ for $n < 0$, $n \geq M$. The system output $y[n]$ is related to the input $x[n]$ via the convolution sum

$$
\begin{aligned}
y[n] &= h[n] * x[n] \\
&= \sum_{k=0}^{M-1} h[k]x[n-k]
\end{aligned}
\tag{4.60}
$$

The goal is to evaluate this sum using a set of periodic convolutions computed via the DTFS.

Consider the N point periodic convolution of $h[n]$ with N consecutive values of the input sequence $x[n]$ and assume $N > M$. Let $\tilde{x}[n]$ and $\tilde{h}[n]$ be N-point periodic versions of $x[n]$ and $h[n]$, respectively,

$$
\begin{aligned}
\tilde{x}[n] &= x[n], & \text{for } 0 \leq n \leq N-1 \\
\tilde{x}[n+mN] &= \tilde{x}[n], & \text{for all integer } m, \; 0 \leq n \leq N-1 \\
\tilde{h}[n] &= h[n], & \text{for } 0 \leq n \leq N-1 \\
\tilde{h}[n+mN] &= \tilde{h}[n], & \text{for all integer } m, \; 0 \leq n \leq N-1
\end{aligned}
$$

The periodic convolution between $\tilde{h}[n]$ and $\tilde{x}[n]$ is

$$
\begin{aligned}
\tilde{y}[n] &= \tilde{h}[n] \circledast \tilde{x}[n], \\
&= \sum_{k=\langle N \rangle} \tilde{h}[k]\tilde{x}[n-k] \\
&= \sum_{k=0}^{N-1} \tilde{h}[k]\tilde{x}[n-k]
\end{aligned}
\tag{4.61}
$$

Using the relationship between $h[n]$ and $\tilde{h}[n]$ to rewrite Eq. (4.61), we have

$$
\tilde{y}[n] = \sum_{k=0}^{M-1} h[k]\tilde{x}[n-k]
\tag{4.62}
$$

Now since $\tilde{x}[n] = x[n]$, $0 \leq n \leq N-1$, we know that

$$
\tilde{x}[n-k] = x[n-k], \qquad 0 \leq n - k \leq N-1
$$

In the sum of Eq. (4.62) k varies from 0 to $M - 1$, and so the condition $0 \leq n - k \leq N - 1$ is always satisfied provided $M - 1 \leq n \leq N - 1$. Substituting $x[n - k] = \tilde{x}[n - k]$, $M - 1 \leq n \leq N - 1$, into Eq. (4.62), we obtain

$$\tilde{y}[n] = \sum_{k=0}^{M-1} h[k]x[n - k], \qquad M - 1 \leq n \leq N - 1$$

$$= y[n], \qquad M - 1 \leq n \leq N - 1$$

Hence the periodic convolution is equal to the linear convolution at $L = N - M + 1$ values of n.

We may obtain values of $y[n]$ other than those on the interval $M - 1 \leq n \leq N - 1$ by shifting $x[n]$ prior to defining $\tilde{x}[n]$. Let

$$\tilde{x}_p[n] = x[n + pL], \qquad 0 \leq n \leq N - 1$$

$$\tilde{x}_p[n + mN] = \tilde{x}_p[n], \qquad \text{for all integer } m, 0 \leq n \leq N - 1$$

and define

$$\tilde{y}_p[n] = \tilde{h}[n] \circledast \tilde{x}_p[n]$$

In this case we have

$$\tilde{y}_p[n] = y[n + pL], \qquad M - 1 \leq n \leq N - 1$$

That is, the last L values in one period of $\tilde{y}_p[n]$ correspond to $y[n]$ for $M - 1 + pL \leq n \leq N - 1 + pL$. Each time we increment p the N-point periodic convolution gives us L new values of the linear convolution.

The relationship between periodic and linear convolution leads to the following procedure, called the *overlap and save method*, for evaluating a linear convolution with the DTFS. We assume that we desire to compute $y[n] = h[n] * x[n]$ for $n \geq 0$ and that $h[n] = 0$, $n < 0$, $n > M - 1$.

Overlap and Save Method of Implementing Convolution

1. Compute the N DTFS coefficients $H[k]$: $h[n] \xleftrightarrow{\text{DTFS; } 2\pi/N} H[k]$.
2. Set $p = 0$ and $L = N - M + 1$.
3. Define $\tilde{x}_p[n] = x[n - (M - 1) + pL]$, $0 \leq n \leq N - 1$.
4. Compute the N DTFS coefficients $\tilde{X}_p[k]$: $\tilde{x}_p[n] \xleftrightarrow{\text{DTFS; } 2\pi/N} \tilde{X}_p[k]$.
5. Compute the product $\tilde{Y}_p[k] = NH[k]\tilde{X}_p[k]$.
6. Compute the time signal $\tilde{y}_p[n]$ from the DTFS coefficients, $\tilde{Y}_p[k]$: $\tilde{y}_p[n] \xleftrightarrow{\text{DTFS; } 2\pi/N} \tilde{Y}_p[k]$.
7. Save L output points: $y[n + pL] = \tilde{y}_p[n + M - 1]$, $0 \leq n \leq L - 1$.
8. Set $p = p + 1$ and return to step 3.

This overlap and save algorithm for computing convolution is so called because there is overlap in the input blocks $x_p[n]$ and a portion of the output $y_p[n]$ is saved. Other algorithms for evaluating linear convolution using circular convolution may also be derived. They are known collectively as *fast convolution algorithms* because they require significantly less computation than direct evaluation of convolution when a fast algorithm is used to compute the required DTFS relationships. The exact savings depends on M and N.

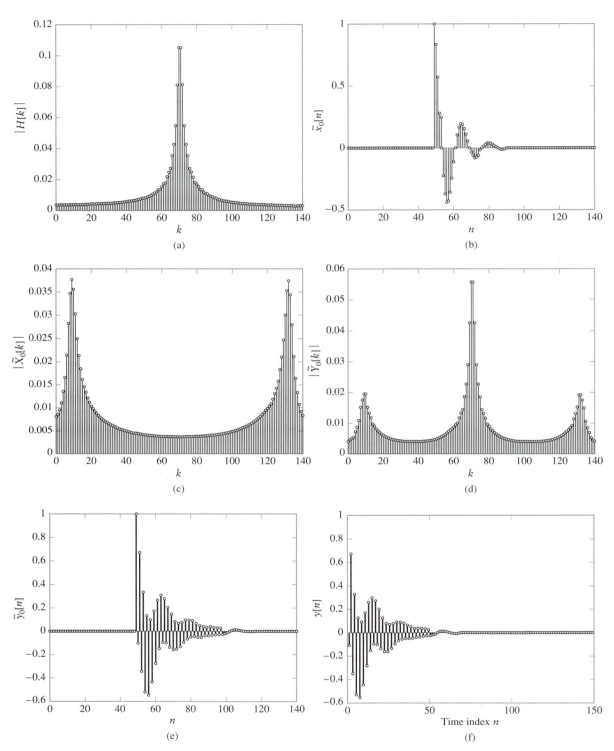

FIGURE 4.53　Example of using periodic convolution to compute linear convolution. (a) $|H[k]|$, (b) $\tilde{x}_0[n]$, (c) $|\tilde{X}_0[k]|$, (d) $|\tilde{Y}_0[k]| = |NH[k]\tilde{X}_0[k]|$, (e) $\tilde{y}_0[n]$, (f) $y[n]$.

EXAMPLE 4.19   Use periodic convolution and the DTFS to determine the output $y[n]$ of the system with impulse response

$$h[n] = \left(\frac{-15}{16}\right)^n (u[n] - u[n - 50])$$

in response to the input

$$x[n] = 0.9^n \cos\left(\frac{\pi}{8} n\right)(u[n] - u[n - 42])$$

Solution: The output is the linear convolution of $h[n]$ and $x[n]$. Both $h[n]$ and $x[n]$ have a finite duration. The linear convolution of two finite-length signals is a finite-length signal whose length is the sum of the lengths of $x[n]$ and $h[n]$. Both $x[n]$ and $h[n]$ are zero for $n < 0$, so $y[n]$ is zero for $n < 0$ and $n \geq 50 + 42$. Our goal is to use the DTFS to determine $y[n]$, $0 \leq n \leq 91$. We know that the periodic convolution is equal to the linear convolution at $L = N - M + 1$ values. Here $M = 50$ and we desire $L \geq 92$ to determine the 92 nonzero values of $y[n]$. We therefore choose the DTFS length $N = 141$ so that all values of interest are determined with one pass through the overlap and save procedure.

In step 1 we use the computer to evaluate $H[k]$ from $h[n]$, where

$$h[n] \xleftarrow{\text{DTFS; } 2\pi/141} H[k]$$

The magnitude of $H[k]$ is depicted in Fig. 4.53(a). Now set $p = 0$ and $L = 92$ to complete step 2. In step 3 we define $\tilde{x}_0[n]$ as a shifted periodic version of $x[n]$. We have

$$\tilde{x}_0[n] = x[n - 49]$$

$$= 0.9^{n-49} \cos\left(\frac{\pi}{8} (n - 49)\right)(u[n - 49] - u[n - 91]), \qquad 0 \leq n \leq 140$$

$$= \begin{cases} 0.9^{n-49} \cos\left(\frac{\pi}{8} (n - 49)\right), & 49 \leq n \leq 90 \\ 0, & 0 \leq n \leq 48 \text{ and } 91 \leq n \leq 140 \end{cases}$$

as one period of $\tilde{x}_0[n]$. This signal is depicted in Fig. 4.53(b). Complete step 4 by using the computer to evaluate $\tilde{X}_0[k]$ described by

$$\tilde{x}_0[n] \xleftarrow{\text{DTFS; } 2\pi/141} \tilde{X}_0[k]$$

Figure 4.53(c) illustrates $|\tilde{X}_0[k]|$. The next step is to compute $\tilde{Y}_0[k] = NH[k]\tilde{X}_0[k]$. Figure 4.53(d) depicts $|\tilde{Y}_0[k]|$. We then compute $\tilde{y}_0[n]$ in step 6, where

$$\tilde{y}_0[n] \xleftarrow{\text{DTFS; } 2\pi/141} \tilde{Y}_0[k]$$

The signal $\tilde{y}_0[n]$ is depicted in Fig. 4.53(e). Finally, set $y[n] = \tilde{y}_0[n + 49]$, $0 \leq n \leq 91$, as illustrated in Fig. 4.53(f).

*4.11 *Efficient Algorithms for Evaluating the DTFS*

The role of the DTFS as a computational tool is greatly expanded by the availability of efficient algorithms for evaluating the forward and the inverse DTFS. Such algorithms are collectively termed *fast Fourier transforms or FFT algorithms*. These fast algorithms operate on the "divide and conquer" principle by splitting the DTFS into a series of lower order DTFS and exploiting the symmetry and periodicity properties of the complex sinusoid $e^{jk2\pi n}$. Less computation is required to evaluate and combine the lower order DTFS than to evaluate the original DTFS. We shall demonstrate the computational savings that accrue from this splitting process.

Recall that the DTFS pair may be evaluated using the expressions

$$X[k] = \frac{1}{N} \sum_{n=0}^{N-1} x[n]e^{-jk\Omega_o n}$$

and

$$x[n] = \sum_{k=0}^{N-1} X[k]e^{jk\Omega_o n} \tag{4.63}$$

These expressions are virtually identical, differing only in the normalization by N and the sign of the complex exponential. Hence the same basic algorithm can be used to compute either relationship; only minor changes are required. We shall consider evaluating Eq. (4.63).

Direct evaluation of Eq. (4.63) for a single value of n requires N complex multiplications and $N - 1$ complex additions. Thus computation of $x[n]$, $0 \leq n \leq N - 1$, requires N^2 complex multiplications and $N^2 - N$ complex additions. In order to demonstrate how this can be reduced, we assume N is even. Split $X[k]$, $0 \leq k \leq N - 1$, into even- and odd-indexed signals, as shown by

$$X_e[k] = X[2k], \qquad 0 \leq k \leq N' - 1$$

and

$$X_o[k] = X[2k + 1], \qquad 0 \leq k \leq N' - 1$$

where $N' = N/2$ and

$$x_e[n] \xleftrightarrow{\text{DTFS; } \Omega_o'} X_e[k], \qquad x_o[n] \xleftrightarrow{\text{DTFS; } \Omega_o'} X_o[k]$$

with $\Omega_o' = 2\pi/N'$. Now express Eq. (4.63) as a combination of the N' DTFS coefficients $X_e[k]$ and $X_o[k]$, as shown by

$$x[n] = \sum_{k=0}^{N-1} X[k]e^{jk\Omega_o n}$$

$$= \sum_{k \text{ even}} X[k]e^{jk\Omega_o n} + \sum_{k \text{ odd}} X[k]e^{jk\Omega_o n}$$

Write the even and odd indices as $2m$ and $2m + 1$, respectively, to obtain

$$x[n] = \sum_{m=0}^{N'-1} X[2m]e^{jm2\Omega_o n} + \sum_{m=0}^{N'-1} X[2m + 1]e^{j(m2\Omega_o n + \Omega_o n)}$$

Substitute the definitions of $X_e[k]$, $X_o[k]$, and $\Omega_o' = 2\Omega_o$ so that

$$x[n] = \sum_{m=0}^{N'-1} X_e[m]e^{jm\Omega_o' n} + e^{j\Omega_o n} \sum_{m=0}^{N'-1} X_o[m]e^{jm\Omega_o' n}$$

$$= x_e[n] + e^{j\Omega_o n}x_o[n], \qquad 0 \leq n \leq N - 1$$

This indicates that $x[n]$ is a weighted combination of $x_e[n]$ and $x_o[n]$.

We may further simplify our result by exploiting the periodicity properties of $x_e[n]$ and $x_o[n]$. Using $x_e[n + N'] = x_e[n]$, $x_o[n + N'] = x_o[n]$, and $e^{j(n+N')\Omega_o} = -e^{j\Omega_o n}$ we obtain

$$x[n] = x_e[n] + e^{j\Omega_o n}x_o[n], \qquad 0 \leq n \leq N' - 1 \tag{4.64}$$

as the first N' DTFS values of $x[n]$ and

$$x[n + N'] = x_e[n] - e^{j\Omega_o n}x_o[n], \qquad 0 \le n \le N' - 1 \tag{4.65}$$

as the second N' values of $x[n]$. Figure 4.54(a) depicts the computation described in Eqs. (4.64) and (4.65) graphically for $N = 8$. We see that we need only multiply by

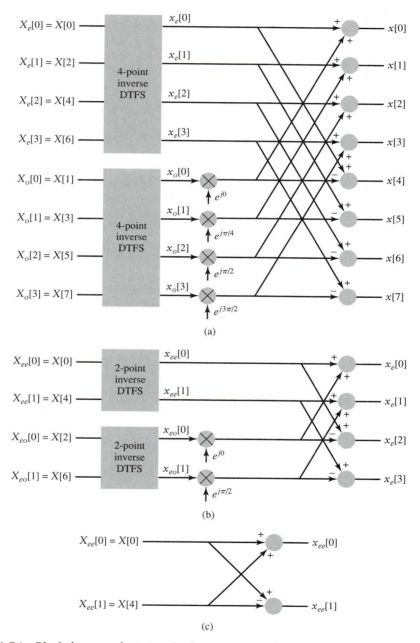

FIGURE 4.54 Block diagrams depicting the decomposition of an inverse DTFS as a combination of lower-order inverse DTFS. (a) Eight-point inverse DTFS represented in terms of two 4-point inverse DTFS. (b) Four-point inverse DTFS represented in terms of 2-point inverse DTFS. (c) Two-point inverse DTFS.

$e^{j\Omega_o n}$ once in computing both equations. The remaining operations are addition and subtraction.

Let us consider the computation required to evaluate Eqs. (4.64) and (4.65). Evaluation of $x_e[n]$ and $x_o[n]$ each requires $(N')^2$ complex multiplications for a total of $N^2/2$ complex multiplications. An additional N' multiplications are required to compute $e^{-jk\Omega_o}X_o[k]$. Thus the total number of complex multiplications is $N^2/2 + N/2$. For large N this is approximately $N^2/2$, about one-half the number of multiplications required to directly evaluate $x[n]$. Additional reductions in computational requirements are obtained if we further split $X_e[k]$ and $X_o[k]$ into even- and odd-indexed sequences. For example, Fig. 4.54(b) illustrates how to split the 4-point inverse DTFS used to calculate $x_e[n]$ into two 2-point inverse DTFSs for $N = 8$. The greatest savings is when N is a power of 2. In that case we can continue subdividing until the size of each inverse DTFS is 2. The 2-point inverse DTFS requires no multiplications, as illustrated in Fig. 4.54(c).

Figure 4.55 illustrates the FFT computation for $N = 8$. The repeated partitioning into even- and odd-indexed sequences permutes the order of the DTFS coefficients at the input. This permutation is termed *bit-reversal*, since the location of $X[k]$ may be determined by reversing the bits in a binary representation for the index k. For example, $X[6]$ has index $k = 6$. Representing $k = 6$ in binary form gives $k = 110_2$. Now reversing the bits gives $k' = 011_2$ or $k' = 3$, so $X[6]$ appears in the third position. The basic two-input, two-output structure depicted in Fig. 4.54(c) that is duplicated in each stage of the FFT (see Fig. 4.55) is termed a *butterfly* because of its appearance.

FFT algorithms for N a power of 2 require on the order of $N \log_2(N)$ complex multiplications. This can represent an extremely large savings in computation relative to N^2 when N is large. For example, if $N = 8192$ or 2^{13}, the direct approach requires approximately 630 times as many arithmetic operations as the FFT algorithm.

Many software packages contain routines that implement FFT algorithms. Unfortunately, the location of the $1/N$ factor is not standardized. Some routines place the $1/N$ in the expression for the DTFS coefficients $X[k]$, as we have here, while others place the

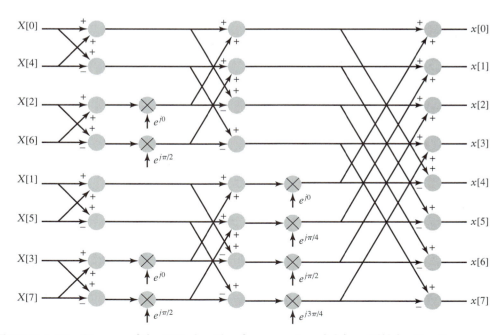

FIGURE 4.55 Diagram of the FFT algorithm for computing $x[n]$ from $X[k]$ for $N = 8$.

$1/N$ in the expression for the time signal $x[n]$. Yet another convention is to place $1/\sqrt{N}$ in both the expressions for $X[k]$ and $x[n]$. The only effect of these alternate conventions is to multiply the DTFS coefficients $X[k]$ by either N or $\sqrt{N}$.

4.12 *Exploring Concepts with MATLAB*

■ FREQUENCY RESPONSE OF LTI SYSTEMS

The MATLAB Signal Processing and Control System Toolboxes contain routines for evaluating the frequency response of LTI systems. Frequency response is a continuous function of frequency. However, numerically we can evaluate only the frequency response at discrete values of frequency. Thus a sufficiently large number of values are normally used to capture the detail in the system's frequency response.

Recall that the impulse and frequency responses of a continuous-time system are related through the FT, while the DTFT relates impulse and frequency responses of discrete-time systems. Hence determining the frequency response from the impulse response description requires approximating either the FT or DTFT using the DTFS, a topic that is discussed in Sections 4.9 and 4.10. We noted in Section 3.7 that the DTFS may be implemented in MATLAB with the `fft` command.

The commands `freqs` and `freqz` evaluate the frequency response for systems described by differential and difference equations, respectively. The command `H = freqs(b,a,w)` returns the values of the continuous-time system frequency response given by Eq. (4.3) at the frequencies specified in the vector `w`. Here we assume that vectors $\mathbf{b} = [b_M, b_{M-1}, \ldots, b_0]$ and $\mathbf{a} = [a_N, a_{N-1}, \ldots, a_0]$ represent the coefficients of the differential equation. The syntax for `freqz` is different in a subtle way. The command `H = freqz(b,a,w)` evaluates the discrete-time system frequency response given by Eq. (4.4) at the frequencies specified in the vector `w`. In the discrete-time case, the entries of `w` must lie between 0 and 2π and the vectors $\mathbf{b} = [b_0, b_1, \ldots, b_M]$ and $\mathbf{a} = [a_0, a_1, \ldots, a_N]$ contain the difference-equation coefficients in the reverse order of that required by `freqs`.

The frequency response of either a continuous- or discrete-time system described in state-variable form (see Eqs. (4.10) and (4.11)) may be computed using `freqresp`. The syntax is `H = freqresp(sys,w)`, where `sys` is the object containing the state-variable description (see Section 2.7) and `w` is a vector containing the frequencies at which to evaluate the frequency response. `freqresp` applies in general to multiple-input, multiple-output systems so the output `H` is a multidimensional array. For the class of single-input, single-output systems considered in this text and `N` frequency points in `w`, `H` is a multidimensional array of size 1 by 1 by `N`. The command `squeeze(H)` converts `H` to a length `N` vector that may be displayed with the `plot` command.

For example, we may numerically evaluate and plot the magnitude response for the system described in Example 4.5 at 500 points evenly spaced on the interval $0 \leq \omega < 10$ by using the commands:

```
>> a = [2, -1; 1, 0];  b = [1; 0];  c = [3, 1];  d = 0;
>> sys = ss(a,b,c,d);
>> w = [0:499]*10/500;
>> H = freqresp(sys,w);
>> Hmag = abs(squeeze(H));
>> plot(w,Hmag)
>> title('System Magnitude Response')
>> xlabel('Frequency (rads/s)');  ylabel('Magnitude')
```

Figure 4.56 depicts the system's magnitude response.

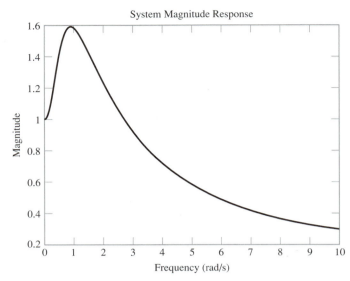

FIGURE 4.56 Magnitude response evaluated from a state-variable description using MATLAB.

■ *DECIMATION AND INTERPOLATION

Recall that decimation reduces the effective sampling rate of a discrete-time signal while interpolation increases the effective sampling rate. Decimation is accomplished by subsampling a lowpass filtered version of the signal, while interpolation is performed by inserting zeros in between samples and then applying a lowpass filter. The Signal Processing Toolbox contains several routines for performing decimation and interpolation. All of them automatically design and apply the lowpass filter required for both decimation and interpolation. The command y = decimate(x,r) decimates the signal represented by x by a positive integer factor r to produce y. The vector y is a factor of r shorter than x. Similarly, y = interp(x,r) interpolates x by a positive integer factor r, producing a vector y that is r times as long as x. The command y = resample(x,p,q) resamples the signal in vector x at p/q times the original sampling rate, where p and q are positive integers. This is conceptually equivalent to first interpolating by a factor p and then decimating by a factor q. The vector y is p/q times the length of x. The values of the resampled sequence may be inaccurate near the beginning and end of y if x contains large deviations from zero at its beginning and end.

Suppose the discrete-time signal

$$x[n] = e^{-n/15} \sin\left(\frac{2\pi}{13} n + \frac{\pi}{8}\right), \qquad 0 \le n \le 59$$

results from sampling a continuous-time signal at a rate of 45 kHz and that we wish to find the discrete-time signal resulting from sampling the underlying continuous-time signal at 30 kHz.

This corresponds to changing the sampling rate by a factor of $\frac{30}{45} = \frac{2}{3}$. The resample command is used to effect this change as follows:

```
>> x = exp(-[0:59]/15).*sin([0:59]*2*pi/13 + pi/8);
>> y = resample(x,2,3);
>> subplot(2,1,1)
```

```
>> stem([0:59],x);
>> title('Signal sampled at 45 kHz);  xlabel('Time');
ylabel('Amplitude')
>> subplot(2,1,2)
>> stem([0:39],y);
>> title('Signal sampled at 30 kHz);  xlabel('Time');
ylabel('Amplitude')
```

The original and resampled signals resulting from these commands are depicted in Fig. 4.57.

■ RELATING THE DTFS TO THE DTFT

Equation (4.48) states that the DTFS coefficients of a finite-duration signal correspond to samples of the DTFT divided by the number of DTFS coefficients, N. As discussed in Section 3.7, the MATLAB command fft calculates N times the DTFS coefficients. Hence fft directly evaluates samples of the DTFT of a finite-duration signal. The process of appending zeros to the finite-duration signal before computing the DTFS is called zero-padding and results in a denser sampling of the underlying DTFT. Zero-padding is easily accomplished with fft by adding an argument that specifies the number of coefficients to compute. If x is a length M vector representing a finite-duration time signal and n is greater than M, then the command $X = fft(x,n)$ evaluates n samples of the DTFT of

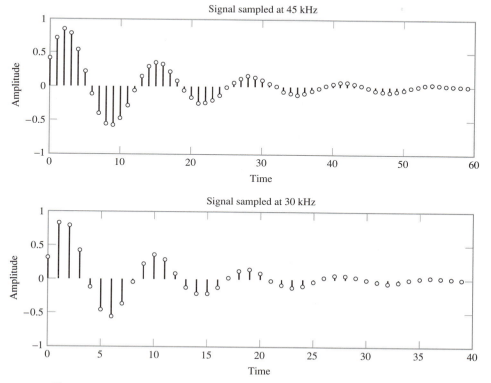

FIGURE 4.57 Original and resampled signals obtained using MATLAB.

x by first padding x with trailing zeros to length n. If n is less than M, then `fft(x,n)` first truncates x to length n.

The frequency values corresponding to the samples in X are represented by a vector n points long with first element zero and remaining entries spaced at intervals of $2\pi/n$. For example, the command `w = [0:(n-1)]*2*pi/n` generates the appropriate vector of frequencies. Note that X describes the DTFT for frequencies on the interval $0 \leq \Omega < 2\pi$. It is sometimes more convenient to view the DTFT over a frequency range centered on zero, that is, $-\pi \leq \Omega < \pi$. The MATLAB command `Y = fftshift(X)` swaps the left and right halves of X in order to put the zero-frequency value in the center. The vector of frequency values corresponding to the values in Y may be generated using `w = [-n/2:(n/2-1)]*2*pi/n`.

Suppose we revisit Example 4.16 using MATLAB to evaluate $|X(e^{j\Omega})|$ at intervals in frequency of (a) $2\pi/32$, (b) $2\pi/60$, and (c) $2\pi/120$. Recall that

$$x[n] = \begin{cases} \cos\left(\dfrac{3\pi n}{8}\right), & 0 \leq n \leq 31 \\ 0, & \text{otherwise} \end{cases}$$

For case (a) we use a 32-point DTFS computed from the 32 nonzero values of the signal. In case (b) and (c) we zero pad to length 60 and 120, respectively, to sample the DTFT at the specified intervals. We evaluate and display the results on $-\pi \leq \Omega < \pi$ using the following commands:

```
>> n = [0:31];
>> x = cos(3*pi*n/8);
>> X32 = abs(fftshift(fft(x))); %magnitude for 32 point
DTFS
>> X60 = abs(fftshift(fft(x,60))); %magnitude for 60
point DTFS
>> X120 = abs(fftshift(fft(x,120))); %magnitude for 120
point DTFS
>> w32 = [-16:15]*2*pi/32;   w60 = [-30:29]*2*pi/60;
w120 = [-60:59]*2*pi/120;
>> stem(w32,X32);   %stem plot for Fig. 4.47(a)
>> stem(w60,X60);   %stem plot for Fig. 4.47(b)
>> stem(w120,X120);   %stem plot for Fig. 4.47(c)
```

The results are depicted as the stem plots in Figs. 4.47(a)–(c).

■ COMPUTATIONAL APPLICATIONS OF THE DTFS

As previously noted, MATLAB's `fft` command may be used to evaluate the DTFS and thus is used for approximating the FT. In particular, the `fft` is used to generate the DTFS approximations in Examples 4.17 and 4.18. To repeat Example 4.18 we use the commands

```
>> ta = 0:0.1:3.9;  %time samples for case (a)
>> tb = 0:0.1:199.9;  %time samples for case (b)
>> xa = cos(0.8*pi*ta) + 0.5*cos(0.9*pi*ta);
>> xb = cos(0.8*pi*tb) + 0.5*cos(0.9*pi*tb);
>> Ya = abs(fft(xa)/40);   Yb = abs(fft(xb)/2000);
>> Ydela = abs(fft(xa,8192)/40); %evaluate 1/M Y_delta
(j omega) for case (a)
>> Ydelb = abs(fft(xa,16000)/2000); %evaluate 1/M
Y_delta(j omega) for case (b)
```

```
≫ fa = [0:19]*5/20;   fb = [0:999]*5/1000;
≫ fdela = [0:4095]*5/4096;   fdelb = [0:7999]*5/8000;
≫ plot(fdela,Ydela(1:4192))   %Fig. 4.52(a)
≫ hold on
≫ stem(fa,Ya(1:20))
≫ xlabel('Frequency (Hz)');   ylabel('Amplitude')
≫ hold off
≫ plot(fdelb(560:800),Ydelb(560:800))   %Fig. 4.52(c)
≫ hold on
≫ stem(fb(71:100),Yb(71:100))
≫ xlabel('Frequency (Hz)');   ylabel('Amplitude')
```

Note that here we evaluated $(1/M)Y_\delta(j\omega)$ using f f t by zero-padding with a large number of zeros relative to the length of $x[n]$. Recall that zero-padding decreases the spacing between the samples of the DTFT that are obtained by the DTFS. Hence by padding with a large number of zeros, we capture sufficient detail such that plot provides a smooth approximation to the underlying DTFT. If plot is used to display the DTFS coefficients without zero-padding, then a much coarser approximation to the underlying DTFT is obtained. Figure 4.58 depicts the DTFS coefficients for case (b) of Example 4.18 using both plot and stem. It is obtained using the commands

```
≫ plot(fb(71:100),Yb(71:100))
≫ hold on
≫ stem(fb(71:100),Yb(71:100))
```

Here the plot command produces triangles centered on the frequencies associated with the sinusoids. The triangles are a consequence of plot drawing straight lines in between the values in Yb.

The f f t command is implemented using a numerically efficient, or fast Fourier transform algorithm based on the divide and conquer principle discussed in Section 4.11. The Signal Processing Toolbox routine f f t f i l t employs a fast Fourier transform-based

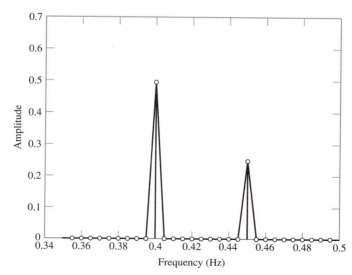

FIGURE 4.58 Use of the MATLAB command plot for displaying the DTFS coefficients in case (b) of Example 4.18.

algorithm for computing convolution with a finite-length sequence. It is based on the *overlap and add* method, which is similar in principle to the overlap and save approach discussed in Section 4.10. Both implement convolution by multiplication of DTFS coefficients. They use different approaches for relating the desired linear convolution to the periodic convolution associated with the DTFS.

4.13 *Summary*

Fourier representations find application in analyzing the interaction between signals and systems and in numerical evaluation of signal characteristics or system behavior. The DTFS is used for numerical computation since it is the only representation that can be evaluated in a computer. The FT and DTFT are most commonly used for analysis. In the course of applying these representations we frequently encounter situations in which there is a mixing of signal classes. This mixing occurs naturally in the interaction between signals and systems and the manipulation of signals. We have established relationships between different Fourier representations in this chapter in order to address mixing of signal classes.

The FT is the most versatile representation for analysis, since there are FT representations for all four signal classes. It is most often used to analyze continuous-time LTI systems, and systems that sample continuous-time signals or reconstruct continuous-time signals from samples. The primary use of the DTFT is to analyze discrete-time systems. We have developed a DTFT representation for discrete-time periodic signals to facilitate this role. The DTFS is used to approximate both the FT and DTFT in computational applications. The existence of computationally efficient or "fast" algorithms for evaluating the DTFS greatly expands the range of problems in which it may be used. We have established the relationships between the DTFS and the FT as well as between the DTFS and DTFT in order to correctly interpret the results of numerical computations.

Fourier methods provide a powerful set of analytic and numerical tools for solving problems involving signals and systems. They provide an important set of tools for the study of communication systems, as we see in the next chapter. They also have extensive applications in the context of filtering, the topic of Chapter 8.

FURTHER READING

1. The topics of sampling, reconstruction, discrete-time signal processing systems, computational applications of the DTFS, and fast algorithms for the DTFS are discussed in greater detail in the following signal processing texts:
   ▶ Porat, B., *A Course in Digital Signal Processing* (Wiley, 1997)
   ▶ Proakis, J. G., and D. G. Manolakis, *Introduction to Digital Signal Processing* (Macmillan, 1988)
   ▶ Oppenhaim, A. V., and R. W. Schafer, *Discrete-Time Signal Processing* Second Edition (Prentice Hall, 1999)
   ▶ Jackson, L. B., *Digital Filters and Signal Processing*, Third Edition (Kluwer, 1996)
   ▶ Roberts, R. A., and C. T. Mullis, *Digital Signal Processing* (Addison-Wesley, 1987)

2. In numerical computation applications, the DTFS is usually termed the discrete Fourier transform, or DFT. We have chosen to retain the DTFS terminology for consistency and to avoid confusion with the DTFT.

3. Rediscovery of the FFT algorithm for evaluating the DTFS is attributed to J. W. Cooley and J. W. Tukey for their 1965 publication "An algorithm for the machine calculation of complex Fourier series," *Mat. Comput.*, vol. 19, pp. 297–301. This paper greatly accelerated the development of a field called digital signal processing, which was in its infancy in the mid-1960s. The availability of a fast algorithm for computing the DTFS opened a tremendous number of new applications for digital signal processing, and resulted in explosive growth of this new field. Indeed, the majority of this chapter and a substantial portion of Chapter 8 concern the field of digital signal processing.

Carl Friedrick Gauss, the eminent German mathematician, has been credited with developing an equivalent, efficient algorithm for computing DTFS coefficients as early as 1805, predating Joseph Fourier's work on harmonic analysis. Additional reading on the history of the FFT and its impact on digital signal processing is found in the following articles:

▶ M. T. Heideman, D. H. Johnson, and C. S. Burrus, "Gauss and the history of the fast Fourier transform," *IEEE ASSP Magazine*, vol. 1, no. 4, pp. 14–21, October 1984.

▶ J. W. Cooley, "How the FFT gained acceptance," *IEEE Signal Processing Magazine*, vol. 9, no. 1, pp. 10–13, January 1992.

PROBLEMS

4.1 Sketch the frequency response of the systems described by the following impulse responses. Characterize each system as lowpass, bandpass, or highpass.

(a) $h(t) = \delta(t) - 2e^{-2t}u(t)$

(b) $h(t) = 4e^{-2t}\cos(20t)u(t)$

(c) $h[n] = \frac{1}{8}(\frac{7}{8})^n u[n]$

(d) $h[n] = \begin{cases} (-1)^n, & |n| \le 10 \\ 0, & \text{otherwise} \end{cases}$

4.2 Find the frequency response and the impulse response of the systems having the output $y(t)$ or $y[n]$ for the input $x(t)$ or $x[n]$.

(a) $x(t) = e^{-t}u(t)$, $y(t) = e^{-2t}u(t) + e^{-3t}u(t)$

(b) $x(t) = e^{-3t}u(t)$, $y(t) = e^{-3(t-2)}u(t-2)$

(c) $x(t) = e^{-2t}u(t)$, $y(t) = 2te^{-2t}u(t)$

(d) $x[n] = (\frac{1}{2})^n u[n]$, $y[n] = \frac{1}{4}(\frac{1}{2})^n u[n] + (\frac{1}{4})^n u[n]$

(e) $x[n] = (\frac{1}{4})^n u[n]$, $y[n] = (\frac{1}{4})^n u[n] - (\frac{1}{4})^{n-1}u[n-1]$

4.3 Determine the frequency response and the impulse response for the systems described by the following differential and difference equations:

(a) $\frac{d}{dt}y(t) + 3y(t) = x(t)$

(b) $\frac{d^2}{dt^2}y(t) + 5\frac{d}{dt}y(t) + 6y(t) = -\frac{d}{dt}x(t)$

(c) $y[n] - \frac{1}{4}y[n-1] - \frac{1}{8}y[n-2] = 3x[n] - \frac{3}{4}x[n-1]$

(d) $y[n] + \frac{1}{2}y[n-1] = x[n] - 2x[n-1]$

4.4 Determine the differential- or difference-equation descriptions for the systems with the following impulse responses:

(a) $h(t) = \frac{1}{a}e^{-t/a}u(t)$

(b) $h(t) = 2e^{-2t}u(t) - 2te^{-2t}u(t)$

(c) $h[n] = \alpha^n u[n]$, $|\alpha| < 1$

(d) $h[n] = \delta[n] + 2(\frac{1}{2})^n u[n] + (-1/2)^n u[n]$

4.5 Determine the differential- or difference-equation descriptions for the systems with the following frequency responses:

(a) $H(j\omega) = \dfrac{2 + 3j\omega - 3(j\omega)^2}{1 + 2j\omega}$

(b) $H(j\omega) = \dfrac{1 - j\omega}{-\omega^2 - 4}$

(c) $H(j\omega) = \dfrac{1 + j\omega}{(j\omega + 2)(j\omega + 1)}$

(d) $H(e^{j\Omega}) = \dfrac{1 + e^{-j\Omega}}{e^{-j2\Omega} + 3}$

(e) $H(e^{j\Omega}) = 1 + \dfrac{e^{-j\Omega}}{(1 - \frac{1}{2}e^{-j\Omega})(1 + \frac{1}{4}e^{-j\Omega})}$

4.6 Determine the frequency response, impulse response, and differential-equation descriptions for the continuous-time systems described by the following state-variable descriptions:

(a) $\mathbf{A} = \begin{bmatrix} -2 & 0 \\ 0 & -1 \end{bmatrix}$, $\mathbf{b} = \begin{bmatrix} 0 \\ 2 \end{bmatrix}$, $\mathbf{c} = [1 \quad 1]$, $D = [0]$

(b) $\mathbf{A} = \begin{bmatrix} 1 & 2 \\ -3 & -4 \end{bmatrix}$, $\mathbf{b} = \begin{bmatrix} 1 \\ 2 \end{bmatrix}$,

     $\mathbf{c} = [0 \quad 1]$, $D = [0]$

4.7 Determine the frequency response, impulse response, and difference-equation descriptions for the discrete-time systems described by the following state-variable descriptions:

(a) $\mathbf{A} = \begin{bmatrix} -\frac{1}{2} & 1 \\ 0 & \frac{1}{4} \end{bmatrix}$, $\mathbf{b} = \begin{bmatrix} 0 \\ 1 \end{bmatrix}$, $\mathbf{c} = [1 \quad 0]$,

     $D = [1]$

(b) $\mathbf{A} = \begin{bmatrix} \frac{1}{4} & \frac{3}{4} \\ \frac{1}{4} & -\frac{1}{4} \end{bmatrix}$, $\mathbf{b} = \begin{bmatrix} 1 \\ 1 \end{bmatrix}$, $\mathbf{c} = [0 \quad 1]$,

     $D = [0]$

4.8 A continuous-time system is described by the state-variable description

$$\mathbf{A} = \begin{bmatrix} -1 & 0 \\ 0 & -3 \end{bmatrix}, \quad \mathbf{b} = \begin{bmatrix} 0 \\ 2 \end{bmatrix},$$

$$\mathbf{c} = [0 \quad 1], \quad D = [0]$$

Transform the state vector associated with this system using the matrix

$$\mathbf{T} = \begin{bmatrix} 1 & -1 \\ 1 & 1 \end{bmatrix}$$

to find a new state-variable description for the system. Show that the frequency responses of the original and transformed systems are equal.

4.9 Find the FT representations for the following periodic signals. Sketch the magnitude and phase spectra.

(a) $x(t) = 2 \sin(\pi t) + \cos(2\pi t)$

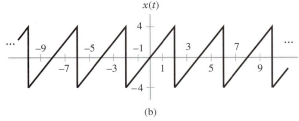

(a)

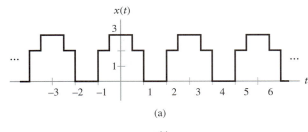

(b)

FIGURE P4.9

(b) $x(t) = \Sigma_{k=0}^{4} [(-1)^k / 2k] \cos((2k + 1)\pi t)$

(c) $x(t) = |\sin(\pi t)|$

(d) $x(t)$ as depicted in Fig. P4.9(a)

(e) $x(t)$ as depicted in Fig. P4.9(b)

4.10 Find the DTFT representations for the following periodic signals. Sketch the magnitude and phase spectra.

(a) $x[n] = \cos(\pi n/4) + \sin(\pi n/5)$

(b) $x[n] = 1 + \Sigma_{m=-\infty}^{\infty} \cos(\pi m/2)\delta[n - m]$

(c) $x[n]$ as depicted in Fig. P4.10(a)

(d) $x[n]$ as depicted in Fig. P4.10(b)

(e) $x[n]$ as depicted in Fig. P4.10(c)

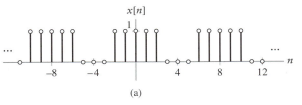

(a)

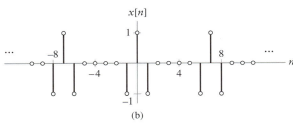

(b)

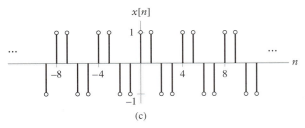

(c)

FIGURE P4.10

4.11 A LTI system has impulse response

$$h(t) = 2\,\frac{\sin(\pi t)}{\pi t}\cos(4\pi t)$$

Use the FT to determine the system output if the input is:

(a) $x(t) = 1 + \cos(\pi t) + \sin(4\pi t)$

(b) $x(t) = \Sigma_{m=-\infty}^{\infty} \delta(t - m)$

(c) $x(t)$ as depicted in Fig. P4.11(a)

(d) $x(t)$ as depicted in Fig. P4.11(b)

(e) $x(t)$ as depicted in Fig. P4.11(c)

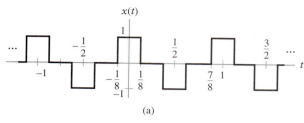

(a)

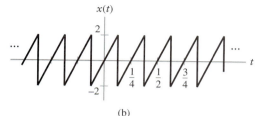

(b)

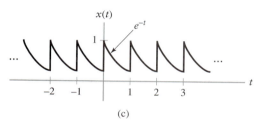

(c)

FIGURE P4.11

4.12 We may design a dc power supply by cascading a full-wave rectifier and an RC circuit as depicted in Fig. P4.12. The full wave rectifier output is given by

$$g(t) = |x(t)|$$

Let $H(j\omega) = Y(j\omega)/G(j\omega)$ be the frequency response of the RC circuit.

(a) Show that

$$H(j\omega) = \frac{1}{j\omega RC + 1}$$

(b) Suppose the input is $x(t) = \cos(120\pi t)$.
   (i) Find the FT representation for $g(t)$.
   (ii) Find the FT representation for $y(t)$.

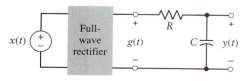

FIGURE P4.12

(iii) Find the range for the time constant RC such that the first harmonic of the ripple in $y(t)$ is less than 1% of the average value.

4.13 Consider the system depicted in Fig. P4.13(a). The FT of the input signal is depicted in Fig. 4.13(b). Let $g(t) \xleftrightarrow{FT} G(j\omega)$ and $y(t) \xleftrightarrow{FT} Y(j\omega)$. Sketch $G(j\omega)$ and $Y(j\omega)$ for the following cases:

(a) $w(t) = \cos(5\pi t)$ and $h(t) = \dfrac{\sin(6\pi t)}{\pi t}$

(b) $w(t) = \cos(5\pi t)$ and $h(t) = \dfrac{\sin(5\pi t)}{\pi t}$

(c) $w(t)$ depicted in Fig. P4.13(c) and $h(t) = \dfrac{\sin(2\pi t)}{\pi t}\cos(5\pi t)$

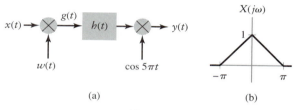

(a)   (b)

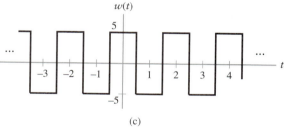

(c)

FIGURE P4.13

4.14 Consider the system depicted in Fig. P4.14. The impulse response $h(t)$ is given by

$$h(t) = \frac{\sin(10\pi t)}{\pi t}$$

and we have

$$x(t) = \sum_{k=1}^{\infty} \frac{1}{k}\cos(k4\pi t)$$

$$g(t) = \sum_{k=1}^{10} \cos(k8\pi t)$$

Use the FT to determine $y(t)$.

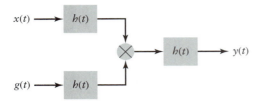

$x(t) \longrightarrow h(t)$

$h(t) \longrightarrow y(t)$

$g(t) \longrightarrow h(t)$

FIGURE P4.14

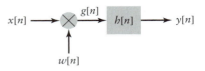

$x[n] \longrightarrow \times \xrightarrow{g[n]} h[n] \longrightarrow y[n]$

$w[n]$

FIGURE P4.16

4.15 The input to a discrete-time system is given by

$$x[n] = \cos\left(\frac{\pi}{4}n\right) + \sin\left(\frac{3\pi}{4}n\right)$$

Use the DTFT to find the output of the system, $y[n]$, if the impulse response is given by

(a) $h[n] = \dfrac{\sin\left(\dfrac{\pi}{2}n\right)}{\pi n}$

(b) $h[n] = (-1)^n \dfrac{\sin\left(\dfrac{\pi}{2}n\right)}{\pi n}$

(c) $h[n] = \cos\left(\dfrac{\pi}{2}n\right)\dfrac{\sin\left(\dfrac{\pi}{8}n\right)}{\pi n}$

4.16 Consider the discrete-time system depicted in Fig. P4.16. Assume $h[n] = [\sin(\pi n/2)]/\pi n$. Use the DTFT to determine the output, $y[n]$, for the following cases. Also sketch $G(e^{j\Omega})$, the DTFT of $g[n]$.

(a) $x[n] = \dfrac{\sin\left(\dfrac{\pi}{2}n\right)}{\pi n}, \quad w[n] = (-1)^n$

(b) $x[n] = \delta[n] - \dfrac{\sin\left(\dfrac{\pi}{2}n\right)}{\pi n}, \quad w[n] = (-1)^n$

(c) $x[n] = \dfrac{\sin\left(\dfrac{\pi}{2}n\right)}{\pi n}, \quad w[n] = \cos\left(\dfrac{\pi}{2}n\right)$

(d) $x[n] = 1 + \sin\left(\dfrac{\pi}{8}n\right) + 2\cos\left(\dfrac{3\pi}{4}n\right)$,

$w[n] = \cos\left(\dfrac{\pi}{2}n\right)$

4.17 Determine and sketch the FT representation, $X_\delta(j\omega)$, for the following discrete-time signals. The sampling interval $\mathcal{T}$ is given for each case.

(a) $x[n] = \dfrac{\sin\left(\dfrac{\pi}{4}n\right)}{\pi n}, \quad \mathcal{T} = 1$

(b) $x[n] = \dfrac{\sin\left(\dfrac{\pi}{4}n\right)}{\pi n}, \quad \mathcal{T} = \frac{1}{4}$

(c) $x[n] = \cos\left(\dfrac{\pi}{2}n\right)\dfrac{\sin\left(\dfrac{\pi}{4}n\right)}{\pi n}, \quad \mathcal{T} = 2$

(d) $x[n]$ depicted in Fig. P4.10(a) with $\mathcal{T} = 4$

(e) $x[n] = \sum_{p=-\infty}^{\infty}\delta[n - 4p], \quad \mathcal{T} = \frac{1}{8}$

4.18 Consider sampling the signal $x(t) = (1/\pi t)\sin(\pi t)$.

(a) Sketch the FT of the sampled signal for the following sampling intervals:

   (i) $\mathcal{T} = \frac{1}{4}$

   (ii) $\mathcal{T} = \frac{1}{2}$

   (iii) $\mathcal{T} = 1$

   (iv) $\mathcal{T} = \frac{4}{3}$

(b) Let $x[n] = x(n\mathcal{T})$. Sketch the DTFT of $x[n]$, $X(e^{j\Omega})$, for each of the sampling intervals given in (a).

4.19 The continuous-time signal $x(t)$ with FT as depicted in Fig. P4.19 is sampled.

(a) Sketch the FT of the sampled signal for the following sampling intervals. Identify in each case if aliasing occurs.

   (i) $\mathcal{T} = \frac{1}{15}$

   (ii) $\mathcal{T} = \frac{2}{15}$

   (iiI) $\mathcal{T} = \frac{1}{5}$

(b) Let $x[n] = x(n\mathcal{T})$. Sketch the DTFT of $x[n]$, $X(e^{j\Omega})$, for each of the sampling intervals given in (a).

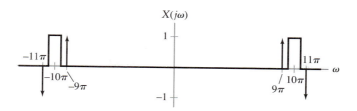

FIGURE P4.19

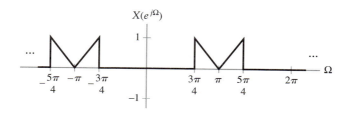

FIGURE P4.21

*4.20 Consider subsampling the signal $x[n] = [\sin(\pi n/6)]/\pi n$ so that $y[n] = x[qn]$. Sketch $Y(e^{j\Omega})$ for the following choices of q:

   (a) $q = 2$
   (b) $q = 4$
   (c) $q = 8$

*4.21 The discrete-time signal $x[n]$ with DTFT depicted in Fig. P4.21 is subsampled to obtain $y[n] = x[qn]$. Sketch $Y(e^{j\Omega})$ for the following choices of q:

   (a) $q = 3$
   (b) $q = 4$
   (c) $q = 8$

4.22 The signals below are sampled with sampling interval $\mathcal{T}$. Determine the bounds on $\mathcal{T}$ that guarantee there will be no aliasing.

   (a) $x(t) = (1/t) \sin \pi t + \cos(2\pi t)$
   (b) $x(t) = \cos(10\pi t)[\sin(\pi t)]/2t$
   (c) $x(t) = e^{-4t}u(t) * [\sin(Wt)]/\pi t$
   (d) $x(t) = w(t)z(t)$, where the FTs $W(j\omega)$ and $Z(j\omega)$ are depicted in Fig. P4.22

*4.23 A continuous-time signal lies in the frequency band $|\omega| < 5\pi$. This signal is contaminated by a large sinusoidal signal of frequency 120π. The contaminated signal is sampled at a sampling rate of $\omega_s = 13\pi$.

   (a) After sampling, at what frequency does the sinusoidal interfering signal appear?

   (b) The contaminated signal is passed through an anti-aliasing filter consisting of the RC circuit depicted in Fig. P4.23. Find the value of the time constant RC required so that the contaminated sinusoid is attenuated by a factor of 1000 prior to sampling.

   (c) Sketch the magnitude response in dB that the anti-aliasing filter presents to the signal of interest for the value of RC identified in (b).

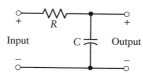

FIGURE P4.23

4.24 Consider the system depicted in Fig. P4.24. Assume $|X(j\omega)| = 0$ for $|\omega| > \omega_m$. Find the largest value of T such that $x(t)$ can be reconstructed from $y(t)$. Determine a system that will perform the reconstruction for this maximum value of T.

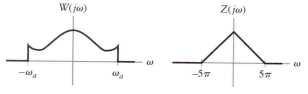

FIGURE P4.22

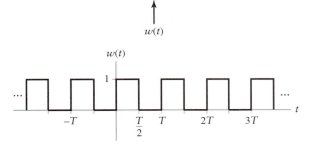

FIGURE P4.24

4.25 Let $|X(j\omega)| = 0$ for $|\omega| > \omega_m$. Form the signal $y(t) = x(t)[\cos(2\pi t) + \sin(10\pi t)]$. Determine the maximum value of ω_m for which $x(t)$ can be reconstructed from $y(t)$ and specify a system that will perform the reconstruction.

*4.26 A bandlimited signal $x(t)$ satisfies $|X(j\omega)| = 0$ for $|\omega| < \omega_1$ and $|\omega| > \omega_2$. Assume $\omega_1 > \omega_2 - \omega_1$. In this case we can sample $x(t)$ at a rate less than that indicated by the sampling theorem and still perform perfect reconstruction by using a bandpass reconstruction filter $H_r(j\omega)$. Let $x[n] = x(n\mathcal{T})$. Determine the maximum sampling interval $\mathcal{T}$ such that $x(t)$ can be perfectly reconstructed from $x[n]$. Sketch the frequency response of the reconstruction filter required for this case.

*4.27 Suppose a periodic signal $x(t)$ has FS coefficients

$$X[k] = \begin{cases} (\frac{3}{4})^k, & |k| \leq 4 \\ 0, & \text{otherwise} \end{cases}$$

The period of this signal is $T = 1$.

(a) Determine the minimum sampling interval for this signal that will prevent aliasing.

(b) The constraints of the sampling theorem can be relaxed somewhat in the case of periodic signals if we allow the reconstructed signal to be a time-scaled version of the original. Suppose we choose a sampling interval $\mathcal{T} = \frac{20}{19}$ and use a reconstruction filter

$$H_r(j\omega) = \begin{cases} 1, & |\omega| < \pi \\ 0, & \text{otherwise} \end{cases}$$

Show that the reconstructed signal is a time-scaled version of $x(t)$ and identify the scaling factor.

(c) Find the constraints on the sampling interval $\mathcal{T}$ so that use of $H_r(j\omega)$ in (b) results in the reconstruction filter being a time-scaled

version of $x(t)$ and determine the relationship between the scaling factor and $\mathcal{T}$.

4.28 A reconstruction system consists of a zero-order hold followed by a continuous-time anti-imaging filter with frequency response $H_c(j\omega)$. The original signal $x(t)$ is bandlimited to ω_m, that is, $X(j\omega) = 0$ for $|\omega| > \omega_m$, and is sampled with a sampling interval of $\mathcal{T}$. Determine the constraints on the magnitude response of the anti-imaging filter so that the overall magnitude response of this reconstruction system is between 0.99 and 1.01 in the signal passband and less than 10^{-4} to the images of the signal spectrum for the following values:

(a) $\omega_m = 10\pi$, $\mathcal{T} = 0.08$

(b) $\omega_m = 10\pi$, $\mathcal{T} = 0.05$

(c) $\omega_m = 10\pi$, $\mathcal{T} = 0.01$

(d) $\omega_m = 2\pi$, $\mathcal{T} = 0.08$

*4.29 In this problem we reconstruct a signal $x(t)$ from its samples $x[n] = x(n\mathcal{T})$ using pulses of width less than $\mathcal{T}$ followed by an anti-imaging filter with frequency response $H_c(j\omega)$. Specifically, we apply

$$x_p(t) = \sum_{n=-\infty}^{\infty} x[n]h_p(t - n\mathcal{T})$$

to the anti-imaging filter, where $h_p(t)$ is a pulse of width T_o as depicted in Fig. P4.29(a). An example of $x_p(t)$ is depicted in Fig. P4.29(b). Determine the constraints on $|H_c(j\omega)|$ so that the overall magnitude response of this reconstruction system is between 0.99 and 1.01 in the signal passband and less than 10^{-4} to the images of the signal spectrum for the following values. Assume $x(t)$ is bandlimited to 10π, that is, $X(j\omega) = 0$ for $|\omega| > 10\pi$.

(a) $\mathcal{T} = 0.08$, $T_o = 0.04$

(b) $\mathcal{T} = 0.08$, $T_o = 0.02$

(c) $\mathcal{T} = 0.04$, $T_o = 0.02$

(d) $\mathcal{T} = 0.04$, $T_o = 0.01$

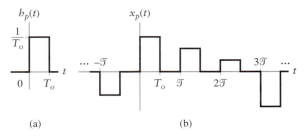

FIGURE P4.29

4.30 The zero-order hold produces a stairstep approximation to the sampled signal $x(t)$ from samples $x[n] = x(n\mathcal{T})$. A device termed a first-order hold linearly interpolates between the samples $x[n]$ and thus produces a smoother approximation to $x(t)$. The output of the first-order hold may be described as

$$x_1(t) = \sum_{n=-\infty}^{\infty} x[n]h_1(t - n\mathcal{T})$$

where $h_1(t)$ is the triangular pulse shown in Fig. P4.30(a). The relationship between $x[n]$ and $x_1(t)$ is depicted in Fig. P4.30(b).

(a) Identify the distortions introduced by the first-order hold and compare them to those introduced by the zero-order hold. *Hint:* $h_1(t) = h_o(t) * h_o(t)$.

(b) Consider a reconstruction system consisting of a first-order hold followed by an anti-imaging filter with frequency response $H_c(j\omega)$. Find $H_c(j\omega)$ so that perfect reconstruction is obtained.

(c) Determine the constraints on $|H_c(j\omega)|$ so that the overall magnitude response of this reconstruction system is between 0.99 and 1.01 in the signal passband and less than 10^{-4} to the images of the signal spectrum for the following values. Assume $x(t)$ is

bandlimited to 10π, that is, $X(j\omega) = 0$ for $|\omega| > 10\pi$.

(i) $\mathcal{T} = 0.08$

(ii) $\mathcal{T} = 0.04$

*4.31 A nonideal sampling operation obtains $x[n]$ from $x(t)$ as

$$x[n] = \int_{(n-1)\mathcal{T}}^{n\mathcal{T}} x(t)\, dt$$

(a) Show that this can be written as ideal sampling of a filtered signal $y(t) = x(t) * h(t)$, that is, $x[n] = y(n\mathcal{T})$, and find $h(t)$.

(b) Express the FT of $x[n]$ in terms of $X(j\omega)$, $H(j\omega)$, and $\mathcal{T}$.

(c) Assume that $x(t)$ is bandlimited to the frequency range $|\omega| < 3\pi/(4\mathcal{T})$. Determine the frequency response of a discrete-time system that will correct the distortion in $x[n]$ introduced by nonideal sampling.

*4.32 The system depicted in Fig. P4.32(a) converts a continuous-time signal $x(t)$ to a discrete-time signal $y[n]$. We have

$$H(e^{j\Omega}) = \begin{cases} 1, & |\Omega| < \pi/4 \\ 0, & \text{otherwise} \end{cases}$$

Find the sampling frequency $\omega_s = 2\pi/\mathcal{T}$ and the constraints on the anti-aliasing filter frequency response $H_a(j\omega)$ so that an input signal with FT $X(j\omega)$ shown in Fig. P4.32(b) results in the output signal with DTFT $Y(e^{j\Omega})$.

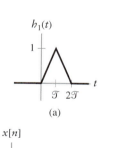

(a)

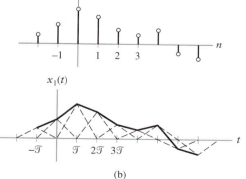

(b)

FIGURE P4.30

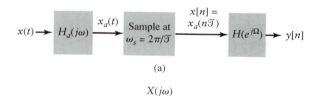

(a)

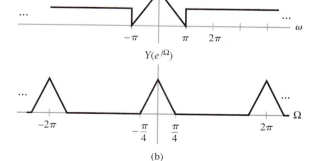

(b)

FIGURE P4.32

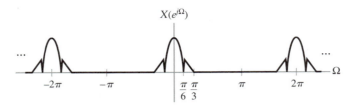

FIGURE P4.33

*4.33 Determine the maximum factor q by which a $x[n]$ with DTFT $X(e^{j\Omega})$ depicted in Fig. P4.33 can be decimated without aliasing. Sketch the DTFT of the sequence that results when $x[n]$ is decimated by this amount.

*4.34 The discrete-time signal $x[n]$ with DTFT $X(e^{j\Omega})$ shown in Fig. P4.34(a) is decimated by first passing $x[n]$ through the filter with frequency response $H(e^{j\Omega})$ shown in Fig. P4.34(b) and then subsampling by a factor of q. For the following values of q and W, determine the minimum value of Ω_p and maximum value of Ω_s so that the subsampling operation does not change the shape of the portion of $X(e^{j\Omega})$ on $|\Omega| < W$. Sketch the DTFT of the subsampled signal.

(a) $q = 2$, $W = \pi/3$
(b) $q = 2$, $W = \pi/4$
(c) $q = 3$, $W = \pi/4$

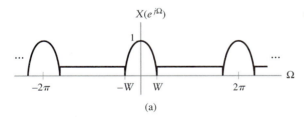

(a)

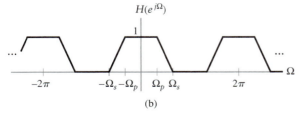

(b)

FIGURE P4.34

*4.35 A signal $x[n]$ is interpolated by a factor of q by first inserting $q - 1$ zeros between each sample and next passing the zero-stuffed sequence through a filter with frequency response $H(e^{j\Omega})$ depicted in Fig. P4.34(b). The DTFT of $x[n]$ is

depicted in Fig. P4.35. Determine the minimum value of Ω_p and maximum value of Ω_s so that ideal interpolation is obtained for the following cases. Also sketch the DTFT of the interpolated signal.

(a) $q = 2$, $W = \pi/2$
(b) $q = 2$, $W = 3\pi/4$
(c) $q = 3$, $W = 3\pi/4$

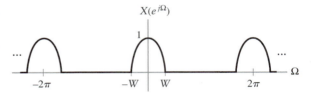

FIGURE P4.35

*4.36 Consider interpolating a signal $x[n]$ by repeating each value q times as depicted in Fig. P4.36. That is, we define $x_o[n] = x[\text{floor}(n/q)]$, where floor($w$) is the largest integer less than or equal to w. Letting $x_z[n]$ be derived from $x[n]$ by inserting $q - 1$ zeros between each value of $x[n]$, that is,

$$x_z[n] = \begin{cases} x[n/q], & n/q \text{ integer} \\ 0, & \text{otherwise} \end{cases}$$

we may write $x_o[n] = x_z[n] * h_o[n]$, where

$$h_o[n] = \begin{cases} 1, & 0 \le n \le q - 1 \\ 0, & \text{otherwise} \end{cases}$$

Note that this is the discrete-time analog of the zero-order hold. The interpolation process is completed by passing $x_o[n]$ through a filter with frequency response $H(e^{j\Omega})$.

(a) Express $X_o(e^{j\Omega})$ in terms of $X(e^{j\Omega})$ and $H_o(e^{j\Omega})$. Sketch $|X_o(e^{j\Omega})|$ if $x[n] = [\sin(3\pi n/4)]/(\pi n)$.

(b) Assume $X(e^{j\Omega})$ is as shown in Fig. P4.35. Specify the constraints on $H(e^{j\Omega})$ so that

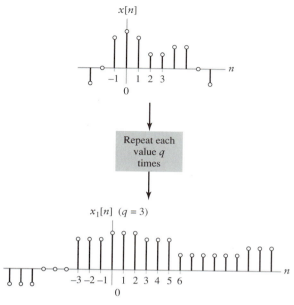

FIGURE P4.36

ideal interpolation is obtained for the following cases:

(i) $q = 2$, $W = 3\pi/4$

(ii) $q = 4$, $W = 3\pi/4$

4.37 A discrete-time system for processing continuous-time signals is shown in Fig. P4.37. Sketch

the frequency response magnitude of an equivalent continuous-time system for the following cases:

(a) $\Omega_1 = \pi/4$, $W_c = 20\pi$

(b) $\Omega_1 = 3\pi/4$, $W_c = 20\pi$

(c) $\Omega_1 = \pi/4$, $W_c = 2\pi$

*4.38 The system shown in Fig. P4.38 is used to implement a bandpass filter. The discrete-time filter $H(e^{j\Omega})$ has frequency response on $-\pi < \Omega \leq \pi$

$$H(e^{j\Omega}) = \begin{cases} 1, & \Omega_a \leq |\Omega| \leq \Omega_b \\ 0, & \text{otherwise} \end{cases}$$

Find the sampling interval $\mathcal{T}$, Ω_a, Ω_b, W_1, W_2, W_3, and W_4, so that the equivalent continuous-time frequency response $G(j\omega)$ satisfies

$$0.9 < |G(j\omega)| < 1.1, \quad \text{for } 100\pi < \omega < 200\pi$$

$$G(j\omega) = 0 \quad \text{elsewhere}$$

In solving this problem, choose W_1 and W_3 as small as possible and choose $\mathcal{T}$, W_2, and W_4 as large as possible.

4.39 Let $X(e^{j\Omega}) = \sin(11\Omega/2)/\sin(\Omega/2)$ and define $\tilde{X}[k] = X(e^{jk\Omega_o})$. Find and sketch $\tilde{x}[n]$, where $\tilde{x}[n] \xleftrightarrow{\ DTFS;\ \Omega_o\ } \tilde{X}[k]$ for the following values of Ω_o:

(a) $\Omega_o = 2\pi/15$

(b) $\Omega_o = \pi/10$

(c) $\Omega_o = \pi/3$

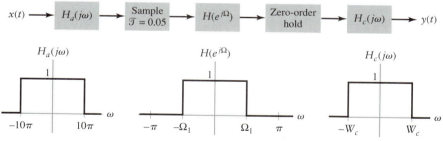

FIGURE P4.37

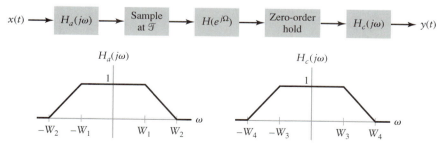

FIGURE P4.38

4.40 Let $X(j\omega) = [\sin(2\omega)]/\omega$ and define $\tilde{X}[k] = X(jk\omega_o)$. Find and sketch $\tilde{x}(t)$, where $\tilde{x}(t) \xleftarrow{FS;\ \omega_o} \tilde{X}[k]$ for the following values of ω_o:

(a) $\omega_o = \pi/8$

(b) $\omega_o = \pi/4$

(c) $\omega_o = \pi/2$

*4.41 The continuous-time representation for a periodic discrete-time signal $x[n] \xleftarrow{DTFS;\ 2\pi/N} X[k]$ is periodic and thus has a FS representation. This FS representation is a function of the DTFS coefficients $X[k]$, as we show in this problem. The result establishes the relationship between the FS and DTFS representations. Let $x[n]$ have period N and let $x_\delta(t) = \sum_{n=-\infty}^{\infty} x[n]\delta(t - n\mathcal{T})$.

(a) Show $x_\delta(t)$ is periodic and find the period, T.

(b) Begin with the definition of the FS coefficients

$$X_\delta[k] = \frac{1}{T}\int_{\langle T\rangle} x_\delta(t)e^{-jk\omega_o t}\ dt$$

Substitute for T, ω_o, and one period of $x_\delta(t)$ to show

$$X_\delta[k] = \frac{1}{\mathcal{T}} X[k]$$

4.42 A signal $x(t)$ is sampled at intervals of $\mathcal{T} = 0.01$. One hundred samples are collected and a 200-point DTFS is taken in an attempt to approximate $X(j\omega)$. Assume $|X(j\omega)| \approx 0$ for $|\omega| > 120\pi$. Determine the frequency range $-\omega_a < \omega < \omega_a$ over which the DTFS offers a reasonable approximation to $X(j\omega)$, the effective resolution of this approximation, ω_r, and the frequency interval between each DTFS coefficient, $\Delta\omega$.

▶ *Computer Experiments*

4.43 Use the MATLAB command `freqs` or `freqz` to plot the magnitude response of the following systems. Determine whether the system has a lowpass, highpass, or bandpass characteristic.

(a) $H(j\omega) = \dfrac{8}{(j\omega)^3 + 4(j\omega)^2 + 8j\omega + 8}$

(b) $H(j\omega) = \dfrac{(j\omega)^3}{(j\omega)^3 + 2(j\omega)^2 + 2j\omega + 1}$

(c) $H(e^{j\Omega}) = \dfrac{1 + 3e^{-j\Omega} + 3e^{-j2\Omega} + e^{-j3\Omega}}{6 + 2e^{-j2\Omega}}$

(d) $H(e^{j\Omega}) = H_1(e^{j\Omega})\ H_2(e^{j\Omega})$

where

$$H_1(e^{j\Omega}) = \frac{0.02426(1 - e^{-j\Omega})^2}{(1 + 1.10416e^{-j\Omega} + 0.4019e^{-j2\Omega})}$$

$$H_2(e^{j\Omega}) = \frac{(1 - e^{-j\Omega})^2}{(1 + 0.56616e^{-j\Omega} + 0.7657e^{-j2\Omega})}$$

4.44 Use the MATLAB command `freqresp` to plot the magnitude and phase response for the systems with state-variable descriptions given in Problems 4.6 and 4.7.

4.45 Repeat Example 4.11 using zero-padding and the MATLAB commands `fft` and `fftshift` to sample and plot $Y(e^{j\Omega})$ at 512 points on $-\pi \leq \Omega < \pi$ for each case.

4.46 The rectangular window is defined as

$$w_r[n] = \begin{cases} 1, & 0 \leq n \leq M \\ 0, & \text{otherwise} \end{cases}$$

We may truncate a signal to the interval $0 \leq n \leq M$ by multiplying the signal with $w[n]$. In the frequency domain we convolve the DTFT of the signal with

$$W_r(e^{j\Omega}) = e^{-j(M/2)\Omega}\ \frac{\sin\left(\dfrac{\Omega(M + 1)}{2}\right)}{\sin\left(\dfrac{\Omega}{2}\right)}$$

The effect of this convolution is to smear detail and introduce ripple in the vicinity of discontinuities. The smearing is proportional to the mainlobe width, while the ripple is proportional to the size of the sidelobes. A variety of alternative windows are used in practice to reduce sidelobe height in return for increased mainlobe width. In this problem we evaluate the effect of windowing time-domain signals on their DTFT. The role of windowing in filter design is explored in Chapter 8.

The Hanning window is defined as

$$w_h[n] = \begin{cases} 0.5 - 0.5\cos(2\pi n/M), & 0 \leq n \leq M \\ 0, & \text{otherwise} \end{cases}$$

(a) Assume $M = 50$ and use the MATLAB command `fft` to evaluate the magnitude spectrum of the rectangular window in dB at frequency intervals of $\pi/50$, $\pi/100$, and $\pi/200$.

(b) Assume $M = 50$ and use the MATLAB command `fft` to evaluate the magnitude spectrum of the Hanning window in dB at frequency intervals of $\pi/50$, $\pi/100$, and $\pi/200$.

(c) Use the results from (a) and (b) to evaluate the mainlobe width and peak sidelobe height in dB for each window.

(d) Let $y_r[n] = x[n]w_r[n]$ and $y_b[n] = x[n]w_b[n]$, where $x[n] = \cos(26\pi n/100) + \cos(29\pi n/100)$ and $M = 50$. Use the MATLAB command fft to evaluate $|Y_r(e^{j\Omega})|$ in dB and $|Y_b(e^{j\Omega})|$ in dB at intervals of $\pi/200$. Does the window choice affect whether you can identify the presence of two sinusoids? Why?

(e) Let $y_r[n] = x[n]w_r[n]$ and $y_b[n] = x[n]w_b[n]$, where $x[n] = \cos(26\pi n/100) + 0.02\cos(51\pi n/100)$ and $M = 50$. Use the MATLAB command fft to evaluate $|Y_r(e^{j\Omega})|$ in dB and $|Y_b(e^{j\Omega})|$ in dB at intervals of $\pi/200$. Does the window choice affect whether you can identify the presence of two sinusoids? Why?

4.47 Let a discrete-time signal $x[n]$ be defined as

$$x[n] = \begin{cases} e^{-(0.1n)^2/2}, & |n| \le 50 \\ 0, & \text{otherwise} \end{cases}$$

Use the MATLAB commands fft and $fftshift$ to numerically evaluate and plot the DTFT of $x[n]$ and the following subsampled signals at 500 values of Ω on the interval $-\pi \le \Omega < \pi$:

(a) $y[n] = x[2n]$
(b) $g[n] = x[4n]$

4.48 Repeat Problem 4.47 assuming

$$x[n] = \begin{cases} \cos(\pi n/2)e^{-(0.1n)^2/2}, & |n| \le 50 \\ 0, & \text{otherwise} \end{cases}$$

4.49 A signal $x(t)$ is defined as

$$x(t) = \cos\left(\frac{3\pi}{2}t\right)e^{-t^2/2}$$

(a) Evaluate the FT $X(j\omega)$ and show that $|X(j\omega)| \approx 0$ for $|\omega| > 3\pi$.

In parts (b)–(d), we compare $X(j\omega)$ to the FT of the sampled signal, $x[n] = x(n\mathcal{T})$, for several sampling intervals. Let $x[n] \overset{FT}{\longleftrightarrow} X_\delta(j\omega)$ be the FT of the sampled version of $x(t)$. Use MATLAB to numerically determine $X_\delta(j\omega)$ by evaluating

$$X_\delta(j\omega) = \sum_{n=-25}^{25} x[n]e^{-j\omega\mathcal{T}n}$$

at 500 values of ω on the interval $-3\pi \le \omega < 3\pi$. In each case, compare $X(j\omega)$ and $X_\delta(j\omega)$ and explain any differences.

(b) $\mathcal{T} = \frac{1}{3}$
(c) $\mathcal{T} = \frac{2}{5}$
(d) $\mathcal{T} = \frac{1}{2}$

4.50 Use the MATLAB command fft to repeat Example 4.16.

4.51 Use the MATLAB command fft to repeat Example 4.17.

4.52 Use the MATLAB command fft to repeat Example 4.18. Also depict the DTFS approximation and the underlying DTFT for $M = 2001$ and $M = 2005$.

4.53 Consider the sum of sinusoids

$$x(t) = \cos(2\pi t) + 2\cos(2\pi(0.8)t) + \tfrac{1}{2}\cos(2\pi(1.1)t)$$

Assume the frequency band of interest is $-5\pi < \omega < 5\pi$.

(a) Determine the sampling interval $\mathcal{T}$ so that the DTFS approximation to the FT of $x(t)$ spans the desired frequency band.

(b) Determine the minimum number of samples M_o so that the DTFS approximation consists of discrete-valued impulses located at the frequency corresponding to each sinusoid.

(c) Use MATLAB to plot $(1/M)|Y_\delta(j\omega)|$ and $|Y[k]|$ for the value of $\mathcal{T}$ chosen in part (a) and $M = M_o$.

(d) Repeat part (c) using $M = M_o + 5$ and $M = M_o + 8$.

4.54 We desire to use the DTFS to approximate the FT of a continuous-time signal $x(t)$ on the band $-\omega_a < \omega < \omega_a$ with resolution ω_r and a maximum sampling interval in frequency of $\Delta\omega$. Find the sampling interval $\mathcal{T}$, number of samples M, and DTFS length N. You may assume that the signal is effectively bandlimited to a frequency ω_m for which $|X(j\omega_a)| \ge 10|X(j\omega)|$, $\omega > \omega_m$. Plot the FT and the DTFS approximation for each of the following cases using the MATLAB command fft. *Hint:* Be sure to sample the pulses in (a) and (b) symmetrically about $t = 0$.

(a) $x(t) = \begin{cases} 1, & |t| < 1 \\ 0, & \text{otherwise} \end{cases}$, $\omega_a = \frac{3\pi}{2}$, $\omega_r = \frac{3\pi}{4}$, and $\Delta\omega = \frac{\pi}{8}$.

(b) $x(t) = 1/(2\pi)\,e^{-t^2/2}$, $\omega_a = 3$, $\omega_r = \frac{1}{2}$, and $\Delta\omega = \frac{1}{8}$.

(c) $x(t) = \cos(20\pi t) + \cos(21\pi t)$, $\omega_a = 40\pi$, $\omega_r = \pi/3$, and $\Delta\omega = \pi/10$.

(d) Repeat case (c) using $\omega_r = \pi/10$.

4.55 Repeat Example 4.19 using MATLAB.

4.56 Write a MATLAB M-file that implements the overlap and save method using fft to evaluate the convolution $y[n] = h[n] * x[n]$ on $0 \leq n < L$ for the following signals:

(a) $h[n] = \frac{1}{5}(u[n] - u[n - 5])$,
$x[n] = \cos(\pi n/6)$, $L = 30$

(b) $h[n] = \frac{1}{5}(u[n] - u[n - 5])$,
$x[n] = (\frac{1}{2})^n u[n]$, $L = 20$

4.57 Plot the ratio of the number of multiplications in the direct method for computing the DTFS coefficients to that of the FFT approach when $N = 2^p$ for $p = 2, 3, 4, \ldots, 16$.

4.58 In this problem we compare the number of multiplications required to evaluate $h[n] * x[n]$ using the overlap and save algorithm to that required for direct evaluation of the convolution sum when the impulse response $h[n] = 0$ for $n < 0$, $n \geq M$. In order to maximize the computational efficiency of the overlap and save algorithm, a FFT algorithm is used to evaluate the DTFS in step 4 and inverse DTFS in step 6.

(a) Show that M multiplications are required per output point if the convolution sum is evaluated directly.

(b) In counting multiplications per output point for the overlap and save algorithm, we need only consider steps 4, 5, and 6, since we assume $H[k]$ is precomputed. Show that

$$\frac{2N \log_2(N) + N}{N - M + 1}$$

multiplications per output point are required.

(c) Let R be the ratio of multiplications required for direct evaluation of the convolution sum to that for the overlap and save algorithm. Evaluate R for the following cases:

   (i) $M = 10$; $N = 16, 256$, and 1024.

   (ii) $M = 20$; $N = 32, 256$, and 1024.

   (iii) $M = 100$; $N = 128, 256$, and 1024.

(d) Show that for $N \gg M$, the overlap and save algorithm requires fewer multiplications than direct evaluation of the convolution sum if $M > 2 \log_2 N + 1$.

4.59 In this experiment we investigate evaluation of the time–bandwidth product with the DTFS. Let $x(t) \overset{FT}{\longleftrightarrow} X(j\omega)$.

(a) Use the Reimann sum approximation to an integral

$$\int_a^b f(u) \, du \approx \sum_{m=m_a}^{m_b} f(m\Delta u) \, \Delta u$$

to show that

$$T_d = \left[\frac{\int_{-\infty}^{\infty} t^2 |x(t)|^2 \, dt}{\int_{-\infty}^{\infty} |x(t)|^2 \, dt} \right]^{1/2}$$

$$\approx \mathcal{T} \left[\frac{\sum_{n=-M}^{M} n^2 |x[n]|^2}{\sum_{n=-M}^{M} |x[n]|^2} \right]^{1/2}$$

provided $x[n] = x(n\mathcal{T})$ represents the samples of $x(t)$ and $x(n\mathcal{T}) \approx 0$ for $|n| > M$.

(b) Use the DTFS approximation to the FT and the Reimann sum approximation to an integral to show that

$$B_w = \left[\frac{\int_{-\infty}^{\infty} \omega^2 |X(j\omega)|^2 \, d\omega}{\int_{-\infty}^{\infty} |X(j\omega)|^2 \, d\omega} \right]^{1/2}$$

$$\approx \frac{\omega_s}{2M + 1} \left[\frac{\sum_{k=-M}^{M} |k|^2 |X[k]|^2}{\sum_{k=-M}^{M} |X[k]|^2} \right]^{1/2}$$

where $x[n] \overset{DTFS;\, 2\pi/(2M + 1)}{\longleftrightarrow} X[k]$, $\omega_s = 2\pi/\mathcal{T}$ is the sampling frequency, and $X(jk\omega_s/(2M + 1)) \approx 0$ for $|k| > M$.

(c) Use the results from (a) and (b) and Eq. (3.58) to show that the time–bandwidth product computed using the DTFS approximation satifies

$$T_d B_w =$$

$$\left[\frac{\sum_{n=-M}^{M} n^2 |x[n]|^2}{\sum_{n=-M}^{M} |x[n]|^2} \right]^{1/2} \left[\frac{\sum_{k=-M}^{M} |k|^2 |X[k]|^2}{\sum_{k=-M}^{M} |X[k]|^2} \right]^{1/2}$$

$$\geq \frac{2M + 1}{4\pi}$$

(d) Repeat Computer Experiment 3.45 to demonstrate that the bound in (c) is satisfied and that Gaussian pulses satisfy the bound with equality.

5 Application to Communication Systems

5.1 Introduction

The purpose of a communication system is to transport a message signal (generated by a source of information) over a channel and deliver an estimate of that message signal to a user. For example, the message signal may be a speech signal. The channel may be a cellular telephone channel or satellite channel. As mentioned in Chapter 1, modulation is basic to the operation of a communication system. Modulation provides the means for shifting the range of frequencies contained in the message signal into another frequency range suitable for transmission over the channel, and a corresponding shift back to the original frequency range after reception. Formally, *modulation* is defined as *the process by which some characteristic of a carrier wave is varied in accordance with the message signal*. The message signal is referred to as the *modulating wave*, and the result of the modulation process is referred to as the *modulated wave*. In the receiver, *demodulation* is used to recover the message signal from the modulated wave. Demodulation is the inverse of the modulation process.

In this chapter we present an introductory treatment of modulation from a system-theoretic viewpoint, building on Fourier analysis as discussed in the previous two chapters. We begin the discussion with a description of the basic types of modulation, followed by the practical benefits derived from their use. This sets the stage for a discussion of the so-called amplitude modulation, which is widely used in practice for analog communications by virtue of its simplicity. One common application of amplitude modulation is in radio broadcasting. We then discuss some important variants of amplitude modulation. The counterpart of amplitude modulation that is used in digital communications is known as pulse-amplitude modulation, which is discussed in the latter part of the chapter. In reality, pulse-amplitude modulation is another manifestation of the sampling process that we studied in Chapter 4.

5.2 Types of Modulation

The specific type of modulation employed in a communication system is determined by the form of carrier wave used to perform the modulation. The two most commonly used forms of carrier are:

- ▶ Sinusoidal wave
- ▶ Periodic pulse train

Correspondingly, we may identify two main classes of modulation as described here.

1. *Continuous-wave (CW) modulation*

Consider the sinusoidal carrier wave

$$c(t) = A_c \cos(\phi(t)) \qquad (5.1)$$

which is uniquely defined by the carrier amplitude A_c and angle $\phi(t)$. Depending on which of these parameters are chosen for modulation, we may identify two subclasses of CW modulation:

▶ *Amplitude modulation*, in which the carrier amplitude is varied with the message signal

▶ *Angle modulation*, in which the angle of the carrier is varied with the message signal

Figure 5.1 shows examples of amplitude-modulated and angle-modulated waves for a sinusoidal modulating wave.

Amplitude modulation can itself be implemented in several different forms. For a given message signal, the frequency content of the modulated wave depends on the form of amplitude modulation used. Specifically, we have:

▶ Full amplitude modulation (double sideband-transmitted carrier)

▶ Double sideband-suppressed carrier modulation

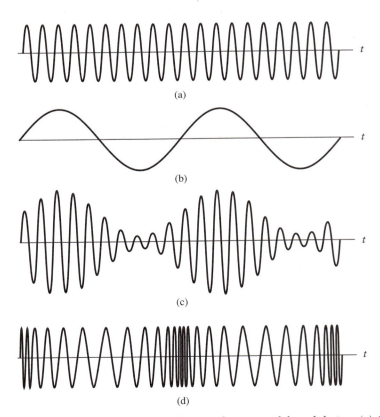

FIGURE 5.1 Amplitude- and angle-modulated waves for sinusoidal modulation. (a) Carrier wave. (b) Sinusoidal modulating wave. (c) Amplitude-modulated wave. (d) Angle-modulated wave.

  ▶ Single sideband modulation
  ▶ Vestigial sideband modulation

The last three types of amplitude modulation are examples of *linear modulation*, in the sense that if the amplitude of the message signal is scaled by a certain factor, then the amplitude of the modulated wave is scaled by exactly the same factor. In this strict sense, full amplitude modulation fails to meet the definition of linear modulation with respect to the message signal for reasons that will become apparent later. Nevertheless, the departure from linearity in the case of full amplitude modulation is of a rather mild sort, such that many of the mathematical procedures applicable to the analysis of linear modulation may be retained. Most importantly from our present perspective, all four different forms of amplitude modulation mentioned here lend themselves to mathematical analysis using the tools presented in this book. Subsequent sections of this chapter develop the details of this analysis.

In contrast, angle modulation is a *nonlinear* modulation process. To describe it in a formal manner, we need to introduce the notion of *instantaneous radian frequency*, denoted by $\omega_i(t)$. It is defined as the derivative of the angle $\phi(t)$ with respect to time t, as shown by

$$\omega_i(t) = \frac{d\phi(t)}{dt} \tag{5.2}$$

Equivalently, we may write (ignoring the constant of integration)

$$\phi(t) = \int_0^t \omega_i(\tau)\,d\tau \tag{5.3}$$

where it is assumed that the initial value

$$\phi(0) = \int_{-\infty}^0 \omega_i(\tau)\,d\tau$$

is zero.

Equation (5.2) includes the usual definition of radian frequency as a special case. Consider the ordinary form of a sinusoidal wave written as

$$c(t) = A_c\cos(\omega_c t + \theta)$$

where A_c is the amplitude, ω_c is the radian frequency, and θ is the phase. For this simple case, the angle $\phi(t)$ is

$$\phi(t) = \omega_c t + \theta$$

in which case the use of Eq. (5.2) yields the expected result

$$\omega_i(t) = \omega_c \quad \text{for all } t$$

Returning to the general definition of Eq. (5.2), when the instantaneous radian frequency $\omega_i(t)$ is varied in accordance with a message signal denoted by $m(t)$, we may write

$$\omega_i(t) = \omega_c + k_f m(t) \tag{5.4}$$

where k_f is the frequency sensitivity factor of the modulator. Hence substituting Eq. (5.4) into (5.3), we get

$$\phi(t) = \omega_c t + k_f \int_0^t m(\tau)\,d\tau$$

The resulting form of angle modulation is known as *frequency modulation* (FM), written as

$$s_{\mathrm{FM}}(t) = A_c \cos\left(\omega_c t + k_f \int_0^t m(\tau)\, d\tau\right) \qquad (5.5)$$

where the carrier amplitude is maintained constant.

When the angle $\phi(t)$ is varied in accordance with the message signal $m(t)$, we may write

$$\phi(t) = \omega_c t + k_p m(t)$$

where k_p is the phase sensitivity factor of the modulator. This time we have a different form of angle modulation known as *phase modulation* (PM), defined by

$$s_{\mathrm{PM}}(t) = A_c \cos(\omega_c t + k_p m(t)) \qquad (5.6)$$

where the carrier amplitude is again maintained constant.

Although the formulas of Eqs. (5.5) and (5.6) for FM and PM signals look different, they are in fact intimately related to each other. For the present, it suffices to say that both of them are nonlinear functions of the message signal $m(t)$, which makes their mathematical analysis more difficult than that of amplitude modulation. Since the primary emphasis in this book is on a linear analysis of signals and systems, we will devote much of the discussion in this chapter to amplitude modulation and its variants.

2. *Pulse modulation*

Consider next a carrier wave that consists of a periodic train of narrow pulses, as shown by

$$c(t) = \sum_{n=-\infty}^{\infty} p(t - n\mathcal{T})$$

where $\mathcal{T}$ is the period, and $p(t)$ denotes a pulse of relatively short duration (compared to the period $\mathcal{T}$) and centered on the origin. When some characteristic parameter of $p(t)$ is

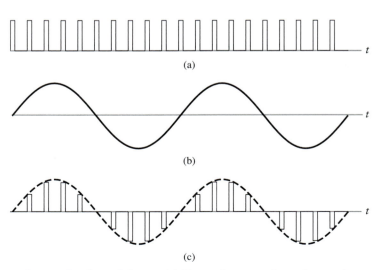

(a)

(b)

(c)

FIGURE 5.2   Pulse-amplitude modulation. (a) Train of rectangular pulses as the carrier wave. (b) Sinusoidal modulating wave. (c) Pulse-amplitude modulated wave.

varied in accordance with the message signal, we have *pulse modulation*. Figure 5.2 shows an example of pulse amplitude modulation for the case of a sinusoidal modulating wave.

Depending on how pulse modulation is actually accomplished, we may distinguish the following two subclasses:

▶ *Analog pulse modulation*, in which a characteristic parameter such as the amplitude, duration, or position of a pulse is varied continuously with the message signal. We thus speak of pulse-amplitude modulation, pulse-duration modulation, and pulse-position modulation as different realizations of analog pulse modulation. This type of pulse modulation may be viewed as the counterpart of CW modulation.

▶ *Digital pulse modulation*, in which the modulated signal is represented in coded form. This representation can be accomplished in a number of different ways. The standard method involves two operations. First, the amplitude of each modulated pulse is approximated by the nearest member of a set of discrete levels that occupies a compatible range of values. This operation is called *quantization*, and the device for performing it is called a *quantizer*. Second, the quantizer output is *coded* (e.g., in binary form). This particular form of digital pulse modulation is known as *pulse-code modulation* (PCM). Quantization is a nonlinear process that results in a loss of information, but the loss is under the designer's control in that it can be made as small as desired simply by using a large enough number of discrete (quantization) levels. In any event, PCM has no CW counterpart. As with angle modulation, a complete discussion of PCM is beyond the scope of this book. Insofar as pulse modulation is concerned, the primary emphasis in this chapter is on pulse-amplitude modulation, which is a linear process.

5.3 Benefits of Modulation

The use of modulation is not confined exclusively to communication systems. Rather, modulation in one form or another is used in signal processing, radiotelemetry, radar, sonar, control systems, and general-purpose instruments such as spectrum analyzers and frequency synthesizers. However, it is in the study of communication systems that we find modulation playing a dominant role.

In the context of communication systems, we may identify three practical benefits that result from the use of modulation:

1. *Modulation is used to shift the spectral content of a message signal so that it lies inside the operating frequency band of a communication channel*
Consider, for example, telephonic communication over a cellular radio channel. For such an application, the frequency components of a speech signal from about 300 to 3100 Hz are considered to be adequate for the purpose of communication. In North America, the band of frequencies assigned to cellular radio systems is 800–900 MHz. The subband 824–849 MHz is used to receive signals from the mobile units, and the subband 869–894 MHz is used for transmitting signals to the mobile units. For this form of telephonic communication to be feasible, we clearly need to do two things: shift the essential spectral content of a speech signal so that it lies inside the prescribed subband for transmission, and shift it back to its original frequency band on reception. The first of these two operations is one of modulation, and the second is one of demodulation.

2. *Modulation provides a mechanism for putting the information content of a message signal into a form that may be less vulnerable to noise or interference*
In a communication system the received signal is ordinarily corrupted by noise generated

at the front end of the receiver or by interference picked up in the course of transmission. Some specific forms of modulation such as frequency modulation and pulse-code modulation have the inherent ability to trade off increased transmission bandwidth for improved system performance in the presence of noise. We are careful here to say that this important property is not shared by all modulation techniques. In particular, those modulation techniques that vary the amplitude of a CW or pulsed carrier provide absolutely no protection against noise or interference in the received signal.

3. *Modulation permits the use of multiplexing*
A communication channel (e.g., telephone channel, mobile radio channel, satellite communications channel) represents a major capital investment and must therefore be deployed in a cost-effective manner. *Multiplexing* is a signal-processing operation that makes this possible. In particular, it permits the simultaneous transmission of information-bearing signals from a number of independent sources over the channel and on to their respective destinations. It can take the form of frequency-division multiplexing for use with CW modulation techniques, or time-division multiplexing for use with digital pulse modulation techniques.

In this chapter we will discuss the frequency-shifting and multiplexing aspects of modulation. However, a study of the issues relating to noise in modulation systems is beyond the scope of this book.

5.4 *Full Amplitude Modulation*

Consider a sinusoidal carrier wave $c(t)$ defined by

$$c(t) = A_c \cos(\omega_c t) \tag{5.7}$$

For convenience of presentation, we have assumed that the phase of the carrier wave is zero in Eq. (5.7). We are justified in making this assumption as the primary emphasis here is on variations imposed on the carrier amplitude. Let $m(t)$ denote a message signal of interest. *Amplitude modulation (AM) is defined as a process in which the amplitude of the carrier is varied proportionately to a message signal* $m(t)$, as shown by

$$s(t) = A_c[1 + k_a m(t)] \cos(\omega_c t) \tag{5.8}$$

where k_a is a constant called the *amplitude sensitivity* factor of the modulator. The modulated wave $s(t)$ so defined is said to be a "full" AM wave for reasons explained later in the section. Note that the radian frequency ω_c of the carrier is maintained constant.

The amplitude of the time function multiplying $\cos(\omega_c t)$ in Eq. (5.8) is called the *envelope* of the AM wave $s(t)$. Using $a(t)$ to denote this envelope, we may thus write

$$a(t) = A_c|1 + k_a m(t)| \tag{5.9}$$

Two cases arise, depending on the magnitude of $k_a m(t)$, compared to unity:

1. $|k_a m(t)| \leq 1$ for all t

Under this condition, the term $1 + k_a m(t)$ is always nonnegative. We may therefore simplify the expression for the envelope of the AM wave by writing

$$a(t) = A_c[1 + k_a m(t)] \quad \text{for all } t \tag{5.10}$$

2. $|k_a m(t)| > 1$ for some t

Under this second condition, we must use Eq. (5.9) for evaluating the envelope of the AM wave.

The maximum absolute value of $k_a m(t)$ multiplied by 100 is referred to as the *percentage modulation*. Accordingly, case 1 corresponds to a percentage modulation less than or equal to 100%, whereas case 2 corresponds to a percentage modulation in excess of 100%.

The waveforms of Fig. 5.3 illustrate the amplitude modulation process. Part (a) of the figure depicts the waveform of a message signal $m(t)$. Part (b) of the figure depicts an AM wave produced by this message signal for a value of k_a for which the percentage modulation is 66.7% (i.e., case 1).

On the other hand, the AM wave shown in Fig. 5.3(c) corresponds to a value of k_a for which the percentage modulation is 166.7% (i.e., case 2). Comparing the waveforms of these two AM waves with that of the message signal, we draw an important conclusion. Specifically, the envelope of the AM wave has a waveform that bears a one-to-one correspondence with that of the message signal if and only if the percentage modulation is less than or equal to 100%. This correspondence is destroyed if the percentage modulation is permitted to exceed 100%. In the latter case, the modulated wave is said to suffer from *envelope distortion*, and the wave itself is said to be *overmodulated*.

▶ **Drill Problem 5.1** For 100% modulation, the envelope $a(t)$ becomes zero for some time t. Why?

Answer: If $k_a m(t) = -1$ for some time t, then $a(t) = 0$. ◀

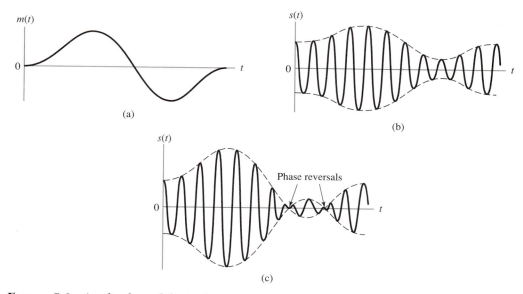

(a)

(b)

Phase reversals

(c)

Figure 5.3 Amplitude modulation for varying percentage of modulation. (a) Message signal $m(t)$. (b) AM wave for $|k_a m(t)| < 1$ for all t, where k_a is the amplitude sensitivity of the modulator. This case represents undermodulation. (c) AM wave for $|k_a m(t)| > 1$ some of the time. This second case represents overmodulation.

■ MORE ON THE TIME-DOMAIN DESCRIPTION OF AN AM WAVE

Earlier we defined linear modulation to be that form of a modulation process in which if the amplitude of the message signal (i.e., modulating wave) is scaled by a certain factor, then the amplitude of the modulated wave is scaled by exactly the same factor. This definition of linear modulation is consistent with the notion of linearity of a system that was introduced in Section 1.8. Amplitude modulation, as defined in Eq. (5.8), fails the linearity test in a strict sense. To demonstrate this, suppose the message signal $m(t)$ consists of the sum of two components, $m_1(t)$ and $m_2(t)$. Let $s_1(t)$ and $s_2(t)$ denote the AM waves produced by these two components acting separately. With the operator H denoting the amplitude modulation process, we may then write

$$H\{m_1(t) + m_2(t)\} = A_c[1 + k_a(m_1(t) + m_2(t))]\cos(\omega_c t)$$
$$\neq s_1(t) + s_2(t)$$

where

$$s_1(t) = A_c[1 + k_a m_1(t)]\cos(\omega_c t)$$

and

$$s_2(t) = A_c[1 + k_a m_2(t)]\cos(\omega_c t)$$

The presence of the carrier wave $A_c \cos(\omega_c t)$ in the AM wave causes the principle of superposition to be violated.

However, as pointed out earlier, the failure of amplitude modulation to meet the criterion for linearity is of a rather mild sort. From the definition given in Eq. (5.8), we see that the AM signal $s(t)$ is, in fact, a linear combination of the carrier component $A_c \cos(\omega_c t)$ and the modulated component $A_c \cos(\omega_c t)m(t)$. Accordingly, amplitude modulation does permit the use of Fourier analysis without difficulty, as discussed next.

■ FREQUENCY-DOMAIN DESCRIPTION OF AMPLITUDE MODULATION

Equation (5.8) defines the full AM wave $s(t)$ as a function of time. To develop the frequency description of this AM wave, we take the Fourier transform of both sides of Eq. (5.8). Let $S(j\omega)$ denote the Fourier transform of $s(t)$, and $M(j\omega)$ denote the Fourier transform of $m(t)$; we refer to $M(j\omega)$ as the *message spectrum*. Accordingly, using the Fourier transform representation of the cosine function $A_c \cos(\omega_c t)$ and the frequency-shifting property of the Fourier transform, we may write

$$S(j\omega) = \pi A_c\,[\delta(\omega - \omega_c) + \delta(\omega + \omega_c)] \\ + \tfrac{1}{2}k_a A_c\,[M(j(\omega - \omega_c)) + M(j(\omega + \omega_c))] \tag{5.11}$$

Let the message signal $m(t)$ be bandlimited to the interval $-\omega_m \leq \omega \leq \omega_m$ as in Fig. 5.4(a). We refer to the highest frequency component ω_m of $m(t)$ as the *message bandwidth*, which is measured in rad/s. The shape of the spectrum shown in this figure is intended for the purpose of illustration only. We find from Eq. (5.11) that the spectrum $S(j\omega)$ of the AM wave is as shown in Fig. 5.4(b) for the case when $\omega_c > \omega_m$. This spectrum consists of two impulse functions weighted by the factor πA_c and occurring at $\pm\omega_c$, and two versions of the message spectrum shifted in frequency by $\pm\omega_c$ and scaled in amplitude by $\tfrac{1}{2}k_a A_c$. The spectrum of Fig. 5.4(b) may be described as follows:

1. For positive frequencies, the portion of the spectrum of the modulated wave lying above the carrier frequency ω_c is called the *upper sideband*, whereas the symmetric portion below ω_c is called the *lower sideband*. For negative frequencies, the image

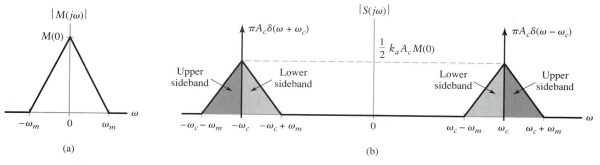

FIGURE 5.4 Spectral content of AM wave. (a) Magnitude spectrum of message signal. (b) Magnitude spectrum of the AM wave, showing the compositions of the carrier, upper and lower sidebands.

of the upper sideband is represented by the portion of the spectrum below $-\omega_c$ and the image of the lower sideband by the portion above $-\omega_c$. The condition $\omega_c > \omega_m$ ensures that the sidebands do not overlap. Otherwise, the modulated wave exhibits *spectral overlap* and, therefore, frequency distortion.

2. For positive frequencies, the highest frequency component of the AM wave is $\omega_c + \omega_m$, and the lowest frequency component is $\omega_c - \omega_m$. The difference between these two frequencies defines the *transmission bandwidth* ω_T for an AM wave, which is exactly twice the message bandwidth ω_m; that is,

$$\omega_T = 2\omega_m \qquad (5.12)$$

The spectrum of the AM wave as depicted in Fig. 5.4(b) is *full* in that the carrier, the upper sideband, and the lower sideband are all completely represented. It is for this reason that we refer to this form of modulation as "full amplitude modulation."

The upper sideband of the AM wave represents the positive frequency components of the message spectrum $M(j\omega)$, shifted upward in frequency by the carrier frequency ω_c. The lower sideband of the AM wave represents the negative frequency components of the message spectrum $M(j\omega)$, also shifted upward in frequency by ω_c. Herein lies the importance of admitting the use of negative frequencies in the Fourier analysis of signals. In particular, the use of amplitude modulation reveals the negative frequency components of $M(j\omega)$ completely, provided that $\omega_c > \omega_m$.

EXAMPLE 5.1 Consider a modulating wave $m(t)$ that consists of a single tone or frequency component, that is,

$$m(t) = A_0 \cos(\omega_0 t)$$

where A_0 is the amplitude of the modulating wave and ω_0 is its radian frequency (see Fig. 5.5(a)). The sinusoidal carrier wave $c(t)$ has amplitude A_c and radian frequency ω_c (see Fig. 5.5(b)). Evaluate the time-domain and frequency-domain characteristics of the AM wave.

Solution: The AM wave is described by

$$s(t) = A_c[1 + \mu \cos(\omega_0 t)] \cos(\omega_c t) \qquad (5.13)$$

where

$$\mu = k_a A_0$$

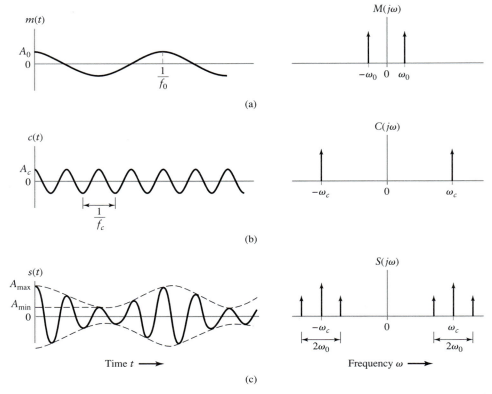

FIGURE 5.5   Time-domain (on the left) and frequency-domain (on the right) characteristics of AM produced by a sinusoidal modulating wave. (a) Modulating wave. (b) Carrier wave. (c) AM wave.

The dimensionless constant μ for a sinusoidal modulating wave is called the *modulation factor*; it equals the percentage modulation when it is expressed numerically as a percentage. To avoid envelope distortion due to overmodulation, the modulation factor μ must be kept below unity. Figure 5.5(c) is a sketch of $s(t)$ for μ less than unity.

Let A_{max} and A_{min} denote the maximum and minimum values of the envelope of the modulated wave. Then, from Eq. (5.13), we get

$$\frac{A_{max}}{A_{min}} = \frac{A_c(1 + \mu)}{A_c(1 - \mu)}$$

Solving for μ:

$$\mu = \frac{A_{max} - A_{min}}{A_{max} + A_{min}}$$

Expressing the product of the two cosines in Eq. (5.13) as the sum of two sinusoidal waves, one having frequency $\omega_c + \omega_0$ and the other having frequency $\omega_c - \omega_0$, we get

$$s(t) = A_c \cos(\omega_c t) + \tfrac{1}{2}\mu A_c \cos[(\omega_c + \omega_0)t]$$
$$+ \tfrac{1}{2}\mu A_c \cos[(\omega_c - \omega_0)t]$$

The Fourier transform of $s(t)$ is therefore

$$S(j\omega) = \pi A_c[\delta(\omega - \omega_c) + \delta(\omega + \omega_c)]$$
$$+ \tfrac{1}{2}\pi\mu A_c[\delta(\omega - \omega_c - \omega_0) + \delta(\omega + \omega_c + \omega_0)]$$
$$+ \tfrac{1}{2}\pi\mu A_c[\delta(\omega - \omega_c + \omega_0) + \delta(\omega + \omega_c - \omega_0)]$$

Thus, in ideal terms, the spectrum of a full AM wave, for the special case of sinusoidal modulation, consists of impulse functions at $\pm\omega_c$, $\omega_c \pm \omega_0$, and $-\omega_c \pm \omega_0$, as depicted in Fig. 5.5(c).

EXAMPLE 5.2 Continuing with Example 5.1, investigate the effect of varying the modulation factor μ on the power content of the AM wave.

Solution: In practice, the AM wave $s(t)$ is a voltage or current signal. In either case, the average power delivered to a 1-ohm load resistor by $s(t)$ is comprised of three components:

$$\text{Carrier power} = \tfrac{1}{2}A_c^2$$
$$\text{Upper side-frequency power} = \tfrac{1}{8}\mu^2 A_c^2$$
$$\text{Lower side-frequency power} = \tfrac{1}{8}\mu^2 A_c^2$$

The ratio of the total sideband power to the total power in the modulated wave is therefore equal to $\mu^2/(2 + \mu^2)$, which depends only on the modulation factor μ. If $\mu = 1$, that is, 100% modulation is used, the total power in the two side-frequencies of the resulting AM wave is only one-third of the total power in the modulated wave.

Figure 5.6 shows the percentage of total power in both side-frequencies and in the carrier plotted versus the percentage modulation.

▶ **Drill Problem 5.2** For a particular case of sinusoidal modulation, the percentage modulation is less than 20%. Show that the power in one side-frequency is less than 1% of the total power in the AM wave. ◀

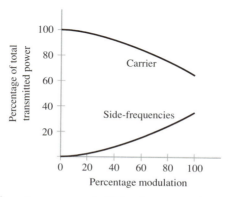

FIGURE 5.6 Variations of carrier power and sideband power with percentage modulation of AM wave for the case of sinusoidal modulation.

▪ GENERATION OF AM WAVE

Various schemes have been devised for the generation of an AM wave. Here we consider a simple circuit that follows from the defining equation (5.8). First, we rewrite this equation in the equivalent form:

$$s(t) = k_a[m(t) + B]A_c \cos(\omega_c t) \tag{5.14}$$

The constant B, equal to $1/k_a$, represents a *bias* that is added to the message signal $m(t)$ before modulation. Equation (5.14) suggests the scheme described in the block diagram of Fig. 5.7 for generating an AM wave. Basically, it consists of two functional blocks:

▶ An adder, which adds the bias B to the incoming message signal $m(t)$.

▶ A multiplier, which multiplies the adder output $(m(t) + B)$ by the carrier wave $A_c \cos(\omega_c t)$, producing the AM wave $s(t)$. The constant k_a is a proportionality constant associated with the multiplier.

The percentage modulation is controlled simply by adjusting the bias B.

▶ **Drill Problem 5.3**   Assuming that M_{max} is the maximum absolute value of the message signal, what is the condition which the bias B must satisfy to avoid overmodulation?

Answer: $B \geq M_{max}$.   ◀

▪ DEMODULATION OF AM WAVE

The so-called *envelope detector* provides a simple and yet effective device for the demodulation of a narrowband AM wave for which the percentage modulation is less than 100%. By "narrowband" we mean that the carrier frequency is large compared with the message bandwidth. Ideally, an envelope detector produces an output signal that follows the envelope of the input signal waveform exactly—hence the name. Some version of this circuit is used in almost all commercial AM radio receivers.

Figure 5.8(a) shows the circuit diagram of an envelope detector that consists of a diode and a resistor–capacitor filter. The operation of this envelope detector is as follows. On the positive half-cycle of the input signal, the diode is forward-biased and the capacitor C charges up rapidly to the peak value of the input signal. When the input signal falls below this value, the diode becomes reverse-biased and the capacitor C discharges slowly through the load resistor R_l. The discharging process continues until the next positive half-cycle. When the input signal becomes greater than the voltage across the capacitor, the diode conducts again and the process is repeated. We assume that the diode is ideal, presenting zero impedance to current flow in the forward-biased region, and infinite impedance in the reverse-biased region. We further assume that the AM wave applied to the

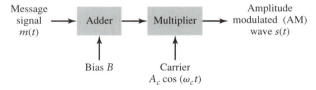

FIGURE 5.7   System involving an adder and multiplier, for generating an AM wave.

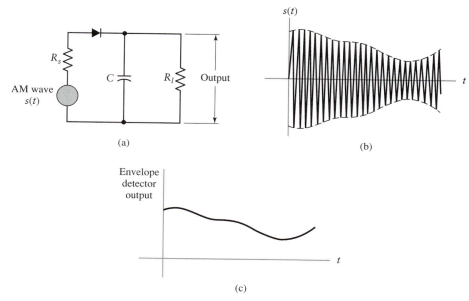

FIGURE 5.8   Envelope detector, illustrated by (a) circuit diagram, (b) AM wave input, and (c) envelope detector output, assuming ideal conditions.

envelope detector is supplied by a voltage source of internal resistance R_s. The charging time constant $R_s C$ must be short compared with the carrier period $2\pi/\omega_c$, that is

$$R_s C \ll \frac{2\pi}{\omega_c}$$

Accordingly, the capacitor C charges rapidly and thereby follows the applied voltage up to the positive peak when the diode is conducting. On the other hand, the discharging time constant $R_l C$ must be long enough to ensure that the capacitor discharges slowly through the load resistor R_l between positive peaks of the carrier wave, but not so long that the capacitor voltage will not discharge at the maximum rate of change of the modulating wave, that is,

$$\frac{2\pi}{\omega_c} \ll R_l C \ll \frac{2\pi}{\omega_m}$$

where ω_m is the message bandwidth. The result is that the capacitor voltage or detector output is very nearly the same as the envelope of the AM wave, as we can see from Figs. 5.8(b) and (c). The detector output usually has a small ripple (not shown in Fig. 5.8(c)) at the carrier frequency; this ripple is easily removed by lowpass filtering.

▶ **Drill Problem 5.4**   An envelope detector has a source resistance $R_s = 75\ \Omega$ and a load resistance $R_l = 10\ \text{k}\Omega$. You are given $\omega_c = 2\pi \times 10^5$ rad/s and $\omega_m = 2\pi \times 10^3$ rad/s. Suggest a suitable value for the capacitor C.

Answer: $C = 0.01\ \mu\text{F}$.   ◀

5.5 Double Sideband-Suppressed Carrier Modulation

In full AM, the carrier wave $c(t)$ is completely independent of the message signal $m(t)$, which means that the transmission of the carrier wave represents a waste of power. This points to a shortcoming of amplitude modulation, namely, that only a fraction of the total transmitted power is affected by $m(t)$, which was well demonstrated in Example 5.2. To overcome this shortcoming, we may suppress the carrier component from the modulated wave, resulting in *double sideband-suppressed carrier* (DSB-SC) *modulation*. By suppressing the carrier, we obtain a modulated wave that is proportional to the product of the carrier wave and the message signal. Thus to describe a DSB-SC modulated wave as a function of time, we simply write

$$s(t) = c(t)m(t) \tag{5.15}$$
$$= A_c \cos(\omega_c t)m(t)$$

This modulated wave undergoes a phase reversal whenever the message signal $m(t)$ crosses zero, as illustrated in Fig. 5.9; part (a) of the figure depicts the waveform of a message signal, and part (b) depicts the corresponding DSB-SC modulated wave. Accordingly, unlike amplitude modulation, the envelope of a DSB-SC modulated wave is entirely different from the message signal.

▶ **Drill Problem 5.5** Sketch the envelope of the DSB-SC modulated wave shown in Fig. 5.9(b) and compare it to the message signal depicted in Fig. 5.9(a). ◀

■ FREQUENCY-DOMAIN DESCRIPTION

The suppression of the carrier from the modulated wave of Eq. (5.15) is well appreciated by examining its spectrum. Specifically, by taking the Fourier transform of both sides of Eq. (5.15), we get

$$S(j\omega) = \tfrac{1}{2}A_c[M(j(\omega - \omega_c)) + M(j(\omega + \omega_c))] \tag{5.16}$$

where, as before, $S(j\omega)$ is the Fourier transform of the modulated wave $s(t)$, and $M(j\omega)$ is the Fourier transform of the message signal $m(t)$. When the message signal $m(t)$ is limited to the interval $-\omega_m \leq \omega \leq \omega_m$ as in Fig. 5.10(a), we find that the spectrum $S(j\omega)$ is as illustrated in part (b) of the figure. Except for a change in scale factor, the modulation process simply translates the spectrum of the message signal by $\pm\omega_c$. Of course, the trans-

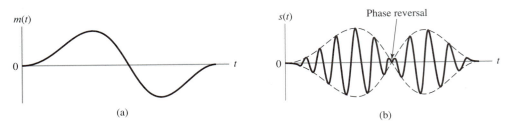

FIGURE 5.9 Double sideband-suppressed carrier modulation. (a) Message signal. (b) DSB-SC modulated wave, resulting from multiplication of the message signal by the sinusoidal carrier wave.

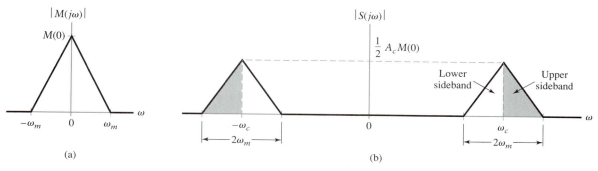

FIGURE 5.10 Spectral content of DSB-SC modulated wave. (a) Magnitude spectrum of message signal. (b) Magnitude spectrum of DSB-SC modulated wave, consisting of upper and lower sidebands only.

mission bandwidth required by DSB-SC modulation is the same as that for full amplitude modulation, namely, $2\omega_m$. However, comparing the spectrum of Fig. 5.10(b) for DSB-SC modulation with that of Fig. 5.4(b) for full AM, we clearly see that the carrier is suppressed in the DSB-SC case, whereas it is present in the full AM case as exemplified by the existence of the pair of impulse functions at $\pm\omega_c$.

The generation of a DSB-SC modulated wave consists simply of the product of the message signal $m(t)$ and the carrier wave $A_c \cos(\omega_c t)$, as indicated in Eq. (5.15). A device for achieving this requirement is called a *product modulator*, which is another term for a straightforward multiplier. Figure 5.11(a) shows the block diagram representation of a product modulator.

■ COHERENT DETECTION

The message signal $m(t)$ may be recovered from a DSB-SC modulated wave $s(t)$ by first multiplying $s(t)$ with a locally generated sinusoidal wave and then lowpass filtering the product, as depicted in Fig. 5.11(b). It is assumed that the local oscillator output is exactly coherent or synchronized, in both frequency and phase, with the carrier wave $c(t)$ used in the product modulator to generate $s(t)$. This method of demodulation is known as *coherent detection* or *synchronous demodulation*.

It is instructive to derive coherent detection as a special case of the more general demodulation process using a local oscillator signal of the same frequency but arbitrary phase difference ϕ, measured with respect to the carrier wave $c(t)$. Thus denoting the local oscillator signal in the receiver by $\cos(\omega_c t + \phi)$, assumed to be of unit amplitude for

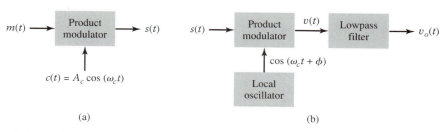

FIGURE 5.11 (a) Product modulator for generating the DSB-SC modulated wave. (b) Coherent detector for demodulation of the DSB-SC modulated wave.

convenience, and using Eq. (5.15) for the DSB-SC modulated wave $s(t)$, we find that the product modulator output in Fig. 5.11(b) is given by

$$
\begin{aligned}
v(t) &= \cos(\omega_c t + \phi)s(t) \\
&= A_c \cos(\omega_c t) \cos(\omega_c t + \phi)m(t) \\
&= \tfrac{1}{2}A_c \cos(\phi)m(t) + \tfrac{1}{2}A_c \cos(2\omega_c t + \phi)m(t)
\end{aligned}
\tag{5.17}
$$

The first term on the right-hand side of Eq. (5.17), namely, $\tfrac{1}{2}A_c \cos(\phi)m(t)$, represents a scaled version of the original message signal $m(t)$. The second term, $\tfrac{1}{2}A_c \cos(2\omega_c t + \phi)m(t)$, represents a new DSB-SC modulated wave with carrier frequency $2\omega_c$. Figure 5.12 shows the magnitude spectrum of $v(t)$. The clear separation between the spectra of the two components of $v(t)$ indicated in Fig. 5.12 hinges on the assumption that the original carrier frequency ω_c satisfies the following condition:

$$
2\omega_c - \omega_m > \omega_m
$$

or, equivalently,

$$
\omega_c > \omega_m
\tag{5.18}
$$

where ω_m is the message bandwidth. Provided that this condition is satisfied, we may then use a lowpass filter to suppress the unwanted second term of $v(t)$. To accomplish this, the passband of the lowpass filter must extend over the entire message spectrum and no more. More precisely, its specifications must satisfy two requirements:

1. Cutoff frequency: ω_m
2. Transition band: $\omega_m \leq \omega \leq 2\omega_c - \omega_m$

Thus the overall output $v_o(t)$ in Fig. 5.11(b) is given by

$$
v_o(t) = \tfrac{1}{2}A_c \cos(\phi)m(t)
\tag{5.19}
$$

The demodulated signal $v_o(t)$ is proportional to $m(t)$ when the phase error ϕ is a constant. The amplitude of this demodulated signal is maximum when $\phi = 0$ and has a minimum of zero when $\phi = \pm\pi/2$. The zero demodulated signal, which occurs for $\phi = \pm\pi/2$, represents the *quadrature null effect* of the coherent detector. The phase error ϕ in the local oscillator causes the detector output to be attenuated by a factor equal to $\cos\phi$. As long as the phase error ϕ is constant, the detector output provides an undistorted version of the original message signal $m(t)$. In practice, however, we usually find that the phase error ϕ varies randomly with time, owing to random variations in the communication channel. The result is that at the detector output, the multiplying factor $\cos\phi$ also varies randomly with time, which is obviously undesirable. Therefore circuitry must be provided

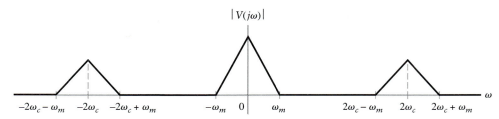

FIGURE 5.12 Magnitude spectrum of the product modulator output $v(t)$ in the coherent detector of Fig. 5.11(b).

in the receiver to maintain the local oscillator in perfect synchronism, in both frequency and phase, with the carrier wave used to generate the DSB-SC modulated wave in the transmitter. The resulting increase in receiver complexity is the price that must be paid for suppressing the carrier wave to save transmitter power.

▶ **Drill Problem 5.6**   For the coherent detector of Fig. 5.11(b) to operate properly, the condition of Eq. (5.18) must be satisfied. What would happen if this condition is violated?

Answer: The lower and upper sidebands overlap, in which case the coherent detector fails to operate properly.   ◀

EXAMPLE 5.3   Consider again the sinusoidal modulating signal

$$m(t) = A_0 \cos(\omega_0 t)$$

with amplitude A_0 and frequency ω_0; see Fig. 5.13(a). The carrier wave is

$$c(t) = A_c \cos(\omega_c t)$$

with amplitude A_c and frequency ω_c; see Fig. 5.13(b). Investigate the time-domain and frequency-domain characteristics of the corresponding DSB-SC modulated wave.

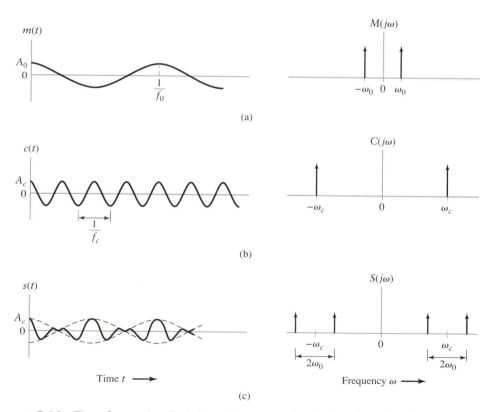

FIGURE 5.13   Time-domain (on the left) and frequency-domain (on the right) characteristics of DSB-SC modulation produced by a sinusoidal modulating wave. (a) Modulating wave. (b) Carrier wave. (c) DSB-SC modulated wave. Note that $\omega = 2\pi f$.

Solution: The DSB-SC modulated wave is defined by

$$s(t) = A_c A_0 \cos(\omega_c t) \cos(\omega_0 t)$$
$$= \tfrac{1}{2} A_c A_0 \cos[(\omega_c + \omega_0)t] + \tfrac{1}{2} A_c A_0 \cos[(\omega_c - \omega_0)t]$$

The Fourier transform of $s(t)$ is given by

$$S(j\omega) = \tfrac{1}{2}\pi A_c A_0 [\delta(\omega - \omega_c - \omega_0) + \delta(\omega + \omega_c + \omega_0)$$
$$+ \; \delta(\omega - \omega_c + \omega_0) + \delta(\omega + \omega_c - \omega_0)]$$

Thus, in ideal terms, the spectrum of the DSB-SC modulated wave, for the special case of a sinusoidal modulating wave, consists of impulse functions located at $\omega_c \pm \omega_0$ and $-\omega_c \pm \omega_0$. Figure 5.13(c) presents a depiction of the modulated wave $s(t)$ and its Fourier transform $S(j\omega)$. Comparison of Fig. 5.13 for DSB-SC modulation with Fig. 5.5 for full AM is noteworthy. Suppression of the carrier has a profound impact on the waveform of the modulated signal and its spectrum.

It is informative to continue the analysis of DSB-SC modulation for a sinusoidal modulating wave. When there is perfect synchronism between the local oscillator in the receiver of Fig. 5.11(b) and the carrier wave $c(t)$ in the transmitter of Fig. 5.11(a), we find that the product modulator output is

$$v(t) = \cos(\omega_c t)\{\tfrac{1}{2} A_c A_0 \cos[(\omega_c - \omega_0)t] + \tfrac{1}{2} A_c A_0 \cos[(\omega_c + \omega_0)t]\}$$
$$= \tfrac{1}{4} A_c A_0 \cos[(2\omega_c - \omega_0)t] + \tfrac{1}{4} A_c A_0 \cos(\omega_0 t)$$
$$+ \; \tfrac{1}{4} A_c A_0 \cos[(2\omega_c + \omega_0)t] + \tfrac{1}{4} A_c A_0 \cos(\omega_0 t)$$

The first two terms of $v(t)$ are produced by the lower side-frequency, and the last two terms are produced by the upper side-frequency. The first and third terms, of frequencies $2\omega_c - \omega_0$ and $2\omega_c + \omega_0$, respectively, are removed by the lowpass filter in Fig. 5.11(b). The coherent detector output thus reproduces the original modulating wave. Note, however, that this detector output appears as two equal terms, one derived from the upper side-frequency and the other from the lower side-frequency. We therefore conclude that, for the transmission of information, only one side-frequency is necessary. This issue is discussed further in Section 5.7.

▶ **Drill Problem 5.7**   For the sinusoidal modulation considered in Example 5.3, what is the average power in the lower or upper side-frequency, expressed as a percentage of the total power in the DSB-SC modulated wave?

Answer: 50%. ◀

5.6 *Quadrature-Carrier Multiplexing*

A *quadrature-carrier multiplexing* or *quadrature-amplitude modulation* (QAM) system enables two DSB-SC modulated waves (resulting from the application of two *independent* message signals) to occupy the same transmission bandwidth, and yet it allows for their separation at the receiver output. It is therefore a *bandwidth-conservation scheme*.

Figure 5.14 is a block diagram of the quadrature-carrier multiplexing system. The transmitter of the system, shown in part (a) of the figure, involves the use of two separate product modulators that are supplied with two carrier waves of the same frequency but

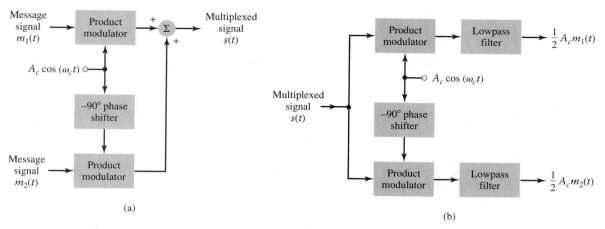

FIGURE 5.14 Quadrature-carrier multiplexing system, exploiting the quadrature null effect. (a) Transmitter. (b) Receiver, assuming perfect synchronization with the transmitter.

differing in phase by $-90°$. The multiplexed signal $s(t)$ consists of the sum of these two product modulator outputs, as shown by

$$s(t) = A_c m_1(t) \cos(\omega_c t) + A_c m_2(t) \sin(\omega_c t) \tag{5.20}$$

where $m_1(t)$ and $m_2(t)$ denote the two different message signals applied to the product modulators. Since each term in Eq. (5.20) has a transmission bandwidth of $2\omega_m$ and is centered on ω_c, we see that the multiplexed signal $s(t)$ occupies a transmission bandwidth of $2\omega_m$ centered on the carrier frequency ω_c, where ω_m is the common message bandwidth of $m_1(t)$ and $m_2(t)$.

The receiver of the system is shown in Fig. 5.14(b). The multiplexed signal $s(t)$ is applied simultaneously to two separate coherent detectors that are supplied with two local carriers of the same frequency, but differing in phase by $-90°$. The output of the top detector is $\frac{1}{2}A_c m_1(t)$, whereas the output of the bottom detector is $\frac{1}{2}A_c m_2(t)$.

For the quadrature-carrier multiplexing system to operate satisfactorily, it is important to maintain the correct phase and frequency relationships between the local oscillators used in the transmitter and receiver parts of the system. This increase in system complexity is the price that must be paid for the practical benefit gained from bandwidth conservation.

▶ **Drill Problem 5.8** Verify that the outputs of the receiver in Fig. 5.14 in response to the $s(t)$ of Eq. (5.20) are as indicated therein. ◀

5.7 *Other Variants of Amplitude Modulation*

The full AM and DSB-SC forms of modulation are wasteful of bandwidth because they both require a transmission bandwidth equal to twice the message bandwidth. In either case, one-half the transmission bandwidth is occupied by the upper sideband of the modulated wave, whereas the other half is occupied by the lower sideband. Indeed, the upper and lower sidebands are uniquely related to each other by virtue of their symmetry about the carrier frequency as illustrated in Example 5.3. That is, given the amplitude and phase spectra of either sideband, we can uniquely determine the other. This means that insofar as the transmission of information is concerned, only one sideband is necessary, and if both the carrier and the other sideband are suppressed at the transmitter, no information is lost. In this way the channel needs to provide only the same bandwidth as the message signal, a conclusion that is intuitively satisfying. When only one sideband is transmitted, the modulation is referred to as *single sideband* (SSB) *modulation.*

■ FREQUENCY-DOMAIN DESCRIPTION OF SSB MODULATION

The precise frequency-domain description of a SSB modulated wave depends on which sideband is transmitted. To investigate this issue, consider a message signal $m(t)$ with a spectrum $M(j\omega)$ limited to the band $\omega_a \leq |\omega| \leq \omega_b$, as in Fig. 5.15(a). The spectrum of the DSB-SC modulated wave, obtained by multiplying $m(t)$ by the carrier wave $A_c \cos(\omega_c t)$, is as shown in Fig. 5.15(b). The upper sideband is represented in duplicate by the frequencies above ω_c and those below $-\omega_c$; and when only the upper sideband is transmitted, the resulting SSB modulated wave has the spectrum shown in Fig. 5.15(c). Likewise, the lower sideband is represented in duplicate by the frequencies below ω_c (for positive frequencies) and those above $-\omega_c$ (for negative frequencies); and when only the lower sideband is transmitted, the spectrum of the corresponding SSB modulated wave is as shown in Fig. 5.15(d). Thus the essential function of SSB modulation is to *translate* the spectrum of the modulating wave, either with or without inversion, to a new location in the frequency domain. Moreover, the transmission bandwidth requirement of a SSB modulation

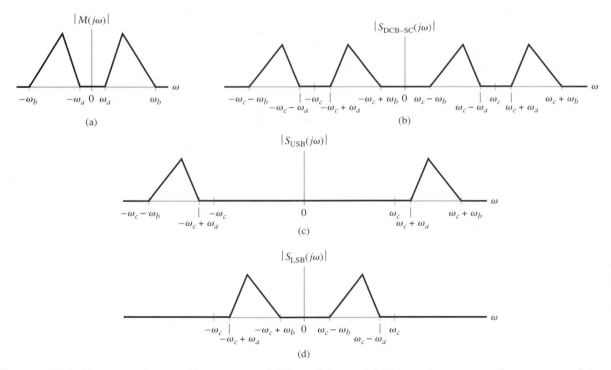

FIGURE 5.15 Frequency-domain characteristics of SSB modulation. (a) Magnitude spectrum of message signal, with energy gap from $-\omega_a$ to ω_a. (b) Magnitude spectrum of DSB-SC signal. (c) Magnitude spectrum of SSB modulated wave, containing upper sideband only. (d) Magnitude spectrum of SSB modulated wave, containing lower sideband only.

system is one-half that of a standard AM or DSB-SC modulation system. The benefit of using SSB modulation is therefore derived principally from the reduced bandwidth requirement and the elimination of the high-power carrier wave, two features that make SSB modulation the *optimum* (and therefore most desired) form of linear CW modulation. The principal disadvantage of SSB modulation, however, is the cost and complexity of implementing both the transmitter and the receiver. Here again we have a tradeoff between increased system complexity and improved system performance.

Using the frequency-domain descriptions in Fig. 5.15, we may readily deduce the *frequency-discrimination scheme* shown in Fig. 5.16 for producing SSB modulation. The scheme consists of a product modulator followed by a bandpass filter. The filter is designed to pass the sideband selected for transmission and suppress the remaining sideband. For a filter to be physically realizable, the transition band separating the passband from the stopband must have a finite width. In the context of the scheme shown in Fig. 5.16, this requirement demands that there be an adequate separation between the lower sideband and upper sideband of the DSB-SC modulated wave produced at the output of the product modulator. Such a requirement can only be satisfied if the message signal $m(t)$ applied to the product modulator has an *energy gap* in its spectrum as indicated in Fig. 5.15(a). Fortunately, speech signals for telephonic communication do exhibit an energy gap extending from -300 to 300 Hz. It is this feature of speech signal that makes SSB modulation well suited for its transmission. Indeed, analog telephony, which was dominant for a good part of the twentieth century, relied on SSB modulation for its transmission needs.

▶ **Drill Problem 5.9**   A SSB modulated wave $s(t)$ is generated using a carrier of frequency ω_c and a sinusoidal modulating wave of frequency ω_0. The carrier amplitude is A_c, and that of the modulating wave is A_0. Define $s(t)$, assuming that (a) only the upper side-frequency is transmitted and (b) only the lower side-frequency is transmitted.

Answer:

(a) $s(t) = \frac{1}{2}A_c A_0 \cos[(\omega_c + \omega_0)t]$

(b) $s(t) = \frac{1}{2}A_c A_0 \cos[(\omega_c - \omega_0)t]$ ◀

▶ **Drill Problem 5.10**   The spectrum of a speech signal lies inside the band $\omega_1 \leq |\omega| \leq \omega_2$. The carrier frequency is ω_c. Specify the passband, transition band, and stopband of the bandpass filter in Fig. 5.16 so as to transmit (a) the lower sideband and (b) the upper sideband.

Answer:

(a) Passband: $\omega_c - \omega_2 \leq |\omega| \leq \omega_c - \omega_1$
     Transition band: $\omega_c - \omega_1 \leq |\omega| \leq \omega_c + \omega_1$
     Stopband: $\omega_c + \omega_1 \leq |\omega| \leq \omega_c + \omega_2$

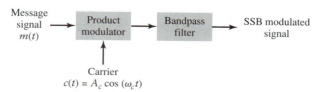

Message signal $m(t)$ → Product modulator → Bandpass filter → SSB modulated signal

Carrier
$c(t) = A_c \cos(\omega_c t)$

FIGURE 5.16   System consisting of product modulator and bandpass filter, for generating SSB modulated wave.

(b) Passband: $\omega_c + \omega_1 \leq |\omega| \leq \omega_c + \omega_2$
       Transition band: $\omega_c - \omega_1 \leq |\omega| \leq \omega_c + \omega_1$
       Stopband: $\omega_c - \omega_2 \leq |\omega| \leq \omega_c - \omega_1$ ◀

* ■ TIME-DOMAIN DESCRIPTION OF SSB MODULATION

The frequency-domain description of SSB modulation depicted in Fig. 5.15 and its generation using the frequency-discrimination scheme shown in Fig. 5.16 build on our knowledge of DSB-SC modulation in a straightforward fashion. However, unlike DSB-SC modulation, the time-domain description of SSB modulation is not as straightforward. To develop the time-domain description of SSB modulation, we need a mathematical tool known as the Hilbert transform. The device used to perform this transformation is known as the *Hilbert transformer*, the frequency response of which is characterized as follows:

▶ The magnitude response is unity for all frequencies, both positive and negative.
▶ The phase response is $-90°$ for positive frequencies and $+90°$ for negative frequencies.

The Hilbert transformer may therefore be viewed as a wideband $-90°$ phase shifter, wideband in the sense that its frequency response occupies a band of frequencies that, in theory, is infinite in extent. Further consideration of the time-domain description of SSB modulation is beyond the scope of this book.

■ VESTIGIAL SIDEBAND MODULATION

Single sideband modulation is well-suited for the transmission of speech because of the energy gap that exists in the spectrum of speech signals between zero and a few hundred hertz. When the message signal contains significant components at extremely low frequencies (as in the case of television signals and wideband data), the upper and lower sidebands meet at the carrier frequency. This means that the use of SSB modulation is inappropriate for the transmission of such message signals owing to the practical difficulty of building a filter to isolate one sideband completely. This difficulty suggests another scheme known as *vestigial sideband (VSB) modulation*, which is a compromise between SSB and DSB-SC forms of modulation. In VSB modulation, one sideband is passed almost completely whereas just a trace, or *vestige*, of the other sideband is retained.

Figure 5.17 illustrates the spectrum of a *VSB modulated* wave $s(t)$ in relation to that of the message signal $m(t)$, assuming that the lower sideband is modified into the vestigial

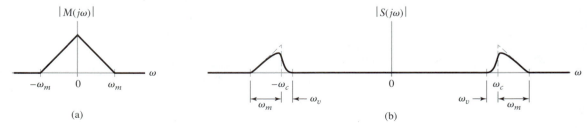

FIGURE 5.17 Spectral content of VSB modulated wave. (a) Magnitude spectrum of message signal. (b) Magnitude spectrum of VSB modulated wave containing a vestige of the lower sideband.

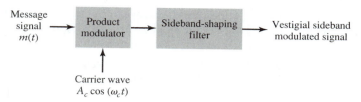

FIGURE 5.18 System consisting of product modulator and sideband shaping filter, for generating VSB modulated wave.

sideband. Specifically, the transmitted vestige of the lower sideband compensates for the amount removed from the upper sideband. The transmission bandwidth required by the VSB modulated wave is therefore given by

$$\omega_T = \omega_m + \omega_v \tag{5.21}$$

where ω_m is the message bandwidth and ω_v is the width of the vestigial sideband.

To generate a VSB modulated wave, we pass a DSB-SC modulated wave through a *sideband shaping filter*, as in Fig. 5.18. Unlike the bandpass filter used for SSB modulation, the filter in Fig. 5.18 does not have a "flat" magnitude response in its passband, because the upper and lower sidebands have to be shaped differently. The filter response is designed so that the original message spectrum $M(j\omega)$ (i.e., the Fourier transform of the message signal $m(t)$) is reproduced on demodulation as a result of the superposition of two spectra:

▶ The positive-frequency part of $S(j\omega)$ (that is, the Fourier transform of the transmitted signal $s(t)$, shifted downward in frequency by ω_c.

▶ The negative-frequency part of $S(j\omega)$, shifted upward in frequency by ω_c.

The magnitudes of these two spectral contributions are illustrated in Figs. 5.19(a) and (b), respectively. In effect, a reflection of the vestige of the lower sideband makes up for the missing part of the upper sideband.

The design requirement described herein makes the implementation of the sideband shaping filter a challenging task.

Vestigial sideband modulation has the virtue of conserving bandwidth almost as efficiently as single sideband modulation, while retaining the excellent low-frequency characteristics of double sideband modulation. Thus VSB modulation has become standard for the analog transmission of television and similar signals, where good phase

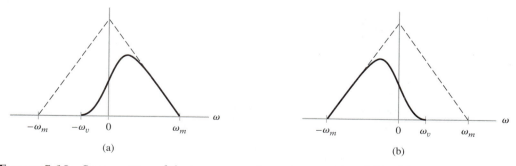

FIGURE 5.19 Superposition of the two spectra shown in parts (a) and (b) of the figure results in the original message spectrum (shown dashed) upon demodulation.

characteristics and transmission of low-frequency components are important, but the bandwidth required for double sideband transmission is unavailable or uneconomical.

In the transmission of television signals in practice, a controlled amount of carrier is added to the VSB modulated signal. This is done to permit the use of an envelope detector for demodulation. The design of the receiver is thereby considerably simplified.

5.8 *Pulse-Amplitude Modulation*

Having familiarized ourselves with continuous-wave AM and its variants, we now turn our attention to *pulse-amplitude modulation* (PAM). PAM represents a widely used form of pulse modulation. Whereas frequency shifting plays a basic role in the operation of AM systems, the basic operation in PAM systems is that of sampling.

■ SAMPLING REVISITED

The sampling process, including a derivation of the sampling theorem and related issues of aliasing and reconstructing the message signal from its sampled version, is covered in detail in Sections 4.6 and 4.7. In this subsection, we tie the discussion of sampling for PAM to the material covered therein. To begin with, we may restate the sampling theorem in the context of PAM in two equivalent parts as follows:

1. *A bandlimited signal of finite energy, which has no radian frequency components higher than ω_m, is uniquely determined by the values of the signal at instants of time separated by π/ω_m seconds.*
2. *A bandlimited signal of finite energy, which has no radian frequency components higher than ω_m, may be completely recovered from a knowledge of its samples taken at the rate of ω_m/π per second.*

Part 1 of the sampling theorem is exploited in the transmitter of a PAM system; part 2 of the theorem is exploited in the receiver of the system. The special value of the sampling rate π/ω_m is referred to as the *Nyquist rate*, in recognition of the pioneering work done by Harry Nyquist on data transmission.

Typically, the spectrum of a message signal is not strictly bandlimited, as required by the sampling theorem. Rather, it approaches zero asymptotically as the frequency approaches infinity, which gives rise to aliasing and therefore signal distortion. Recall that aliasing refers to a high-frequency component in the spectrum of the message signal apparently taking on the identity of a lower frequency in the spectrum of a sampled version of the message signal. To combat the effects of aliasing in practice, we use two corrective measures:

▸ Prior to sampling, a lowpass anti-aliasing filter is used to attenuate those high-frequency components of the signal that lie outside the band of interest.
▸ The filtered signal is sampled at a rate higher than the Nyquist rate.

On this basis, the generation of a PAM signal as a sequence of flat-topped pulses, whose amplitudes are determined by the corresponding signal samples, follows the block diagram shown in Fig. 5.20.

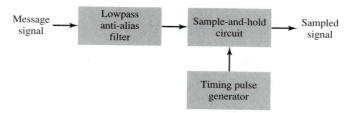

FIGURE 5.20   System consisting of anti-alias filter and sample-and-hold circuit, for converting a message signal into flat-top PAM signal.

EXAMPLE 5.4   The highest frequency component of a speech signal needed for telephonic communications is about 3.1 kHz. Suggest a suitable value for the sampling rate.

Solution: The highest frequency component of 3.1 kHz corresponds to

$$\omega_m = 6.2\pi \times 10^3 \text{ rad/s}$$

Correspondingly, the Nyquist rate is

$$\frac{\omega_m}{\pi} = 6.2 \text{ kHz}$$

For a suitable value for the sampling rate, slightly higher than the Nyquist rate, we may suggest 8 kHz. Indeed, this sampling rate is the international standard for speech signals.

■ MATHEMATICAL DESCRIPTION OF PAM

The carrier wave used in PAM consists of a sequence of short pulses of fixed duration, in terms of which PAM is formally defined as follows. PAM is a *form of pulse modulation in which the amplitude of the pulsed carrier is varied in accordance with instantaneous sample values of the message signal*; the duration of the pulsed carrier is maintained constant throughout. Figure 5.21 illustrates the waveform of such a PAM signal. Note that the fundamental frequency of the carrier wave (i.e., the pulse repetition frequency) is the same as the sampling rate.

For a mathematical representation of the PAM signal $s(t)$ for a message signal $m(t)$, we may write

$$s(t) = \sum_{n=-\infty}^{\infty} m[n]h(t - n\mathcal{T}) \tag{5.22}$$

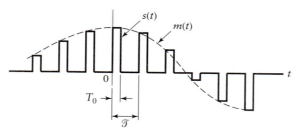

FIGURE 5.21   Waveform of flat-top PAM signal with pulse duration T_0 and sampling period $\mathcal{T}$.

where $\mathcal{T}$ is the sampling period. The term $m[n]$ is the value of the message signal $m(t)$ at time $t = n\mathcal{T}$. The term $h(t)$ is a rectangular pulse of unit amplitude and duration T_0, defined as follows (see Fig. 5.22(a)):

$$h(t) = \begin{cases} 1, & 0 < t < T_0 \\ 0, & \text{otherwise} \end{cases} \tag{5.23}$$

In physical terms, Eq. (5.22) represents a *sample-and-hold operation*. This operation is analogous to the zero-order hold-based reconstruction described in Section 4.7. These two operations differ from each other in that the impulse response $h(t)$ in Eq. (5.22) is T_0 wide instead of $\mathcal{T}$. Bearing this difference in mind, we may follow the material presented in Section 4.7 to derive the spectrum of the PAM signal $s(t)$.

The impulse-sampled version of the message signal $m(t)$ is given by

$$m_\delta(t) = \sum_{n=-\infty}^{\infty} m[n]\, \delta(t - n\mathcal{T}) \tag{5.24}$$

The PAM signal $s(t)$ is itself expressed as

$$\begin{aligned} s(t) &= \sum_{n=-\infty}^{\infty} m[n]\, h(t - n\mathcal{T}) \\ &= m_\delta(t) * h(t) \end{aligned} \tag{5.25}$$

Equation (5.25) states that $s(t)$ is mathematically equivalent to the convolution of $m_\delta(t)$, the impulse-sampled version of $m(t)$, and the pulse $h(t)$.

Taking the Fourier transform of both sides of Eq. (5.25) and recognizing that the convolution of two time functions is transformed into the multiplication of their respective Fourier transforms, we get

$$S(j\omega) = M_\delta(j\omega)H(j\omega) \tag{5.26}$$

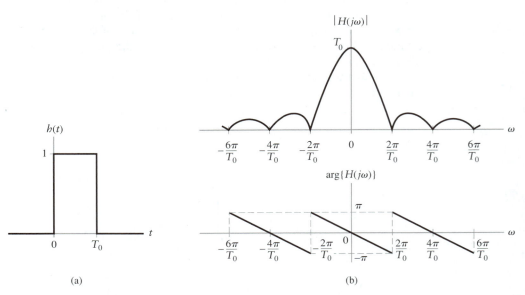

(a)   (b)

FIGURE 5.22   (a) Rectangular pulse $h(t)$ of unit amplitude and duration T_0. (b) Magnitude spectrum $|H(j\omega)|$ and phase spectrum $\arg\{H(j\omega)\}$ of pulse $h(t)$.

where $S(j\omega) \overset{FT}{\longleftrightarrow} s(t)$, $M_\delta(j\omega) \overset{FT}{\longleftrightarrow} m_\delta(t)$, and $H(j\omega) \overset{FT}{\longleftrightarrow} h(t)$. From Chapter 4 we recall that impulse sampling of the message signal $m(t)$ introduces periodicity into the spectrum as shown by

$$M_\delta(j\omega) = \frac{1}{\mathcal{T}} \sum_{k=-\infty}^{\infty} M\left(j\left(\omega - \frac{2\pi k}{\mathcal{T}}\right)\right) \tag{5.27}$$

where $1/\mathcal{T}$ is the sampling rate. Therefore substitution of Eq. (5.27) into (5.26) yields

$$S(j\omega) = \frac{1}{\mathcal{T}} \sum_{k=-\infty}^{\infty} M\left(j\left(\omega - \frac{2\pi k}{\mathcal{T}}\right)\right)H(j\omega) \tag{5.28}$$

where $M(j\omega) \overset{FT}{\longleftrightarrow} m(t)$.

Finally, suppose that $m(t)$ is strictly bandlimited and that the sampling rate $1/\mathcal{T}$ is greater than the Nyquist rate. Then passing $s(t)$ through a reconstruction filter chosen as an ideal lowpass filter with cutoff frequency ω_m, we find that the spectrum of the resulting filter output is equal to $M(j\omega)H(j\omega)$. This result is equivalent to that which would be obtained by passing the original message signal $m(t)$ through a lowpass filter of frequency response $H(j\omega)$.

From Eq. (5.23) we find that

$$H(j\omega) = T_0 \operatorname{sinc}(\omega T_0/2\pi)e^{-j\omega T_0/2} \tag{5.29}$$

whose magnitude and phase components are plotted in Fig. 5.22(b). Hence in light of Eqs. (5.26) and (5.29) we see that by using PAM to represent a continuous-time message signal, we introduce amplitude distortion as well as a delay of $T_0/2$. Both of these effects are also present in the sample-and-hold reconstruction scheme described in Section 4.7. A similar form of amplitude distortion is caused by the finite size of the scanning aperture in television and facsimile. Accordingly, the frequency distortion caused by the use of flat-top samples in the generation of a PAM wave, as in Fig. 5.22(b), is referred to as the *aperture effect*.

▶ **Drill Problem 5.11**   What happens to the scaled frequency response $H(j\omega)/T_0$ of Eq. (5.29) as the pulse duration T_0 approaches zero?

Answer: $\displaystyle \lim_{T_0 \to 0} \frac{H(j\omega)}{T_0} = 1.$ ◀

■ Demodulation of PAM Signal

Given a sequence of flat-topped samples, $s(t)$, we may reconstruct the original message signal $m(t)$ using the scheme shown in Fig. 5.23. It consists of two components connected in cascade. The first component is a lowpass filter with a cutoff frequency that equals the highest frequency component ω_m of the message signal. The second component is an *equal-*

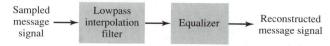

FIGURE 5.23   System consisting of lowpass interpolation filter and equalizer, for reconstructing a message signal from its flat-top sampled version.

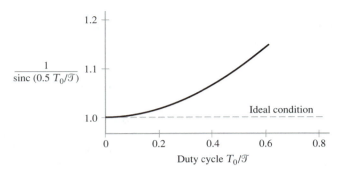

FIGURE 5.24   Normalized equalization (to compensate for aperture effect) plotted versus the duty cycle $T_0/\mathcal{T}$.

izer that corrects for the aperture effect due to flat-top sampling in the sample-and-hold circuit. The equalizer has the effect of decreasing the in-band loss of the interpolation filter as the frequency increases in such a manner as to compensate for the aperture effect. Ideally, the amplitude response of the equalizer is given by

$$\frac{1}{|H(j\omega)|} = \frac{1}{T_0 \, \mathrm{sinc}(\omega T_0/2)} = \frac{1}{2T_0} \frac{\omega T_0}{\sin(\omega T_0/2)}$$

where $H(j\omega)$ is the frequency response defined in Eq. (5.29). The amount of equalization needed in practice is usually small.

EXAMPLE 5.5   The duty cycle in a PAM signal, namely, $T_0/\mathcal{T}$, is 10%. Evaluate the equalization required at $\omega = \omega_m$, where ω_m is the modulation frequency.

Solution: At $\omega_m = \pi/\mathcal{T}$, which corresponds to the highest frequency component of the message signal for a sampling rate equal to the Nyquist rate, we find from Eq. (5.29) that the magnitude response of the equalizer at ω_m, normalized to that at zero frequency, is equal to

$$\frac{1}{\mathrm{sinc}(0.5T_0/T)} = \frac{(\pi/2)(T_0/\mathcal{T})}{\sin[(\pi/2)T_0/\mathcal{T}]}$$

where the ratio $T_0/\mathcal{T}$ is equal to the duty cycle of the sampling pulses. In Fig. 5.24, this result is plotted as a function of $T_0\mathcal{T}$. Ideally, it should be equal to 1 for all values of $T_0/\mathcal{T}$. For a duty cycle of 10%, it is equal to 1.0041. It follows therefore that for duty cycles of less than 10%, the magnitude equalization required is less than 1.0041, and the aperture effect is usually considered to be negligible.

5.9 *Multiplexing*

In Section 5.3 we pointed out that modulation provides a method for multiplexing, whereby message signals derived from independent sources are combined into a composite signal suitable for transmission over a common channel. In a telephone system, for example, multiplexing is used to transmit multiple conversations over a single long-distance line. The signals associated with different speakers are combined in such a way as to not interfere with each other during transmission and so that they can be separated at the receiving end of the system. Multiplexing can be accomplished by separating the different message signals either in frequency or in time, or through the use of coding techniques. We thus have three basic types of multiplexing:

1. *Frequency-division multiplexing*, where the signals are separated by allocating them to different frequency bands. This is illustrated in Fig. 5.25(a) for the case of six different message signals. Frequency-division multiplexing favors the use of CW modulation, where each message signal is able to use the channel on a continuous-time basis.

2. *Time-division multiplexing*, where the signals are separated by allocating them different time slots within a sampling interval. This second type of multiplexing is illustrated in Fig. 5.25(b) for the case of six different message signals. Time-division multiplexing favors the use of pulse modulation, where each message signal has access to the complete frequency response of the channel.

3. *Code-division multiplexing*, which relies on the assignment of different codes to the individual users of the channel.

The first two methods of multiplexing are described in the sequel; discussion of code-division multiplexing is beyond the scope of this book.

■ FREQUENCY-DIVISION MULTIPLEXING (FDM)

A block diagram of a FDM system is shown in Fig. 5.26. The incoming message signals are assumed to be of the lowpass type, but their spectra do not necessarily have nonzero values all the way down to zero frequency. Following each signal input, we have shown a lowpass filter, which is designed to remove high-frequency components that do not contribute significantly to signal representation but are capable of disturbing other message signals that share the common channel. These lowpass filters may be omitted only if the input signals are sufficiently bandlimited initially. The filtered signals are applied to modulators that shift the frequency ranges of the signals so as to occupy mutually exclusive frequency intervals. The necessary carrier frequencies needed to perform these frequency

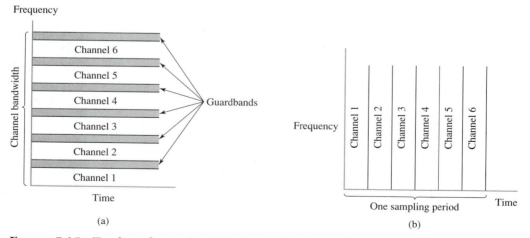

FIGURE 5.25 Two basic forms of multiplexing. (a) Frequency-division multiplexing (with guardbands). (b) Time-division multiplexing; no provision is made here for synchronizing pulses.

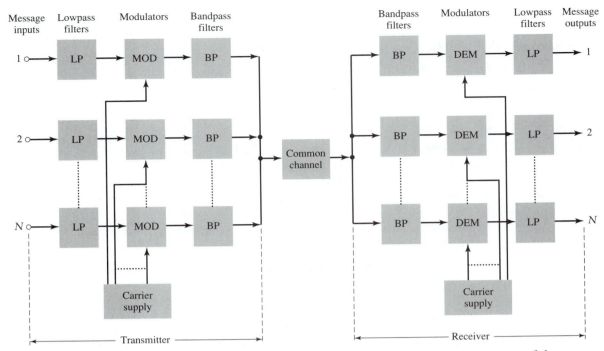

Figure 5.26 Block diagram of FDM system, showing the important constituents of the transmitter and receiver.

translations are obtained from a carrier supply. For the modulation, we may use any one of the methods described in previous sections of this chapter. However, the most widely used method of modulation in frequency-division multiplexing is single sideband modulation, which, in the case of voice signals, requires a bandwidth that is approximately equal to that of the original voice signal. In practice, each voice input is usually assigned a bandwidth of 4 kHz. The bandpass filters following the modulators are used to restrict the band of each modulated wave to its prescribed range. The resulting bandpass filter outputs are next combined in parallel to form the input to the common channel. At the receiving terminal, a bank of bandpass filters, with their inputs connected in parallel, is used to separate the message signals on a frequency-occupancy basis. Finally, the original message signals are recovered by individual demodulators. Note that the FDM system shown in Fig. 5.26 operates in only one direction. To provide for two-way transmission, as in telephony, for example, we have to completely duplicate the multiplexing facilities, with the components connected in reverse order and with the signal waves proceeding from right to left.

Example 5.6 A FDM system is used to multiplex 24 independent voice signals. SSB modulation is used for the transmission. Given that each voice signal is allotted a bandwidth of 4 kHz, calculate the overall transmission bandwidth of the channel.

Solution: With each voice signal allotted a bandwidth of 4 kHz, the use of SSB modulation requires a bandwidth of 4 kHz for its transmission. Accordingly, the overall transmission bandwidth provided by the channel is $24 \times 4 = 96$ kHz.

■ TIME-DIVISION MULTIPLEXING (TDM)

Basic to the operation of a TDM system is the sampling theorem. It states that we can transmit all the information contained in a bandlimited message signal by using samples of the message signal taken uniformly at a rate that is usually slightly higher than the Nyquist rate. An important feature of the sampling process is a conservation of time. That is, the transmission of the message samples engages the transmission channel for only a fraction of the sampling interval on a periodic basis, equal to the width T_0 of the PAM modulating wave. In this way some of the time interval between adjacent samples is cleared for use by other independent message sources on a time-shared basis.

The concept of TDM is illustrated by the block diagram shown in Fig. 5.27. Each input message signal is first restricted in bandwidth by a lowpass filter to remove the frequencies that are nonessential to an adequate signal representation. The lowpass filter outputs are then applied to a *commutator* that is usually implemented using electronic switching circuitry. The function of the commutator is twofold: (a) to take a narrow sample of each of the N input message signals at a rate $1/\mathcal{T}$ that is slightly higher than ω_c/π, where ω_c is the cutoff frequency of the input lowpass filter; and (2) to sequentially interleave these N samples inside a sampling interval $\mathcal{T}$. Indeed, this latter function is the essence of the time-division multiplexing operation. Following the commutation process, the multiplexed signal is applied to a *pulse modulator* (e.g., pulse-amplitude modulator), the purpose of which is to transform the multiplexed signal into a form suitable for transmission over the common channel. The use of time-division multiplexing introduces a *bandwidth expansion factor N*, because the scheme must squeeze N samples derived from N independent message sources into a time slot equal to one sampling interval. At the receiving end of the system, the received signal is applied to a *pulse demodulator*, which performs the inverse operation of the pulse modulator. The narrow samples produced at the pulse demodulator output are distributed to the appropriate lowpass reconstruction filters by means of a *decommutator*, which operates in synchronism with the commutator in the transmitter.

Synchronization between the timing operations of the transmitter and receiver in a TDM system is essential for satisfactory performance of the system. In the case of a TDM system using PAM, synchronization may be achieved by inserting an extra pulse into each sampling interval on a regular basis. The combination of N PAM signals and synchroni-

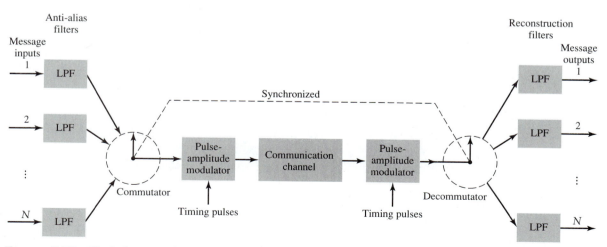

FIGURE 5.27 Block diagram of TDM system, showing the important constituents of the transmitter and receiver.

zation pulse contained in a single sampling period is referred to as a *frame*. In PAM, the feature of a message signal that is used for modulation is its amplitude. Accordingly, a simple way of identifying the synchronizing pulse train at the receiver is to make sure that its constant amplitude is large enough to stand above every one of the PAM signals. On this basis, the synchronizing pulse train is identified at the receiver by using a threshold device set at the appropriate level. Note that the use of time synchronization in the manner described here increases the bandwidth expansion factor to $N + 1$, where N is the number of message signals being multiplexed.

The TDM system is highly sensitive to dispersion in the common transmission channel, that is, to variations of amplitude with frequency or nonlinear phase response. Accordingly, accurate equalization of both the amplitude and phase responses of the channel is necessary to ensure a satisfactory operation of the system. Equalization of a communication channel is discussed in Chapter 8.

EXAMPLE 5.7   A TDM system is used to multiplex four independent voice signals using PAM. Each voice signal is sampled at the rate of 8 kHz. The system incorporates a synchronizing pulse train for its proper operation.

  (a) Determine the timing relationships between the synchronizing pulse train and the impulse trains used to sample the four voice signals.

  (b) Calculate the transmission bandwidth of the channel for the TDM system, and compare the result with a corresponding FDM system using SSB modulation.

Solution:

  (a) The sampling period is

$$\mathcal{T} = \frac{1}{8 \times 10^3}\ \text{s} = 125\ \mu\text{s}$$

In this example, the number of voice signals is $N = 4$. Hence dividing the sampling period of 125 μs among these voice signals and the synchronizing pulse train, the time slot allocated to each one of them is

$$T_0 = \frac{\mathcal{T}}{N + 1}$$

$$= \frac{125}{5} = 25\ \mu\text{s}$$

Figure 5.28 shows the timing relationships between the synchronizing pulse train and the four impulse trains used to sample the different voice signals in a single frame. Each frame includes time slots of common duration $T_0 = 25\ \mu\text{s}$, which are allocated to corresponding PAM signals or synchronizing pulse train.

  (b) As a consequence of the time bandwidth product discussion in Section 3.6, there is an inverse relationship between the duration of a pulse and bandwidth (i.e., cuttoff frequency) of the channel needed for its transmission. Accordingly, the overall transmission bandwidth of the channel is

$$f_T = \frac{\omega_T}{2\pi}$$

$$= \frac{1}{T_0}$$

$$= \frac{1}{25}\ \text{MHz} = 40\ \text{kHz}$$

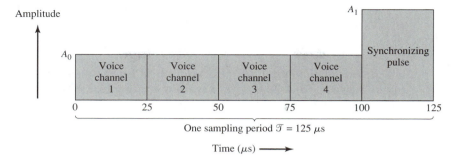

FIGURE 5.28 Composition of one frame of a multiplexed PAM signal, incorporating four voice signals and synchronizing pulse.

In contrast, the use of a FDM system based on SSB modulation requires a channel bandwidth equal to N times that of a single voice signal, that is, $4 \times 4 = 16$ kHz. Thus the use of PAM-TDM requires a channel bandwidth that is $40/16 = 2.5$ times that of SSB-FDM.

In practice, pulse-code modulation is commonly used as the method of modulation for TDM; this results in a further increase in channel bandwidth, depending on the codeword length used in the digital representation of each pulse in the PAM signal.

*5.10 *Phase and Group Delays*

Whenever a signal is transmitted through a dispersive (i.e., frequency-selective) system, such as a communication channel, some *delay* is introduced into the output signal in relation to the input signal. The delay is determined by the phase response of the system, as discussed here.

For convenience of presentation, let $\phi(\omega)$ denote the phase response of a dispersive communication channel, as shown by

$$\phi(\omega) = \arg\{H(j\omega)\} \tag{5.30}$$

where $H(j\omega)$ is the frequency response of the channel. Suppose that a sinusoidal signal at frequency ω_c is transmitted through the channel. The signal received at the channel output lags the transmitted signal by $\phi(\omega_c)$ radians. The time delay corresponding to this phase lag is simply equal to $-\phi(\omega_c)/\omega_c$, where the minus sign accounts for the phase lag. This time delay is called the *phase delay* of the channel. The phase delay, denoted by τ_p, is formally defined by

$$\tau_p = -\frac{\phi(\omega_c)}{\omega_c} \tag{5.31}$$

It is important to realize, however, that the phase delay is *not* necessarily the true signal delay. This follows from the fact that a sinusoidal signal has infinite duration, with each cycle exactly like the preceding cycle. Such a signal does not convey information except for the fact that it is there, so to speak. It would therefore be incorrect to deduce from the above reasoning that the phase delay is the true signal delay. In actual fact, as we have seen from the material presented in this chapter, information can only be transmitted through a channel by applying some form of modulation to a carrier.

Assume that we have a transmitted signal that consists of a DSB-SC modulated wave with carrier frequency ω_c and sinusoidal modulation frequency ω_0, as shown by

$$s(t) = A \cos(\omega_c t) \cos(\omega_0 t) \tag{5.32}$$

which corresponds to the signal considered in Example 5.3. (For convenience of presentation, we have set $A = A_c A_0$.) Expressing the modulated signal $s(t)$ in terms of its upper and lower side-frequencies, we may write

$$s(t) = \tfrac{1}{2}A \cos(\omega_1 t) + \tfrac{1}{2}A \cos(\omega_2 t)$$

where

$$\omega_1 = \omega_c + \omega_0 \tag{5.33}$$

and

$$\omega_2 = \omega_c - \omega_0 \tag{5.34}$$

Now let the signal $s(t)$ be transmitted through the channel with phase response $\phi(\omega)$. For illustrative purposes, we assume that the magnitude response of the channel is essentially constant (equal to unity) over the frequency range from ω_1 to ω_2. Accordingly, the signal received at the channel output is

$$y(t) = \tfrac{1}{2}A \cos(\omega_1 t + \phi(\omega_1)) + \tfrac{1}{2}A \cos(\omega_2 t + \phi(\omega_2))$$

where $\phi(\omega_1)$ and $\phi(\omega_2)$ are the phase shifts produced by the channel at frequencies ω_1 and ω_2, respectively. Equivalently, we may express $y(t)$ as

$$y(t) = A \cos\left(\omega_c t + \frac{\phi(\omega_1) + \phi(\omega_2)}{2}\right) \cos\left(\omega_0 t + \frac{\phi(\omega_1) - \phi(\omega_2)}{2}\right) \tag{5.35}$$

where we have invoked the definitions of ω_1 and ω_2 given in Eqs. (5.33) and (5.34), respectively. Comparing the sinusoidal carrier and message components of the received signal $y(t)$ in Eq. (5.35) with those of the transmitted signal $s(t)$ in Eq. (5.32), we can make the following two statements:

1. The carrier component at frequency ω_c in $y(t)$ lags its counterpart in $s(t)$ by $\tfrac{1}{2}(\phi(\omega_1) + \phi(\omega_2))$, which represents a time delay equal to

$$-\frac{\phi(\omega_1) + \phi(\omega_2)}{2\omega_c} = -\frac{\phi(\omega_1) + \phi(\omega_2)}{\omega_1 + \omega_2} \tag{5.36}$$

2. The message component at frequency ω_0 in $y(t)$ lags its counterpart in $s(t)$ by $\tfrac{1}{2}(\phi(\omega_1) - \phi(\omega_2))$, which represents a time delay equal to

$$-\frac{\phi(\omega_1) - \phi(\omega_2)}{2\omega_0} = -\frac{\phi(\omega_1) - \phi(\omega_2)}{\omega_1 - \omega_2} \tag{5.37}$$

Suppose that the modulation frequency ω_0 is small compared with the carrier frequency ω_c, which implies that the side-frequencies ω_1 and ω_2 are close together with ω_c between them. Such a modulated signal is said to be a *narrowband signal*. Then we may approximate the phase response $\phi(\omega)$ in the vicinity of $\omega = \omega_c$, using the two-term Taylor series expansion

$$\phi(\omega) = \phi(\omega_c) + \left.\frac{d\phi(\omega)}{d\omega}\right|_{\omega = \omega_c} (\omega - \omega_c) \tag{5.38}$$

Using this expansion to evaluate $\phi(\omega_1)$ and $\phi(\omega_2)$ for substitution in Eq. (5.36), we see that the *carrier delay* is equal to $-\phi(\omega_c)/\omega_c$, which is identical to the formula given in Eq. (5.31) for phase delay. Treating Eq. (5.37) in a similar way, we find that the time delay incurred by the message signal (i.e., the "envelope" of the modulated signal) is given by

$$\tau_g = -\frac{d\phi(\omega)}{d\omega}\Bigg|_{\omega=\omega_c} \tag{5.39}$$

The time delay τ_g is called the *envelope delay* or *group delay*. Thus group delay is defined as the negative of the derivative of the phase response $\phi(\omega)$ of the channel with respect to ω, evaluated at the carrier frequency ω_c.

In general, we thus find that when a modulated signal is transmitted through a communication channel, there are two different delays to be considered:

1. Carrier or phase delay, τ_p, defined by Eq. (5.31)
2. Envelope or group delay, τ_g, defined by Eq. (5.39)

The group delay is the true signal delay.

▶ **Drill Problem 5.12** What are the conditions for which the phase delay and group delay assume a common value?

Answer: The phase response $\phi(\omega)$ must be linear in ω, and $\phi(\omega_c) = 0$. ◀

EXAMPLE 5.8 The phase response of a bandpass communication channel is defined by

$$\phi(\omega) = -\tan^{-1}\left(\frac{\omega^2 - \omega_c^2}{\omega\omega_c}\right)$$

The signal $s(t)$ defined in Eq. (5.32) is transmitted through this channel with

$$\omega_c = 4.75 \text{ rad/s} \quad \text{and} \quad \omega_0 = 0.25 \text{ rad/s}$$

Calculate (a) the phase delay and (b) the group delay.

Solution:

(a) At $\omega = \omega_c$, $\phi(\omega_c) = 0$. According to Eq. (5.31), the phase delay τ_p is zero.
(b) Differentiating $\phi(\omega)$ with respect to ω, we get

$$\frac{d\phi(\omega)}{d\omega} = -\frac{\omega_c(\omega^2 + \omega_c^2)}{\omega_c^2\omega^2 + (\omega^2 - \omega_c^2)^2}$$

Using this result in Eq. (5.39), we find that the group delay is

$$\tau_g = \frac{2}{\omega_c} = \frac{2}{4.75} = 0.4211 \text{ s}$$

To display the results obtained in parts (a) and (b) in graphical form, Fig. 5.29 shows a superposition of two waveforms obtained as follows:

1. One waveform, shown as a solid curve, was obtained by multiplying the transmitted signal $s(t)$ by the carrier wave $\cos(\omega_c t)$.
2. The second waveform, shown as a dotted curve, was obtained by multiplying the received signal $y(t)$ by the carrier wave $\cos(\omega_c t)$.

Figure 5.29 clearly shows that the carrier (phase) delay τ_p is zero, and the envelope of the received signal $y(t)$ is lagging behind that of the transmitted signal by τ_g seconds. For the

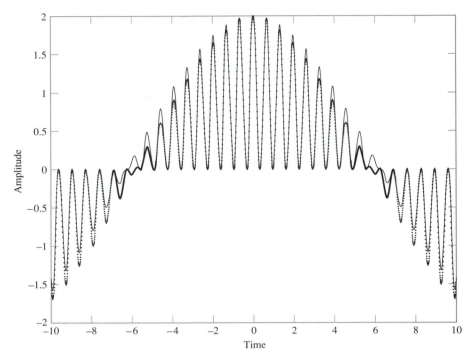

FIGURE 5.29   Highlighting the zero carrier delay (solid curve) and group delay τ_g (dotted curve), which are determined in accordance with Example 5.9.

presentation of waveforms in this figure, we purposely did not use a filter to suppress the high-frequency components resulting from the multiplications described under points 1 and 2 because of the desire to retain a contribution due to the carrier for display.

Note also that the separation between the upper side-frequency $\omega_1 = \omega_c + \omega_0 = 5.00$ rad/s and the lower side-frequency $\omega_2 = \omega_c - \omega_0 = 4.50$ rad/s is about 10% of the carrier frequency $\omega_c = 4.75$ rad/s, which justifies referring to the modulated signal in this example as a narrowband signal.

■ SOME PRACTICAL CONSIDERATIONS

Having established that group delay is the true signal delay when a modulated signal is transmitted through a communication channel, we now need to address the following question: What is the practical importance of group delay? To deal with this question, we first have to realize that the formula of Eq. (5.39) for determining group delay applies strictly to modulated signals that are narrowband, that is, the bandwidth of the message signal is small compared to the carrier frequency. It is only when this condition is satisfied that we would be justified to use the two-term approximation of Eq. (5.38) for the phase response $\phi(\omega)$, on the basis of which Eq. (5.39) was derived.

However, there are many practical situations where this narrowband assumption is not satisfied because the message bandwidth is comparable to the carrier frequency. In

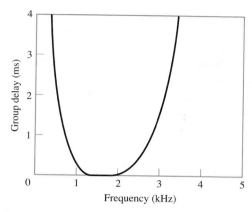

FIGURE 5.30   Group delay response of voice-grade telephone channel. (Adapted from J. C. Bellamy, *Digital Telephony*, Wiley, 1982.)

situations of this kind, the group delay is formulated as a *frequency-dependent parameter*, as shown by

$$\tau_g(\omega) = -\frac{d\phi(\omega)}{d\omega} \tag{5.40}$$

which includes Eq. (5.39) as a special case. Now we begin to see the real importance of group delay. Specifically, when a *wideband* modulated signal is transmitted through a dispersive channel, the frequency components of the message signal are delayed by different amounts at the channel output. Consequently, the message signal experiences a form of linear distortion known as *delay distortion*. To reconstruct a faithful version of the original message signal in the receiver, we have to use a *delay equalizer*. This equalizer has to be designed in such a way that when it is connected in cascade with the channel, the overall group delay is constant (i.e., the overall phase is linear with frequency).

As an illustrative example, consider the ubiquitous telephone channel, the useful frequency band of which extends from about 0.1 to 3.1 kHz. Over this band of frequencies, the magnitude response of the channel is considered to be essentially constant, so that there is little amplitude distortion. In contrast, the group delay of the channel is highly dependent on frequency, as shown in Fig. 5.30. Insofar as telephonic communication is concerned, the variation of group delay in the channel with frequency is of no real consequence because our ears are relatively insensitive to delay distortion. The story is dramatically different, however, when wideband data are transmitted over a telephone channel. For example, for a data rate of 16 kilobits per second, the bit duration is about 60 μs. From Fig. 5.30 we see that over the useful frequency band of the telephone channel, the group delay varies from zero to several milliseconds. Accordingly, delay distortion is extremely harmful to wideband data transmission over a telephone channel. In such an application, delay equalization is essential for satisfactory operation.

5.11 *Exploring Concepts with MATLAB*

In this chapter, we discussed the idea of modulation for the transmission of a message signal over a bandpass channel. To illustrate this idea, we used a sinusoidal wave as the message (modulating) signal. In particular, we used Examples 5.1 and 5.3 to illustrate the spectra of sinusoidally modulated waves based on full AM and DSB-SC modulation, assuming ideal conditions. In this section, we use MATLAB to expand on those examples by considering modulated waves of finite duration, which is how they always are in real-life situations. In particular, we build on the results presented in Example 4.18, where we used the DTFS to approximate the Fourier transform of a finite-duration signal consisting of a pair of sinusoidal components.

■ FULL AM

In the time-domain description of amplitude modulation, the modulated wave consists of the carrier plus a product of the message signal (i.e., the modulating wave) and the carrier. Thus for the case of sinusoidal modulation considered in Example 5.1, we have

$$s(t) = A_c[1 + \mu \cos(\omega_0 t)] \cos(\omega_c t)$$

where μ is the modulation factor. The term $1 + \mu \cos(\omega_0 t)$ is a modified version of the modulating signal and $A_c \cos(\omega_c t)$ is the carrier.

For the AM experiment described here, we have

Carrier amplitude, $A_c = 1$

Carrier frequency, $\omega_c = 0.8\pi$ rad/s

Modulation frequency, $\omega_0 = 0.1\pi$ rad/s

We wish to display and analyze 10 full cycles of the AM wave. This corresponds to a total duration of 200 s. Choosing a sampling rate $1/\mathcal{T} = 10$ Hz, we have a total of $N = 2000$ time samples. The frequency band of interest is $-10\pi \leq \omega \leq 10\pi$. Since the separation between the carrier and either side-frequency is equal to the modulation frequency $\omega_0 = 0.1\pi$ rad/s, we would like to have a frequency resolution $\omega_r = 0.01\pi$ rad/s. Accordingly, to achieve this resolution, we require the following number of frequency samples (see Eq. (4.54)):

$$M \geq \frac{\omega_s}{\omega_r} = \frac{20\pi}{0.01\pi} = 2000$$

We therefore choose $M = 2000$. To approximate the Fourier transform of the AM wave $s(t)$, we may use a 2000-point DTFS. The only variable in the AM experiment is the modulation factor μ. Specifically, we wish to investigate three different situations:

▶ $\mu = 0.5$, corresponding to undermodulation

▶ $\mu = 1.0$, for which the AM system is on the verge of overmodulation

▶ $\mu = 2.0$, corresponding to overmodulation

Putting all of these points together, we may now formulate the MATLAB commands for generating the AM wave and analyzing its frequency content as follows:

```
Ac = 1;  % carrier amplitude
wc = 0.8*pi;  % carrier frequency
w0 = 0.1*pi;  % modulation frequency
mu = 0.5;  % modulation factor
t = 0:0.1:199.9;
s = Ac*(1 + mu*cos(wo*t)).*cos(wc*t);
plot(t,s)
```

```
Smag = abs(fftshift(fft,2000)))/2000;
% Smag denotes the magnitude spectrum of the AM wave
w = 10*[-1000:999]*2*pi/2000;
plot(w,Smag)
```

The fourth command is written for $\mu = 0.5$. The computations are repeated for $\mu = 1, 2$.

In what follows, we describe the effect of varying the modulation factor μ on the time-domain and frequency-domain characteristics of the AM wave:

1. *$\mu = 0.5$.*

Figure 5.31(a) shows 10 cycles of the full AM wave $s(t)$ corresponding to $\mu = 0.5$. The envelope of $s(t)$ is clearly seen to faithfully follow the sinusoidal modulating wave. This

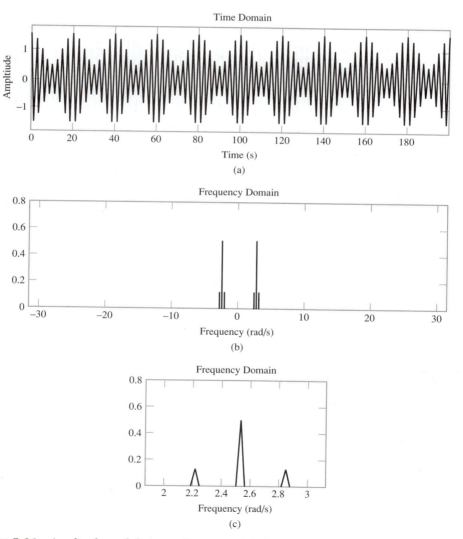

FIGURE 5.31 Amplitude modulation with 50% modulation. (a) AM wave, (b) magnitude spectrum of the AM wave, and (c) expanded spectrum around the carrier frequency.

means that we can use an envelope detector for demodulation. Figure 5.31(b) shows the magnitude spectrum of $s(t)$. In Fig. 5.31(c), we have zoomed in on the fine structure of the spectrum of $s(t)$ around the carrier frequency. This latter figure clearly displays the exact relationships between the side-frequencies and the carrier in accordance with modulation theory. In particular, the lower side-frequency, the carrier, and the upper side-frequency in Fig. 5.31(c) are located at $\omega_c - \omega_0 = \pm 0.7\pi$ rad/s, $\omega_c = \pm 0.8\pi$ rad/s, and $\omega_c + \omega_0 = \pm 0.9\pi$ rad/s, respectively. Moreover, the amplitude of both sidebands is $(\mu/2) = 0.25$ times that of the carrier; see Fig. 5.5 for comparison.

2. $\mu = 1.0$.

Figure 5.32(a) shows 10 cycles of the AM wave $s(t)$ with the same parameters as in Fig. 5.31(a), except for the fact that $\mu = 1.0$. This figure shows that the AM wave is now on

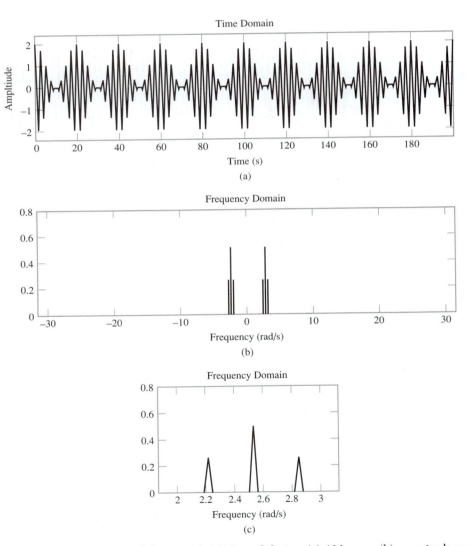

FIGURE 5.32 Amplitude modulation with 100% modulation. (a) AM wave, (b) magnitude spectrum of the AM wave, and (c) expanded spectrum around the carrier frequency.

the verge of overmodulation. The magnitude spectrum of $s(t)$ is shown in Fig. 5.32(b), and its zoomed version (around the carrier frequency) is shown in Fig. 5.32(c). Here again we see that the basic structure of the magnitude spectrum of the full AM wave is in perfect accord with the theory.

3. $\mu = 2.0$.

Figure 5.33(a) demonstrates the effect of overmodulation by using a modulation factor of $\mu = 2$. Here we see that there is no clear relationship between the envelope of the over-modulated wave $s(t)$ and the sinusoidal modulating wave. This implies that an envelope detector will not work, and we must use a coherent detector to perform the process of demodulation. Note, however, that the basic spectral content of the AM wave displayed in Figs. 5.33(b) and (c) follows exactly what the theory predicts.

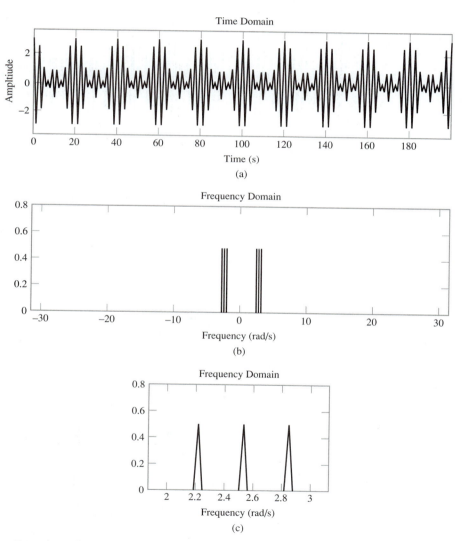

FIGURE 5.33 Amplitude modulation with 200% modulation. (a) AM wave, (b) magnitude spectrum of the AM wave, and (c) expanded spectrum around the carrier frequency.

■ DSB-SC MODULATION

In a DSB-SC modulated wave, the carrier is suppressed and both sidebands are transmitted in full. It is produced simply by multiplying the modulating wave by the carrier wave. Thus for the case of sinusoidal modulation, we have

$$s(t) = A_c A_0 \cos(\omega_c t) \cos(\omega_0 t)$$

The MATLAB commands for generating $s(t)$ and analyzing its frequency content are as follows:

```
Ac = 1;   % carrier amplitude
wc = 0.8*pi;   % carrier frequency in rad/s
```

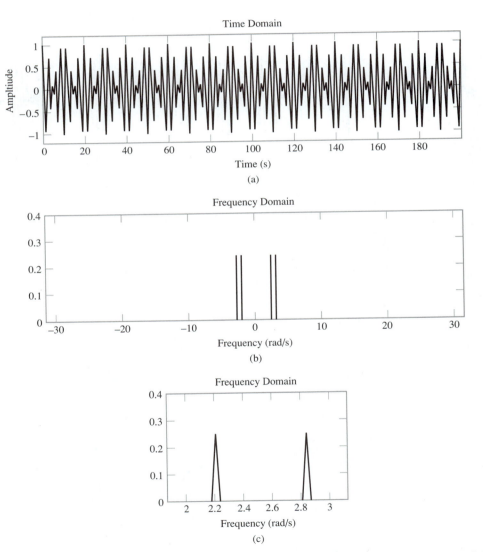

FIGURE 5.34   DSB-SC modulation. (a) DSB-SC modulated wave, (b) magnitude spectrum of the modulated wave, and (c) expanded spectrum around the carrier frequency.

```
A0 = 1;   % amplitude of modulating signal
w0 = 0.1*pi;   % frequency of modulating signal
t = 0:.1:199.9;
s = Ac*A0*cos(wc*t).*cos(w0*t);
plot(t,s)
Smag = abs(fftshift(fft(s,2000)))/2000;
w = 10*[-1000:999]*2*pi/2000;
plot(w,Smag)
```

These commands were used to investigate different aspects of DSB-SC modulation, as described here.

1. Figure 5.34(a) shows 10 cycles of the DSB-SC modulated wave $s(t)$ for the sinusoidal modulating wave. As expected, the envelope of the modulated wave bears no clear relationship to the sinusoidal modulating wave. Accordingly, we must use coherent detection for demodulation, which is discussed further under point 2. Figure 5.34(b) shows the magnitude spectrum of $s(t)$. An expanded view of the spectrum around the carrier frequency is shown in Fig. 5.34(c). These two figures clearly show that the carrier is indeed suppressed, and that the upper and lower side-frequencies are located exactly where they should be, namely, 0.9π and 0.7π rad/s, respectively.

2. To perform coherent detection, we multiply the DSB-SC modulated wave $s(t)$ by a replica of the carrier, and then pass the result through a lowpass filter, as described in Section 5.5. The output of the product modulator in Fig. 5.11(b) is defined by (assuming perfect synchronism between the transmitter and receiver)

$$v(t) = s(t)\ \cos(\omega_c t)$$

Correspondingly, the MATLAB command is

```
v = s.*cos(wc*t);
```

where s is itself as computed previously. Figure 5.35(a) shows the waveform of $v(t)$. Applying the fft command to v and taking the absolute value of the result, we obtain the magnitude spectrum shown in Fig. 5.35(b). This latter figure readily shows that $v(t)$ consists of the following components:

▶ A sinusoidal component with frequency 0.1π rad/s, representing the modulating wave.

▶ A new DSB-SC modulated wave with double carrier frequency of 1.6π rad/s; in actual fact, the side frequencies of this modulated wave are located at 1.5π and 1.7π rad/s.

Accordingly, we may recover the sinusoidal modulating signal by passing $v(t)$ through a lowpass filter with the following requirements:

▶ The frequency of the modulating wave lies inside the passband of the filter.

▶ The upper and lower side-frequencies of the new DSB-SC modulated wave lie inside the stopband of the filter.

The issue of how to design a filter with these requirements will be considered in detail in Chapter 8. For the present, it suffices to say that the above requirements can be met by using the MATLAB commands:

```
[b,a] = butter(3,0.025);
output = filter(b,a,v);
```

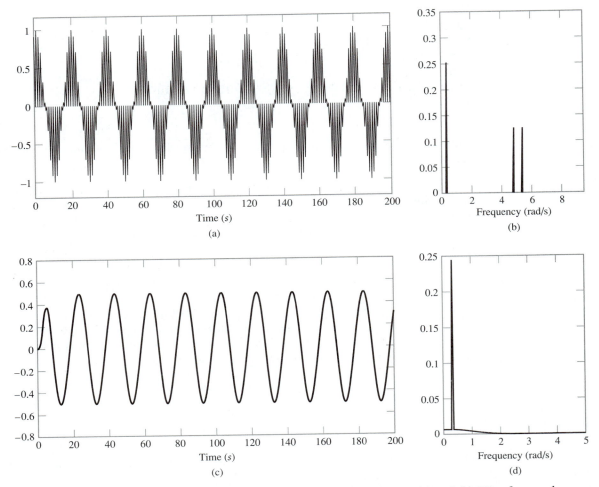

FIGURE 5.35 Coherent detection of DSB-SC modulated wave. (a) and (b): Waveform and magnitude spectrum of signal produced at the output of product modulator; (c) and (d) waveform and magnitude spectrum of lowpass filter output.

The first command produces a special type of filter called a Butterworth filter. For the experiment considered here, the filter order is 3 and its *normalized cutoff frequency* of 0.025 is calculated as follows:

$$\frac{\text{Actual cutoff frequency of filter}}{\text{Half the sampling rate}} = \frac{0.25\pi \text{ rad/s}}{10\pi \text{ rad/s}}$$

$$= 0.025$$

The second command computes the filter's output in response to the product modulator output $v(t)$. (We will revisit the design of this filter in Chapter 8.) Figure 5.35(c) displays the waveform of the lowpass filter output; it represents a sinusoidal signal of frequency 0.05 Hz. This observation is confirmed by using the **fft** command to approximate the spectrum of the filter output; the result of the computation is shown in Fig. 5.35(d).

3. In Fig. 5.36, we explore another aspect of DSB-SC modulation, namely, the effect of varying the modulation frequency. Figure 5.36(a) shows five cycles of a DSB-SC modulated wave that has the same carrier frequency as that in Fig. 5.34(a), but the modulation frequency has been reduced to 0.025 Hz (i.e., radian frequency of 0.05π). Figure 5.36(b) shows the magnitude spectrum of this second DSB-SC modulated wave. Its zoomed-in version is shown in Fig. 5.36(c). Comparing this latter figure with Fig. 5.34(c), we clearly see that decreasing the modulation frequency has the effect of moving the upper and lower side-frequencies closer together, which is exactly consistent with modulation theory.

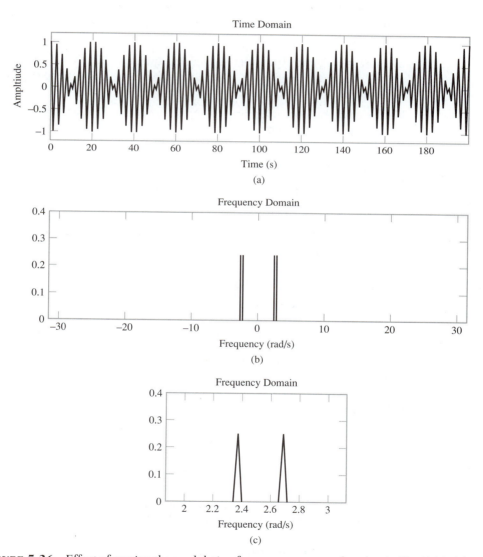

FIGURE 5.36 Effect of varying the modulation frequency, compared to that in Fig. 5.35. (a) and (b): Waveform and magnitude spectrum of DSB-SC modulated wave with a modulation frequency one-half that used in Fig. 5.34; (c) expanded spectrum around the carrier frequency.

▶ **Drill Problem 5.13** A *radiofrequency (RF) pulse* is defined as the product of a rectangular pulse and a sinusoidal carrier wave. Using MATLAB, plot the waveform of this pulse for each of the following two cases:

(a) Pulse duration = 1 s
    Carrier frequency = 5 Hz

(b) Pulse duration = 1 s
    Carrier frequency = 25 Hz

Use a sampling frequency of 1 kHz.   ◀

▶ **Drill Problem 5.14** Using the `fft` command, plot the magnitude spectrum of the RF pulse for each of the two cases described in Drill Problem 5.13. Hence, demonstrate the following:

(a) For case (a), corresponding to carrier frequency 5 Hz, the lower sidebands for positive and negative frequencies overlap each other. This effect is known as *sideband overlap*.

(b) For case (b), corresponding to carrier frequency 25 Hz, the spectrum is essentially free from sideband overlap.   ◀

■ *PHASE AND GROUP DELAYS

In Example 5.8, we studied the phase and group delays for a bandpass channel with phase response

$$\phi(\omega) = -\tan^{-1}\left(\frac{\omega^2 - \omega_c^2}{\omega\omega_c}\right)$$

At $\omega = \omega_c$, the phase delay is $\tau_p = 0$ and the group delay is $\tau_g = 0.4211$s. The two waveforms displayed in Fig. 5.29 are defined as follows:

(1)
$$x_1(t) = s(t)\cos(\omega_c t)$$

$$= \frac{A}{2}[\cos((\omega_c + \omega_0)t) + \cos((\omega_c - \omega_0)t)]$$

$$= \frac{A}{2}[\cos(\omega_1 t) + \cos(\omega_2 t)]$$

where $\omega_1 = \omega_c + \omega_0$ and $\omega_2 = \omega_c - \omega_0$. The waveform shown in Fig. 5.29 as a solid curve is a plot of $x_1(t)$.

(2)
$$x_2(t) = y(t)\cos(\omega_c t)$$

$$= \frac{A}{2}[\cos(\omega_1 t + \phi(\omega_1)) + \cos(\omega_2 t + \phi(\omega_2))]\cos(\omega_c t)$$

where the angles $\phi(\omega_1)$ and $\phi(\omega_2)$ are the values of the phase response $\phi(\omega)$ at $\omega = \omega_1$ and $\omega = \omega_2$, respectively. The waveform shown as a dotted curve in Fig. 5.29 is a plot of $x_2(t)$.

The generation of $x_1(t)$ and $x_2(t)$ in MATLAB is achieved using the following commands:

```
o1=-atan((w1^2-wc^2)/(w1*wc));
o2=-atan((w2^2-wc^2)/(w2*wc));
```

```
s=cos(w1*t)+cos(w2*t);
y=cos(w1*t+o1)+cos(w2*t+o2);
x1=s.*cos(4.75*t);
x2=y.*cos(4.75*t);
```

where we have set $(A/2) = 1$ for convenience of presentation. The function **atan** in the first two commands returns the arctangent. Note also that both **x1** and **x2** involve element-by-element multiplications, hence the use of a period followed by an asterisk.

5.12 *Summary*

In this chapter we presented a discussion of linear modulation techniques for the transmission of a message signal over a communication channel.

In particular, we described amplitude modulation (AM) and its variants, as summarized here:

▶ In full AM, the spectrum consists of two sidebands (one termed the upper sideband and the other termed the lower sideband) and the carrier. The primary advantage of full AM is the simplicity of its implementation, which explains its popular use for radio broadcasting. Its disadvantages include a wastage of transmission bandwidth and transmit power.

▶ In double sideband-suppressed carrier (DSB-SC) modulation, the carrier is suppressed, saving transmit power. However, the transmission bandwidth for DSB-SC modulation is the same as that of full AM, that is, twice the message bandwidth.

▶ In single sideband (SSB) modulation, only one of the sidebands is transmitted. SSB modulation is therefore the optimum form of continuous-wave (CW) modulation, in that it requires the least amount of channel bandwidth and power for its transmission. The use of SSB modulation requires the presence of an energy gap in the spectrum of the message signal around zero frequency.

▶ In vestigial sideband (VSB) modulation, one sideband and a vestige of the other sideband are transmitted. It is well suited for transmission of wideband signals whose spectra extend down to zero frequency. VSB modulation is the standard method for the transmission of television signals.

The other form of linear modulation discussed in the chapter was that of pulse-amplitude modulation (PAM). PAM represents the simplest form of pulse modulation. It may be viewed as a direct manifestation of the sampling process. As such, PAM is commonly used as a method of modulation in its own right. Moreover, it constitutes an operation that is basic to all the other forms of pulse modulation, including pulse-code modulation.

We then discussed the notion of multiplexing, which permits the sharing of a common communication channel among a number of independent users. In frequency-division multiplexing (FDM), the sharing is performed in the frequency domain. In time-division multiplexing (TDM), the sharing is performed in the time domain.

The other topic discussed in the chapter was that of phase (carrier) delay and group (envelope) delay, both of which are defined in terms of the phase response of a channel over which a modulated signal is transmitted. The group delay is the true signal delay; it becomes of paramount importance when a wideband modulated signal is transmitted over the channel.

One final comment is in order. In discussing the modulation systems presented in this chapter, we made use of two functional blocks:

▶ Filters for the suppression of spurious signals
▶ Equalizers for correcting signal distortion produced by physical transmission systems

The approach taken herein was from a system-theoretic viewpoint, and we did not concern ourselves with the design of these functional blocks. Design considerations of filters and equalizers are taken up in Chapter 8.

FURTHER READING

1. Communications technology has an extensive history that dates back to the invention of the telegraph (the predecessor to digital communications) by Samuel Morse in 1837. This was followed by the invention of the telephone by Alexander Graham Bell in 1875, in whose honor the decibel is named. Other notable contributors to the subject include Harry Nyquist, who published a classic paper on the theory of signal transmission in telegraphy in 1928, and Claude Shannon who laid down the foundations of *information theory* in 1948. Information theory is a broad subject, encompassing the transmission, processing, and utilization of information.

   For a historical account of communication systems, see Chapter 1 of the book:

   ▶ Haykin, S., *Communication Systems*, Third Edition (Wiley, 1994)

2. For a more complete treatment of modulation theory, see the books:

   ▶ Carlson, A. B., *Communication Systems: An Introduction to Signals and Noise in Electrical Communications*, Third Edition (McGraw-Hill, 1986)
   ▶ Couch, L. W. III, *Digital and Analog Communication Systems*, Third Edition (Prentice Hall, 1990)
   ▶ Haykin, S., *Communication Systems*, Third Edition (Wiley, 1994)
   ▶ Schwartz, M., *Information Transmission Modulation and Noise: A Unified Approach*, Third Edition (McGraw-Hill, 1980)
   ▶ Stremler, F. G., *Introduction to Communication Systems*, Third Edition (Addison-Wesley, 1990)
   ▶ Ziemer, R. E., and W. H. Tranter, *Principles of Communication Systems*, Third Edition (Houghton Mifflin, 1990)

   These books cover both continuous-wave modulation and pulse modulation techniques.

*3. The Hilbert transform of a signal $x(t)$ is defined by

$$\hat{x}(t) = \frac{1}{\pi} \int_{-\infty}^{\infty} \frac{x(\tau)}{t - \tau} \, d\tau$$

Equivalently, we may define the Hilbert transform $\hat{x}(t)$ as the convolution of $x(t)$ with $1/\pi t$. The Fourier transform of $1/\pi t$ is j times the signum function, where

$$\text{sgn}(j\omega) = \begin{cases} +1, & \text{for } \omega > 0 \\ 0, & \text{for } \omega = 0 \\ -1, & \text{for } \omega < 0 \end{cases}$$

Passing $x(t)$ through a Hilbert transformer is therefore equivalent to the combination of the following two operations in the frequency-domain:

▶ Keeping $|X(j\omega)|$ (i.e., the magnitude spectrum of $x(t)$) unchanged for all ω
▶ Shifting $\arg\{X(j\omega)\}$ (i.e., the phase spectrum of $x(t)$) by $+90°$ for negative frequencies and $-90°$ for positive frequencies

For a more complete discussion of the Hilbert transform and its use in the time-domain description of single sideband modulation, see Chapters 2 and 3 of the book:

▶ Haykin, S., *Communication Systems*, Third Edition (Wiley, 1994)

4. The books cited above also include the study of how noise affects the performance of modulation systems.

5. For an advanced treatment of phase delay and group delay, see Chapter 2 of the book:

▶ Haykin, S., *Communication Systems*, Third Edition (Wiley, 1994)

PROBLEMS

5.1 You are given a nonlinear device whose input–output relation is described by

$$i_o = a_1 v_i + a_2 v_i^2$$

where a_1 and a_2 are constants, v_i is the input voltage, and i_o is the output current. Let

$$v_i(t) = A_c \cos(\omega_c t) + A_m \cos(\omega_m t)$$

where the first term represents a sinusoidal carrier and the second term represents a sinusoidal modulating signal.

(a) Determine the frequency content of $i_o(t)$.

(b) The output current $i_o(t)$ contains an AM signal produced by the two components of $v_i(t)$. Describe the specification of a filter that extracts this AM signal from $i_o(t)$.

5.2 Consider a message signal $m(t)$ with the spectrum shown in Fig. P5.2. The message bandwidth $\omega_m = 2\pi \times 10^3$ rad/s. This signal is applied to a product modulator, together with a carrier wave $A_c \cos(\omega_c t)$, producing the DSB-SC modulated signal $s(t)$. The modulated signal is next applied to a coherent detector. Assuming perfect synchronism between the carrier waves in the modulator and detector, determine the spectrum of the detector output when (a) the carrier frequency $\omega_c = 2.5\pi \times 10^3$ rad/s and (b) the carrier frequency $\omega_c = 1.5\pi \times 10^3$ rad/s. What is the lowest carrier frequency for which

each component of the modulated signal $s(t)$ is uniquely determined by $m(t)$?

5.3 Figure P5.3 shows the circuit diagram of a balanced modulator. The input applied to the top AM modulator is $m(t)$, whereas that applied to the lower AM modulator is $-m(t)$; these two modulators have the same amplitude sensitivity. Show that the output $s(t)$ of the balanced modulator consists of a DSB-SC modulated signal.

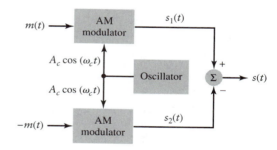

FIGURE P5.3

5.4 Phase-shift keying (PSK) is a form of phase modulation widely used for digital communications. The simplest form of PSK, known as binary PSK, arises when the following representation for the modulation signal $s(t)$ is adopted:

$$s(t) = \begin{cases} A_c \cos(\omega_c t), & \text{symbol 1} \\ A_c \cos(\omega_c t + \pi), & \text{symbol 0} \end{cases}$$

Binary PSK may also be viewed as double sideband-suppressed carrier (DSB-SC) modulation. Specifically, we may redefine $s(t)$ as

$$s(t) = A_c \cos(\omega_c t) b(t)$$

where $b(t)$ represents the sequence of 1's and 0's; symbol 1 is represented by $+1$ and symbol 0 is represented by -1.

Consider then a binary sequence $b(t)$ consisting of a square wave with fundamental pe-

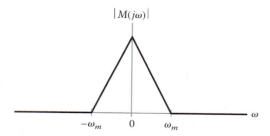

FIGURE P5.2

riod $2T_0$; that is, symbols 1 and 0 alternate. For this special sequence, do the following:

(a) Sketch the waveform of the binary PSK signal $s(t)$ for $\omega_c T_0 = 4\pi$.

(b) Derive a general formula for the spectrum of $s(t)$.

(c) Plot the magnitude spectrum of $s(t)$ for $\omega_c T_0 = 4\pi, 10\pi, 20\pi$.

5.5 The so-called radiofrequency (RF) pulse $s(t)$ is defined by

$$s(t) = \begin{cases} A_c \cos(\omega_c t), & -T/2 \leq t \leq T/2 \\ 0, & \text{otherwise} \end{cases}$$

(a) Derive a formula for the spectrum of $s(t)$, assuming that $\omega_c T_0 \gg 2\pi$.

(b) Sketch the magnitude spectrum of $s(t)$ for $\omega_c T_0 = 20\pi$.

5.6 The transmitted signal $s(t)$ of a radar system consists of a periodic sequence of short RF pulses. The fundamental period of the sequence is T_0. Each RF pulse has duration T_1 and frequency ω_c. Typical values are:

$$T_0 = 1 \text{ ms}$$
$$T_1 = 1 \text{ } \mu\text{s}$$
$$\omega_c = 2\pi \times 10^9 \text{ rad/s}$$

Using the results of Problem 5.5, sketch the magnitude spectrum of $s(t)$.

5.7 A DSB-SC modulated signal is demodulated by applying it to a coherent detector. Evaluate the effect of a frequency error $\Delta\omega$ in the local carrier frequency of the detector, measured with respect to the carrier frequency of the incoming DSB-SC signal.

*5.8 Consider the quadrature-carrier multiplex system of Fig. 5.14. The multiplexed signal $s(t)$ produced at the transmitter input in Fig. 5.14(a) is applied to a communication channel with frequency response $H(j\omega)$. The output of this channel is in turn applied to the receiver input in Fig. 5.14(b). Prove that the condition

$$H(j\omega_c + j\omega) = H^*(j\omega_c - j\omega), \quad 0 < \omega < \omega_m$$

is necessary for recovery of the message signals $m_1(t)$ at the receiver outputs; ω_c is the carrier frequency, and ω_m is the message bandwidth. *Hint:* Evaluate the spectra of the two receiver outputs.

5.9 Using the message signal

$$m(t) = \frac{1}{1 + t^2}$$

sketch the modulated waves for the following methods of modulation:

(a) Amplitude modulation with 50% modulation

(b) Double sideband-suppressed carrier modulation

*5.10 The spectrum of a voice signal $m(t)$ is zero outside the interval $\omega_a \leq |\omega| \leq \omega_b$. In order to ensure communication privacy, this signal is applied to a *scrambler* that consists of the following cascade of components: a product modulator, a highpass filter, a second product modulator, and a lowpass filter. The carrier wave applied to the first product modulator has a frequency equal to ω_c, whereas that applied to the second product modulator has a frequency equal to $\omega_b + \omega_c$; both of them have unit amplitude. The highpass and lowpass filters have the same cutoff frequency at ω_c. Assume that $\omega_c > \omega_b$.

(a) Derive an expression for the scrambler output $s(t)$, and sketch its spectrum.

(b) Show that the original voice signal $m(t)$ may be recovered from $s(t)$ by using an *unscrambler* that is identical to the unit described above.

5.11 A single sideband modulated wave $s(t)$ is applied to the coherent detector shown in Fig. P5.11. The cutoff frequency of the lowpass filter is set equal to the highest frequency component of the message signal. Using frequency-domain ideas, show that this detector produces an output that is a scaled version of the original message signal. You may assume that the carrier frequency ω_c satisfies the condition $\omega_c > \omega_m$.

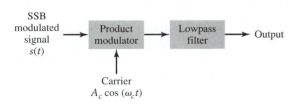

FIGURE P5.11

5.12 Figure P5.12 shows the block diagram of a *frequency synthesizer*, which enables the generation of many frequencies, each with the same high accuracy as the *master oscillator*. The master oscillator of frequency 1 MHz feeds two *spectrum generators*, one directly and the other through a *frequency divider*. Spectrum generator 1 produces a signal rich in the following har-

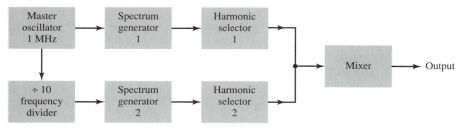

FIGURE P5.12

monics: 1, 2, 3, 4, 5, 6, 7, 8, and 9 MHz. The frequency divider provides a 100-kHz output, in response to which spectrum generator 2 produces a second signal rich in the following harmonics: 100, 200, 300, 400, 500, 600, 700, 800, and 900 kHz. The harmonic selectors are designed to feed two signals into the mixer, one from spectrum generator 1 and the other from spectrum generator 2. Find the range of possible frequency outputs of this synthesizer and its resolution (i.e., the separation between adjacent frequency outputs).

*5.13 Consider a multiplex system in which four input signals $m_1(t)$, $m_2(t)$, $m_3(t)$, and $m_4(t)$ are respectively multiplied by the carrier waves

$$[\cos(\omega_a t) + \cos(\omega_b t)]$$
$$[\cos(\omega_a t + \alpha_1) + \cos(\omega_b t + \beta_1)]$$
$$[\cos(\omega_a t + \alpha_2) + \cos(\omega_b t + \beta_2)]$$
$$[\cos(\omega_a t + \alpha_3) + \cos(\omega_b t + \beta_3)]$$

and the resulting DSB-SC signals are summed and then transmitted over a common channel. In the receiver, demodulation is achieved by multiplying the sum of the DSB-SC signals by the four carrier waves separately and then using filtering to remove the unwanted components. Determine the conditions that the phase angles α_1, α_2, α_3 and β_1, β_2, β_3 must satisfy in order that the output of the kth demodulator is $m_k(t)$, where $k = 1, 2, 3, 4$.

*5.14 In this problem we study the idea of mixing used in a *superheterodyne receiver*. To be specific, consider the block diagram of the *mixer* shown in Fig. P5.14 that consists of a product modulator with a local oscillator of variable frequency, followed by a bandpass filter. The input signal is an AM wave of bandwidth 10 kHz and carrier frequency that may lie anywhere in the range 0.535–1.605 MHz; these parameters are typical of AM radio broadcasting. It is required to translate this signal to a frequency band centered at a fixed *intermediate frequency* (IF) of

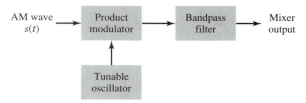

FIGURE P5.14

0.455 MHz. Find the range of tuning that must be provided in the local oscillator in order to achieve this requirement.

5.15 Compare full AM with PAM, emphasizing their similarities and differences.

5.16 In *natural sampling*, an analog signal $g(t)$ is multiplied by a periodic train of rectangular pulses $c(t)$. You are given that the pulse repetition frequency of this periodic train is ω_s and the duration of each rectangular pulse is T (with $\omega_s T \gg 2\pi$). Find the spectrum of the signal $s(t)$ that results from the use of natural sampling; you may assume that time $t = 0$ corresponds to the midpoint of a rectangular pulse in $c(t)$.

5.17 Specify the Nyquist rate for each of the following signals:

(a) $g(t) = \text{sinc}(200t)$

(b) $g(t) = \text{sinc}^2(200t)$

(c) $g(t) = \text{sinc}(200t) + \text{sinc}^2(200t)$

5.18 Twenty-four voice signals are sampled uniformly and then time-division multiplexed using PAM. The PAM signal is reconstructed from flat-top pulses with 1-μs duration. The multiplexing operation includes provision for synchronization by adding an extra pulse of sufficient amplitude and also 1-μs duration. The highest frequency component of each voice signal is 3.4 kHz.

(a) Assuming a sampling rate of 8 kHz, calculate the spacing between successive pulses of the multiplexed signal.

(b) Repeat your calculation assuming the use of Nyquist rate sampling.

5.19 Twelve different message signals, each with a bandwidth of 10 kHz, are to be multiplexed and transmitted. Determine the minimum bandwidth required for each method if the multiplexing/modulation method used is

(a) FDM, SSB

(b) TDM, PAM

5.20 A PAM *telemetry* system involves the multiplexing of four input signals $s_i(t)$, $i = 1, 2, 3, 4$. Two of the signals $s_1(t)$ and $s_2(t)$ have bandwidths of 80 Hz each, whereas the remaining two signals $s_3(t)$ and $s_4(t)$ have bandwidths of 1 kHz each. The signals $s_3(t)$ and $s_4(t)$ are each sampled at the rate of 2400 samples per second. This sampling rate is divided by 2^R (i.e., an integer power of 2) in order to derive the sampling rate for $s_1(t)$ and $s_2(t)$.

(a) Find the maximum value of R.

(b) Using the value of R found in part (a), design a multiplexing system that first multiplexes $s_1(t)$ and $s_2(t)$ into a new sequence, $s_5(t)$, and then multiplexes $s_3(t)$, $s_4(t)$ and $s_5(t)$.

▶ *Computer Experiments*

Note: The reader is expected to choose sampling rates for the computer experiments described here. Thorough understanding of the material presented in Chapter 4 is needed.

5.21 Use MATLAB to generate and display an AM wave with the following specifications:

Modulating wave	Sinusoidal
Modulation frequency	1 kHz
Carrier frequency	20 kHz
Percentage modulation	75%

Compute and display the magnitude spectrum of the AM wave.

5.22 (a) Generate a symmetric triangular wave $m(t)$ with fundamental frequency of 1 Hz, alternating between -1 and $+1$.

(b) Use $m(t)$ to modulate a carrier of frequency $f_c = 25$ Hz, generating a full AM wave with 80% modulation. Compute the magnitude spectrum of the AM wave.

5.23 Continuing with Problem 5.22, investigate the effect of varying the carrier frequency f_c on the spectrum of the AM wave. Determine the minimum value of f_c that is necessary to ensure that there is no overlap between the lower and upper sidebands of the AM wave.

5.24 The triangular wave described in Problem 5.22(a) is used to perform DSB-SC modulation on a carrier of frequency $f_c = 25$ Hz.

(a) Generate and display the DSB-SC modulated wave so produced.

(b) Compute and display the spectrum of the modulated wave. Investigate the use of coherent detection for demodulation.

5.25 Use MATLAB to do the following:

(a) Generate a PAM wave using a sinusoidal modulating signal of frequency $\omega_m = 0.5\pi$ rad/s, sampling period $\mathcal{T} = 1$s, and pulse duration $T = 0.05$s.

(b) Compute and display the magnitude spectrum of the PAM wave.

(c) Repeat the experiment for pulse duration $T = 0.1, 0.2, 0.3, 0.4, 0.5$s.

Comment on the results of your experiment.

5.26 *Natural sampling* involves the multiplication of a message signal by a rectangular pulse train, as discussed in Problem 5.16. The fundamental period of the pulse train is T_c and the pulse duration is T.

(a) Generate and display the modulated wave for a sinusoidal modulating wave, with the following specifications:

Modulation frequency	1 kHz
Pulse-repetition frequency	$(1/T_c) = 10$ kHz
Pulse duration	$T = 10$ μs

(b) Compute and display the spectrum of the modulated wave. Hence verify that the original modulating wave can be recovered without distortion by passing the modulated wave through a lowpass filter. Specify the requirements that this filter must satisfy.

6 Representation of Signals Using Continuous-Time Complex Exponentials: The Laplace Transform

6.1 Introduction

In this chapter we generalize the complex sinusoidal representation of a continuous-time signal offered by the FT to a representation in terms of complex exponential signals, which is termed the *Laplace transform*. The Laplace transform provides a broader characterization of systems and their interaction with signals than is possible with Fourier methods. For example, the Laplace transform can be used to analyze a large class of continuous-time problems involving signals that are not absolutely integrable, such as the impulse response of an unstable system. The FT does not exist for signals that are not absolutely integrable, so FT-based methods cannot be used in this class of problems.

The Laplace transform possesses a powerful set of properties for analysis of signals and systems. Many of these properties parallel those of the FT. For example, we shall see that continuous-time complex exponentials are eigenfunctions of LTI systems. As with complex sinusoids, this results in convolution of time signals transforming to multiplication of the associated Laplace transforms. Hence the output of a LTI system is obtained by multiplying the Laplace transform of the input with the Laplace transform of the impulse response. The Laplace transform of the impulse response is defined as the transfer function of the system. The transfer function generalizes the frequency response characterization of a system's input–output behavior and offers new insights into system characteristics.

The Laplace transform comes in two varieties, (1) unilateral or one-sided and (2) bilateral or two-sided. The unilateral Laplace transform is a convenient tool for solving differential equations with initial conditions. The bilateral Laplace transform offers insight into the nature of system characteristics such as stability, causality, and frequency response. The primary role of the Laplace transform in engineering is transient and stability analysis of causal LTI systems described by differential equations. We shall develop the Laplace transform with these roles in mind throughout this chapter.

6.2 The Laplace Transform

Let e^{st} be a complex exponential with complex frequency $s = \sigma + j\omega$. We may write e^{st} as the complex-valued signal

$$e^{st} = e^{\sigma t} \cos(\omega t) + j e^{\sigma t} \sin(\omega t) \qquad (6.1)$$

The real part of e^{st} is an exponentially damped cosine and the imaginary part is an exponentially damped sine, as depicted in Fig. 6.1. In this figure it is assumed that σ is negative. The real part of s is the exponential damping factor σ, and the imaginary part of s is the frequency of the cosine and sine factor, ω.

Consider applying an input of the form $x(t) = e^{st}$ to a LTI system with impulse response $h(t)$. The system output $y(t)$ is given by

$$y(t) = H\{e^{st}\}$$
$$= h(t) * x(t)$$
$$= \int_{-\infty}^{\infty} h(\tau)x(t - \tau)\, d\tau$$

Substitute for $x(t)$ to obtain

$$y(t) = \int_{-\infty}^{\infty} h(\tau)e^{s(t-\tau)}\, d\tau$$
$$= e^{st}\int_{-\infty}^{\infty} h(\tau)e^{-s\tau}\, d\tau$$

Define the transfer function

$$H(s) = \int_{-\infty}^{\infty} h(\tau)e^{-s\tau}\, d\tau \tag{6.2}$$

so that we may write

$$H\{e^{st}\} = H(s)e^{st}$$

The action of the system on an input e^{st} is multiplication by $H(s)$. Hence we identify e^{st} as an eigenfunction of the system and $H(s)$ as an eigenvalue.

Express the complex number $H(s)$ in polar form as $H(s) = |H(s)|e^{j\phi(s)}$, where $|H(s)|$ and $\phi(s)$ are the magnitude and phase of $H(s)$, respectively. Now rewrite the system output as

$$y(t) = |H(s)|e^{j\phi(s)}e^{st}$$

Substitute $s = \sigma + j\omega$ to obtain

$$y(t) = |H(\sigma + j\omega)|e^{\sigma t}e^{j\omega t + \phi(\sigma + j\omega)}$$
$$= |H(\sigma + j\omega)|e^{\sigma t}\cos(\omega t + \phi(\sigma + j\omega)) + j|H(\sigma + j\omega)|e^{\sigma t}\sin(\omega t + \phi(\sigma + j\omega))$$

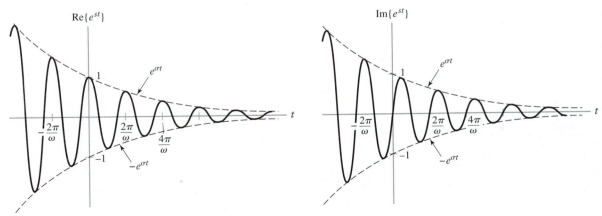

FIGURE 6.1   Real and imaginary parts of the complex exponential e^{st}.

Since the input $x(t)$ has the form given in Eq. (6.1), we see that the system changes the amplitude of the input by $|H(\sigma + j\omega)|$ and shifts the phase of the sinusoidal components by $\phi(\sigma + j\omega)$. The system does not change the damping factor σ of the input or the sinusoidal frequency ω.

Given the simplicity of describing the action of the system on inputs of the form e^{st}, we now seek a representation for arbitrary signals as a weighted superposition of eigenfunctions e^{st}. Substituting $s = \sigma + j\omega$ in Eq. (6.2) and using t as the variable of integration, we obtain

$$H(\sigma + j\omega) = \int_{-\infty}^{\infty} h(t)e^{-(\sigma+j\omega)t}\, dt$$

$$= \int_{-\infty}^{\infty} [h(t)e^{-\sigma t}]e^{-j\omega t}\, dt$$

This indicates that $H(\sigma + j\omega)$ is the Fourier transform of $h(t)e^{-\sigma t}$. Hence the inverse Fourier transform of $H(\sigma + j\omega)$ must be $h(t)e^{-\sigma t}$, that is,

$$h(t)e^{-\sigma t} = \frac{1}{2\pi}\int_{-\infty}^{\infty} H(\sigma + j\omega)e^{j\omega t}\, d\omega$$

We may recover $h(t)$ by multiplying both sides of this equation by $e^{\sigma t}$, as shown by

$$h(t) = e^{\sigma t}\frac{1}{2\pi}\int_{-\infty}^{\infty} H(\sigma + j\omega)e^{j\omega t}\, d\omega$$

$$= \frac{1}{2\pi}\int_{-\infty}^{\infty} H(\sigma + j\omega)e^{(\sigma+j\omega)t}\, d\omega \qquad (6.3)$$

Now substituting $s = \sigma + j\omega$ and $d\omega = ds/j$ into Eq. (6.3), we get

$$h(t) = \frac{1}{2\pi j}\int_{\sigma-j\infty}^{\sigma+j\infty} H(s)e^{st}\, ds \qquad (6.4)$$

The limits on the integral are also a result of the substitution $s = \sigma + j\omega$. Equation (6.2) indicates how to determine $H(s)$ from $h(t)$, while Eq. (6.4) expresses $h(t)$ as a function of $H(s)$. We say that $H(s)$ is the *Laplace transform* of $h(t)$, and that $h(t)$ is the *inverse Laplace transform* of $H(s)$.

We have obtained the Laplace transform for the impulse response of a system. This relationship holds for an arbitrary signal. The Laplace transform of $x(t)$ is

$$X(s) = \int_{-\infty}^{\infty} x(t)e^{-st}\, dt \qquad (6.5)$$

and the inverse Laplace transform of $X(s)$ is

$$x(t) = \frac{1}{2\pi j}\int_{\sigma-j\infty}^{\sigma+j\infty} X(s)e^{st}\, ds \qquad (6.6)$$

We express this relationship with the notation

$$x(t) \xleftarrow{\quad\mathcal{L}\quad} X(s)$$

Note that Eq. (6.6) represents the signal $x(t)$ as a weighted superposition of complex exponentials e^{st}. The weights are proportional to $X(s)$. In practice, we usually do not evaluate this integral directly since it requires techniques of contour integration. Instead, we determine inverse Laplace transforms by inspection utilizing the one-to-one relationship between $x(t)$ and $X(s)$.

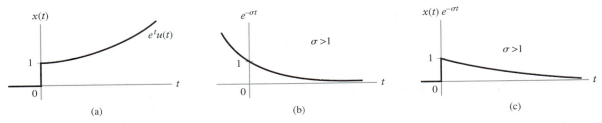

FIGURE 6.2   The Laplace transform applies to more general signals than the Fourier transform. (a) Signal for which the Fourier transform does not exist. (b) Attenuating factor associated with Laplace transform. (c) The modified signal $x(t)e^{-\sigma t}$ is absolutely integrable for $\sigma > 1$.

Our development indicates that the Laplace transform is the Fourier transform of $x(t)e^{-\sigma t}$. Hence a necessary condition for convergence of the Laplace transform is absolute integrability of $x(t)e^{-\sigma t}$. That is, we must have

$$\int_{-\infty}^{\infty} |x(t)e^{-\sigma t}| \, dt < \infty$$

The range of σ for which the Laplace transform converges is termed the *region of convergence* (ROC).

Note that the Laplace transform exists for signals that do not have a Fourier transform. By limiting ourselves to a certain range of σ, we may ensure that $x(t)e^{-\sigma t}$ is absolutely integrable even though $x(t)$ is not absolutely integrable by itself. For example, the Fourier transform of $x(t) = e^{t}u(t)$ does not exist since $x(t)$ is an increasing real exponential signal and is thus not absolutely integrable. However, if $\sigma > 1$, then $x(t)e^{-\sigma t} = e^{(1-\sigma)t}u(t)$ is absolutely integrable, and so the Laplace transform, which is the Fourier transform of $x(t)e^{-\sigma t}$, does exist for some values of σ. This scenario is illustrated in Fig. 6.2. The existence of the Laplace transform for signals that have no Fourier transform is a significant advantage of using the complex exponential representation.

It is convenient to graphically represent the complex frequency s in terms of a complex plane termed the s-plane as depicted in Fig. 6.3. The horizontal axis represents the real part of s, that is, the exponential damping factor σ, and the vertical axis represents the imaginary part of s, or the sinusoidal frequency ω. Note that if $x(t)$ is absolutely integrable, then we may obtain the Fourier transform from the Laplace transform by setting $\sigma = 0$, as shown here:

$$X(j\omega) = X(s)\big|_{\sigma=0}$$

In the s-plane, $\sigma = 0$ corresponds to the imaginary axis. We thus say that the Fourier transform is given by the Laplace transform evaluated along the imaginary axis.

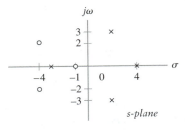

FIGURE 6.3   The s-plane. The horizontal axis is Re$\{s\}$ and the vertical axis is Im$\{s\}$. Zeros are depicted at $s = -1$, $s = -4 \pm 2j$ and poles are depicted at $s = -3$, $s = 2 \pm 3j$, $s = 4$.

The $j\omega$-axis divides the s-plane in half. The region of the s-plane to the left of the $j\omega$-axis is termed the *left half of the s-plane*, while the region to the right of the $j\omega$-axis is termed the *right half of the s-plane*. The real part of s is negative in the left half of the s-plane, and positive in the right half of the s-plane.

The most commonly encountered form of the Laplace transform in engineering is a ratio of two polynomials in s. That is,

$$X(s) = \frac{b_M s^M + b_{M-1} s^{M-1} + \cdots + b_1 s + b_0}{s^N + a_{N-1} s^{N-1} + \cdots + a_1 s + a_0}$$

It is useful to rewrite $X(s)$ as a product of terms involving the roots of the denominator and numerator polynomials, as shown by

$$X(s) = \frac{b_M \Pi_{k=1}^{M}(s - c_k)}{\Pi_{k=1}^{N}(s - d_k)}$$

The c_k are the roots of the numerator polynomial and are termed the *zeros* of $X(s)$. The d_k are the roots of the denominator polynomial and are termed the *poles* of $X(s)$. We denote the locations of zeros in the s-plane with the "$\circ$" symbol and the pole locations with the "$\times$" symbol, as illustrated in Fig. 6.3. The pole and zero locations in the s-plane completely specify $X(s)$, except for the constant gain factor b_M.

EXAMPLE 6.1   Determine the Laplace transform of

$$x(t) = e^{at} u(t)$$

and depict the ROC and pole and zero locations in the s-plane. Assume a is real.

Solution: Substitute $x(t)$ into Eq. (6.5), obtaining

$$X(s) = \int_{-\infty}^{\infty} e^{at} u(t) e^{-st}\, dt$$

$$= \int_{0}^{\infty} e^{-(s-a)t}\, dt$$

$$= \frac{-1}{s - a} e^{-(s-a)t} \Big|_{0}^{\infty}$$

To evaluate $e^{-(s-a)t}$ at the limits, we substitute $s = \sigma + j\omega$ and so write

$$X(s) = \frac{-1}{\sigma + j\omega - a} e^{-(\sigma-a)t} e^{-j\omega t} \Big|_{0}^{\infty}$$

Now if $\sigma > a$, then $e^{-(\sigma-a)t}$ goes to zero as t approaches infinity and

$$X(s) = \frac{-1}{\sigma + j\omega - a}(0 - 1), \qquad \sigma > a$$

$$= \frac{1}{s - a}, \qquad \mathrm{Re}(s) > a$$

(6.7)

The Laplace transform $X(s)$ does not exist for $\sigma \leq a$ since the integral is unbounded. The ROC for this signal is thus $\sigma > a$, or equivalently $\mathrm{Re}(s) > a$. This ROC is depicted as the shaded region of the s-plane in Fig. 6.4. The pole is located at $s = a$.

The expression for the Laplace transform does not uniquely correspond to a signal $x(t)$ if the ROC is not specified. That is, two different signals may have identical Laplace transforms but different ROCs. We demonstrate this with the following example.

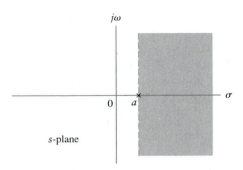

FIGURE 6.4   The ROC for $e^{at}u(t)$ is depicted by the shaded region. A pole is located at $s = a$.

EXAMPLE 6.2   Determine the Laplace transform and ROC for the signal

$$y(t) = -e^{at}u(-t)$$

Solution: Substitute for $x(t)$ in Eq. (6.5), obtaining

$$Y(s) = \int_{-\infty}^{\infty} -e^{at}u(-t)e^{-st} \, dt$$

$$= -\int_{-\infty}^{0} e^{-(s-a)t} \, dt$$

$$= \frac{1}{s-a} e^{-(s-a)t} \Big|_{-\infty}^{0} \tag{6.8}$$

$$= \frac{1}{s-a}, \qquad \text{Re}(s) < a$$

The ROC and pole and zero locations are depicted in Fig. 6.5.

Examples 6.1 and 6.2 reveal that the Laplace transforms $X(s)$ and $Y(s)$ are equal, even though the signals $x(t)$ and $y(t)$ are clearly different. However, their ROCs are different. This ambiguity occurs in general with signals that are one sided. To see this, let $x(t) = g(t)u(t)$ and $y(t) = -g(t)u(-t)$. We may thus write

$$X(s) = \int_{0}^{\infty} g(t)e^{-st} \, dt$$

$$= G(s, \infty) - G(s, 0)$$

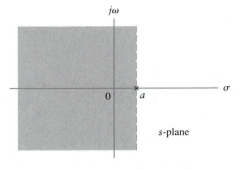

FIGURE 6.5   The ROC for $-e^{at}u(-t)$ is depicted by the shaded region. A pole is located at $s = a$.

where $G(s, t)$ is the indefinite integral

$$G(s, t) = \int g(t)e^{-st}\,dt$$

Next, we write

$$
\begin{aligned}
Y(s) &= -\int_{-\infty}^{0} g(t)e^{-st}\,dt \\
&= \int_{0}^{-\infty} g(t)e^{-st}\,dt \\
&= G(s, -\infty) - G(s, 0)
\end{aligned}
$$

We see that $X(s) = Y(s)$ whenever $G(s, \infty) = G(s, -\infty)$. In Examples 6.1 and 6.2 we have $G(s, -\infty) = G(s, \infty) = 0$. The values of s for which $G(s, \infty)$ is finite are different from those for which $G(s, -\infty)$ is finite, and thus the ROCs are different. *The ROC must be specified for the Laplace transform to be unique.*

▶ **Drill Problem 6.1**   Determine the Laplace transform and ROC of

$$x(t) = u(t - 5)$$

Answer:

$$X(s) = \frac{e^{-5s}}{s}, \qquad \text{Re}(s) > 0 \qquad\qquad\qquad ◀$$

▶ **Drill Problem 6.2**   Determine the Laplace transform, ROC, and pole and zero locations of $X(s)$ for

$$x(t) = e^{j\omega_o t}u(t)$$

Answer:

$$X(s) = \frac{1}{s - j\omega_o}, \qquad \text{Re}(s) > 0$$

Pole at $s = j\omega_o$.   ◀

6.3   *The Unilateral Laplace Transform*

There are many applications of Laplace transforms in which it is reasonable to assume that the signals involved are causal, that is, zero for times $t < 0$. For example, if we apply an input that is zero for time $t < 0$ to a causal system, the output will also be zero for $t < 0$. Also, the choice of time origin is arbitrary in many system problems. Thus time $t = 0$ is often chosen as the time at which an input is presented to the system, and the behavior of the system at time $t \geq 0$ is of interest. In such problems it is advantageous to define the unilateral or one-sided Laplace transform, which is based on only the positive time $(t > 0)$ portions of a signal. By working with causal signals, we remove the ambiguity inherent in the bilateral transform and thus do not need to consider the ROC. Also, the differentiation property for the unilateral Laplace transform may be used to analyze the behavior of a causal system described by a differential equation with initial conditions. This is the most common use for the unilateral transform in engineering applications.

■ DEFINITION

The *unilateral Laplace transform* of a signal $x(t)$ is defined by

$$X(s) = \int_{0^+}^{\infty} x(t)e^{-st}\, dt \tag{6.9}$$

The lower limit of 0^+ implies that we do not include the point $t = 0$ in the integral. Hence $X(s)$ depends only on $x(t)$ for $t > 0$. Discontinuities and impulses at $t = 0$ are excluded. The inverse unilateral transform is given by Eq. (6.6). We shall denote the relationship between $X(s)$ and $x(t)$ as

$$x(t) \overset{\mathscr{L}_u}{\longleftrightarrow} X(s)$$

where the subscript u in $\mathscr{L}_u$ denotes the unilateral transform. The unilateral and bilateral Laplace transforms are naturally equivalent for signals that are zero for times $t < 0$. For example,

$$e^{at}u(t) \overset{\mathscr{L}_u}{\longleftrightarrow} \frac{1}{s - a}$$

is equivalent to

$$e^{at}u(t) \overset{\mathscr{L}}{\longleftrightarrow} \frac{1}{s - a} \quad \text{with ROC Re}\{s\} > a$$

■ PROPERTIES

The properties of the Laplace transform are similar to those of the Fourier transform. Hence we simply state many of the properties. Proofs of some of these are given as problems at the end of the chapter. The properties described in this subsection specifically apply to the unilateral Laplace transform. The unilateral and bilateral transforms have many properties in common, although there are important differences. These differences are discussed in Section 6.6. In the properties given below, we assume that

$$x(t) \overset{\mathscr{L}_u}{\longleftrightarrow} X(s)$$

and

$$y(t) \overset{\mathscr{L}_u}{\longleftrightarrow} Y(s)$$

Linearity

$$ax(t) + by(t) \overset{\mathscr{L}_u}{\longleftrightarrow} aX(s) + bY(s) \tag{6.10}$$

The linearity of the Laplace transform follows from its definition as an integral and the fact that integration is a linear operation.

Scaling

$$x(at) \overset{\mathscr{L}_u}{\longleftrightarrow} \frac{1}{|a|} X\!\left(\frac{s}{a}\right) \tag{6.11}$$

Scaling in time introduces the inverse scaling in s.

Time Shift

$$x(t - \tau) \xleftrightarrow{\mathscr{L}_u} e^{-s\tau}X(s) \quad \text{for all } \tau \text{ such that } x(t - \tau)u(t) = x(t - \tau)u(t - \tau) \quad (6.12)$$

A shift of τ in time corresponds to multiplication by the complex exponential $e^{-s\tau}$. The restriction on the shift arises because the unilateral transform is defined solely in terms of the positive-time portions of the signal. Hence this property applies only if the shift does not move a nonzero positive-time component of the signal to negative time, as depicted in Fig. 6.6(a), or move a nonzero negative-time portion of the signal to positive time, as depicted in Fig. 6.6(b). The time-shift property is most commonly applied to causal signals $x(t)$ with shifts $\tau > 0$, in which case the shift restriction is always satisfied.

s-Domain Shift

$$e^{s_o t}x(t) \xleftrightarrow{\mathscr{L}_u} X(s - s_o) \quad (6.13)$$

Multiplication by a complex exponential in time introduces a shift in complex frequency s to the Laplace transform.

Convolution

$$x(t) * y(t) \xleftrightarrow{\mathscr{L}_u} X(s)Y(s) \quad (6.14)$$

Convolution in time corresponds to multiplication of Laplace transforms.

Differentiation in the s-Domain

$$-tx(t) \xleftrightarrow{\mathscr{L}_u} \frac{d}{ds} X(s) \quad (6.15)$$

Differentiation in the s-domain corresponds to multiplication by $-t$ in the time domain.

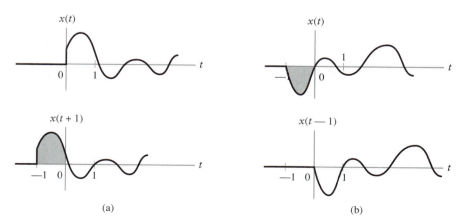

(a)   (b)

FIGURE 6.6   Time shifts for which the unilateral Laplace transform time-shift property does not apply. (a) A nonzero portion of $x(t)$ that occurs at times $t > 0$ is shifted to times $t < 0$. (b) A nonzero portion of $x(t)$ that occurs at times $t < 0$ is shifted to times $t > 0$.

EXAMPLE 6.3   Find the unilateral Laplace transform of

$$x(t) = (-e^{3t}u(t)) * (tu(t))$$

Solution: First we note that

$$-e^{3t}u(t) \xleftrightarrow{\mathcal{L}_u} \frac{-1}{s-3}$$

and

$$u(t) \xleftrightarrow{\mathcal{L}_u} \frac{1}{s}$$

Apply the s-domain differentiation property to obtain

$$tu(t) \xleftrightarrow{\mathcal{L}_u} \frac{1}{s^2}$$

Now use the convolution property to obtain

$$x(t) = (-e^{3t}u(t)) * (tu(t)) \xleftrightarrow{\mathcal{L}_u} X(s) = \frac{-1}{s^2(s-3)}$$

▶ **Drill Problem 6.3**   Find the unilateral Laplace transform of

$$x(t) = e^{-t}(t-2)u(t-2)$$

Answer:

$$X(s) = \frac{e^{-2(s+1)}}{(s+1)^2}$$

◀

Differentiation in the Time Domain

Consider the unilateral Laplace transform of $dx(t)/dt$. By definition, we have

$$\frac{d}{dt}x(t) \xleftrightarrow{\mathcal{L}_u} \int_{0^+}^{\infty} \left(\frac{d}{dt}x(t)\right)e^{-st}\,dt$$

Integrating by parts, we obtain

$$\frac{d}{dt}x(t) \xleftrightarrow{\mathcal{L}_u} x(t)e^{-st}\Big|_{0^+}^{\infty} + s\int_{0^+}^{\infty} x(t)e^{-st}\,dt$$

Since $X(s)$ exists, we know that $x(t)e^{-st}|_{t=\infty} = 0$. Furthermore, the integral corresponds to the unilateral Laplace transform definition of Eq. (6.9), and so we have

$$\frac{d}{dt}x(t) \xleftrightarrow{\mathcal{L}_u} sX(s) - x(0^+) \qquad (6.16)$$

EXAMPLE 6.4   Verify the differentiation property of Eq. (6.16) for the signal $x(t) = e^{at}u(t)$.

Solution: The derivative of $x(t)$ is

$$\frac{d}{dt}e^{at} = ae^{at}, \; t > 0$$

The unilateral Laplace transform of $ae^{at}u(t)$ is a times the unilateral Laplace transform of $e^{at}u(t)$; that is,

$$\frac{d}{dt}x(t) = ae^{at}u(t) \overset{\mathcal{L}_u}{\longleftrightarrow} \frac{a}{s-a}$$

Next, let us rederive this result using the differentiation property. We have

$$\frac{d}{dt}x(t) \overset{\mathcal{L}_u}{\longleftrightarrow} sX(s) - x(0^+) = s\frac{1}{s-a} - 1$$

$$= \frac{s-(s-a)}{s-a}$$

$$= \frac{a}{s-a}$$

The general form for the differentiation property is

$$\frac{d^n}{dt^n}x(t) \overset{\mathcal{L}_u}{\longleftrightarrow} s^nX(s) - \left.\frac{d^{n-1}}{dt^{n-1}}x(t)\right|_{t=0^+} - s\left.\frac{d^{n-2}}{dt^{n-2}}x(t)\right|_{t=0^+}$$

$$- \cdots - s^{n-2}\left.\frac{d}{dt}x(t)\right|_{t=0^+} - s^{n-1}x(0^+) \tag{6.17}$$

Integration Property

$$\int_{-\infty}^{t}x(\tau)\,d\tau \overset{\mathcal{L}_u}{\longleftrightarrow} \frac{x^{(-1)}(0^+)}{s} + \frac{X(s)}{s} \tag{6.18}$$

where

$$x^{(-1)}(0^+) = \int_{-\infty}^{0^+}x(\tau)\,d\tau$$

is the area under $x(t)$ from $t = -\infty$ to $t = 0^+$.

▶ **Drill Problem 6.4** Use the integration property to show that the unilateral Laplace transform of

$$tu(t) = \int_{-\infty}^{t}u(\tau)\,d\tau$$

is given by $1/s^2$. ◀

Initial and Final Value Theorems

The initial and final value theorems allow us to determine the initial value of $x(t)$, $x(0^+)$, and the final value, $x(\infty)$, directly from $X(s)$. These theorems are most often used to evaluate either the initial or final values of a system output without explicitly determining the entire time response of the system. The initial value theorem states that

$$\lim_{s\to\infty} sX(s) = x(0^+) \tag{6.19}$$

The initial value theorem does not apply to rational functions $X(s)$ whose numerator polynomial order is greater than the denominator polynomial order. The final value theorem states that

$$\lim_{s \to 0} sX(s) = x(\infty) \qquad (6.20)$$

The final value theorem applies only if all the poles of $X(s)$ are in the left half of the s-plane, with at most a single pole at $s = 0$.

EXAMPLE 6.5   Determine the initial and final values of a signal $x(t)$ whose unilateral Laplace transform is

$$X(s) = \frac{7s + 10}{s(s + 2)}$$

Solution: We may apply the initial value theorem to obtain

$$x(0^+) = \lim_{s \to \infty} s \frac{7s + 10}{s(s + 2)}$$

$$= \lim_{s \to \infty} \frac{7s + 10}{s + 2}$$

$$= 7$$

The final value theorem is applicable since $X(s)$ has only a single pole at $s = 0$ and the remaining poles are in the left half of the s-plane. We have

$$x(\infty) = \lim_{s \to 0} s \frac{7s + 10}{s(s + 2)}$$

$$= \lim_{s \to 0} \frac{7s + 10}{s + 2}$$

$$= 5$$

The reader may verify these results by showing the inverse Laplace transform of $X(s)$ is $x(t) = 5u(t) + 2e^{-2t}u(t)$.

▶ **Drill Problem 6.5**   Find the initial and final values of $x(t)$ if

$$X(s) = e^{-5s}\left(\frac{-2}{s(s + 2)}\right)$$

Answer:

$$x(0^+) = 0$$
$$x(\infty) = -1 \qquad \blacktriangleleft$$

6.4 *Inversion of the Laplace Transform*

Direct inversion of the inverse Laplace transform given in Eq. (6.6) requires an understanding of contour integration and will not be pursued here. We shall determine inverse Laplace transforms using the one-to-one relationship between a signal and its unilateral Laplace transform. Given knowledge of several basic transform pairs and the Laplace transform properties, we are able to invert a very large class of Laplace transforms in this manner. A table of basic Laplace transform pairs is given in Appendix D.1.

In the study of LTI systems, we frequently encounter Laplace transforms that are a ratio of polynomials in s. In this case, the inverse transform is obtained by expressing $X(s)$ as a sum of terms for which we already know the time function. Suppose

$$X(s) = \frac{B(s)}{A(s)}$$

$$= \frac{b_M s^M + b_{M-1} s^{M-1} + \cdots + b_1 s + b_0}{s^N + a_{N-1} s^{N-1} + \cdots + a_1 s + a_0}$$

where $M < N$. The case $M \geq N$ does not apply to the unilateral Laplace transform, since that implies terms of the form $c_k s^k$, $k = 0, 1, \ldots, M - N$, are contained in $X(s)$. Such terms correspond to impulses and their derivatives located at time $t = 0$, which are excluded from the unilateral transform as defined in Eq. (6.9).

We begin by factoring the denominator polynomial as a product of poles to obtain

$$X(s) = \frac{b_M s^M + b_{M-1} s^{M-1} + \cdots + b_1 s + b_0}{\Pi_{k=1}^N (s - d_k)}$$

If all the poles d_k are distinct, then we may rewrite $X(s)$ as a sum of simple terms using a partial fraction expansion as follows:

$$X(s) = \sum_{k=1}^N \frac{A_k}{s - d_k}$$

The A_k are determined using the method of residues or by solving a system of linear equations as described in Appendix B. The inverse Laplace transform of each term in the sum may now be found using the pair

$$A_k e^{d_k t} u(t) \xleftarrow{\quad \mathcal{L}_u \quad} \frac{A_k}{s - d_k}$$

If a pole d_i is repeated r times, then there are r terms in the partial fraction expansion associated with this pole. They are

$$\frac{A_{i_1}}{s - d_i}, \frac{A_{i_2}}{(s - d_i)^2}, \ldots, \frac{A_{i_r}}{(s - d_i)^r}$$

The inverse Laplace transform of each term is found using the pair

$$\frac{A t^{n-1}}{(n - 1)!} e^{d_k t} u(t) \xleftarrow{\quad \mathcal{L}_u \quad} \frac{A}{(s - d_k)^n}$$

EXAMPLE 6.6 Find the inverse Laplace transform of

$$X(s) = \frac{3s + 4}{(s + 1)(s + 2)^2}$$

Solution: Use a partial fraction expansion of $X(s)$ to write

$$X(s) = \frac{A_1}{s + 1} + \frac{A_2}{s + 2} + \frac{A_3}{(s + 2)^2}$$

Solving for A_1, A_2, and A_3 using the method of residues, we obtain

$$X(s) = \frac{1}{s + 1} - \frac{1}{s + 2} + \frac{2}{(s + 2)^2}$$

We may construct $x(t)$ from the inverse Laplace transform of each term in the partial fraction expansion:

► The pole of the first term is at $s = -1$:

$$e^{-t}u(t) \xleftrightarrow{\mathscr{L}_u} \frac{1}{s+1}$$

► The second term has a pole at $s = -2$:

$$-e^{-2t}u(t) \xleftrightarrow{\mathscr{L}_u} -\frac{1}{s+2}$$

► The double pole in the last term is at $s = -2$:

$$2te^{-2t}u(t) \xleftrightarrow{\mathscr{L}_u} \frac{2}{(s+2)^2}$$

Combining these three terms, we obtain

$$x(t) = e^{-t}u(t) - e^{-2t}u(t) + 2te^{-2t}u(t)$$

► **Drill Problem 6.6** Find the inverse Laplace transform of

$$X(s) = \frac{-5s - 7}{(s+1)(s-1)(s+2)}$$

Answer:

$$x(t) = e^{-t}u(t) - 2e^{t}u(t) + e^{-2t}u(t) \qquad \blacktriangleleft$$

If the coefficients in the denominator polynomial are real, then all the complex poles occur in complex-conjugate pairs. The partial fraction expansion procedure described above is applicable with either real or complex poles. A complex pole usually results in complex-valued expansion coefficients and a complex exponential function of time. In cases where $X(s)$ has real-valued coefficients and thus corresponds to a real-valued time signal, we may simplify the algebra by combining complex-conjugate poles in the partial fraction expansion in such a way as to ensure real-valued expansion coefficients and a real-valued inverse transform. This is accomplished by combining all pairs of complex-conjugate poles into quadratic terms with real coefficients. The inverse Laplace transforms of these quadratic terms are exponentially damped sinusoids.

Suppose $\alpha + j\omega_o$ and $\alpha - j\omega_o$ are a pair of complex-conjugate poles. The two first-order terms associated with these in the partial fraction expansion are written as

$$\frac{A_1}{s - \alpha - j\omega_o} + \frac{A_2}{s - \alpha + j\omega_o}$$

In order for this sum to represent a real-valued signal, A_1 and A_2 must be the complex conjugate of each other. Hence we may replace these two terms with the single quadratic term

$$\frac{B_1 s + B_2}{(s - \alpha - j\omega_o)(s - \alpha + j\omega_o)} = \frac{B_1 s + B_2}{(s - \alpha)^2 + \omega_o^2}$$

having real-valued B_1 and B_2. We then solve for B_1 and B_2 and factor the result into the sum of two quadratic terms for which the inverse Laplace transforms are known. That is, we write

$$\frac{B_1 s + B_2}{(s - \alpha)^2 + \omega_o^2} = \frac{C_1(s - \alpha)}{(s - \alpha)^2 + \omega_o^2} + \frac{C_2\omega_o}{(s - \alpha)^2 + \omega_o^2}$$

where $C_1 = B_1$ and $C_2 = (B_1\alpha + B_2)/\omega_o$. The inverse Laplace transform of the term

$$\frac{C_1(s - \alpha)}{(s - \alpha)^2 + \omega_o^2}$$

is

$$C_1 e^{\alpha t} \cos(\omega_o t)u(t)$$

Likewise, the inverse Laplace transform of

$$\frac{C_2\omega_o}{(s - \alpha)^2 + \omega_o^2}$$

is

$$C_2 e^{\alpha t} \sin(\omega_o t)u(t)$$

The following example illustrates this approach.

EXAMPLE 6.7 Find the inverse Laplace transform of

$$X(s) = \frac{4s^2 + 6}{s^3 + s^2 - 2}$$

Solution: There are three poles in $X(s)$. By trial and error we find that $s = 1$ is a pole. We factor $s - 1$ out of $s^3 + s^2 - 2$ to obtain $s^2 + 2s + 2 = 0$ as the equation defining the remaining two poles. Rooting this quadratic equation gives the complex-conjugate poles $s = -1 \pm j$.

We may write the quadratic equation $s^2 + 2s + 2$ in terms of the perfect square $(s^2 + 2s + 1) + 1 = (s + 1)^2 + 1$, and so the partial fraction expansion for $X(s)$ takes the form

$$X(s) = \frac{A}{s - 1} + \frac{B_1 s + B_2}{(s + 1)^2 + 1} \tag{6.21}$$

The expansion coefficient A is easily obtained by the method of residues. That is, multiply both sides of Eq. (6.21) by $(s - 1)$ and evaluate at $s = 1$ to obtain

$$A = X(s)(s - 1)|_{s=1}$$

$$= \frac{4s^2 + 6}{(s + 1)^2 + 1}\bigg|_{s=1}$$

$$= 2$$

The remaining expansion coefficients B_1 and B_2 are obtained by placing both terms in Eq. (6.21) over a common denominator and equating the numerator to the numerator of $X(s)$. We may thus write

$$4s^2 + 6 = 2((s + 1)^2 + 1) + (B_1 s + B_2)(s - 1)$$

$$= (2 + B_1)s^2 + (4 - B_1 + B_2)s + (4 - B_2)$$

Equating coefficients on s^2 gives $B_1 = 2$, and equating the coefficients of s^0 gives $B_2 = -2$. Thus

$$X(s) = \frac{2}{s-1} + \frac{2s-2}{(s+1)^2 + 1}$$

$$= \frac{2}{s-1} + 2\frac{s+1}{(s+1)^2 + 1} - 4\frac{1}{(s+1)^2 + 1}$$

In arriving at the second equation we have factored $2s - 2$ as $2(s + 1) - 4$. Now take the inverse Laplace transform of each term. Putting these results together, we obtain

$$x(t) = 2e^t u(t) + 2e^{-t}\cos(t)u(t) - 4e^{-t}\sin(t)u(t)$$

▶ **Drill Problem 6.7** Find the inverse Laplace transform of

$$X(s) = \frac{s^2 + s - 2}{s^3 + 3s^2 + 5s + 3}$$

Answer:

$$x(t) = -e^{-t}u(t) + 2e^{-t}\cos(\sqrt{2}t)u(t) - \frac{1}{\sqrt{2}}e^{-t}\sin(\sqrt{2}t)u(t) \qquad ◀$$

The poles of $X(s)$ determine the inherent characteristics of the signal $x(t)$. A pole at $s = d_k$ results in an exponentially damped sinusoidal term of the form $e^{d_k t}u(t)$. Letting $d_k = \sigma_k + j\omega_k$, we may write this term as $e^{\sigma_k t}e^{j\omega_k t}u(t)$. Hence the real part of the pole determines the exponential damping factor, and the imaginary part determines the sinusoidal frequency. If $\mathrm{Re}\{d_k\}$ is large and negative, then the term decays very quickly. If $\mathrm{Im}\{d_k\}$ is large, then the term oscillates rapidly. With these properties in mind, we can infer much about the characteristics of a signal from its pole locations in the s-plane.

6.5 *Solving Differential Equations with Initial Conditions*

The primary application of the unilateral Laplace transform in systems analysis is solving differential equations with nonzero initial conditions. The initial conditions are incorporated into the solution as the values of the signal and its derivatives that occur at time zero in the differentiation property of Eq. (6.17). This is illustrated by way of an example.

EXAMPLE 6.8 Use the Laplace transform to find the output of the system described by the differential equation

$$\frac{d}{dt}y(t) + 5y(t) = x(t)$$

in response to the input $x(t) = 3e^{-2t}u(t)$ and initial condition $y(0^+) = -2$.

Solution: Take the unilateral Laplace transform of each side of the differential equation and apply the differentiation property of Eq. (6.16) to obtain

$$sY(s) - y(0^+) + 5Y(s) = X(s)$$

Solving for $Y(s)$, we get

$$Y(s) = \frac{1}{s + 5} [X(s) + y(0^+)]$$

Now substitute $X(s) = 3/(s + 2)$ and the initial condition $y(0^+) = -2$, obtaining

$$Y(s) = \frac{3}{(s + 2)(s + 5)} + \frac{-2}{s + 5}$$

Expanding $Y(s)$ in partial fractions,

$$Y(s) = \frac{1}{s + 2} + \frac{-1}{s + 5} + \frac{-2}{s + 5}$$

and taking the inverse unilateral Laplace transform, we get

$$y(t) = e^{-2t}u(t) - 3e^{-5t}u(t)$$

The Laplace transform method for solving differential equations offers clear separation between the natural response of the system to initial conditions and the forced response of the system associated with the input. Taking the unilateral Laplace transform of both sides of the general differential equation

$$a_N \frac{d^N}{dt^N} y(t) + a_{N-1} \frac{d^{N-1}}{dt^{N-1}} y(t) + \cdots + a_1 \frac{d}{dt} y(t) + a_0 y(t)$$

$$= b_M \frac{d^M}{dt^M} x(t) + b_{M-1} \frac{d^{M-1}}{dt^{M-1}} x(t) + \cdots + b_1 \frac{d}{dt} x(t) + b_0 x(t)$$

we obtain

$$A(s)Y(s) - C(s) = B(s)X(s) - D(s)$$

where

$$A(s) = a_N s^N + a_{N-1} s^{N-1} + \cdots + a_1 s + a_0$$
$$B(s) = b_M s^M + b_{M-1} s^{M-1} + \cdots + b_1 s + b_0$$
$$C(s) = \sum_{k=1}^{N} \sum_{l=0}^{k-1} a_k s^{k-1-l} \frac{d^l}{dt^l} y(t)\big|_{t=0^+}$$

and

$$D(s) = \sum_{k=1}^{M} \sum_{l=0}^{k-1} b_k s^{k-1-l} \frac{d^l}{dt^l} x(t)\big|_{t=0^+}$$

We note that $C(s) = 0$ if all the initial conditions are zero and $B(s)X(s) - D(s) = 0$ if the input is zero. We may now write $Y(s)$ as

$$Y(s) = \frac{B(s)X(s) - D(s)}{A(s)} + \frac{C(s)}{A(s)}$$
$$= Y^{(f)}(s) + Y^{(n)}(s)$$

where

$$Y^{(f)}(s) = \frac{B(s)X(s) - D(s)}{A(s)} \quad \text{and} \quad Y^{(n)}(s) = \frac{C(s)}{A(s)}$$

The first term, $Y^{(f)}(s)$, represents the component of the response associated entirely with the input, or the *forced response* of the system. It represents the output when the initial conditions are zero. The second term, $Y^{(n)}(s)$, represents the component of the output due to the initial conditions, or the *natural response* of the system. It represents the system output when the input is zero.

EXAMPLE 6.9   Use the Laplace transform to determine the output of a system represented by the differential equation

$$\frac{d^2}{dt^2}\,y(t) + 5\,\frac{d}{dt}\,y(t) + 6y(t) = 2\,\frac{d}{dt}\,x(t) + x(t)$$

in response to the input $x(t) = u(t)$. Assume that the initial conditions on the system are

$$y(0^+) = 1 \quad \text{and} \quad \frac{d}{dt}\,y(t)\big|_{t=0^+} = 2$$

Identify the forced response of the system, $y^{(f)}(t)$, and the natural response, $y^{(n)}(t)$.

Solution: Taking the unilateral Laplace transform of both sides of the differential equation, we obtain

$$(s^2 + 5s + 6)Y(s) - sy(0^+) - \frac{d}{dt}\,y(t)\big|_{t=0^+} - 5y(0^+) = (2s + 1)X(s) - 2x(0^+)$$

Solving for $Y(s)$, we get

$$Y(s) = \frac{(2s + 1)X(s) - 2x(0^+)}{s^2 + 5s + 6} + \frac{sy(0^+) + \dfrac{d}{dt}\,y(t)\big|_{t=0^+} + 5y(0^+)}{s^2 + 5s + 6}$$

The first term is associated with the forced response of the system, $Y^{(f)}(s)$. The second term corresponds to the natural response, $Y^{(n)}(s)$. Substituting for $X(s)$, $x(0^+)$, and the initial conditions, we obtain

$$Y^{(f)}(s) = \frac{1}{s(s + 2)(s + 3)}$$

and

$$Y^{(n)}(s) = \frac{s + 7}{(s + 2)(s + 3)}$$

Expanding both of these in a partial fraction expansion yields

$$Y^{(f)}(s) = \frac{\frac{1}{6}}{s} + \frac{-\frac{1}{2}}{s + 2} + \frac{\frac{1}{3}}{s + 3}$$

$$Y^{(n)}(s) = \frac{5}{s + 2} + \frac{-4}{s + 3}$$

Next, taking the inverse unilateral Laplace transforms of $Y^{(f)}(s)$ and $Y^{(n)}(s)$, we obtain

$$y^{(f)}(t) = \tfrac{1}{6}u(t) - \tfrac{1}{2}e^{-2t}u(t) + \tfrac{1}{3}e^{-3t}u(t)$$

and

$$y^{(n)}(t) = 5e^{-2t}u(t) - 4e^{-3t}u(t)$$

The output of the system is $y(t) = y^{(f)}(t) + y^{(n)}(t)$. Figures 6.7(a), (b), and (c) depict the forced response, natural response, and system output, respectively.

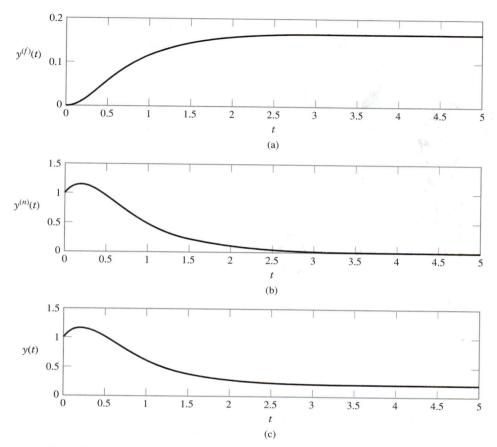

FIGURE 6.7   Solution to Example 6.9. (a) Forced response of the system, $y^f(t)$. (b) Natural response of the system, $y^n(t)$. (c) Overall system output.

EXAMPLE 6.10   The mechanical system depicted in Fig. 6.8 is governed by the differential equation

$$m \frac{d^2}{dt^2} y(t) + f \frac{d}{dt} y(t) + ky(t) = x(t)$$

Here $x(t)$ is the applied force and $y(t)$ is the position of the mass. Find the forced and natural response of this system for $m = 2$ kg, $f = 4 \frac{N \cdot s}{m}$, and $k = 202 \frac{N}{m}$ assuming the initial position is $y(0^+) = 5$ m, the initial velocity is $\left. \frac{dy(t)}{dt} \right|_{t=0^+} = -5 \frac{m}{s}$, and the input is $x(t) = 2[u(t) - u(t-1)]$ N.

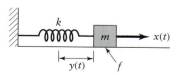

FIGURE 6.8   Mechanical system with input $x(t)$ and output $y(t)$.

Solution: Take the unilateral Laplace transform of both sides of the differential equation and rearrange terms to identify

$$Y^{(f)}(s) = \frac{X(s)}{2s^2 + 4s + 202}$$

$$Y^{(n)}(s) = \frac{(2s + 4)y(0^+) + 2\frac{d}{dt}y(t)|_{t=0^+}}{2s^2 + 4s + 202}$$

Substituting for the initial conditions and simplifying, we obtain

$$Y^{(n)}(s) = \frac{5(s + 1)}{(s + 1)^2 + 10^2}$$

Now take the inverse unilateral transform of $Y^{(n)}(s)$ to obtain the natural response as shown by

$$y^{(n)}(t) = 5e^{-t}\cos(10t)u(t)$$

Next we substitute for $X(s)$ to obtain the Laplace transform of the forced response as

$$Y^{(f)}(s) = (1 - e^{-s})\frac{1}{s(s^2 + 2s + 101)}$$

Perform a partial fraction expansion of $Y^{(f)}(s)$ to obtain

$$Y^{(f)}(s) = (1 - e^{-s})\frac{1}{101}\left[\frac{1}{s} - \frac{(s + 1)}{(s + 1)^2 + 10^2} - \frac{1}{10}\frac{10}{(s + 1)^2 + 10^2}\right]$$

The e^{-s} term introduces a time delay of 1 s to each term in the partial fraction expansion. Taking the inverse unilateral Laplace transform, we obtain

$$y^{(f)}(t) = \frac{1}{101}\left[u(t) - u(t - 1) - e^{-t}\cos(10t)u(t) + e^{-(t-1)}\cos(10(t - 1))u(t - 1)\right.$$

$$\left. - \frac{1}{10}e^{-t}\sin(10t)u(t) + \frac{1}{10}e^{-(t-1)}\sin(10(t - 1))u(t - 1)\right]$$

▶ **Drill Problem 6.8** Determine the forced response, natural response, and output of a system described by the differential equation

$$\frac{d}{dt}y(t) + 3y(t) = 4x(t)$$

in response to the input $x(t) = \cos(2t)u(t)$ and initial condition $y(0^+) = -2$.

Answer:

$$y^{(f)}(t) = -\tfrac{12}{13}e^{-3t}u(t) + \tfrac{12}{13}\cos(2t)u(t) + \tfrac{8}{13}\sin(2t)u(t)$$

$$y^{(n)}(t) = -2e^{-3t}u(t)$$

and $y(t) = y^{(f)}(t) + y^{(n)}(t)$.   ◀

The natural response of the system is obtained from a partial fraction expansion using the poles of $C(s)/A(s)$, which are the roots of $A(s)$. For this reason, these roots are sometimes termed the *natural frequencies* of the system. Note that if the system is stable, then the roots of $A(s)$ must have negative real parts; that is, they must lie in the left half of the s-plane.

The differentiation and integration properties may also be used to transform circuits involving capacitive and inductive elements so that the circuit may be solved directly in terms of Laplace transforms rather than by first writing the differential equation in the time domain. This is accomplished by replacing resistive, capacitive, and inductive elements by their Laplace transform equivalents.

A resistance R with corresponding voltage $v_R(t)$ and current $i_R(t)$ satisfies the relation

$$v_R(t) = Ri_R(t)$$

Transforming this equation, we write

$$V_R(s) = RI_R(s) \tag{6.22}$$

which is represented by the transformed resistive element of Fig. 6.9(a). Next, we consider an inductor, for which

$$v_L(t) = L \frac{d}{dt} i_L(t)$$

Transforming this relationship and using the differentiation property yields

$$V_L(s) = sLI_L(s) - Li_L(0^+) \tag{6.23}$$

This relationship is represented by the transformed inductive element of Fig. 6.9(b). We now consider a capacitor, which may be described by

$$v_C(t) = \frac{1}{C} \int_{0^+}^{t} i_C(\tau) \, d\tau + v_C(0^+)$$

Transforming and using the integration property we obtain

$$V_C(s) = \frac{1}{sC} I_C(s) + \frac{v_C(0^+)}{s} \tag{6.24}$$

Figure 6.9(c) depicts the transformed capacitive element described by this equation.

The circuit models corresponding to Eqs. (6.22), (6.23), and (6.24) are most useful when applying Kirchhoff's voltage law to solve a circuit. If Kirchhoff's current law is to be used, then it is more convenient to rewrite Eqs. (6.22), (6.23), and (6.24) to express current as a function of voltage. This results in the transformed circuit elements depicted in Fig. 6.10. The following example illustrates the Laplace transform method for solving an electrical circuit.

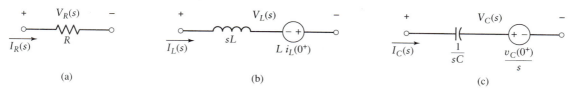

FIGURE 6.9   Laplace transform circuit models for use with Kirchhoff's voltage law. (a) Resistor. (b) Inductor with initial current $i_L(0^+)$. (c) Capacitor with initial voltage $v_C(0^+)$.

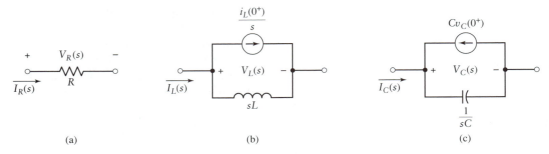

FIGURE 6.10   Laplace transform circuit models for use with Kirchhoff's current law. (a) Resistor. (b) Inductor with initial current $i_L(0^+)$. (c) Capacitor with initial voltage $v_C(0^+)$.

EXAMPLE 6.11   Use Laplace transform circuit models to determine the voltage $y(t)$ in the circuit of Fig. 6.11(a) for an applied voltage $x(t) = 2e^{-10t}u(t)$ V. The voltage across the capacitor at time $t = 0^+$ is 5 V.

Solution: The transformed circuit is drawn in Fig. 6.11(b) with symbols $I_1(s)$ and $I_2(s)$ representing the current through each branch. We may write the following equations to describe the circuit:

$$Y(s) = 1000(I_1(s) + I_2(s))$$

$$X(s) = Y(s) + \frac{1}{s10^{-4}} I_1(s) + \frac{5}{s}$$

and

$$X(s) = Y(s) + 1000I_2(s)$$

Combining these three equations to eliminate $I_1(s)$ and $I_2(s)$, we may write

$$Y(s) = X(s)\frac{s + 10}{s + 20} - \frac{5}{s + 20}$$

Substituting for $X(s)$ we obtain

$$Y(s) = \frac{-3}{s + 20}$$

and thus we conclude that

$$y(t) = -3e^{-20t}u(t) \quad \text{V}$$

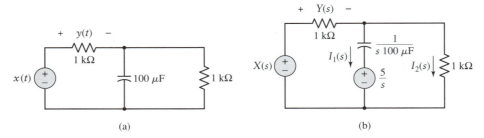

FIGURE 6.11   Electrical circuit for Example 6.11. (a) Original circuit. (b) Transformed circuit.

The natural and forced responses of a circuit are easily determined using the transformed circuit representation. The natural response is obtained by setting the voltage or current source associated with the input equal to zero. In this case the only voltage or current sources in the circuit are those associated with the initial conditions in the transformed capacitive and inductive circuit models. The forced response is obtained by setting the initial conditions equal to zero, which eliminates the voltage or current sources present in the transformed capacitive and inductive circuit models.

6.6   *The Bilateral Laplace Transform*

The bilateral Laplace transform involves the values of the signal $x(t)$ for both $t > 0$ and $t \leq 0$, as shown by

$$x(t) \xleftrightarrow{\mathscr{L}} X(s) = \int_{-\infty}^{\infty} x(t) e^{-st}\, dt$$

Hence the bilateral Laplace transform is well suited to problems involving noncausal signals and systems. These applications are studied in the following section. As we discovered in Section 6.2, the bilateral Laplace transform is not unique unless the ROC is specified. In this section we examine the properties of the ROC, note important differences between unilateral and bilateral Laplace transform properties, and discuss inversion of the bilateral Laplace transform.

■   *PROPERTIES OF THE REGION OF CONVERGENCE

In this subsection we show how the ROC is related to the characteristics of a signal $x(t)$. We develop these properties using intuitive arguments rather than rigorous proofs. Once we know the ROC properties, we can often identify the ROC from knowledge of the Laplace transform $X(s)$ and limited knowledge of the characteristics of $x(t)$.

First, we note that the ROC cannot contain any poles. If the Laplace transform converges, then $X(s)$ is finite over the entire ROC. Suppose d is a pole of $X(s)$. This implies $X(d) = \pm\infty$, and so the Laplace transform does not converge at d. Thus, d cannot lie in the ROC.

Next, convergence of the Laplace transform for a signal $x(t)$ implies that

$$I(\sigma) = \int_{-\infty}^{\infty} |x(t)| e^{-\sigma t}\, dt < \infty$$

for some values of σ. The set of σ for which this integral is finite determines the ROC. The quantity σ is the real part of s, so the ROC depends only on the real part of s; the imaginary component of s does not affect convergence. This implies that the ROC consists of strips parallel to the $j\omega$-axis in the s-plane.

Suppose $x(t)$ is a finite-duration signal, that is, $x(t) = 0$ for $t < a$ and $t > b$. If we can find a finite bounding constant A such that $|x(t)| \leq A$, then

$$I(\sigma) \leq \int_{a}^{b} A e^{-\sigma t}\, dt$$

$$\leq \begin{cases} (-A/\sigma) e^{-\sigma t} \Big|_{a}^{b}, & \sigma \neq 0 \\ A(b - a), & \sigma = 0 \end{cases}$$

In this case we see that $I(\sigma)$ is finite for all finite values of σ and we conclude that the ROC for a finite-duration signal includes the entire s-plane.

Now separate $I(\sigma)$ into positive- and negative-time sections, as shown by

$$I(\sigma) = I_-(\sigma) + I_+(\sigma)$$

where

$$I_-(\sigma) = \int_{-\infty}^0 |x(t)| e^{-\sigma t}\, dt$$

and

$$I_+(\sigma) = \int_0^\infty |x(t)| e^{-\sigma t}\, dt$$

In order for $I(\sigma)$ to be finite, both of these integrals must be finite. This implies that $|x(t)|$ must be bounded in some sense.

Suppose we can bound $|x(t)|$ for both positive and negative t by finding the smallest constants A and σ_p such that

$$|x(t)| \leq A e^{\sigma_p t}, \qquad t > 0$$

and the largest constant σ_n such that

$$|x(t)| \leq A e^{\sigma_n t}, \qquad t < 0$$

A signal $x(t)$ that satisfies these bounds is said to be of *exponential order*. These bounds imply that $|x(t)|$ grows no faster than $e^{\sigma_p t}$ for positive t and $e^{\sigma_n t}$ for negative t. There are signals that are not of exponential order, such as e^{t^2} or t^{3t}, but such signals generally do not arise in the study of physical systems.

Using the exponential-order bounds on $|x(t)|$, we may write

$$I_-(\sigma) \leq A \int_{-\infty}^0 e^{(\sigma_n - \sigma)t}\, dt$$
$$\leq \frac{A}{\sigma_n - \sigma} e^{(\sigma_n - \sigma)t} \Big|_{-\infty}^0$$

and

$$I_+(\sigma) \leq A \int_0^\infty e^{(\sigma_p - \sigma)t}\, dt$$
$$\leq \frac{A}{\sigma_p - \sigma} e^{(\sigma_p - \sigma)t} \Big|_0^\infty$$

We note that $I_-(\sigma)$ is finite whenever $\sigma < \sigma_n$, and $I_+(\sigma)$ is finite whenever $\sigma > \sigma_p$. The quantity $I(\sigma)$ is finite at values σ for which both $I_-(\sigma)$ and $I_+(\sigma)$ are finite. Hence the Laplace transform converges for $\sigma_p < \sigma < \sigma_n$. Note that if $\sigma_p > \sigma_n$, then there are no values σ for which the Laplace transform converges.

We may draw the following conclusions about the ROC for infinite duration signals from the analysis just presented. Define a *left-sided signal* as one for which $x(t) = 0$ for $t > b$, a *right-sided signal* as one for which $x(t) = 0$ for $t < a$, and a *two-sided signal* as one that is infinite in extent in both directions. If $x(t)$ is of exponential order, we may then make the following statements:

▶ The ROC of a left-sided signal is of the form $\sigma < \sigma_n$.
▶ The ROC of a right-sided signal is of the form $\sigma > \sigma_p$.
▶ The ROC of a two-sided signal is of the form $\sigma_p < \sigma < \sigma_n$.

Each of these cases is illustrated in Fig. 6.12.

Exponential signals are frequently encountered in physical problems. In this case there is a clear relationship between the ROC and the signal. Specifically, the real part of one or more poles determines the ROC boundaries σ_n and σ_p. Suppose we have the right-sided signal, $x(t) = e^{at}u(t)$, where in general a is complex. This signal is of exponential order with the smallest exponential bounding signal $e^{\mathrm{Re}(a)t}$. Hence $\sigma_p = \mathrm{Re}(a)$ and the ROC is $\sigma > \mathrm{Re}(a)$. The Laplace transform of $x(t)$ has a pole at $s = a$, and so the ROC is the region of the s-plane that lies to the right of the pole. Likewise, if $x(t) = e^{at}u(-t)$, then the ROC is $\sigma < \mathrm{Re}(a)$, that is, the region of the s-plane to the left of the pole. If a signal $x(t)$ consists of a sum of exponentials, then the ROC is the intersection of the ROCs associated with each term in the sum. This is demonstrated in the following example.

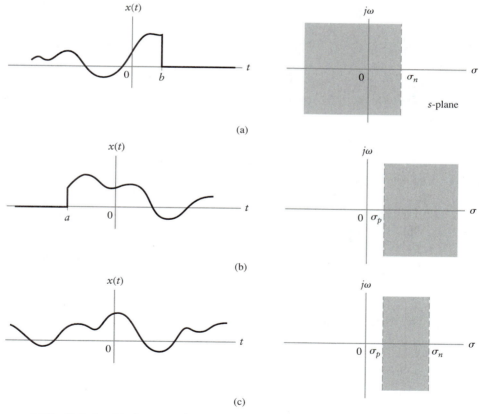

(a)

(b)

(c)

FIGURE 6.12   Relationship between the time extent of a signal and the ROC. (a) A left-sided signal has ROC to the left of a vertical line in the s-plane. (b) A right-sided signal has ROC to the right of a vertical line in the s-plane. (c) A two-sided signal has ROC given by a vertical strip in the s-plane of finite width.

EXAMPLE 6.12   Consider the two signals

$$x_1(t) = e^{-2t}u(t) + e^{-t}u(-t)$$

and

$$x_2(t) = e^{-t}u(t) + e^{-2t}u(-t)$$

Identify the ROC for each signal.

Solution: Check the absolute integrability of $|x_1(t)|e^{-\sigma t}$ by writing

$$
\begin{aligned}
I_1(\sigma) &= \int_{-\infty}^{\infty} |x_1(t)|e^{-\sigma t}\, dt \\
&= \int_{-\infty}^{0} e^{-(1+\sigma)t}\, dt + \int_{0}^{\infty} e^{-(2+\sigma)t}\, dt \\
&= \left. \frac{-1}{1+\sigma} e^{-(1+\sigma)t} \right|_{-\infty}^{0} + \left. \frac{-1}{2+\sigma} e^{-(2+\sigma)t} \right|_{0}^{\infty}
\end{aligned}
$$

The first term converges for $\sigma < -1$, while the second term converges for $\sigma > -2$. Hence both terms converge for $-2 < \sigma < -1$. This is the intersection of the ROC for each term. The ROC for each term is depicted in Fig. 6.13(a). The intersection of ROCs is shown as the doubly shaded region. The reader may verify that the Laplace transform of $x_1(t)$ is

$$
\begin{aligned}
X_1(s) &= \frac{-1}{s+1} + \frac{1}{s+2} \\
&= \frac{-1}{(s+1)(s+2)}
\end{aligned}
$$

which has poles at $s = -1$ and $s = -2$. We see that the ROC is the strip of the s-plane located between the poles.

For the second signal $x_2(t)$, we have

$$
\begin{aligned}
I_1(\sigma) &= \int_{-\infty}^{\infty} |x_2(t)|e^{-\sigma t}\, dt \\
&= \int_{-\infty}^{0} e^{-(2+\sigma)t}\, dt + \int_{0}^{\infty} e^{-(1+\sigma)t}\, dt \\
&= \left. \frac{-1}{2+\sigma} e^{-(2+\sigma)t} \right|_{-\infty}^{0} + \left. \frac{-1}{1+\sigma} e^{-(1+\sigma)t} \right|_{0}^{\infty}
\end{aligned}
$$

The first term converges for $\sigma < -2$ and the second term for $\sigma > -1$. Here there is no value of σ for which both terms converge, and so the intersection is empty. Hence there are no values of s for which $X_2(s)$ converges. This situation is illustrated in Fig. 6.13(b).

▶ **Drill Problem 6.9**   Describe the ROC of the signal

$$x(t) = e^{-b|t|}$$

for $b > 0$ and $b \leq 0$.

Answer: For $b > 0$ the ROC is the region $-b < \sigma < b$. For $b \leq 0$ the ROC is the empty set.   ◀

■ **PROPERTIES OF THE BILATERAL LAPLACE TRANSFORM**

The linearity, scaling, s-domain shift, convolution, and differentiation in the s-domain properties are identical for the bilateral and unilateral Laplace transforms, although the

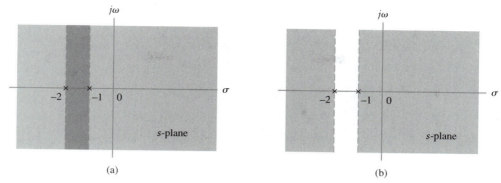

FIGURE 6.13   ROCs for signals in Example 6.12. (a) The shaded regions denote the ROCs of each individual term, $e^{-2t}u(t)$ and $e^{-t}u(-t)$. The doubly shaded region is the intersection of the individual ROCs and represents the ROC of the sum. (b) The shaded regions represent the individual ROCs of $e^{-2t}u(-t)$ and $e^{-t}u(t)$. In this case there is no intersection and the Laplace transform of the sum does not converge for any value of s.

operations associated with these properties may change the ROC. The effect of each of these operations on the ROC is given in the table of Laplace transform properties in Appendix D.2.

As an example illustrating the change in ROC that may occur, consider the linearity property. If $x(t) \overset{\mathcal{L}}{\longleftrightarrow} X(s)$ with ROC R_x and $y(t) \overset{\mathcal{L}}{\longleftrightarrow} Y(s)$ with ROC R_y, then $ax(t) + by(t) \overset{\mathcal{L}}{\longleftrightarrow} aX(s) + bY(s)$ with ROC at least $R_x \cap R_y$, where the symbol $\cap$ indicates intersection. In general, the ROC for a sum of signals is just the intersection of the individual ROCs. The ROC may be larger than the intersection of the individual ROCs if a pole and zero cancel in the sum $aX(s) + bY(s)$. This is illustrated in the following example.

EXAMPLE 6.13   Suppose

$$x(t) = e^{-2t}u(t) \overset{\mathcal{L}}{\longleftrightarrow} X(s) = \frac{1}{s + 2} \quad \text{with ROC Re}(s) > -2$$

and

$$y(t) = e^{-2t}u(t) + e^{-3t}u(t) \overset{\mathcal{L}}{\longleftrightarrow} Y(s) = \frac{1}{(s + 2)(s + 3)} \quad \text{with ROC Re}(s) > -2$$

The s-plane representations of the ROCs are shown in Fig. 6.14. The intersection of the ROCs is Re$(s) > -2$. However, if we choose $a = 1$ and $b = -1$, then the sum $x(t) - y(t) = e^{-3t}u(t)$ has ROC Re$(s) > -3$, which is larger than the intersection of ROCs. Here the subtraction eliminates the signal $e^{-2t}u(t)$ in the time domain, and consequently the ROC is enlarged. This corresponds to a pole–zero cancellation in the s-domain since

$$X(s) - Y(s) = \frac{1}{s + 2} - \frac{1}{(s + 2)(s + 3)}$$

$$= \frac{(s + 3) - 1}{(s + 2)(s + 3)}$$

$$= \frac{(s + 2)}{(s + 2)(s + 3)}$$

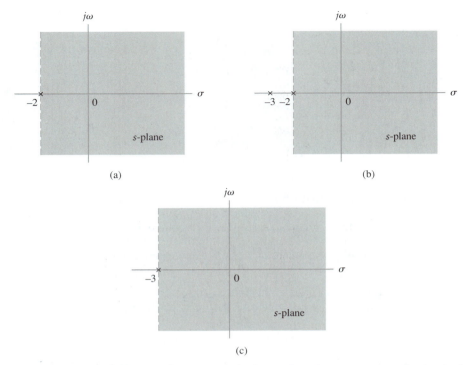

FIGURE 6.14   The ROC of a sum of signals may be larger than the intersection of individual ROCs when pole–zero cancellation occurs. (a) ROC for $x(t) = e^{-2t}u(t)$; (b) ROC for $y(t) = e^{-3t}u(t) + e^{-2t}u(t)$; (c) ROC for $x(t) - y(t)$.

The zero of $(X(s) - Y(s))$ located at $s = -2$ cancels the pole at $s = -2$, and so we have

$$X(s) - Y(s) = \frac{1}{s + 3}$$

If the intersection of the ROCs is the empty set and pole–zero cancellation does not occur, then the Laplace transform of $ax(t) + by(t)$ does not exist. Note that the ROC for the convolution of two signals may also be larger than the intersection of individual ROCs if pole–zero cancellation occurs.

The bilateral Laplace transform properties involving time shifts, differentiation in the time domain, and time integration differ slightly from their unilateral counterparts. We state them as follows:

Time Shift

$$x(t - \tau) \xleftrightarrow{\;\mathscr{L}\;} e^{-s\tau}X(s) \tag{6.25}$$

The restriction on the shift that is present in the unilateral case is removed because the bilateral Laplace transform is evaluated over both positive and negative values of time.

Differentiation in the Time Domain

$$\frac{d}{dt} x(t) \xleftrightarrow{\;\mathscr{L}\;} sX(s) \quad \text{with ROC at least } R_x \qquad (6.26)$$

Differentiation in time corresponds to multiplication by s. Here again the ROC may be larger than R_x if $X(s)$ has a single pole at $s = 0$ on the ROC boundary. Multiplication by s, corresponding to differentiation, cancels this pole and so eliminates the dc component in $x(t)$.

EXAMPLE 6.14   Find the Laplace transform of

$$x(t) = \frac{d^2}{dt^2} \left(e^{-3(t-2)} u(t - 2) \right)$$

Solution: We know that

$$e^{-3t} u(t) \xleftrightarrow{\;\mathscr{L}\;} \frac{1}{s + 3} \quad \text{with ROC Re}(s) > -3$$

The time-shift property implies

$$e^{-3(t-2)} u(t - 2) \xleftrightarrow{\;\mathscr{L}\;} \frac{1}{s + 3} e^{-2s} \quad \text{with ROC Re}(s) > -3$$

Now apply the time differentiation property twice, as shown by

$$x(t) = \frac{d^2}{dt^2} \left(e^{-3(t-2)} u(t - 2) \right) \xleftrightarrow{\;\mathscr{L}\;} X(s) = \frac{s^2}{s + 3} e^{-2s} \quad \text{with ROC Re}(s) > -3$$

Time Integration

$$\int_{-\infty}^{t} x(\tau)\, d\tau \xleftrightarrow{\;\mathscr{L}\;} \frac{X(s)}{s} \quad \text{with ROC } R_x \cap \text{Re}(s) > 0 \qquad (6.27)$$

Integration corresponds to division by s. Since this introduces a pole at $s = 0$ and we are integrating to the right, the ROC must lie to the right of $s = 0$.

The initial and final value theorems apply to the bilateral Laplace transform with the additional restriction that $x(t) = 0$ for $t < 0$.

■ INVERSION OF THE BILATERAL LAPLACE TRANSFORM

As in the unilateral case, we consider inversion of bilateral Laplace transforms that are expressed as ratios of polynomials in s. The primary difference between inversion of bilateral and unilateral Laplace transforms is that we must use the ROC to determine a unique inverse transform in the bilateral case. Recall that the ROC of a right-sided exponential signal lies to the right of a pole, while the ROC of a left-sided exponential signal lies to the left of a pole.

We begin by assuming that the transform to be inverted is expressed as a partial fraction expansion in terms of nonrepeated poles, as shown by

$$X(s) = \sum_{k=1}^{N} \frac{A_k}{s - d_k}$$

In the bilateral case, there are two possibilities for the inverse Laplace transform of each term in the sum. We may use either the right-sided transform pair

$$A_k e^{d_k t} u(t) \overset{\mathcal{L}}{\longleftrightarrow} \frac{A_k}{s - d_k} \quad \text{with ROC } \text{Re}(s) > d_k$$

or the left-sided transform pair

$$-A_k e^{d_k t} u(-t) \overset{\mathcal{L}}{\longleftrightarrow} \frac{A_k}{s - d_k} \quad \text{with ROC } \text{Re}(s) < d_k$$

The ROC associated with $X(s)$ determines whether the left-sided or right-sided inverse transform is chosen.

The linearity property states that the ROC of $X(s)$ is the intersection of the ROCs of the individual terms in the partial fraction expansion. In order to find the inverse transform of each term, we must infer the ROC of each term from the given ROC of $X(s)$. This is easily accomplished by comparing each pole location with the ROC of $X(s)$. If the ROC of $X(s)$ lies to the left of a particular pole, we choose the left-sided inverse Laplace transform for that pole. If the ROC of $X(s)$ lies to the right of a particular pole, we choose the right-sided inverse Laplace transform for that pole. This procedure is illustrated in the following example.

EXAMPLE 6.15   Find the inverse Laplace transform of

$$X(s) = \frac{-5s - 7}{(s + 1)(s - 1)(s + 2)} \quad \text{with ROC } -1 < \text{Re}(s) < 1$$

Solution: Use a partial fraction expansion

$$X(s) = \frac{A_1}{s + 1} + \frac{A_2}{s - 1} + \frac{A_3}{s + 2}$$

Solving for A_1, A_2, and A_3 gives

$$X(s) = \frac{1}{s + 1} - \frac{2}{s - 1} + \frac{1}{s + 2}$$

The ROC and pole locations are depicted in Fig. 6.15. We find the inverse Laplace transform of each term using the relationship between the pole location and ROC:

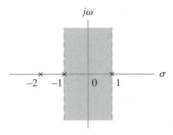

FIGURE 6.15   The ROC for Example 6.15.

▶ The pole of the first term is at $s = -1$. The ROC lies to the right of this pole, so we choose the right-sided inverse Laplace transform

$$e^{-t}u(t) \xleftrightarrow{\ \mathcal{L}\ } \frac{1}{s+1}$$

▶ The second term has a pole at $s = 1$. Here the ROC is to the left of the pole, so we choose the left-sided inverse Laplace transform

$$2e^{t}u(-t) \xleftrightarrow{\ \mathcal{L}\ } -\frac{2}{s-1}$$

▶ The pole in the last term is at $s = -2$. The ROC is to the right of this pole, so we choose the right-sided inverse Laplace transform

$$e^{-2t}u(t) \xleftrightarrow{\ \mathcal{L}\ } \frac{1}{s+2}$$

Combining these three terms, we obtain

$$x(t) = e^{-t}u(t) + 2e^{t}u(-t) + e^{-2t}u(t)$$

▶ **Drill Problem 6.10**   Repeat the previous example if the ROC is $-2 < \text{Re}(s) < -1$.

Answer:
$$x(t) = -e^{-t}u(-t) + 2e^{t}u(-t) + e^{-2t}u(t)$$
◀

The relationship between pole and ROC locations in the s-plane also determines the inverse transform for the other terms that can occur in a partial fraction expansion. For example, the inverse Laplace transform of the term

$$\frac{A}{(s-d_k)^n}$$

is given by the right-sided signal

$$\frac{At^{n-1}}{(n-1)!}e^{d_k t}u(t)$$

if the ROC lies to the right of the poles at $s = d_k$. If the ROC lies to the left of the poles at $s = d_k$, then the inverse Laplace transform is

$$\frac{-At^{n-1}}{(n-1)!}e^{d_k t}u(-t)$$

Similarly, the inverse Laplace transform of the term

$$\frac{C_1(s-\alpha)}{(s-\alpha)^2 + \omega_o^2}$$

is the right-sided signal

$$C_1 e^{\alpha t} \cos(\omega_o t)u(t)$$

if the ROC lies to the right of the poles at $s = \alpha \pm j\omega_o$. If the ROC lies to the left of the poles at $s = \alpha \pm j\omega_o$, then the inverse Laplace transform is

$$-C_1 e^{\alpha t} \cos(\omega_o t) u(-t)$$

▶ **Drill Problem 6.11**   Find the inverse Laplace transform of

$$X(s) = \frac{4s^2 + 6}{s^3 + s^2 - 2} \quad \text{with ROC } -1 < \text{Re}(s) < 1$$

Answer:

$$x(t) = -2e^t u(-t) + 2e^{-t} \cos(t) u(t) - 4e^{-t} \sin(t) u(t) \qquad ◀$$

Note that knowledge other than the ROC can be used to determine a unique inverse Laplace transform. The most common form of other knowledge is causality, stability, or existence of the Fourier transform. If the signal is known to be causal, then we choose the right-sided inverse transform for each term. This is the approach followed with the unilateral Laplace transform. A stable signal is absolutely integrable and thus has a Fourier transform. Hence stability and existence of the Fourier transform are equivalent conditions. In both of these cases, the ROC includes the $j\omega$-axis in the s-plane, or $\text{Re}(s) = 0$. The inverse Laplace transform is obtained by comparing pole locations to the $j\omega$-axis. If a pole lies to the right of the $j\omega$-axis, then the right-sided inverse transform is chosen. If it lies to the left of the $j\omega$-axis, then the left-sided inverse transform is chosen.

▶ **Drill Problem 6.12**   Find the inverse Laplace transform of

$$X(s) = \frac{4s^2 + 15s + 8}{(s + 2)^2(s - 1)}$$

assuming (a) $x(t)$ is causal, and (b) the Fourier transform of $x(t)$ exists.

Answer:

(a) $x(t) = e^{-2t}u(t) + 2te^{-2t}u(t) + 3e^t u(t)$

(b) $x(t) = e^{-2t}u(t) + 2te^{-2t}u(t) - 3e^t u(-t) \qquad ◀$

6.7 *Transform Analysis of Systems*

The transfer function of a system was defined in Section 6.2 as the Laplace transform of the impulse response. The output of a LTI system is related to the input in terms of the impulse response via the convolution

$$y(t) = h(t) * x(t)$$

In general, this equation applies to $h(t)$ and $x(t)$ that are either causal or noncausal. Hence if we take the bilateral Laplace transform of both sides of this equation and use the convolution property, then we see that

$$Y(s) = H(s)X(s) \tag{6.28}$$

The Laplace transform of the system output is the product of the transfer function and the Laplace transform of the input. Hence the transfer function provides yet another description for the input–output behavior of a LTI system.

Note that Eq. (6.28) implies

$$H(s) = \frac{Y(s)}{X(s)} \tag{6.29}$$

That is, the transfer function is the ratio of the Laplace transform of the output to the Laplace transform of the input. This definition applies at values of s for which $X(s)$ is nonzero. We now examine how the system characteristics are related to the transfer function.

■ THE TRANSFER FUNCTION AND DIFFERENTIAL EQUATIONS

The transfer function may be related directly to the differential-equation description for a LTI system using the bilateral Laplace transform. Recall that the relationship between the input and output of an Nth order LTI system is described by the differential equation

$$\sum_{k=0}^{N} a_k \frac{d^k}{dt^k} y(t) = \sum_{k=0}^{M} b_k \frac{d^k}{dt^k} x(t)$$

In Section 6.2 we showed that the input e^{st} is an eigenfunction of the system with eigenvalue equal to transfer function $H(s)$. That is, if $x(t) = e^{st}$, then $y(t) = e^{st}H(s)$. Substitution of e^{st} for $x(t)$ and $e^{st}H(s)$ for $y(t)$ in the differential equation gives

$$\sum_{k=0}^{N} a_k \frac{d^k}{dt^k} \{e^{st}\} H(s) = \sum_{k=0}^{M} b_k \frac{d^k}{dt^k} \{e^{st}\}$$

We now substitute the relationship

$$\frac{d^k}{dt^k} \{e^{st}\} = s^k e^{st}$$

and solve for $H(s)$ to obtain

$$H(s) = \frac{\sum_{k=0}^{M} b_k s^k}{\sum_{k=0}^{N} a_k s^k} \tag{6.30}$$

The system transfer function is a ratio of polynomials in s and is thus termed a *rational transfer function*. The coefficient of s^k in the numerator polynomial corresponds to the coefficient of the kth derivative of $x(t)$. The coefficient of s^k in the denominator polynomial corresponds to the coefficient of the kth derivative of $y(t)$. Hence we may obtain the transfer function from the differential-equation description of a system. Conversely, we may determine the differential-equation description of a system from its transfer function.

EXAMPLE 6.16 Find the transfer function of the LTI system described by the differential equation

$$\frac{d^2}{dt^2} y(t) + 3 \frac{d}{dt} y(t) + 2y(t) = 2 \frac{d}{dt} x(t) - 3x(t)$$

Solution: Apply Eq. (6.30), obtaining

$$H(s) = \frac{2s - 3}{s^2 + 3s + 2}$$

The poles and zeros of a rational transfer function offer much insight into system characteristics, as we shall see in the following subsections. The transfer function is expressed in pole–zero form by factoring the numerator and denominator polynomials in Eq. (6.30) as follows:

$$H(s) = \frac{(b_M/a_N)\Pi_{k=1}^{M}(s - c_k)}{\Pi_{k=1}^{N}(s - d_k)} \tag{6.31}$$

The c_k and d_k are the zeros and poles of the system, respectively. Knowledge of the system poles, zeros, and the gain factor (b_M/a_N) completely determines the system transfer function and thus offers yet another description for a LTI system.

EXAMPLE 6.17   An electromechanical system consisting of a dc motor and a load is depicted in Fig. 6.16(a). The input is the applied voltage $x(t)$ and output is the angular position of the load $y(t)$. The rotational inertia of the load is given by J. Under ideal circumstances the torque produced by the motor is directly proportional to the input current; that is,

$$\tau(t) = K_1 i(t)$$

Rotation of the motor results in a back electromotive force $v(t)$ that is proportional to the angular velocity, as shown by

$$v(t) = K_2 \frac{d}{dt} y(t)$$

The circuit diagram in Fig. 6.16(b) depicts the relationship between the input current, applied voltage, back electromotive force, and armature resistance R. Express the transfer function of this system in pole–zero form.

Solution: The torque experienced by the load is given by the rotational inertia times the angular acceleration. Equating the torque produced by the motor and that experienced by the load results in the following relationship

$$J \frac{d^2}{dt^2} y(t) = K_1 i(t)$$

Application of Ohm's law to the circuit in Fig. 6.16(b) indicates that the current is expressed in terms of the input and back electromotive force as shown by

$$i(t) = \frac{1}{R}[x(t) - v(t)]$$

Hence, substituting for $i(t)$, we have

$$J \frac{d^2}{dt^2} y(t) = \frac{K_1}{R}[x(t) - v(t)]$$

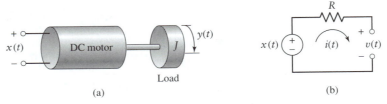

(a)   (b)

FIGURE 6.16   (a) Electromechanical system in which a motor is used to position a load. (b) Circuit diagram relating applied voltage to back electromotive force, armature resistance, and input current.

Next we express $v(t)$ in terms of the angular velocity to obtain the differential equation relating the applied voltage to position, as shown by

$$J \frac{d^2}{dt^2} y(t) + \frac{K_1 K_2}{R} \frac{d}{dt} y(t) = \frac{K_1}{R} x(t)$$

Application of Eq. (6.30) shows that the transfer function is given by

$$H(s) = \frac{\dfrac{K_1}{R}}{Js^2 + \dfrac{K_1 K_2}{R} s}$$

Hence the transfer function is expressed in pole–zero form as

$$H(s) = \frac{\dfrac{K_1}{RJ}}{s\left(s + \dfrac{K_1 K_2}{RJ}\right)}$$

■ TRANSFER FUNCTION AND STATE-VARIABLE DESCRIPTION

The transfer function is also directly related to the state-variable description

$$\frac{d}{dt} \mathbf{q}(t) = \mathbf{A}\mathbf{q}(t) + \mathbf{b}x(t) \tag{6.32}$$

$$y(t) = \mathbf{c}\mathbf{q}(t) + Dx(t) \tag{6.33}$$

Taking the Laplace transform of Eq. (6.32) and applying the differentiation property of Eq. (6.26), we obtain

$$s\tilde{\mathbf{q}}(s) = \mathbf{A}\tilde{\mathbf{q}}(s) + \mathbf{b}X(s) \tag{6.34}$$

Here we have

$$\tilde{\mathbf{q}}(s) = \begin{bmatrix} Q_1(s) \\ Q_2(s) \\ \vdots \\ Q_N(s) \end{bmatrix}$$

where $Q_i(s)$ is the Laplace transform of the ith element of $\mathbf{q}(t)$. Equation (6.34) may be rewritten as

$$s\tilde{\mathbf{q}}(s) - \mathbf{A}\tilde{\mathbf{q}}(s) = \mathbf{b}X(s)$$

or, equivalently,

$$(s\mathbf{I} - \mathbf{A})\tilde{\mathbf{q}}(s) = \mathbf{b}X(s)$$

The latter result implies that

$$\tilde{\mathbf{q}}(s) = (s\mathbf{I} - \mathbf{A})^{-1}\mathbf{b}X(s)$$

Next, taking the Laplace transform of Eq. (6.33) yields

$$Y(s) = \mathbf{c}\tilde{\mathbf{q}}(s) + DX(s)$$

Substitute for $\tilde{\mathbf{q}}(s)$ to obtain

$$Y(s) = [\mathbf{c}(s\mathbf{I} - \mathbf{A})^{-1}\mathbf{b} + D]X(s)$$

Hence we may identify the transfer function of the system as

$$H(s) = \mathbf{c}(s\mathbf{I} - \mathbf{A})^{-1}\mathbf{b} + D \qquad (6.35)$$

EXAMPLE 6.18   Find the transfer function for a system with state-variable description matrices

$$\mathbf{A} = \begin{bmatrix} 1 & 2 \\ 3 & -2 \end{bmatrix}, \qquad \mathbf{b} = \begin{bmatrix} 0 \\ 2 \end{bmatrix}$$
$$\mathbf{c} = [1 \quad 1], \qquad D = 0$$

Solution: Substituting $(\mathbf{A}, \mathbf{b}, \mathbf{c}, D)$ into Eq. (6.35) we get

$$s\mathbf{I} - \mathbf{A} = \begin{bmatrix} s - 1 & -2 \\ -3 & s + 2 \end{bmatrix}$$

Accordingly, we may write

$$(s\mathbf{I} - \mathbf{A})^{-1} = \frac{1}{s^2 + s - 8} \begin{bmatrix} s + 2 & 2 \\ 3 & s - 1 \end{bmatrix}$$

and

$$(s\mathbf{I} - \mathbf{A})^{-1}\mathbf{b} = \frac{1}{s^2 + s - 8} \begin{bmatrix} 4 \\ 2s - 2 \end{bmatrix}$$

Thus the use of Eq. (6.35) yields

$$H(s) = \mathbf{c}(s\mathbf{I} - \mathbf{A})^{-1}\mathbf{b} + D$$
$$= \frac{2s + 2}{s^2 + s - 8}$$

The transfer function is an input–output description for a system. Transformations of the state vector change the state-variable description but do not change the input–output characteristics of the system. Hence it follows that Eq. (6.35) is invariant to transformations of the state vector.

▪ CAUSALITY AND STABILITY

The impulse response is the inverse Laplace transform of the transfer function. In order to obtain a unique inverse transform, we must know the ROC or have other knowledge of the impulse response. The differential-equation description for a system does not contain this information. Hence to obtain the impulse response we must have additional knowledge of the system characteristics. The relationships between the poles, zeros, and system characteristics can provide this additional knowledge.

The impulse response of a causal system is zero for $t < 0$. Therefore if we know that a system is causal, the impulse response is determined from the transfer function by using

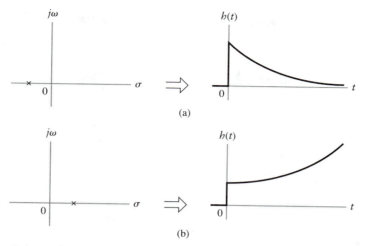

(a)

(b)

FIGURE 6.17 Relationship between the pole location and impulse response for a causal system. (a) A pole in the left half of the s-plane corresponds to an exponentially decaying impulse response. (b) A pole in the right half of the s-plane corresponds to an exponentially increasing impulse response.

right-sided inverse Laplace transforms. A pole at $s = d_k$ in the left half of the s-plane $[\mathrm{Re}(d_k) < 0]$ contributes an exponentially decaying term to the impulse response, while a pole in the right half of the s-plane $[\mathrm{Re}(d_k) > 0]$ contributes an increasing exponential term to the impulse response. These relationships are illustrated in Fig. 6.17.

Alternatively, if we know a system is stable, then the impulse response is absolutely integrable. This implies that the Fourier transform exists and thus the ROC includes the $j\omega$-axis in the s-plane. This knowledge is sufficient to uniquely determine the inverse La-

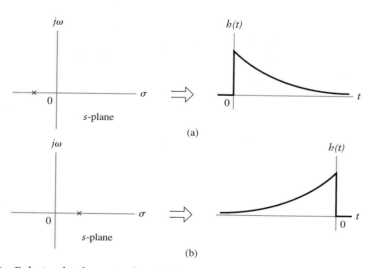

(a)

(b)

FIGURE 6.18 Relationship between the pole location and impulse response for a stable system. (a) A pole in the left half of the s-plane corresponds to a right-sided impulse response. (b) A pole in the right half of the s-plane corresponds to a left-sided impulse response.

place transform of the transfer function. A pole that is in the right half of the s-plane contributes a left-sided decaying exponential term to the impulse response, while a pole in the left half of the s-plane contributes a right-sided decaying exponential term to the impulse response as depicted in Fig. 6.18. Note that a stable impulse response cannot contain any increasing exponential terms, since an increasing exponential is not absolutely integrable.

Suppose a system is known to be both causal and stable. A pole that is in the left half of the s-plane contributes a right-sided decaying exponential term to the impulse response. We cannot have a pole in the right half of the s-plane, however. A pole in the right half will contribute either a left-sided decaying exponential that is not causal, or a right-sided increasing exponential that results in an unstable impulse response. That is, the inverse Laplace transform of a pole in the right half of the s-plane is either stable or causal, but cannot be both stable and causal. Systems that are stable and causal must have all their poles in the left half of the s-plane. Such a system is illustrated in Fig. 6.19.

EXAMPLE 6.19   A system has the transfer function

$$H(s) = \frac{2}{s + 3} + \frac{1}{s - 2}$$

Find the impulse response assuming (a) the system is stable and (b) the system is causal. Can this system be both stable and causal?

Solution: This system has a pole at $s = -3$ and at $s = 2$. If the system is stable, then the pole at $s = -3$ contributes a right-sided term to the impulse response, while the pole at $s = 2$ contributes a left-sided term. We have

$$h(t) = 2e^{-3t}u(t) - e^{2t}u(-t)$$

If the system is causal, then both poles must contribute right-sided terms to the impulse response and we have

$$h(t) = 2e^{-3t}u(t) + e^{2t}u(t)$$

Note that this system is not stable since the term $e^{2t}u(t)$ is not absolutely integrable. In fact, this system cannot be both stable and causal since the pole at $s = 2$ is in the right half of the s-plane.

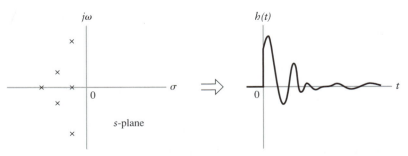

FIGURE 6.19   A system that is both stable and causal must have all of its poles in the left half of the s-plane, as shown here.

▶ **Drill Problem 6.13**   A stable and causal system is described by the differential equation

$$\frac{d^2}{dt^2}\, y(t) + 5\, \frac{d}{dt}\, y(t) + 6y(t) = \frac{d^2}{dt^2}\, x(t) + 8\, \frac{d}{dt}\, x(t) + 13x(t)$$

Find the system impulse response.

Answer:

$$h(t) = 2e^{-3t}u(t) + e^{-2t}u(t) + \delta(t)$$ ◀

■ **INVERSE SYSTEMS**

Given a LTI system with impulse response $h(t)$, the impulse response of the inverse system, $h^{-1}(t)$, satisfies the condition

$$h^{-1}(t) * h(t) = \delta(t)$$

If we take the Laplace transform of both sides of this equation, we find that the inverse system transfer function $H^{-1}(s)$ satisfies

$$H^{-1}(s)H(s) = 1$$

or

$$H^{-1}(s) = \frac{1}{H(s)}$$

The inverse system's transfer function is therefore the inverse of the transfer function of the original system. If $H(s)$ is written in pole–zero form, as in Eq. (6.31), then we have

$$H^{-1}(s) = \frac{\Pi_{k=1}^{N}(s - d_k)}{(b_M/a_N)\Pi_{k=1}^{M}(s - c_k)} \tag{6.36}$$

The zeros of the inverse system are the poles of $H(s)$, and the poles of the inverse system are the zeros of $H(s)$. We conclude that any system with a rational transfer function has an inverse system.

Often we are interested in inverse systems that are both stable and causal. In the previous subsection we concluded that a stable, causal system must have all of its poles in the left half of the s-plane. Since the poles of the inverse system $H^{-1}(s)$ are the zeros of $H(s)$, a stable and causal inverse system exists only if all of the zeros of $H(s)$ are in the left half of the s-plane. A system whose transfer function $H(s)$ has all of its poles and zeros in the left half of the s-plane is said to be *minimum phase*. A nonminimum phase system cannot have a stable and causal inverse system as it has zeros in the right half of the s-plane.

EXAMPLE 6.20   Consider a system H described by the differential equation

$$\frac{d}{dt}\, y(t) + 2y(t) = \frac{d^2}{dt^2}\, x(t) + 2\, \frac{d}{dt}\, x(t) - 3x(t)$$

Find the transfer function of the inverse system for H. Does a stable and causal inverse system exist?

Solution: First find the system transfer function $H(s)$ by taking the Laplace transform of both sides of the given differential equation, obtaining

$$Y(s)(s + 2) = X(s)(s^2 + 2s - 3)$$

Hence the transfer function of the system is

$$H(s) = \frac{Y(s)}{X(s)}$$

$$= \frac{s^2 + 2s - 3}{s + 2}$$

and the inverse system has the transfer function

$$H^{-1}(s) = \frac{1}{H(s)}$$

$$= \frac{s + 2}{s^2 + 2s - 3}$$

$$= \frac{s + 2}{(s - 1)(s + 3)}$$

The inverse system has poles at $s = 1$ and $s = -2$. The pole at $s = 1$ is in the right half of the s-plane. Therefore the inverse system cannot be both stable and causal.

▶ **Drill Problem 6.14**   Consider the system with impulse response

$$h(t) = \delta(t) + e^{-3t}u(t) + 2e^{-t}u(t)$$

Find the transfer function of the inverse system. Does a stable and causal inverse system exist?

Answer:

$$H^{-1}(s) = \frac{s^2 + 4s + 3}{(s + 2)(s + 5)}$$

A stable and causal system does exist.   ◀

■ **DETERMINING THE FREQUENCY RESPONSE FROM POLES AND ZEROS**

The locations of the poles and zeros in the s-plane provide insight into the frequency response of a system. Recall that the frequency response is obtained from the transfer function by substituting $j\omega$ for s, that is, by evaluating the transfer function along the $j\omega$-axis in the s-plane. This assumes the $j\omega$ axis is in the ROC. Substituting $j\omega$ for s in Eq. (6.31) yields

$$H(j\omega) = \frac{(b_M/a_N)\Pi_{k=1}^{M}(j\omega - c_k)}{\Pi_{k=1}^{N}(j\omega - d_k)} \tag{6.37}$$

We shall examine both the magnitude and phase of $H(j\omega)$. Beginning with the magnitude at some fixed value of ω, say, ω_o, we may write

$$|H(j\omega_o)| = \frac{|b_M/a_N|\Pi_{k=1}^{M}|j\omega_o - c_k|}{\Pi_{k=1}^{N}|j\omega_o - d_k|}$$

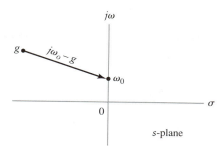

FIGURE 6.20 The quantity $j\omega_o - g$ as a vector from g to $j\omega_o$ in the s-plane.

This expression involves a ratio of products of terms having the form $|j\omega_o - g|$, where g is either a pole or a zero. The zero contributions are in the numerator, while the pole contributions are in the denominator. The factor $(j\omega_o - g)$ is a complex number that may be represented in the s-plane as a vector from the point g to the point $j\omega_o$ as illustrated in Fig. 6.20. The length of this vector is $|j\omega_o - g|$. By examining the length of this vector as ω_o changes, we may assess the contribution of each pole or zero to the overall magnitude response.

Figure 6.21(a) depicts the vector $j\omega - g$ for several different values of ω and Fig. 6.21(b) depicts $|j\omega - g|$ as a continuous function of frequency. Note that when $\omega = \text{Im}\{g\}$, $|j\omega - g| = |\text{Re}\{g\}|$. Hence if g is close to the $j\omega$-axis ($\text{Re}\{g\} \approx 0$), then $|j\omega - g|$ will become very small for $\omega = \text{Im}\{g\}$. Also, if g is close to the $j\omega$-axis, then the most rapid change in $|j\omega - g|$ occurs at frequencies closest to g.

If g represents a zero, then $|j\omega - g|$ contributes to the numerator of $|H(j\omega)|$. Hence at frequencies close to a zero, $|H(j\omega)|$ tends to decrease. How far $|H(j\omega)|$ decreases depends on how close the zero is to the $j\omega$-axis. If the zero is on the $j\omega$-axis, then $|H(j\omega)|$ goes to zero at the frequency corresponding to the zero location. At frequencies far from a zero, $|\omega| \gg |g|$, $|j\omega - g|$ is approximately equal to $|\omega|$. The component of the magnitude response due to a zero is illustrated in Fig. 6.22(a). On the other hand, if g corresponds to a pole, then $|j\omega - g|$ contributes to the denominator of $|H(j\omega)|$ and when $|j\omega - g|$ decreases, $|H(j\omega)|$ increases. How far $|H(j\omega)|$ increases depends

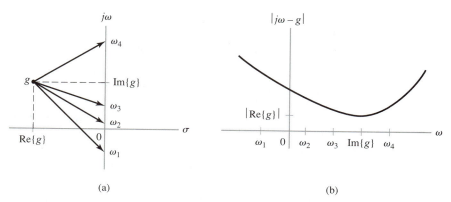

(a)   (b)

FIGURE 6.21 The function $|j\omega - g|$ corresponds to the length of vectors from g to the $j\omega$-axis in the s-plane. (a) Vectors from g to $j\omega$ for several frequencies. (b) $|j\omega - g|$ as a function of $j\omega$.

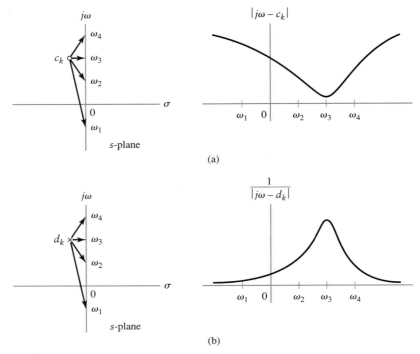

FIGURE 6.22   Components of the magnitude response. (a) Magnitude response associated with a zero. (b) Magnitude response associated with a pole.

on how close the pole is to the $j\omega$-axis. A pole that is close to the $j\omega$-axis will result in a large peak in $|H(j\omega)|$. The component of the magnitude response associated with a pole is illustrated in Fig. 6.22(b). Hence zeros near the $j\omega$-axis tend to pull the response magnitude down, while poles near the $j\omega$-axis tend to push the response magnitude up. Note that a pole cannot lie on the $j\omega$-axis since we have assumed the ROC includes the $j\omega$-axis.

EXAMPLE 6.21   Sketch the magnitude response of the system having the transfer function

$$H(s) = \frac{(s - 0.5)}{(s + 0.1 - 5j)(s + 0.1 + 5j)}$$

Solution: The system has a zero at $s = 0.5$, and poles at $s = -0.1 \pm 5j$ as depicted in Fig. 6.23(a). Hence the response will tend to decrease near $\omega = 0$ and increase near $\omega = \pm 5$. At $\omega = 0$ we have

$$|H(j0)| = \frac{0.5}{|0.1 - 5j||0.1 + 5j|}$$

$$\approx \frac{0.5}{5^2}$$

$$\approx 0.02$$

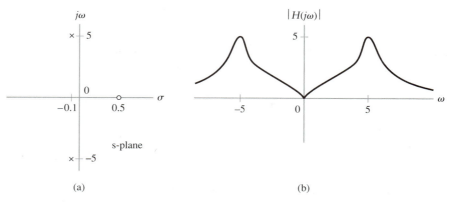

FIGURE 6.23   Solution to Example 6.20. (a) Pole–zero plot. (b) Approximate magnitude response.

At $\omega = 5$, we have

$$|H(j5)| = \frac{|j5 - 0.5|}{|0.1||j10 + 0.1|}$$

$$\approx \frac{5}{0.1(10)}$$

$$\approx 5$$

For $\omega \gg 5$, the length of the vector from $j\omega$ to one of the poles is approximately equal to the length of the vector from $j\omega$ to the zero, and so the zero is cancelled by one of the poles. The distance from $j\omega$ to the remaining pole increases as frequency increases and thus the magnitude response goes to zero. The magnitude response is sketched in Fig. 6.23(b).

The phase of $H(j\omega)$ may also be evaluated in terms of the phase associated with each pole and zero. Using Eq. (6.37), we may evaluate $\arg\{H(j\omega)\}$ as

$$\arg\{H(j\omega)\} = \arg\{b_M/a_N\} + \sum_{k=1}^{M} \arg\{j\omega - c_k\} - \sum_{k=1}^{N} \arg\{j\omega - d_k\} \qquad (6.38)$$

In this case the phase of $H(j\omega)$ is the sum of the phase angles due to all the zeros minus the sum of the phase angles due to all the poles. The first term, $\arg\{b_M/a_N\}$, is independent of frequency. The phase associated with each zero and pole is evaluated by considering a term of the form $\arg\{j\omega_o - g\}$. This is the angle of a vector pointing from g to $j\omega_o$ in the s-plane. The angle of the vector is measured relative to the horizontal line through g as illustrated in Fig. 6.24. By examining the phase of this vector as ω changes, we may assess the contribution of each pole or zero to the overall phase response.

Figure 6.25(a) depicts the phase of this vector for several different frequencies and Fig. 6.25(b) illustrates the phase as a continuous function of frequency. Note that since g is in the left half of the s-plane, the phase is $-\pi/2$ for ω large and negative, increases to zero when $\omega = \text{Im}\{g\}$, and continues increasing to $\pi/2$ for ω large and positive. If g is in the right half of the s-plane, then the phase begins at $-\pi/2$ for ω large and negative, decreases to $-\pi$ when $\omega = \text{Im}\{g\}$, and then decreases to $-3\pi/2$ for ω large and positive. If g is close to the $j\omega$-axis, then the change from $-\pi/2$ to $\pi/2$ (or $-3\pi/2$) occurs rapidly in the vicinity of $\omega = \text{Im}\{g\}$. If g corresponds to a zero, then the phase associated with g

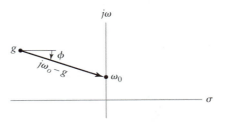

FIGURE 6.24 The quantity $j\omega_o - g$ as a vector from g to $j\omega_o$ in the s-plane. The phase angle of the vector is ϕ, its angle with respect to a horizontal line through g.

adds to the phase of $H(j\omega)$, while if g is a pole, the phase associated with g subtracts from the phase of $H(j\omega)$.

EXAMPLE 6.22 Sketch the phase response of a system described by the transfer function

$$H(s) = \frac{(s - 0.5)}{(s + 0.1 - 5j)(s + 0.1 + 5j)}$$

Solution: The pole and zero locations of this system in the s-plane are depicted in Fig. 6.23(a). The phase response associated with the zero at $s = 0.5$ is illustrated in Fig. 6.26(a), the phase response associated with the pole at $s = -0.1 + j5$ in Fig. 6.26(b), and that associated with the pole at $s = -0.1 - j5$ in Fig. 6.26(c). The phase response of the system is obtained by subtracting the phase contributions of the poles from that of the zero. The result is shown in Fig. 6.26(d).

▶ **Drill Problem 6.15** Sketch the magnitude response and phase response for a system with the transfer function

$$H(s) = \frac{-2}{(s + 0.2)(s^2 + 2s + 5)}$$

Answer: Poles at $s = -0.2$, $s = -1 \pm j2$. See Figs. 6.27(a) and (b). ◀

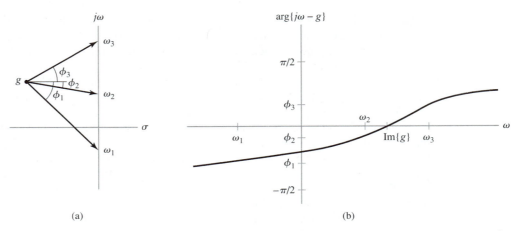

(a)   (b)

FIGURE 6.25 The phase angle of $j\omega - g$. (a) Vectors from g to $j\omega$ for several different values of ω. (b) Plot of $\arg\{j\omega - g\}$ as a continuous function of ω.

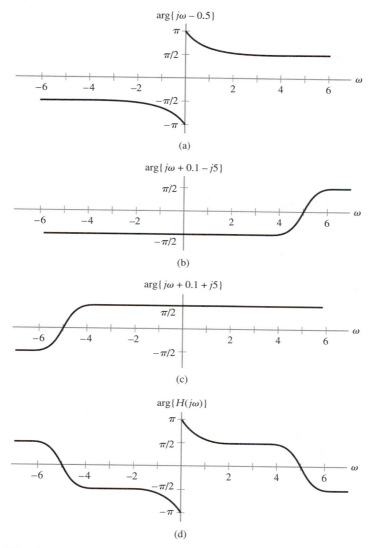

FIGURE 6.26 The phase response of the system in Example 6.22. (a) Phase of the zero at $s = 0.5$. (b) Phase of the pole at $s = -0.1 + j5$. (c) Phase of the pole at $s = -0.1 - j5$. (d) Phase response of the system.

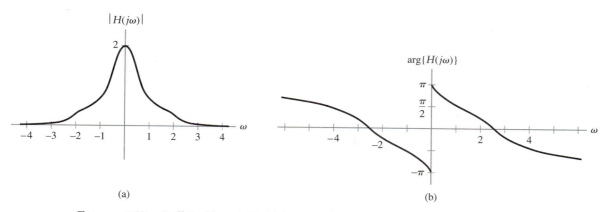

FIGURE 6.27 Drill Problem 6.15. (a) Magnitude response. (b) Phase response.

6.8 Exploring Concepts with MATLAB

The MATLAB Control System Toolbox contains numerous commands that are useful for working with Laplace transforms and continuous-time LTI systems described in terms of transfer functions, poles and zeros, or state-variable descriptions.

■ POLES AND ZEROS

The command `r = roots(a)` finds the roots of a polynomial described by the vector `a` and thus may be used to determine the zeros and poles of a Laplace transform expressed as a ratio of polynomials in s. The elements of the vector `a` correspond to descending powers of s.

For example, we may find the poles and zeros of the Laplace transform in Example 6.7,

$$X(s) = \frac{4s^2 + 6}{s^3 + s^2 - 2}$$

using the commands

```
>> z = roots([4, 0, 6])
z =

     0 + 1.2247i
     0 - 1.2247i
>> p = roots([1, 1, 0, -2])
p =

    -1.0000 + 1.0000i
    -1.0000 - 1.0000i
     1.0000
```

Hence we identify a pair of complex-conjugate zeros at $s = \pm j1.2247$, a pole at $s = 1$, and a pair of complex-conjugate poles at $s = -1 \pm j$.

The command `poly(r)` uses the poles or zeros specified in the vector `r` to determine the coefficients of the corresponding polynomial.

■ PARTIAL FRACTION EXPANSIONS

The `residue` command finds the partial fraction expansion for a ratio of two polynomials. The syntax is `[r,p,k] = residue(b,a)`, where `b` represents the numerator polynomial, `a` represents the denominator polynomial, `r` represents the partial fraction expansion coefficients or residues, `p` represents the poles, and `k` is a vector describing any terms in powers of s. If the numerator order is less than the denominator order, as assumed in Section 6.4, then `k` is an empty matrix.

To illustrate this command, we find the partial fraction expansion for the Laplace transform considered in Example 6.6,

$$X(s) = \frac{3s + 4}{s^3 + 5s^2 + 8s + 4}$$

using the commands

```
>> [r,p,k] = residue([3, 4], [1, 5, 8, 4])
r =

    -1.0000
     2.0000
     1.0000
p =

    -2.0000
    -2.0000
    -1.0000

k =

    [ ]
```

Hence the residue `r(1) = -1` corresponds to the pole at $s = -2$ given by `p(1)`, the residue `r(2) = 2` corresponds to the double pole at $s = -2$ given by `p(2)`, and the residue `r(3) = 1` corresponds to the pole at $s = -1$ given by `p(3)`. The partial fraction expansion is therefore

$$X(s) = \frac{-1}{s + 2} + \frac{2}{(s + 2)^2} + \frac{1}{s + 1}$$

This result agrees with Example 6.6.

▶ **Drill Problem 6.16**   Use `residue` to solve Drill Problem 6.6.

■ RELATING SYSTEM DESCRIPTIONS

Recall that a LTI system may be described in terms of a differential equation, impulse response, transfer function, poles and zeros, frequency response, or a state-variable description. The Control System Toolbox contains routines for relating transfer function, pole–zero, and state-variable representations of LTI systems. All of these are based on LTI objects that represent the different forms of the system description. Definition of state-space objects is accomplished with the MATLAB command `ss` as discussed in Section 2.7. The command `H = tf(b, a)` creates a LTI object `H` representing a transfer function with numerator and denominator polynomials defined by the coefficients in `b` and `a`, respectively. The coefficients in `b` and `a` are ordered in descending powers of s. The command `H = zpk(z, p, k)` creates a LTI object representing the pole–zero–gain form of system description. The zeros and poles are described by the vectors `z` and `p`, respectively, and the gain is represented by the scalar `k`.

The commands `ss`, `tf`, and `zpk` also perform model conversion when applied to a LTI object of a different form. For example, if `syszpk` is a LTI object representing a system in pole–zero–gain form, then the command `sysss = ss(syszpk)` generates a state-space object `sysss` representing the same system.

The commands `tzero(sys)` and `pole(sys)` find the zeros and poles of the LTI object `sys`, while `pzmap(sys)` produces a pole–zero plot. Additional commands that apply directly to LTI objects include `freqresp` for determining frequency response (see Section 4.13), `step` for determining step response, and `lsim` for simulating the system output in response to a specified input.

Consider a system containing zeros at $s = 0$, $s = \pm j10$ and poles at $s = -0.5 \pm j5$, $s = -3$, and $s = -4$ with gain 2. We may determine the transfer function representation for this system, plot the pole and zero locations in the s-plane, and plot the system's magnitude response using the MATLAB commands:

```
>> z = [0, j*10, -j*10];
p = [-0.5+j*5, -0.5-j*5, -3, -4];   k = 2;
>> syszpk = zpk(z, p, k)
Zero/pole/gain
    2 s (s ^ 2 + 100)
---------------------------
(s+4) (s+3) (s ^ 2 + s + 25.25)
>> systf = tf(syszpk) % convert to transfer function form
Transfer function:
    2 s ^ 3 + 200 s
-----------------------------------------------
s ^ 4 + 8 s ^ 3 + 44.25 s ^ 2 + 188.8 s + 303
>> pzmap(systf)   % generate pole-zero plot
>> w = [0:499]*20/500; % Frequencies from 0 to 20 rads/sec
>> H = freqresp(systf,w);
>> Hmag = abs(squeeze(H));   plot(w,Hmag)
```

Figure 6.28 depicts the pole–zero plot resulting from these commands while Fig. 6.29 illustrates the magnitude response of this system for $0 \leq \omega < 20$. Note that the magnitude response is zero at frequencies corresponding to the locations of zeros on the $j\omega$-axis, at $\omega = 0$ and $\omega = 10$. Similarly, the magnitude response is large at the frequency corresponding to the pole location near the $j\omega$-axis, at $\omega = 5$.

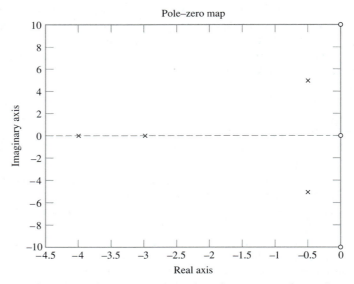

FIGURE 6.28   Pole and zero locations in the s-plane for a system obtained using MATLAB.

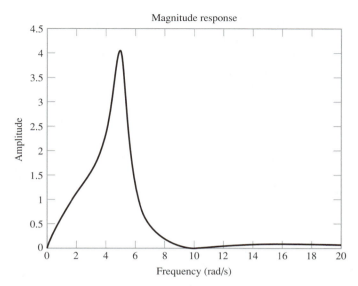

FIGURE 6.29 Magnitude response obtained using MATLAB.

6.9 Summary

The Laplace transform represents continuous-time signals as a weighted superposition of complex exponentials. Complex exponentials are more general signals than complex sinusoids; they include complex sinusoids as a special case. Correspondingly, the Laplace transform represents a more general class of signals than the Fourier transform, including signals that are not absolutely integrable. Hence we may use the Laplace transform to analyze signals and systems that are not stable. The transfer function is the Laplace transform of the impulse response and offers another description of the input–output characteristics of a system. The Laplace transform converts convolution of time signals to multiplication of Laplace transforms, so the Laplace transform of a system output is the product of the Laplace transform of the input and the transfer function. The locations of the poles and zeros of a transfer function in the s-plane offer yet another characterization for a LTI system. They provide information regarding system stability, causality, invertibility, and frequency response.

Complex exponentials are parameterized by a complex variable, s. The Laplace transform is a function of s and is represented in a complex plane termed the s-plane. The Fourier transform is obtained by evaluating the Laplace transform on the $j\omega$-axis, or by setting $s = j\omega$. The properties of the Laplace transform are analogous to those of the Fourier transform.

The unilateral or one-sided Laplace transform applies to causal signals and provides a convenient tool for solving system problems involving differential equations with initial conditions. The bilateral Laplace transform applies to two-sided signals. It is not unique unless the ROC is specified. The relative positions of the ROC and Laplace transform poles in the s-plane determine whether a signal is left-sided, right-sided or two-sided.

The Fourier and Laplace transforms have many similarities and can often be used interchangeably, yet have distinctly different roles in signal and system analysis. The Laplace transform is most often used to solve problems involving transient and stability analysis of systems. Problems of this type occur frequently in control system applications, where we are interested in the manner in which the system output tracks a desired system output. The system poles provide essential information about a LTI system's stability and transient response characteristics, while the unilateral transform can be used to obtain a LTI system's response given the input and initial conditions. There is no Fourier transform counterpart with these capabilities. The role of the Laplace transform in transient and stability analysis of control systems is further explored in Chapter 9.

In contrast, the Fourier transform is usually employed as a signal representation tool and for systems problems in which steady-state characteristics are of interest. The Fourier transform is easier to visualize and use than the Laplace transform in such problems because it is a function of the real-valued frequency ω, while the Laplace transform is a function of the complex frequency $s = \sigma + j\omega$. Examples of the signal representation and steady-state system analysis problems for which the Fourier transform is used were given in Chapters 4 and 5 and are further developed in the context of filtering in Chapter 8.

FURTHER READING

1. The Laplace transform is named after Pierre-Simon de Laplace (1749–1827). Laplace studied a wide variety of natural phenomena, including hydrodynamics, propagation of sound, heat, tides, and the liquid state of matter, although the majority of his life was devoted to celestial mechanics. Laplace presented complete analytic solutions to the mechanical problems of the solar system in a series of five volumes titled *Mécanique céleste*. A famous passage in another of his books asserted that the future of the world is determined entirely by the past, that if one possessed knowledge of the "state" of the world at any instant, then one could predict the future. Although Laplace made many important discoveries in mathematics, his primary interest was the study of nature. Mathematics was simply a means to this end.

2. The following texts are devoted specifically to the Laplace transform and applications thereof:

   ▶ Holbrook, J. G., *Laplace Transforms for Electronic Engineers*, Second Edition (Pergamon Press, 1966)

   ▶ Kuhfittig, P. K. F., *Introduction to the Laplace Transform* (Plenum Press, 1978)

   ▶ Thomson, W. T., *Laplace Transformation*, Second Edition (Prentice Hall, 1960)

PROBLEMS

6.1 A signal $x(t)$ has Laplace transform $X(s)$ as given below. Plot the poles and zeros in the s-plane and determine the Fourier transform of $x(t)$ without inverting $X(s)$. Assume $x(t)$ is absolutely integrable.

(a) $X(s) = \dfrac{s^2 - 1}{s^2 + 3s + 2}$

(b) $X(s) = \dfrac{2s^2}{s^2 + 2\sqrt{2}s + 4}$

(c) $X(s) = \dfrac{1}{s - 3} + \dfrac{2}{s + 2}$

6.2 Determine the bilateral Laplace transform and ROC for each of the following signals:

(a) $x(t) = u(t - 2)$

(b) $x(t) = e^{2t}u(-t + 2)$

(c) $x(t) = \delta(t - t_o)$

(d) $x(t) = \cos(2t)u(t)$

6.3 Determine the unilateral Laplace transform of each of the following signals using the defining equation:

(a) $x(t) = u(t - 1)$

(b) $x(t) = u(t + 1)$

(c) $x(t) = e^{-t+2}u(t)$

(d) $x(t) = \cos(\omega_o t)u(t - 3)$

(e) $x(t) = \sin(\omega_o t)u(t + 2)$

(f) $x(t) = e^{2t}[u(t) - u(t - 4)]$

(g) $x(t) = \begin{cases} \sin(\pi t), & 0 < t < 1 \\ 0, & \text{otherwise} \end{cases}$

6.4 Use the Laplace transform properties and the table of basic Laplace transforms to determine the unilateral Laplace transform of each of the following signals:

(a) $x(t) = t^2 e^{-2t}u(t)$

(b) $x(t) = e^{-t}u(t) * \sin(3\pi t)u(t)$

(c) $x(t) = \dfrac{d}{dt}\{tu(t)\}$

(d) $x(t) = tu(t) - (t - 1)u(t - 1) - (t - 2)u(t - 2) + (t - 3)u(t - 3)$

(e) $x(t) = \int_0^t e^{-2\tau}\cos(3\tau)\,d\tau$

(f) $x(t) = 2t\dfrac{d}{dt}(e^{-t}\sin(t)u(t))$

6.5 Use the Laplace transform properties and the table of basic Laplace transforms to determine the time signals corresponding to the following unilateral Laplace transforms:

(a) $X(s) = \left(\dfrac{1}{s}\right)\left(\dfrac{1}{s+1}\right)$

(b) $X(s) = \dfrac{d}{ds}\left(e^{-2s}\dfrac{1}{(s+2)^2}\right)$

(c) $X(s) = \dfrac{1}{(2s+1)^2 + 4}$

(d) $X(s) = s\dfrac{d^2}{ds^2}\left(\dfrac{4}{s^2+4}\right)$

6.6 Given the transform pair $\cos(2t)u(t) \overset{\mathcal{L}_u}{\longleftrightarrow} X(s)$, determine the time signals corresponding to the following Laplace transforms:

(a) $sX(s) - 1$

(b) $X(2s)$

(c) $X(s + 1)$

(d) $s^{-1}X(s)$

(e) $\dfrac{d}{ds}(e^{-2s}X(s))$

6.7 Given the transform pair $x(t) \overset{\mathcal{L}_u}{\longleftrightarrow} 2s/(s^2 - 2)$, determine the Laplace transform of each of the following time signals:

(a) $x(2t)$

(b) $x(t - 3)$

(c) $x(t) * tx(t)$

(d) $e^{-2t}x(t)$

(e) $2\dfrac{d}{dt}x(t)$

(f) $\int_0^t x(2\tau)\,d\tau$

6.8 Use the s-domain shift property and the transform pair $e^{-at}u(t) \overset{\mathcal{L}_u}{\longleftrightarrow} 1/(s + a)$ to derive the unilateral Laplace transform of $x(t) = e^{-at}\cos(\omega_1 t)u(t)$.

6.9 Prove the following properties of the unilateral Laplace transform:

(a) Linearity

(b) Scaling

(c) Time shift

(d) s-domain shift

(e) Convolution

(f) Differentiation in the s-domain

6.10 Determine the initial value $x(0^+)$ given the following Laplace transforms:

(a) $X(s) = \dfrac{3}{s^2 + 5s - 1}$

(b) $X(s) = \dfrac{2s + 3}{s^2 + 5s - 6}$

(c) $X(s) = e^{-5s}\dfrac{3s^2 + 2s}{s^2 + s - 1}$

6.11 Determine the final value $x(\infty)$ given the following Laplace transforms:

(a) $X(s) = \dfrac{2s^2 + 3}{s^2 + 5s + 1}$

(b) $X(s) = \dfrac{2s + 3}{s^3 + 5s^2 + 6s}$

(c) $X(s) = e^{-3s}\dfrac{3s^2 + 2}{s(s + 1)(s + 2)}$

6.12 Use the method of partial fractions to find the time signals corresponding to the following unilateral Laplace transforms:

(a) $X(s) = \dfrac{-s - 4}{s^2 + 3s + 2}$

(b) $X(s) = \dfrac{s}{s^2 + 5s + 6}$

(c) $X(s) = \dfrac{2s - 1}{s^2 + 2s + 1}$

(d) $X(s) = \dfrac{5s + 4}{s^3 + 3s^2 + 2s}$

(e) $X(s) = \dfrac{3s^2 + 8s + 5}{(s + 2)(s^2 + 2s + 1)}$

(f) $X(s) = \dfrac{3s + 2}{s^2 + 4s + 5}$

(g) $X(s) = \dfrac{4s^2 + 8s + 10}{(s + 2)(s^2 + 2s + 5)}$

(h) $X(s) = \dfrac{-9}{(s + 1)(s^2 + 2s + 10)}$

(i) $X(s) = \dfrac{s + 4 + e^{-2s}}{s^2 + 5s + 6}$

6.13 Determine the forced and natural responses for the systems described by the following differential equations with the specified input and initial conditions:

(a) $5\dfrac{d}{dt}y(t) + 10y(t) = 2x(t)$, $y(0^+) = 2$,

$x(t) = u(t)$

(b) $\dfrac{d^2}{dt^2}y(t) + 5\dfrac{d}{dt}y(t) + 6y(t) =$

$-4x(t) - 3\dfrac{d}{dt}x(t)$, $y(0^+) = -1$,

$\dfrac{d}{dt}y(t)|_{t=0^+} = 5$, $x(t) = e^{-t}u(t)$

(c) $\dfrac{d^2}{dt^2}y(t) + 4y(t) = 8x(t)$, $y(0^+) = 1$,

$\dfrac{d}{dt}y(t)|_{t=0^+} = 2$, $x(t) = u(t)$

(d) $\dfrac{d^2}{dt^2}y(t) + 2\dfrac{d}{dt}y(t) + 5y(t) = \dfrac{d}{dt}x(t)$,

$y(0^+) = 2$, $\dfrac{d}{dt}y(t)|_{t=0^+} = 0$, $x(t) = e^{-t}u(t)$

6.14 Use Laplace transform circuit models to determine the current $y(t)$ in the circuit of Fig. P6.14 in response to an applied voltage $x(t) = e^{-t}u(t)$ V, assuming normalized values $R = 1\ \Omega$ and $L = \frac{1}{2}$ H. The current through the inductor at time $t = 0^+$ is 2 A.

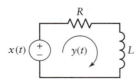

FIGURE P6.14

6.15 The circuit in Fig. P6.15 represents a system with input $x(t)$ and output $y(t)$. Determine the forced response and the natural response of this system under the specified conditions.

(a) $R = 3\ \Omega, L = 1$ H, $C = \frac{1}{2}$ F, $x(t) = e^{-3t}u(t)$ V, current through the inductor at $t = 0^+$ is 2 A, and the voltage across the capacitor at $t = 0^+$ is 1 V.

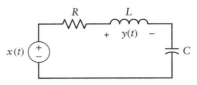

FIGURE P6.15

(b) $R = 2\ \Omega, L = 1$ H, $C = \frac{1}{5}$ F, $x(t) = 2e^{-t}u(t)$ V, current through the inductor at $t = 0^+$ is 2 A, and the voltage across the capacitor at $t = 0^+$ is 1 V.

6.16 Determine the bilateral Laplace transform and the corresponding ROC for each of the following signals:

(a) $x(t) = e^{-2t}u(t) + e^{-t}u(t) + e^t u(-t)$

(b) $x(t) = e^{2t}\cos(2t)u(-t) + e^{-t}u(t) + e^t u(t)$

(c) $x(t) = e^{2t+4}u(t + 2)$

(d) $x(t) = \cos(3t)u(-t) * e^{-t}u(t)$

(e) $x(t) = e^t \sin(2t + 4)u(t + 2)$

(f) $x(t) = e^{-t}\dfrac{d}{dt}(e^{-t}u(t + 1))$

(g) $x(t) = \int_{-\infty}^{t} e^{2\tau}\sin(\tau)u(-\tau)\,d\tau$

6.17 Use the tables of transforms and properties to determine the time signals that correspond to the following bilateral Laplace transforms:

(a) $X(s) = e^{5s}\dfrac{1}{s + 2}$ with ROC Re$(s) > -2$

(b) $X(s) = \dfrac{d^2}{ds^2}\left(\dfrac{1}{s - 1}\right)$ with ROC Re$(s) < 1$

(c) $X(s) = s\left(\dfrac{1}{s^2} - \dfrac{e^{-s}}{s^2} - \dfrac{e^{-2s}}{s}\right)$ with ROC Re$(s) < 0$

(d) $X(s) = s^{-1}\dfrac{d}{ds}\left(\dfrac{e^{-2s}}{s}\right)$ with ROC Re$(s) > 0$

6.18 Use the method of partial fractions to determine the time signals corresponding to the following bilateral Laplace transforms:

(a) $X(s) = \dfrac{s + 3}{s^2 + 3s + 2}$

(i) with ROC Re$(s) < -2$

(ii) with ROC Re$(s) > -1$

(iii) with ROC $-2 <$ Re$(s) < -1$

(b) $X(s) = \dfrac{s^2 + 7}{(s + 1)(s^2 - 2s + 4)}$

(i) with ROC Re$(s) < -1$

(ii) with ROC Re$(s) > 1$

(iii) with ROC $-1 <$ Re$(s) < 1$

(c) $X(s) = \dfrac{2s^2 + 4s + 2}{s^2 + 2s}$

(i) with ROC $\text{Re}(s) < -2$

(ii) with ROC $\text{Re}(s) > 0$

(iii) with ROC $-2 < \text{Re}(s) < 0$

(d) $X(s) = \dfrac{s^2 + 3s + 4}{s^2 + 2s + 1}$ with ROC

$\text{Re}(s) < -1$

6.19 A system has transfer function $H(s)$ as given below. Determine the impulse response assuming

(i) that the system is causal

(ii) that the system is stable.

(a) $H(s) = \dfrac{3s - 1}{s^2 - 1}$

(b) $H(s) = \dfrac{5s + 7}{s^2 + 3s + 2}$

(c) $H(s) = \dfrac{s^2 + 5s - 9}{(s + 1)(s^2 - 2s + 10)}$

6.20 A stable system has input $x(t)$ and output $y(t)$ as given below. Use Laplace transforms to determine the transfer function and impulse response of the system.

(a) $x(t) = u(t)$, $y(t) = e^{-t}\cos(2t)u(t)$

(b) $x(t) = e^{-2t}u(t)$,
$y(t) = -2e^{-t}u(t) + 2e^{-3t}u(t)$

6.21 The relationship between the input $x(t)$ and output $y(t)$ of a causal system is described by each of the differential equations given below. Use Laplace transforms to determine the transfer function and impulse response of each system.

(a) $5\dfrac{d}{dt}y(t) + 10y(t) = 2x(t)$

(b) $\dfrac{d^2}{dt^2}y(t) + 5\dfrac{d}{dt}y(t) + 6y(t) = x(t) + \dfrac{d}{dt}x(t)$

(c) $\dfrac{d^2}{dt^2}y(t) - 2\dfrac{d}{dt}y(t) + 10y(t) = x(t) +$

$2\dfrac{d}{dt}x(t)$

6.22 Determine the differential-equation description for a system with each of the following transfer functions:

(a) $H(s) = \dfrac{2s + 1}{s(s + 2)}$

(b) $H(s) = \dfrac{3s}{s^2 + 2s + 10}$

(c) $H(s) = \dfrac{2(s + 1)(s - 2)}{(s + 1)(s + 2)(s + 3)}$

6.23 Determine the transfer function, impulse response, and differential-equation descriptions

for a stable system represented by the following state-variable descriptions:

(a) $\mathbf{A} = \begin{bmatrix} -1 & 0 \\ 0 & -3 \end{bmatrix}$, $\mathbf{b} = \begin{bmatrix} 1 \\ 2 \end{bmatrix}$,

$\mathbf{c} = [1 \quad -1]$, $D = [0]$

(b) $\mathbf{A} = \begin{bmatrix} 1 & 2 \\ 1 & -6 \end{bmatrix}$, $\mathbf{b} = \begin{bmatrix} 1 \\ 2 \end{bmatrix}$,

$\mathbf{c} = [0 \quad 1]$, $D = [0]$

6.24 Determine whether the systems described by the following transfer functions can be

(i) both stable and causal

(ii) whether a stable and causal inverse system exists.

(a) $H(s) = \dfrac{(s + 1)(s + 2)}{(s + 1)(s^2 + 2s + 10)}$

(b) $H(s) = \dfrac{s^2 - 2s - 3}{(s + 2)(s^2 + 4s + 5)}$

(c) $H(s) = \dfrac{s^2 + 3s + 2}{(s + 2)(s^2 - 2s + 8)}$

(d) $H(s) = \dfrac{s^2 + 2s}{(s^2 + 3s + 2)(s^2 + s - 2)}$

6.25 The relationship between the input $x(t)$ and output $y(t)$ of a system is described by the differential equation

$\dfrac{d^2}{dt^2}y(t) + \dfrac{d}{dt}y(t) - 6y(t)$

$= \dfrac{d^2}{dt^2}x(t) - \dfrac{d}{dt}x(t) - 2x(t)$

(a) Does this system have a stable and causal inverse? Why?

(b) Find a differential-equation description for the inverse system.

6.26 Sketch the magnitude responses for the systems described by the following transfer functions using the relationship between the pole and zero locations and the $j\omega$-axis in the s-plane:

(a) $H(s) = \dfrac{s}{s^2 + 2s + 101}$

(b) $H(s) = \dfrac{s^2 + 25}{s + 1}$

(c) $H(s) = \dfrac{s - 2}{s + 2}$

*6.27 Suppose a system has M poles at $d_k = \alpha_k + j\beta_k$ and M zeros at $c_k = -\alpha_k + j\beta_k$. That is, the pole and zero locations are symmetric about the $j\omega$-axis.

(a) Show that the magnitude response of any system that satisfies this condition is unity. Such a system is termed an *allpass* system since it passes all frequencies with unit gain.

(b) Evaluate the phase response of a single real pole–zero pair; that is, sketch the phase response of $(s - \alpha)/(s + \alpha)$, where $\alpha > 0$.

6.28 Sketch the phase responses for the systems described by the following transfer functions using the relationship between the pole and zero locations and the $j\omega$-axis in the s-plane:

(a) $H(s) = \dfrac{s - 1}{s + 2}$

(b) $H(s) = \dfrac{s + 1}{s + 2}$

(c) $H(s) = \dfrac{1}{s^2 + 2s + 101}$

(d) $H(s) = s^2$

*6.29 Consider the nonminimum phase system described by the transfer function

$$H(s) = \frac{(s + 2)(s - 1)}{(s + 4)(s + 3)(s + 5)}$$

(a) Does this system have a stable and causal inverse system?

(b) Express $H(s)$ as the product of a minimum phase system, $H_{min}(s)$, and an allpass system, $H_{ap}(s)$, containing a single pole and zero. (See Problem 6.27 for the definition of an all-pass system.)

(c) Let $H_{min}^{-1}(s)$ be the inverse system for $H_{min}(s)$. Find $H_{min}^{-1}(s)$. Can it be both stable and causal?

(d) Sketch the magnitude response and phase response of the system $H(s)H_{min}^{-1}(s)$.

(e) Generalize your results from parts (b) and (c) to an arbitrary nonminimum phase system $H(s)$ and determine the magnitude response of the system $H(s)H_{min}^{-1}(s)$.

▶ *Computer Experiments*

6.30 Use the MATLAB command `roots` to determine the poles and zeros of the following systems:

(a) $H(s) = \dfrac{s^2 + 2}{s^3 + 2s^2 - s + 1}$

(b) $H(s) = \dfrac{s^3 + 1}{s^4 + 2s^2 + 1}$

(c) $H(s) = \dfrac{4s^2 + 8s + 10}{2s^3 + 8s^2 + 18s + 20}$

6.31 Use the MATLAB command `pzmap` to plot the poles and zeros for the following systems:

(a) $H(s) = \dfrac{s^3 + 1}{s^4 + 2s^2 + 1}$

(b) $\mathbf{A} = \begin{bmatrix} 1 & 2 \\ 1 & -6 \end{bmatrix}, \quad \mathbf{b} = \begin{bmatrix} 1 \\ 2 \end{bmatrix},$
$\mathbf{c} = [0 \quad 1], \quad \mathbf{D} = [0]$

6.32 Use the MATLAB command `freqresp` to evaluate and plot the magnitude and phase responses for Examples 6.21 and 6.22.

6.33 Use the MATLAB command `freqresp` to evaluate and plot the magnitude and phase responses for Problem 6.26.

6.34 Use your knowledge of the effect of poles and zeros on the magnitude response to design systems having the magnitude response specified below. Place poles and zeros in the s-plane, and evaluate the corresponding magnitude response using the MATLAB command `freqresp`. Repeat this process until you find pole and zero locations that satisfy the specifications.

(a) Design a highpass filter with two poles and two zeros that satisfies $|H(j0)| = 0$, $0.8 \leq |H(j\omega)| \leq 1.2$ for $|\omega| > 100\pi$, and has real-valued coefficients.

(b) Design a lowpass filter with real-valued coefficients that satisfies $0.8 \leq |H(j\omega)| \leq 1.2$ for $|\omega| < \pi$, and $|H(j\omega)| < 0.1$ for $|\omega| > 10\pi$.

6.35 Use the MATLAB command `ss` to find state-variable descriptions for the systems in Problem 6.22.

6.36 Use the MATLAB command `tf` to find transfer function descriptions for the systems in Problem 6.23.

7 Representation of Signals Using Discrete-Time Complex Exponentials: The z-Transform

7.1 Introduction

In this chapter we generalize the complex sinusoidal representation of a discrete-time signal offered by the DTFT to a representation in terms of complex exponential signals, which is termed the *z-transform*. The z-transform is the discrete-time counterpart to the Laplace transform. By using this more general representation for discrete-time signals, we are able to obtain a broader characterization of systems and their interaction with signals than possible with the DTFT. For example, the DTFT can be applied only to stable systems, since the DTFT exists only if the impulse response is absolutely summable. In contrast, the z-transform of the impulse response exists for unstable systems, and thus the z-transform can be used to study a much larger class of systems and signals.

As in continuous time, we shall see that discrete-time complex exponentials are eigenfunctions of LTI systems. This characteristic endows the z-transform with a powerful set of properties for analysis of signals and systems. Many of these properties parallel those of the DTFT; for example, convolution of time signals corresponds to multiplication of z-transforms. Hence the output of a LTI system is obtained by multiplying the z-transform of the input with the z-transform of the impulse response. We define the z-transform of the impulse response as the transfer function of a system. The transfer function generalizes the frequency response characterization of a system's input–output behavior and offers new insights into system characteristics.

The primary roles of the z-transform in engineering practice are the study of system characteristics and derivation of computational structures for implementing discrete-time systems on computers. The unilateral z-transform is also used to solve difference equations subject to initial conditions. We shall explore such problems in this chapter as we study the z-transform.

7.2 The z-Transform

We shall derive the z-transform by examining the effect of applying a complex exponential input to a LTI system. Let $z = re^{j\Omega}$ be a complex number with magnitude r and phase Ω. The signal $x[n] = z^n$ is a complex exponential signal. Substitute $re^{j\Omega}$ for z to write $x[n]$ as

$$x[n] = r^n \cos(\Omega n) + jr^n \sin(\Omega n) \tag{7.1}$$

As illustrated in Fig. 7.1, the real part of $x[n]$ is an exponentially damped cosine and the imaginary part is an exponentially damped sine. The positive number r determines the damping factor and Ω is the sinusoidal frequency. Note that $x[n]$ is a complex sinusoid if $r = 1$.

Consider applying $x[n] = z^n$ to a LTI system with impulse response $h[n]$. The system output $y[n]$ is given by

$$
\begin{aligned}
y[n] &= H\{x[n]\} \\
&= h[n] * x[n] \\
&= \sum_{k=-\infty}^{\infty} h[k]x[n-k]
\end{aligned}
$$

Substitute $x[n] = z^n$ to obtain

$$
\begin{aligned}
y[n] &= \sum_{k=-\infty}^{\infty} h[k]z^{n-k} \\
&= z^n \left(\sum_{k=-\infty}^{\infty} h[k]z^{-k} \right)
\end{aligned}
$$

Define the *transfer function*

$$
H(z) = \sum_{k=-\infty}^{\infty} h[k]z^{-k} \tag{7.2}
$$

so that we may write

$$
H\{z^n\} = H(z)z^n
$$

This equation has the form of an eigenrelation, where z^n is the eigenfunction and $H(z)$ is the eigenvalue. The action of the system on an input z^n is equivalent to multiplication

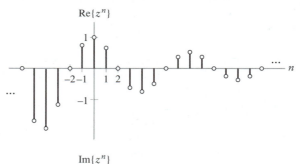

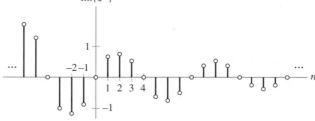

FIGURE 7.1   Real and imaginary parts of the signal z^n.

of the input by the complex number $H(z)$. If we express $H(z)$ in polar form as $H(z) = |H(z)|e^{j\phi(z)}$, then the system output is written as

$$y[n] = |H(z)|e^{j\phi(z)}z^n$$

Substitute $z = re^{j\Omega}$ to obtain

$$y[n] = |H(re^{j\Omega})|r^n \cos(\Omega n + \phi(re^{j\Omega})) + j|H(re^{j\Omega})|r^n \sin(\Omega n + \phi(re^{j\Omega}))$$

Comparing $y[n]$ to $x[n]$ in Eq. (7.1), we see that the system modifies the amplitude of the input by $|H(re^{j\Omega})|$ and shifts the phase of the sinusoidal components by $\phi(re^{j\Omega})$.

We now seek to represent arbitrary signals as a weighted superposition of the eigenfunctions z^n. Substituting $z = re^{j\Omega}$ into Eq. (7.2) yields

$$H(re^{j\Omega}) = \sum_{n=-\infty}^{\infty} h[n](re^{j\Omega})^{-n}$$

$$= \sum_{n=-\infty}^{\infty} (h[n]r^{-n})e^{-j\Omega n}$$

We see that $H(re^{j\Omega})$ corresponds to the DTFT of a signal $h[n]r^{-n}$. Hence the inverse DTFT of $H(re^{j\Omega})$ must be $h[n]r^{-n}$, and so we may write

$$h[n]r^{-n} = \frac{1}{2\pi} \int_{-\pi}^{\pi} H(re^{j\Omega})e^{j\Omega n} \, d\Omega$$

Multiplying this result by r^n gives

$$h[n] = \frac{r^n}{2\pi} \int_{-\pi}^{\pi} H(re^{j\Omega})e^{j\Omega n} \, d\Omega$$

$$= \frac{1}{2\pi} \int_{-\pi}^{\pi} H(re^{j\Omega})(re^{j\Omega})^n \, d\Omega$$

We may convert this to an integral over z by subtituting $re^{j\Omega} = z$. Since the integration is performed only over Ω, r may be considered a constant and we have $dz = jre^{j\Omega} \, d\Omega$. Hence we identify $d\Omega = (1/j)z^{-1} \, dz$. Lastly, consider the limits on the integral. As Ω goes from $-\pi$ to π, z traverses a circle of radius r in a counterclockwise direction. Thus we write

$$h[n] = \frac{1}{2\pi j} \oint H(z)z^{n-1} \, dz \tag{7.3}$$

where the symbol $\oint$ denotes integration around a circle of radius $|z| = r$ in a counterclockwise direction. Equation (7.2) indicates how to determine $H(z)$ from $h[n]$, while Eq. (7.3) expresses $h[n]$ as a function of $H(z)$. We say that the transfer function $H(z)$ is the *z-transform* of the impulse response $h[n]$.

More generally, the *z*-transform of an arbitrary signal $x[n]$ is

$$X[z] = \sum_{n=-\infty}^{\infty} x[n]z^{-n} \tag{7.4}$$

and the inverse *z*-transform is

$$x[n] = \frac{1}{2\pi j} \oint X(z)z^{n-1} \, dz \tag{7.5}$$

We express the relationship between $x[n]$ and $X(z)$ with the notation

$$x[n] \overset{z}{\longleftrightarrow} X(z)$$

Note that Eq. (7.5) expresses the signal $x[n]$ as a weighted superposition of complex exponentials z^n. The weights are $(1/2\pi j)X(z)z^{-1}\,dz$. In practice we will not evaluate this integral directly since this requires knowledge of complex variable theory, which is beyond the scope of this book. Instead, we evaluate inverse z-transforms by inspection, using the one-to-one relationship between $x[n]$ and $X(z)$.

The z-transform exists when the infinite sum in Eq. (7.4) converges. A necessary condition for convergence is absolute summability of $x[n]z^{-n}$. Since $|x[n]z^{-n}| = |x[n]r^{-n}|$, we must have

$$\sum_{n=-\infty}^{\infty} |x[n]r^{-n}| < \infty$$

The range of r for which this condition is satisfied is termed the *region of convergence* (ROC).

The z-transform exists for signals that do not have a DTFT. Recall that existence of the DTFT requires absolute summability of $x[n]$. By limiting ourselves to restricted values of r, we may ensure that $x[n]r^{-n}$ is absolutely summable even though $x[n]$ is not. For example, the DTFT of $x[n] = \alpha^n u[n]$ does not exist for $|\alpha| > 1$ since then $x[n]$ is an increasing exponential signal, as illustrated in Fig. 7.2(a). However, if $r > \alpha$, then r^{-n}, depicted in Fig. 7.2(b), decays faster than $x[n]$ grows. Hence the signal $x[n]r^{-n}$, depicted in Fig. 7.2(c), is absolutely summable and the z-transform exists. The ability to work with signals that do not have a DTFT is a significant advantage offered by the z-transform.

It is convenient to represent the complex number z as a location in a complex plane termed the *z-plane* and depicted graphically in Fig. 7.3. The point $z = re^{j\Omega}$ is located at a distance r from the origin and angle Ω from the real axis. Note that if $x[n]$ is absolutely

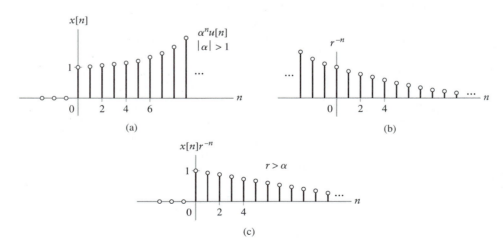

FIGURE 7.2 Illustration of a signal that has a z-transform but does not have a DTFT. (a) An increasing exponential signal for which the DTFT does not exist. (b) The attenuating factor r^{-n} associated with the z-transform. (c) The modified signal $x[n]r^{-n}$ is absolutely summable provided $r > \alpha$ and thus the z-transform of $x[n]$ exists.

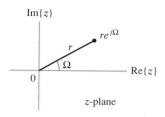

FIGURE 7.3   The z-plane. A point $z = re^{j\Omega}$ is located at a distance r from the origin and angle Ω relative to the real axis.

summable, then the DTFT is obtained from the z-transform by setting $r = 1$, or substituting $z = e^{j\Omega}$ into Eq. (7.4). That is,

$$X(e^{j\Omega}) = X(z)|_{z=e^{j\Omega}}$$

The equation $z = e^{j\Omega}$ describes a circle of unit radius centered on the origin in the z-plane as illustrated in Fig. 7.4. This contour is termed the *unit circle* in the z-plane. Frequency Ω in the DTFT corresponds to the point on the unit circle at an angle of Ω with respect to the real axis. As discrete-time frequency Ω goes from $-\pi$ to π, we take one trip around the unit circle. We say that the DTFT corresponds to the z-transform evaluated on the unit circle.

EXAMPLE 7.1   Determine the z-transform of the signal

$$x[n] = \begin{cases} 1, & n = -1 \\ 2, & n = 0 \\ -1, & n = 1 \\ 1, & n = 2 \\ 0, & \text{otherwise} \end{cases}$$

Use the z-transform to determine the DTFT of $x[n]$.

Solution: Substitute $x[n]$ into Eq. (7.4) to obtain

$$X(z) = z + 2 - z^{-1} + z^{-2}$$

We obtain the DTFT from $X(z)$ by substituting $z = e^{j\Omega}$:

$$X(e^{j\Omega}) = e^{j\Omega} + 2 - e^{-j\Omega} + e^{-j2\Omega}$$

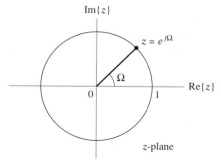

FIGURE 7.4   The unit circle, $z = e^{j\Omega}$, in the z-plane.

The most commonly encountered form of the z-transform in engineering applications is a ratio of two polynomials in z^{-1}, as shown by the rational function

$$X(z) = \frac{b_0 + b_1 z^{-1} + \cdots + b_M z^{-M}}{1 + a_1 z^{-1} + \cdots + a_N z^{-N}}$$

It is useful to rewrite $X(z)$ as a product of terms involving the roots of the numerator and denominator polynomials; that is,

$$X(z) = \frac{b_0 \Pi_{k=1}^{M}(1 - c_k z^{-1})}{\Pi_{k=1}^{N}(1 - d_k z^{-1})}$$

The c_k are the roots of the numerator polynomial and are termed the *zeros* of $X(z)$. The d_k are the roots of the denominator polynomial and are termed the *poles* of $X(z)$. Zero locations are denoted with the "○" symbol and pole locations with the "×" symbol in the z-plane. The pole and zero locations completely specify $X(z)$ except for the gain factor b_0.

EXAMPLE 7.2   Determine the z-transform of the signal

$$x[n] = \alpha^n u[n]$$

Depict the ROC and pole and zero locations of $X(z)$ in the z-plane.

Solution: Substitute $x[n]$ into Eq. (7.4), obtaining

$$X(z) = \sum_{n=-\infty}^{\infty} \alpha^n u[n] z^{-n}$$
$$= \sum_{n=0}^{\infty} \left(\frac{\alpha}{z}\right)^n$$

This is a geometric series of infinite length in the ratio α/z; the sum converges provided $|\alpha/z| < 1$, or $|z| > |\alpha|$. Hence

$$X(z) = \frac{1}{1 - \alpha z^{-1}}, \qquad |z| > |\alpha|$$
$$= \frac{z}{z - \alpha}, \qquad |z| > |\alpha| \tag{7.6}$$

There is a pole at $z = \alpha$ and a zero at $z = 0$ as illustrated in Fig. 7.5. The ROC is depicted as the shaded region of the z-plane.

As with the Laplace transform, the expression for $X(z)$ does not correspond to a unique time signal unless the ROC is specified. This means that two different time signals may have identical z-transforms but different ROCs, as demonstrated in the following example.

EXAMPLE 7.3   Determine the z-transform of the signal

$$y[n] = -\alpha^n u[-n - 1]$$

Depict the ROC and pole and zero locations in the z-plane.

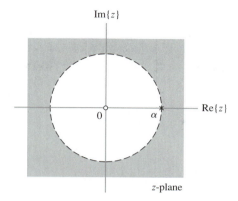

FIGURE 7.5   Pole and zero locations in the z-plane for $x[n] = \alpha^n u[n]$. The ROC is the shaded area.

Solution: Substitute $y[n]$ into Eq. (7.4) and write

$$Y(z) = \sum_{n=-\infty}^{\infty} -\alpha^n u[-n-1]z^{-n}$$

$$= -\sum_{n=-\infty}^{-1} \left(\frac{\alpha}{z}\right)^n$$

$$= -\sum_{k=1}^{\infty} \left(\frac{z}{\alpha}\right)^k$$

$$= 1 - \sum_{k=0}^{\infty} \left(\frac{z}{\alpha}\right)^k$$

The sum converges provided $|z/\alpha| < 1$, or $|z| < |\alpha|$. Hence

$$Y(z) = 1 - \frac{1}{1 - z\alpha^{-1}}, \qquad |z| < |\alpha|$$

$$= \frac{z}{z - \alpha}, \qquad |z| < |\alpha| \tag{7.7}$$

The ROC and pole and zero locations are depicted in Fig. 7.6.

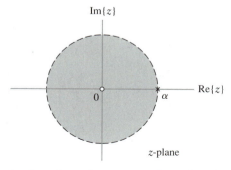

FIGURE 7.6   ROC and pole and zero locations in the z-plane for $y[n] = -\alpha^n u[-n-1]$.

Note that $Y(z)$ in Eq. (7.7) is identical to $X(z)$ in Eq. (7.6) even though the time signals are quite different. Only the ROC differentiates them. We must know the ROC to determine the correct inverse z-transform. This ambiguity generally occurs with signals that are one sided.

EXAMPLE 7.4   Determine the z-transform of

$$x[n] = -u[-n-1] + (\tfrac{1}{2})^n u[n]$$

Depict the ROC and pole and zero locations of $X(z)$ in the z-plane.

Solution: Substitute for $x[n]$ in Eq. (7.4), obtaining

$$X(z) = \sum_{n=-\infty}^{\infty} \left((\tfrac{1}{2})^n u[n] z^{-n} - u[-n-1] z^{-n} \right)$$

$$= \sum_{n=0}^{\infty} \left(\frac{1}{2z} \right)^n - \sum_{n=-\infty}^{-1} \left(\frac{1}{z} \right)^n$$

$$= \sum_{n=0}^{\infty} \left(\frac{1}{2z} \right)^n + 1 - \sum_{k=0}^{\infty} z^k$$

Both of the sums in this example must converge in order for $X(z)$ to converge. This implies we must have $|z| > \tfrac{1}{2}$ and $|z| < 1$. Hence

$$X(z) = \frac{1}{1 - \tfrac{1}{2}z^{-1}} + 1 - \frac{1}{1-z}, \qquad \tfrac{1}{2} < |z| < 1$$

$$= \frac{z(2z - \tfrac{3}{2})}{(z - \tfrac{1}{2})(z - 1)}, \qquad \tfrac{1}{2} < |z| < 1$$

The ROC and pole and zero locations are depicted in Fig. 7.7. In this case the ROC is a ring in the z-plane.

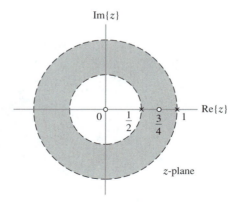

FIGURE 7.7   ROC and pole and zero locations in the z-plane for Example 7.4.

▶ **Drill Problem 7.1**   Determine the z-transform, ROC, and pole and zero locations of $X(z)$ for

$$x[n] = (\tfrac{1}{2})^n u[n] + (-\tfrac{1}{3})^n u[n]$$

Answer:

$$X(z) = \frac{z(2z - \tfrac{1}{6})}{(z - \tfrac{1}{2})(z + \tfrac{1}{3})}, \qquad |z| > \tfrac{1}{2}$$

Poles at $z = \tfrac{1}{2}$, $z = -\tfrac{1}{3}$ and zeros at $z = 0$, $z = \tfrac{1}{12}$.   ◀

▶ **Drill Problem 7.2**   Determine the z-transform, ROC, and pole and zero locations of $X(z)$ for

$$x[n] = e^{j\Omega_o n} u[n]$$

Answer:

$$X(z) = \frac{z}{z - e^{j\Omega_o}}, \qquad |z| > 1$$

Pole at $z = e^{j\Omega_o}$, zero at $z = 0$.   ◀

*7.3 Properties of the Region of Convergence

The basic properties of the ROC are examined in this section. In particular, we show how the ROC is related to the characteristics of a signal, $x[n]$. Given the ROC properties, we can often identify the ROC from $X(z)$ and limited knowledge of the characteristics of $x[n]$. The relationship between the ROC and the characteristics of the time-domain signal is used to find inverse z-transforms in Section 7.5. The results presented here are derived using intuitive arguments rather than rigorous proofs.

First, we note that the ROC cannot contain any poles. This is because the ROC is defined as the set of z for which the z-transform converges. Hence $X(z)$ must be finite for all z in the ROC. However, if d is a pole, then $|X(d)| = \infty$ and the z-transform does not converge at the pole. Thus the pole cannot lie in the ROC.

Second, the ROC for a finite-duration signal includes the entire z-plane, except possibly $z = 0$ and/or $|z| = \infty$. To see this, suppose $x[n]$ is nonzero only on the interval $n_1 \leq n \leq n_2$. We have

$$X(z) = \sum_{n=n_1}^{n_2} x[n]z^{-n}$$

This sum will converge provided that each term in the sum is finite. If a signal has any nonzero causal components ($n_2 > 0$), then the expression for $X(z)$ will have a term involving z^{-1} and thus the ROC cannot include $z = 0$. If a signal has any nonzero noncausal components ($n_1 < 0$), then the expression for $X(z)$ will have a term involving z and thus the ROC cannot include $|z| = \infty$. Conversely, if a signal does not have any nonzero causal components ($n_2 \leq 0$), then the ROC will include $z = 0$. If a signal has no nonzero noncausal components ($n_1 \geq 0$), then the ROC will include $|z| = \infty$. This also indicates that $x[n] = c\delta[n]$ is the only signal whose ROC is the entire z-plane.

Now consider infinite-duration signals. The condition for convergence is $|X(z)| < \infty$. We may thus write

$$|X(z)| = \left| \sum_{n=-\infty}^{\infty} x[n]z^{-n} \right|$$

$$\leq \sum_{n=-\infty}^{\infty} |x[n]z^{-n}|$$

$$= \sum_{n=-\infty}^{\infty} |x[n]| \cdot |z|^{-n}$$

The second line follows from the fact that the magnitude of a sum of numbers is less than or equal to the sum of the magnitudes. We obtain the third line by noting that the magnitude of a product equals the product of magnitudes. Splitting the infinite sum into negative- and positive-time portions, we may write

$$I_-(z) = \sum_{n=-\infty}^{-1} |x[n]| \cdot |z|^{-n}$$

and

$$I_+(z) = \sum_{n=0}^{\infty} |x[n]| \cdot |z|^{-n}$$

We note that $|X(z)| \leq I_-(z) + I_+(z)$. If both $I_-(z)$ and $I_+(z)$ are finite, then $|X(z)|$ is guaranteed to be finite. This clearly requires that $|x[n]|$ be bounded in some way.

Suppose we can bound $|x[n]|$ by finding the smallest positive constants A_-, A_+, r_-, and r_+ such that

$$|x[n]| \leq A_-(r_-)^n, \qquad n < 0 \tag{7.8}$$

and

$$|x[n]| \leq A_+(r_+)^n, \qquad n \geq 0 \tag{7.9}$$

A signal that satisfies these bounds grows no faster than $(r_+)^n$ for positive n and $(r_-)^n$ for negative n. While we can construct signals that do not satisfy these bounds, such as a^{n^2}, such signals do not generally occur in engineering problems.

If bound (7.8) is satisfied, then

$$I_-(z) \leq A_- \sum_{n=-\infty}^{-1} (r_-)^n |z|^{-n}$$

$$= A_- \sum_{n=-\infty}^{-1} \left(\frac{r_-}{|z|} \right)^n$$

$$= A_- \sum_{k=1}^{\infty} \left(\frac{|z|}{r_-} \right)^k$$

This last sum converges if and only if $|z| < r_-$. Now consider the positive-time portion. If bound (7.9) is satisfied, then

$$I_+(z) \leq A_+ \sum_{n=0}^{\infty} (r_+)^n |z|^{-n}$$

$$= A_+ \sum_{n=0}^{\infty} \left(\frac{r_+}{|z|} \right)^n$$

This sum converges if and only if $|z| > r_+$. Hence if $r_+ < |z| < r_-$, then both $I_+(z)$ and $I_-(z)$ converge and $|X(z)|$ also converges. Note that if $r_+ > r_-$, then there are no values of z for which convergence is guaranteed.

This analysis supports the following conclusions for infinite duration signals. Define a *left-sided signal* as one for which $x[n] = 0$ for $n > n_2$, a *right-sided signal* as one for which $x[n] = 0$ for $n < n_1$, and a *two-sided signal* as one that has infinite duration in both positive and negative directions. For signals $x[n]$ that satisfy the exponential bounds in Eqs. (7.8) and (7.9), we have:

▶ The ROC of a right-sided signal is of the form $|z| > r_+$.
▶ The ROC of a left-sided signal is of the form $|z| < r_-$.
▶ The ROC of a two-sided signal is of the form $r_+ < |z| < r_-$.

Each of these cases is illustrated in Fig. 7.8.

One-sided exponential signals are frequently encountered in engineering problems. In this case the magnitude of one or more poles determines the ROC boundaries r_- and

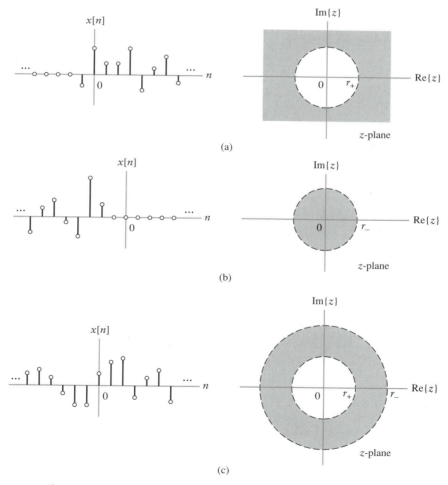

FIGURE 7.8   The relationship between ROC and time extent of a signal. (a) Right-sided signal has a ROC of the form $|z| > r_+$. (b) Left-sided signal has a ROC of the form $|z| < r_-$. (c) Two-sided signal has a ROC of the form $r_+ < |z| < r_-$.

r_+. Suppose we have a right-sided signal $x[n] = \alpha^n u[n]$, where α is complex in general. The z-transform of $x[n]$ has a pole at $z = \alpha$ and the ROC is $|z| > |\alpha|$. Hence the ROC is the region of the z-plane with a radius greater than the radius of the pole. Likewise, if $x[n]$ is the left-sided signal $x[n] = \alpha^n u[-n - 1]$, then the ROC is $|z| < |\alpha|$, the region of the z-plane with radius less than the radius of the pole. If a signal consists of a sum of exponentials, then the ROC is the intersection of the ROCs associated with each term. It will have radius greater than that of the pole of largest radius associated with right-sided terms and radius less than that of the pole of smallest radius associated with left-sided terms. Some of these aspects are illustrated in the following example.

EXAMPLE 7.5   Identify the ROC associated with the z-transform for each of the following signals:

$$x[n] = (-\tfrac{1}{2})^n u[-n] + 2(\tfrac{1}{4})^n u[n]$$
$$y[n] = (-\tfrac{1}{2})^n u[n] + 2(\tfrac{1}{4})^n u[n]$$
$$w[n] = (-\tfrac{1}{2})^n u[-n] + 2(\tfrac{1}{4})^n u[-n]$$

Solution: Beginning with $x[n]$, we write

$$X(z) = \sum_{n=-\infty}^{0} \left(-\frac{1}{2z}\right)^n + 2\sum_{n=0}^{\infty} \left(\frac{1}{4z}\right)^n$$
$$= \sum_{k=0}^{\infty} (-2z)^k + 2\sum_{n=0}^{\infty} \left(\frac{1}{4z}\right)^n$$

The first series converges for $|z| < \tfrac{1}{2}$, while the second converges for $|z| > \tfrac{1}{4}$. Both series must converge for $X(z)$ to converge, so the ROC is $\tfrac{1}{4} < |z| < \tfrac{1}{2}$. The ROC of this two-sided signal is depicted in Fig. 7.9(a). Summing the two geometric series, we obtain

$$X(z) = \frac{1}{1 + 2z} + \frac{2z}{z - \frac{1}{4}}$$

which has poles at $z = -\tfrac{1}{2}$ and $z = \tfrac{1}{4}$. Note that the ROC is in the ring-shaped region located between the poles.

Next, $y[n]$ is a right-sided signal, and

$$Y(z) = \sum_{n=0}^{\infty} \left(\frac{-1}{2z}\right)^n + 2\sum_{n=0}^{\infty} \left(\frac{1}{4z}\right)^n$$

The first series converges for $|z| > \tfrac{1}{2}$, while the second converges for $|z| > \tfrac{1}{4}$. Hence the combined ROC is $|z| > \tfrac{1}{2}$ as depicted in Fig. 7.9(b). In this case, we write

$$Y(z) = \frac{z}{z + \frac{1}{2}} + \frac{2z}{z - \frac{1}{4}}$$

for which the poles are again at $z = -\tfrac{1}{2}$ and $z = \tfrac{1}{4}$. The ROC is outside a circle containing the pole of largest radius, $z = -\tfrac{1}{2}$.

The last signal, $w[n]$, is left sided, and

$$W(z) = \sum_{n=-\infty}^{0} \left(-\frac{1}{2z}\right)^n + 2\sum_{n=-\infty}^{0} \left(\frac{1}{4z}\right)^n$$
$$= \sum_{k=0}^{\infty} (-2z)^k + 2\sum_{k=0}^{\infty} (4z)^k$$

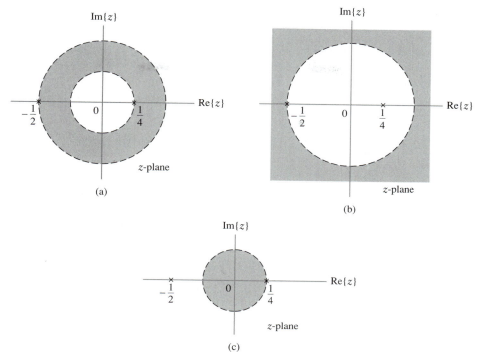

FIGURE 7.9   ROCs for Example 7.5. (a) Two-sided signal $x[n]$ has ROC in between the poles. (b) Right-sided signal $y[n]$ has ROC outside the circle containing the largest pole. (c) Left-sided signal $w[n]$ has ROC inside the circle containing the smallest pole.

Here the first series converges for $|z| < \frac{1}{2}$, while the second series converges for $|z| < \frac{1}{4}$, giving a combined ROC of $|z| < \frac{1}{4}$ as depicted in Fig. 7.9(c). Here we have

$$W(z) = \frac{1}{1 + 2z} + \frac{2}{1 - 4z}$$

where the poles are at $z = -\frac{1}{2}$ and $z = \frac{1}{4}$. The ROC is inside a circle containing the pole of smallest radius, $z = \frac{1}{4}$.

Hence this example illustrates that the ROC of a two-sided signal is a ring, the ROC of a right-sided signal is the exterior of a circle, and the ROC of a left-sided signal is the interior of a circle. In each case the poles define the ROC boundaries.

▶ **Drill Problem 7.3**   Determine the z-transform and ROC for the two-sided signal

$$x[n] = \alpha^{|n|}$$

assuming $|\alpha| < 1$. Repeat for $|\alpha| > 1$.

Answer: For $|\alpha| < 1$

$$X(z) = \frac{z}{z - \alpha} - \frac{z}{z - 1/\alpha}, \qquad |\alpha| < |z| < 1/|\alpha|$$

For $|\alpha| > 1$ the ROC is the empty set.   ◀

7.4 Properties of the z-Transform

Most of the z-transform properties are analogous to those of the DTFT. Hence in this section we state the properties and defer the proofs to the exercises. In the properties given below we assume that

$$x[n] \overset{z}{\longleftrightarrow} X(z) \quad \text{with ROC } R_x$$

and

$$y[n] \overset{z}{\longleftrightarrow} Y(z) \quad \text{with ROC } R_y$$

The ROC is changed by certain operations. In the previous section we established that the general form of the ROC is a ring in the z-plane. Thus, the effect of an operation on the ROC is described by a change in the radii of the ROC boundaries.

■ LINEARITY

The linearity property states that the z-transform of a sum of signals is just the sum of the individual z-transforms. That is,

$$ax[n] + by[n] \overset{z}{\longleftrightarrow} aX(z) + bY(z) \quad \text{with ROC at least } R_x \cap R_y \qquad (7.10)$$

The ROC is the intersection of the individual ROCs because the z-transform of the sum is valid only where both $X(z)$ and $Y(z)$ converge. The ROC can be larger than the intersection if one or more terms in $x[n]$ or $y[n]$ cancel each other in the sum. In the z-plane this corresponds to a zero cancelling a pole that defines one of the ROC boundaries. This phenomenon is illustrated in the following example.

EXAMPLE 7.6   Suppose

$$x[n] = (\tfrac{1}{2})^n u[n] - (\tfrac{3}{2})^n u[-n-1] \overset{z}{\longleftrightarrow} X(z) = \frac{-z}{(z-\tfrac{1}{2})(z-\tfrac{3}{2})} \quad \text{with ROC } \tfrac{1}{2} < |z| < \tfrac{3}{2}$$

and

$$y[n] = (\tfrac{1}{4})^n u[n] - (\tfrac{1}{2})^n u[n] \overset{z}{\longleftrightarrow} Y(z) = \frac{-\tfrac{1}{4}z}{(z-\tfrac{1}{4})(z-\tfrac{1}{2})} \quad \text{with ROC } |z| > \tfrac{1}{2}$$

Evaluate the z-transform of $ax[n] + by[n]$.

Solution: The pole–zero plots and ROCs for $x[n]$ and $y[n]$ are illustrated in Fig. 7.10(a) and (b), respectively. The linearity property indicates that

$$ax[n] + by[n] \overset{z}{\longleftrightarrow} a\frac{-z}{(z-\tfrac{1}{2})(z-\tfrac{3}{2})} + b\frac{-\tfrac{1}{4}z}{(z-\tfrac{1}{4})(z-\tfrac{1}{2})}$$

In general, the ROC is the intersection of individual ROCs, or $\tfrac{1}{2} < |z| < \tfrac{3}{2}$, which corresponds to the ROC depicted in Fig. 7.10(a). Note, however, what happens when $a = b$. In this case, we have

$$ax[n] + ay[n] = a(-(\tfrac{3}{2})^n u[-n-1] + (\tfrac{1}{4})^n u[n])$$

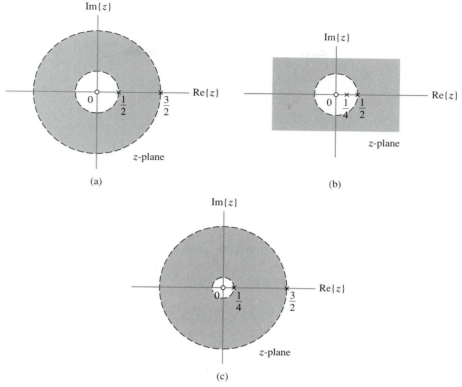

FIGURE 7.10 ROCs for Example 7.6. (a) ROC and pole–zero plot for $X(z)$. (b) ROC and pole–zero plot for $Y(z)$. (c) ROC and pole–zero plot for $a(X(z) + Y(z))$.

and we see the term $(\frac{1}{2})^n u[n]$ has been cancelled in the time-domain signal. The ROC is now easily verified to be $\frac{1}{4} < |z| < \frac{3}{2}$, as shown in Fig. 7.10(c). This ROC is larger than the intersection of the individual ROCs, because the term $(\frac{1}{2})^n u[n]$ is no longer present. Combining z-transforms and using the linearity property gives

$$aX(z) + aY(z) = a\left(\frac{-z}{(z - \frac{1}{2})(z - \frac{3}{2})} + \frac{-\frac{1}{4}z}{(z - \frac{1}{4})(z - \frac{1}{2})}\right)$$

$$= a\frac{-\frac{1}{4}z(z - \frac{3}{2}) - z(z - \frac{1}{4})}{(z - \frac{1}{4})(z - \frac{1}{2})(z - \frac{3}{2})}$$

$$= a\frac{-\frac{5}{4}z(z - \frac{1}{2})}{(z - \frac{1}{4})(z - \frac{1}{2})(z - \frac{3}{2})}$$

The zero at $z = \frac{1}{2}$ cancels the pole at $z = \frac{1}{2}$, and so we have

$$aX(z) + aY(z) = a\frac{-\frac{5}{4}z}{(z - \frac{1}{4})(z - \frac{3}{2})}$$

Hence cancellation of the $(\frac{1}{2})^n u[n]$ term in the time domain corresponds to cancellation of the pole at $z = \frac{1}{2}$ by a zero in the z-domain. This pole defined the ROC boundary, so the ROC enlarges when it is removed.

■ TIME REVERSAL

$$x[-n] \overset{z}{\longleftrightarrow} X\left(\frac{1}{z}\right) \quad \text{with ROC } \frac{1}{R_x} \tag{7.11}$$

Time reversal, or reflection, corresponds to replacing z by z^{-1}. Hence if R_x is of the form $a < |z| < b$, the ROC of the reflected signal is $a < 1/|z| < b$ or $1/b < |z| < 1/a$.

■ TIME SHIFT

$$x[n - n_o] \overset{z}{\longleftrightarrow} z^{-n_o}X(z) \quad \text{with ROC } R_x \text{ except possibly } z = 0 \text{ or } |z| = \infty \tag{7.12}$$

Multiplication by z^{-n_o} introduces a pole of order n_o at $z = 0$ if $n_o > 0$. In this case the ROC cannot include $z = 0$, even if R_x does include $z = 0$, unless $X(z)$ has a zero of at least order n_o at $z = 0$ that cancels all of the new poles. If $n_o < 0$, then multiplication by z^{-n_o} introduces n_o poles at infinity. If these poles are not cancelled by zeros at infinity in $X(z)$, then the ROC of $z^{-n_o}X(z)$ cannot include $|z| = \infty$.

■ MULTIPLICATION BY EXPONENTIAL SEQUENCE

Let α be a complex number. We have

$$\alpha^n x[n] \overset{z}{\longleftrightarrow} X\left(\frac{z}{\alpha}\right) \quad \text{with ROC } |\alpha|R_x \tag{7.13}$$

The notation $|\alpha|R_x$ implies the ROC boundaries are multiplied by $|\alpha|$. If R_x is $a < |z| < b$, then the new ROC is $|\alpha|a < |z| < |\alpha|b$. If $X(z)$ contains a factor $1 - dz^{-1}$ in the denominator so that d is a pole, then $X(z/\alpha)$ has a factor $1 - \alpha dz^{-1}$ in the denominator and thus has a pole at αd. Similarly, if c is a zero of $X(z)$, then $X(z/\alpha)$ has a zero at αc. This indicates that the poles and zeros of $X(z)$ have their radii changed by $|\alpha|$ and their angles are changed by $\arg\{\alpha\}$. This is depicted in Fig. 7.11. If α has unit magnitude then the radius is unchanged, while if α is a positive real number then the angle is unchanged.

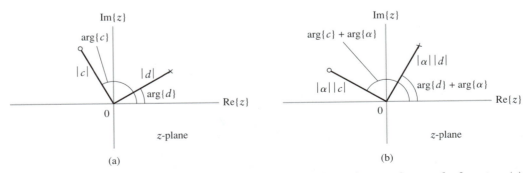

FIGURE 7.11   The effect of multiplication by α^n on the poles and zeros of a transfer function. (a) Pole and zero locations of $X(z)$. (b) Pole and zero locations of $X(z/\alpha)$.

■ CONVOLUTION

$$x[n] * y[n] \overset{z}{\longleftrightarrow} X(z)Y(z) \quad \text{with ROC at least } R_x \cap R_y \qquad (7.14)$$

Convolution of time signals corresponds to multiplication of z-transforms. As in the linearity property, the ROC may be larger than the intersection if a pole–zero cancellation occurs in the product $X(z)Y(z)$.

■ DIFFERENTIATION IN THE z-DOMAIN

$$nx[n] \overset{z}{\longleftrightarrow} -z\frac{d}{dz}X(z) \quad \text{with ROC } R_x \qquad (7.15)$$

Multiplication by n in time corresponds to differentiation with respect to z and multiplication of the result by $-z$ in the z-domain. This operation does not change the ROC.

EXAMPLE 7.7   Find the z-transform of the signal

$$x[n] = (n(-\tfrac{1}{2})^n u[n]) * (\tfrac{1}{4})^{-n} u[-n]$$

Solution: First, we find the z-transform of $w[n] = n(-\tfrac{1}{2})^n u[n]$. We know that

$$(-\tfrac{1}{2})^n u[n] \overset{z}{\longleftrightarrow} \frac{z}{z + \tfrac{1}{2}} \quad \text{with ROC } |z| > \tfrac{1}{2}$$

Thus the z-domain differentiation property of Eq. (7.15) implies

$$w[n] = n(-\tfrac{1}{2})^n u[n] \overset{z}{\longleftrightarrow} W(z) = -z\frac{d}{dz}\left(\frac{z}{z + \tfrac{1}{2}}\right) \quad \text{with ROC } |z| > \tfrac{1}{2}$$

$$= -z\left(\frac{z + \tfrac{1}{2} - z}{(z + \tfrac{1}{2})^2}\right)$$

$$= \frac{-\tfrac{1}{2}z}{(z + \tfrac{1}{2})^2} \quad \text{with ROC } |z| > \tfrac{1}{2}$$

Next, we find the z-transform of $y[n] = (\tfrac{1}{4})^{-n} u[-n]$. We do this by using the time reversal property. Note that

$$(\tfrac{1}{4})^n u[n] \overset{z}{\longleftrightarrow} \frac{z}{z - \tfrac{1}{4}} \quad \text{with ROC } |z| > \tfrac{1}{4}$$

Hence Eq. (7.11) implies

$$y[n] \overset{z}{\longleftrightarrow} Y(z) = \frac{z^{-1}}{z^{-1} - \tfrac{1}{4}} \quad \text{with ROC } \frac{1}{|z|} > \frac{1}{4}$$

$$= \frac{-4z}{z - 4} \quad \text{with ROC } |z| < 4$$

Lastly, we apply the convolution property to obtain $X(z)$ and thus write

$$x[n] = w[n] * y[n] \overset{z}{\longleftrightarrow} X(z) = W(z)Y(z) \quad \text{with ROC } R_w \cap R_y$$

$$= \frac{2z^2}{(z - 4)(z + \tfrac{1}{2})^2} \quad \text{with ROC } \tfrac{1}{2} < |z| < 4$$

EXAMPLE 7.8   Use the properties of linearity and multiplication by a complex exponential to find the z-transform of

$$x[n] = a^n \cos(\Omega_o n) u[n]$$

where a is real and positive.

Solution: First note that $y[n] = a^n u[n]$ has the z-transform

$$Y(z) = \frac{1}{1 - az^{-1}} \quad \text{with ROC } |z| > a$$

Now rewrite $x[n]$ as the sum

$$x[n] = \tfrac{1}{2} e^{j\Omega_o n} y[n] + \tfrac{1}{2} e^{-j\Omega_o n} y[n]$$

and apply the multiplication by a complex exponential property to each term, obtaining

$$X(z) = \tfrac{1}{2} Y(e^{-j\Omega_o} z) + \tfrac{1}{2} Y(e^{j\Omega_o} z) \quad \text{with ROC } |z| > a$$

$$= \frac{1}{2} \frac{1}{(1 - ae^{j\Omega_o} z^{-1})} + \frac{1}{2} \frac{1}{(1 - ae^{-j\Omega_o} z^{-1})}$$

$$= \frac{1}{2} \left(\frac{1 - ae^{-j\Omega_o} z^{-1} + 1 - ae^{j\Omega_o} z^{-1}}{(1 - ae^{j\Omega_o} z^{-1})(1 - ae^{-j\Omega_o} z^{-1})} \right)$$

$$= \frac{1 - a \cos(\Omega_o) z^{-1}}{1 - 2a \cos(\Omega_o) z^{-1} + a^2 z^{-2}} \quad \text{with ROC } |z| > a$$

▶ **Drill Problem 7.4**   Find the z-transform of

$$x[n] = u[n - 2] * (\tfrac{2}{3})^n u[n]$$

Answer:

$$X(z) = \frac{1}{(z - 1)(z - \tfrac{2}{3})} \quad \text{with ROC } |z| > 1 \qquad \blacktriangleleft$$

▌7.5 *Inversion of the z-Transform*

We now turn our attention to the problem of recovering a time-domain signal from its z-transform. Direct evaluation of the inversion integral defined in Eq. (7.5) requires an understanding of contour integration and is not discussed further here. Two alternative methods for determining inverse z-transforms are presented. The method of partial fractions uses knowledge of several basic z-transform pairs and the z-transform properties to invert a large class of z-transforms. This approach also relies on an important property of the ROC. A right-sided time signal has a ROC that lies outside the pole radius, while a left-sided time signal has a ROC that lies inside the pole radius. The second inversion method expresses $X(z)$ as a power series in z^{-1} of the form Eq. (7.4), so that the values of the signal may be determined by inspection.

■ PARTIAL FRACTION EXPANSIONS

In the study of LTI systems, we often encounter z-transforms that are a rational function of z^{-1}. Let

$$X(z) = \frac{B(z)}{A(z)}$$

$$= \frac{b_0 + b_1 z^{-1} + \cdots + b_M z^{-M}}{1 + a_1 z^{-1} + \cdots + a_N z^{-N}} \tag{7.16}$$

and assume that $M < N$. If $M \geq N$, then we may use long division to express $X(z)$ in the form

$$X(z) = \sum_{k=0}^{M-N} f_k z^{-k} + \frac{\tilde{B}(z)}{A(z)}$$

The numerator polynomial $\tilde{B}(z)$ now has order one less than that of the denominator polynomial, and the partial fraction expansion method is applied to determine the inverse transform of $\tilde{B}(z)/A(z)$. The inverse z-transform of the terms in the sum are obtained from the pair $\delta[n] \xleftarrow{\quad z \quad} 1$ and the time-shift property.

In some problems $X(z)$ may be expressed as a ratio of polynomials in z rather than z^{-1}. In this case, we may use the partial fraction expansion method described here if we first convert $X(z)$ to a ratio of polynomials in z^{-1} as described by Eq. (7.16). This conversion is accomplished by factoring out the highest power of z present in the numerator and the term with the highest power of z present in the denominator. This ensures that the remainder has the form described by Eq. (7.16). For example, if

$$X(z) = \frac{2z^2 - 2z + 10}{3z^3 - 6z + 9}$$

then we factor z^2 from the numerator and $3z^3$ from the denominator, and thus write

$$X(z) = \frac{z^2}{3z^3} \left(\frac{2 - 2z^{-1} + 10z^{-2}}{1 - 2z^{-2} + 3z^{-3}} \right)$$

$$= \frac{1}{3} z^{-1} \left(\frac{2 - 2z^{-1} + 10z^{-2}}{1 - 2z^{-2} + 3z^{-3}} \right)$$

The partial fraction expansion is applied to the term in parentheses, and the factor $\frac{1}{3} z^{-1}$ is incorporated later using the time-shift property.

The partial fraction expansion of Eq. (7.16) is obtained by factoring the denominator polynomial into a product of first-order terms, as shown by

$$X(z) = \frac{b_0 + b_1 z^{-1} + \cdots + b_M z^{-M}}{\Pi_{k=1}^{N}(1 - d_k z^{-1})}$$

where the d_k are the poles of $X(z)$. If none of the poles are repeated, then we may rewrite $X(z)$ as a sum of first-order terms using the partial fraction expansion:

$$X(z) = \sum_{k=1}^{N} \frac{A_k}{1 - d_k z^{-1}}$$

The inverse z-transform associated with each term is then determined using the transform pair:

$$A_k(d_k)^n u[n] \xleftrightarrow{\ z\ } \frac{A_k}{1 - d_k z^{-1}} \quad \text{with ROC } |z| > d_k$$

or

$$-A_k(d_k)^n u[-n-1] \xleftrightarrow{\ z\ } \frac{A_k}{1 - d_k z^{-1}} \quad \text{with ROC } |z| < d_k$$

The relationship between the ROC associated with $X(z)$ and each pole determines whether the right-sided or left-sided inverse transform is chosen for each term.

If a pole d_i is repeated r times, then there are r terms in the partial fraction expansion associated with this pole. They are

$$\frac{A_{i_1}}{1 - d_i z^{-1}}, \frac{A_{i_2}}{(1 - d_i z^{-1})^2}, \cdots, \frac{A_{i_r}}{(1 - d_i z^{-1})^r}$$

The inverse z transform of each term here is determined using either

$$A \frac{(n+1) \cdots (n+m-1)}{(m-1)!} (d_i)^n u[n] \xleftrightarrow{\ z\ } \frac{A}{(1 - d_i)^m} \quad \text{with ROC } |z| > d_i$$

or

$$-A \frac{(n+1) \cdots (n+m-1)}{(m-1)!} (d_i)^n u[-n-1] \xleftrightarrow{\ z\ } \frac{A}{(1 - d_i)^m} \quad \text{with ROC } |z| < d_i$$

Again, the ROC of $X(z)$ determines whether the right- or left-sided inverse transform is chosen.

The linearity property indicates that the ROC of $X(z)$ is the intersection of the ROCs associated with the individual terms in the partial fraction expansion. In order to choose the correct inverse transform, we must infer the ROC of each term from the ROC of $X(z)$. This is accomplished by comparing each pole location to the ROC of $X(z)$. If the ROC of $X(z)$ has radius greater than that of the pole associated with a given term, we choose the right-sided inverse transform. If the ROC of $X(z)$ has radius less than that of the pole, we choose the left-sided inverse transform for this term. The following example illustrates this procedure.

EXAMPLE 7.9   Find the inverse z-transform of

$$X(z) = \frac{1 - z^{-1} + z^{-2}}{(1 - \frac{1}{2}z^{-1})(1 - 2z^{-1})(1 - z^{-1})} \quad \text{with ROC } 1 < |z| < 2$$

Solution: Use a partial fraction expansion to write

$$X(z) = \frac{A_1}{1 - \frac{1}{2}z^{-1}} + \frac{A_2}{1 - 2z^{-1}} + \frac{A_3}{1 - z^{-1}}$$

Solving for A_1, A_2, and A_3 gives

$$X(z) = \frac{1}{1 - \frac{1}{2}z^{-1}} + \frac{2}{1 - 2z^{-1}} - \frac{2}{1 - z^{-1}}$$

Now we find the inverse z-transform of each term using the relationship between the pole location and the ROC of $X(z)$. The ROC and pole locations are depicted in Fig. 7.12. The ROC has a radius greater than the pole at $z = \frac{1}{2}$, so this term has the right-sided inverse transform

$$(\tfrac{1}{2})^n u[n] \xleftrightarrow{\;\;z\;\;} \frac{1}{1 - \tfrac{1}{2}z^{-1}}$$

The ROC has a radius less than the pole at $z = 2$, so this term has the left-sided inverse transform

$$-2(2)^n u[-n-1] \xleftrightarrow{\;\;z\;\;} \frac{2}{1 - 2z^{-1}}$$

Lastly, the ROC has a radius greater than the pole at $z = 1$, so this term has the right-sided inverse z-transform

$$-2u[n] \xleftrightarrow{\;\;z\;\;} -\frac{2}{1 - z^{-1}}$$

Combining the individual terms gives

$$x[n] = (\tfrac{1}{2})^n u[n] - 2(2)^n u[-n-1] - 2u[n]$$

▶ **Drill Problem 7.5**   Repeat the previous example if the ROC is changed to $\frac{1}{2} < |z| < 1$.

Answer:

$$x[n] = (\tfrac{1}{2})^n u[n] - 2(2)^n u[-n-1] + 2u[-n-1]$$ ◀

EXAMPLE 7.10   Find the inverse z-transform of

$$X(z) = \frac{z^3 - 10z^2 - 4z + 4}{2z^2 - 2z - 4} \quad \text{with ROC } |z| < 1$$

Solution: The poles at $z = -1$ and $z = 2$ are found by rooting the denominator polynomial. The ROC and pole locations in the z-plane are illustrated in Fig. 7.13. Now convert $X(z)$ into

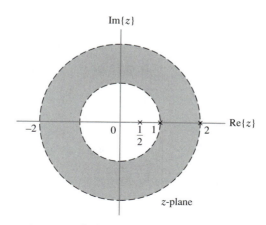

FIGURE 7.12   Pole locations and ROC for Example 7.9.

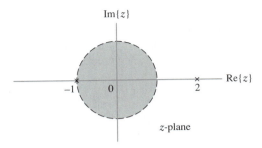

Figure 7.13   Pole locations and ROC for Example 7.10.

a ratio of polynomials in z^{-1} in accordance with Eq. (7.16). We do this by factoring z^3 from the numerator and $2z^2$ from the denominator, yielding

$$X(z) = \frac{1}{2} z \left(\frac{1 - 10z^{-1} - 4z^{-2} + 4z^{-3}}{1 - z^{-1} - 2z^{-2}} \right)$$

The factor $\frac{1}{2}z$ is easily incorporated later using the time-shift property, so we focus on the ratio of polynomials in parentheses. Using long division to reduce the order of the numerator polynomial, we have

$$
\begin{array}{r}
-2z^{-1} + 3 \\
-2z^{-2} - z^{-1} + 1 \overline{)\, 4z^{-3} - 4z^{-2} - 10z^{-1} + 1} \\
4z^{-3} + 2z^{-2} - 2z^{-1} \\
\hline
-6z^{-2} - 8z^{-1} + 1 \\
-6z^{-2} - 3z^{-1} + 3 \\
\hline
-5z^{-1} - 2
\end{array}
$$

Thus we may write

$$\frac{1 - 10z^{-1} - 4z^{-2} + 4z^{-3}}{1 - z^{-1} - 2z^{-2}} = -2z^{-1} + 3 + \frac{-5z^{-1} - 2}{1 - z^{-1} - 2z^{-2}}$$

$$= -2z^{-1} + 3 + \frac{-5z^{-1} - 2}{(1 + z^{-1})(1 - 2z^{-1})}$$

Next, using a partial fraction expansion, we write

$$\frac{-5z^{-1} - 2}{(1 + z^{-1})(1 - 2z^{-1})} = \frac{1}{1 + z^{-1}} - \frac{3}{1 - 2z^{-1}}$$

and thus define

$$X(z) = \tfrac{1}{2} z W(z)$$

where

$$W(z) = -2z^{-1} + 3 + \frac{1}{1 + z^{-1}} - \frac{3}{1 - 2z^{-1}} \quad \text{with ROC } |z| < 1$$

The ROC has a smaller radius than either pole, and so the inverse z-transform of $W(z)$ is

$$w[n] = -2\delta[n - 1] + 3\delta[n] - (-1)^n u[-n - 1] + 3(2)^n u[-n - 1]$$

The time-shift property indicates that

$$x[n] = \tfrac{1}{2} w[n + 1]$$

and thus

$$x[n] = -\delta[n] + \tfrac{3}{2}\delta[n + 1] - \tfrac{1}{2}(-1)^{n+1} u[-n - 2] + \tfrac{3}{2}(2)^{n+1} u[-n - 2]$$

▶ **Drill Problem 7.6**   Find the inverse z-transform of

$$X(z) = \frac{16z^2 - 4z + 1}{8z^2 + 2z - 1} \quad \text{with ROC } |z| > \tfrac{1}{2}$$

Answer:

$$x[n] = -\delta[n] + \tfrac{2}{3}(\tfrac{1}{4})^n u[n] + \tfrac{7}{3}(-\tfrac{1}{2})^n u[n]$$ ◀

The method of partial fractions also applies when the poles are complex valued. In this case the expansion coefficients will generally also be complex valued. However, if the coefficients in $X(z)$ are real valued, then the expansion coefficients corresponding to complex-conjugate poles will be complex conjugates of each other.

Note that information other than the ROC can be used to establish a unique inverse transform. For example, causality, stability, or existence of the DTFT is sufficient to determine the inverse transform. If a signal is known to be causal, then right-sided inverse transforms are chosen. If a signal is stable, then it is absolutely summable and has a DTFT. Hence stability and existence of the DTFT are equivalent conditions. In both cases the ROC includes the unit circle in the z-plane, $|z| = 1$. The inverse z-transform is determined by comparing the pole locations to the unit circle. If a pole is inside the unit circle, then the right-sided inverse z-transform is chosen. If a pole is outside the unit circle, then the left-sided inverse z-transform is chosen.

▶ **Drill Problem 7.7**   Find the inverse z-transform of

$$X(z) = \frac{1}{1 - \tfrac{1}{2}z^{-1}} + \frac{2}{1 - 2z^{-1}}$$

assuming (a) the signal is causal and (b) the signal has a DTFT.

Answer:
  (a) $x[n] = (\tfrac{1}{2})^n u[n] + 2(2)^n u[n]$
  (b) $x[n] = (\tfrac{1}{2})^n u[n] - 2(2)^n u[-n - 1]$ ◀

■ POWER SERIES EXPANSION

Here we seek to express $X(z)$ as a power series in z^{-1} or z of the form defined in Eq. (7.4). The value of the signal $x[n]$ is then given by the coefficient associated with z^{-n}. This inversion method is limited to signals that are one sided; that is, with ROCs of the form $|z| < a$ or $|z| > a$. If the ROC is $|z| > a$, then we express $X(z)$ as a power series in z^{-1} so that we obtain a right-sided signal. If the ROC is $|z| < a$, then we express $X(z)$ as a power series in z and obtain a left-sided inverse transform.

EXAMPLE 7.11   Find the inverse z-transform of

$$X(z) = \frac{2 + z^{-1}}{1 - \tfrac{1}{2}z^{-1}} \quad \text{with ROC } |z| > \tfrac{1}{2}$$

using a power series expansion.

Solution: We use long division to write $X(z)$ as a power series in z^{-1}, since the ROC indicates that $x[n]$ is right sided. We have

$$
\begin{array}{r}
2 + 2z^{-1} + z^{-2} + \frac{1}{2}z^{-3} + \cdots \\
1 - \frac{1}{2}z^{-1}\overline{)2 + \ z^{-1}} \\
\underline{2 - \ z^{-1}} \\
2z^{-1} \\
\underline{2z^{-1} - z^{-2}} \\
z^{-2} \\
\underline{z^{-2} - \frac{1}{2}z^{-3}} \\
\frac{1}{2}z^{-3}
\end{array}
$$

That is,

$$X(z) = 2 + 2z^{-1} + z^{-2} + \tfrac{1}{2}z^{-3} + \cdots$$

Thus we conclude that

$$
\begin{aligned}
x[n] &= 0, \qquad n < 0 \\
x[0] &= 2 \\
x[1] &= 2 \\
x[2] &= 1 \\
x[3] &= \tfrac{1}{2} \\
&\ \ \vdots
\end{aligned}
$$

If the ROC is changed to $|z| < \frac{1}{2}$, then we expand $X(z)$ as a power series in z:

$$
\begin{array}{r}
-2 - 8z - 16z^2 - 32z^3 + \cdots \\
-\frac{1}{2}z^{-1} + 1\overline{)z^{-1} + 2} \\
\underline{z^{-1} - 2} \\
4 \\
\underline{4 - 8z} \\
8z \\
\underline{8z - 16z^2} \\
16z^2
\end{array}
$$

That is,

$$X(z) = -2 - 8z - 16z^2 - 32z^3 + \cdots$$

Thus in this case we have

$$
\begin{aligned}
x[n] &= 0, \qquad n > 0 \\
x[0] &= -2 \\
x[-1] &= -8 \\
x[-2] &= -16 \\
x[-3] &= -32 \\
&\ \ \vdots
\end{aligned}
$$

Long division may be used to obtain the power series whenever $X(z)$ is a ratio of polynomials and is simple to perform. However, long division may not lead to a closed-form expression for $x[n]$.

An advantage of the power series approach is the ability to find inverse z-transforms for signals that are not a ratio of polynomials in z. This is illustrated in the following example.

EXAMPLE 7.12   Find the inverse z-transform of

$$X(z) = e^{z^2} \quad \text{with ROC all } z \text{ except } |z| = \infty$$

Solution: Using the power series representation for e^a,

$$e^a = \sum_{k=0}^{\infty} \frac{a^k}{k!}$$

we write

$$X(z) = \sum_{k=0}^{\infty} \frac{(z^2)^k}{k!}$$

$$= \sum_{k=0}^{\infty} \frac{z^{2k}}{k!}$$

Thus

$$x[n] = \begin{cases} 0, & n > 0 \text{ and } n \text{ odd} \\ \dfrac{1}{(n/2)!}, & \text{otherwise} \end{cases}$$

7.6 Transform Analysis of LTI Systems

In this section we examine the relationship between the transfer function and input–output characteristics of LTI discrete-time systems. In Section 7.2 we defined the transfer function as the z-transform of the impulse response. The output $y[n]$ of a LTI system is expressed in terms of the impulse response $h[n]$ and the input $x[n]$ using convolution

$$y[n] = h[n] * x[n]$$

If we take the z-transform of both sides of this equation using the convolution property, then we may express the transformed output $Y(z)$ as the product of the transfer function $H(z)$ and the transformed input, $X(z)$, as shown by

$$Y(z) = H(z)X(z) \tag{7.17}$$

The z-transform has converted convolution of time sequences into multiplication of transforms. Thus the transfer function offers yet another description for the input–output characteristics of a LTI discrete-time system.

Note that Eq. (7.17) implies that the transfer function may also be viewed as the ratio of the z-transform of the output to that of the input, that is,

$$H(z) = \frac{Y(z)}{X(z)} \tag{7.18}$$

This definition applies to all values of z for which $X(z)$ is nonzero.

■ **RELATING THE TRANSFER FUNCTION AND THE DIFFERENCE EQUATION**

The transfer function may be obtained directly from the difference-equation description of a system. Recall that an Nth order difference equation relates the input $x[n]$ to the output $y[n]$ as

$$\sum_{k=0}^{N} a_k y[n-k] = \sum_{k=0}^{M} b_k x[n-k]$$

In Section 7.2 we showed that the transfer function $H(z)$ is an eigenvalue of the system associated with the eigenfunction z^n. That is, if $x[n] = z^n$, then the output of a LTI system is $y[n] = z^n H(z)$. Substituting $x[n-k] = z^{n-k}$ and $y[n-k] = z^{n-k} H(z)$ into the difference equation gives the relationship

$$z^n \sum_{k=0}^{N} a_k z^{-k} H(z) = z^n \sum_{k=0}^{M} b_k z^{-k}$$

We may now solve for $H(z)$, obtaining

$$H(z) = \frac{\displaystyle\sum_{k=0}^{M} b_k z^{-k}}{\displaystyle\sum_{k=0}^{N} a_k z^{-k}} \tag{7.19}$$

The transfer function of a system described by a difference equation is a ratio of polynomials in z^{-1} and is thus termed a *rational transfer function*. The coefficient of z^{-k} in the numerator polynomial is the coefficient associated with $x[n-k]$ in the difference equation. The coefficient of z^{-k} in the denominator polynomial is the coefficient associated with $y[n-k]$ in the difference equation. This correspondence allows us not only to find the transfer function given the difference equation, but also to find a difference-equation description for a system given a rational transfer function.

EXAMPLE 7.13   Find the difference-equation description for a system with transfer function

$$H(z) = \frac{5z + 2}{z^2 + 3z + 2}$$

Solution: First we rewrite $H(z)$ as a ratio of polynomials in z^{-1}. Dividing both the numerator and denominator by z^2, we obtain

$$H(z) = \frac{5z^{-1} + 2z^{-2}}{1 + 3z^{-1} + 2z^{-2}}$$

Comparing this to Eq. (7.19) we conclude that $M = 2$, $N = 2$, $b_0 = 0$, $b_1 = 5$, $b_2 = 2$, $a_0 = 1$, $a_1 = 3$, and $a_2 = 2$. Hence this system is described by the difference equation

$$y[n] + 3y[n-1] + 2y[n-2] = 5x[n-1] + 2x[n-2]$$

The poles and zeros of a rational transfer function offer much insight into the system characteristics, as we shall see in the following subsections. The transfer function is expressed in pole–zero form by factoring the numerator and denominator polynomials in Eq. (7.19). We write

$$H(z) = \frac{\tilde{b}\Pi_{k=1}^{M}(1 - c_k z^{-1})}{\Pi_{k=1}^{N}(1 - d_k z^{-1})} \tag{7.20}$$

where the c_k and the d_k are the poles and zeros of the system and $\tilde{b} = b_0/a_0$ is the gain factor. This form assumes that there are no poles or zeros at $z = 0$. A pth order pole at $z = 0$ occurs when $b_0 = b_1 = \cdots = b_{p-1} = 0$, while an lth order zero at $z = 0$ occurs when $a_0 = a_1 = \cdots = a_{l-1} = 0$. In this case we write

$$H(z) = \frac{\tilde{b}z^{-p}\Pi_{k=1}^{M-p}(1 - c_k z^{-1})}{z^{-l}\Pi_{k=1}^{N-l}(1 - d_k z^{-1})} \tag{7.21}$$

where $\tilde{b} = b_p/a_l$. The system in the previous example had a first-order pole at $z = 0$. The system's poles, zeros, and gain factor $\tilde{b}$ uniquely determine the system transfer function and thus provide another description for the input–output behavior of the system.

The impulse response is the inverse z-transform of the transfer function. In order to uniquely determine the impulse response from the transfer function, we must know the ROC. However, the difference equation does not provide ROC information, so other system characteristics, such as stability or causality, must be known in order to uniquely determine the impulse response.

■ **RELATING THE TRANSFER FUNCTION AND THE STATE-VARIABLE DESCRIPTION**

The transfer function may also be obtained from the state-variable description

$$\mathbf{q}[n + 1] = \mathbf{A}\mathbf{q}[n] + \mathbf{b}x[n] \tag{7.22}$$
$$y[n] = \mathbf{c}\mathbf{q}[n] + Dx[n] \tag{7.23}$$

Let

$$\tilde{\mathbf{q}}(z) = \begin{bmatrix} Q_1(z) \\ Q_2(z) \\ \vdots \\ Q_N(z) \end{bmatrix}$$

be the z-transform of $\mathbf{q}[n]$. Here $Q_i(z)$ is the z-transform of the ith element of $\mathbf{q}[n]$, $q_i[n]$. Taking the z-transform of Eq. (7.22) using the time-shift property, and solving for $\tilde{\mathbf{q}}(z)$ in a manner analogous to that used to find the frequency response in Section 4.2, we obtain

$$\tilde{\mathbf{q}}(z) = (z\mathbf{I} - \mathbf{A})^{-1}\mathbf{b}X(z)$$

Now take the z-transform of Eq. (7.23), obtaining

$$Y(z) = \mathbf{c}\tilde{\mathbf{q}}(z) + DX(z)$$

Substitute the expression for $\tilde{\mathbf{q}}(z)$ in this z-transform, yielding

$$Y(z) = [\mathbf{c}(z\mathbf{I} - \mathbf{A})^{-1}\mathbf{b} + D]X(z)$$

Hence the system's transfer function is

$$H(z) = \mathbf{c}(z\mathbf{I} - \mathbf{A})^{-1}\mathbf{b} + D \tag{7.24}$$

EXAMPLE 7.14   Determine the transfer function for a LTI system with state-variable description

$$\mathbf{A} = \begin{bmatrix} 0 & 1 \\ -1 & 1 \end{bmatrix}, \quad \mathbf{b} = \begin{bmatrix} 0 \\ 2 \end{bmatrix}$$
$$\mathbf{c} = [3 \quad 0], \quad D = [0]$$

Solution: We begin with

$$z\mathbf{I} - \mathbf{A} = \begin{bmatrix} z & -1 \\ 1 & z-1 \end{bmatrix}$$

This implies

$$(z\mathbf{I} - \mathbf{A})^{-1} = \frac{1}{z^2 - z + 1} \begin{bmatrix} z-1 & 1 \\ -1 & z \end{bmatrix}$$

Hence the use of Eq. (7.24) yields

$$H(z) = \frac{1}{z^2 - z + 1} [3 \quad 0] \begin{bmatrix} z-1 & 1 \\ -1 & z \end{bmatrix} \begin{bmatrix} 0 \\ 2 \end{bmatrix} + 0$$
$$= \frac{1}{z^2 - z + 1} [3 \quad 0] \begin{bmatrix} 2 \\ 2z \end{bmatrix}$$
$$= \frac{6}{z^2 - z + 1}$$

Note that the expression for $H(z)$ has the same form as the frequency response defined in Eq. (4.11). Here $e^{j\Omega}$ is replaced by z. Hence as an input–output system description, the transfer function is invariant to a transformation of the system's state vector.

■ STABILITY AND CAUSALITY

The impulse response of a causal system is zero for $n < 0$. Therefore the impulse response of a causal system is determined from the transfer function by using right-sided inverse transforms. A pole that is inside the unit circle in the z-plane ($|d_k| < 1$) contributes an exponentially decaying term to the impulse response, while a pole that is outside the unit circle ($|d_k| > 1$) contributes an exponentially increasing term. These relationships are illustrated in Fig. 7.14.

Alternatively, if we know a system is stable, then the impulse response is absolutely summable and the DTFT of the impulse response exists. It follows that the ROC must include the unit circle in the z-plane. Hence the relationship between a pole location and the unit circle determines the component of the impulse response associated with the pole. A pole inside the unit circle contributes a right-sided decaying exponential term to the impulse response, while a pole outside the unit circle contributes a left-sided decaying exponential term to the impulse response as depicted in Fig. 7.15. Note that a stable impulse response cannot contain any increasing exponential terms since then the impulse response is not absolutely summable.

Systems that are both stable and causal must have all their poles inside the unit circle. A pole that is inside the unit circle in the z-plane contributes a right-sided or causal decaying exponential term to the impulse response. We cannot have a pole outside the unit circle, since the inverse transform of a pole located outside the circle will contribute either a right-sided increasing exponential term, which is not stable, or a left-sided decaying exponential

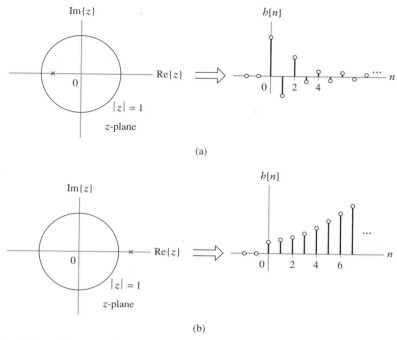

(a)

(b)

FIGURE 7.14 The relationship between the pole location and impulse response characteristics for a causal system. (a) A pole inside the unit circle contributes an exponentially decaying term to the impulse response. (b) A pole outside the unit circle contributes an exponentially increasing term to the impulse response.

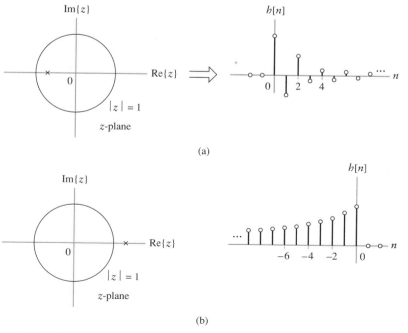

(a)

(b)

FIGURE 7.15 The relationship between the pole location and impulse response characteristics for a stable system. (a) A pole inside the unit circle contributes a right-sided term to the impulse response. (b) A pole outside the unit circle contributes a left-sided term to the impulse response.

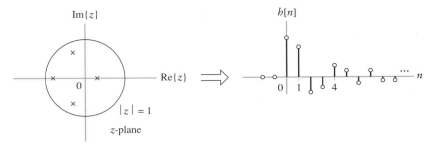

FIGURE 7.16   A system that is both stable and causal must have all its poles inside the unit circle in the z-plane, as illustrated here.

term, which is not causal. An example of a stable and causal system is depicted in Fig. 7.16.

EXAMPLE 7.15   A system has the transfer function

$$H(z) = \frac{2}{1 - 0.9e^{j\pi/4}z^{-1}} + \frac{2}{1 - 0.9e^{-j\pi/4}z^{-1}} + \frac{3}{1 + 2z^{-1}}$$

Find the impulse response assuming the system is (a) stable and (b) causal. Can this system be both stable and causal?

Solution: This system has poles at $z = 0.9e^{j\pi/4}$, $z = 0.9e^{-j\pi/4}$, and $z = -2$ as depicted in Fig. 7.17. If the system is stable, then the ROC includes the unit circle. The two poles inside the unit circle contribute right-sided terms to the impulse response, while the pole outside the unit circle contributes a left-sided term. Hence

$$h[n] = 2(0.9e^{j\pi/4})^n u[n] + 2(0.9e^{-j\pi/4})^n u[n] - 3(-2)^n u[-n - 1]$$

$$= 4(0.9)^n \cos\left(\frac{\pi}{4} n\right) u[n] - 3(-2)^n u[-n - 1]$$

for case (a). If the system is assumed causal, then all the poles contribute right-sided terms to the impulse response and so we have

$$h[n] = 2(0.9e^{j\pi/4})^n u[n] + 2(0.9e^{-j\pi/4})^n u[n] + 3(-2)^n u[n]$$

$$= 4(0.9)^n \cos\left(\frac{\pi}{4} n\right) u[n] + 3(-2)^n u[n]$$

as the solution to case (b). Note that this system cannot be both stable and causal since there is a pole outside the unit circle.

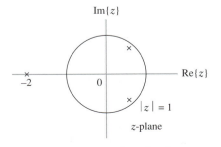

FIGURE 7.17   Pole locations in the z-plane for the system in Example 7.15.

▶ **Drill Problem 7.8** A stable and causal system is described by the difference equation

$$y[n] + \tfrac{1}{4}y[n-1] - \tfrac{1}{8}y[n-2] = -2x[n] + \tfrac{5}{4}x[n-1]$$

Find the system impulse response.

Answer:

$$h[n] = (\tfrac{1}{4})^n u[n] - 3(-\tfrac{1}{2})^n u[n]$$

◀

■ INVERSE SYSTEMS

Recall that the impulse response of an inverse system, $h^{-1}[n]$, satisfies

$$h^{-1}[n] * h[n] = \delta[n]$$

where $h[n]$ is the impulse response of the system to be inverted. Taking the z-transform of both sides of this equation we find that the transfer function of the inverse system must satisfy

$$H^{-1}(z)H(z) = 1$$

That is,

$$H^{-1}(z) = \frac{1}{H(z)}$$

The transfer function of an inverse system is the inverse of the transfer function of the system that we desire to invert. If $H(z)$ is written in the pole–zero form shown in Eq. (7.21), then

$$H^{-1}(z) = \frac{z^{-l}\prod_{k=1}^{N-l}(1 - d_k z^{-1})}{\tilde{b}z^{-p}\prod_{k=1}^{M-p}(1 - c_k z^{-1})} \tag{7.25}$$

The zeros of $H(z)$ are the poles of $H^{-1}(z)$, and the poles of $H(z)$ are the zeros of $H^{-1}(z)$. Any system described by a rational transfer function has an inverse system of this form.

We are often interested in inverse systems that are both stable and causal. $H^{-1}(z)$ is both stable and causal if all of its poles are inside the unit circle. Since the poles of $H^{-1}(z)$ are the zeros of $H(z)$, we conclude that a stable and causal inverse for a system $H(z)$ exists if and only if all the zeros of $H(z)$ are inside the unit circle. If $H(z)$ has any zeros outside the unit circle, then a stable and causal inverse system does not exist. A system with all its poles and zeros inside the unit circle, as illustrated in Fig. 7.18, is termed a *minimum phase* system.

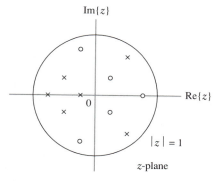

FIGURE 7.18 A system that has a causal and stable inverse must have all its poles and zeros inside the unit circle, as illustrated here.

EXAMPLE 7.16   A system is described by the difference equation

$$y[n] - y[n-1] + \tfrac{1}{4}y[n-2] = x[n] + \tfrac{1}{4}x[n-1] - \tfrac{1}{8}x[n-2]$$

Find the transfer function of the inverse system. Does a stable and causal inverse system exist?

Solution: We find the transfer function of this system by applying Eq. (7.19) to obtain

$$H(z) = \frac{1 + \tfrac{1}{4}z^{-1} - \tfrac{1}{8}z^{-2}}{1 - z^{-1} + \tfrac{1}{4}z^{-2}}$$

$$= \frac{(1 - \tfrac{1}{4}z^{-1})(1 + \tfrac{1}{2}z^{-1})}{(1 - \tfrac{1}{2}z^{-1})^2}$$

and the inverse system has the transfer function

$$H^{-1}(z) = \frac{(1 - \tfrac{1}{2}z^{-1})^2}{(1 - \tfrac{1}{4}z^{-1})(1 + \tfrac{1}{2}z^{-1})}$$

The poles of the inverse system are at $z = \tfrac{1}{4}$ and $z = -\tfrac{1}{2}$. Both of these poles are inside the unit circle and therefore the inverse system can be both stable and causal.

▶ **Drill Problem 7.9**   A system has the impulse response

$$h[n] = 2\delta[n] + \tfrac{5}{2}(\tfrac{1}{2})^n u[n] - \tfrac{7}{2}(-\tfrac{1}{4})^n u[n]$$

Find the transfer function of the inverse system. Does a stable and causal inverse system exist?

Answer:

$$H^{-1}(z) = \frac{(1 - \tfrac{1}{2}z^{-1})(1 + \tfrac{1}{4}z^{-1})}{(1 - \tfrac{1}{8}z^{-1})(1 + 2z^{-1})}$$

The inverse system cannot be both stable and causal.   ◀

■ **DETERMINING THE FREQUENCY RESPONSE FROM POLES AND ZEROS**

We now explore the relationship between pole and zero locations in the z-plane and the frequency response of the system. Recall that the frequency response is obtained from the transfer function by substituting $e^{j\Omega}$ for z in $H(z)$. That is, the frequency response corresponds to the transfer function evaluated on the unit circle in the z-plane. This assumes that the ROC includes the unit circle. Substituting $z = e^{j\Omega}$ into Eq. (7.21) gives

$$H(e^{j\Omega}) = \frac{\tilde{b}e^{-jp\Omega}\prod_{k=1}^{M-p}(1 - c_k e^{-j\Omega})}{e^{-jl\Omega}\prod_{k=1}^{N-l}(1 - d_k e^{-j\Omega})}$$

Multiplying both the numerator and denominator by $e^{jN\Omega}$ yields

$$H(e^{j\Omega}) = \frac{\tilde{b}e^{j(N-M)\Omega}\prod_{k=1}^{M-p}(e^{j\Omega} - c_k)}{\prod_{k=1}^{N-l}(e^{j\Omega} - d_k)} \tag{7.26}$$

We shall examine both the magnitude and phase of $H(e^{j\Omega})$ using Eq. (7.26).

The magnitude of $H(e^{j\Omega})$ at some fixed value of Ω, say, Ω_o, is defined by

$$|H(e^{j\Omega_o})| = \frac{|\tilde{b}|\prod_{k=1}^{M-p}|e^{j\Omega_o} - c_k|}{\prod_{k=1}^{N-l}|e^{j\Omega_o} - d_k|}$$

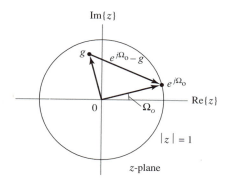

FIGURE 7.19   Vector interpretation of $e^{j\Omega_o} - g$ in the z-plane.

This expression involves a ratio of products of terms of the form $|e^{j\Omega_o} - g|$, where g represents either a pole or a zero. The terms involving zeros are in the numerator, while those involving poles are in the denominator. If we use vectors to represent complex numbers in the z-plane, then $e^{j\Omega_o}$ is a vector from the origin to the point $e^{j\Omega_o}$ and g is a vector from the origin to g. Hence $e^{j\Omega_o} - g$ is represented as a vector from the point g to the point $e^{j\Omega_o}$ as illustrated in Fig. 7.19. The length of this vector is $|e^{j\Omega_o} - g|$. We assess the contribution of each pole and zero to the overall frequency response by examining the length of this vector as Ω_o changes.

Figure 7.20(a) depicts the vector $e^{j\Omega} - g$ for several different values of Ω while Fig. 7.20(b) depicts $|e^{j\Omega} - g|$ as a continuous function of frequency. Note that if $\Omega = \arg\{g\}$, then $|e^{j\Omega} - g|$ attains its minimum value of $1 - |g|$ when g is inside the unit circle and $|g| - 1$ when g is outside the unit circle. Hence if g is close to the unit circle ($|g| \approx 1$), then $|e^{j\Omega} - g|$ becomes very small when $\Omega = \arg\{g\}$.

If g represents a zero, then $|e^{j\Omega} - g|$ contributes to the numerator of $|H(e^{j\Omega})|$. Hence at frequencies near $\arg\{g\}$, $|H(e^{j\Omega})|$ tends to decrease. How far $|H(e^{j\Omega})|$ decreases depends on how close the zero is to the unit circle. If the zero is on the unit circle, then $|H(e^{j\Omega})|$ goes to zero at the frequency corresponding to the zero. On the other hand, if g represents a pole, then $|e^{j\Omega} - g|$ contributes to the denominator of $|H(e^{j\Omega})|$. When $|e^{j\Omega} - g|$

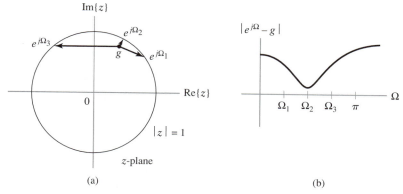

(a)   (b)

FIGURE 7.20   The quantity $|e^{j\Omega} - g|$ is the length of vectors from g to $e^{j\Omega}$ in the z-plane. (a) Vectors from g to $e^{j\Omega}$ at several frequencies. (b) The function $|e^{j\Omega} - g|$.

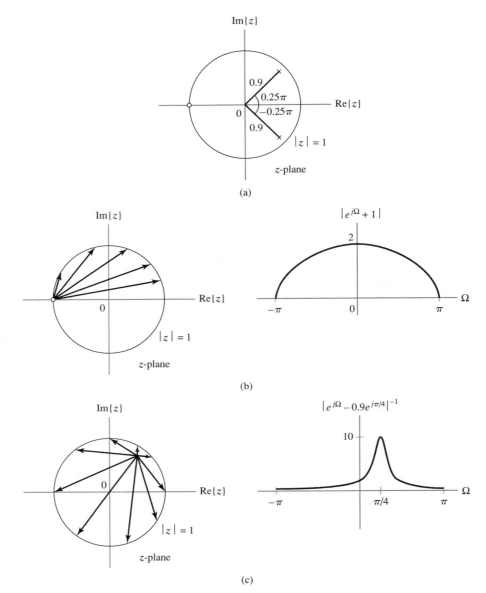

FIGURE 7.21 Solution for Example 7.17. (a) Pole and zero locations in the z-plane. (b) The component of the magnitude response associated with a zero is given by the length of a vector from the zero to $e^{j\Omega}$. (c) The component of the magnitude response associated with the pole at $z = 0.9e^{j\pi/4}$ is the inverse of the length of a vector from the pole to $e^{j\Omega}$.

decreases, $|H(e^{j\Omega})|$ will increase, with the size of the increase dependent on how far the pole is from the unit circle. A pole that is very close to the unit circle will cause a large peak in $|H(e^{j\Omega})|$ at the frequency corresponding to the phase angle of the pole. Hence zeros tend to pull the frequency response magnitude down, while poles tend to push it up.

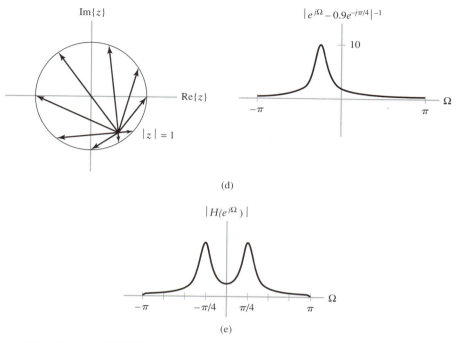

(d)

(e)

FIGURE 7.21   Continued (d) The component of the magnitude response associated with the pole at $z = 0.9e^{-j\pi/4}$ is the inverse of the length of a vector from the pole to $e^{j\Omega}$. (e) The system magnitude response is the product of the response in parts (b)–(d).

EXAMPLE 7.17   Sketch the magnitude response for the system having the transfer function

$$H(z) = \frac{1 + z^{-1}}{(1 - 0.9e^{j\pi/4}z^{-1})(1 - 0.9e^{-j\pi/4}z^{-1})}$$

Solution: The system has a zero at $z = -1$ and poles at $z = 0.9e^{j\pi/4}$ and $z = 0.9e^{-j\pi/4}$ as depicted in Fig. 7.21(a). Hence the magnitude response will be zero at $\Omega = \pi$ and large at $\Omega = \pm\pi/4$ because the poles are close to the unit circle. Figures 7.21(b)–(d) depict the component of the magnitude response associated with the zero and each pole. Multiplication of these contributions gives the magnitude response sketched in Fig. 7.21(e).

The phase of $H(e^{j\Omega})$ may also be evaluated in terms of the phase associated with each pole and zero. Using Eq. (7.26), we evaluate $\arg\{H(e^{j\Omega})\}$ as

$$\arg\{H(e^{j\Omega})\} = \arg\{\tilde{b}\} + (N - M)\Omega + \sum_{k=1}^{M-p} \arg\{e^{j\Omega} - c_k)\} - \sum_{k=1}^{N-l} \arg\{e^{j\Omega} - d_k\}$$

The phase of $H(e^{j\Omega})$ involves the sum of the phase angles due to each zero minus the phase angle due to each pole. The first term, $\arg\{\tilde{b}\}$, is independent of frequency. The phase associated with each zero and pole is evaluated by considering a term of the form $\arg\{e^{j\Omega} - g\}$. This is the angle associated with a vector pointing from g to $e^{j\Omega}$. The angle of this vector is measured with respect to a horizontal line passing through g as illustrated in Fig. 7.22. The contribution of any pole or zero to the overall phase response is determined by the angle of this vector as frequency changes.

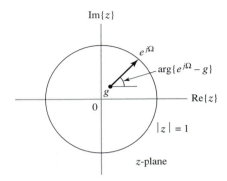

FIGURE 7.22   The quantity $\arg\{e^{j\Omega} - g\}$ is the angle of the vector from g to $e^{j\Omega}$ with respect to a horizontal line through g, as shown here.

Exact evaluation of frequency response is best performed numerically. However, we can often obtain a rough approximation from the pole and zero locations, and thus develop insight into the nature of the frequency response from the pole and zero locations.

7.7 *Computational Structures for Implementing Discrete-Time Systems*

Discrete-time systems are often implemented on a computer. In order to write the computer program that determines the system output from the input, we must first specify the order in which each computation is to be performed. The z-transform is often used to develop such computational structures for implementing discrete-time systems that have a given transfer function. We recall from Chapter 2 that there are many different block diagram implementations corresponding to a system with a given input–output characteristic. The freedom to choose between alternative implementations can be used to optimize some criteria associated with the computation, such as the number of numerical operations or the sensitivity of the system to numerical rounding of computations. A detailed study of such issues is beyond the scope of this book. Here we illustrate the role of the z-transform in obtaining alternative computational structures.

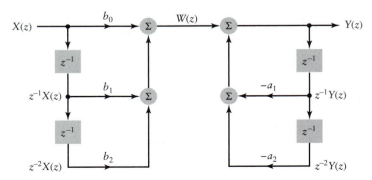

FIGURE 7.23   Block diagram of the transfer function corresponding to Fig. 2.27.

Several block diagrams for implementing systems described by difference equations were derived in Section 2.5. They consist of time-shift operations, denoted by the operator S, multiplication by constants, and summing junctions. We may represent rational transfer function descriptions for systems with analogous block diagrams by taking the z-transform of the block diagram representing the difference equation. The time-shift operator corresponds to multiplication by z^{-1} in the z-domain. Scalar multiplication and addition are linear operations and are thus not modified by taking the z-transform. Hence the block diagrams representing rational transfer functions use z^{-1} in place of the shift operator. For example, the block diagram depicted in Fig. 2.27 represents a system described by the difference equation

$$y[n] + a_1y[n-1] + a_2y[n-2] = b_0x[n] + b_1x[n-1] + b_2x[n-2] \quad (7.27)$$

Taking the z-transform of the difference equation gives

$$Y(z)(1 + a_1z^{-1} + a_2z^{-2}) = X(z)(b_0 + b_1z^{-1} + b_2z^{-2})$$

The block diagram depicted in Fig. 7.23 implements this relationship. It is obtained by replacing the shift operators in Fig. 2.27 with z^{-1}. The transfer function of the system in Fig. 7.23 is given by

$$\begin{aligned} H(z) &= \frac{Y(z)}{X(z)} \\ &= \frac{b_0 + b_1z^{-1} + b_2z^{-2}}{1 + a_1z^{-1} + a_2z^{-2}} \end{aligned} \quad (7.28)$$

The direct form II representation for a system was derived in Section 2.5 by writing the difference equation described by Eq. (7.27) as two coupled difference equations involving an intermediate signal $f[n]$. We may also derive the direct form II directly from the system transfer function. The transfer function of the system described by Eq. (7.27) is given by $H(z)$ in Eq. (7.28). Now suppose we write $H(z) = H_1(z)H_2(z)$, where

$$H_1(z) = b_0 + b_1z^{-1} + b_2z^{-2}$$

and

$$H_2(z) = \frac{1}{1 + a_1z^{-1} + a_2z^{-2}}$$

The direct form II implementation for $H(z)$ is obtained by writing

$$Y(z) = H_1(z)F(z) \quad (7.29)$$

where

$$F(z) = H_2(z)X(z) \quad (7.30)$$

The block diagram depicted in Fig. 7.24(a) implements Eqs. (7.29) and (7.30). The z^{-1} blocks in $H_1(z)$ and $H_2(z)$ generate identical quantities and thus may be combined to obtain the direct form II block diagram depicted in Fig. 7.24(b).

The pole–zero form of the transfer function leads to two alternative system implementations, the cascade and parallel forms. In these forms the transfer function is represented as an interconnection of lower order transfer functions, or sections. In the cascade form we write

$$H(z) = \prod_{i=1}^{P} H_i(z)$$

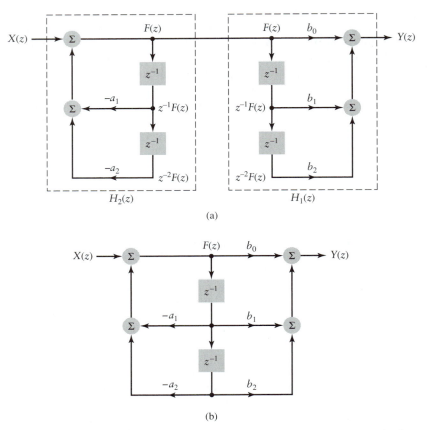

(a)

(b)

FIGURE 7.24   Development of the direct form II. (a) Representation of the transfer function $H(z)$ as $H_1(z)H_2(z)$. (b) Direct form II implementation of the transfer function $H(z)$ obtained from (a) by collapsing the two sets of z^{-1} blocks.

where the $H_i(z)$ contain distinct subsets of the poles and zeros of $H(z)$. Usually one or two of the poles and zeros of $H(z)$ are assigned to each $H_i(z)$. We say that the system is represented as a cascade of first- or second-order sections in this case. Poles and zeros that occur in complex-conjugate pairs are usually placed into the same section so that the coefficients of the section are real valued. In the parallel form we use a partial fraction expansion to write

$$H(z) = \sum_{i=1}^{P} H_i(z)$$

where each $H_i(z)$ contains a distinct set of the poles of $H(z)$. Here again usually one or two poles are assigned to each section, and we say that the system is represented by a parallel connection of first- or second-order sections. The following example and drill problem illustrate both the parallel and cascade forms.

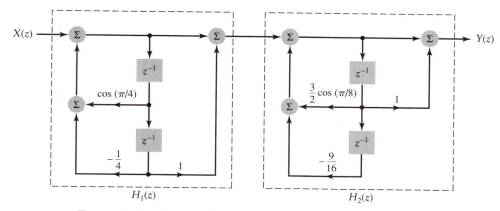

FIGURE 7.25   Cascade form of implementation for Example 7.18.

EXAMPLE 7.18   Consider the system represented by the transfer function

$$H(z) = \frac{(1 + jz^{-1})(1 - jz^{-1})(1 + z^{-1})}{(1 - \frac{1}{2}e^{j\pi/4}z^{-1})(1 - \frac{1}{2}e^{-j\pi/4}z^{-1})(1 - \frac{3}{4}e^{j\pi/8}z^{-1})(1 - \frac{3}{4}e^{-j\pi/8}z^{-1})}$$

Depict the cascade form for this system using real-valued second-order sections. Assume each second-order section is implemented as a direct form II.

Solution: We combine complex-conjugate poles and zeros into the sections, as shown by

$$H_1(z) = \frac{1 + z^{-2}}{1 - \cos(\pi/4)z^{-1} + \frac{1}{4}z^{-2}}$$

$$H_2(z) = \frac{1 + z^{-1}}{1 - \frac{3}{2}\cos(\pi/8)z^{-1} + \frac{9}{16}z^{-2}}$$

The block diagram corresponding to $H_1(z)H_2(z)$ is depicted in Fig. 7.25. Note that this solution is not unique, since we could have interchanged the order of $H_1(z)$ and $H_2(z)$ or interchanged the pairing of poles and zeros.

▶ **Drill Problem 7.10**   Depict the parallel form representation for the transfer function

$$H(z) = \frac{4 - \frac{1}{2}z^{-1} - \frac{1}{2}z^{-2}}{(1 - \frac{1}{2}z^{-1})(1 + \frac{1}{2}z^{-1})(1 - \frac{1}{4}z^{-1})}$$

using first-order sections implemented as a direct form II.

Answer:

$$H(z) = \frac{1}{1 - \frac{1}{2}z^{-1}} + \frac{1}{1 + \frac{1}{2}z^{-1}} + \frac{2}{1 - \frac{1}{4}z^{-1}}$$

See Fig. 7.26.   ◀

7.8   *The Unilateral z-Transform*

The *unilateral* or *one-sided z-transform* is evaluated using the portion of a signal associated with nonnegative values of the time index ($n \geq 0$). This form of the z-transform is appro-

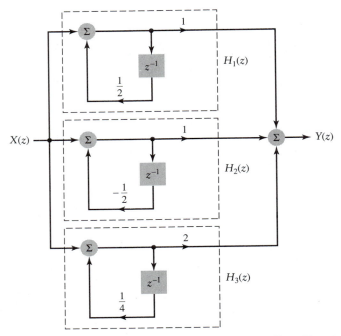

FIGURE 7.26   Parallel form of implementation for Drill Problem 7.10.

priate for problems involving causal signals and systems. It is reasonable to assume causality in many applications of z-transforms. For example, we are often interested in the
response of a causal system to an input signal. The choice of time origin is usually arbitrary,
so we may choose $n = 0$ as the time at which the input is applied and then study the
response for times $n \geq 0$. There are several advantages to using the unilateral transform
in such problems. First, we do not need to use ROCs. Second, and perhaps most important,
the unilateral transform allows us to study systems described by difference equations with
initial conditions.

■ **DEFINITION AND PROPERTIES**

The unilateral z-transform of a signal $x[n]$ is defined as

$$X(z) = \sum_{n=0}^{\infty} x[n]z^{-n} \tag{7.31}$$

which depends only on $x[n]$ for $n \geq 0$. The inverse z-transform may be obtained by evaluating Eq. (7.5) for $n \geq 0$. We denote the relationship between $x[n]$ and $X(z)$ as

$$x[n] \xleftrightarrow{\;z_u\;} X(z)$$

The unilateral and bilateral z-transforms are equivalent for causal signals. For example,

$$\alpha^n u[n] \xleftrightarrow{\;z_u\;} \frac{1}{1 - \alpha z^{-1}}$$

and

$$a^n \cos(\Omega_o n)u[n] \xleftrightarrow{\;z_u\;} \frac{1 - a\cos(\Omega_o)z^{-1}}{1 - 2a\cos(\Omega_o)z^{-1} + a^2 z^{-2}}$$

It is straightforward to show that the unilateral z-transform satisfies the same properties as the bilateral z-transform with one important exception, the time-shift property. In order to develop the time-shift property, let $w[n] = x[n - 1]$. Now from Eq. (7.31) we have

$$X(z) = \sum_{n=0}^{\infty} x[n]z^{-n}$$

The unilateral z-transform of $w[n]$ is defined similarly,

$$W(z) = \sum_{n=0}^{\infty} w[n]z^{-n}$$

We express $W(z)$ as a function of $X(z)$. Substituting $w[n] = x[n - 1]$, we obtain

$$
\begin{aligned}
W(z) &= \sum_{n=0}^{\infty} x[n - 1]z^{-n} \\
&= x[-1] + \sum_{n=1}^{\infty} x[n - 1]z^{-n} \\
&= x[-1] + \sum_{m=0}^{\infty} x[m]z^{-(m+1)} \\
&= x[-1] + z^{-1} \sum_{m=0}^{\infty} x[m]z^{-m} \\
&= x[-1] + z^{-1}X(z)
\end{aligned}
$$

Hence a unit time shift results in multiplication by z^{-1} and addition of the constant $x[-1]$. We obtain the time-shift property for delays greater than one in an identical manner. If

$$x[n] \xleftrightarrow{\ z_u\ } X(z)$$

then we have

$$
x[n - k] \xleftrightarrow{\ z_u\ } x[-k] + x[-k + 1]z^{-1} \tag{7.32}
$$
$$
+ \cdots + x[-1]z^{-k+1} + z^{-k}X(z) \quad \text{for } k > 0
$$

In the case of a time advance, the shift property changes somewhat. Here we obtain

$$
x[n + k] \xleftrightarrow{\ z_u\ } -x[0]z^k - x[1]z^{k-1} - \cdots - x[k - 1]z + z^kX(z) \quad \text{for } k > 0 \tag{7.33}
$$

Both shift properties correspond to the bilateral shift property with additional terms that account for values of the sequence that are shifted into or out of the nonnegative time portion of the signal.

■ SOLVING DIFFERENCE EQUATIONS WITH INITIAL CONDITIONS

The primary application of the unilateral z-transform is in solving difference equations subject to nonzero initial conditions. The difference equation is solved by taking the unilateral z-transform of both sides, using algebra to obtain the z-transform of the solution, and then inverse z-transforming. The initial conditions are incorporated naturally into the problem as a consequence of the time-shift property, Eq. (7.32). We begin with an example.

EXAMPLE 7.19 Consider the system described by the difference equation

$$y[n] - 0.9y[n - 1] = x[n]$$

Find the output if the input is $x[n] = u[n]$ and the initial condition on the output is $y[-1] = 2$.

Solution: Take the unilateral z-transform of both sides of the difference equation and use the time-shift property, obtaining

$$Y(z) - 0.9(y[-1] + z^{-1}Y(z)) = X(z)$$

Now rearrange this equation to determine $Y(z)$, as shown by

$$(1 - 0.9z^{-1})Y(z) = X(z) + 0.9y[-1]$$

or

$$Y(z) = \frac{X(z)}{1 - 0.9z^{-1}} + \frac{0.9y[-1]}{1 - 0.9z^{-1}}$$

Note that $Y(z)$ is given as the sum of two terms: one that depends on the input and another that depends on the initial condition. The input-dependent term represents the forced response of the system. The initial condition term represents the natural response of the system. Substitute $X(z) = 1/(1 - z^{-1})$ and $y[-1] = 2$ to obtain

$$Y(z) = \frac{1}{(1 - 0.9z^{-1})(1 - z^{-1})} + \frac{1.8}{1 - 0.9z^{-1}}$$

Now perform a partial fraction expansion on the first term of $Y(z)$, obtaining

$$Y(z) = \frac{-9}{1 - 0.9z^{-1}} + \frac{10}{1 - z^{-1}} + \frac{1.8}{1 - 0.9z^{-1}}$$

The system output is obtained by inverse z-transforming $Y(z)$, as shown by

$$y[n] = -9(0.9)^n u[n] + 10u[n] + 1.8(0.9)^n u[n]$$

The last term, $1.8(0.9)^n u[n]$, is the natural response of the system, while the first two terms represent the forced response. The natural response, forced response, and overall response are illustrated in Fig. 7.27.

In the previous example there was a natural separation between the forced and natural response of the system. This separation occurs in general. Consider taking the unilateral z-transform of both sides of the following difference equation:

$$\sum_{k=0}^{N} a_k y[n - k] = \sum_{k=0}^{M} b_k x[n - k]$$

We may write the z-transform as shown by

$$A(z)Y(z) + C(z) = B(z)X(z)$$

where

$$A(z) = \sum_{k=0}^{N} a_k z^{-k}$$

$$B(z) = \sum_{k=0}^{M} b_k z^{-k}$$

and

$$C(z) = \sum_{m=0}^{N-1} \sum_{k=m+1}^{N} a_k y[-k + m]z^{-m}$$

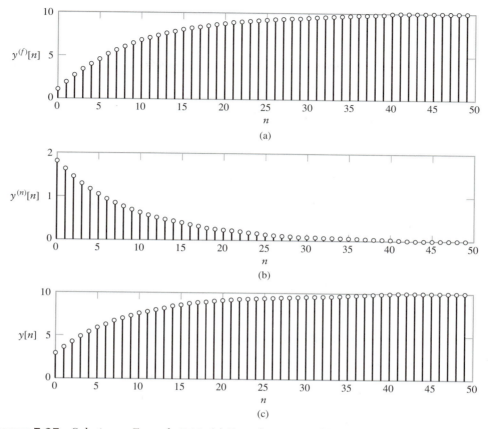

FIGURE 7.27   Solution to Example 7.19. (a) Forced response. (b) Natural response. (c) Overall system output.

Here we have assumed $x[n]$ is causal, so that $x[n - k] \xleftrightarrow{\;z_u\;} z^{-k}X(z)$. The term $C(z)$ depends on the N initial conditions $y[-1], y[-2], \ldots, y[-N]$ and the a_k. It is zero if all the initial conditions are zero. Solving for $Y(z)$ yields

$$Y(z) = \frac{B(z)}{A(z)} X(z) - \frac{C(z)}{A(z)}$$

The output is the sum of the forced response due to the input, $[B(z)/A(z)]X(z)$, and the natural response induced by the initial conditions, $C(z)/A(z)$. Since $C(z)$ is a polynomial, the poles of the natural response are the roots of $A(z)$, which are also the poles of the transfer function. Hence the form of the natural response depends only on the poles of the system. Note that if the system is stable, then the poles must lie inside the unit circle.

▶ **Drill Problem 7.11**   Determine the forced response $y^{(f)}[n]$, natural response $y^{(n)}[n]$, and output $y[n]$ of the system described by the difference equation

$$y[n] + 3y[n - 1] = x[n] + x[n - 1]$$

if the input is $x[n] = (\frac{1}{2})^n u[n]$ and $y[-1] = 2$ is the initial condition.

Answer:

$$y^{(f)}[n] = \tfrac{4}{7}(-3)^n u[n] + \tfrac{3}{7}(\tfrac{1}{2})^n u[n]$$
$$y^{(n)}[n] = -6(-3)^n u[n]$$
$$y[n] = y^{(f)}[n] + y^{(n)}[n]$$

◀

7.9 Exploring Concepts with MATLAB

The MATLAB Signal Processing Toolbox contains routines for working with z-transforms.

■ POLES AND ZEROS

The poles and zeros of a system may be determined by applying `roots` to the respective polynomial. For example, to find the roots of $1 + 4z^{-1} + 3z^{-2}$, we give the command `roots([1, 4, 3])`. The poles and zeros may be displayed in the z-plane using `zplane(b,a)`. If b and a are row vectors, then `zplane` roots the numerator and denominator polynomials represented by b and a, respectively, to find the poles and zeros before displaying them. If b and a are column vectors, then `zplane` assumes b and a contain the zero and pole locations, respectively.

■ INVERSION OF THE z-TRANSFORM

The `residuez` command computes partial fraction expansions for z-transforms expressed as a ratio of two polynomials in z^{-1}. The syntax is `[r, p, k] = residuez(b,a)`, where b and a are vectors representing the numerator and denominator polynomial coefficients ordered in descending powers of z. The vector r represents the partial fraction expansion coefficients corresponding to the poles given in p. The vector k contains the coefficients associated with powers of z^{-1} that result from long division when the numerator order equals or exceeds the denominator order.

For example, we may use MATLAB to find the partial fraction expansion for the z-transform given in Example 7.10:

$$X(z) = \frac{z^3 - 10z^2 - 4z + 4}{2z^2 - 2z - 4}$$

Since `residuez` assumes the numerator and denominator polynomials are expressed in powers of z^{-1}, we first write $X(z) = zY(z)$, where

$$Y(z) = \frac{1 - 10z^{-1} - 4z^{-2} + 4z^{-3}}{2 - 2z^{-1} - 4z^{-2}}$$

Now use `residuez` to find the partial fraction expansion for $Y(z)$ as follows:

```
>> [r, p, k] = residuez([1, -10, -4, 4], [2, -2, -4])

r =
      -1.5000
       0.5000

p =
       2
      -1

k =
       1.5000    -1.0000
```

This implies a partial fraction expansion of the form

$$Y(z) = \frac{-1.5}{1 - 2z^{-1}} + \frac{0.5}{1 + z^{-1}} + 1.5 - z^{-1}$$

which, as expected, corresponds to $\frac{1}{2}W(z)$ in Example 7.10.

▶ **Drill Problem 7.12** Solve Drill Problem 7.6 using MATLAB and the `residuez` command.

■ TRANSFORM ANALYSIS OF LTI SYSTEMS

Recall that the impulse response, difference equation, transfer function, poles and zeros, frequency response, and state-variable description offer different, yet equivalent, representations for the input–output characteristics of an LTI system. The MATLAB Signal Processing Toolbox contains several routines for converting between different system descriptions. If `b` and `a` contain the coefficients of the transfer function numerator and denominator polynomials, respectively, ordered in descending powers of z, then `tf2ss(b,a)` determines a state-variable description for the system and `tf2zp(b,a)` determines the pole–zero–gain description for the system. Similarly, `zp2ss` and `zp2tf` convert from pole–zero–gain descriptions to state-variable and transfer function descriptions, respectively, while `ss2tf` and `ss2zp` convert from state-variable description to transfer function and pole–zero–gain forms, respectively. As noted in Section 4.12, the frequency response is evaluated from the transfer function using `freqz`.

Consider a system with transfer function

$$H(z) = \frac{0.094(1 + 4z^{-1} + 6z^{-2} + 4z^{-3} + z^{-4})}{1 + 0.4860z^{-2} + 0.0177z^{-4}} \tag{7.34}$$

We may depict the poles and zeros of $H(z)$ in the z-plane and plot the system's magnitude response with the following commands:

```
>> b = .094*[1, 4, 6, 4, 1];
>> a = [1, 0, 0.486, 0, 0.0177];
>> zplane(b,a)
>> [H,w] = freqz(b,a,250);
>> plot(w,abs(H))
```

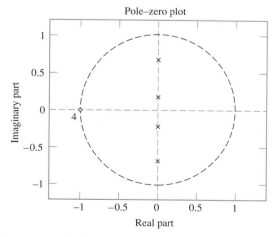

FIGURE 7.28 Locations of poles and zeros in the z-plane obtained using MATLAB.

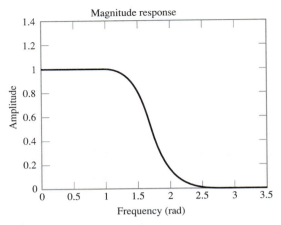

FIGURE 7.29   Magnitude response evaluated using MATLAB.

Figure 7.28 indicates that this system has a zero of order four at $z = -1$ and four poles on the imaginary axis. The magnitude response is depicted in Fig. 7.29. Note that the zeros at $z = -1$ force the magnitude response to be small at high frequencies.

■ COMPUTATIONAL STRUCTURES FOR IMPLEMENTING DISCRETE-TIME SYSTEMS

One useful means for implementing a discrete-time system is as a cascade of second-order sections. The MATLAB Signal Processing Toolbox contains routines for converting a state-variable or pole–zero–gain description of a system to a cascade connection of second-order sections. This is accomplished using $\mathtt{ss2sos}$ and $\mathtt{zp2sos}$, respectively. The syntax for $\mathtt{zp2sos}$ is $\mathtt{sos = zp2sos(z,p,k)}$, where $\mathtt{z}$ and $\mathtt{p}$ are vectors containing zeros and poles, respectively, and $\mathtt{k}$ is the gain. The matrix $\mathtt{sos}$ is L by 6, where each row contains the coefficients of the transfer function for that section. The first three elements of the row contain the numerator coefficients while the last three elements contain the denominator coefficients. The commands $\mathtt{sos2zp}$, $\mathtt{sos2ss}$, and $\mathtt{sos2tf}$ convert from a cascade of second-order sections to pole–zero–gain, state-variable, and transfer function descriptions, respectively.

Suppose we repeat Example 7.18 using MATLAB to obtain a representation of the system as a cascade of second-order sections. The transfer function is given in pole–zero–gain form

$$H(z) = \frac{(1 + jz^{-1})(1 - jz^{-1})(1 + z^{-1})}{(1 - \frac{1}{2}e^{j\pi/4}z^{-1})(1 - \frac{1}{2}e^{-j\pi/4}z^{-1})(1 - \frac{3}{4}e^{j\pi/8}z^{-1})(1 - \frac{3}{4}e^{-j\pi/8}z^{-1})}$$

The system has zeros at $z = \pm j$ and $z = -1$, while the poles are at $z = \frac{1}{2}e^{\pm j\pi/4}$ and $z = \frac{3}{4}e^{\pm j\pi/8}$. We employ $\mathtt{zp2sos}$ to convert from pole–zero–gain form to second-order sections as follows:

```
≫ z = [-1, -j, j]; ...
≫ p = [0.5*exp(j*pi/4), 0.5*exp(-j*pi/4), ...
0.75*exp(j*pi/8), 0.75exp(-j*pi/8)];
≫ k = 1;
≫ sos = zp2sos(z,p,k)
```

```
SOS =

    0.2706    0.2706         0    1.0000   -0.7071    0.2500
    3.6955         0    3.6955    1.0000   -1.3858    0.5625
```

Hence the system is described as the cascade of the second-order sections

$$F_1(z) = \frac{0.2706 + 0.2706z^{-1}}{1 - 0.7071z^{-1} + 0.25z^{-2}} \qquad F_2(z) = \frac{3.6955 + 3.6955z^{-2}}{1 - 1.3858z^{-1} + 0.5625z^{-2}}$$

Note that this solution differs from that of Example 7.18 in that the pairing of zeros and poles are interchanged. A scaling factor is also introduced into each section by zp2sos. The overall gain is unchanged, however, since the product of scaling factors is unity. The procedures employed by zp2sos for scaling and pairing poles with zeros are chosen to minimize the effect of numerical errors when such systems are implemented with fixed point arithmetic.

7.10 *Summary*

The z-transform represents discrete-time signals as a weighted superposition of complex exponentials. Complex exponentials are a more general signal class than complex sinusoids, so the z-transform can represent a broader class of signals than the DTFT, including signals that are not absolutely summable. Thus we may use the z-transform to analyze signals and systems that are not stable. The transfer function of a system is the z-transform of its impulse response. It offers another description for the input–output characteristics of a LTI system. The z-transform converts convolution of time signals into multiplication of z-transforms, so the z-transform of a system's output is the product of the z-transform of the input and the system's transfer function.

A complex exponential is described by a complex number. Hence the z-transform is a function of a complex variable z represented in a complex plane. The DTFT is obtained by evaluating the z-transform on the unit circle, $|z| = 1$, by setting $z = e^{j\Omega}$. The properties of the z-transform are analogous to those of the DTFT. The ROC defines the values of z for which the z-transform converges. The ROC must be specified in order to have a unique relationship between the time signal and its z-transform. The relative locations of the ROC and z-transform poles determine whether the corresponding time signal is right sided, left sided, or both. The locations of z-transform poles and zeros offer another representation for the input–output characteristics of the system. They provide information regarding system stability, causality, invertibility, and frequency response.

The z-transform and DTFT have many common features. However, they have distinct roles in signal and system analysis. The z-transform is generally used to study system characteristics, such as stability and causality, and to develop computational structures for implementing discrete-time systems. The z-transform is also used for transient and stability analysis of sampled-data control systems, a topic we visit in Chapter 9. The unilateral z-transform applies to causal signals and offers a convenient tool for solving systems problems defined by difference equations with nonzero initial conditions. None of these problems are addressed with the DTFT. Instead, the DTFT is usually used as a signal representation tool and to study the steady-state characteristics of systems, as we illustrated in Chapters 3 and 4. In these problems the DTFT is easier to visualize than the z-transform, since it is a function of real-valued frequency Ω, while the z-transform is a function of a complex number $z = re^{j\Omega}$.

FURTHER READING

1. The following text is devoted entirely to z-transforms:
   ▶ Vich, R., *Z Transform Theory and Applications* (D. Reidel Publishing, 1987)

2. The z-transform is also discussed in most texts on signal processing, including:
   ▶ Oppenheim, A. V., and R. W. Schafer, *Discrete-Time Signal Processing* Second Edition (Prentice Hall, 1999)
   ▶ Proakis, J. G., and D. G. Manolakis, *Introduction to Digital Signal Processing* (Macmillan, 1988)
   ▶ Roberts, R. A., and C. T. Mullis, *Digital Signal Processing* (Addison-Wesley, 1987)

3. One interesting property of minimum phase systems is the unique relationship between their magnitude and phase response. If we are given one component, then we can uniquely determine the other. This and other properties of minimum phase systems are discussed in
   ▶ Oppenheim, A. V., and R. W. Schafer, *Discrete-Time Signal Processing* Second Edition (Prentice Hall, 1999)

PROBLEMS

7.1 Determine the z-transform and ROC for the following time signals. Sketch the ROC, poles, and zeros in the z-plane.
(a) $x[n] = \delta[n]$
(b) $x[n] = \delta[n - k], \quad k > 0$
(c) $x[n] = \delta[n + k], \quad k > 0$
(d) $x[n] = u[n]$
(e) $x[n] = (\frac{1}{2})^n(u[n] - u[n - 10])$
(f) $x[n] = (\frac{1}{2})^n u[-n]$
(g) $x[n] = 2^n u[-n - 1]$
(h) $x[n] = (\frac{1}{2})^{|n|}$
(i) $x[n] = (\frac{1}{2})^n u[n - 2]$
(j) $x[n] = (\frac{1}{2})^n u[n] + (\frac{1}{3})^n u[-n - 1]$

7.2 Given the following z-transforms, determine whether the DTFT of the corresponding time signal exists without determining the time signal. Identify the DTFT in those cases where it exists.
(a) $X(z) = \dfrac{10}{1 + \frac{1}{2}z^{-1}}, \quad |z| > \frac{1}{2}$
(b) $X(z) = \dfrac{10}{1 + \frac{1}{2}z^{-1}}, \quad |z| < \frac{1}{2}$
(c) $X(z) = z^{-5}, \quad |z| > 0$
(d) $X(z) = z^5, \quad |z| < \infty$
(e) $X(z) = \dfrac{z^{-1}}{(1 - \frac{1}{3}z^{-1})(1 + 3z^{-1})}, \quad |z| < \frac{1}{3}$
(f) $X(z) = \dfrac{z^{-1}}{(1 - \frac{1}{3}z^{-1})(1 + 3z^{-1})}, \quad \frac{1}{3} < |z| < 3$

7.3 The pole and zero locations of $X(z)$ are depicted in the z-plane on the following figures. In each case, identify all valid ROCs for $X(z)$ and specify the characteristics of the time signal corresponding to each ROC.
(a) Figure P7.3(a)
(b) Figure P7.3(b)
(c) Figure P7.3(c)

7.4 Prove the following properties of the z-transform:
(a) Time shift
(b) Time reversal
(c) Multiplication by exponential sequence
(d) Differentiation in the z-domain

7.5 Use the tables of z-transforms and z-transform properties to determine the z-transforms of the following signals:
(a) $x[n] = (\frac{1}{2})^n u[n] * (\frac{1}{3})^n u[n]$
(b) $x[n] = n((\frac{1}{2})^n u[n] * (\frac{1}{2})^n u[n])$
(c) $x[n] = u[-n]$
(d) $x[n] = \sin\left(\dfrac{\pi}{8} n - \dfrac{\pi}{4}\right)u[n - 2]$
(e) $x[n] = n \sin\left(\dfrac{\pi}{2} n\right)u[-n]$

7.6 Given the z-transform pair $x[n] \xleftrightarrow{z} z/(z^2 + 4)$ with ROC $|z| < 2$, use the z-transform properties to determine the z-transform of the following signals:
(a) $y[n] = x[n - 4]$
(b) $y[n] = 2^n x[n]$
(c) $y[n] = x[-n]$
(d) $y[n] = nx[n]$

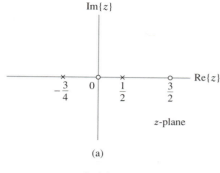

(a)

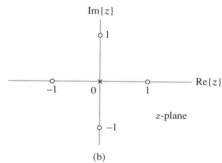

(b)

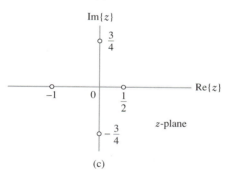

(c)

FIGURE P7.3

(e) $y[n] = x[n + 1] + x[n - 1]$

(f) $y[n] = \underbrace{x[n] * x[n] * \cdots * x[n]}_{m \text{ times}}$

(g) $y[n] = (n - 3)x[n - 2]$

7.7 Given the z-transform pair $3^n u[n] \xleftrightarrow{z} X(z)$, use the z-transform properties to determine the time-domain signals corresponding to the following z-transforms:

(a) $Y(z) = X(3z)$

(b) $Y(z) = X(z^{-1})$

(c) $Y(z) = \dfrac{d}{dz} X(z)$

(d) $Y(z) = \dfrac{z + z^{-1}}{2} X(z)$

(e) $Y(z) = X(z)X(9z)$

7.8 Use the method of partial fractions to obtain the time-domain signals corresponding to the following z-transforms:

(a) $X(z) = \dfrac{\frac{1}{4}z^{-1}}{(1 - \frac{1}{2}z^{-1})(1 - \frac{1}{4}z^{-1})}, \quad |z| > \frac{1}{2}$

(b) $X(z) = \dfrac{\frac{1}{4}z^{-1}}{(1 - \frac{1}{2}z^{-1})(1 - \frac{1}{4}z^{-1})}, \quad |z| < \frac{1}{4}$

(c) $X(z) = \dfrac{\frac{1}{4}z^{-1}}{(1 - \frac{1}{2}z^{-1})(1 - \frac{1}{4}z^{-1})}, \quad \frac{1}{4} < |z| < \frac{1}{2}$

(d) $X(z) = \dfrac{z^2 - 3z}{z^2 + \frac{3}{2}z - 1}, \quad \frac{1}{2} < |z| < 2$

(e) $X(z) = \dfrac{12(11z^2 - 3z)}{12z^2 - 7z + 1}, \quad |z| > \frac{1}{3}$

(f) $X(z) = \dfrac{8z^2 + 4z}{4z^2 - 4z + 1}, \quad |z| > \frac{1}{2}$

(g) $X(z) = \dfrac{z^3 + z^2 + \frac{3}{2}z + \frac{1}{2}}{z^3 + \frac{3}{2}z^2 + \frac{1}{2}z}, \quad |z| < \frac{1}{2}$

(h) $X(z) = \dfrac{z^3 + z^2 + \frac{3}{2}z + \frac{1}{2}}{z^2 + \frac{3}{2}z + \frac{1}{2}}, \quad |z| > 1$

(i) $X(z) = \dfrac{2z^4 - 2z^3 - 2z^2}{z^2 - 1}, \quad |z| > 1$

7.9 Determine the time-domain signals corresponding to the following z-transforms:

(a) $X(z) = 1 + 2z^{-2} + 4z^{-4}, \quad |z| > 0$

(b) $X(z) = \sum_{k=5}^{10} (1/k)z^{-k}, \quad |z| > 0$

(c) $X(z) = (1 + z^{-1})^4, \quad |z| > 0$

(d) $X(z) = z^4 + 2z^2 + 3 + 2z^{-2} + z^{-4}, \quad 0 < |z| < \infty$

7.10 Determine the impulse response corresponding to the following transfer functions if (i) the system is stable; (ii) the system is causal:

(a) $H(z) = \dfrac{3z^{-1}}{(1 - 2z^{-1})^2}$

(b) $H(z) = \dfrac{12z^2 + 24z}{12z^2 + 13z + 3}$

(c) $H(z) = \dfrac{4z}{z^2 - \frac{1}{2}z + \frac{1}{16}}$

7.11 Use a power series expansion to determine the time-domain signal corresponding to the following z-transforms:

(a) $X(z) = \dfrac{1}{1 - z^{-2}}, \quad |z| > 1$

(b) $X(z) = \dfrac{1}{1 - z^{-2}}, \quad |z| < 1$

(c) $X(z) = \cos(2z), \quad |z| < \infty$

(d) $X(z) = \cos(z^{-2}), \quad |z| > 0$

(e) $X(z) = \ln(1 + z^{-1}), \quad |z| > 0$

7.12 A causal system has input $x[n]$ and output $y[n]$. Use the transfer function to determine the impulse response of this system.

(a) $x[n] = \delta[n] + \frac{1}{4}\delta[n-1] - \frac{1}{8}\delta[n-2]$,
$y[n] = \delta[n] - \frac{3}{4}\delta[n-1]$

(b) $x[n] = (-\frac{1}{3})^n u[n]$,
$y[n] = 3(-1)^n u[n] + (\frac{1}{3})^n u[n]$

(c) $x[n] = (-3)^n u[n]$,
$y[n] = 4(2)^n u[n] - (\frac{1}{2})^n u[n]$

7.13 A system has impulse response $h[n] = (\frac{1}{2})^n u[n]$. Determine the input to the system if the output is given by:

(a) $y[n] = \delta[n-2]$

(b) $y[n] = (\frac{1}{2})^n u[n] + (-\frac{1}{2})^n u[n]$

(c) $y[n] = \frac{1}{3}u[n] + \frac{2}{3}(-\frac{1}{2})^n u[n]$

7.14 Determine (i) the transfer function and (ii) the impulse response of the systems described by the following difference equations:

(a) $y[n] - \frac{1}{2}y[n-1] = 2x[n-1]$

(b) $y[n] = x[n] - x[n-2] + x[n-4] - x[n-6]$

(c) $y[n] - \frac{1}{4}y[n-1] - \frac{3}{8}y[n-2] = -x[n] + 2x[n-1]$

(d) $y[n] - \frac{4}{5}y[n-1] - \frac{16}{25}y[n-2] = 2x[n] + x[n-1]$

7.15 Determine (i) the transfer function and (ii) a difference-equation representation for the systems with the following impulse responses:

(a) $h[n] = 3(\frac{1}{4})^n u[n-1]$

(b) $h[n] = (\frac{1}{3})^n u[n] + (\frac{1}{2})^{n-2} u[n-1]$

(c) $h[n] = 2(\frac{2}{3})^n u[n-1] + (\frac{1}{4})^n [\cos(\pi n/6) - 2\sin(\pi n/6)] u[n]$

(d) $h[n] = \delta[n] - \delta[n-5]$

7.16 Determine (i) the transfer function and (ii) a difference-equation representation for the systems described by the following state-variable descriptions. Plot the pole and zero locations in the z-plane.

(a) $\mathbf{A} = \begin{bmatrix} -\frac{1}{2} & 0 \\ 0 & \frac{1}{2} \end{bmatrix}$, $\mathbf{b} = \begin{bmatrix} 0 \\ 2 \end{bmatrix}$,
$\mathbf{c} = \begin{bmatrix} 1 & -1 \end{bmatrix}$, $D = [1]$

(b) $\mathbf{A} = \begin{bmatrix} \frac{1}{2} & -\frac{1}{2} \\ -\frac{1}{2} & -\frac{1}{4} \end{bmatrix}$, $\mathbf{b} = \begin{bmatrix} 1 \\ 0 \end{bmatrix}$,
$\mathbf{c} = \begin{bmatrix} 2 & 1 \end{bmatrix}$, $D = [0]$

(c) $\mathbf{A} = \begin{bmatrix} -\frac{1}{4} & \frac{1}{8} \\ -\frac{7}{2} & \frac{3}{4} \end{bmatrix}$, $\mathbf{b} = \begin{bmatrix} 2 \\ 2 \end{bmatrix}$,
$\mathbf{c} = \begin{bmatrix} 0 & 1 \end{bmatrix}$, $D = [0]$

7.17 Determine whether each of the systems described below is (i) causal and stable and (ii) minimum phase.

(a) $H(z) = \dfrac{2z+1}{z^2 + z - \frac{5}{16}}$

(b) $H(z) = \dfrac{1 + 2z^{-1}}{1 + \frac{14}{8}z^{-1} + \frac{49}{64}z^{-2}}$

(c) $y[n] - \frac{6}{5}y[n-1] - \frac{16}{25}y[n-2] = 2x[n] + x[n-1]$

(d) $y[n] - 2y[n-2] = x[n] - \frac{1}{2}x[n-1]$

(e) $\mathbf{A} = \begin{bmatrix} -\frac{1}{4} & \frac{1}{8} \\ -\frac{7}{2} & \frac{3}{4} \end{bmatrix}$, $\mathbf{b} = \begin{bmatrix} 0 \\ 2 \end{bmatrix}$,
$\mathbf{c} = \begin{bmatrix} 1 & 1 \end{bmatrix}$, $D = [0]$

7.18 For each system described below, identify the transfer function of the inverse system, and determine whether it can be both causal and stable.

(a) $H(z) = \dfrac{1 - 4z^{-1} + 4z^{-2}}{1 - \frac{1}{2}z^{-1} + \frac{1}{4}z^{-2}}$

(b) $H(z) = \dfrac{z^2 - \frac{49}{64}}{z^2 - 4}$

(c) $h[n] = 10(-\frac{1}{2})^n u[n] - 9(-\frac{1}{4})^n u[n]$

(d) $h[n] = 24(\frac{1}{2})^n u[n-1] - 30(\frac{1}{3})^n u[n-1]$

(e) $y[n] - \frac{1}{4}y[n-2] = 6x[n] - 7x[n-1] + 3x[n-2]$

(f) $y[n] - 2y[n-1] = x[n]$

7.19 Use the graphical method to sketch the magnitude response of the systems having the following transfer functions:

(a) $H(z) = \dfrac{z-1}{z+0.9}$

(b) $H(z) = \dfrac{z^{-2}}{1 + \frac{49}{64}z^{-2}}$

(c) $H(z) = \dfrac{1 + z^{-1} + z^{-2} + z^{-3}}{4}$

__*__**7.20** Consider a system with transfer function

$$H(z) = \frac{1 - a^* z}{z - a}, \qquad |a| < 1$$

Here the pole and zero are a conjugate reciprocal pair.

(a) Sketch a pole–zero plot for this system in the z-plane.

(b) Use the graphical method to show that the magnitude response of this system is unity for all frequencies. A system with this characteristic is termed an *allpass* system.

(c) Use the graphical method to sketch the phase response of this system for $a = \frac{1}{2}$.

(d) Use the result from (b) to prove that any system with a transfer function of the form

$$H(z) = \prod_{k=1}^{P} \frac{1 - a_k^* z}{z - a_k}, \qquad |a_k| < 1$$

corresponds to a stable and causal allpass system.

(e) Can a stable and causal allpass system also be minimum phase? Explain.

7.21 Draw block diagram implementations of the following systems as a cascade of second-order sections with real-valued coefficients:

(a) $H(z) = \left[\dfrac{(1 - \frac{1}{2}e^{j\pi/3}z^{-1})(1 - \frac{1}{2}e^{-j\pi/3}z^{-1})}{(1 - \frac{1}{2}e^{j\pi/8}z^{-1})(1 - \frac{1}{2}e^{-j\pi/8}z^{-1})}\right]$
$\cdot \left[\dfrac{(1 + \frac{1}{2}e^{j\pi/3}z^{-1})(1 + \frac{1}{2}e^{-j\pi/3}z^{-1})}{(1 - \frac{3}{4}e^{j7\pi/8}z^{-1})(1 - \frac{3}{4}e^{-j7\pi/8}z^{-1})}\right]$

(b) $H(z) = \left[\dfrac{(1 + z^{-1})^2}{(1 - \frac{3}{4}z^{-1})(1 - \frac{3}{4}e^{j\pi/4}z^{-1})}\right]$
$\cdot \left[\dfrac{(1 - \frac{1}{2}e^{j\pi/2}z^{-1})(1 - \frac{1}{2}e^{-j\pi/2}z^{-1})}{(1 - \frac{3}{4}e^{-j\pi/4}z^{-1})(1 + \frac{3}{4}z^{-1})}\right]$

7.22 Draw block diagram implementations of the following systems as a parallel combination of second-order sections with real-valued coefficients:

(a) $h[n] = 2\left(\dfrac{1}{2}\right)^n u[n] + \left(\dfrac{j}{2}\right)^n u[n] +$
$\left(\dfrac{-j}{2}\right)^n u[n] + \left(-\dfrac{1}{2}\right)^n u[n]$

(b) $h[n] = 2(\frac{1}{2}e^{j\pi/4})^n u[n] + (\frac{1}{4}e^{j\pi/3})^n u[n] + (\frac{1}{4}e^{-j\pi/3})^n u[n] + 2(\frac{1}{2}e^{-j\pi/4})^n u[n]$

7.23 Determine the transfer function of the system depicted in Fig. P7.23.

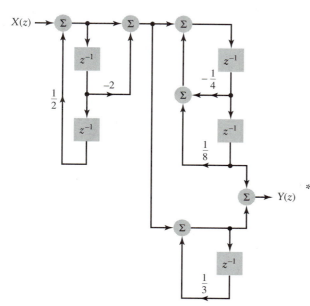

FIGURE P7.23

*7.24 A very useful structure for implementing non-recursive systems in the so-called lattice structure. The lattice is constructed as a cascade of two-input, two-output sections of the form depicted in Fig. P7.24(a). An Mth order lattice structure is depicted in Fig. P7.24(b).

(a) Find the transfer function of a second-order ($M = 2$) lattice having $c_1 = \frac{1}{2}$ and $c_2 = -\frac{1}{4}$.

(b) We may determine the relationship between the transfer function and lattice structure by examining the effect of adding a section on the transfer function, as depicted in Fig. P7.24(c). Here we have defined $H_i(z)$ as the transfer function between the input and the output of the lower branch in the ith section and $\tilde{H}_i(z)$ as the transfer function between the input and the output of the upper branch in the ith section. Write the relationship between the transfer functions to the $(i - 1)$st and ith stages as

$$\begin{bmatrix} \tilde{H}_i(z) \\ H_i(z) \end{bmatrix} = \mathbf{T}(z) \begin{bmatrix} \tilde{H}_{i-1}(z) \\ H_{i-1}(z) \end{bmatrix}$$

where $\mathbf{T}(z)$ is a 2 by 2 matrix. Express $\mathbf{T}(z)$ in terms of c_i and z^{-1}.

(c) Use induction to prove that $\tilde{H}_i(z) = z^{-i}H_i(z^{-1})$.

(d) Show that the coefficient of z^{-i} in $H_i(z)$ is given by c_i.

(e) By combining the results of (b)–(d) we may derive an algorithm for finding the c_i required by the lattice structure to implement an arbitrary order M nonrecursive transfer function $H(z)$. Start with $i = M$ so that $H_M(z) = H(z)$. The result of (d) implies that c_M is the coefficient of z^{-M} in $H(z)$. By decreasing i, continue this algorithm to find the remaining c_i. *Hint:* Use the result of (b) to find a 2 by 2 matrix $\mathbf{A}(z)$ such that

$$\begin{bmatrix} \tilde{H}_{i-1}(z) \\ H_{i-1}(z) \end{bmatrix} = \mathbf{A}(z) \begin{bmatrix} \tilde{H}_i(z) \\ H_i(z) \end{bmatrix}$$

*7.25 Causal filters always have a nonzero phase response. One technique for attaining zero phase response from a causal filter involves filtering the signal twice, once in the forward direction and the second time in the reverse direction. We may describe this operation in terms of the input $x[n]$ and filter impulse response $h[n]$ as follows. Let $y_1[n] = x[n] * h[n]$ represent filtering the signal in the forward direction. Now filter $y_1[n]$ backward to obtain $y_2[n] = y_1[-n] * h[n]$. The

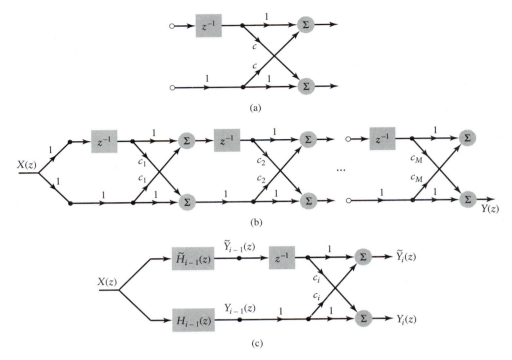

(a)

(b)

(c)

FIGURE P7.24

output is then given by reversing $y_2[n]$ to obtain $y[n] = y_2[-n]$.

(a) Show that this set of operations is equivalently represented by a filter with impulse response $h_o[n]$ as $y[n] = x[n] * h_o[n]$ and express $h_o[n]$ in terms of $h[n]$.

(b) Show that $h_o[n]$ is an even signal and that the phase response of any system with an even impulse response is zero.

(c) Show for every pole or zero at $z = \beta$ in $h[n]$ that $h_o[n]$ has a pair of poles or zeros at $z = \beta$ and $z = 1/\beta$.

7.26 Let $x[n] = \delta[n - 2] + \delta[n + 2]$.

(a) Determine the unilateral z-transform of $x[n]$.

(b) Use the unilateral z-transform time-shift property and the result of (a) to determine the unilateral z-transform of the following signals:

   (i) $w[n] = x[n - 1]$

   (ii) $w[n] = x[n - 3]$

7.27 Use the unilateral z-transform to determine the forced response, the natural response, and the complete response of the systems described by

the following difference equations with the given inputs and initial conditions:

(a) $y[n] - \frac{1}{2}y[n - 1] = 2x[n]$,
$y[-1] = 3$, $x[n] = 2(-\frac{1}{2})^n u[n]$

(b) $y[n] - \frac{1}{9}y[n - 2] = x[n - 1]$,
$y[-1] = 0$, $y[-2] = 1$, $x[n] = 3u[n]$

(c) $y[n] - \frac{1}{4}y[n - 1] - \frac{1}{8}y[n - 2] = x[n] + x[n - 1]$, $y[-1] = 1$,
$y[-2] = -1$, $x[n] = 3^n u[n]$

(d) $y[n] - \frac{3}{4}y[n - 1] + \frac{1}{8}y[n - 2] = 2x[n]$, $y[-1] = 1$, $y[-2] = -1$,
$x[n] = u[n]$

7.28 Use the unilateral z-transform to determine the impulse response of the systems described by the following difference equations by setting $x[n] = \delta[n]$. *Hint*: The impulse response is defined in terms of a system that is initially at rest.

(a) $y[n] + \frac{1}{2}y[n - 1] = 2x[n]$

(b) $y[n] - \frac{1}{4}y[n - 2] = x[n - 1]$

(c) $y[n] + y[n - 1] + \frac{1}{4}y[n - 2] = x[n] - x[n - 1]$

7.29 The present value of a loan with interest compounded monthly may be described in terms of the first-order difference equation

$$y[n] = \left(1 + \frac{r}{12}\right)y[n-1] - x[n]$$

where r is the annual interest rate expressed as a decimal number, $x[n]$ is the payment credited at the end of the nth month, and $y[n]$ is the loan balance at the beginning of the $(n+1)$st month. The beginning loan balance is the initial condition $y[-1]$. If uniform payments of $\$c$ are made for L consecutive months, then $x[n] = c\{u[n] - u[n-L]\}$.

(a) Is the system represented by this difference equation stable? Why?

(b) Find the natural response of the system. What happens as $n \to \infty$?

(c) Use the unilateral z-transform to show that

$$Y(z) = \frac{y[-1](1 + r/12) - c\sum_{n=0}^{L-1}z^{-n}}{1 - (1 + r/12)z^{-1}}$$

Hint: Use long division to show that

$$\frac{1 - z^{-L}}{1 - z^{-1}} = \sum_{n=0}^{L-1} z^{-n}$$

(d) Show that $(1 + r/12)$ must be a zero of $Y(z)$ if the loan is to have zero balance after L payments.

(e) Find the monthly payment $\$c$ as a function of the initial loan value $y[-1]$ and the interest rate r assuming the loan has zero balance after L payments.

▶ *Computer Experiments*

7.30 Use the MATLAB command `zplane` to obtain a pole–zero plot for the following systems:

(a) $H(z) = \dfrac{1 + z^{-2}}{2 + z^{-1} - \frac{1}{2}z^{-2} + \frac{1}{4}z^{-3}}$

(b) $H(z) = \dfrac{1 + z^{-1} + \frac{3}{2}z^{-2} + \frac{1}{2}z^{-3}}{1 + \frac{3}{2}z^{-1} + \frac{1}{2}z^{-2}}$

7.31 Use the MATLAB command `residuez` to obtain the partial fraction expansions required to solve Problem 7.8(d)–(i).

7.32 Use the MATLAB command `tf2ss` to find state-variable descriptions for the systems in Problem 7.14.

7.33 Use the MATLAB command `ss2tf` to find the transfer functions in Problem 7.16.

7.34 Use the MATLAB command `zplane` to solve Problem 7.17.

7.35 Use the MATLAB command `freqz` to evaluate and plot the magnitude and phase response of the systems given in Example 7.17.

7.36 Use the MATLAB command `freqz` to evaluate and plot the magnitude and phase response of the systems given in Problem 7.19.

7.37 Use the MATLAB commands `filter` and `filtic` to plot the loan balance at the start of each month $n = 0, 1, \ldots, L + 1$ for Problem 7.29. Assume that $y[-1] = \$10,000$, $L = 60$, $r = 0.12$, and the monthly payment is chosen to bring the loan balance to zero after 60 payments.

7.38 Use the MATLAB command `zp2sos` to determine a cascade connection of second-order sections for implementing the systems in Problem 7.21.

7.39 A causal discrete-time system has transfer function given by

$$H(z) = \left[\frac{0.0976(z-1)^2}{(z - 0.357 - j0.589)(z - 0.357 + j0.589)}\right]$$
$$\cdot \left[\frac{(z+1)^2}{(z - 0.769 - j0.334)(z - 0.769 + j0.334)}\right]$$

(a) Use the pole and zero locations to sketch the magnitude response.

(b) Use the MATLAB commands `zp2tf` and `freqz` to evaluate and plot the magnitude and phase response.

(c) Use the MATLAB command `zp2sos` to obtain a representation for this filter as a cascade of two second-order sections with real-valued coefficients.

(d) Use the MATLAB command `freqz` to evaluate and plot the magnitude response of each section in (c).

(e) Use the MATLAB command `filter` to determine the impulse response of this system by obtaining the output for an input $x[n] = \delta[n]$.

(f) Use the MATLAB command `filter` to determine the system output for an input of the form

$$x[n] = \left(1 + \cos\left(\frac{\pi}{4}n\right) + \cos\left(\frac{\pi}{2}n\right)\right.$$
$$\left. + \cos\left(\frac{3\pi}{4}n\right) + \cos(\pi n)\right)u[n]$$

Plot the first 250 points of the input and output.

8 | Application to Filters and Equalizers

8.1 *Introduction*

In Chapters 4 and 5 we made use of *filters* as functional blocks to suppress spurious signals by exploiting the fact that the frequency content of these signals is separated from the frequency content of wanted signals. In Chapter 5 we made use of *equalizers* as functional blocks to compensate for signal distortion that arises when a signal is transmitted through a physical system such as a telephone channel. The treatments of both filters and equalizers presented in those chapters were from a system-theoretic viewpoint. Now that we have the Laplace and *z*-transforms at our disposal, we are ready to describe procedures for the *design* of these two important functional blocks.

We begin the discussion by considering the issue of distortionless transmission, which is basic to the study of linear filters and equalizers. This leads naturally to a discussion of an idealized framework for filtering, which, in turn, provides the basis for the design of practical filters. The design of a filter can be accomplished using continuous-time concepts, in which case we speak of *analog filters*. Alternatively, the filter design can be accomplished using discrete-time concepts, in which case we speak of *digital filters*. Analog and digital filters have their own advantages and disadvantages. Both of these filter types are discussed in this chapter. The topic of equalization is covered toward the end of the chapter.

8.2 *Conditions for Distortionless Transmission*

Consider a continuous-time LTI system with impulse response $h(t)$. Equivalently, the system may be described in terms of its frequency response $H(j\omega)$, defined as the Fourier transform of $h(t)$. Let a signal $x(t)$ with Fourier transform $X(j\omega)$ be applied to the input of the system. Let the signal $y(t)$ with Fourier transform $Y(j\omega)$ denote the output of the system. We wish to know conditions for *distortionless transmission* through the system. By "distortionless transmission" we mean that the output signal of the system is an exact replica of the input signal, except for two minor modifications:

- ▶ A possible scaling of amplitude
- ▶ A constant time delay

On this basis, we say that a signal $x(t)$ is transmitted through the system without distortion if the output signal $y(t)$ is defined by (see Fig. 8.1)

$$y(t) = Kx(t - t_0) \tag{8.1}$$

where the constant K accounts for a change in amplitude and the constant t_0 accounts for a delay in transmission.

Applying the Fourier transform to Eq. (8.1), and using the time-shifting property of the Fourier transform, we get

$$Y(j\omega) = KX(j\omega)e^{-j\omega t_0} \tag{8.2}$$

The frequency response of a distortionless system is therefore

$$\begin{aligned} H(j\omega) &= \frac{Y(j\omega)}{X(j\omega)} \\ &= Ke^{-j\omega t_0} \end{aligned} \tag{8.3}$$

Correspondingly, the impulse response of the system is given by

$$h(t) = K\delta(t - t_0) \tag{8.4}$$

Equations (8.4) and (8.3) describe the time-domain and frequency-domain conditions, respectively, which a LTI system has to satisfy for distortionless transmission. From a practical viewpoint, however, Eq. (8.3) is the more revealing one of the two, as discussed next.

Equation (8.3) indicates that in order to achieve the distortionless transmission of a signal with some finite frequency content through a continuous-time LTI system, the frequency response of the system must satisfy two conditions:

1. The magnitude response $|H(j\omega)|$ is constant for all frequencies of interest, as shown by

$$|H(j\omega)| = K \tag{8.5}$$

2. For the same frequencies of interest, the phase response $\arg\{H(\omega)\}$ is linear in frequency, with slope $-t_0$ and intercept zero, as shown by

$$\arg\{H(j\omega)\} = -\omega t_0 \tag{8.6}$$

These two conditions are illustrated in Figs. 8.2(a) and (b), respectively.

Consider next the case of a discrete-time LTI system with transfer function $H(e^{j\Omega})$. Following a procedure similar to that described above, we may show that the conditions for ideal transmission through such a system are as follows:

1. The magnitude response $|H(e^{j\Omega})|$ is constant for all frequencies of interest, as shown by

$$|H(e^{j\Omega})| = K \tag{8.7}$$

where K is a constant.

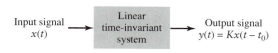

FIGURE 8.1 Time-domain condition for distortionless transmission of a signal through a linear time-invariant system.

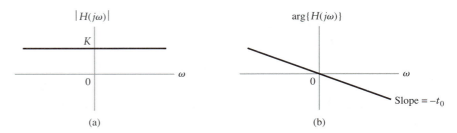

FIGURE 8.2   Frequency response for distortionless transmission through a linear time-invariant system. (a) Magnitude response. (b) Phase response.

2. For the same frequencies of interest, the phase response $\arg\{H(e^{j\Omega})\}$ is linear in frequency, as shown by

$$\arg\{H(e^{j\Omega})\} = -\Omega n_0 \tag{8.8}$$

where n_0 accounts for an integer delay in transmission through the system.

▶ **Drill Problem 8.1**   Using the impulse response of Eq. (8.4) in the convolution integral, show that the input–output relation of a distortionless system is as defined in Eq. (8.1).   ◀

EXAMPLE 8.1   Suppose that the condition of Eq. (8.6) on the phase response $\arg\{H(j\omega)\}$ for distortionless response is modified by adding a constant equal to a positive or negative integer multiple of π radians (i.e., 180°). What is the effect of this modification?

Solution: We begin by rewriting Eq. (8.6) as

$$\arg\{H(j\omega)\} = -\omega t_0 + k\pi$$

where k is an integer. Correspondingly, the frequency response of the system given in Eq. (8.3) takes the new form

$$H(j\omega) = Ke^{-j\omega t_0 + jk\pi}$$

But

$$e^{+jk\pi} = \begin{cases} -1, & k = \text{odd integer} \\ +1, & k = \text{even integer} \end{cases}$$

It follows therefore that

$$H(j\omega) = \pm Ke^{-j\omega t_0}$$

which is of exactly the same form as that described in Eq. (8.3), except for a possible change in the algebraic sign of the scaling factor K. We conclude therefore that the conditions for distortionless transmission through a linear time-invariant system remain unchanged when the phase response of the system is changed by a constant amount equal to a positive or negative integer multiple of 180°.

8.3 *Ideal Lowpass Filters*

Typically, the spectral content of an information-bearing signal occupies a frequency band of some finite extent. For example, the spectral content of a speech signal essential for telephonic communications lies in the frequency band from 300 to 3100 Hz. To extract the essential information content of a speech signal for such an application, we need a frequency-selective system, that is, a filter that limits the spectrum of the signal to the desired band of frequencies. Indeed, filters are basic to the study of signals and systems, in the sense that every system used to process signals contains a filter of some kind in its composition.

As noted in Chapter 4, the frequency response of a filter is characterized by a passband and a stopband, which are separated by a transition band or guardband. Signals with frequencies inside the passband are transmitted with little or no distortion, whereas those with frequencies inside the stopband are effectively rejected. The filter may thus be of the lowpass, highpass, bandpass, or bandstop type, depending on whether it transmits low, high, intermediate, or all but intermediate frequencies, respectively.

Consider then an *ideal* lowpass filter, which transmits all frequencies inside the passband without any distortion and rejects all frequencies inside the stopband. The transition from the passband to the stopband is assumed to occupy zero width. Insofar as lowpass filtering is concerned, the primary interest is in the faithful transmission of an information-bearing signal where spectral content is confined to some frequency band defined by $0 \leq \omega \leq \omega_c$. Accordingly, in such an application the conditions for distortionless transmission need only be satisfied inside the passband of the filter, as illustrated in Fig. 8.3. Specifically, the frequency response of an ideal lowpass filter with cutoff frequency ω_c is defined by

$$H(j\omega) = \begin{cases} e^{-j\omega t_0}, & -\omega_c \leq \omega \leq \omega_c \\ 0, & |\omega| > \omega_c \end{cases} \tag{8.9}$$

where, for convenience of presentation, we have set the constant $K = 1$. For a finite delay t_0, the ideal lowpass filter is noncausal, which is confirmed by examining the impulse response $h(t)$ of the filter.

To evaluate $h(t)$, we take the inverse Fourier transform of Eq. (8.9), obtaining

$$h(t) = \frac{1}{2\pi} \int_{-\omega_c}^{\omega_c} e^{j\omega(t-t_0)} \, d\omega \tag{8.10}$$

From the definition of the sinc function given in Eq. (3.21), we have

$$\text{sinc}(t) = \frac{1}{2\pi} \int_{-\pi}^{\pi} e^{j\omega t} \, d\omega$$
$$= \frac{\sin(\pi t)}{\pi t} \tag{8.11}$$

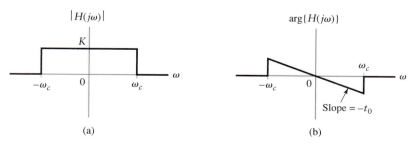

(a)   (b)

FIGURE 8.3 Frequency response of ideal lowpass filter. (a) Magnitude response. (b) Phase response.

Accordingly, we may rewrite Eq. (8.10) in the compact form

$$h(t) = \frac{\omega_c}{\pi} \text{sinc}\left(\frac{\omega_c}{\pi}(t - t_0)\right) \tag{8.12}$$

This impulse response has a peak amplitude of ω_c/π centered on time t_0, as shown in Fig. 8.4 for $\omega_c = 1$ and $t_0 = 8$. The duration of the mainlobe of the impulse response is $2\pi/\omega_c$, and the rise time from the zero at the beginning of the mainlobe to the peak is π/ω_c. We see from Fig. 8.4 that, for any finite value of t_0, there is some response from the filter before the time $t = 0$, at which the unit impulse is applied to the input of the filter. This confirms that the ideal lowpass filter is noncausal.

Despite the noncausal nature of the ideal lowpass filter, it is a useful theoretical concept. In particular, it provides the ideal framework for the design of practical (i.e., causal) filters.

■ TRANSMISSION OF A RECTANGULAR PULSE THROUGH AN IDEAL LOWPASS FILTER

A rectangular pulse plays a key role in digital communications. For example, for the electrical representation of a binary sequence transmitted through a channel, we may use the following:

▶ Transmit a rectangular pulse for symbol 1.
▶ Switch off the pulse for symbol 0.

Consider then a rectangular pulse $x(t)$ of unit amplitude and duration T, written as

$$x(t) = \begin{cases} 1, & -\dfrac{T}{2} \leq t \leq \dfrac{T}{2} \\ 0, & |t| > \dfrac{T}{2} \end{cases} \tag{8.13}$$

This pulse is applied to a communication channel modeled as an ideal lowpass filter whose frequency response is defined by Eq. (8.9). The issue of interest is that of determining the response of the channel to the pulse input.

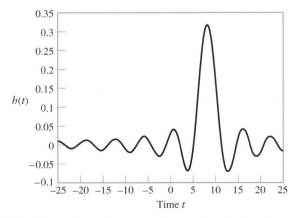

FIGURE 8.4 Time-shifted form of sinc function, representing the impulse response of ideal (but noncausal) lowpass filter.

The impulse response of the filter representing the channel is given by Eq. (8.12), rewritten here in the expanded form

$$h(t) = \frac{\omega_c}{\pi} \frac{\sin(\omega_c(t - t_0))}{\omega_c(t - t_0)} \tag{8.14}$$

We may express the response of the filter using the convolution integral as follows:

$$y(t) = \int_{-\infty}^{\infty} x(\tau)h(t - \tau)\, d\tau \tag{8.15}$$

Substituting Eqs. (8.13) and (8.14) in (8.15), we get

$$y(t) = \frac{\omega_c}{\pi} \int_{-T/2}^{T/2} \frac{\sin(\omega_c(t - t_0 - \tau))}{\omega_c(t - t_0 - \tau)}\, d\tau$$

Let

$$\lambda = \omega_c(t - t_0 - \tau)$$

Then, changing the variable of integration from τ to λ, we may rewrite $y(t)$ as

$$\begin{aligned} y(t) &= \frac{1}{\pi} \int_b^a \frac{\sin \lambda}{\lambda}\, d\lambda \\ &= \frac{1}{\pi} \left[\int_0^a \frac{\sin \lambda}{\lambda}\, d\lambda - \int_0^b \frac{\sin \lambda}{\lambda}\, d\lambda \right] \end{aligned} \tag{8.16}$$

where the limits of integration, a and b, are defined by

$$a = \omega_c \left(t - t_0 + \frac{T}{2} \right) \tag{8.17}$$

and

$$b = \omega_c \left(t - t_0 - \frac{T}{2} \right) \tag{8.18}$$

To rewrite Eq. (8.16) in a compact form, we introduce the *sine integral*, defined by

$$\text{Si}(u) = \int_0^u \frac{\sin \lambda}{\lambda}\, d\lambda \tag{8.19}$$

The sine integral cannot be evaluated in closed form in terms of elementary functions, but it can be integrated using a power series. It is shown plotted in Fig. 8.5. From this figure we see that:

- ▶ The sine integral $\text{Si}(u)$ has odd symmetry about the origin $u = 0$.
- ▶ It has maxima and minima at multiples of π.
- ▶ It approaches the limiting value of $\pm \pi/2$ for large values of $|u|$.

Using the definition of the sine integral in Eq. (8.19), we may rewrite the response $y(t)$ defined in Eq. (8.16) in compact form:

$$y(t) = \frac{1}{\pi} [\text{Si}(a) - \text{Si}(b)] \tag{8.20}$$

where a and b are themselves defined in Eqs. (8.17) and (8.18), respectively.

Figure 8.6 depicts the response $y(t)$ for three different values of cutoff frequency ω_c, assuming that the transmission delay t_0 is zero. In each case, we see that the response $y(t)$

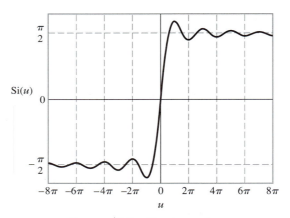

Figure 8.5 Sine integral.

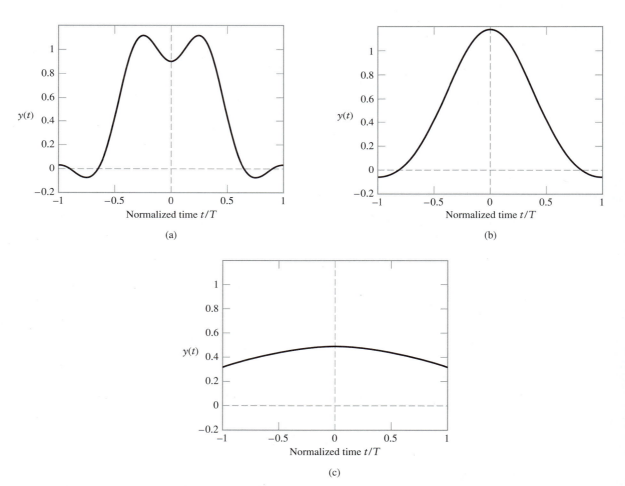

Figure 8.6 Pulse response of an ideal lowpass filter for pulse duration $T = 1$s and varying filter cutoff frequency ω_c: (a) $\omega_c = 4\pi$ rad/s; (b) $\omega_c = 2\pi$ rad/s; and (c) $\omega_c = 0.4\pi$ rad/s.

is symmetric about $t = 0$. We further observe that the shape of the response $y(t)$ is markedly dependent on the cutoff frequency. In particular, we note the following points:

1. When ω_c is larger than $2\pi/T$, as in Fig. 8.6(a), the response $y(t)$ has approximately the same duration as the rectangular pulse $x(t)$ applied to the filter input. However, it differs from $x(t)$ in two major respects:

   ▶ Unlike the input $x(t)$, the response $y(t)$ has nonzero rise and fall times that are inversely proportional to the cutoff frequency ω_c.

   ▶ The response $y(t)$ exhibits *ringing* at both the leading and trailing edges.

2. When $\omega_c = 2\pi/T$, as in Fig. 8.6(b), the response $y(t)$ is recognizable as a pulse. However, the rise and fall times of $y(t)$ are now significant compared to the duration of the input rectangular pulse $x(t)$.

3. When the cutoff frequency ω_c is smaller than $2\pi/T$, as in Fig. 8.6(c), the response $y(t)$ is a grossly distorted version of the input $x(t)$.

These observations point to the inverse relationship that exists between the two parameters: (1) the duration of a rectangular pulse acting as the input signal to an ideal lowpass filter, and (2) the cutoff frequency of the filter. This inverse relationship is a manifestation of the constancy of the time–bandwidth product discussed in Chapter 3. From a practical perspective, the inverse relationship between pulse duration and filter cutoff frequency has a simple interpretation, as illustrated here, in the context of digital communications. Specifically, if the requirement is merely that of recognizing the response of a lowpass channel to be due to the transmission of symbol 1 represented by a rectangular pulse of duration T, it is adequate to set the cutoff frequency of the channel at $\omega_c = 2\pi/T$.

EXAMPLE 8.2   The response $y(t)$ shown in Fig. 8.6(a), corresponding to cutoff frequency $\omega_c = 4\pi/T$, exhibits an overshoot of approximately 9%. Investigate what happens to this overshoot when the cutoff frequency ω_c is allowed to approach infinity.

Solution: In Figs. 8.7(a) and (b) we show the pulse response of the ideal lowpass filter for cutoff frequency $\omega_c = 10\pi/T$ and $\omega_c = 40\pi/T$. These two graphs illustrate that the overshoot remains approximately equal to 9% in a manner that is practically independent of how large the cutoff frequency ω_c is. This result is in fact another manifestation of the *Gibbs phenomenon* that was discussed in Chapter 3.

To provide an analytic proof for this result, we observe from Fig. 8.5 that the sine integral $\text{Si}(u)$ defined in Eq. (8.19) oscillates at a frequency of $1/2\pi$. The implication of this observation is that the filter response $y(t)$ will oscillate at a frequency equal to $\omega_c/2\pi$, where ω_c is the cutoff frequency of the filter. The maximum value of $\text{Si}(u)$ occurs at $u_{\max} = \pi$ and is equal to 1.8519, which we may write as $(1.179)(\pi/2)$. The filter response $y(t)$ thus has its first maximum at

$$t_{\max} = \frac{T}{2} - \frac{\pi}{\omega_c} \tag{8.21}$$

Correspondingly, the integration limits a and b, defined in Eqs. (8.17) and (8.18), take on the following values (assuming $t_0 = 0$):

$$a_{\max} = \omega_c\left(t_{\max} + \frac{T}{2}\right)$$

$$= \omega_c\left(\frac{T}{2} - \frac{\pi}{\omega_c} + \frac{T}{2}\right) \tag{8.22}$$

$$= \omega_c T - \pi$$

$$b_{max} = \omega_c\left(t_{max} - \frac{T}{2}\right)$$

$$= \omega_c\left(\frac{T}{2} - \frac{\pi}{\omega_c} - \frac{T}{2}\right) \tag{8.23}$$

$$= -\pi$$

Substituting Eqs. (8.22) and (8.23) in (8.20) yields

$$y(t_{max}) = \frac{1}{\pi}\left[\text{Si}(a_{max}) - \text{Si}(b_{max})\right]$$

$$= \frac{1}{\pi}\left[\text{Si}(\omega_c T - \pi) - \text{Si}(-\pi)\right] \tag{8.24}$$

$$= \frac{1}{\pi}\left[\text{Si}(\omega_c T - \pi) + \text{Si}(\pi)\right]$$

Let

$$\text{Si}(\omega_c T - \pi) = \frac{\pi}{2}(1 + \Delta) \tag{8.25}$$

where Δ is the absolute value of the deviation in the value of $\text{Si}(\omega_c T - \pi)$, expressed as a fraction of the final value $+\pi/2$. Thus recognizing that

$$\text{Si}(\pi) = (1.179)\left(\frac{\pi}{2}\right)$$

we may rewrite Eq. (8.24) as

$$y(t_{max}) = \tfrac{1}{2}(1.179 + 1 + \Delta)$$

$$= 1.09 + \tfrac{1}{2}\Delta \tag{8.26}$$

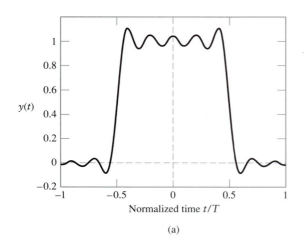

(a)

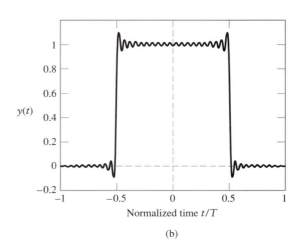

(b)

FIGURE 8.7   The Gibbs phenomenon, exemplified by pulse response of an ideal lowpass filter. The overshoot remains essentially the same despite a significant increase in cutoff frequency ω_c: (a) $\omega_c = 10\pi$ rad/s and (b) $\omega_c = 40\pi$ rad/s. The pulse duration T is maintained constant at 1 s.

Viewing ω_c as a measure of the filter's bandwidth, we note that for a time–bandwidth product $\omega_c T \gg 1$ the fractional deviation Δ has a very small value. We may thus write the approximation

$$y(t_{max}) \approx 1.09 \quad \text{for } \omega_c \gg 2\pi/T \tag{8.27}$$

which shows that the overshoot in the filter response is approximately 9%, a result that is practically independent of the cutoff frequency ω_c.

8.4 *Design of Filters*

The lowpass filter with frequency response shown in Fig. 8.3 is "ideal" in that it passes all frequency components lying inside the passband with no distortion and rejects all frequency components lying inside the stopband, and the transition from the passband to the stopband is abrupt. Recall that these characteristics result in a nonimplementable filter. Therefore, from a practical perspective the prudent approach is to tolerate an acceptable level of distortion by permitting prescribed "deviations" from these ideal conditions, as described here for the case of continuous-time or analog filters:

▶ Inside the passband, the magnitude response of the filter should lie between 1 and $1 - \epsilon$:

$$1 - \epsilon \leq |H(j\omega)| \leq 1 \quad \text{for } 0 \leq |\omega| \leq \omega_p \tag{8.28}$$

where ω_p is the *passband cutoff frequency* and ϵ is a *tolerance parameter*.

▶ Inside the stopband, the magnitude response of the filter should not exceed δ:

$$|H(j\omega)| \leq \delta \quad \text{for } |\omega| \geq \omega_s \tag{8.29}$$

where ω_s is the *stopband cutoff frequency* and δ is another tolerance parameter.

▶ The transition bandwidth has a finite width equal to $\omega_s - \omega_p$.

The tolerance diagram of Fig. 8.8 presents a portrayal of these filter specifications. Analogous specifications are used for discrete-time filters, with the added provision that the

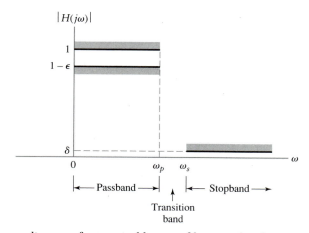

FIGURE 8.8 Tolerance diagram of a practical lowpass filter: passband, transition band, and stopband are shown for positive frequencies.

response is always 2π periodic in Ω. So long as these specifications meet the goal for the filtering problem at hand and the filter design is accomplished at a reasonable cost, the job is satisfactorily done. Indeed, this is the very nature of engineering design.

The specifications just described favor an approach that focuses on the design of the filter based on its frequency response rather than its impulse response. This is in recognition of the fact that the application of a filter usually involves the separation of signals on the basis of their frequency content.

Having formulated a set of specifications describing the desired properties of the frequency-selective filter, there are two distinct steps involved in the design of the filter, which are pursued in the following order:

1. The *approximation* of a prescribed frequency response (i.e., magnitude response, phase response, or both) by a rational transfer function that represents a system that is both causal and stable.
2. The *realization* of the approximating transfer function by a physical system.

Both of these steps can be implemented in a variety of different ways, with the result that there is no unique solution to the filter design problem for a prescribed set of specifications.

Nevertheless, we may mention three different approaches to the design of analog and digital filters, as summarized here:

1. *Analog approach*, which applies to the class of analog filters.
2. *Analog-to-digital approach*, where the motivation is to design a digital filter by building on an analog filter design.
3. *Direct digital approach*, which applies to the class of digital filters.

In what follows, the basic ideas of these approaches, particularly the first two, are presented and illustrated with different design examples.

8.5 *Approximating Functions*

The choice of a transfer function for solving the approximation problem is the transition step from a set of design specifications to the realization of the transfer function by means of a specific filter structure. As such, it is the most fundamental step in filter design: the choice of the transfer function determines the filter performance. At the outset, however, it must be emphasized that there is no unique solution to the approximation problem. Rather, we have a set of possible solutions, each of which has its own distinctive properties.

Basically, the approximation problem is an *optimization problem*, which can only be solved in the context of a specific *criterion of optimality*. In other words, before we proceed to solve the approximation problem we have to specify a criterion of optimality in an implicit or explicit sense. Moreover, the choice of that criterion uniquely determines the solution. Two optimality criteria commonly used in filter design are as follows:

1. *Maximally flat magnitude response.*
Let $|H(j\omega)|$ denote the magnitude response of an analog lowpass filter of order N. The magnitude response $|H(j\omega)|$ is said to be *maximally flat* at the origin if its $(2N - 1)$ derivatives with respect to ω vanish at $\omega = 0$; that is,

$$\frac{\partial^{2N-1}}{\partial \omega^{2N-1}} |H(j\omega)| = 0 \quad \text{at } \omega = 0$$

2. *Equiripple magnitude response.*

Let the squared value of the magnitude response $|H(j\omega)|$ of an analog lowpass filter be expressed in the form

$$|H(j\omega)|^2 = \frac{1}{1 + \delta^2 F^2(\omega)}$$

where δ is related to the passband tolerance parameter ϵ, and $F(\omega)$ is some function of ω. The magnitude response $|H(j\omega)|$ is said to be *equiripple in the passband* if $F^2(\omega)$ oscillates between maxima and minima of equal amplitude over the entire passband. Here we must distinguish between two cases, depending on whether the filter order N is odd or even. We illustrate the formulation of this second optimality criterion for two cases, $N = 3$ and $N = 4$, as follows:

> *Case (a):* $N = 3$, and $\omega_e = 1$

(i) $F(\omega) = 0$ if $\omega = 0, \pm\omega_a$

(ii) $F^2(\omega) = 1$ if $\omega = \pm\omega_b, \pm1$

(iii) $\dfrac{\partial}{\partial\omega} F^2(\omega) = 0$ if $\omega = 0, \pm\omega_b, \pm\omega_a, \pm1$

> where $0 < \omega_b < \omega_a < 1$.

> *Case (b):* $N = 4$, and $\omega_e = 1$

(i) $F(\omega) = 0$ if $\omega = \pm\omega_{a1}, \pm\omega_{a2}$

(ii) $F^2(\omega) = 1$ if $\omega = 0, \pm\omega_b, \pm1$

(iii) $\dfrac{\partial}{\partial\omega} F^2(\omega) = 0$ if $\omega = 0, \pm\omega_{a1}, \pm\omega_b, \pm\omega_{a2}, \pm1$

> where $0 < \omega_{a1} < \omega_b < \omega_{a2} < 1$.

The two optimality criteria described under points 1 and 2 are satisfied by two classes of filters known as Butterworth filters and Chebyshev filters, respectively. They are both described in what follows.

■ **BUTTERWORTH FILTERS**

A Butterworth function of order N is defined by

$$|H(j\omega)|^2 = \frac{1}{1 + \left(\dfrac{\omega}{\omega_c}\right)^{2N}} \tag{8.30}$$

and a filter so designed is referred to as a *Butterworth filter of order N.*

The approximating function of Eq. (8.30) satisfies the requirement that $H(j\omega)$ must be an even function of ω. The parameter ω_c is the *cutoff frequency* of the filter. For prescribed values of tolerance parameters ϵ and δ defined in Fig. 8.8, we readily find from Eq. (8.30) that the passband and stopband cutoff frequencies are as follows, respectively:

$$\omega_p = \omega_c\left(\frac{\epsilon}{1 - \epsilon}\right)^{1/2N} \tag{8.31}$$

$$\omega_s = \omega_c\left(\frac{1 - \delta}{\delta}\right)^{1/2N} \tag{8.32}$$

The squared magnitude response $|H(j\omega)|^2$ obtained by using the approximating function of Eq. (8.30) is shown plotted in Fig. 8.9 for four different values of filter order N. All these curves pass through the half-power point at $\omega = \omega_c$.

A Butterworth function is monotonic throughout the passband and stopband. In particular, in the vicinity of $\omega = 0$, we may expand the magnitude of $H(j\omega)$ as a power series as follows:

$$|H(j\omega)| = 1 - \frac{1}{2}\left(\frac{\omega}{\omega_c}\right)^{2N} + \frac{3}{8}\left(\frac{\omega}{\omega_c}\right)^{4N} - \frac{5}{16}\left(\frac{\omega}{\omega_c}\right)^{6N} + \cdots \qquad (8.33)$$

which implies that the first $2N - 1$ derivatives of $|H(j\omega)|$ with respect to ω are zero at the origin. It follows therefore that the Butterworth function is indeed *maximally flat* at $\omega = 0$.

To design an analog filter, we need to know the transfer function $H(s)$, expressed as a function of the complex variable s. Given the Butterworth function $|H(j\omega)|^2$, how do we find the corresponding transfer function $H(s)$? To address this issue, we put $j\omega = s$ and recognize that

$$H(s)H(-s)\big|_{s=j\omega} = |H(j\omega)|^2 \qquad (8.34)$$

Hence we may rewrite Eq. (8.30) in the equivalent form

$$H(s)H(-s) = \frac{1}{1 + \left(\dfrac{s}{j\omega_c}\right)^{2N}} \qquad (8.35)$$

The roots of the denominator polynomial are located at the following points in the s plane:

$$\begin{aligned} s &= j\omega_c(-1)^{1/2N} \\ &= \omega_c e^{j\pi(2k+N-1)/2N} \quad \text{for } k = 0, 1, \ldots, 2N - 1 \end{aligned} \qquad (8.36)$$

That is, the poles of $H(s)H(-s)$ form symmetrical patterns on a circle of radius ω_c, as illustrated in Fig. 8.10 for $N = 3$ and $N = 4$. Note that, for any N, none of the poles fall on the imaginary axis of the s-plane.

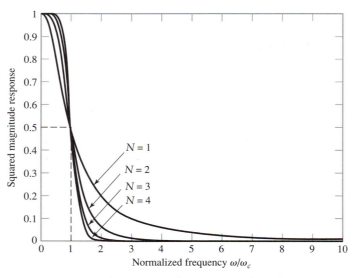

FIGURE 8.9   Magnitude response of Butterworth filter for varying orders.

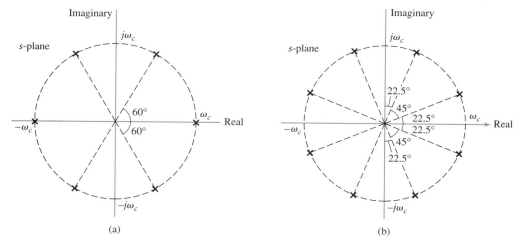

(a)   (b)

FIGURE 8.10 Distribution of poles of $H(s)H(-s)$ in the s-plane for two different filter orders: (a) $N = 3$ and (b) $N = 4$.

Which of these $2N$ poles belong to $H(s)$? To answer this fundamental question, we recall from Chapter 6 that for the transfer function $H(s)$ to represent a stable and causal filter, all of its poles must lie in the left half of the s-plane. Accordingly, those N poles of $H(s)H(-s)$ that lie in the left half of the s-plane are allocated to $H(s)$, and the remaining right-half poles are allocated to $H(-s)$.

EXAMPLE 8.3   Determine the transfer function of a Butterworth filter of the lowpass type with order $N = 3$. Assume that the 3-dB cutoff frequency $\omega_c = 1$.

Solution: For $N = 3$, the $2N = 6$ poles of $H(s)H(-s)$ are located on a circle of unit radius with angular spacing $60°$, as shown in Fig. 8.10(a). Hence allocating the left-half plane poles to $H(s)$, we may define them as

$$s = -\frac{1}{2} + j\frac{\sqrt{3}}{2}$$

$$s = -1$$

$$s = -\frac{1}{2} - j\frac{\sqrt{3}}{2}$$

The transfer function of a Butterworth filter of order 3 is therefore

$$H(s) = \frac{1}{(s + 1)\left(s + \frac{1}{2} - j\frac{\sqrt{3}}{2}\right)\left(s + \frac{1}{2} + j\frac{\sqrt{3}}{2}\right)}$$

$$= \frac{1}{(s + 1)(s^2 + s + 1)} \tag{8.37}$$

$$= \frac{1}{s^3 + 2s^2 + 2s + 1}$$

▶ **Drill Problem 8.2**   How is the transfer function of Eq. (8.37) modified for a Butterworth filter of order 3 and cutoff frequency ω_c?

Answer:

$$H(s) = \cfrac{1}{\left(\cfrac{s}{\omega_c}\right)^3 + 2\left(\cfrac{s}{\omega_c}\right)^2 + 2\left(\cfrac{s}{\omega_c}\right) + 1}$$

◀

▶ **Drill Problem 8.3**   Find the transfer function of a Butterworth filter with cutoff frequency $\omega_c = 1$ and filter order (a) $N = 1$ and (b) $N = 2$.

Answers:

(a) $H(s) = \cfrac{1}{s + 1}$

(b) $H(s) = \cfrac{1}{s^2 + \sqrt{2}\,s + 1}$

◀

In Table 8.1 we present a summary of the transfer functions of Butterworth filters of cutoff frequency $\omega_c = 1$ for up to order $N = 6$.

■ **CHEBYSHEV FILTERS**

The tolerance diagram of Fig. 8.8 calls for an approximating function that lies between 1 and $1 - \epsilon$ inside the passband range $0 \le |\omega| < \omega_p$. The Butterworth function meets this requirement but concentrates its approximating ability near $\omega = 0$. For a given filter order, we can obtain a filter with a reduced transition bandwidth by using an approximating function that exhibits an *equiripple* characteristic in the passband (i.e., it oscillates uniformly between 1 and $1 - \epsilon$ for $0 \le \omega \le \omega_p$) as illustrated in Figs. 8.11(a) and (b) for $N = 3, 4$, respectively, and 0.5-dB ripple in the passband. The magnitude responses plotted here satisfy the equiripple criteria described earlier for N odd and N even, respectively. Approximating functions with an equiripple magnitude response are known collectively as *Chebyshev functions*. A filter designed on this basis is called a *Chebyshev filter*. The poles of transfer function $H(s)$ pertaining to a Chebyshev filter lie on an ellipse in the s-plane in a manner closely related to those of the corresponding Butterworth filter.

TABLE 8.1   *Summary of Butterworth filters*

$$H(s) = \frac{1}{Q(s)}$$

Filter Order N	Polynomial $Q(s)$
1	$s + 1$
2	$s^2 + \sqrt{2}\,s + 1$
3	$s^3 + 2s^2 + 2s + 1$
4	$s^4 + 2.6131s^3 + 3.4142s^2 + 2.6131s + 1$
5	$s^5 + 3.2361s^4 + 5.2361s^3 + 5.2361s^2 + 3.2361s + 1$
6	$s^6 + 3.8637s^5 + 7.4641s^4 + 9.1416s^3 + 7.4641s^2 + 3.8637s + 1$

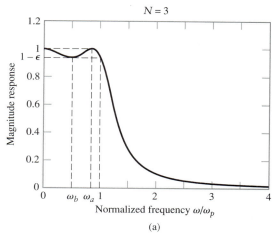

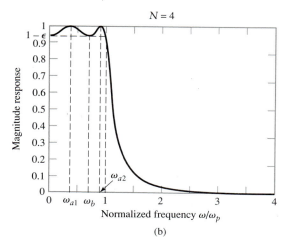

FIGURE 8.11   Magnitude response of Chebyshev filter for order (a) $N = 3$, and (b) $N = 4$, and passband ripple = 0.5 dB. The frequencies ω_b and ω_a in case (a) and the frequencies ω_{a1}, ω_b and ω_{a2} in case (b) are defined in accordance with the optimality criteria for equiripple amplitude response.

The Chebyshev functions shown in Fig. 8.11 exhibit a monotonic behavior in the stopband. Alternatively, we may use another class of Chebyshev functions that exhibit a monotonic response in the passband but an equiripple response in the stopband, as illustrated in Figs. 8.12(a) and (b) for $N = 3, 4$, respectively, and 30-dB stopband ripple. A filter designed on this latter basis is called an *inverse Chebyshev filter*. Unlike a Chebyshev filter, the transfer function of an inverse Chebyshev filter has zeros on the $j\omega$-axis of the s-plane.

The ideas embodied in Chebyshev and inverse Chebyshev filters can be combined to further reduce the transition bandwidth by making the approximating function equiripple in both the passband and the stopband. Such an approximating function is called an *elliptic*

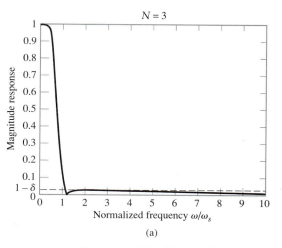

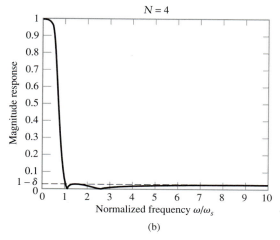

FIGURE 8.12   Magnitude response of inverse Chebyshev filter for order (a) $N = 3$, and (b) $N = 4$, and stopband ripple = 30 dB.

function, and a filter resulting from its use is called an *elliptic filter*. An elliptic filter is optimum in the sense that for a prescribed set of design specifications, the width of the transition band is the smallest that we can achieve. This permits the smallest possible separation between the passband and stopband of the filter. From the standpoint of analysis, however, the determination of transfer function $H(s)$ is simplest for a Butterworth filter and most challenging for an elliptic filter. The elliptic filter is able to achieve its optimum behavior by virtue of the fact that its transfer function $H(s)$ has finite zeros in the s-plane, the number of which is uniquely determined by filter order N. In contrast, the transfer function $H(s)$ for a Butterworth filter or that of a Chebyshev filter has all of its zeros located at $s = \infty$.

8.6 *Frequency Transformations*

Up to this point in the discussion, we have considered the issue of solving the approximation problem for lowpass filters. In this context, it is common practice to speak of a lowpass "prototype" filter, by which we mean a lowpass filter whose cutoff frequency ω_c is normalized to unity. Given that we have found the transfer function of a lowpass prototype filter, we may use it to derive the transfer function of a lowpass filter with an arbitrary cutoff frequency, a highpass, bandpass, or bandstop filter by means of an appropriate transformation of the independent variable. Such a transformation has no effect on the tolerances within which the ideal characteristic of interest is approximated. In what follows, we consider two specific *frequency transformations*: lowpass to highpass and lowpass to bandpass. Other frequency transformations follow the principles described herein.

■ LOWPASS TO HIGHPASS TRANSFORMATION

The points $s = 0$ and $s = \infty$ in the s-plane are of particular interest here. In the case of a lowpass filter, $s = 0$ defines the midpoint of the passband (defined for both positive and negative frequencies), and $s = \infty$ defines the vicinity where the transfer function of the filter behaves in an asymptotic sense. The roles of these two points are interchanged in a highpass filter. Accordingly, the lowpass to highpass transformation is described by

$$s \to \frac{\omega_c}{s} \tag{8.38}$$

where ω_c is the desired cutoff frequency of the highpass filter. This notation implies that we replace s in the transfer function of the lowpass prototype with ω_c/s to obtain the transfer function of the corresponding highpass filter with cutoff frequency ω_c.

To be more precise, let $(s - d_j)$ denote a pole factor of the transfer function $H(s)$ of a lowpass prototype. Using the formula (8.38), we may thus write

$$\frac{1}{s - d_j} \to \frac{-s/d_j}{s - D_j} \tag{8.39}$$

where $D_j = \omega_c/d_j$. The transformation of Eq. (8.39) results in a zero at $s = 0$ and a pole at $s = D_j$ for a pole at $s = d_j$ in the original transfer function $H(s)$.

EXAMPLE 8.4 Equation (8.37) defines the transfer function of a Butterworth lowpass filter of order 3 and unity cutoff frequency. Determine the transfer function of the corresponding highpass filter with cutoff frequency $\omega_c = 1$.

Solution: Applying the frequency transformation of Eq. (8.38) to the lowpass transfer function of Eq. (8.37) yields the transfer function of the corresponding highpass filter with $\omega_c = 1$ as

$$H(s) = \frac{1}{\left(\dfrac{1}{s} + 1\right) + \left(\dfrac{1}{s^2} + \dfrac{1}{s} + 1\right)}$$

$$= \frac{s^3}{(s + 1)(s^2 + s + 1)}$$

▶ **Drill Problem 8.4** Given the transfer function

$$H(s) = \frac{1}{s^2 + \sqrt{2}\, s + 1}$$

pertaining to a Butterworth lowpass filter, find the transfer function of the corresponding highpass filter with cutoff frequency ω_c.

Answer:

$$\frac{s^2}{s^2 + \sqrt{2}\, \omega_c s + \omega_c^2}$$ ◀

▶ **Drill Problem 8.5** Let $(s - c_j)$ denote a zero factor in the transfer function of a lowpass prototype. How is this factor transformed by the use of Eq. (8.38)?

Answer:

$$\frac{s - C_j}{-s/c_j}, \quad \text{where } C_j = \omega_c/c_j$$ ◀

■ LOWPASS TO BANDPASS TRANSFORMATION

Consider next the transformation of a lowpass prototype filter into a bandpass filter. By definition, a bandpass filter rejects both low- and high-frequency components and passes a certain band of frequencies somewhere between them. Thus the frequency response $H(j\omega)$ of a bandpass filter has the following properties:

1. $H(j\omega) = 0$ at both $\omega = 0$ and $\omega = \infty$.
2. $|H(j\omega)| \approx 1$ for a frequency band centered on ω_0, where ω_0 is the midband frequency of the filter.

Accordingly, we want to create a transfer function with zeros at $s = 0$ and $s = \infty$ and poles near $s = \pm j\omega_0$ on the $j\omega$-axis in the s-plane. A lowpass to bandpass transformation that meets these requirements is described by

$$s \rightarrow \frac{s^2 + \omega_0^2}{Bs} \tag{8.40}$$

where B is the bandwidth of the bandpass filter. Both ω_0 and B are measured in radians per second. According to Eq. (8.40), the point $s = 0$ for the lowpass prototype filter is transformed into $s = \pm j\omega_0$ for the bandpass filter, and the point $s = \infty$ for the lowpass prototype filter is transformed into $s = 0$ and $s = \infty$ for the bandpass filter.

Thus a pole factor $(s - d_j)$ in the transfer function of a lowpass prototype filter is transformed as follows:

$$\frac{1}{s - d_j} \rightarrow \frac{Bs}{(s - p_1)(s - p_2)} \tag{8.41}$$

where the poles p_1 and p_2 are defined by

$$p_1, p_2 = \tfrac{1}{2}(Bd_j \pm \sqrt{B^2 d_j^2 - 4\omega_0^2}) \tag{8.42}$$

An important point to note is that the frequency transformations described in Eqs. (8.38) and (8.40) are *reactance functions*. By "reactance function" we mean the driving-point impedance of a network composed entirely of inductors and capacitors. Indeed, we may generalize this result by saying that all frequency transformations, regardless of complications in the passband specifications of interest, are in the form of reactance functions.

▶ **Drill Problem 8.6**   Consider a lowpass filter whose transfer function is

$$H(s) = \frac{1}{s + 1}$$

Find the transfer function of the corresponding bandpass filter with midband frequency $\omega_0 = 1$ and bandwidth $B = 0.1$.

Answer:

$$\frac{0.1s}{s^2 + 0.1s + 1} \qquad \blacktriangleleft$$

▶ **Drill Problem 8.7**   Here, we revisit the phase response of a bandpass channel considered in Example 5.8, namely

$$\phi(\omega) = -\tan^{-1}\left(\frac{\omega^2 - \omega_c^2}{\omega\omega_c}\right)$$

where ω_c is the carrier frequency of the modulated signal applied to the filter. Show that $\phi(\omega)$ is the phase response of the filter obtained by applying the lowpass to bandpass transformation to the Butterworth lowpass filter of order 1:

$$H(s) = \frac{1}{s + 1} \qquad \blacktriangleleft$$

Having familiarized ourselves with the notion of approximating functions and the fundamental role of lowpass prototype filters, we consider the implementation of passive analog filters in the next section, followed by the design of digital filters in Section 8.8.

8.7 *Passive Filters*

A filter is said to be *passive* when its composition is made up entirely of passive circuit elements (i.e., inductors, capacitors, and resistors). However, the design of highly

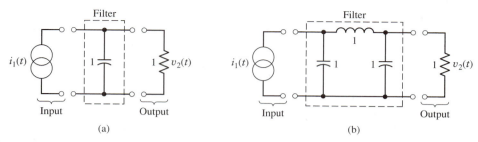

FIGURE 8.13 Lowpass Butterworth filters driven from ideal current source: (a) order $N = 1$ and (b) order $N = 3$.

frequency-selective passive filters is based exclusively on reactive elements (i.e., inductors and capacitors). Resistive elements enter the design only as source resistance and/or load resistance. The order N of the filter is usually determined by the number of reactive elements contained in the filter.

Figure 8.13(a) shows a Butterworth filter of the lowpass type of order $N = 1$ and 3-dB cutoff frequency $\omega_c = 1$. The filter is driven from an ideal current source. The resistance $R_l = 1$ represents the load resistance. The capacitor $C = 1$ represents the only reactive element of the filter.

Figure 8.13(b) shows a Butterworth filter of the lowpass type of order $N = 3$ and 3-dB cutoff frequency $\omega_c = 1$. As with the previous configuration, the filter is driven from a current source, and $R_l = 1$ represents the load resistance. In this case, the filter is made up of three reactive elements: two shunt capacitors and a single series inductor.

Note that in both Figs. 8.13(a) and (b) the transfer function $H(s)$ is in actual fact in the form of a transfer impedance, defined by the Laplace transform of the output voltage $v_2(t)$ divided by the Laplace transform of the current source $i_1(t)$.

▶ **Drill Problem 8.8**   Show that the transfer function of the filter in Fig. 8.13(b) is equal to the Butterworth function given in Eq. (8.37).   ◀

The determination of the elements of a filter, starting from a particular transfer function $H(s)$, is referred to as *network synthesis*. It encompasses a number of highly advanced procedures that are beyond the scope of the present book. Indeed, passive filters occupied a dominant role in the design of communication and other systems for several decades until the advent of active filters and digital filters in the 1960s. Active filters (using operational amplifiers) are discussed in Chapter 9, and digital filters are discussed in the next section.

8.8 Digital Filters

A digital filter uses *computation* to implement the filtering action that is to be performed on a continuous-time signal. Figure 8.14 shows the block diagram of the operations in-

FIGURE 8.14   System for filtering a continuous-time signal, built around a digital filter.

volved in such an approach for designing a frequency-selective filter; the ideas behind these operations were discussed in Section 4.8. The block labeled analog-to-digital (A/D) converter is used to convert the continuous-time signal $x(t)$ into a sequence $x[n]$ of numbers. The digital filter processes the sequence of numbers $x[n]$ on a sample-by-sample basis to produce a new sequence of numbers, $y[n]$, which is then converted into a corresponding continuous-time signal by the digital-to-analog (D/A) converter. Finally, the reconstruction (lowpass) filter at the output of the system produces a continuous-time signal $y(t)$, representing the filtered version of the original input signal $x(t)$.

Two important points should be carefully noted in the study of digital filters:

1. The underlying design procedures are usually based on the use of an analog or infinite precision model for the samples of input data and all internal calculations; this is done in order to take advantage of well-understood discrete-time but continuous-amplitude mathematics. The resulting *discrete-time filter* provides the designer with a "theoretical" framework for the task at hand.

2. When the discrete-time filter is implemented in digital form for practical use, the input data and internal calculations are all quantized to a finite precision. In so doing, *roundoff errors* are introduced into the operation of the digital filter, causing its performance to deviate from that of the discrete-time filter from which it is derived.

In this section we confine ourselves to matters relating to point 1. Although in light of this point the filters considered herein should in reality be referred to as discrete-time filters, we will refer to them as digital filters to conform to commonly used terminology.

Analog filters, exemplified by the passive filters discussed in Section 8.7, are characterized by an impulse response of infinite duration. In contrast, there are two classes of digital filters, depending on the duration of the impulse response:

1. *Finite-duration impulse response (FIR) digital filters*, the operation of which is governed by linear constant-coefficient difference equations of a nonrecursive nature. The transfer function of a FIR digital filter is a polynomial in z^{-1}. Consequently, FIR digital filters exhibit three important properties:
   ▶ They have finite memory and, therefore, any transient start-up is of limited duration.
   ▶ They are always BIBO stable.
   ▶ They can realize a desired magnitude response with exactly linear phase response (i.e., with no phase distortion), as explained in the sequel.

2. *Infinite-duration impulse response (IIR) digital filters*, whose input–output characteristics are governed by linear constant-coefficient difference equations of a recursive nature. The transfer function of an IIR digital filter is a rational function in z^{-1}. Consequently, for a prescribed frequency response the use of an IIR digital filter usually results in a shorter filter length than the corresponding FIR digital filter. However, this improvement is achieved at the expense of phase distortion and a transient start-up that is not limited to a finite-time interval.

In the sequel, examples of both FIR and IIR digital filters are discussed.

8.9 *FIR Digital Filters*

An inherent property of FIR digital filters is that they can realize a frequency response with *linear phase*. Recognizing that a linear phase response corresponds to a constant delay, the approximation problem in the design of FIR digital filters is therefore greatly simplified. Specifically, the design simplifies to that of approximating a desired magnitude response.

Let $h[n]$ denote the impulse response of an FIR digital filter, defined as the inverse discrete-time Fourier transform of the frequency response $H(e^{j\Omega})$. Let M denote the filter order, corresponding to a filter length of $M + 1$. To design the filter, we are required to determine the filter coefficients, $h[n]$, $n = 0, 1, \ldots, M$, so that the actual frequency response of the filter, namely, $H(e^{j\Omega})$, provides a good approximation to a desired frequency response $H_d(e^{j\Omega})$ over the frequency interval $-\pi < \Omega \le \pi$. As a measure of the goodness of this approximation, we define the *mean-square error*

$$E = \frac{1}{2\pi} \int_{-\pi}^{\pi} |H_d(e^{j\Omega}) - H(e^{j\Omega})|^2 \, d\Omega \tag{8.43}$$

Let $h_d[n]$ denote the inverse discrete-time Fourier transform of $H_d(e^{j\Omega})$. Then, invoking Parseval's theorem that was discussed in Chapter 3, we may redefine the error measure E in the equivalent form

$$E = \sum_{n=-\infty}^{\infty} |h_d[n] - h[n]|^2 \tag{8.44}$$

The only adjustable parameters in this equation are the filter coefficients $h[n]$. Accordingly, the error measure is minimized by setting

$$h[n] = \begin{cases} h_d[n], & 0 \le n \le M \\ 0, & \text{otherwise} \end{cases} \tag{8.45}$$

The approximation of Eq. (8.45) is equivalent to the use of a *rectangular window* defined by

$$w(n) = \begin{cases} 1, & 0 \le n \le M \\ 0, & \text{otherwise} \end{cases} \tag{8.46}$$

We may therefore rewrite Eq. (8.45) in the equivalent form:

$$h[n] = w[n]h_d[n] \tag{8.47}$$

It is for this reason that the design of a FIR filter based on Eq. (8.45) is called the *window method*. The mean-square error resulting from the use of the window method is:

$$E = \sum_{n=-\infty}^{-1} h_d^2[n] + \sum_{n=M+1}^{\infty} h_d^2[n] \tag{8.47a}$$

Since the multiplication of two sequences is equivalent to the convolution of their DTFTs, we may express the frequency response of the FIR filter with impulse response $h[n]$ as follows:

$$\begin{aligned} H(e^{j\Omega}) &= \sum_{n=0}^{M} h[n]e^{-jn\Omega} \\ &= \frac{1}{2\pi} \int_{-\pi}^{\pi} W(e^{j\Omega}) H_d(e^{j(\Omega-\Lambda)}) \, d\Lambda \end{aligned} \tag{8.48}$$

The function $W(e^{j\Omega})$ is the frequency response of the rectangular window $w[n]$; it is defined by

$$W(e^{j\Omega}) = \frac{\sin[\Omega(M + 1)/2]}{\sin(\Omega/2)} e^{-j\Omega(M+1)/2}, \qquad -\pi < \Omega \le \pi \tag{8.49}$$

In Fig. 8.15 we have plotted the magnitude response $|W(e^{j\Omega})|$ of the rectangular window for filter order $M = 12$. For the actual frequency response $H(e^{j\Omega})$ of the FIR digital filter to equal the ideal frequency response $H_d(e^{j\Omega})$, one period of the function $W(e^{j\Omega})$ would have to be an impulse. The frequency response $W(e^{j\Omega})$ of the rectangular window $w[n]$ can only approximate this ideal condition in an oscillatory manner.

The *mainlobe* of a window $w[n]$ is defined as the frequency band between the first zero crossing of its frequency response $W(e^{j\Omega})$ on either side of the origin. The individual transition regions that lie on either side of the mainlobe are referred to as *sidelobes*. The width of the mainlobe and amplitudes of the sidelobes provide measures of the extent by which the frequency response $W(e^{j\Omega})$ deviates from an impulse function.

▶ **Drill Problem 8.9**   Referring to Fig. 8.15 describing the frequency response of a rectangular window, verify that the width of the mainlobe is

$$\Delta\Omega_{\text{mainlobe}} = \frac{4\pi}{M + 2}$$

where M is the order of the filter.   ◀

▶ **Drill Problem 8.10**   Again referring to the frequency response of Fig. 8.15, verify that for a rectangular window (a) all the sidelobes have a common width equal to $2\pi/(M + 1)$, and (b) the first sidelobes have a peak amplitude that is 13 dB below that of the mainlobe.   ◀

From the discussion presented in Chapter 3 we recall that convolution with $W(e^{j\Omega})$ described in Eq. (8.48) results in an oscillatory approximation of the desired frequency response $H_d(e^{j\Omega})$ by the frequency response $H(e^{j\Omega})$ of the FIR filter. The oscillations are a consequence of the sidelobes in $W(e^{j\Omega})$. They may be reduced using a window with smaller sidelobes. A window commonly used for this purpose is the *Hamming window*, defined by

$$w[n] = \begin{cases} 0.54 - 0.46 \cos\left(\dfrac{2\pi n}{M}\right), & 0 \leq n \leq M \\ 0, & \text{otherwise} \end{cases} \tag{8.50}$$

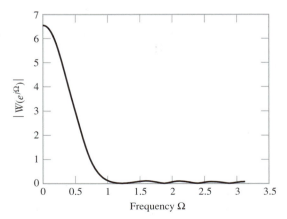

FIGURE 8.15   Magnitude response of rectangular window for FIR filter order $M = 12$, depicted on $0 \leq \Omega \leq \pi$.

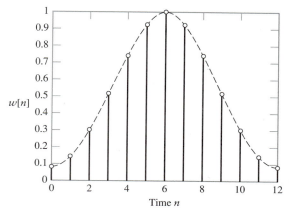

FIGURE 8.16   Impulse response of Hamming window for FIR filter order $M = 12$.

With M assumed even, $w[n]$ is *symmetric* about the point $n = M/2$. In Fig. 8.16, we have plotted $w[n]$ for the Hamming window with $M = 12$.

To further compare the frequency response of the Hamming window with that of the rectangular window, we have chosen to plot $20 \log_{10} |W(e^{j\Omega})|$ for these two windows in Fig. 8.17 for $M = 12$. From this figure we may make two important observations:

▶ The mainlobe of the rectangular window is about half the width of the mainlobe of the Hamming window.

▶ The sidelobes of the Hamming window, relative to the mainlobe, are greatly reduced compared to those of the rectangular window. Specifically, the peak amplitude of the first sidelobe of the rectangular window is only about 13 dB below that of the mainlobe, whereas the corresponding value for the Hamming window is about 40 dB.

It is because of the latter point that the Hamming window reduces oscillations in the frequency response of a FIR digital filter as illustrated in the following two examples. However, there is a price to be paid for this improvement, namely a wider transition band.

In order to obtain the best possible approximation of the desired response, the window must preserve as much of the energy in $h_d[n]$ as possible. Since the windows are

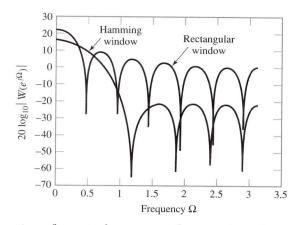

FIGURE 8.17   Comparison of magnitude responses of rectangular and Hamming windows for filter order $M = 12$, plotted in decibels.

symmetric about $n = M/2$, we desire to concentrate the maximum values of $h_d[n]$ about $n = M/2$. This is accomplished by choosing the phase response $\arg\{H_d(e^{j\Omega})\}$ to be linear with zero intercept and a slope equal to $-M/2$. This point is illustrated in the next example.

EXAMPLE 8.5   Consider the desired frequency response

$$H_d(e^{j\Omega}) = \begin{cases} e^{-jM\Omega/2}, & |\Omega| < \Omega_c \\ 0, & \Omega_c < |\Omega| \leq \pi \end{cases} \tag{8.51}$$

which represents the frequency response of an ideal lowpass filter with linear phase. Investigate the frequency response of a FIR digital filter of length $M = 12$, using (a) a rectangular window and (b) a Hamming window. Assume $\Omega_c = 0.2\pi$ radians.

Solution: The desired response is

$$\begin{aligned} h_d[n] &= \frac{1}{2\pi} \int_{-\pi}^{\pi} H_d(e^{j\Omega}) e^{jn\Omega} \, d\Omega \\ &= \frac{1}{2\pi} \int_{-\Omega_c}^{\Omega_c} e^{j\Omega(n-M/2)} \, d\Omega \end{aligned} \tag{8.52}$$

Invoking the definition of the sinc function, we may express $h_d[n]$ in the compact form

$$h_d[n] = \frac{\Omega_c}{\pi} \text{sinc}\left[\frac{\Omega_c}{\pi}\left(n - \frac{M}{2}\right)\right], \qquad -\infty < n < \infty \tag{8.53}$$

This impulse response is symmetric about $n = M/2$, for M even, at which point we have

$$h_d\left[\frac{M}{2}\right] = \frac{\Omega_c}{\pi} \tag{8.54}$$

(a) *Rectangular window.* For the case of a rectangular window, the use of Eq. (8.53) yields

$$h[n] = \begin{cases} \dfrac{\Omega_c}{\pi} \text{sinc}\left[\dfrac{\Omega_c}{\pi}\left(n - \dfrac{M}{2}\right)\right], & 0 \leq n \leq M \\ 0, & \text{otherwise} \end{cases} \tag{8.55}$$

the value of which is given in the first column of Table 8.1 for $\Omega_c = 0.2\pi$ and $M = 12$. The corresponding magnitude response $|H(e^{j\Omega})|$ is plotted in Fig. 8.18. The $|H(e^{j\Omega})|$

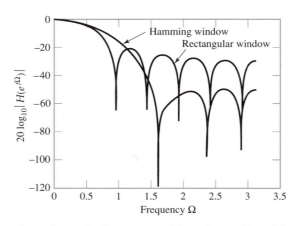

FIGURE 8.18   Comparison of magnitude responses (plotted on a dB scale) of two lowpass FIR digital filters of order $M = 12$, one filter using the rectangular window and the other using the Hamming window.

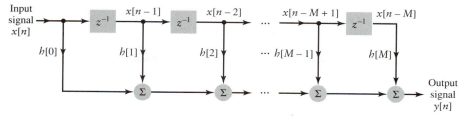

FIGURE 8.19   Structure for implementing an FIR digital filter.

oscillations due to windowing the ideal impulse response are clearly evident at frequencies greater than $\Omega_c = 0.2\pi$.

(b) *Hamming window.* For the case of a Hamming window, the use of Eqs. (8.50) and (8.53) yields

$$h[n] = \begin{cases} \left[\dfrac{\Omega_c}{\pi} \operatorname{sinc}\left[\dfrac{\Omega_c}{\pi}\left(n - \dfrac{M}{2}\right)\right]\right]\left(0.54 - 0.46 \cos\left(2\pi \dfrac{n}{M}\right)\right), & 0 \le n \le M \\ 0, & \text{otherwise} \end{cases} \tag{8.56}$$

the value of which is given in the second column of Table 8.1 for $\Omega_c = 0.2\pi$ and $M = 12$. The corresponding magnitude response $|H(e^{j\Omega})|$ is also plotted in Fig. 8.18. We see that the oscillations due to windowing have been greatly reduced in amplitude. However, this improvement has been achieved at the expense of a wider transition band compared to that attained using a rectangular window.

Note that the filter coefficients in Table 8.2 have been *scaled* so that the magnitude response of the filter at $\omega = 0$ is exactly one after windowing. This explains the deviation of the coefficient $h[M/2]$ from the theoretical value $\Omega_c/\pi = 0.2$.

The structure of the FIR digital filter for both windows is the same as shown in Fig. 8.19. The filter coefficients for the two windows are of course different, taking the respective values given in Table 8.2.

TABLE 8.2   *Filter Coefficients of Rectangular and Hamming Windows for Lowpass Filter*

	$h[n]$	
n	*Rectangular Window*	*Hamming Window*
0	−0.0281	−0.0027
1	0.0000	0.0000
2	0.0421	0.0158
3	0.0909	0.0594
4	0.1364	0.1271
5	0.1686	0.1914
6	0.1802	0.2180
7	0.1686	0.1914
8	0.1364	0.1271
9	0.0909	0.0594
10	0.0421	0.0158
11	0.0000	0.0000
12	−0.0281	−0.0027

EXAMPLE 8.6   Consider next the design of a discrete-time *differentiator*, the frequency response of which is defined by

$$H_d(e^{j\Omega}) = j\Omega e^{-jM\Omega/2}, \qquad -\pi < \Omega \leq \pi \tag{8.57}$$

Design a FIR digital filter that approximates this desired frequency response for $M = 12$ using (a) a recentangular window and (b) a Hamming window.

Solution: The desired impulse response is

$$
\begin{aligned}
h_d[n] &= \frac{1}{2\pi} \int_{-\pi}^{\pi} H_d(e^{j\Omega}) e^{jn\Omega} \, d\Omega \\
&= \frac{1}{2\pi} \int_{-\pi}^{\pi} j\Omega e^{j\Omega(n-M/2)} \, d\Omega
\end{aligned}
\tag{8.58}
$$

Using integration by parts, we get

$$h_d[n] = \frac{\cos[\pi(n - M/2)]}{(n - M/2)} - \frac{\sin[\pi(n - M/2)]}{\pi(n - M/2)^2}, \qquad -\infty < n < \infty \tag{8.59}$$

(a) *Rectangular window*. Multiplying this impulse response by the rectangular window of Eq. (8.46), we get

$$h[n] = \frac{\cos[\pi(n - M/2)]}{(n - M/2)} - \frac{\sin[\pi(n - M/2)]}{\pi(n - M/2)^2}, \qquad 0 \leq n \leq M \tag{8.60}$$

This impulse response is antisymmetric in that $h[M - n] = -h[n]$. Note also that for M even, $h[n]$ is zero at $n = M/2$. The value of $h[n]$ is given in the second column of Table 8.3 for $M = 12$. which clearly demonstrates the antisymmetric property of $h[n]$. The corresponding magnitude response $|H(e^{j\Omega})|$ is plotted in Fig. 8.20(a). The oscillatory deviations from the ideal frequency response are manifestations of windowing the ideal impulse response in Eq. (8.59).

TABLE 8.3   Filter Coefficients of Rectangular and Hamming Windows for a Differentiator

	$h[n]$	
n	Rectangular Window	Hamming Window
0	−0.1667	−0.0133
1	0.2000	0.0283
2	−0.2500	−0.0775
3	0.3333	0.1800
4	−0.5000	−0.3850
5	1.0000	0.9384
6	0	0
7	−1.0000	−0.9384
8	0.5000	0.3850
9	−0.3333	−0.1800
10	0.2500	0.0775
11	−0.2000	−0.0283
12	0.1667	0.0133

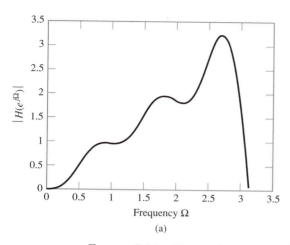
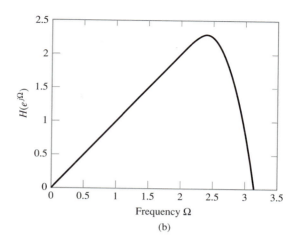

(a)

(b)

FIGURE 8.20 Magnitude response of FIR digital filter as differentiator, designed using (a) a rectangular window and (b) a Hamming window. In both cases, the filter order M is 12.

(b) *Hamming window.* Multiplying the impulse response $h_d[n]$ of Eq. (8.59) by the Hamming window of Eq. (8.50), we get the impulse response $h[n]$ given in the last column of Table 8.3. The corresponding magnitude response $|H(e^{j\Omega})|$ is plotted in Fig. 8.20(b). Comparing this response with that of Fig. 8.20(a) we see that the oscillations have been greatly reduced in amplitude, but the bandwidth over which $|H(e^{j\Omega})|$ is linear with Ω has been reduced, implying less usable bandwidth for the operation of differentiation.

Note that there are many other windows besides the Hamming window that allow different trade-offs between mainlobe width and sidelobe height.

▶ **Drill Problem 8.11** Starting with Eq. (8.58), derive the formula for the impulse response $h_d[n]$ given in Eq. (8.59), and show that $h_d[M/2] = 0$. ◀

■ FILTERING OF SPEECH SIGNALS

The preprocessing of speech signals is fundamental to many applications such as the digital transmission and storage of speech, automatic speech recognition, and automatic speaker recognition systems. FIR digital filters are well suited for the preprocessing of speech signals for two important reasons:

1. In speech processing applications, it is essential to maintain precise time alignment. The *exact* linear phase property inherent to a FIR digital filter caters to this requirement in a natural way.

2. The approximation problem in the filter design is greatly simplified by the exact linear phase property of a FIR digital filter. In particular, in not having to deal with delay (phase) distortion, our only concern is that of approximating a desired magnitude response.

However, there is a price to be paid for achieving these two desirable features. To design a FIR digital filter with a sharp cutoff characteristic, the length of the filter has to be large. This, in turn, produces an impulse response with a long duration.

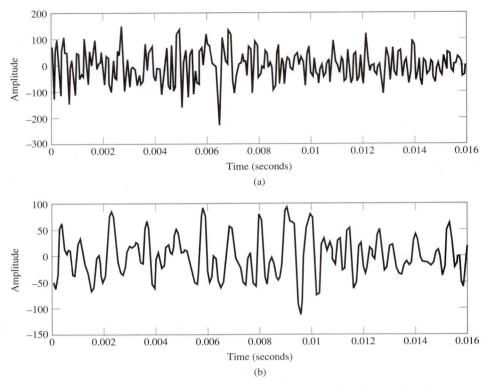

FIGURE 8.21   (a) Waveform of raw speech signal, containing an abundance of high-frequency noise. (b) Waveform of speech signal after passing it through a lowpass FIR digital filter of order $M = 98$ and cutoff frequency $\omega_c = 2\pi \times 3.1 \times 10^3$ rad/s.

In this subsection, we will illustrate the use of a FIR digital filter for the preprocessing of a real-life speech signal, so that it would be suitable for transmission over a telephone channel. Figure 8.21(a) shows a short portion of the waveform of a speech signal produced by a female speaker saying the phrase: "This was easy for us." The original sampling rate of this speech signal was 16 kHz, and the total number of samples contained in the whole sequence was 27,751. The magnitude spectrum of the speech signal, computed by means of a fast Fourier transform (FFT) algorithm, is shown in Fig. 8.22(a).

Before transmission, the speech signal is applied to a FIR digital lowpass filter with the following specifications:

Length of filter, $M + 1 = 99$

Symmetric about midpoint to obtain a linear phase response

Cutoff frequency $\omega_c = 2\pi \times 3.1 \times 10^3$ rad/s

The design of the filter was based on the window method, using the *Hanning* or *raised-cosine window*, which is not to be confused with the Hamming window. This new window is defined by

$$
w[n] = \begin{cases} \dfrac{1}{2}\left[1 - \cos\left(\dfrac{2\pi n}{M}\right)\right], & 0 \le n \le M \\ 0, & \text{otherwise} \end{cases} \qquad (8.61)
$$

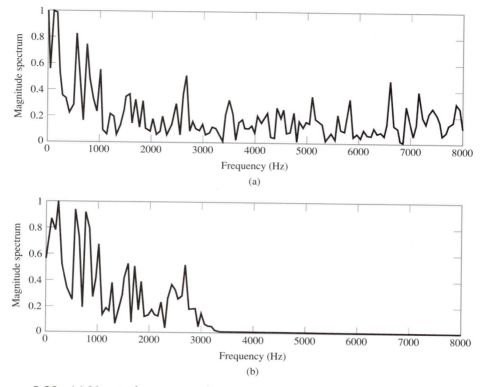

FIGURE 8.22   (a) Magnitude spectrum of prefiltered speech signal. (b) Magnitude spectrum of postfiltered speech signal. Note the sharp cutoff of the spectrum around 3100 Hz.

The Hanning window goes to zero with zero slope at the edges of the window [i.e., $n = 0$ and $n = M$].

Figure 8.21(b) shows the waveform of the postfiltered speech signal. Its magnitude spectrum, again computed using the FFT algorithm, is shown in Fig. 8.22(b). In comparing this spectrum with that of Fig. 8.22(a), we clearly see the relatively sharp cutoff produced by the FIR lowpass filter around 3.1 kHz.

In listening to the prefiltered and postfiltered versions of the speech signal, the following observations were made:

1. The prefiltered speech signal was harsh, with an abundance of high-frequency noise such as clicks, pops, and hissing sounds.
2. The postfiltered signal, in contrast, was found to be much softer, more smooth, and natural sounding.

The essence of these observations can be confirmed by examining 10 milliseconds of the speech waveforms and their spectra shown in Figs. 8.21 and 8.22, respectively.

As mentioned previously, the original speech signal was sampled at the rate of 16 kHz, which corresponds to a sampling period $\mathcal{T} = 62.5$ μs. The structure used to implement the FIR filter was similar to that described in Fig. 8.19. In passing the speech signal through this filter with $M + 1 = 99$ coefficients, a delay of

$$\mathcal{T}\left(\frac{M}{2}\right) = 62.5 \times 49 = 3.0625 \text{ ms}$$

is introduced into the postfiltered speech signal. This time delay is clearly discernible in comparing the waveform of the postfiltered speech signal in Fig. 8.21(b) with that of the prefiltered speech signal in Fig. 8.21(a).

8.10 *IIR Digital Filters*

Various techniques have been developed for the design of IIR digital filters. In this section we describe a popular method for converting analog transfer functions to digital transfer functions. The method is based on the *bilinear transform*, which provides a unique mapping between points in the s-plane and those in the z-plane.

The bilinear transform is defined by

$$s = \left(\frac{2}{\mathcal{T}}\right)\left(\frac{z-1}{z+1}\right) \tag{8.62}$$

where $\mathcal{T}$ is the implied sampling interval associated with conversion from the s-domain to the z-domain. To simplify matters, we shall set $\mathcal{T} = 2$, henceforth. The resulting filter design is independent of the actual choice of $\mathcal{T}$. Let $H_a(s)$ denote the transfer function of an analog (continuous-time) filter. The transfer function of the corresponding digital filter is obtained by substituting the bilinear transformation of Eq. (8.62) in $H_a(s)$ as shown by

$$H(z) = H_a(s)\big|_{s=((z-1)/(z+1))} \tag{8.63}$$

What can we say about the properties of the transfer function $H(z)$ derived from Eq. (8.63)? To answer this question, we rewrite Eq. (8.62) in the form (with $\mathcal{T} = 2$)

$$z = \frac{1+s}{1-s}$$

Putting $s = \sigma + j\omega$ in this equation, we may express the complex variable z in the polar form

$$z = re^{j\theta}$$

where the radius r and angle θ are defined by, respectively,

$$\begin{aligned} r &= |z| \\ &= \left[\frac{(1+\sigma)^2 + \omega^2}{(1-\sigma)^2 + \omega^2}\right]^{1/2} \end{aligned} \tag{8.64}$$

and

$$\begin{aligned} \theta &= \arg\{z\} \\ &= \tan^{-1}\left(\frac{\omega}{1+\sigma}\right) + \tan^{-1}\left(\frac{\omega}{1-\sigma}\right) \end{aligned} \tag{8.65}$$

From Eqs. (8.64) and (8.65) we readily see that

▶ $r < 1$ for $\sigma < 0$.
▶ $r = 1$ for $\sigma = 0$.
▶ $r > 1$ for $\sigma > 0$.
▶ $\theta = 2\tan^{-1}(\omega)$ for $\sigma = 0$.

Accordingly, we may state the properties of the bilinear transform as follows:

1. The left-half of the s-plane is mapped onto the interior of the unit circle in the z-plane.
2. The entire $j\omega$-axis of the s-plane is mapped onto one complete revolution of the unit circle in the z-plane.
3. The right-half of the s-plane is mapped onto the exterior of the unit circle in the z-plane.

These properties are illustrated in Fig. 8.23.

An immediate implication of property 1 is that if the analog filter represented by transfer function $H_a(s)$ is stable and causal, then the digital filter derived from it by using the bilinear transform is guaranteed to be stable and causal also. Since the bilinear transform has real coefficients, it follows that $H(z)$ will have real coefficients if $H_a(s)$ has real coefficients. Hence the transfer function $H(z)$ resulting from the use of Eq. (8.63) is indeed physically realizable.

▶ **Drill Problem 8.12**   What do the points $s = 0$ and $s = \pm j\infty$ in the s-plane map onto in the z-plane using the bilinear transform?

Answer: $s = 0$ is mapped onto $z = +1$. The points $s = j\infty$ and $s = -j\infty$ are mapped onto just above and just below $z = -1$, respectively.   ◀

For $\sigma = 0$ and $\theta = \Omega$, Eq. (8.65) reduces to

$$\Omega = 2 \tan^{-1}(\omega) \qquad (8.66)$$

which is plotted in Fig. 8.24 for $\omega > 0$. Note that Eq. (8.66) has odd symmetry. The infinitely long range of frequency variations $-\infty < \omega < \infty$ for an analog (continuous-time) filter is nonlinearly compressed into the finite frequency range $-\pi < \Omega < \pi$ for a digital (discrete-time) filter. This form of nonlinear distortion is known as *warping*. In the design of frequency-selective filters where the emphasis is on the approximation of a piecewise magnitude response, we must compensate for this nonlinear distortion by *prewarping* the design specifications of the analog filter. Specifically, the critical frequencies (i.e., the pre-

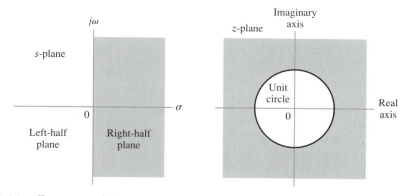

FIGURE 8.23   Illustration of the properties of the bilinear transform. The left half of the s-plane (shown on the left) is mapped onto the interior of the unit circle in the z-plane (shown on the right). Likewise, the right half of the s-plane is mapped onto the exterior of the unit circle in the z-plane; these two corresponding regions are shown shaded.

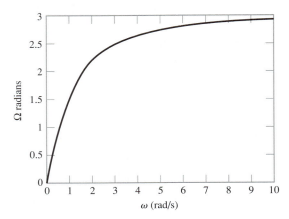

FIGURE 8.24   Graphical plot of the relation between the frequency Ω pertaining to the discrete-time domain and the frequency ω pertaining to the continuous-time domain: $\Omega = 2 \tan^{-1}(\omega)$.

scribed passband cutoff and stopband cutoff frequencies) are prewarped in accordance with the formula

$$\omega = \tan\left(\frac{\Omega}{2}\right) \tag{8.67}$$

which is the inverse of Eq. (8.66). To illustrate the prewarping procedure, let Ω'_k, $k = 1$, $2, \ldots$, denote the critical frequencies that a digital filter is required to realize. Before applying the bilinear transform, the corresponding critical frequencies of the continuous-time filter are prewarped using Eq. (8.67) to obtain

$$\omega_k = \tan\left(\frac{\Omega'_k}{2}\right), \qquad k = 1, 2, \ldots \tag{8.68}$$

Then, when the bilinear transform is applied to the transfer function of the analog filter designed using the prewarped frequencies in Eq. (8.68), we find from Eq. (8.66) that

$$\Omega_k = \Omega'_k, \qquad k = 1, 2, \ldots \tag{8.69}$$

That is, the prewarping procedure ensures that the digital filter will meet the prescribed design specifications exactly.

EXAMPLE 8.7   Using an analog filter with a Butterworth response of order 3, design a digital IIR lowpass filter with a 3-dB cutoff frequency $\Omega_c = 0.2\pi$.

Solution: The prewarping formula of Eq. (8.68) indicates that the cutoff frequency of the analog filter should be

$$\omega_c = \tan(0.1\pi) = 0.3429$$

Adapting Eq. (8.37) to the problem at hand in light of Drill Problem 8.2, the transfer function of the analog filter is

$$H_a(s) = \frac{1}{\left(\dfrac{s}{\omega_c} + 1\right)\left(\dfrac{s^2}{\omega_c^2} + \dfrac{s}{\omega_c} + 1\right)} \tag{8.70}$$

$$= \frac{0.0403}{(s + 0.3429)(s^2 + 0.3429s + 0.1176)}$$

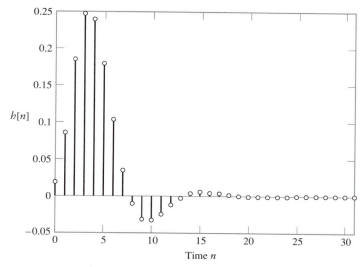

FIGURE 8.25 Impulse response of digital IIR lowpass filter with Butterworth response of order 3 and 3-dB cutoff frequency $\Omega_c = 0.2\pi$.

Hence, using Eq. (8.63), we get

$$H(z) = \frac{0.0205(z + 1)^3}{(z - 0.4893)(z^2 - 1.208z + 0.5304)} \tag{8.71}$$

Figure 8.25 shows the impulse response $h[n]$ of the filter (i.e., the inverse z-transform of the $H(z)$ given in Eq. (8.71)).

In Section 7.7 we discussed different computational structures (i.e., cascade and parallel forms) for implementing discrete-time systems. In light of the material covered therein, we readily see that the transform function of Eq. (8.71) can be realized using a cascade of two sections, as shown in Fig. 8.26. The section resulting from the bilinear transformation of

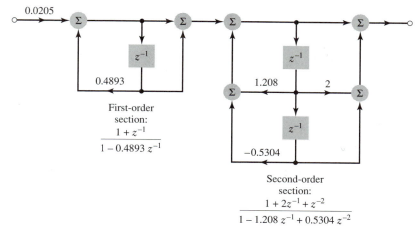

FIGURE 8.26 Cascade implementation of IIR lowpass digital filter, made up of first-order section followed by second-order section.

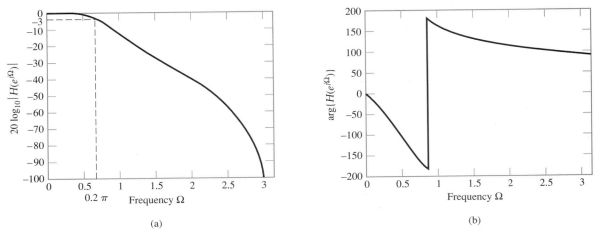

FIGURE 8.27 (a) Magnitude response of IIR lowpass digital filter characterized by the impulse response shown in Fig. 8.25, plotted in decibels. (b) Phase response of the filter.

the simple pole factor $((s/\omega_c) + 1)$ in $H_a(s)$ is referred to as a *first-order section*. By the same token, the section resulting from the bilinear transformation of the quadratic pole factor $((s/\omega_c)^2 + (s/\omega_c) + 1)$ in $H_a(s)$ is referred to as a *second-order section*. Indeed, this result may be generalized to say that the application of the bilinear transform to $H_s(s)$ in factored form results in a realization for $H(z)$ that consists of a cascade of first-order and second-order sections. From a practical point of view, this kind of structure for implementing a digital filter has intuitive appeal.

Putting $z = e^{j\Omega}$ in Eq. (8.71) and plotting $H(e^{j\Omega})$ versus Ω, we get the magnitude and phase responses shown in Fig. 8.27. We see that the passband of the filter extends up to 0.2π, as prescribed.

8.11 Linear Distortion

In practice, the conditions for distortionless transmission described in Section 8.2 can only be satisfied approximately; the material presented in the previous sections testifies to this statement. That is to say, there is always a certain amount of distortion present in the output signal of a physical LTI system, be that of a continuous-time or discrete-time type, due to deviations in the frequency response of the system from the ideal conditions described in Eqs. (8.5) and (8.6). In particular, we may distinguish two components of linear distortion produced by signal transmission through a LTI system:

1. *Amplitude distortion.* When the magnitude response of the system is not constant inside the frequency band of interest, the frequency components of the input signal are transmitted through the system with different amounts of gain or attenuation. This effect is called *amplitude distortion*. The most common form of amplitude distortion is excess gain or attenuation at one or both ends of the frequency band of interest.

2. *Phase distortion.* The second form of linear distortion arises when the phase response of the system is not linear with frequency inside the frequency band of interest. If the input signal is divided into a set of components, each one of which occupies a

narrow band of frequencies, we find that each such component is subject to a different delay in passing through the system, with the result that the output signal emerges with a waveform different from that of the input signal. This form of linear distortion is called *phase* or *delay distortion*.

We emphasize the distinction between a constant delay and a constant phase shift. In the case of a continuous-time LTI system, constant delay means a linear phase response, that is, $\arg\{H(j\omega)\} = -t_0\omega$, where t_0 is the constant delay. On the other hand, constant phase shift means that $\arg\{H(j\omega)\}$ equals some constant for all ω. These two conditions have different implications. Constant delay is a requirement for distortionless transmission. Constant phase shift, on the other hand, causes signal distortion.

A LTI system that suffers from linear distortion is said to be *dispersive*, in that the frequency components of the input signal emerge with amplitude and phase characteristics that are different from those of the original input signal after transmission through the system. The telephone channel is an example of a dispersive system.

8.12 *Equalization*

To compensate for linear distortion, we may use a network known as an *equalizer* connected in cascade with the system in question, as illustrated in Fig. 8.28. The equalizer is designed in such a way that, inside the frequency band of interest, the overall magnitude and phase responses of this cascade connection approximate the conditions for distortionless transmission to within prescribed limits.

Consider, for example, a communication channel with frequency response $H_c(j\omega)$. Let an equalizer of frequency response $H_{eq}(j\omega)$ be connected in cascade with the channel, as in Fig. 8.28. The overall frequency response of this combination is equal to $H_c(j\omega)H_{eq}(j\omega)$. For overall transmission through the cascade connection of Fig. 8.28 to be distortionless, we require that (see Eq. (8.3))

$$H_c(j\omega)H_{eq}(j\omega) = e^{-j\omega t_0} \tag{8.72}$$

where t_0 is a constant time delay; for convenience of presentation, we have set the scaling factor K in Eq. (8.3) equal to unity. Ideally, therefore, the frequency response of the equalizer is *inversely related* to that of the channel, as shown by

$$H_{eq}(j\omega) = \frac{e^{-j\omega t_0}}{H_c(j\omega)} \tag{8.73}$$

In practice, the equalizer is designed such that its frequency response approximates the ideal value of Eq. (8.73) closely enough for the linear distortion to be reduced to a satisfactory level.

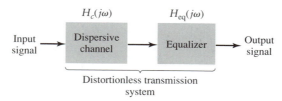

FIGURE 8.28   Cascade connection of a dispersive (LTI) channel and an equalizer for distortionless transmission.

The frequency response $H_{eq}(j\omega)$ of the equalizer in Eq. (8.73) is formulated in continuous time. Although indeed it is possible to design the equalizer using an analog filter, the preferred method is to do the design in discrete time using a digital filter. With a discrete-time approach, the channel output is sampled prior to equalization. Depending on the application, the equalizer output may be converted back to a continuous-time signal or left in discrete-time form.

A system that is well suited for equalization is the FIR digital filter, also referred to as *tapped-delay-line equalizer*. The structure of such a filter is depicted in Fig. 8.19. Since the channel frequency response is represented in terms of the Fourier transform, we shall employ the Fourier transform representation for the FIR filter frequency response. If the sampling interval is $\mathcal{T}$ seconds, then we see from Eq. (4.27) that the equalizer frequency response is

$$H_{\delta,eq}(j\omega) = \sum_{n=0}^{M} h[n]\exp(-jn\omega\mathcal{T}) \tag{8.74}$$

The subscript δ in $H_{\delta,eq}(j\omega)$ is intended to distinguish it from its continuous-time counterpart $H_{eq}(j\omega)$. For convenience of analysis it is assumed that the number of filter coefficients $M + 1$ in the equalizer is odd; that is, M is even.

The goal of equalizer design is to determine the filter coefficients $h[0]$, $h[1]$, . . . , $h[M]$, so that $H_{\delta,eq}(j\omega)$ approximates $H_{eq}(j\omega)$ in Eq. (8.73) over a frequency band of interest, say, $-\omega_c \leq \omega \leq \omega_c$. Note that $H_{\delta,eq}(j\omega)$ is periodic, with one period occupying the frequency range $-\pi/\mathcal{T} \leq \omega \leq \pi/\mathcal{T}$. Hence we choose $\mathcal{T} = \pi/\omega_c$, so that one period of $H_{\delta,eq}(j\omega)$ corresponds to the frequency band of interest. Let

$$H_d(j\omega) = \begin{cases} e^{-j\omega t_0}/H_c(j\omega), & -\omega_c \leq \omega \leq \omega_c \\ 0, & \text{otherwise} \end{cases} \tag{8.75}$$

be the frequency response we seek to approximate with $H_{\delta,eq}(j\omega)$. We accomplish this task by using a variation of the window method of FIR filter design, as described here:

1. Set $t_0 = (M/2)\mathcal{T}$.
2. Take the inverse Fourier transform of $H_d(j\omega)$ to obtain a desired impulse response $h_d(t)$.
3. Set $h[n] = w[n]h_d(n\mathcal{T})$, where $w[n]$ is a window of length $(M + 1)$. Note that the sampling operation does not cause aliasing of the desired response in the band $-\omega_c \leq \omega \leq \omega_c$ since we chose $\mathcal{T} = \pi/\omega_c$.

Typically $H_c(j\omega)$ is given numerically in terms of its magnitude and phase, in which case numerical integration is used to evaluate $h_d(n\mathcal{T})$. The number of terms, $M + 1$, is chosen just big enough to produce a satisfactory approximation to $H_d(j\omega)$.

EXAMPLE 8.8   Consider a simple channel whose frequency response is described by the first-order Butterworth response

$$H_c(j\omega) = \frac{1}{1 + j\omega/\pi}$$

Design a FIR filter with 13 coefficients (i.e., $M = 12$) for equalizing this channel over the frequency band $-\pi \leq \omega \leq \pi$.

Solution: In this example, the channel equalization problem is simple enough for us to solve it without having to resort to the use of numerical integration. With $\omega_c = \pi$, the sampling interval is $\mathcal{T} = 1$ s. Now, from Eq. (8.75), we have

$$H_d(j\omega) = \begin{cases} \left(1 + \dfrac{j\omega}{\pi}\right)e^{-j6\omega}, & -\pi \leq \omega \leq \pi \\ 0, & \text{otherwise} \end{cases}$$

The nonzero part of the frequency response $H_d(j\omega)$ consists of the sum of two terms: unity and $j\omega/\pi$, except for a linear phase term. These two terms are approximated as follows, respectively:

▶ The term $j\omega/\pi$ represents a scaled form of differentiation. The design of a differentiator using a FIR filter was discussed in Example 8.6. Indeed, evaluating the inverse Fourier transform of $j\omega/\pi$, and then setting $t = n$ for a sampling period $\mathcal{T} = 1$ s, we get the result of Eq. (8.59), scaled by $1/\pi$. Thus using the result obtained in that example, which incorporated the Hamming window of length 13, and scaling it by $1/\pi$, we get the values listed in the second column of Table 8.4.

▶ The inverse Fourier transform of the unity term is sinc(t). Setting $t = n\mathcal{T} = n$, and weighting it with the Hamming window of length 13, we get the set of values listed in column 4 of Table 8.4.

Adding these two sets of values, we get the Hamming-windowed FIR filter coefficients for the equalizer listed in the last column of Table 8.4. Note that this filter is antisymmetric about the midpoint, $n = 6$.

Figure 8.29 displays superposition of the magnitude responses of the channel, the FIR equalizer, and the equalized channel. The responses are plotted for the band $0 \leq \omega \leq \pi$. From Fig. 8.29, we see that the magnitude response of the equalized channel is essentially flat over the band $0 \leq \omega \leq 2.5$. In other words, an FIR filter with $m = 12$ equalizes the channel for over 80% of its passband.

TABLE 8.4 *Filter Coefficients for Example 8.8 on Equalization*

n	Hamming Windowed Inverse Fourier Transform of $j\omega/\pi$	Hamming Windowed Inverse Fourier Transform of 1	$h_d[n]$
0	-0.0042	0	-0.0042
1	0.0090	0	0.0090
2	-0.0247	0	-0.0247
3	0.0573	0	0.0573
4	-0.1225	0	-0.1225
5	0.2987	0	0.2987
6	0	1	1.0000
7	-0.2987	0	-0.2987
8	0.1225	0	0.1225
9	-0.0573	0	-0.0573
10	0.0247	0	0.0247
11	-0.0090	0	-0.0090
12	0.0042	0	0.0042

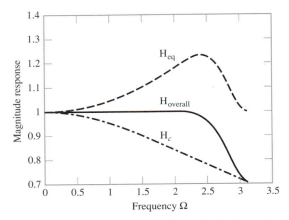

FIGURE 8.29   Magnitude response of Butterworth channel of order 1: dashed and dotted (—·—·) curve. Magnitude response of FIR equalizer of order $M = 12$: dashed (——) curve. Magnitude response of equalized channel: continuous curve. The flat region of the overall (equalized) magnitude response is extended up to about $\omega = 2.5$.

8.13 *Exploring Concepts with MATLAB*

In this chapter, we studied the design of linear filters and equalizers. While these two systems act on input signals of their own, their purposes are entirely different. The goal of filtering is to produce an output signal with a specified frequency content. An equalizer, on the other hand, is used to compensate for some form of linear distortion contained in the input signal.

The MATLAB Signal Processing Toolbox possesses a rich collection of functions that are tailor-made for the analysis and design of linear filters and equalizers. In this section, we explore the use of some of those functions as tools for consolidating ideas and design procedures described in previous sections of this chapter.

■ TRANSMISSION OF A RECTANGULAR PULSE THROUGH AN IDEAL LOWPASS FILTER

In Example 8.1, we studied the response of an ideal lowpass filter to an input rectangular pulse. This response, denoted by $y(t)$, is given in terms of the sine integral by Eq. (8.20); that is,

$$y(t) = \frac{1}{\pi} \left[\text{Si}(a) - \text{Si}(b) \right]$$

where

$$a = \omega_c \left(t - t_0 + \frac{T}{2} \right)$$

$$b = \omega_c \left(t - t_0 - \frac{T}{2} \right)$$

where T is the pulse duration, ω_c is the cutoff frequency of the filter, and t_0 is the transmission delay through the filter. For convenience of presentation, we set $t_0 = 0$.

The expression for $y(t)$ shows that the sine integral Si(u), defined in Eq. (8.19), plays a dominant role in determining the response of an ideal lowpass filter to a rectangular pulse input. Unfortunately, there is no analytic solution for this integral. We therefore have to resort to the use of numerical integration for its evaluation. A common procedure for numerical integration is to compute an estimate of the area under the integrand between the limits of integration. Such a procedure is referred to as a *quadrature technique*. The MATLAB function

```
quad('function_name,'a,b)
```

returns the area under the integrand between the limits of integration, a and b. The function quad uses a form of Simpson's rule, in which the integrand is uniformly sampled across [a,b]. For the sine integral plotted in Fig. 8.5, we used the following commands:

```
≫ x = -20:.1:20;
   For u = 1:length(x),
       z(u) = quad('sincnopi,'0,x(u));
   end
```

which incorporate the M-file called 'sincnopi.m' described here:

```
Function y = sincnopi(w)
y = ones(size(w));
i = find(w);
y(i) = sin(w(i))./w(i);
```

Returning to the issue at hand, the MATLAB M-file for computing the pulse response $y(t)$ is as follows:

```
function [y]=sin_pr(wc, r)
```

%r is a user-specified resolution parameter

```
≫ T=1;
≫ to=0;              % transmission delay = 0
≫ t=-T*1.01:r:T*1.01;
≫ ta=wc*(t-to+T/2);
≫ tb=wc*(t-to-T/2);
≫ for q=1:length(ta),
     z1(q)=quad('sincnopi',0,ta(q));
end
≫ for q=1:length(tb),
     z2(q)=quad('sincnopi',0,tb(q));
end
≫ plot(t,(z1-z2)/pi)
   axis(gca,'YLim',[-.2 1.2])
   axis(gca,'XLim',[-1 1])
```

■ FIR DIGITAL FILTERS

The MATLAB Signal Processing Toolbox has two types of routines, namely, fir1 and fir2, for designing FIR filters based on the window methods. Their functions are summarized here.

1. The command

   ```
   b=fir1(M,wc)
   ```

   designs an *M*th order lowpass digital filter and returns the filter coefficients in vector **b** of length **M+1**. The cutoff frequency **wc** is normalized so that it lies in the interval [0, 1], with 1 corresponding to one-half the sampling rate, or $\Omega = \pi$ in discrete-time frequency. By default, **fir1** uses a Hamming window. It allows the use of several other windows, including the rectangular and Hanning windows; in MATLAB the rectangular window is referred to as the *boxcar*. The use of a desired window can be specified with an optional trailing argument. For example **fir1(M,wc,boxcar(M+1))** uses a rectangular window. Note that by default the filter is *scaled* so that the center of the first pass band has a magnitude of exactly one after windowing.

2. **fir2** designs a FIR filter with arbitrary frequency response. The command

   ```
   b=fir2(M,F,K)
   ```

   designs a filter of order **M** with frequency response specified by vectors **F** and **K**. The vector **F** is a vector of frequency points in the range [0, 1], where 1 corresponds to one-half the sampling rate or $\Omega = \pi$. The vector **K** is a vector containing the desired magnitude response at the points specified in **F**. The vectors **F** and **K** must have the same length. As with **fir1**, by default **fir2** uses a Hamming window; other windows can be specified with an optional trailing argument.

   fir1 was used to design the FIR digital filters considered in Examples 8.5 and 8.6. In particular, in Example 8.5 we studied the window method for the design of a lowpass filter of order **M** = 12, using (a) a rectangular (boxcar) window and (b) a Hamming window. We used the following MATLAB commands for designing these filters and evaluating their frequency responses:

   (a) *Rectangular window*

   ```
   ≫ b=fir1(12,0.2,boxcar(13));
   ≫ [H,w]=freqz(b,1,512);
   ≫ db=20*log10(abs(H))
   ≫ plot(w,db)
   ```

   (b) *Hamming window*

   ```
   ≫ b=fir1(12,0.2,hamming(13));
   ≫ [H,w]=freqz(b,1,512);
   ≫ db=20*log10(abs(H));
   ≫ plot(w,db)
   ```

   In Example 8.6, we studied the design of a discrete-time differentiator whose frequency response is defined by

   $$H_d(e^{j\Omega}) = j\Omega e^{-jM\Omega/2}$$

   Here again we examined the use of a rectangular window and a Hamming window as the basis for filter design. The respective MATLAB commands for designing these filters are as follows:

   ```
   ≫ taps=13;  M=taps-1;  %M - filter length
   ≫ n=0:M;  f=n-M/2;
   ≫ a = cos(pi*f) ./ f;    % integration by parts as done
   ≫ b = sin(pi*f) ./ (pi*f.^2);  % in equation 8.59
   ```

```
>> h=a-b;    % impulse response for rectangular windowing
>> k=isnan(h); h(k)=0;  % get rid of NaN
>> [H,w]=freqz(h,1,512,2*pi);
>> hh=hamming(taps)'.*h;        % apply Hamming window
>> [HH,w]=freqz(hh,1,512,2*pi);
```

■ PROCESSING OF SPEECH SIGNALS

Filtering of speech signals was used as an illustration in Section 8.9. The filter considered therein was a FIR lowpass filter designed using the Hanning window. Insight into the effect of filtering was achieved by comparing the spectra of the raw and filtered speech signals. Since the speech data represent a continuous-time signal, the Fourier transform is the appropriate Fourier representation. We shall approximate the Fourier transform by evaluating the discrete-time Fourier series of a finite-duration sampled section of speech using the `fft` command as discussed in Chapter 4. Thus the MATLAB commands for studying the effect of filtering on speech signals are as follows:

```
>> load spk-sam  % load speech data
>> b=fir1(98,31/80,hanning(99));
>> filt_sp=filter(b,1,speech);
>> f=0:8000/(127):8000;  % sampling rate = 16,000 Hz
>> subplot(2,1,1)
>> spect=fft(speech,256);  % speech refers to speech data
>> plot(f,abs(spect(:128))/max(abs(spect(1:128))))
>> xlabel('frequency')
>> subplot(2,1,2)
>> filt_spect=fft(filt_sp,256);
>> plot(f,abs(filt_spect(1:128))/max(abs(filt_spect(1:128))))
```

■ IIR DIGITAL FILTERS

In Example 8.7, we used an analog filter as the basis for the design of an IIR lowpass filter with cutoff frequency Ω_c. It is a simple matter to design such a digital filter, using the Signal Processing Toolbox. For the problem posed in Example 8.8, the requirement is to design an IIR digital lowpass filter with a Butterworth response of order 3.

The MATLAB command

```
[b,a]=butter(N,w)
```

designs a lowpass digital IIR filter of Butterworth response and order `N` and returns the coefficients of the transfer function's numerator and denominator polynomials in vectors `b` and `a`, respectively, of length $N + 1$. The cutoff frequency `w` of the filter must be normalized so that it lies in the interval [0, 1], with 1 corresponding to $\Omega = \pi$.

Thus the commands for designing the IIR digital filter in Example 8.7 and the evaluation of its frequency response are as follows:

```
>> [b,a]=butter(3,0.2);
>> [H,w]=freqz(b,a,512);
>> mag=20*log10(abs(H));
>> plot(w,mag)
>> phi=angle(H);
>> phi=(180/pi)*phi;  % convert from radians to degrees
>> plot(w,phi)
```

▶ **Drill Problem 8.13** In the experiment on double sideband-suppressed carrier modulation described in Section 5.11, we used a Butterworth lowpass digital filter with the following specifications:

Filter order 3
Cutoff frequency 0.125 Hz
Sampling rate 10 Hz

Use the MATLAB command `butter` to design this filter. Plot the frequency response of the filter, demonstrating that it satisfies the specifications of the above-mentioned experiment.

■ EQUALIZATION

In Example 8.9, we considered the design of a FIR digital filter to equalize a channel with frequency response

$$H_c(j\omega) = \frac{1}{1 + (j\omega/\pi)}$$

The desired frequency response of the equalizer is

$$H_d(j\omega) = \begin{cases} \left(1 + \dfrac{j\omega}{\pi}\right)e^{-jM\omega/2}, & -\pi < \omega \leq \pi \\ 0, & \text{otherwise} \end{cases}$$

where M is the length of the equalizer. Following the procedure described in Example 8.9, we note that the equalizer consists of two components connected in parallel: an ideal lowpass filter and a differentiator. Assuming the use of a Hamming window of length $M + 1$, we may build on the MATLAB commands used for Examples 8.1 and 8.6. The corresponding set of commands for designing the equalizer and evaluating its frequency response may thus be formulated as follows:

```
>> % get a response of the channel
>> w=0:pi/(length(hh)-1):pi;
>> den=sqrt(1+(w/pi).^2);
>> Hchan=1./den;

>> % get response of the equalizer
>> taps=13;  M=taps-1;  % M - filter length
>> n=0:M;   f=n-M/2;
>> a = cos(pi*f) ./ f;          % integration by parts as done
>> b = sin(pi*f) ./ (pi*f.^2);  % in equation 8.59
>> h=a-b;                  % impulse response
>> k=isnan(h);   h(k)=0;   % get rid of NaN
>> hh = (hamming(taps)'.*h)/pi;
>> k=fftshift(ifft(ones(13,1))).*hamming(13);
>> [Heq,w]=freqz(hh+k',1,512,2*pi);

>> % get magnitude response of the equalized channel
>> Hcheq=abs(Heq).*Hchan';
>> hold on
```

```
>> plot(w,abs(Heq))
   plot(w,abs(Hchan))
   plot(w,abs(Hcheq))
   hold off
```

8.14 *Summary*

In this chapter we discussed procedures for the design of two important building blocks for systems and processing signals—linear filters and equalizers—and explored them using MATLAB. The purpose of a filter is to separate signals on the basis of their frequency content. The purpose of an equalizer is to compensate for linear distortion produced by signal transmission through a dispersive channel.

Frequency-selective analog filters may be realized using inductors and capacitors. The resulting networks are referred to as passive filters; their design is based on continuous-time ideas. Alternatively, we may use digital filters whose design is based on discrete-time concepts. Digital filters are of two kinds: finite-duration impulse response (FIR) and infinite-duration impulse response (IIR) filters.

FIR digital filters are characterized by finite memory and BIBO stability; they can be designed to have linear phase response. IIR digital filters have infinite memory; they are therefore able to realize a prescribed magnitude response with a shorter filter length than is possible with FIR filters.

For the design of FIR digital filters, we may use the window method, where a window is used to provide a trade-off between transition bandwidth and passband/stopband ripple. For the design of IIR digital filters, we may start with a suitable continuous-time transfer function (e.g., Butterworth or Chebyshev function) and then apply the bilinear transform. Both of these digital filters can be designed directly from the prescribed specifications, using computer-aided procedures. Here, algorithmic computational complexity is traded off for a more efficient design.

Turning finally to the issue of equalization, the method most commonly used in practice involves a FIR digital filter. The central problem here is to evaluate the filter coefficients such that when the equalizer is connected in cascade with a communication channel, say, the combination of the two approximates a distortionless filter.

FURTHER READING

1. The classic texts for the synthesis of passive filters include:
   - ▶ Guillemin, E. A., *Synthesis of Passive Networks* (Wiley, 1957)
   - ▶ Tuttle, D. F. Jr., *Network Synthesis* (Wiley, 1958)
   - ▶ Weinberg, L., *Network Analysis and Synthesis* (McGraw-Hill, 1962)

2. The Hamming window and Hanning window (also referred to as the Hann window) are named after their respective originators: Richard W. Hamming and Julius von Hann. The term "Hanning" window was introduced by Blackman and Tukey in their book:
   - ▶ Blackman, R. B., and J. W. Tukey, *The Measurement of Power Spectra* (Dover Publications, 1958)

   A discussion of the window method for the design of FIR digital filters would be incomplete without mentioning the *Kaiser window*, named after James F. Kaiser. It is defined in terms of an adjustable parameter, denoted by *a*, that controls the trade-off between mainlobe width and sidelobe level. When a goes to zero the Kaiser window becomes simply the rectangular window. For a succinct description of the Kaiser window, see:

▶ Kaiser, J. F., "Nonrecursive digital filter design using the $I_0 - \sinh$ window function," *Selected Papers in Digital Signal Processing, II*, edited by The Digital Signal Processing Committee, IEEE Acoustics, Speech, and Signal Processing Society, pp. 123–126 (IEEE Press, 1975)

3. Digital filters were first described in the following books:
   ▶ Gold, B., and C. M. Rader, *Digital Processing of Signals* (McGraw-Hill, 1969)
   ▶ Kuo, F., and J. F. Kaiser, editors, *System Analysis by Digital Computer* (Wiley, 1966)

4. For an advanced treatment of digital filters, see the following books:
   ▶ Antoniou, A., *Digital Filters: Analysis, Design, and Applications*, Second Edition (McGraw-Hill, 1993)
   ▶ Mitra, S. K., *Digital Signal Processing: A Computer-Based Approach* (McGraw-Hill, 1998)
   ▶ Oppenheim, A. V., and R. W. Schafer, *Discrete-Time Signal Processing*, Second edition (Prentice Hall, 1999)
   ▶ Parks, T. W., and C. S. Burrus, *Digital Filter Design* (Wiley, 1987)
   ▶ Rabiner, L. R., and B. Gold, *Theory and Application of Digital Signal Processing* (Prentice Hall, 1975)

5. For books on speech processing using digital filter techniques, see the books:
   ▶ Rabiner, L. R., and R. W. Schafer, *Digital Processing of Speech Signals* (Prentice Hall, 1978)
   ▶ Deller, J., J. G. Proakis, and J. H. L. Hanson, *Discrete-Time Processing of Speech Signals* (Prentice Hall, 1993)

6. For a discussion of equalization, see the classic book:
   ▶ Lucky, R. W., J. Salz, and E. J. Weldon, Jr., *Principles of Data Communication* (McGraw-Hill, 1968)

   Equalization is also discussed in the following books:
   ▶ Haykin, S., *Communications Systems*, Third Edition (Wiley, 1994)
   ▶ Proakis, J. G., *Digital Communications*, Third Edition (McGraw-Hill, 1995)

| PROBLEMS

8.1 A rectangular pulse of 1-μs duration is transmitted through a lowpass channel. Suggest a small enough value for the cutoff frequency of the channel, such that the pulse is recognizable at the output of the filter.

8.2 Suppose that, for a given signal $x(t)$, the integrated value of the signal over an interval T is required, as shown by

$$y(t) = \int_{t-T}^{t} x(\tau)\, d\tau$$

(a) Show that $y(t)$ can be obtained by transmitting $x(t)$ through a filter with transfer function given by

$$H(j\omega) = T \operatorname{sinc}(\omega T/2\pi) \exp(-j\omega T/2)$$

(b) Assuming the use of an ideal lowpass filter, determine the filter output at time $t = T$ due

to a step function applied to the filter at $t = 0$. Compare the result with the corresponding output of the ideal integrator.

Note: Si(π) = 1.85 and Si(∞) = $\pi/2$.

8.3 Derive Eqs. (8.31) and (8.32), defining the passband and stopband cutoff frequencies of a Butterworth filter of order N.

8.4 Consider a Butterworth lowpass filter of order $N = 5$ and cutoff frequency $\omega_c = 1$.

(a) Find the $2N$ poles of $H(s)H(-s)$.

(b) Determine $H(s)$.

8.5 Show that for a Butterworth lowpass filter, the following properties are satisfied:

(a) The transfer function $H(s)$ has a pole at $s = -\omega_c$ for N odd.

(b) All the poles of the transfer function $H(s)$ appear in complex-conjugate form for N even.

8.6 The denominator polynomial of the transfer function of a Butterworth lowpass prototype filter of order $N = 5$ is defined by

$$(s + 1)(s^2 + 0.618s + 1)(s^2 + 1.618s + 1)$$

Find the transfer function of the corresponding highpass filter with cutoff frequency $\omega_c = 1$. Plot the magnitude response of the filter.

8.7 Consider again the lowpass prototype filter of order $N = 5$ described in Problem 8.6. The requirement is to modify the cutoff frequency of the filter to some arbitrary value ω_c. Find the transfer function of this filter.

8.8 For the lowpass transfer function $H(s)$ specified in Example 8.3, find the transfer function of the corresponding bandpass filter with midband frequency $\omega_0 = 1$ and bandwidth $B = 0.1$. Plot the magnitude response of the filter.

8.9 A lowpass prototype filter is to be transformed into a bandstop filter with midband rejection frequency ω_0. Suggest a suitable frequency transformation for doing this transformation.

8.10 A FIR digital filter has a total of $M + 1$ coefficients, where M is an even integer. The impulse response of the filter is symmetric with respect to the $(M/2)$th point:

$$h[n] = h[M - n], \qquad 0 \le n \le M$$

(a) Find the magnitude response of the filter.

(b) Show that this filter has a linear phase response.

Hence show that there are no restrictions on the frequency response $H(e^{j\Omega})$ at both $\Omega = 0$ and $\Omega = \pi$. This filter is labeled type I.

8.11 Suppose that in Problem 8.10 the $M + 1$ coefficients of the FIR digital filter satisfy the antisymmetry condition with respect to the $(M/2)$th point:

$$h[n] = -h[M - n], \qquad 0 \le n \le \frac{M}{2} - 1$$

In this case, show that the frequency response $H(e^{j\Omega})$ of the filter must satisfy the following two conditions: $H(e^{j0}) = 0$ and $H(e^{j\pi}) = 0$. Also show that the filter has a linear phase response. This filter is labeled type III.

8.12 In Problems 8.10 and 8.11, the filter length M is an even integer. In this problem and the next one, the filter length M is an odd integer. Suppose the impulse response $h[n]$ of the filter is

symmetric about the noninteger point $n = M/2$. Let

$$b[k] = 2h[(M + 1)/2 - k],$$
$$k = 1, 2, \ldots, (M + 1)/2$$

Find the frequency response $H(e^{j\Omega})$ of the filter in terms of $b[k]$. Hence show that:

(a) The phase response of the filter is linear.

(b) There is no restriction on $H(e^{j0})$, but $H(e^{j\pi}) = 0$.

The filter considered in this problem is labeled type II.

8.13 Continuing with Problem 8.12 involving a FIR digital filter of length M that is an odd integer, suppose that the impulse response $h[n]$ of the filter is antisymmetric about the noninteger point $n = M/2$. Let

$$c[k] = 2h[(M + 1)/2 - k],$$
$$k = 1, 2, \ldots, (M + 1)/2$$

Find the frequency response $H(e^{j\Omega})$ of the filter in terms of $c[k]$. Hence show that:

(a) The phase response of the filter is linear.

(b) $H(e^{j0}) = 0$, but there is no restriction on $H(e^{j\pi})$.

The filter considered in this problem is labeled type IV.

8.14 A FIR digital filter is required to have a zero at $z = 1$ for its transfer function $H(z)$. Find the condition that the impulse response $h[n]$ of the filter must satisfy for this requirement to be realized.

8.15 In Section 8.9 we presented one procedure for deriving a FIR digital filter using the window method. In this problem, we want to proceed in two steps, as described here:

(a) Define

$$h[n] = \begin{cases} h_d[n], & -M/2 \le n \le M/2 \\ 0, & \text{otherwise} \end{cases}$$

where $h_d[n]$ is the desired impulse response corresponding to a frequency response with zero phase. The phase response of $h[n]$ is also zero.

(b) Having determined $h[n]$, shift it to the right by $M/2$ samples. This second step makes the filter causal.

Show that this procedure is equivalent to that described in Section 8.9.

8.16 Equation (8.59) defines the impulse response $h_d[n]$ of a FIR digital filter used as a differentiator with a rectangular window. Show that $h_d[n]$ is antisymmetric, that is,

$$h[n - M] = -h[n], \qquad 0 \le n \le M/2 - 1$$

where M is the length of the filter, assumed to be even. In light of Problem 8.11, what can you say about the frequency response of this particular differentiator? Check your answer against the magnitude responses shown in Fig. 8.20.

8.17 Equations (8.64) and (8.65) pertain to the bilinear transform

$$s = \frac{z - 1}{z + 1}$$

How are these equations modified for

$$s = \frac{2}{\mathcal{T}} \frac{z - 1}{z + 1}$$

where $\mathcal{T}$ is the sampling interval?

8.18 It is possible for a digital IIR filter to be unstable. How can such a condition arise? Assuming the bilinear transform is used, where would some of the poles of the corresponding analog transfer function have to lie for this to happen?

8.19 Figure 8.26 shows a cascade realization of the digital IIR filter specified in Eq. (8.71). Formulate a direct form II for realizing this transfer function.

8.20 In Section 8.10 we described the method of bilinear transform for the design of IIR digital filters. Here, we consider another method called the *method of impulse invariance* for digital filter design. In this procedure for transforming a continuous-time (analog) filter into a discrete-time (digital) filter, the impulse response $h[n]$ of the discrete-time filter is chosen as equally spaced samples of the continuous-time filter's impulse response $h_a(t)$; that is,

$$h[n] = \mathcal{T} h_a(n\mathcal{T})$$

where $\mathcal{T}$ is the sampling interval. Let

$$H_a(s) = \sum_{k=1}^{N} \frac{A_k}{s - d_k}$$

denote the transfer function of the continuous-time filter. Show that the transfer function of the corresponding discrete-time filter obtained using the method of impulse invariance is given by

$$H(z) = \sum_{k=1}^{N} \frac{\mathcal{T} A_k}{1 - e^{d_k \mathcal{T}} z^{-1}}$$

8.21 Equation (8.73) defines the frequency response of an equalizer for dealing with linear distortion

produced by a continuous-time LTI system. Formulate the corresponding relation for an equalizer used to deal with linear distortion produced by a discrete-time LTI system.

8.22 Consider a tapped-delay-line equalizer whose frequency response is specified in Eq. (8.84). In theory, this equalizer can compensate for any linear distortion simply by making the number of coefficients, $M + 1$, in the equalizer large enough. What is the penalty for making M large? Justify your answer.

8.23 Some radio systems suffer from multipath distortion, which is caused by the existence of more than one propagation path between the transmitter and receiver. Consider a channel governed by the input–output relation

$$y(t) = K_1 x(t - t_{01}) + K_2 x(t - t_{02})$$

where $x(t)$ is the input, $y(t)$ is the output, K_1 and K_2 are constants, and t_{01} and t_{02} represent transmission delays. It is proposed to use the FIR digital filter of Fig. 8.19 to equalize the multipath distortion produced by this channel.

(a) Evaluate the transfer function of the channel.

(b) Evaluate the coefficients of the equalizer assuming that $K_2 \ll K_1$ and $t_{02} > t_{01}$.

▶ **Computer Experiments**

8.24 Design a FIR digital lowpass filter with a total of 23 coefficients. Use a Hamming window for the design. The cutoff frequency of the filter is $\omega_c = \pi/3$.

(a) Plot the impulse response of the filter.

(b) Plot the magnitude response of the filter.

8.25 Design a differentiator using a FIR digital filter of order $M = 100$. For this design, use (a) a rectangular window and (b) a Hamming window. In each case, plot the impulse response and magnitude response of the filter.

8.26 You are given a data sequence with sampling rate of $2\pi \times 8000$ rad/s. A lowpass digital IIR filter is required for processing this data sequence to meet the following specifications:

Cutoff frequency $\omega_c = 2\pi \times 800$ rad/s

Attenuation at $2\pi \times 1200$ rad/s = 15 dB

(a) Assuming a Butterworth response, determine a suitable value for filter order N.

(b) Using the bilinear transform, design the filter.

(c) Plot the magnitude response and phase response of the filter.

8.27 Design a highpass digital IIR filter with Butterworth response. The filter specifications are as follows: filter order $N = 5$ and cutoff frequency $\omega_c = 0.6$.

8.28 Consider a channel whose frequency response is described by the second-order Butterworth response

$$H_c(j\omega) = \frac{1}{(1 + j\omega/\pi)^2}$$

Design a FIR filter with 95 coefficients for equalizing this channel over the frequency band $-\pi < \omega \le \pi$.

9 Application to Feedback Systems

9.1 *Introduction*

Feedback is defined as the *returning of a fraction of the output signal of a system to its input*. It can arise in a system in two ways.

- ▶ Feedback is *naturally* present in the system, in that it is an integral part of the underlying physical mechanism responsible for the operation of the system.
- ▶ Feedback is applied *externally* to the system with a specific purpose in mind.

It is the latter form of feedback that we shall be concerned with here.

The application of feedback is motivated by several desirable properties:

- ▶ Reduction in the sensitivity of the gain of a system to parameter variations due to changes in supply voltage, environmental conditions, and so on
- ▶ Reduction in the effect of external disturbances (e.g., noise) on performance of a system
- ▶ Improvement in the linear behavior of a system by reducing distortion due to nonlinear effects in the system

These benefits are of profound engineering importance. However, they are achieved at the cost of a more complicated system behavior. There is also the potential possibility that a feedback system may become unstable, unless special precautions are taken in the design of the system.

We begin the study of linear feedback systems in this chapter by describing some basic feedback concepts, followed by discussions of sensitivity analysis, the effect of feedback on external disturbances, and distortion analysis. This introductory material on linear feedback theory provides the motivation for two important applications: operational amplifiers and feedback control systems, which are discussed in that order. The next topic discussed is the stability problem, which features prominently in the study of feedback systems. Two approaches are taken here: one approach is based on the natural response of a feedback system, and the other is based on the frequency response of the system. One other important topic covered toward the end of the chapter is that of a sampled-data system, which is a feedback control system that uses a computer for control. The study of sampled-data systems is important not only in an engineering context but also a theoretical one: it combines the use of the z-transform and Laplace transform under one umbrella.

9.2 *Basic Feedback Concepts*

Figure 9.1(a) shows the block diagram of a feedback system in its most basic form. The system consists of three components connected together to form a single feedback loop:

- ▶ A *plant*, which acts on an *error signal e(t)* to produce the output signal $y(t)$
- ▶ A *sensor*, which measures the output signal $y(t)$ to produce a *feedback signal r(t)*
- ▶ A *comparator*, which calculates the difference between the externally applied input (reference) signal $x(t)$ and the feedback signal $r(t)$ to produce the error signal $e(t)$ as shown by

$$e(t) = x(t) - r(t) \tag{9.1}$$

The terminology used here pertains more closely to a control system but can readily be adapted to deal with a feedback amplifier.

In what follows, we assume that the plant dynamics and sensor dynamics in Fig. 9.1(a) are both modeled as LTI systems. Given the time-domain descriptions of both of them, we may proceed to relate the output signal $y(t)$ to the input signal $x(t)$. However, we find it more convenient to work with Laplace transforms and do the formulation in the complex s-domain, as described in Fig. 9.1(b). Let $X(s)$, $Y(s)$, $R(s)$, and $E(s)$ denote the Laplace transforms of $x(t)$, $y(t)$, $r(t)$, and $e(t)$, respectively. We may then transform Eq. (9.1) into the equivalent form

$$E(s) = X(s) - R(s) \tag{9.2}$$

Let $G(s)$ denote the transfer function of the plant and $H(s)$ denote the transfer function of the sensor. Then by definition, we may write

$$G(s) = \frac{Y(s)}{E(s)} \tag{9.3}$$

and

$$H(s) = \frac{R(s)}{Y(s)} \tag{9.4}$$

Using Eq. (9.3) to eliminate $E(s)$ and Eq. (9.4) to eliminate $R(s)$ from Eq. (9.2), we get

$$\frac{Y(s)}{G(s)} = X(s) - H(s)Y(s)$$

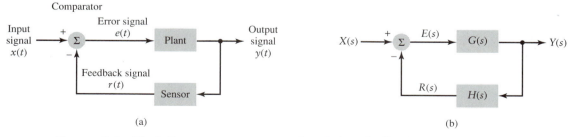

(a)   (b)

FIGURE 9.1   Block diagram representations of single-loop feedback system: (a) time-domain representation and (b) complex s-domain representation.

Collecting terms and solving for the ratio $Y(s)/X(s)$, we find that the *closed-loop transfer function* of the feedback system in Fig. 9.1(b) is

$$T(s) = \frac{Y(s)}{X(s)}$$

$$= \frac{G(s)}{1 + G(s)H(s)} \tag{9.5}$$

The term "closed-loop" is used here to emphasize the fact that there is a closed signal-transmission loop, around which signals may flow in the system.

The quantity $1 + G(s)H(s)$ in the denominator of Eq. (9.5) provides a measure of the feedback acting around $G(s)$. For a physical interpretation of this quantity, consider the configuration of Fig. 9.2. In this figure we have made two changes to the feedback system of Fig. 9.1(b):

▶ The input signal $x(t)$ is reduced to zero.
▶ The feedback loop around $G(s)$ is opened.

Suppose that a test signal with unit Laplace transform is applied to $G(s)$ (i.e., the plant) in Fig. 9.2. The signal returned to the other open end of the loop is $-G(s)H(s)$. The difference between the unit test signal and the returned signal is equal to $1 + G(s)H(s)$. Accordingly, this quantity is called the *return difference*. Denoting it by $F(s)$, we may thus write

$$F(s) = 1 + G(s)H(s) \tag{9.6}$$

The product term $G(s)H(s)$ is called the *loop transfer function* of the system. It is simply the transfer function of the plant and the sensor connected in cascade, as shown in Fig. 9.3. This latter configuration is the same as that of Fig. 9.2 with the comparator removed. Denoting the loop transfer function by $L(s)$, we may thus write

$$L(s) = G(s)H(s) \tag{9.7}$$

and so relate the return difference $F(s)$ to $L(s)$ as follows:

$$F(s) = 1 + L(s) \tag{9.8}$$

In the sequel we use $G(s)H(s)$ and $L(s)$ interchangeably, when referring to the loop transfer function.

■ NEGATIVE AND POSITIVE FEEDBACK

Consider an operating range of frequencies for which G and H, pertaining respectively to the plant and sensor, may be treated as essentially independent of the complex frequency s. In such a situation, the feedback in Fig. 9.1(a) is said to be *negative*. When the comparator in Fig. 9.1(a) is replaced with an adder, the feedback is said to be *positive*.

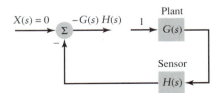

FIGURE 9.2   Complex s-domain representation of a scheme for measuring the return difference $F(s)$.

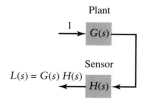

Plant

$$\frac{1}{\longrightarrow} \boxed{G(s)}$$

Sensor

$$L(s) = G(s)\,H(s)$$

$$\longleftarrow \boxed{H(s)} \longleftarrow$$

Figure 9.3   Complex s-domain representation of a scheme for measuring the loop transfer function $L(s)$.

These terms, however, are of limited value. We say this because in the general setting depicted in Fig. 9.1(b) the loop transfer function $G(s)H(s)$ is dependent on the complex frequency s. For $s = j\omega$, we find that $G(j\omega)H(j\omega)$ has a phase that varies with the frequency ω. When the phase of $G(j\omega)H(j\omega)$ is zero, the situation in Fig. 9.1(b) corresponds to negative feedback. When the phase of $G(j\omega)H(j\omega)$ is 180°, the same configuration behaves like a positive feedback system. Thus for the single-loop feedback system of Fig. 9.1(b), there will be different frequency bands for which the feedback is alternately negative and positive. Care must therefore be exercised in the use of the terms negative feedback and positive feedback.

9.3 *Sensitivity Analysis*

A primary motivation for the use of feedback is to reduce the sensitivity of the closed-loop transfer function of the system in Fig. 9.1 to changes in the transfer function of the plant. For the purpose of this discussion, we ignore the dependencies on the complex frequency s in Eq. (9.5) and treat G and H as "scalar" parameters. We may thus write

$$T = \frac{G}{1 + GH} \tag{9.9}$$

In Eq. (9.9), we refer to G as the *gain* of the plant and to T as the *closed-loop gain* of the feedback system.

Suppose now the gain G is changed by a small amount ΔG. Then, differentiating Eq. (9.9) with respect to G, we find that the corresponding change in T is

$$\begin{aligned}
\Delta T &= \frac{\partial T}{\partial G}\,\Delta G \\
&= \frac{1}{(1 + GH)^2}\,\Delta G
\end{aligned} \tag{9.10}$$

The *sensitivity* of T with respect to changes in G is formally defined by

$$S_G^T = \frac{\Delta T/T}{\Delta G/G} \tag{9.11}$$

In words, the sensitivity of T with respect to G is the percentage change in T divided by the percentage change in G. Using Eqs. (9.5) and (9.10) in (9.11) yields

$$\begin{aligned}
S_G^T &= \frac{1}{1 + GH} \\
&= \frac{1}{F}
\end{aligned} \tag{9.12}$$

which shows that the sensitivity of T with respect to G is equal to the reciprocal of the return difference F.

With the availability of two degrees of freedom represented by the parameters G and H pertaining to the plant and sensor, respectively, the use of feedback permits a system designer to simultaneously realize prescribed values for the closed-loop gain T and sensitivity S_G^T. This is achieved through the use of Eqs. (9.9) and (9.12), respectively.

▶ **Drill Problem 9.1**   To make the sensitivity S_G^T, small compared to unity, the loop gain GH must be large compared to unity. What are the approximate values of the closed-loop gain T and sensitivity S_G^T under this condition?

Answer:

$$T \simeq \frac{1}{H} \quad \text{and} \quad S_G^T \simeq \frac{1}{GH}$$

◀

EXAMPLE 9.1   Consider a single-loop feedback amplifier, the block diagram of which is shown in Fig. 9.4. It consists of a linear amplifier with gain A and a passive feedback network; the latter component feeds a controllable fraction β of the output signal back to the input. We are given that the gain $A = 1000$.

(a) Determine the value of β that will result in a closed-loop gain $T = 10$.

(b) Suppose that the gain A changes by 10%. What is the corresponding percentage change in the closed-loop gain T?

Solution

(a) For the problem at hand, the plant and sensor are represented by the amplifier and feedback network, respectively. We may thus put $G = A$ and $H = \beta$ and so rewrite Eq. (9.9) in the form

$$T = \frac{A}{1 + \beta A}$$

Solving for β, we get

$$\beta = \frac{1}{A}\left(\frac{A}{T} - 1\right)$$

With $A = 1000$ and $T = 10$, we therefore get

$$\beta = \frac{1}{1000}\left(\frac{1000}{10} - 1\right) = 0.099$$

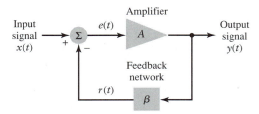

FIGURE 9.4   Block diagram of single-loop feedback amplifier.

(b) From Eq. (9.12), the sensitivity of the closed-loop gain T with respect to A is

$$S_A^T = \frac{1}{1 + \beta A}$$

$$= \frac{1}{1 + 0.099 \times 1000} = \frac{1}{100}$$

Hence, with a 10% change in A, the corresponding percentage change in T is

$$\Delta T = S_A^T \frac{\Delta A}{A}$$

$$= \frac{1}{100} \times 10\% = 0.1\%$$

which indicates that, for this example, the feedback amplifier of Fig. 9.4 is relatively insensitive to variations in the gain A of the internal amplifier.

9.4 *Effect of Feedback on Disturbance or Noise*

The use of feedback has another beneficial effect on system performance: it reduces the effect of a disturbance or noise generated inside the feedback loop.

Consider the single-loop feedback system depicted in Fig. 9.5. It differs from the basic configuration of Fig. 9.1 in two respects: G and H are both treated as scalar parameters, and the system includes a disturbance denoted by ν inside the loop. Since the system is linear, we may use the principle of superposition to calculate the effects of the externally applied input signal x and the disturbance ν separately, and then add the results. In light of Eq. (9.9), the output signal resulting from x acting alone is

$$y|_{\nu=0} = \frac{G}{1 + GH} x$$

To evaluate the output signal resulting from ν acting alone, we readily find from Fig. 9.5 that

$$y|_{x=0} = \frac{1}{1 + GH} \nu$$

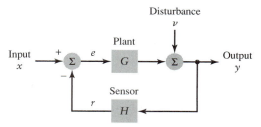

FIGURE 9.5 Block diagram of a single-loop feedback system including a disturbance inside the loop.

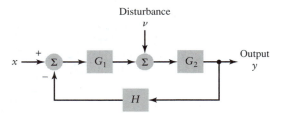

FIGURE 9.6   Feedback system for Drill Problem 9.2.

Adding these two contributions, the output in Fig. 9.5 due to the combined action of x and ν is obtained to be

$$y = \frac{G}{1 + GH} x + \frac{1}{1 + GH} \nu \qquad (9.13)$$

where the first term represents the desired output and the second term represents the unwanted output. Equation (9.13) clearly shows that the use of feedback in Fig. 9.5 has the effect of reducing the disturbance ν by the factor $1 + GH$ (i.e., the return difference F).

▶ **Drill Problem 9.2**   Consider the system configuration of Fig. 9.6. Determine the effect of disturbance ν on the output of this system.

Answer:

$$\frac{G_2}{1 + G_1 G_2 H} \nu \qquad \blacktriangleleft$$

9.5 *Distortion Analysis*

Nonlinearity arises in a physical system whenever it is driven outside its linear range of operation. We may improve the linearity of such a system by applying feedback around it. To investigate this important effect, we may proceed in one of two ways:

▸ The output of the system is expressed as a nonlinear function of the input, and a pure sine wave is used as the input signal.

▸ The input to the system is expressed as a nonlinear function of the output.

The latter approach may seem strange at first sight; however, it is more general in formulation and provides a more intuitively satisfying description of how feedback affects the nonlinear behavior of a system. It is therefore the approach that is pursued in the sequel.

Consider then a feedback system in which the input–output relationship of the plant is represented by

$$e = a_1 y + a_2 y^2 \qquad (9.14)$$

where the error e is shown to be dependent on the output y, and a_1 and a_2 are constants. The linear term $a_1 y$ in Eq. (9.14) represents the desired behavior of the plant, and the parabolic term $a_2 y^2$ accounts for its deviation from linearity. Let the parameter H determine the fraction of the plant output y that is fed back to the input. With x denoting the input applied to the feedback system, we may thus write

$$e = x - Hy \qquad (9.15)$$

Eliminating e between Eqs. (9.14) and (9.15), and rearranging terms, we get

$$x = (a_1 + H)y + a_2 y^2$$

Differentiating x with respect to y yields

$$\frac{dx}{dy} = a_1 + H + 2a_2 y$$
$$= (a_1 + H)\left(1 + \frac{2a_2}{a_1 + H}\, y\right) \qquad (9.16)$$

which holds in the presence of feedback.

In the absence of feedback, the plant operates by itself, as shown by

$$x = a_1 y + a_2 y^2 \qquad (9.17)$$

which is a rewrite of Eq. (9.14) with the input x used in place of the error e. Differentiating Eq. (9.17) with respect to y yields

$$\frac{dx}{dy} = a_1 + 2a_2 y$$
$$= a_1\left(1 + \frac{2a_2}{a_1}\, y\right) \qquad (9.18)$$

The derivatives in Eqs. (9.16) and (9.18) have both been *normalized* to their respective linear terms to make for a fair comparison between them. In the presence of feedback, the term $2a_2 y/(a_1 + H)$ in Eq. (9.16) provides a measure of *distortion* due to the parabolic term $a_2 y^2$ in the input–output relationship of the plant. The corresponding measure of distortion in the absence of feedback is represented by the term $2a_2 y/a_1$ in Eq. (9.18). Accordingly, the application of feedback has reduced distortion, due to deviation of the plant from linearity, by the following factor:

$$D = \frac{2a_2 y/(a_1 + H)}{2a_2 y/a_1} = \frac{a_1}{a_1 + H}$$

From Eq. (9.17), we readily see that the coefficient a_1 is the reciprocal of the gain G of the plant. Hence we may rewrite this result as

$$D = \frac{1/G}{(1/G) + H} = \frac{1}{1 + GH} = \frac{1}{F} \qquad (9.19)$$

which shows that the distortion is reduced by an amount equal to the return difference F.

From the analysis presented in this section and that in the preceding two sections, we now see that the return difference F plays a central role in the study of feedback systems in three important respects:

1. The control of sensitivity
2. The control of the effect of an internal disturbance
3. The control of distortion in a nonlinear system

In case 1, feedback reduces the sensitivity by an amount equal to F. In case 2, the transmission of a disturbance from some point inside the feedback loop to the output is also reduced by an amount equal to F. Finally, in case 3, distortion due to nonlinear effects is again reduced by an amount equal to F.

▶ **Drill Problem 9.3** Suppose the nonlinear input–output relation of Eq. (9.14) is expanded to include a cubic term, as shown by

$$e = a_1 y + a_2 y^2 + a_3 y^3$$

Show that the application of feedback also reduces the effect of distortion due to this cubic term by a factor equal to the return difference F.   ◀

9.6 Cost of Feedback

In the previous three sections we discussed the benefits that are gained by the application of feedback to a control system. It is to be expected that there is an attendant *cost* to these benefits as described here.

1. *Increased complexity.* Naturally, the application of feedback to a control system requires the addition of new components. Thus there is the cost of increased system complexity.

2. *Reduced gain.* In the absence of feedback, the transfer function of a plant is $G(s)$. When feedback is applied to the plant, the transfer function of the system is modified to $G(s)/F(s)$, where $F(s)$ is the return difference. Now the benefits of feedback are realized only when $F(s)$ is greater than 1. It follows therefore that the application of feedback results in reduced gain.

3. *Possible instability.* Often, an open-loop system (i.e., the plant operating on its own) is stable. However, when feedback is applied to the system, there is a real possibility that the closed-loop system may become unstable. To guard against such a possibility, we have to take precautionary measures in the design of the feedback control system.

In general, the advantages of feedback outweigh the disadvantages. It is therefore necessary that we consider the increased complexity in designing a control system and pay particular attention to the stability problem.

9.7 Operational Amplifiers

An important application of feedback is in operational amplifiers. An *operational amplifier*, or an *op amp* as it is often referred to, provides the basis for realizing a transfer function with prescribed poles and zeros in a relatively straightforward manner. Ordinarily, it has two input terminals, one inverting and the other noninverting, and an output terminal. Figure 9.7(a) shows the conventional symbol used to represent an operational amplifier; only the principal signal terminals are included in this symbol.

The *ideal model* for an operational amplifier may be described as follows (see Fig. 9.7(b)):

1. It is a voltage-controlled voltage source, described by the input–output relation

$$v_o = A(v_2 - v_1) \tag{9.20}$$

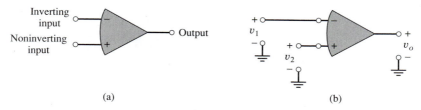

(a)   (b)

FIGURE 9.7   (a) Conventional symbol for operational amplifier. (b) Operational amplifier with input and output voltages.

where v_1 and v_2 are the signals applied to the inverting and noninverting input terminals, respectively, and v_o is the output signal. All these signals are measured in volts.

2. The open-loop voltage gain A has a constant value that is very large compared to unity, which means that for a finite output signal v_o we must have $v_1 = v_2$. This property is referred to as *virtual ground*.

3. The impedance between the two input terminals is infinitely large, and so is the impedance between each one of them and the ground, which means that the input terminal currents are zero.

4. The output impedance is zero.

An operational amplifier is not used in an open-loop fashion. Rather, it is normally used as the amplifier component of a feedback circuit, in which the feedback controls the closed-loop transfer function of the circuit. Figure 9.8 shows one such circuit, where the noninverting input terminal of the operational amplifier is grounded, and the impedances $Z_1(s)$ and $Z_2(s)$ represent the input element and feedback element of the circuit, respectively. Let $V_{in}(s)$ and $V_{out}(s)$ denote the Laplace transforms of the input and output voltage signals, respectively. Then, using the ideal model to describe the operational amplifier, we may in a corresponding way construct the model shown in Fig. 9.9 for the feedback circuit of Fig. 9.8. The following condition readily follows from properties 2 and 3 of the operational amplifier:

$$\frac{V_{in}(s)}{Z_1(s)} \simeq -\frac{V_{out}(s)}{Z_2(s)}$$

The closed-loop transfer function of the feedback circuit in Fig. 9.8 is therefore

$$\begin{aligned} T(s) &= \frac{V_{out}(s)}{V_{in}(s)} \\ &\simeq -\frac{Z_2(s)}{Z_1(s)} \end{aligned} \qquad (9.21)$$

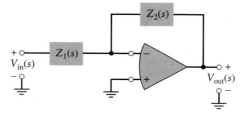

FIGURE 9.8   Operational amplifier embedded in a single-loop feedback circuit.

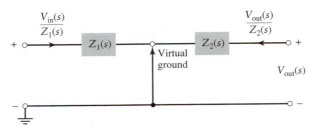

FIGURE 9.9   Ideal model for the feedback circuit of Fig. 9.8.

We derived this result without recourse to the feedback theory developed in Section 9.2. How then do we interpret the result in light of the general feedback formula of Eq. (9.5)? To answer this question, we have to understand the way in which feedback manifests itself in the operational amplifier circuit of Fig. 9.8. The feedback element $Z_2(s)$ is connected in parallel to the amplifier at both its input and output ports. This would therefore suggest the use of currents as the basis for representing the input signal $x(t)$ and feedback signal $r(t)$. The application of feedback in the system of Fig. 9.8 has the effect of making the input impedance measured looking into the operational amplifier small compared to both $Z_1(s)$ and $Z_2(s)$, but nevertheless of some finite value. Let $Z_{in}(s)$ denote this input impedance. We may then use current signals to represent the Laplace transforms of $x(t)$ and $r(t)$ in terms of the Laplace transforms of $v_{in}(t)$ and $v_{out}(t)$, respectively, as follows:

$$X(s) = \frac{V_{in}(s)}{Z_1(s)} \tag{9.22}$$

and

$$R(s) = -\frac{V_{out}(s)}{Z_2(s)} \tag{9.23}$$

The error signal $e(t)$, defined as the difference between $x(t)$ and $r(t)$, is applied across the input terminals of the operational amplifier to produce an output voltage equal to $v_{out}(t)$. With $e(t)$ viewed as a current signal, we may express the Laplace transform of the output signal $y(t)$ as follows:

$$\begin{aligned} Y(s) &= V_{out}(s) \\ &= -AZ_{in}(s)E(s) \end{aligned} \tag{9.24}$$

By definition (see Eq. (9.3)), the transfer function of the operational amplifier (viewed as the plant) is

$$G(s) = \frac{Y(s)}{E(s)}$$

For the problem at hand, we therefore deduce from Eq. (9.24) that

$$G(s) = -AZ_{in}(s) \tag{9.25}$$

Also, by definition (see Eq. (9.4)), the transfer function of the feedback path comprising Z_2 (viewed as the sensor) is

$$H(s) = \frac{R(s)}{Y(s)}$$

Hence, with $V_{\text{out}}(s) = Y(s)$, we deduce from Eq. (9.23) that

$$H(s) = -\frac{1}{Z_2(s)} \qquad (9.26)$$

Using Eqs. (9.22) and (9.24), we may now reformulate the feedback circuit of Fig. 9.8 as depicted in Fig. 9.10, where $G(s)$ and $H(s)$ are defined by Eqs. (9.25) and (9.26), respectively. Figure 9.10 is configured in the same way as the basic feedback system shown in Fig. 9.1(b).

From Fig. 9.10 we readily find that

$$\frac{Y(s)}{X(s)} = \frac{G(s)}{1 + G(s)H(s)}$$

$$= \frac{-AZ_{\text{in}}(s)}{1 + \dfrac{AZ_{\text{in}}(s)}{Z_2(s)}}$$

In light of Eq. (9.22) and the first line of Eq. (9.24), we may rewrite this result in the equivalent form:

$$\frac{V_{\text{out}}(s)}{V_{\text{in}}(s)} = \frac{-AZ_{\text{in}}(s)}{Z_1(s)\left[1 + \dfrac{AZ_{\text{in}}(s)}{Z_2(s)}\right]} \qquad (9.27)$$

Since for an operational amplifier the gain A is very large compared to unity, we may approximate Eq. (9.27) as

$$\frac{V_{\text{out}}(s)}{V_{\text{in}}(s)} \simeq -\frac{Z_2(s)}{Z_1(s)}$$

which is the result that we derived earlier.

EXAMPLE 9.2 Consider the operational amplifier circuit of Fig. 9.11, in which the input element consists of a resistor R_1 and the feedback element consists of a capacitor C_2. Show that this circuit operates as an integrator.

Solution: The impedances $Z_1(s)$ and $Z_2(s)$ are

$$Z_1(s) = R_1$$

and

$$Z_2(s) = \frac{1}{sC_2}$$

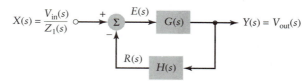

FIGURE 9.10 Reformulation of the feedback circuit of Fig. 9.8 so as to correspond to the basic feedback system of Fig. 9.1(b).

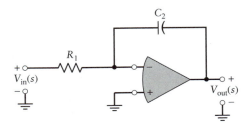

FIGURE 9.11   Operational amplifier circuit used as integrator.

Thus, substituting these values into Eq. (9.21), we get

$$T(s) = -\frac{1}{sC_2R_1}$$

which shows that the closed-loop transfer function of Fig. 9.11 has a pole at the origin. Since division by the complex variable s corresponds to integration in time, we conclude that this circuit performs integration on the input signal.

▶ **Drill Problem 9.4**   The circuit elements of the integrator in Fig. 9.11 have the following values: $R_1 = 100$ kΩ and $C_2 = \mu$F. The initial value of the output voltage is $v_{out}(0)$. Determine the output voltage $v_{out}(t)$ for varying time t.

Answer: $v_{out}(t) = -\int_0^t 10v_{in}(\tau)\,d\tau + v_{out}(0)$, where time t is measured in seconds.   ◀

EXAMPLE 9.3   Consider the operational amplifier circuit of Fig. 9.12. Determine the closed-loop transfer function of this circuit.

Solution: The input element consists of the parallel combination of resistor R_1 and capacitor C_1; hence

$$Z_1(s) = \frac{R_1}{1 + sC_1R_1}$$

The feedback element consists of the parallel combination of resistor R_2 and capacitor C_2; hence

$$Z_2(s) = \frac{R_2}{1 + sC_2R_2}$$

Substituting these values into Eq. (9.21) yields the closed-loop transfer function

$$T(s) = -\frac{R_2}{R_1}\frac{1 + sC_1R_1}{1 + sC_2R_2}$$

which has a zero at $s = -1/C_1R_1$ and a pole at $s = -1/C_2R_2$.

Note that by cascading different versions of the basic circuit of Fig. 9.12, it is possible to synthesize an overall transfer function with arbitrary real poles and arbitrary real zeros. Indeed, with more elaborate forms of $Z_1(s)$ and $Z_2(s)$, we can realize a transfer function

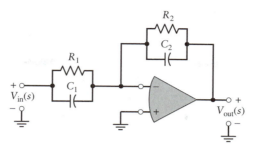

FIGURE 9.12 Operational amplifier circuit for Example 9.3.

with arbitrary complex poles and zeros. Filters synthesized using operational amplifiers are referred to as *active filters*.

9.8 Control Systems

Consider a *plant* that is controllable. The function of a control system is to *obtain accurate control over the plant, such that the output of the plant remains close to a target (desired) response.* This is accomplished through proper modification of the plant input. We may identify two basic types of control systems:

▶ *Open-loop control*, in which the modification to the plant input is derived directly from the target response
▶ *Closed-loop control*, in which feedback is used around the plant

In both cases, the target response acts as the input to the control system. In the sequel, these two types of control are considered in that order.

■ OPEN-LOOP CONTROL

Figure 9.13 shows the block diagram of an open-loop control system. The plant dynamics are represented by the transfer function $G(s)$. The controller, represented by the transfer function $H(s)$, acts on the target response $y_d(t)$ to produce the desired control signal $c(t)$. The disturbance $\nu(t)$ is included to account for noise and distortion produced at the output of the plant. The dashed part of the figure has nothing to do with the operation of the system; it is included here merely to depict the error $e(t)$ as the difference between the target response $y_d(t)$ and the actual output $y(t)$ of the system, as shown by

$$e(t) = y_d(t) - y(t) \tag{9.28}$$

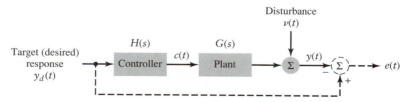

FIGURE 9.13 Block diagram of open-loop control system; the dashed line from the input to the output is included merely to illustrate calculation of the error signal $e(t)$.

Let $Y_d(s)$, $Y(s)$, and $E(s)$ denote the Laplace transforms of $y_d(t)$, $y(t)$, and $e(t)$, respectively. We may therefore rewrite Eq. (9.28) in the complex s-domain as follows:

$$E(s) = Y_d(s) - Y(s) \tag{9.29}$$

From Fig. 9.13 we also readily find that

$$Y(s) = G(s)H(s)Y_d(s) + N(s) \tag{9.30}$$

where $N(s)$ is the Laplace transform of the disturbance $v(t)$. Eliminating $Y(s)$ between Eqs. (9.29) and (9.30) yields

$$E(s) = [1 - G(s)H(s)]Y_d(s) - N(s) \tag{9.31}$$

The error $e(t)$ is minimized by setting

$$1 - G(s)H(s) = 0$$

For this condition to be satisfied, the controller must act as the inverse of the plant as shown by

$$H(s) = \frac{1}{G(s)} \tag{9.32}$$

From Fig. 9.13 we readily see that with $y_d(t) = 0$, the plant output $y(t)$ is equal to $v(t)$. Therefore the best that an open-loop control system can do is to leave the disturbance $v(t)$ unchanged.

The overall transfer function of the system (in the absence of the disturbance $v(t)$) is simply

$$T(s) = \frac{Y(s)}{Y_d(s)}$$
$$= G(s)H(s) \tag{9.33}$$

Ignoring the dependence on s, the sensitivity of T with respect to changes in G is therefore

$$S_G^T = \frac{\Delta T/T}{\Delta G/G}$$
$$= \frac{H\,\Delta G/(GH)}{\Delta G/G} \tag{9.34}$$
$$= 1$$

The implication of $S_G^T = 1$ is that a percentage change in G is translated into an equal percentage change in T.

The conclusion to be drawn from this analysis is that an open-loop control system leaves both the sensitivity and the effect of a disturbance unchanged.

■ CLOSED-LOOP CONTROL

Consider next the closed-loop control system shown in Fig. 9.14. As before, the plant and controller are represented by the transfer functions $G(s)$ and $H(s)$, respectively. The controller or compensator in the forward path preceding the plant is the only "free" part of the system that is available for adjustment by the system designer. Accordingly, the closed-loop control system of Fig. 9.14 is referred to as a *single-degree-of-freedom (1-DOF) structure*.

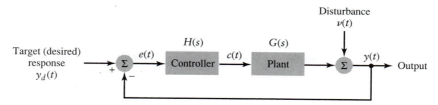

FIGURE 9.14   Control system with unity feedback.

To simplify matters, Fig. 9.14 assumes that the sensor (measuring the output signal to produce a feedback signal) is perfect. That is, the transfer function of the sensor is unity, and noise produced by the sensor is zero. Under this assumption, the actual output $y(t)$ of the plant is fed back directly to the input of the system. The system of Fig. 9.14 is therefore said to be a *unity-feedback system*. The controller in this system is actuated by the "measured" error $e(t)$, defined as the difference between the target (desired) response $y_d(t)$ (acting as the input $x(t)$) and the feedback (output) signal $y(t)$.

For the purpose of analysis, we may recast the closed-loop control system of Fig. 9.14 into the equivalent form shown in Fig. 9.15. Here we have made use of the equivalence between the two block diagrams shown in Fig. 9.16. Except for the block labeled $H(s)$ at the input end of Fig. 9.15, the single-loop feedback system shown in this figure is of exactly the same form as that of Fig. 9.1(b). By transforming the original closed-loop control system of Fig. 9.14 into the equivalent form shown in Fig. 9.15, we may make full use of the results developed in Section 9.3. Specifically, we note from Fig. 9.15 that

$$X(s) = H(s)Y_d(s) \tag{9.35}$$

Hence, using Eq. (9.35) in Eq. (9.5), we readily find that the closed-loop transfer function of the 1-DOF system of Fig. 9.14 is given by

$$
\begin{aligned}
T(s) &= \frac{Y(s)}{Y_d(s)} \\
&= \frac{Y(s)}{X(s)} \cdot \frac{X(s)}{Y_d(s)} \\
&= \frac{G(s)H(s)}{1 + G(s)H(s)}
\end{aligned}
\tag{9.36}
$$

Assuming that $G(s)H(s)$ is large compared to unity for all values of s of interest, Eq. (9.36) reduces to

$$T(s) \simeq 1$$

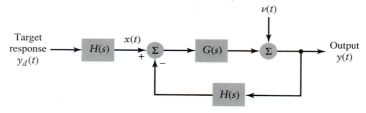

FIGURE 9.15   Reformulation of the feedback control system of Fig. 9.14.

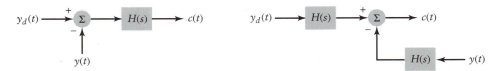

FIGURE 9.16 A pair of equivalent block diagrams used to change Fig. 9.14 into the equivalent form shown in Fig. 9.15.

That is,

$$y(t) \simeq y_d(t) \tag{9.37}$$

It is therefore desirable to have a large loop gain $G(s)H(s)$. Under this condition, the system of Fig. 9.14 has the potential to achieve the desired goal of accurate control, exemplified by the actual output $y(t)$ of the system closely approximating the target response $y_d(t)$.

There are other good reasons for using a large loop gain. Specifically, in light of the results derived in Sections 9.3 to 9.5, we may state that:

▶ The sensitivity of the closed-loop control system $T(s)$ is reduced by a factor equal to the return difference

$$F(s) = 1 + G(s)H(s)$$

▶ The disturbance $v(t)$ inside the feedback loop is reduced by the same factor $F(s)$.

▶ The effect of distortion due to nonlinear behavior of the plant is also reduced by $F(s)$.

■ **STEADY-STATE ERROR SPECIFICATIONS**

The error signal $e(t)$ is defined as the difference between the actuating or target signal $y_d(t)$ and the controlled signal (i.e., actual response) $y(t)$. Expressing these signals in terms of their respective Laplace transforms, we may write

$$
\begin{aligned}
E(s) &= Y_d(s) - Y(s) \\
&= [1 - T(s)]Y_d(s)
\end{aligned}
\tag{9.38}
$$

Substituting Eq. (9.36) into (9.38) yields

$$
\begin{aligned}
E(s) &= \left[1 - \frac{G(s)H(s)}{1 + G(s)H(s)}\right]Y_d(s) \\
&= \frac{1}{1 + G(s)H(s)}Y_d(s)
\end{aligned}
\tag{9.39}
$$

The *steady-state error* of a feedback control system is defined as the value of the error signal $e(t)$ when time t approaches infinity. Denoting this quantity by ϵ_{ss}, we may thus write

$$\epsilon_{ss} = \lim_{t \to \infty} e(t) \tag{9.40}$$

Using the final value theorem of Laplace transform theory described in Chapter 6, we may redefine the steady-state error ϵ_{ss} in the equivalent form

$$\epsilon_{ss} = \lim_{s \to 0} sE(s) \tag{9.41}$$

Hence, substituting Eq. (9.39) in (9.41), we get

$$\epsilon_{ss} = \lim_{s \to 0} \frac{sY_d(s)}{1 + G(s)H(s)} \tag{9.42}$$

Equation (9.42) shows that the steady-state error ϵ_{ss} of a feedback control system depends on two quantities:

▶ The open-loop transfer function $G(s)H(s)$ of the system
▶ The Laplace transform $Y_d(s)$ of the target signal $y_d(t)$

However, for Eq. (9.42) to be valid, the closed-loop control system of Fig. 9.14 must be stable.

In general, $G(s)H(s)$ may be written in the form of a rational function as follows:

$$G(s)H(s) = \frac{P(s)}{s^p Q_1(s)} \tag{9.43}$$

where neither the polynomial $P(s)$ nor $Q_1(s)$ has a zero at $s = 0$. Since $1/s$ is the transfer function of an integrator, it follows that p is the number of free integrators in the loop transfer function $G(s)H(s)$. The order p is referred to as the *type* of the feedback control system. We thus speak of a feedback control system being of type 0, type 1, type 2, and so on for $p = 0, 1, 2, \ldots$, respectively. In light of this classification, we next consider the steady-state error for three different input functions.

Step Input

For the step input $y_d(t) = u(t)$, we have $Y_d(s) = 1/s$. Hence Eq. (9.42) yields the steady-state error

$$\epsilon_{ss} = \lim_{s \to 0} \frac{1}{1 + G(s)H(s)}$$
$$= \frac{1}{1 + K_p} \tag{9.44}$$

where K_p is called the *position error constant*, defined by

$$K_p = \lim_{s \to 0} G(s)H(s)$$
$$= \lim_{s \to 0} \frac{P(s)}{s^p Q_1(s)} \tag{9.45}$$

For $p \geq 1$, K_p is unbounded and therefore $\epsilon_{ss} = 0$. For $p = 0$, K_p is finite and therefore $\epsilon_{ss} \neq 0$. Accordingly, we may state that the steady-state error for a step input is zero for a feedback control system of type 1 or higher. On the other hand, for a system of type 0, the steady-state error is not zero, and its value is given by Eq. (9.44).

Ramp Input

For the ramp input $y_d(t) = tu(t)$, we have $Y_d = 1/s^2$. In this case, Eq. (9.42) yields

$$\epsilon_{ss} = \lim_{s \to 0} \frac{1}{s + sG(s)H(s)}$$
$$= \frac{1}{K_v} \tag{9.46}$$

where K_v is the *velocity error constant*, defined by

$$
\begin{aligned}
K_v &= \lim_{s \to 0} sG(s)H(s) \\
&= \lim_{s \to 0} \frac{P(s)}{s^{p-1}Q_1(s)}
\end{aligned}
\tag{9.47}
$$

For $p \geq 2$, K_v is unbounded and therefore $\epsilon_{ss} = 0$. For $p = 1$, K_v is finite and $\epsilon_{ss} \neq 0$. For $p = 0$, K_v is zero and $\epsilon_{ss} = \infty$. Accordingly, we may state that the steady-state error for a ramp input is zero for a feedback control system of type 2 or higher. For a system of type 1, the steady-state error is not zero, and its value is given by Eq. (9.46). For a system of type 0, the steady-state error is unbounded.

Parabolic Input

For the parabolic input $y_d(t) = (t^2/2)u(t)$, we have $Y_d = 1/s^3$. The use of Eq. (9.42) yields

$$
\begin{aligned}
\epsilon_{ss} &= \lim_{s \to 0} \frac{1}{s^2 + s^2 G(s)H(s)} \\
&= \frac{1}{K_a}
\end{aligned}
\tag{9.48}
$$

where K_a is called the *acceleration error constant*, defined by

$$
K_a = \lim_{s \to 0} \frac{P(s)}{s^{p-2}Q_1(s)}
\tag{9.49}
$$

For $p \geq 3$, K_a is unbounded and therefore $\epsilon_{ss} = 0$. For $p = 2$, K_a is finite and therefore $\epsilon_{ss} = 0$. For $p = 0, 1$, K_a is zero and therefore $\epsilon_{ss} = \infty$. Accordingly, we may state that for a parabolic input, the steady-state error is zero for a type 3 system or higher. For a system of type 2, the steady-state error is not zero, and its value is given by Eq. (9.48). For a system of type 0 or type 1, the steady-state error is unbounded.

In Table 9.1 we present a summary of the steady-state errors according to system type as determined above.

A type 0 system is referred to as a *regulator*. The object of such a system is to maintain a physical variable of interest at some prescribed constant value despite the presence of external disturbances. Examples of regulator systems include:

▶ Control of the moisture content of paper, a problem that arises in the paper-making process

▶ Control of the chemical composition of the material outlet produced by a reactor

▶ Biological control system, which maintains the temperature of the human body at approximately 37°C despite variations in the temperature of the surrounding environment

TABLE 9.1   *Steady-State Errors According to System Type*

	Step	Ramp	Parabolic
Type 0	$1/(1 + K_p)$	∞	∞
Type 1	0	$1/K_v$	∞
Type 2	0	0	$1/K_a$

Type 1 and higher systems are referred to as *servomechanisms*. The objective here is to make a physical variable follow, or track, some desired time-varying function. Examples of servomechanisms include:

▶ Control of a robot, in which the robot manipulator is made to follow a preset trajectory in space
▶ Control of a missile, guiding it to follow a specified flight path
▶ Tracking of a maneuvering target (e.g., enemy aircraft) by a radar system

EXAMPLE 9.4   Figure 9.17 shows the block diagram of a feedback control system involving a *dc motor*. The dc motor is an electromechanical device that converts dc electrical energy into rotational mechanical energy. Its transfer function is approximately defined by

$$G(s) = \frac{K}{s(\tau_L s + 1)}$$

where K is the gain and τ_L is the load time constant. Here it is assumed that the field time constant of the dc motor is small compared to the load time constant. The transfer function of the controller is

$$H(s) = \frac{\alpha \tau s + 1}{\tau s + 1}$$

Find the steady-state errors of this control system. What is the effect of making the gain K assume a large value?

Solution: The loop transfer function of the control system in Fig. 9.17 is

$$G(s)H(s) = \frac{K(\alpha \tau s + 1)}{s(\tau_L s + 1)(\tau s + 1)} \tag{9.50}$$

which represents a type 1 system. Comparing Eq. (9.50) with (9.43), we have for the example at hand

$$P(s) = \frac{\alpha K}{\tau_L}\left(s + \frac{1}{\alpha \tau}\right)$$

$$Q_1(s) = \left(s + \frac{1}{\tau_L}\right)\left(s + \frac{1}{\tau}\right)$$

$$p = 1$$

Hence the use of these values in Eqs. (9.45), (9.47), and (9.49) yields

$$K_p = \infty$$
$$K_v = K$$
$$K_a = 0$$

Correspondingly, substituting these error constants into Eqs. (9.44), (9.46), and (9.48) yields the following steady-state errors for the control system of Fig. 9.17:

(a) Step input: $\epsilon_{ss} = 0$.

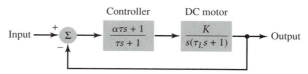

FIGURE 9.17   Type 1 feedback control system considered in Example 9.4.

(b) Ramp input: $\epsilon_{ss} = 1/K = $ constant.

(c) Parabolic input: $\epsilon_{ss} = \infty$.

The effect of making the gain K large is to reduce the steady-state error of the system due to a ramp input. This, in turn, improves the behavior of the system as a velocity control system.

▶ **Drill Problem 9.5**   Consider a type 2 feedback control system whose loop transfer function is defined by

$$G(s)H(s) = \frac{0.25K(s + 2)}{s^2(s + 1)(s + \frac{1}{2})}$$

Calculate the steady-state error of the system for each of the following three inputs:

(a) Step function of unit amplitude.

(b) Ramp of unit slope.

(c) Parabolic input of unit second-order derivative.

Answer:

(a) $\epsilon_{ss} = 0$.

(b) $\epsilon_{ss} = 0$.

(c) $\epsilon_{ss} = 1/K = $ constant.   ◀

9.9 *Transient Response of Low-Order Systems*

To set the stage for material on the stability analysis of feedback control systems presented in subsequent sections of this chapter, we find it informative to examine the transient response of first-order and second-order systems. Although feedback control systems of such low order are indeed rare in practice, their transient analysis forms the basis for a better understanding of higher-order systems.

■ FIRST-ORDER SYSTEM

Using the notation of Chapter 6, the transfer function of a first-order system is defined by

$$T(s) = \frac{b_0}{s + a_0}$$

In order to give physical meaning to the coefficients of the transfer function $T(s)$, we find it more convenient to rewrite it in the *standard form*:

$$T(s) = \frac{T(0)}{\tau s + 1} \tag{9.51}$$

where $T(0) = b_0/a_0$ and $\tau = 1/a_0$. The parameter $T(0)$ is the gain of the system at $s = 0$. The parameter τ is measured in units of time and is therefore referred to as the *time constant* of the system. According to Eq. (9.51), the single pole of $T(s)$ is located at $s = -1/\tau$.

For a step input (i.e., $Y_d(s) = 1/s$), the response of the system has the Laplace transform

$$Y(s) = \frac{T(0)}{s(\tau s + 1)} \tag{9.52}$$

Expanding $Y(s)$ in partial fractions and using the table of Laplace transform pairs in Appendix D, we find that the step response of the system is as follows (assuming that $T(0) = 1$:

$$y(t) = 1 - e^{-t/\tau} \tag{9.53}$$

which is plotted in Fig. 9.18. At $t = \tau$, the response $y(t)$ reaches 63.21% of its final value, hence the "time-constant" terminology.

■ SECOND-ORDER SYSTEM

Again using the notation of Chapter 6, the transfer function of a second-order system is defined by

$$T(s) = \frac{b_0}{s^2 + a_1 s + a_0}$$

However, as in the first-order case, we find it more convenient to reformulate the transfer function $T(s)$ so that its coefficients have physical meaning. Specifically, we redefine $T(s)$ in the *standard form*:

$$T(s) = \frac{T(0)\omega_n^2}{s^2 + 2\zeta\omega_n s + \omega_n^2} \tag{9.54}$$

where $T(0) = b_0/a_0$, $\omega_n^2 = a_0$, and $2\zeta\omega_n = a_1$. The parameter $T(0)$ is the gain of the system at $s = 0$. The dimensionless parameter ζ is called the *damping ratio*, and ω_n is called the *natural frequency* or *undamped frequency* of the system. These three parameters characterize the system in their own individual ways. The poles of the system are located at

$$s = -\zeta\omega_n \pm j\omega_n\sqrt{1 - \zeta^2} \tag{9.55}$$

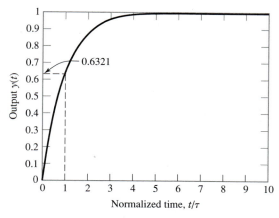

FIGURE 9.18 Transient response of first-order system, plotted versus the normalized time t/τ, where τ is the time constant of the system.

For a step input, the response of the system has the Laplace transform

$$Y(s) = \frac{T(0)\omega_n^2}{s(s^2 + 2\zeta\omega_n s + \omega_n^2)}$$

For the moment, we assume that the poles of $T(s)$ are complex with negative real parts. This implies $0 < \zeta < 1$. Then, expanding $Y(s)$ in partial fractions and using the table of Laplace transform pairs in Appendix D, we may express the step response of the system as the following exponentially damped sinusoidal signal (assuming that $T(0) = 1$):

$$y(t) = \left[1 - \frac{1}{\sqrt{1-\zeta^2}} e^{-\zeta\omega_n t} \sin\left(\omega_n\sqrt{1-\zeta^2}t + \tan^{-1}\left(\frac{\sqrt{1-\zeta^2}}{\zeta}\right)\right)\right]u(t) \quad (9.56)$$

The *time constant* of the exponentially damped sinusoid is defined by

$$\tau = \frac{1}{\zeta\omega_n} \quad (9.57)$$

which is measured in seconds. The frequency of the exponentially damped sinusoid is $\omega_n\sqrt{1-\zeta^2}$.

Depending on the value of ζ, we may now formally identify three regimes of operation:

1. $0 < \zeta < 1$. In this case, the two poles of $T(s)$ constitute a complex-conjugate pair, and the step response of the system is defined by Eq. (9.56). The system is said to be *underdamped*.

2. $\zeta > 1$. In this second case, the two poles of $T(s)$ are both real. The step response $y(t)$ now involves two exponential functions, as shown by

$$y(t) = (1 + k_1 e^{-t/\tau_1} + k_2 e^{-t/\tau_2})u(t) \quad (9.58)$$

where the time constants τ_1 and τ_2 are defined by, respectively,

$$\tau_1 = \frac{1}{\zeta\omega_n - \omega_n\sqrt{\zeta^2-1}}$$

$$\tau_2 = \frac{1}{\zeta\omega_n + \omega_n\sqrt{\zeta^2-1}}$$

and the scaling factors k_1 and k_2 are defined by, respectively,

$$k_1 = \frac{1}{2}\left(1 + \frac{\zeta}{\sqrt{\zeta^2-1}}\right)$$

$$k_2 = \frac{1}{2}\left(1 - \frac{\zeta}{\sqrt{\zeta^2-1}}\right)$$

Thus, for $\zeta > 1$, the system is said to be *overdamped*.

3. $\zeta = 1$. In this final case, the two poles are coincident at $s = -\omega_n$, and the step response of the system is defined by

$$y(t) = (1 - e^{-t/\tau} - te^{-t/\tau})u(t) \quad (9.59)$$

where $\tau = 1/\omega_n$ is the only time constant of the system. The system is said to be *critically damped*.

Figure 9.19 shows the step response $y(t)$ plotted versus time t for $\omega_n = 1$ and three different damping ratios: $\zeta = 2$, $\zeta = 1$, and $\zeta = 0.1$. These three values of ζ correspond to cases 2, 3, and 1, respectively.

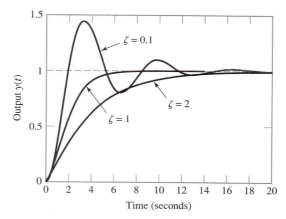

FIGURE 9.19   Transient response of second-order system for three different values of damping ratio ζ: overdamped, $\zeta = 2$; critically damped, $\zeta = 1$; underdamped, $\zeta = 0.1$.

With the above material on first-order and second-order systems at our disposal, we are ready to resume our study of feedback control systems.

▶ **Drill Problem 9.6**   Using the table of Laplace transform pairs in Appendix D, derive the following:

(a) The step response of an underdamped system as defined in Eq. (9.56).

(b) The step response of an overdamped system as defined in Eq. (9.58).

(c) The step response of a critically damped system as defined in Eq. (9.59).   ◀

9.10 Time-Domain Specifications

For linear control systems, the transient response is usually measured in terms of the step response. Typically, the step response, denoted by $y(t)$, is oscillatory as illustrated in Fig. 9.20. In describing such a response, we have two conflicting criteria: swiftness of the response, and closeness of the response to the desired response. Swiftness of the response is measured in terms of the rise time and peak time. Closeness of the response to the desired response is measured in terms of the percentage overshoot and settling time. These four quantities are defined as follows:

▶ *Rise time*, T_r, is defined as the time taken by the step response to rise from 10% to 90% of its final value $y(\infty)$.

▶ *Peak time*, T_p, is defined as the time taken by the step response to reach the overshoot (overall) maximum value y_{max}.

▶ *Percentage overshoot*, P.O., is defined in terms of the maximum value y_{max} and final value $y(\infty)$ by

$$\text{P.O.} = \frac{y_{max} - y(\infty)}{y(\infty)} \times 100$$

▶ *Settling time*, T_s, is defined as the time required by the step response to settle within $\pm\delta\%$ of the final value $y(\infty)$, where δ is user specified.

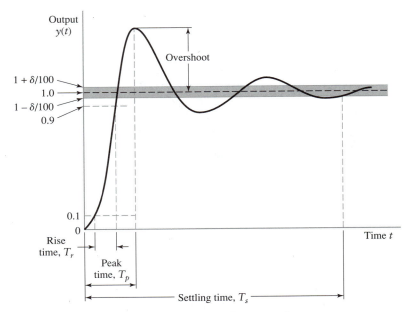

FIGURE 9.20   Time-domain specifications of a step response.

Figure 9.20 illustrates the definitions of these four quantities, assuming that $y(\infty) = 1.0$. They provide an adequate description of the step response $y(t)$. Most importantly, they lend themselves to measurement. Note that in the case of an overdamped system, the peak time and percentage overshoot are not defined. In such a case, the step response of the system is specified simply in terms of the rise time and settling time.

For reasons that will become apparent later, the underdamped response of a second-order system to a step input often provides an adequate approximation to the step response of a linear feedback control system. Accordingly, it is of particular interest to relate the above-mentioned quantities to the parameters of a second-order system.

EXAMPLE 9.5   Consider an underdamped second-order system of damping ratio ζ and natural frequency ω_n. Determine the rise time, peak time, percentage overshoot, and settling time of the system. For the settling time, use $\delta = 1$.

Solution: Unfortunately, it is difficult to obtain an explicit expression for the rise time T_r in terms of the damping ratio ζ and natural frequency ω_n. Nevertheless, it can be determined by simulation. Table 9.2 presents the results of simulation for the range $0.1 \leq \zeta \leq 0.9$. In this table we have also included the results obtained by using the approximate formula:

$$T_r \simeq \frac{1}{\omega_n}(0.60 + 2.16\zeta)$$

This formula yields fairly accurate results for $0.3 \leq \zeta \leq 0.8$, as can be seen from Table 9.2.

To determine the peak time T_p, we may differentiate Eq. (9.56) with respect to time t and then set the result equal to zero. We thus obtain the solutions $t = \infty$ and

$$t = \frac{n\pi}{\omega_n\sqrt{1 - \zeta^2}}, \quad n = 0, 1, 2, \ldots$$

TABLE 9.2 *Normalized Rise Time as a Function of the Damping Ratio ζ for a Second-Order System*

ζ	$\omega_n T_r$	
	Simulation	*Approximate Value*
0.1	1.1	0.82
0.2	1.2	1.03
0.3	1.3	1.25
0.4	1.45	1.46
0.5	1.65	1.68
0.6	1.83	1.90
0.7	2.13	2.11
0.8	2.5	2.33
0.9	2.83	2.54

The solution $t = \infty$ defines the maximum of the step response $y(t)$ only when $\zeta \geq 1$ (i.e., the critically damped or overdamped case). We are interested in the underdamped case. The first maximum of $y(t)$ occurs for $n = 1$; hence the peak time is

$$T_p = \frac{\pi}{\omega_n \sqrt{1 - \zeta^2}}$$

The first maximum of $y(t)$ also defines the percentage overshoot. Thus, putting $t = T_p$ in Eq. (9.56) and simplifying the result, we get

$$y_{\max} = 1 + e^{-\pi\zeta/\sqrt{1 - \zeta^2}}$$

The percentage overshoot is therefore

$$\text{P.O.} = 100e^{-\pi\zeta/\sqrt{1 - \zeta^2}}$$

Finally, to determine the settling time T_s we seek the time t for which the step response $y(t)$ defined in Eq. (9.56) decreases and stays within $\pm\delta\%$ of the final value $y(\infty) = 1.0$. For $\delta = 1$, this time is closely approximated by the decaying exponential factor $e^{-\zeta\omega_n t}$, as shown by

$$e^{-\zeta\omega_n T_s} \simeq 0.01$$

or

$$T_s \simeq \frac{4.6}{\zeta\omega_n}$$

9.11 *The Stability Problem*

In Sections 9.3 to 9.5 we showed that a large loop gain $G(s)H(s)$ is required to make the closed-loop transfer function $T(s)$ of a feedback system less sensitive to parameter variations, reduce the effects of disturbance or noise, and reduce nonlinear distortion. Indeed, based on the findings presented there, it would be tempting to propose the following recipe for improving the performance of a feedback system: make the loop gain $G(s)H(s)$ of the system as large as possible in the passband of the system. Unfortunately, the utility of this simple recipe is limited by a stability problem that is known to arise in feedback systems under certain conditions. If the number of poles contained in $G(s)H(s)$ is three or higher, then the system becomes more prone to instability and therefore more difficult to control as the loop gain is increased. In the design of a feedback system, the task is therefore not only to meet the various performance requirements imposed on the system for satisfactory operation inside a prescribed passband, but also to ensure that the system is stable and remains stable under all possible operating conditions.

The stability of a feedback system, like any other LTI system, is completely determined by the location of its poles or natural frequencies in the s-plane. The *natural frequencies* of a linear feedback system with closed-loop transfer function $T(s)$ are defined as the roots of the *characteristic equation*

$$A(s) = 0 \qquad (9.60)$$

where $A(s)$ is the denominator polynomial of $T(s)$. The feedback system is stable if the roots of this characteristic equation are all confined to the left half of the s-plane.

It would therefore seem appropriate for us to begin a detailed study of the stability problem by discussing how the natural frequencies of a feedback system are modified by the application of feedback. We now examine this issue by using three simple feedback systems.

■ FIRST-ORDER FEEDBACK SYSTEM

Consider a first-order feedback system of type 0, with unity feedback. The loop transfer function of the system is defined by

$$G(s)H(s) = \frac{K}{\tau_0 s + 1} \qquad (9.61)$$

where τ_0 is the *open-loop time constant* of the system, and K is an adjustable loop gain. The loop transfer function $G(s)H(s)$ has a single pole at $s = -1/\tau_0$. Using Eq. (9.61) in (9.36), we find that the closed-loop transfer function of the system is

$$\begin{aligned}
T(s) &= \frac{G(s)H(s)}{1 + G(s)H(s)} \\
&= \frac{K}{\tau_0 s + K + 1}
\end{aligned}$$

The characteristic equation of the system is therefore

$$\tau_0 s + K + 1 = 0 \qquad (9.62)$$

which has a single root at $s = -(K + 1)/\tau_0$. As K is increased, this root moves along the real axis of the s-plane, tracing the locus shown in Fig. 9.21. Indeed, it remains confined to the left half of the s-plane for $K > -1$. We may therefore state that the first-order feedback system with a loop transfer function described by Eq. (9.61) is stable for all $K > -1$.

■ SECOND-ORDER FEEDBACK SYSTEM

Consider next a second-order feedback system of type 1, with unity feedback. The loop transfer function of the system is defined by

$$G(s)H(s) = \frac{K}{s(\tau s + 1)} \qquad (9.63)$$

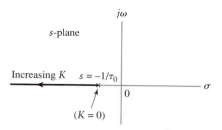

FIGURE 9.21   Effect of feedback on the pole location of a first-order system.

where K is an adjustable loop gain measured in rad/s, and $G(s)H(s)$ has simple poles at $s = 0$ and $s = -1/\tau$. Using Eq. (9.63) in (9.36), we find that the closed-loop transfer function of the system is

$$T(s) = \frac{G(s)H(s)}{1 + G(s)H(s)}$$

$$= \frac{K}{\tau s^2 + s + K}$$

The characteristic equation of the system is therefore

$$\tau s^2 + s + K = 0 \qquad (9.64)$$

This is a quadratic equation in s with a pair of roots defined by

$$s = -\frac{1}{2\tau} \pm \sqrt{\frac{1}{4\tau^2} - \frac{K}{\tau}} \qquad (9.65)$$

Figure 9.22 shows the locus traced by the two roots of Eq. (9.65) as the loop gain K is varied, starting from zero. We see that for $K = 0$, the characteristic equation has a root at $s = 0$ and one other at $s = -1/\tau$. As K is increased, the two roots move toward each other along the real axis, until they meet at $s = -1/2\tau$ for $K = 1/4\tau$. When K is increased further, the two roots separate from each other along a line parallel to the $j\omega$-axis, and which passes through the point $s = -1/2\tau$. This point is called the *breakaway point*, where the root loci break away from the real axis of the s-plane.

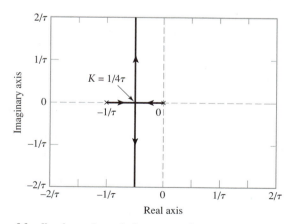

FIGURE 9.22   Effect of feedback on the pole locations of a second-order system. The loop transfer function has poles at $s = 0$ and $s = -1/\tau$.

When there is no feedback applied to the system (i.e., $K = 0$), the characteristic equation of the system has a root at $s = 0$, and the system is therefore on the verge of instability. When K is assigned a value greater than zero, the two roots of the characteristic equation are both confined to the left half of the s-plane. It follows therefore that the second-order feedback system with a loop transfer function described by Eq. (9.63) is stable for all positive values of K.

▶ **Drill Problem 9.7**   Referring to the second-order feedback system of Eq. (9.63), identify the values of K that result in the following forms of step response for the system: (a) underdamped, (b) overdamped, and (c) critically damped.

Answer: (a) $K > 0.25/\tau$.   (b) $K < 0.25/\tau$.   (c) $K = 0.25/\tau$.   ◀

▶ **Drill Problem 9.8**   For the case when the loop gain K is large enough to produce an underdamped step response, show that the damping ratio and natural frequency of the second-order feedback system are defined in terms of the loop gain K and time constant τ as follows, respectively:

$$\zeta = \frac{1}{2\sqrt{\tau K}} \quad \text{and} \quad \omega_n = \sqrt{\frac{K}{\tau}}$$

◀

▶ **Drill Problem 9.9**   The characteristic equation of a second-order feedback system may, in general, be written in the form

$$s^2 + as + b = 0$$

Show that such a system is stable provided the coefficients a and b are both positive.   ◀

■ THIRD-ORDER FEEDBACK SYSTEM

From the analysis just presented, we see that first-order and second-order feedback systems, as described therein, do not pose a stability problem. In both cases, the feedback system is stable for all positive values of the loop gain K. To probe further into the stability problem, we now consider a third-order feedback system whose loop transfer function is described by

$$G(s)H(s) = \frac{K}{(s + 1)^3} \tag{9.66}$$

Correspondingly, the closed-loop transfer function of the system is

$$T(s) = \frac{G(s)H(s)}{1 + G(s)H(s)}$$

$$= \frac{K}{s^3 + 3s^2 + 3s + K + 1}$$

The characteristic equation of the system is therefore

$$s^3 + 3s^2 + 3s + K + 1 = 0 \tag{9.67}$$

This cubic characteristic equation is more difficult to handle than the lower-order characteristic equations (9.62) and (9.64). So we resort to the use of a computer in order to gain some insight into how variations in the loop gain K affect the stability performance of the system.

TABLE 9.3 *Roots of the Characteristic Equation $s^3 + 3s^2 + 3s + K + 1 = 0$*	

K	Roots
0	Third-order root at $s = -1$
5	$s = -2.71$ $s = -0.1450 \pm j1.4809$
10	$s = -3.1544$ $s = 0.0772 \pm j1.8658$

Table 9.3 presents the roots of the characteristic equation (9.67) for three different values of K. For $K = 0$, we have a third-order root as $s = -1$. For $K = 5$, the characteristic equation has a simple root and a pair of complex-conjugate roots, all of which have negative real parts (i.e., they are located in the left half of the s-plane). Hence, for $K = 5$, the system is stable. For $K = 10$, the pair of complex-conjugate roots moves into the right half of the s-plane, and the system is therefore unstable. Thus, in the case of a third-order feedback system with a loop transfer function described by Eq. (9.66), the loop gain K has a profound influence on the stability of the system.

The majority of feedback systems used in practice are of order 3 or higher. The stability of such systems is therefore an issue of paramount importance. Much of the material presented in the rest of this chapter is devoted to a study of this problem.

9.12   *Routh–Hurwitz Criterion*

The *Routh–Hurwitz criterion* provides a simple procedure to ascertain whether or not all the roots of a polynomial $A(s)$ have negative real parts (i.e., lie in the left-half of the s-plane); it does so *without* having to compute the roots of $A(s)$. Let the polynomial $A(s)$ be expressed in the expanded form

$$A(s) = a_n s^n + a_{n-1} s^{n-1} + \cdots + a_1 s + a_0 \qquad (9.68)$$

where $a_n \neq 0$. The procedure begins by arranging all the coefficients of $A(s)$ in the form of two rows as follows:

Row n: $a_n \quad a_{n-2} \quad a_{n-4} \quad \cdots$
Row $n - 1$: $a_{n-1} \quad a_{n-3} \quad a_{n-5} \quad \cdots$

If the order n of polynomial $A(s)$ is even and therefore coefficient a_0 belongs to row n, then a zero is placed under a_0 in row $n - 1$. The next step is to construct row $n - 2$ by using the entries of rows n and $n - 1$ as follows:

Row $n - 2$: $\dfrac{a_{n-1}a_{n-2} - a_n a_{n-3}}{a_{n-1}} \qquad \dfrac{a_{n-1}a_{n-4} - a_n a_{n-5}}{a_{n-1}} \qquad \cdots$

Note that the entries in this row have determinant-like quantities for their numerators. That is, $a_{n-1}a_{n-2} - a_n a_{n-3}$ corresponds to the negative of the determinant of the 2 by 2 matrix

$$\begin{bmatrix} a_n & a_{n-2} \\ a_{n-1} & a_{n-3} \end{bmatrix}$$

A similar formulation applies to the numerators of the other entries in row $n - 2$. Next, the entries of rows $n - 1$ and $n - 2$ are used to construct row $n - 3$ following a procedure similar to that described above, and the process is continued until we reach row 0. The resulting array of $(n + 1)$ rows is called the *Routh array*.

We may now state the Routh–Hurwitz criterion: *All the roots of the polynomial A(s) lie in the left half of the s-plane if all the entries in the leftmost column of the Routh array are nonzero and have the same sign. If sign changes are encountered in scanning the leftmost column, the number of sign changes is the number of roots of A(s) in the right half of the s-plane.*

EXAMPLE 9.6   The characteristic polynomial of a fourth-order feedback system is given by

$$A(s) = s^4 + 3s^3 + 7s^2 + 2s + 10$$

Construct the Routh array, and hence determine whether or not the system is stable.

Solution: Constructing the Routh array for $n = 4$, we obtain the following:

Row 4:	1	7	10
Row 3:	3	3	0
Row 2:	$\dfrac{3 \times 7 - 3 \times 1}{3} = 6$	$\dfrac{3 \times 10 - 0 \times 1}{3} = 10$	0
Row 1:	$\dfrac{6 \times 3 - 10 \times 3}{6} = -2$	0	0
Row 0:	$\dfrac{-2 \times 10 - 0 \times 6}{-2} = 10$	0	0

There are two sign changes in the entries in the leftmost column of the Routh array. We therefore conclude that (1) the system is unstable, and (2) the characteristic equation of the system has two roots in the right half of the *s*-plane.

The Routh–Hurwitz criterion may be used to determine the critical value of the loop gain K for which the polynomial $A(s)$ has a pair of roots on the $j\omega$-axis of the s-plane by exploiting a special case of this criterion. If $A(s)$ has a pair of roots on the axis, the Routh–Hurwitz test terminates prematurely in that an entire row (always an odd-numbered row) of zeros is encountered in constructing the Routh array. When this happens, the feedback system is said to be *on the verge of instability*. The critical value of K is deduced from the entries of the particular row in question. The corresponding pair of roots on the $j\omega$-axis is found in the auxiliary polynomial formed from the entries of the preceding row, as illustrated in the following example.

EXAMPLE 9.7   Consider again a third-order feedback system whose loop transfer function $L(s) = G(s)H(s)$ is defined by Eq. (9.66), that is,

$$L(s) = \frac{K}{(s + 1)^3}$$

Find (a) the value of K for which the system is on the verge of instability and (b) the corresponding pair of roots on the $j\omega$-axis of the s-plane.

Solution: The characteristic polynomial $A(s)$ of the system is defined by

$$A(s) = (s + 1)^3 + K$$
$$= s^3 + 3s^2 + 3s + 1 + K$$

Constructing the Routh array, we obtain the following:

Row 3:	1	3
Row 2:	3	$1 + K$
Row 1:	$\dfrac{9 - (1 + K)}{3}$	0
Row 0:	$1 + K$	0

(a) For the only nonzero entry of row 1 to become zero, we require

$$9 - (1 + K) = 0$$

which yields $K = 8$.

(b) For this value of K, the auxiliary polynomial is obtained from row 2:

$$3s^2 + 9 = 0$$

which has a pair of roots at $s = \pm j\sqrt{3}$.

▶ **Drill Problem 9.10**   Consider a linear feedback system with its loop transfer function defined by

$$L(s) = \frac{0.2K(s + 5)}{(s + 1)^3}$$

Find (a) the critical value of loop gain K for which the system is on the verge of instability and (b) the corresponding pair of roots on the $j\omega$-axis of the s-plane.

Answer: (a) $K = 20$. (b) $s = \pm j\sqrt{7}$.   ◀

■ **SINUSOIDAL OSCILLATORS**

In the design of *sinusoidal oscillators*, feedback is applied to an amplifier with the specific objective of making the system *unstable*. In such an application, the oscillator consists of an amplifier and a frequency-determining network, forming a closed-loop feedback system. The amplifier sets the necessary condition for oscillation. To avoid distorting the output signal, the degree of nonlinearity in the amplifier is maintained at a very low level. In the next example, we show how the Routh–Hurwitz criterion may be used for such an application.

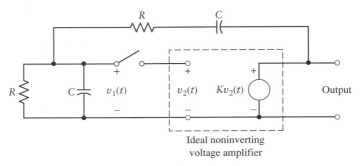

FIGURE 9.23   *RC* audio oscillator.

EXAMPLE 9.8   Figure 9.23 shows the simplified circuit diagram of an *RC audio oscillator*. Determine the frequency of oscillation and the condition for oscillation.

Solution: With the switch open, we find from Fig. 9.23 that the loop transfer function is (in accordance with the terminology of Fig. 9.2)

$$L(s) = -\frac{V_1(s)}{V_2(s)}$$

$$= -\frac{K\left(\dfrac{R}{sC}\right)\Big/\left(R + \dfrac{1}{sC}\right)}{\left(\dfrac{R}{sC}\Big/\left(R + \dfrac{1}{sC}\right)\right) + R + \dfrac{1}{sC}}$$

$$= -\frac{KRCs}{(RCs)^2 + 3(RCs) + 1}$$

The characteristic equation of the feedback circuit is therefore

$$(RCs)^2 + (3 - K)RCs + 1 = 0$$

A quadratic characteristic equation is simple enough for us to determine the condition for instability without having to set up the Routh array. For the problem at hand, we see that when the switch in Fig. 9.23 is closed, the circuit will be on the verge of instability provided that the voltage gain K of the amplifier is 3. The natural frequencies of the circuit will then lie on the $j\omega$-axis at $s = \pm j/RC$. In practice, the gain K is chosen to be slightly larger than 3 so that the two roots of the characteristic equation lie just to the right of the $j\omega$-axis. This is done in order to make sure that the oscillator is self-starting. As the oscillations build up in amplitude, a resistive component of the amplifier (not shown in Fig. 9.23) is modified slightly, helping to stabilize the gain K at the desired value of 3.

▶ **Drill Problem 9.11**   The element values in the oscillator circuit of Fig. 9.23 are $R = 10\ \text{k}\Omega$ and $C = 0.01\ \mu\text{F}$. Find the oscillation frequency.

Answer: 1.5915 kHz.   ◀

9.13 *Root Locus Method*

The *root locus method* is an analytical tool for the design of a linear feedback system, with emphasis on the pole locations of its closed-loop transfer function. Recall that the poles of a system's transfer function determine its transient response. Hence by knowing the closed-loop pole locations, we can deduce considerable information about the transient response of the feedback system. The method derives its name from the fact that a "root locus" is the geometric path or locus traced out by the roots of the system's characteristic equation in the complex s-plane as some parameter (usually, but not necessarily, the loop gain) is varied from zero to infinity. Such a root locus is exemplified by the plots shown in Fig. 9.21 for a first-order feedback system and Fig. 9.22 for a second-order feedback system of type 1.

In a general setting, construction of the root locus begins with the loop transfer function of the system, expressed in factored form as follows:

$$L(s) = G(s)H(s)$$
$$= K \frac{\Pi_{i=1}^{M}(1 - s/c_i)}{\Pi_{j=1}^{N}(1 - s/d_j)} \tag{9.69}$$

where K is the loop gain, and d_j and c_i are the poles and zeros of $L(s)$, respectively. These poles and zeros are fixed numbers, independent of K. In a linear feedback system, they may be determined directly from the block diagram of the system, since the system is usually made up of a cascade of first- and second-order components.

Traditionally, the term "root locus" refers to a situation where the loop gain is nonnegative, that is, $0 \leq K \leq \infty$. This is the case treated in the sequel.

■ ROOT LOCUS CRITERIA

Let the numerator and denominator polynomials of the loop transfer function $L(s)$ be defined by

$$P(s) = \prod_{i=1}^{M} \left(1 - \frac{s}{c_i}\right) \tag{9.70}$$

and

$$Q(s) = \prod_{j=1}^{N} \left(1 - \frac{s}{d_j}\right) \tag{9.71}$$

The characteristic equation of the system is defined by

$$A(s) = Q(s) + KP(s) = 0 \tag{9.72}$$

Equivalently, we may write

$$L(s) = K \frac{P(s)}{Q(s)} = -1 \tag{9.73}$$

Since $s = \sigma + j\omega$ is complex valued, we may express the polynomial $P(s)$ in terms of its magnitude and phase components as follows:

$$P(s) = |P(s)|e^{j \arg\{P(s)\}} \tag{9.74}$$

where

$$|P(s)| = \prod_{i=1}^{M} \left|1 - \frac{s}{c_i}\right| \tag{9.75}$$

and

$$\arg\{P(s)\} = \sum_{i=1}^{M} \arg\left\{1 - \frac{s}{c_i}\right\} \tag{9.76}$$

Similarly, the polynomial $Q(s)$ may be expressed in terms of its magnitude and phase components as follows:

$$Q(s) = |Q(s)|e^{j \arg\{Q(s)\}} \tag{9.77}$$

where

$$|Q(s)| = \prod_{j=1}^{N} \left|1 - \frac{s}{d_j}\right| \tag{9.78}$$

and

$$\arg\{Q(s)\} = \sum_{j=1}^{N} \arg\left\{1 - \frac{s}{d_j}\right\} \tag{9.79}$$

Substituting Eqs. (9.75), (9.76), (9.78), and (9.79) into (9.69), we may readily establish two basic criteria for a root locus (assuming that K is nonnegative):

▶ *Angle criterion.* For a point s_l to lie on a root locus, the angle criterion

$$\arg\{P(s)\} - \arg\{Q(s)\} = (2k + 1)\pi, \qquad k = 0, \pm 1, \pm 2, \ldots \tag{9.80}$$

must be satisfied for $s = s_l$. The angles $\arg\{Q(s)\}$ and $\arg\{P(s)\}$ are themselves determined by the angles of the pole and zero factors of $L(s)$ as in Eqs. (9.76) and (9.79).

▶ *Magnitude criterion.* Once a root locus is constructed, the value of loop gain K corresponding to the point s_l is determined from the magnitude criterion

$$K = \frac{|Q(s)|}{|P(s)|} \tag{9.81}$$

evaluated at $s = s_l$. The magnitudes $|Q(s)|$ and $|P(s)|$ are themselves determined by the magnitudes of the pole and zero factors of $L(s)$ as in Eqs. (9.75) and (9.78).

To illustrate the use of the angle and magnitude criteria for the construction of root loci, consider the loop transfer function

$$L(s) = \frac{K(1 - s/c)}{s(1 - s/d)(1 - s/d^*)}$$

which has a zero at $s = c$, a simple pole at $s = 0$, and a pair of complex-conjugate poles at $s = d, d^*$. Select an arbitrary trial point g in the s-plane, and construct vectors from the poles and zeros of $L(s)$ to this point, as depicted in Fig. 9.24. For the angle criterion of Eq. (9.80) and the magnitude criterion of Eq. (9.81) to be both satisfied by the choice of point g, we should find that

$$\theta_{z_1} - \theta_{p_1} - \theta_{p_2} - \theta_{p_3} = (2k + 1)\pi, \qquad k = 0, \pm 1, \ldots$$

and

$$K = \frac{BCD}{A}$$

where the angles and lengths of the vectors are defined in Fig. 9.24.

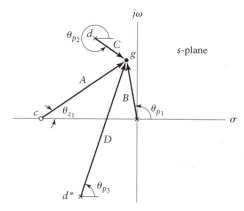

FIGURE 9.24 Illustrating the angle criterion of Eq. (9.80) and the magnitude criterion of Eq. (9.81) for the loop transfer function:

$$L(s) = \frac{K(1 - s/c)}{s(1 - s/d)(1 - s/d^*)}$$

The various angles and lengths of vectors drawn from the poles and zeros of $L(s)$ to point g in the complex s-plane are defined as follows:

$$\theta_{z_1} = \arg\left\{1 - \frac{g}{c}\right\}, \qquad A = \left|1 - \frac{g}{c}\right|$$

$$\theta_{p_1} = \arg\{g\}, \qquad B = |g|$$

$$\theta_{p_2} = \arg\left\{1 - \frac{g}{d}\right\}, \qquad C = \left|1 - \frac{g}{d}\right|$$

$$\theta_{p_3} = \arg\left\{1 - \frac{g}{d^*}\right\}, \qquad D = \left|1 - \frac{g}{d^*}\right|$$

■ **PROPERTIES OF ROOT LOCUS**

Given the poles and zeros of the loop transfer function as described in Eq. (9.69), we may construct an approximate form of the root locus for a linear feedback system by exploiting some basic properties of the root locus, as described next.

Property 1. *The root locus has a number of branches equal to N or M, whichever one is the greater.*
A *branch* of the root locus refers to the locus of one root of the characteristic equation $A(s) = 0$ when K varies from zero to infinity. Property 1 follows from Eq. (9.72), bearing in mind that the polynomials $P(s)$ and $Q(s)$ are themselves defined by Eqs. (9.70) and (9.71).

Property 2. *The root locus starts at the poles of the loop transfer function.*
For $K = 0$, the characteristic equation (9.72) reduces to

$$Q(s) = 0$$

The roots of this equation are the same as the poles of the loop transfer function $L(s)$, which proves Property 2.

Property 3. *The root locus terminates on the zeros of the loop transfer function, including those zeros that lie at infinity.*

For $K = \infty$, the characteristic equation (9.72) reduces to

$$P(s) = 0$$

The roots of this equation are the same as the zeros of the loop transfer function $L(s)$, which proves Property 3.

Property 4. *The root locus is symmetrical about the real axis of the s-plane.*

The poles and zeros of the loop transfer function $L(s)$ are real or else they occur in complex-conjugate pairs. The roots of the characteristic equation (9.72) must therefore be real or complex-conjugate pairs, from which Property 4 follows immediately.

Property 5. *As the complex frequency s approaches infinity, the branches of the root locus tend to straight-line asymptotes with angles given by*

$$\theta_k = \frac{(2k + 1)\pi}{N - M}, \qquad k = 0, 1, 2, \ldots, |N - M| - 1 \tag{9.82}$$

The asymptotes intersect at a common point on the real axis of the s-plane, the location of which is defined by

$$\sigma_0 = \frac{\sum_{j=1}^{N} d_j - \sum_{i=1}^{M} c_i}{N - M} \tag{9.83}$$

That is,

$$\sigma_0 = \frac{(\text{sum of finite poles}) - (\text{sum of finite zeros})}{(\text{number of finite poles}) - (\text{number of finite zeros})}$$

The intersection point $s = \sigma_0$ is called the *centroid* of the root locus.

▶ **Drill Problem 9.12**   The loop transfer function of a linear feedback system is defined by

$$L(s) = \frac{0.2K(s + 5)}{(s + 1)^3}$$

Find (a) the asymptotes of the root locus of the system and (b) its centroid.

Answer: (a) $\theta = 90°, 270°$. (b) $\sigma_0 = 1$.   ◀

Property 6. *The intersection points of the root locus with the imaginary axis of the s-plane, and the corresponding values of loop gain K, may be determined from the Routh–Hurwitz criterion.*

This property was discussed in Section 9.12.

Property 7. *The breakaway points, where the branches of the root locus intersect, must satisfy the condition*

$$\frac{d}{ds}\left(\frac{1}{L(s)}\right) = 0 \tag{9.84}$$

where L(s) is the loop transfer function.

The condition of Eq. (9.84) is a necessary but not sufficient condition for a breakaway point. In other words, all breakaway points satisfy Eq. (9.84), but not all solutions of this equation are breakaway points.

EXAMPLE 9.9   Consider again the second-order feedback system of Eq. (9.63), assuming that $\tau = 1$. The loop transfer function of the system is

$$L(s) = \frac{K}{s(1 + s)}$$

Find the breakaway point of the root locus for this system.

Solution: The use of Eq. (9.84) yields

$$\frac{d}{ds} = [s(1 + s)] = 0$$

That is,

$$1 + 2s = 0$$

from which we readily see that the breakaway point is at $s = -\frac{1}{2}$. This agrees with the result displayed in Fig. 9.22 for $\tau = 1$.

The seven properties described above are usually adequate to construct a reasonably accurate root locus, starting from the factored form of the loop transfer function of a linear feedback system. The following two examples illustrate how this is done.

EXAMPLE 9.10   Consider a linear feedback amplifier, involving three transistor stages. The loop transfer function of the amplifier is defined by

$$L(s) = \frac{6K}{(s + 1)(s + 2)(s + 3)}$$

Sketch the root locus of this feedback amplifier.

Solution: The loop transfer function $L(s)$ has poles at $s = -1$, $s = -2$, and $s = -3$. All three zeros of $L(s)$ occur at infinity. Thus the root locus has three branches that start at the above-mentioned poles and terminate at infinity.

From Eq. (9.82), we find that the angles made by the three asymptotes are 60°, 180°, and 300°. Moreover, the intersection point of these asymptotes (i.e., centroid of the root locus) is obtained from Eq. (9.83) as

$$\sigma_0 = \frac{-1 - 2 - 3}{3} = -2$$

The asymptotes are depicted in Fig. 9.25.

To find the intersection points of the root locus with the imaginary axis of the s-plane, we first form the characteristic polynomial:

$$A(s) = (s + 1)(s + 2)(s + 3) + 6K$$
$$= s^3 + 6s^2 + 11s + 6(K + 1)$$

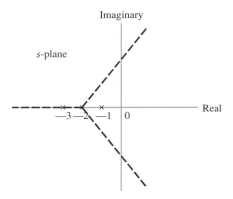

FIGURE 9.25 Diagram showing the intersection point of the three asymptotes (i.e., centroid of the root locus) for the feedback system of Example 9.10.

Next, we construct the Routh array:

Row 3:   1   11

Row 2:   6 $6(K + 1)$

Row 1: $\dfrac{66 - 6(K + 1)}{6}$ 0

Row 0: $6(K + 1)$ 0

Setting the only nonzero entry of row 1 equal to zero in accordance with Property 6, we find that the critical value of K for which the system is on the verge of instability is

$$K = 10$$

Using row 2 to construct the auxiliary polynomial with $K = 10$, we write

$$6s^2 + 66 = 0$$

Hence the intersection points of the root locus with the imaginary axis are at $s = \pm j\sqrt{11}$.
     Finally, using Eq. (9.84), we find that the breakaway point must satisfy the condition

$$\frac{d}{ds}[(s + 1)(s + 2)(s + 3)] = 0$$

That is,

$$3s^2 + 12s + 11 = 0$$

The roots of this quadratic equation are

$$s = -1.423 \quad \text{and} \quad s = -2.577$$

Examining the real-axis segments of the root locus, we find that the first point is on it and is therefore a breakaway point of the root locus, but the second point is not on the root locus. Moreover, for $s = -1.423$, the use of Eq. (9.73) yields

$$K = |1 - 1.423| \cdot |2 - 1.423| \cdot |3 - 1.423|$$
$$= 0.385$$

     Finally, putting all of these results together, we may sketch the root locus of the feedback amplifier as shown in Fig. 9.26.

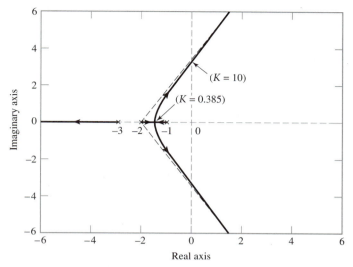

FIGURE 9.26   Root locus of third-order feedback system with loop transfer function

$$L(s) = \frac{6K}{(s+1)(s+2)(s+3)}$$

EXAMPLE 9.11   Consider the unity-feedback system of Fig. 9.27. The plant is unstable, with a transfer function defined by

$$G(s) = \frac{0.5K}{(s+5)(s-4)}$$

The controller has a transfer function defined by

$$H(s) = \frac{(s+2)(s+5)}{s(s+12)}$$

Sketch the root locus of the system, and determine the values of K for which the system is stable.

Solution: The plant has two poles, one at $s = -5$ and the other at $s = 4$. The latter pole is responsible for instability of the plant. The controller has a pair of zeros at $s = -2$ and $s = -5$, and a pair of poles at $s = 0$ and $s = -12$. When the controller is connected in cascade with the plant, a *pole–zero cancellation* takes place, yielding the loop transfer function

$$L(s) = G(s)H(s)$$
$$= \frac{0.5K(s+2)}{s(s+12)(s-4)}$$

which is indicative of a type 1 feedback system.

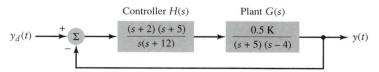

FIGURE 9.27   Unity-feedback system for Example 9.11.

The root locus has three branches. One branch starts at the pole $s = -12$ and terminates at the zero $s = -2$. The other two branches start at the poles $s = 0$ and $s = 4$ and terminate at infinity.

With $L(s)$ having three poles and one finite zero, we find from Eq. (9.82) that the root locus has two asymptotes defined by $\theta = 90°$ and $180°$. The centroid of the root locus is obtained from Eq. (9.83) as

$$\sigma_0 = \frac{(-12 + 0 + 4) - (-2)}{3 - 1}$$

$$= -3$$

Next, the characteristic polynomial of the feedback system is

$$A(s) = s^3 + 8s^2 + (0.5K - 48)s + K$$

Constructing the Routh array, we obtain the following:

Row 3:	1	$0.5K - 48$
Row 2:	8	K
Row 1:	$\dfrac{8(0.5K - 48) - K}{8}$	0
Row 0:	K	0

Setting the only nonzero entry of row 1 to zero, we obtain

$$8(0.5K - 48) - K = 0$$

which yields the critical value of loop gain K as

$$K = 128$$

Next, using the entries of row 2 with $K = 128$, we get the auxiliary polynomial:

$$8s^2 + 128 = 0$$

which has roots at $s = \pm j4$. Thus the root locus intersects the imaginary axis of the s-plane at $s = \pm j4$, and the corresponding value of K is 128.

Finally, applying Eq. (9.84), we find that the breakaway point of the root locus must satisfy the condition

$$\frac{d}{ds}\left(\frac{s(s + 12)(s - 4)}{0.5K(s + 2)}\right) = 0$$

that is,

$$s^3 + 7s^2 + 16s - 48 = 0$$

Using the computer, we find that this cubic equation has a single real root at $s = 1.6083$. The corresponding value of K is 5.5611.

Putting these results together, we may construct the root locus shown in Fig. 9.28. Here we see that the feedback system is unstable for $0 \le K \le 128$. When $K > 128$, all three roots of the characteristic equation become confined to the left half of the s-plane. We thus see that the application of feedback has the beneficial effect of *stabilizing* an unstable plant, provided that the loop gain is large enough.

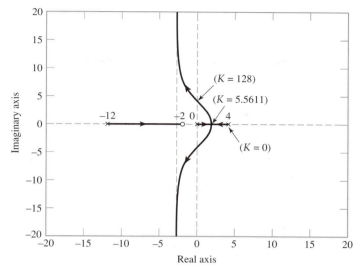

FIGURE 9.28 Root locus of closed-loop control system with loop transfer function

$$L(s) = \frac{0.5K(s + 2)}{s(s - 4)(s + 12)}$$

▶ **Drill Problem 9.13** Sketch the root locus of a linear feedback system whose loop transfer function is defined by

$$L(s) = \frac{0.2K(s + 5)}{(s + 1)^3}$$

Hint: Use the answers to Drill Problems 9.10 and 9.12.   ◀

9.14 *Reduced-Order Models*

We often find that the poles and zeros of the closed-loop transfer function $T(s)$ of a feedback system are grouped in the complex s-plane roughly in the manner illustrated in Fig. 9.29. In particular, depending on how close the poles and zeros are to the $j\omega$-axis we may identify two groupings:

1. *Dominant poles and zeros*, which are those poles and zeros of $T(s)$ that lie close to the $j\omega$-axis. They are said to be dominant because they exert a profound influence on the frequency response of the system. Another way of viewing this situation is to recognize that poles close to the $j\omega$-axis correspond to large time constants of the system. The contributions made by these poles to the transient response of the system are slow and therefore dominant.

2. *Insignificant poles and zeros*, which are those poles and zeros of $T(s)$ that are far removed from the $j\omega$-axis. They are said to be insignificant because they have relatively little influence on the frequency response of the system. In terms of time-domain behavior, poles that are far away from the $j\omega$-axis correspond to small time constants.

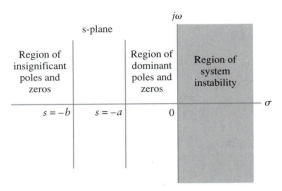

FIGURE 9.29   Different regions of the complex *s*-plane.

The contributions of such poles to the transient response of the system are much faster and therefore insignificant.

Let $s = -a$ and $s = -b$ define the boundaries of the dominant poles and the insignificant poles, as indicated in Fig. 9.29. As a rule of thumb, the grouping of poles into dominant poles and insignificant poles is justified if the ratio b/a is greater than 4.

Given that we have a high-order feedback system whose closed-loop transfer function fits the picture portrayed in Fig. 9.29, we may then approximate the system by a *reduced-order model* simply by retaining the dominant poles and zeros of $T(s)$. The use of a reduced-order model in place of the original system is motivated by the following considerations:

▶ Low-order models are simple; they are therefore intuitively appealing in system analysis and design.

▶ Low-order models are less demanding in computational terms than high-order ones.

A case of particular interest is when the use of a first-order or second-order model as the reduced-order model is justified, for then we can exploit the wealth of information available on such low-order models.

When discarding poles and zeros in the derivation of a reduced-order model, it is important to rescale the gain of the system. Specifically, the reduced-order model and the original system should have the same gain at zero frequency.

EXAMPLE 9.12   Consider again the linear feedback amplifier of Example 9.10. Assuming that $K = 8$, approximate this system using a second-order model. Use the step response to assess the quality of the approximation.

Solution: For $K = 8$, the characteristic equation of the feedback amplifier is given by

$$s^3 + 6s^2 + 11s + 54 = 0$$

Using the computer, the roots of this equation are found to be

$$s = -5.7259, \quad s = -0.1370 \pm j3.0679$$

The locations of these three roots are plotted in Fig. 9.30. We immediately observe from the pole–zero map of Fig. 9.30 that the poles at $s = -0.1370 \pm j3.0679$ are the dominant poles, and the pole at $s = -5.7259$ is an insignificant pole. The numerator of the closed-loop transfer function $T(s)$ consists simply of $6K$. Accordingly, the closed-loop gain of the feedback amplifier may be approximated as

$$T'(s) \simeq \frac{6K'}{(s + 0.1370 + j3.0679)(s + 0.1370 - j3.0679)}$$

$$= \frac{6K'}{s^2 + 0.2740s + 9.4308}$$

To make sure that $T'(s)$ is scaled properly relative to the original $T(s)$, the gain K' is chosen as follows:

$$K' = \frac{9.4308}{54} \times 8 = 1.3972$$

for which we thus have $T'(0) = T(0)$.

In light of Eq. (9.54), we set

$$2\zeta\omega_n = 0.2740$$
$$\omega_n^2 = 9.4308$$

from which we readily find that

$$\zeta = 0.0446$$
$$\omega_n = 3.0710$$

The step response of the feedback amplifier is therefore underdamped. The time constant of the exponentially damped response is from Eq. (9.57):

$$\tau = \frac{1}{\zeta\omega_n} = \frac{1}{0.137} = 8.2993 \text{ s}$$

The frequency of the exponentially damped response is

$$\omega_n\sqrt{1 - \zeta^2} = 3.0710\sqrt{1 - (0.0446)^2}$$
$$= 3.0679 \text{ rad/s}$$

Figure 9.31 shows two plots, one displaying the step response of the original third-order feedback amplifier with $K = 8$, and the other displaying the step response of the approximating second-order model with $K' = 1.3972$. The two plots are very similar, which indicates that the reduced-order model is adequate for the example at hand.

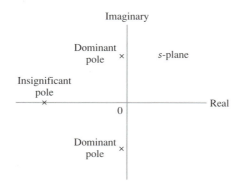

FIGURE 9.30   Pole–zero map for feedback amplifier of Example 9.10 with gain $K = 8$.

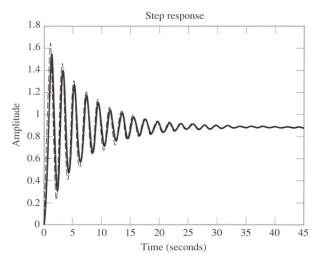

FIGURE 9.31   Step response for third-order system shown as dashed curve, and step response for (reduced) second-order system shown as solid curve.

*9.15 *Nyquist Stability Criterion*

The root locus method provides information on the roots of the characteristic equation of a linear feedback system (i.e., poles of the system's closed-loop transfer function) as the loop gain is varied. This information may, in turn, be used to assess not only the stability of the system but also matters relating to its transient response, as discussed in the previous two sections. For the method to work, we require knowledge of the poles and zeros of the system's loop transfer function. However, there are situations where this requirement may be difficult to meet. For example, it could be that the only way of assessing the stability of a feedback system is by experimental means, or the feedback loop includes a time delay, in which case the loop transfer function is not a rational function. In such situations, we may look to the Nyquist criterion as an alternative method for evaluating the stability performance of the system. In any event, the Nyquist criterion is important enough to be considered in its own right.

The *Nyquist stability criterion* is a frequency-domain method that is based on a plot (in polar coordinates) of the loop transfer function $L(s)$ for $s = j\omega$. It has a number of desirable features that make it a useful tool for the analysis and design of a linear feedback system:

1. It provides information on the absolute stability of the system, degree of stability, and how to stabilize the system if it is unstable.

2. It provides information on the frequency-domain response of the system.

3. It can be used to study the stability of a linear feedback system with time delay, which may arise due to the presence of distributed components.

A limitation of the Nyquist criterion, however, is that, unlike the root locus technique, it does not give the exact location of the roots of the system's characteristic equation.

■ ENCLOSURES AND ENCIRCLEMENTS

To prepare the way for a statement of the Nyquist stability criterion, we need to understand what is meant by the terms "enclosure" and "encirclement," which arise in the context of contour mapping.

To be specific, consider some function $F(s)$ of the complex variable s. We are accustomed to the representation of matters relating to s in a complex plane, referred to as the *s*-plane. Suppose the function $F(s)$ is represented in a complex plane of its own, hereafter referred to as the *F*-plane. Let C denote a *closed contour* traversed by the complex variable s in the *s*-plane. A contour is said to be closed if it terminates onto itself and does not intersect itself as it is transversed by the complex variable s. Let Γ denote the corresponding contour traversed by the function $F(s)$ in the *F*-plane. If $F(s)$ is a *single-valued* function of s, then Γ is also a closed contour. The customary practice is to traverse the contour C in a counterclockwise direction, as indicated in Fig. 9.32(a). Two different situations may arise in the *F*-plane, as described here:

▶ The interior of contour C in the *s*-plane is mapped onto the interior of contour Γ in the *F*-plane, as illustrated in Fig. 9.32(b). In this case, contour Γ is traversed in the counterclockwise direction (i.e., in the same direction as contour C).

▶ The interior of contour C in the *s*-plane is mapped onto the exterior of contour Γ in the *F*-plane, as illustrated in Fig. 9.32(c). In this second case, contour Γ is traversed in the clockwise direction (i.e., in the opposite direction to contour C).

On the basis of this figure, we may thus offer the following definition: *A region or point in a plane is said to be "enclosed" by a closed contour if it is encircled in the counterclockwise direction.* For example, point s_A inside contour C in Fig. 9.32(a) is mapped onto point F_A inside contour Γ in Fig. 9.32(b) but outside contour Γ in Fig. 9.32(c). Thus point F_A is enclosed by Γ in Fig. 9.32(b) but not in Fig. 9.32(c).

The notion of enclosure as defined herein should be carefully distinguished from that of encirclement. For the latter, we may offer the following definition: *A point is said to be encircled by a closed contour if it lies inside the contour.* It is possible for a point of interest in the *F*-plane to be encircled more than once in a positive or negative direction. In particular, the contour Γ in the *F*-plane makes a total of m positive encirclements of a point A, say, if the phasor (i.e., the line drawn from point A to a moving point $F(s_1)$ on the

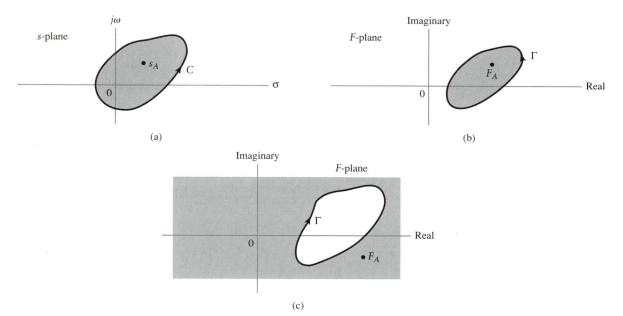

FIGURE 9.32 (a) Contour C traversed in counterclockwise direction in *s*-plane. (b) and (c) Two possible ways in which contour C is mapped onto *F*-plane.

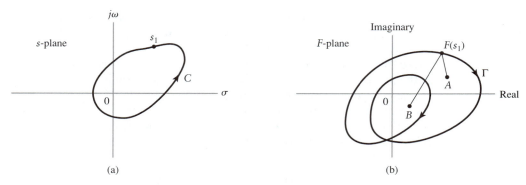

(a)   (b)

FIGURE 9.33   Illustration of the definition of encirclement. As point s_1 traverses contour C in the s-plane in the counterclockwise direction as shown in part (a) of the figure, point A is encircled by contour Γ only once and point B is encircled twice, both in the clockwise direction in the F-plane, as shown in part (b) of the figure.

contour Γ) rotates through $2\pi m$ in a counterclockwise direction as the point s_1 traverses contour C in the s-plane once in the same counterclockwise direction. Thus, in the situation described in Fig. 9.33, we find that as the point s_1 traverses the contour C in the s-plane once in the counterclockwise direction, point A is encircled by contour Γ in the F-plane only once, whereas point B is encircled by contour Γ twice, both in the clockwise direction. Thus, in the case of point A, we have $m = -1$; and in the case of point B, we have $m = -2$.

▶ **Drill Problem 9.14**   Consider the situations described in Fig. 9.34. How many times are points A and B encircled by the locus Γ in this figure?

Answer: For point A, the number of encirclements is 2, and for point B it is 1.   ◀

■ **PRINCIPLE OF THE ARGUMENT**

Assume that a function $F(s)$ is a single-valued rational function of s that satisfies the following two requirements:

   1. The function $F(s)$ is *analytic* in the interior of a closed contour C in the s-plane, except at a finite number of poles. The analytic requirement means that at every

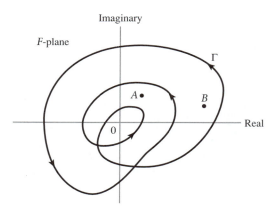

FIGURE 9.34   Diagram for Drill Problem 9.14.

point $s = s_0$ inside the contour C, excluding the points where the poles are located, the function $F(s)$ has a derivative at $s = s_0$ and at every point in the neighborhood of s_0.

2. The function $F(s)$ has neither poles nor zeros on the contour C.

We may then state the *principle of the argument* in complex-variable theory as follows:

$$\frac{1}{2\pi} \arg\{F(s)\}_C = Z - P \tag{9.85}$$

where $\arg\{F(s)\}_C$ is the change in the argument (angle) of the function $F(s)$ as the contour C is traversed once in the counterclockwise direction; and Z and P are the number of zeros and poles, respectively, of the function $F(s)$ inside the contour C. Note that the change in the magnitude of $F(s)$, as s moves on the contour C once, is zero because $F(s)$ is single valued and the contour C is closed; hence $\arg\{F(s)\}_C$ is the only term representing the change in $F(s)$ on the left-hand side of Eq. (9.85) as s traverses the contour C once. Suppose now that the origin in the F-plane is encircled a total of m times as the contour C is traversed once in the counterclockwise direction. We may then write

$$\arg\{F(s)\}_C = 2\pi m \tag{9.86}$$

in light of which Eq. (9.85) reduces to

$$m = Z - P \tag{9.87}$$

As mentioned previously, m may be positive or negative. Accordingly, we may identify three distinct cases, given that the contour C is traversed in the s-plane once in the counterclockwise direction:

1. $Z > P$, in which case the contour Γ encircles the origin of the F-plane m times in the counterclockwise direction.
2. $Z = P$, in which case the origin of the F-plane is not encircled by the contour Γ.
3. $Z < P$, in which case the contour Γ encircles the origin of the F-plane m times in the clockwise direction.

■ NYQUIST CONTOUR

We are now equipped with the tools we need to return to the issue at hand: evaluation of the stability of a linear feedback system. From Eq. (9.6), we know that the characteristic equation of such a system is defined in terms of its loop transfer function $L(s) = G(s)H(s)$ as follows:

$$1 + L(s) = 0$$

or, equivalently,

$$F(s) = 0 \tag{9.88}$$

where $F(s)$ is the return difference. With $F(s)$ as the function of interest, the Nyquist stability criterion is basically an application of the principle of the argument, as described here: determine the number of roots of the characteristic equation (9.88) that lie in the right half of the s-plane. With this part of the s-plane as the domain of interest, we may solve the stability problem by considering the contour C shown in Fig. 9.35, which is constructed so as to satisfy the requirements of the principle of the argument:

▶ The semicircle has a radius R that tends to infinity; hence the contour C encompasses the entire right half of the s-plane as $R \to \infty$.

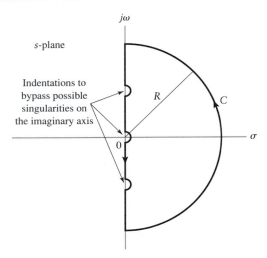

FIGURE 9.35   Nyquist contour.

▶ The small semicircles shown along the imaginary axis are included to bypass the singularities (i.e., poles and zeros) of $F(s)$ that are located at the centers of the semi-circles. This would ensure that the return difference $F(s)$ has no poles or zeros on the contour C.

The contour C of Fig. 9.35 is referred to as the *Nyquist contour.*

Let Γ be the closed contour traced by the return difference $F(s)$ in the F-plane as the Nyquist contour C of Fig. 9.35 is traversed once in the s-plane in the counterclockwise direction. If Z is the (unknown) number of the zeros of $F(s)$ in the right half of the s-plane, then from Eq. (9.87) we readily see that

$$Z = m - P \tag{9.89}$$

where P is the number of poles of $F(s)$ in the right half of the s-plane, and m is the net number of counterclockwise encirclements of the origin in the F-plane by the contour Γ. Recognizing that the zeros of $F(s)$ are the same as the roots of the system's characteristic equation, we may now formally state the Nyquist stability criterion as follows: *A linear feedback system is absolutely stable, provided that its characteristic equation has no roots in the right half of the s-plane or on the jω-axis, that is,*

$$m - P = 0 \tag{9.90}$$

The Nyquist stability criterion may be simplified for a large class of linear feedback systems as described here. By definition, the return difference $F(s)$ is related to the loop transfer function $L(s)$ as follows:

$$F(s) = 1 + L(s) \tag{9.91}$$

The poles of $F(s)$ are therefore the same as the poles of $L(s)$. If $L(s)$ has no poles in the right half of the s-plane (i.e., the system is stable in the absence of feedback), then $P = 0$, and Eq. (9.90) reduces to $m = 0$. That is, the feedback system is absolutely stable provided that the contour Γ does not encircle the origin in the F-plane.

From Eq. (9.91) we also note that the origin in the *F*-plane corresponds to the point $(-1, 0)$ in the *L*-plane. For the case when $L(s)$ has no poles in the right half of the *s*-plane, we may therefore reformulate the Nyquist stability criterion as follows: *A linear feedback system with loop transfer function L(s) is absolutely stable, provided that the locus traced by L(s) in the L-plane does not encircle the point $(-1, 0)$ as s traverses the Nyquist contour once in the s-plane.* The point $(-1, 0)$ in the *L*-plane is called the *critical point*.

Typically, the loop transfer function $L(s)$ has more poles than zeros, which means that $L(s)$ approaches zero as *s* approaches infinity. Hence the contribution of the semicircular part of the Nyquist contour *C* to the $L(s)$-locus approaches zero as the radius *R* approaches infinity. That is, the $L(s)$-locus reduces simply to a plot of $L(j\omega)$ for $-\infty < \omega < \infty$ (i.e., the values of *s* on the imaginary axis of the *s*-plane). It is also helpful to view the locus as a *polar* plot of $L(j\omega)$ for varying ω, with $|L(j\omega)|$ denoting the magnitude and arg$\{L(j\omega)\}$ denoting the phase angle. The resulting plot is called the *Nyquist locus* or *Nyquist diagram*.

Construction of the Nyquist locus is simplified by recognizing that

$$|L(-j\omega)| = |L(j\omega)|$$

and

$$\text{arg}\{L(-j\omega)\} = -\text{arg}\{L(j\omega)\}$$

Accordingly, it is only necessary to plot the Nyquist locus for positive frequencies $0 \leq \omega < \infty$. The locus for negative frequencies is inserted simply by reflecting the locus for positive frequencies about the real axis of the *L*-plane, as illustrated in Fig. 9.36 for a system whose loop transfer function has a pole at $s = 0$. Figure 9.36(a) represents a stable system, whereas Fig. 9.36(b) represents an unstable system whose characteristic equation has two roots in the right-half plane, and for which the Nyquist locus encircles the critical point $(-1, 0)$ twice in the counterclockwise direction. Note that in Fig. 9.36, both Nyquist loci start at $\omega = \infty$ and terminate at $\omega = 0$, so as to be consistent with the fact that the Nyquist locus in Fig. 9.35 is traversed in the counterclockwise direction.

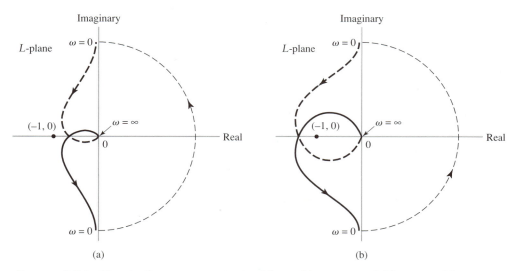

FIGURE 9.36 Nyquist diagrams representing (a) a stable system and (b) an unstable system.

EXAMPLE 9.13   Using the Nyquist stability criterion, investigate the stability performance of the three-stage transistor feedback amplifier considered in Example 9.10. Putting $s = j\omega$ in $L(s)$, we get the loop frequency response

$$L(j\omega) = \frac{6K}{(j\omega + 1)(j\omega + 2)(j\omega + 3)}$$

Show that the amplifier is stable with $K = 6$.

Solution: The magnitude and phase of $L(j\omega)$ are given by (with $K = 6$)

$$|L(j\omega)| = \frac{36}{(\omega^2 + 1)^{1/2}(\omega^2 + 4)^{1/2}(\omega^2 + 9)^{1/2}}$$

and

$$\arg\{L(j\omega)\} = -\tan^{-1}(\omega) - \tan^{-1}\left(\frac{\omega}{2}\right) - \tan^{-1}\left(\frac{\omega}{3}\right)$$

Figure 9.37 shows a plot of the Nyquist locus, which is seen not to encircle the critical point $(-1, 0)$. The amplifier is therefore stable.

▶ **Drill Problem 9.15**   Consider a feedback amplifier described by the loop frequency response

$$L(j\omega) = \frac{K}{(1 + j\omega)^3}$$

Using the Nyquist stability criterion, show that the amplifier is on the verge of instability for $K = 8$.   ◀

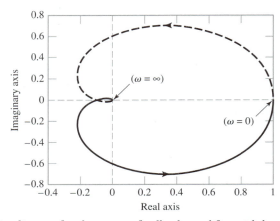

FIGURE 9.37   Nyquist diagram for three-stage feedback amplifier with loop frequency response

$$L(j\omega) = \frac{6K}{(j\omega + 1)(j\omega + 2)(j\omega + 3)} \quad \text{with } K = 6.$$

9.16 Bode Diagram

Another method for studying the stability performance of a feedback system involves plotting the loop transfer function $L(s)$ of the system for $s = j\omega$ in the form of two separate graphs. In one graph the magnitude of $L(j\omega)$ is plotted in decibels versus the logarithm of ω. In the other graph, the phase of $L(j\omega)$ in degrees is plotted versus the logarithm of ω. The combination of these two graphs is called the *Bode diagram* or *Bode plot*. The attractive feature of this method is twofold:

▶ The relative ease and speed with which the necessary calculations for different frequencies can be performed, making the Bode diagram a useful design tool.

▶ The concepts learned from the Bode diagram are very helpful in developing engineering intuition regarding the effect of pole–zero placement on the system frequency response.

To proceed with a description of the Bode diagram, let $L(s)$ be expressed in the factored form shown in Eq. (9.69). Setting $s = j\omega$, and multiplying the logarithm of $L(j\omega)$ by 20, we get (assuming that the loop gain K is positive)

$$|L(j\omega)|_{dB} = 20 \log_{10}K + 20 \sum_{i=1}^{M} \log_{10}\left|1 - \frac{j\omega}{c_i}\right| - 20 \sum_{k=1}^{N} \log_{10}\left|1 - \frac{j\omega}{d_k}\right| \quad (9.92)$$

and

$$\arg\{L(j\omega)\} = \sum_{i=1}^{M} \arg\left\{1 - \frac{j\omega}{c_i}\right\} - \sum_{k=1}^{N} \arg\left\{1 - \frac{j\omega}{d_k}\right\} \quad (9.93)$$

(In Eqs. (9.92) and (9.93) we have used the index k in place of j so as not to cause confusion with j, the square root of -1.) Hence, in computing the *gain component* $|L(j\omega)|_{dB}$, the product and division factors in the magnitude response $|L(j\omega)|$ are replaced by additions and subtractions, respectively. Moreover, the individual contributions of the zero and pole factors to the phase response $\arg\{L(j\omega)\}$ involve additions and subtractions in a natural way. The computation of $L(j\omega)$ for varying ω is thereby made relatively easy.

The intuitive appeal of the Bode diagram comes from the fact that the computation of $|L(j\omega)|_{dB}$ may readily be approximated by straight-line segments. The nature of the approximation depends on whether the pole/zero factor in question is a simple or quadratic factor, as described next.

1. *Simple factor.* Consider the case of a pole factor $(1 - s/d_0)$ for which $d_0 = -\sigma_0$, where σ_0 is some real number. The contribution of this pole factor to the gain component $|L(j\omega)|_{dB}$ is written as

$$-20 \log_{10}\left|1 + \frac{j\omega}{\sigma_0}\right| = -10 \log_{10}\left(1 + \frac{\omega^2}{\sigma_0^2}\right) \quad (9.94)$$

We may obtain asymptotic approximations of this contribution by considering both very small and very large values of ω, compared to σ_0, as described here:

▶ *Low-frequency asymptote.* For $\omega \ll \sigma_0$, Eq. (9.94) approximates to

$$-20 \log_{10}\left|1 + \frac{j\omega}{\sigma_0}\right| \simeq -10 \log_{10}(1) = 0 \text{ dB} \quad (9.95)$$

which consists simply of the 0-dB line.

▶ *High-frequency asymptote.* For $\omega \gg \sigma_0$, Eq. (9.94) approximates to

$$-20 \log_{10}\left|1 + \frac{j\omega}{\sigma_0}\right| \simeq -10 \log_{10}\left(\frac{\omega}{\sigma_0}\right)^2 = -20 \log_{10}\left(\frac{\omega}{\sigma_0}\right) \text{ dB} \qquad (9.96)$$

which represents a straight line with a slope of -20 dB/decade.

These two asymptotes intersect at $\omega = \sigma_0$. Accordingly, the contribution of the pole factor $(1 + s/\sigma_0)$ to $|L(j\omega)|_{\text{dB}}$ may be approximated by a pair of straight-line segments, as illustrated in Fig. 9.38(a). The intersection frequency σ_0 is called a *corner* or *break frequency* of the Bode diagram. Figure 9.38(a) also includes the actual magnitude characteristic of a simple pole factor. The approximation error, that is, the difference between the actual magnitude characteristic and its approximate form, attains its maximum value of 3 dB at the corner frequency σ_0. The table included in Fig. 9.38(a) presents a listing of the approximation errors for a logarithmically spaced set of frequencies normalized with respect to σ_0.

The phase response of the simple pole factor is defined by

$$\arg\left\{\frac{1}{1 + j\omega/\sigma_0}\right\} = -\tan^{-1}\left(\frac{\omega}{\sigma_0}\right) \qquad (9.97)$$

which is shown plotted exactly in Fig. 9.38(b). Figure 9.38(b) also includes a piece-wise linear approximation to the phase response; the approximation is shown dashed.

2. *Quadratic factor.* Consider next the quadratic pole factor represented by a pair of complex-conjugate poles, written as follows:

$$Q(s) = \frac{1}{1 + (2\zeta/\omega_n)s + s^2/\omega_n^2}$$

$$= \frac{\omega_n^2}{(s + \zeta\omega_n - j\omega_n\sqrt{1 - \zeta^2})(s + \zeta\omega_n + j\omega_n\sqrt{1 - \zeta^2})} \qquad (9.98)$$

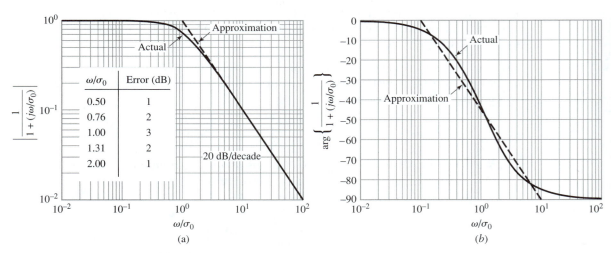

FIGURE 9.38 (a) Gain component of Bode diagram for first-order pole factor: $1/(1 + s/\sigma_0)$. (b) Phase component of Bode diagram.

where we have assumed the damping factor $\zeta \leq 1$. For $s = j\omega$, the magnitude of $Q(j\omega)$ in decibels is

$$|Q(j\omega)|_{dB} = -10 \log_{10}\left[\left(1 - \left(\frac{\omega}{\omega_n}\right)^2\right)^2 + 4\zeta^2\left(\frac{\omega}{\omega_n}\right)^2\right] \qquad (9.99)$$

For $\omega \ll \omega_n$, Eq. (9.99) is approximated as

$$|Q(j\omega)|_{dB} \simeq -10 \log_{10}(1) = 0 \text{ dB}$$

For $\omega \gg \omega_n$, it is approximated as

$$|Q(j\omega)|_{dB} \simeq -10 \log_{10}\left(\frac{\omega}{\omega_n}\right)^4 = -40 \log_{10}\left(\frac{\omega}{\omega_n}\right)$$

Thus the gain component $|Q(j\omega)|_{dB}$ may be approximated by a pair of straight-line segments, one represented by the 0-dB line and the other having a slope of -40 dB/decade, as shown in Fig. 9.39. The two asymptotes intersect at $\omega = \omega_n$, which is referred to as the corner frequency of the quadratic factor. However, unlike the case of a simple pole factor, the actual magnitude of the quadratic pole factor $Q(j\omega)$ may differ markedly from its asymptotic approximation, depending on how small the damping factor ζ is compared to unity. Figure 9.40(a) shows the exact plot of $|Q(j\omega)|_{dB}$ for three different values of ζ in the range $0 < \zeta < 1$. The difference between the exact curve and the asymptotic approximation of $|Q(j\omega)|_{dB}$ defines the approximation error. Evaluating Eq. (9.99) at $\omega = \omega_n$ and noting that the corresponding value of the asymptotic approximation is 0 dB, we find that the value of the error at $\omega = \omega_n$ is given by

$$(\text{Error})_{\omega=\omega_n} = -20 \log_{10}(2\zeta) \text{ dB} \qquad (9.100)$$

This error is zero for $\zeta = 0.5$, positive for $\zeta < 0.5$, and negative for $\zeta > 0.5$.

From Eq. (9.98), we find that the phase component of $Q(j\omega)$ is given by

$$\arg\{Q(j\omega)\} = -\tan^{-1}\left(\frac{2\zeta(\omega/\omega_n)}{1 - (\omega/\omega_n)^2}\right) \qquad (9.101)$$

Figure 9.40(b) shows exact plots of $\arg\{Q(j\omega)\}$ for the same values of ζ used in Fig. 9.40(a). At $\omega = \omega_n$, we have

$$\arg\{Q(j\omega_n)\} = -90°$$

Note also the change in the algebraic sign of $\arg\{Q(j\omega)\}$ at $\omega = \omega_n$.

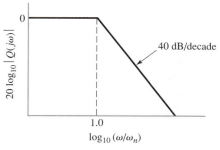

FIGURE 9.39   Asymptotic approximation to $20 \log_{10}|Q(j\omega)|$, where

$$Q(s) = \frac{1}{1 + (2\zeta/\omega_n)s + s^2/\omega_n^2}$$

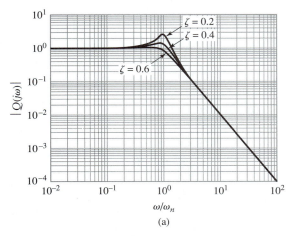
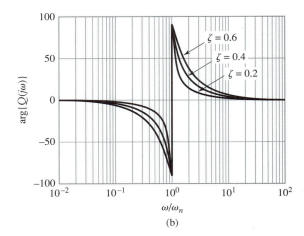

FIGURE 9.40   Bode diagram of second-order pole factor:

$$Q(s) = \frac{1}{1 + (2\zeta/\omega_n)s + s^2/\omega_n^2}$$

for varying ζ. (a) Gain component. (b) Phase component.

We can now see the practical merit of the Bode diagram. Specifically, using the approximations described above for simple and quadratic pole/zero factors of the loop transfer function $L(s)$, we can quickly make a sketch of $|L(j\omega)|_{dB}$ and investigate the effect of pole–zero placement on $|L(j\omega)|$.

EXAMPLE 9.14   Consider a three-stage feedback amplifier having the loop frequency response

$$L(j\omega) = \frac{6K}{(j\omega + 1)(j\omega + 2)(j\omega + 3)}$$

$$= \frac{K}{(1 + j\omega)(1 + j\omega/2)(1 + j\omega/3)}$$

Construct the Bode diagram for $K = 6$.

Solution: The numerator in the second line of $L(j\omega)$ is a constant equal to 6 for $K = 6$. Expressed in decibels, this numerator contributes a constant gain equal to

$$20 \log_{10} 6 = 15.56 \text{ dB}$$

The denominator is made up of three simple pole factors with corner frequencies equal to 1, 2, and 3 rad/s. Putting the contributions of the numerator and denominator terms together, we get the straight-line approximation to the gain component of $L(j\omega)$ shown in Fig. 9.41.

Figures 9.42(a) and (b) show the exact gain and phase components of $L(j\omega)$, respectively.

■ RELATIVE STABILITY OF A FEEDBACK SYSTEM

Now that we have familiarized ourselves with the construction of the Bode diagram, we are ready to consider its use in studying the stability problem. The *relative stability* of a

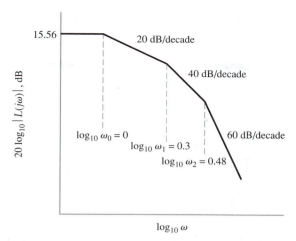

FIGURE 9.41   Straight-line approximation to gain component of Bode diagram for open-loop response

$$L(j\omega) = \frac{6K}{(j\omega + 1)(j\omega + 2)(j\omega + 3)} \quad \text{for } K = 6.$$

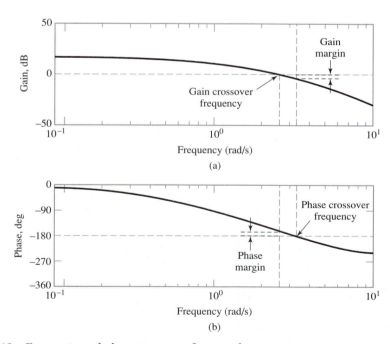

FIGURE 9.42   Exact gain and phase responses for open-loop response

$$L(j\omega) = \frac{6K}{(j\omega + 1)(j\omega + 2)(j\omega + 3)} \quad \text{for } K = 6.$$

feedback system is determined by how close a plot of the loop transfer function $L(s)$ of the system is to the critical point $L(s) = -1$ for $s = j\omega$. With the Bode diagram consisting of two graphs, one pertaining to $20 \log_{10}|L(j\omega)|$ and the other to $\arg\{L(j\omega)\}$, there are two commonly used measures of relative stability, as illustrated in Fig. 9.43.

The first of these two measures is the *gain margin*, expressed in decibels. For a stable feedback system, the gain margin is defined as the number of decibels by which $20 \log_{10}|L(j\omega)|$ must be changed to bring the system to the verge of instability. Assume that when the phase angle of the loop frequency response $L(j\omega)$ equals $-180°$, its magnitude $|L(j\omega)|$ equals $1/K_m$, where $K_m > 1$. The quantity $20 \log_{10} K_m$ is equal to the gain margin of the system, as indicated in Fig. 9.43(a). The frequency ω_p at which $\arg\{L(j\omega_p)\} = -180°$ is called the *phase crossover frequency*.

The second measure of relative stability is the *phase margin*, expressed in degrees. Again for a stable feedback system, the phase margin is defined as the magnitude of the minimum angle by which $\arg\{L(j\omega)\}$ must be changed in order to intersect the critical point $L(j\omega) = -1$. Assume that when the magnitude $|L(j\omega)|$ equals unity, the phase angle $\arg\{L(j\omega)\}$ equals $-180° + \phi_m$. The angle ϕ_m is called the phase margin of the system, as indicated in Fig. 9.43(b). The frequency ω_g at which $|L(j\omega_g)| = 1$ is called the *gain crossover frequency*.

On the basis of these definitions, we can make the following observations on the stability of a feedback system:

1. For a stable feedback system, both the gain margin and phase margin must be positive. By implication, the phase crossover frequency must be larger than the gain crossover frequency.
2. The system is unstable if the gain margin is negative, the phase margin is negative, or both are negative.

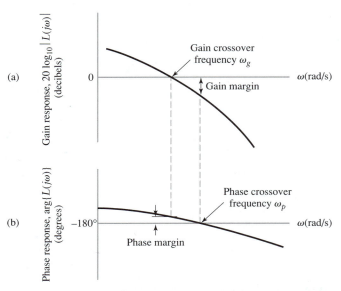

FIGURE 9.43 Illustration of the definitions of (a) gain margin and gain crossover frequency and (b) phase margin and phase crossover frequency.

EXAMPLE 9.15   Calculate the gain and phase margins for the loop frequency response of Example 9.14 for $K = 6$.

Solution: Figure 9.42 also includes the locations of the gain and phase crossover frequencies:

$$\omega_p = \text{phase crossover frequency} = 3.317 \text{ rad/s}$$
$$\omega_g = \text{gain crossover frequency} = 2.59 \text{ rad/s}$$

With $\omega_p > \omega_g$, we have further confirmation that the three-stage feedback amplifier described by the loop frequency response $L(j\omega)$ of Examples 9.13 and 9.14 is stable for $K = 6$.

At $\omega = \omega_p$, we have, by definition, $\arg\{L(j\omega_p)\} = -180°$. At this frequency, we find from Fig. 9.42(a) that

$$20 \log_{10}|L(j\omega_p)| = -4.437 \text{ dB}$$

The gain margin is therefore equal to 4.437 dB.

At $\omega = \omega_g$, we have, by definition, $|L(j\omega_g)| = 1$. At this frequency we find from Fig. 9.42(b) that

$$\arg\{L(j\omega_p)\} = -162.01°$$

The phase margin is therefore equal to

$$180 - 162.01 = 17.99°$$

These stability margins are included in the Bode diagram of Fig. 9.42.

■ RELATION BETWEEN PHASE MARGIN AND DAMPING RATIO

The guidelines used in the classical (frequency-domain) approach to the design of a linear feedback control system are usually derived from the analysis of second-order servomechanism dynamics. Such an approach is justified on the following grounds. First, when the loop gain is large the closed-loop transfer function of the system develops a pair of dominant complex-conjugate poles. Second, a second-order model provides an adequate approximation to a higher-order model whose transfer function is dominated by a pair of complex-conjugate poles (see Section 9.14).

Consider then a second-order system whose loop transfer function is given by

$$T(s) = \frac{K}{s(\tau s + 1)}$$

This system was studied in Section 9.11. When the loop gain K is large enough to produce an underdamped step response (i.e., the closed-loop poles form a complex-conjugate pair), the damping ratio and natural frequency of the system are defined by, respectively (see Drill Problem 9.8),

$$\zeta = \frac{1}{2\sqrt{\tau K}} \quad \text{and} \quad \omega_n = \sqrt{\frac{K}{\tau}}$$

Accordingly, we may redefine the loop transfer function in terms of ζ and ω_n as follows:

$$L(s) = \frac{\omega_n^2}{s(s + 2\zeta\omega_n)} \tag{9.102}$$

By definition, the gain crossover frequency ω_g is determined from the relation

$$|L(j\omega_g)| = 1$$

Therefore, putting $s = j\omega_g$ in Eq. (9.102) and solving for ω_g, we get

$$\omega_g = \omega_n\sqrt{\sqrt{4\zeta^4 + 1} - 2\zeta^2} \tag{9.103}$$

The phase margin, measured in degrees, is therefore

$$
\begin{aligned}
\phi_m &= 180° - 90° - \tan^{-1}\left(\frac{\omega_g}{2\zeta\omega_n}\right) \\
&= \tan^{-1}\left(\frac{2\zeta\omega_n}{\omega_g}\right) \\
&= \tan^{-1}\left(\frac{2\zeta}{\sqrt{\sqrt{4\zeta^4 + 1} - 2\zeta^2}}\right)
\end{aligned}
\tag{9.104}
$$

Equation (9.104) provides an exact relationship between the phase margin ϕ_m, a quantity pertaining to the open-loop frequency response, and the damping ratio ζ, a quantity pertaining to the closed-loop step response. This exact relationship is plotted in Fig. 9.44. For values of the damping ratio in the range $0 \leq \zeta \leq 0.6$, Eq. (9.104) is closely approximated by

$$\phi_m \approx 100\zeta, \quad 0 \leq \zeta \leq 0.6$$

Equivalently, we may write

$$\zeta = \frac{\phi_m}{100}, \quad 0 \leq \zeta \leq 0.6 \tag{9.105}$$

where, as mentioned previously, the phase margin ϕ_m is measured in degrees. This approximation is included as the dashed line in Fig. 9.44.

Once we have obtained a value for the damping ratio ζ, we can use Eq. (9.103) to determine the corresponding value of the natural frequency ω_n in terms of the gain crossover frequency ω_g. With the values of ζ and ω_n at hand, we can then go on to determine

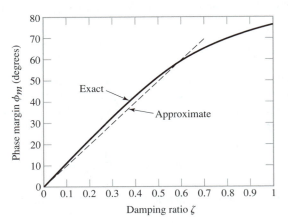

FIGURE 9.44 Graph of the relationship between damping ratio and phase margin computed for a second-order system.

the rise time, peak time, percentage overshoot, and settling time as descriptors of the step response using the formulas derived in Example 9.5.

■ *RELATION BETWEEN THE BODE DIAGRAM AND NYQUIST CRITERION

The Bode diagram discussed in this section and the Nyquist diagram discussed in the previous section are frequency-domain techniques that offer different perspectives on the stability performance of a linear feedback system. The Bode diagram consists of two separate graphs, one for displaying the gain response and the other for displaying the phase response. On the other hand, the Nyquist diagram combines the magnitude and phase responses in a single polar plot.

The Bode diagram illustrates the frequency response of the system. It uses straight-line approximations that can be sketched with little effort, thereby providing an easy-to-use method for assessing the absolute stability and relative stability of the system. Accordingly, a great deal of insight can be derived from using the Bode diagram to design a feedback system by frequency-domain techniques.

The Nyquist criterion is important for two reasons:

1. It provides the theoretical basis for using the loop frequency response to determine the stability of a closed-loop system.
2. It may be used to assess stability from experimental data describing the system.

The Nyquist criterion is the ultimate test for stability, in the sense that the determination of stability may be misleading unless it is used in conjunction with the Nyquist criterion. This is particularly so when the system is *conditionally stable*, which means that the system goes through stable/unstable conditions as the loop gain is varied. Such a phenomenon is illustrated in Fig. 9.45, where we see that there are *two* phase crossover frequencies, namely, ω_{p1} and ω_{p2}. For $\omega_{p1} \leq \omega \leq \omega_{p2}$, the magnitude response $|L(j\omega)|$ is greater than unity. Moreover, we see that the gain crossover frequency ω_g is greater than both ω_{p1} and ω_{p2}. Based on these superficial observations, it would be tempting to conclude that a closed-loop feedback system represented by Fig. 9.45 is unstable. In reality, however, the system is stable since the Nyquist locus shown therein does *not* encircle the critical point $(-1, 0)$.

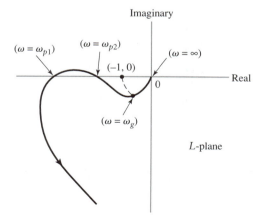

FIGURE 9.45 Nyquist diagram illustrating the notion of conditional stability.

A closed-loop system characterized by a Nyquist locus such as that shown in Fig. 9.45 is said to be conditionally stable for the following reason: a reduced loop gain or an increased loop gain will make the system unstable.

▶ **Drill Problem 9.16**   Verify that the Nyquist locus shown in Fig. 9.45 does not encircle the critical point $(-1, 0)$.   ◀

*9.17   *Sampled-Data Systems*

In the treatment of feedback control systems discussed thus far, we have assumed that the whole system behaves in a continuous-time fashion. However, in many applications of control theory, a digital computer is included as an integral part of the control system. Examples of digital control of dynamic systems include such important applications as aircraft autopilots, mass-transit vehicles, oil refineries, and paper-making machines. A distinct advantage of using a digital computer for control is increased *flexibility* of the control program and decision-making.

The use of a digital computer to calculate the control action for a continuous-time system introduces two effects: sampling and quantization. Sampling is made necessary by virtue of the fact that a digital computer can only manipulate discrete-time signals. Thus samples are taken from physical signals such as position or velocity, and these samples are used in the computer to calculate the appropriate control. As for quantization, it arises because the digital computer operates with finite arithmetic. The computer takes in numbers, stores them, performs calculations on them, and then returns them with some finite accuracy. In other words, quantization introduces roundoff errors into the calculations performed by the computer. In this section, we confine our attention to the effects of sampling in feedback control systems.

Feedback control systems using digital computers are "hybrid" systems, hybrid in the sense that continuous-time signals appear in some places and discrete-time signals appear in other places. Such systems are commonly referred to as *sampled-data systems*. The hybrid nature of such a system makes its analysis somewhat less straightforward than that of a purely continuous-time system or a purely discrete-time system, since it requires the combined use of both continuous-time and discrete-time analysis methods.

■ SYSTEM DESCRIPTION

Consider, for example, the feedback control system of Fig. 9.46, in which the digital computer (controller) performs the controlling action. The analog-to-digital (A/D) converter, at the front end of the system, acts on the continuous-time error signal and converts it into a stream of numbers for processing in the computer. The control calculated by the

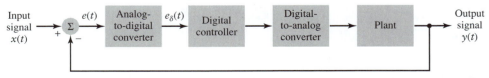

FIGURE 9.46   Block diagram of sampled-data feedback control system, which includes both discrete-time and continuous-time components.

computer is a second stream of numbers, which is converted by the digital-to-analog (D/A) converter back into a continuous-time signal applied to the plant.

For the purpose of analysis, the various components of the sampled-data system of Fig. 9.46 are modeled as follows:

1. *A/D converter.* This component is represented simply by an impulse sampler. Let $e(t)$ denote the error signal, defined as the difference between the system input $x(t)$ and system output $y(t)$. Let $e[n] = e(n\mathcal{T})$ be the samples of $e(t)$, where $\mathcal{T}$ is the sampling period. Recall from Chapter 4 that the discrete-time signal $e[n]$ can be represented by the continuous-time signal

$$e_\delta(t) = \sum_{n=-\infty}^{\infty} e[n]\delta(t - n\mathcal{T}) \tag{9.106}$$

2. *Digital controller.* The computer program responsible for the control is viewed as a difference equation, whose input–output effect is represented by the z-transform $D(z)$ or, equivalently, the impulse response $d[n]$:

$$D(z) = \sum_{n=-\infty}^{\infty} d[n]z^{-n} \tag{9.107}$$

Alternatively, we may represent the computer program by the continuous-time transfer function $D_\delta(s)$, where s is the complex frequency in the Laplace transform. This follows from the continuous-time representation of the signal $d[n]$ given by

$$d_\delta(t) = \sum_{n=-\infty}^{\infty} d[n]\delta(t - n\mathcal{T})$$

Taking the Laplace transform of $d_\delta(t)$ gives

$$D_\delta(s) = \sum_{n=-\infty}^{\infty} d[n]e^{-ns\mathcal{T}} \tag{9.108}$$

From Eqs. (9.107) and (9.108) we see that, given the transfer function $D_\delta(s)$, we may determine the corresponding z-transform $D(z)$ by letting $z = e^{\mathcal{T}s}$, as shown by

$$D(z) = D_\delta(s)\big|_{e^{\mathcal{T}s}=z} \tag{9.109}$$

Conversely, given $D(z)$, we may determine $D_\delta(s)$ by writing

$$D_\delta(s) = D(z)\big|_{z=e^{\mathcal{T}s}} \tag{9.110}$$

The inverse z-transform of $D(z)$ is a sequence of numbers whose individual values are equal to the impulse response $d[n]$. On the other hand, the inverse Laplace transform of $D_\delta(s)$ is a sequence of impulses whose individual strengths are weighted by the impulse response $d[n]$. Note also that $D_\delta(s)$ is periodic in s with a period equal to $2\pi/\mathcal{T}$.

3. *D/A converter.* A commonly used type of D/A converter is the *zero-order hold*, which simply holds the amplitude of an incoming sample constant for the entire sampling period, until the next sample arrives. The impulse response of the zero-order hold, denoted by $h_o(t)$, may thus be described as shown in Fig. 9.47 (see Section 4.7); that is,

$$h_o(t) = \begin{cases} 1, & 0 < t < \mathcal{T} \\ 0, & \text{otherwise} \end{cases}$$

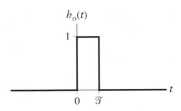

FIGURE 9.47   Impulse response of zero-order hold.

The transfer function of the zero-order hold is therefore

$$H_o(s) = \int_o^{\mathcal{T}} e^{-st}\, dt$$
$$= \frac{1 - e^{-\mathcal{T}s}}{s} \tag{9.111}$$

4. *Plant.* The plant operates on the continuous-time control delivered by the zero-order hold to produce the overall system output. The plant, as usual, is represented by the transfer function $G(s)$.

On the basis of these representations, we may model the digital control system of Fig. 9.46 as depicted in Fig. 9.48.

■ PROPERTIES OF LAPLACE TRANSFORMS OF SAMPLED SIGNALS

To prepare the way for determining the closed-loop transfer function of the sampled-data system modeled in Fig. 9.48, we need to introduce some properties of the Laplace transforms of sampled signals. Let $a_\delta(t)$ denote the impulse-sampled version of a continuous-time signal $a(t)$, that is,

$$a_\delta(t) = \sum_{n=-\infty}^{\infty} a(n\mathcal{T})\delta(t - n\mathcal{T})$$

Let $A_\delta(s)$ denote the Laplace transform of $a_\delta(t)$. (In the control literature, $a^*(t)$ and $A^*(s)$ are commonly used to denote the impulse-sampled version of $a(t)$ and its Laplace transform, respectively, and so $A^*(s)$ is referred to as a *starred transform*. We have not used this terminology largely because the asterisk is used to denote complex conjugation in this book.) The Laplace transform $A_\delta(s)$ has two important properties that follow from the material on impulse sampling presented in Chapter 4:

1. *The Laplace transform $A_\delta(s)$ of a sampled signal $a_\delta(t)$ is periodic in the complex variable s with period $j\omega_s$, where $\omega_s = 2\pi/\mathcal{T}$, and $\mathcal{T}$ is the sampling period. This*

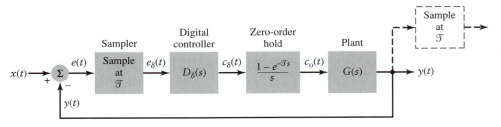

FIGURE 9.48   Model of sampled-data feedback control system shown in Fig. 9.46.

property follows directly from Eq. (4.32). Specifically, using s in place of $j\omega$ in that equation, we may write

$$A_\delta(s) = \frac{1}{\mathcal{T}} \sum_{k=-\infty}^{\infty} A(s - jk\omega_s) \tag{9.112}$$

from which we readily find that

$$A_\delta(s) = A_\delta(s + j\omega_s) \tag{9.113}$$

2. *If the Laplace transform $A(s)$ of the original continuous-time signal $a(t)$ has a pole at $s = s_1$, then the Laplace transform $A_\delta(s)$ of the sampled signal $a_\delta(t)$ has poles at $s = s_1 + jm\omega_s$, where $m = 0, \pm 1, \pm 2, \ldots$.* This property follows directly from Eq. (9.112) by rewriting it in the expanded form

$$A_\delta(s) = \frac{1}{\mathcal{T}} [A(s) + A(s + j\omega_s) + A(s - j\omega_s) + A(s + j2\omega_s) + A(s + j2\omega_s) + \cdots]$$

Here we clearly see that if $A(s)$ has a pole at $s = s_1$, then each term of the form $A(s - jm\omega_s)$ contributes a pole at $s = s_1 + jm\omega_s$, because

$$A(s - jm\omega_s)\big|_{s=s_1+jm\omega_s} = A(s_1), \qquad m = 0, \pm 1, \pm 2, \ldots$$

Property 2 of $A_\delta(s)$ is illustrated in Fig. 9.49.

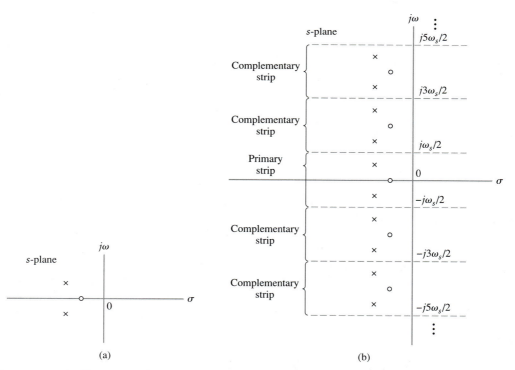

(a)   (b)

FIGURE 9.49   Illustration of Property 2 of the Laplace transform of sampled signal. (a) Pole–zero map of $A(s)$. (b) Pole–zero map of $A_\delta(s)$ defined by

$$A_\delta(s) = \frac{1}{\mathcal{T}} \sum_{k=-\infty}^{\infty} A(s - jk\omega_s)$$

where ω_s is the sampling frequency.

Examining Eq. (9.112), we see that because of the summation involving terms of the form $A(s - jk\omega_s)$, both the poles and zeros of $A(s)$ contribute to the zeros of $A_\delta(s)$. Accordingly, no statement equivalent to Property 2 can be made regarding the zeros of $A_\delta(s)$. Nevertheless, we can say that the zeros of $A_\delta(s)$ exhibit periodicity with period $j\omega_s$, as illustrated in Fig. 9.49.

Thus far in this subsection, we have discussed only discrete-time signals. However, in a sampled-data system we have a mixture of continuous-time and discrete-time signals. The issue we discuss next concerns such a situation. Suppose we have a signal $l(t)$ that is the result of convolving a discrete-time signal $a_\delta(t)$ with a continuous-time signal $b(t)$, as shown by

$$l(t) = a_\delta(t) * b(t)$$

We now sample $l(t)$ at the same rate as $a_\delta(t)$, and so write

$$l_\delta(t) = [a_\delta(t) * b(t)]_\delta$$

Transforming this relation into the complex s-domain, we may equivalently write

$$L_\delta(s) = [A_\delta(s)B(s)]_\delta$$

where $a_\delta(t) \xleftrightarrow{\mathscr{L}} A_\delta(s)$, $b(t) \xleftrightarrow{\mathscr{L}} B(s)$, and $l_\delta(t) \xleftrightarrow{\mathscr{L}} L_\delta(s)$. Adapting Eq. (9.112) to this new situation, we have

$$L_\delta(s) = \frac{1}{\mathscr{T}} \sum_{k=-\infty}^{\infty} A_\delta(s - jk\omega_s)B(s - jk\omega_s) \qquad (9.114)$$

where, as before, $\omega_s = 2\pi/\mathscr{T}$. However, by definition, the Laplace transform $A_\delta(s)$ is periodic in s with period $j\omega_s$. It follows therefore that

$$A_\delta(s - jk\omega_s) = A_\delta(s) \quad \text{for } k = 0, \pm1, \pm2, \ldots$$

Hence we may simplify Eq. (9.114) to

$$
\begin{aligned}
L_\delta(s) &= A_\delta(s) \cdot \frac{1}{\mathscr{T}} \sum_{k=-\infty}^{\infty} B(s - jk\omega_s) \\
&= A_\delta(s)B_\delta(s)
\end{aligned}
\qquad (9.115)
$$

where $b_\delta(t) \xleftrightarrow{\mathscr{L}} B_\delta(s)$ and $b_\delta(t)$ is the impulse-sampled version of $b(t)$; that is,

$$B_\delta(s) = \frac{1}{\mathscr{T}} \sum_{k=-\infty}^{\infty} B(s - jk\omega_s)$$

In light of Eq. (9.115), we may now state another property of impulse sampling. *If the Laplace transform of a signal to be sampled at the rate $1/\mathscr{T}$ is the product of a Laplace transform that is already periodic in s with period $j\omega_s = j2\pi/\mathscr{T}$ and another Laplace transform that is not, then the periodic Laplace transform comes out as a factor of the result.*

■ CLOSED-LOOP TRANSFER FUNCTION

Returning to the issue at hand, namely, that of determining the closed-loop transfer function of the sampled-data system in Fig. 9.46, we note that each one of the functional blocks in the model of Fig. 9.48 is characterized by a transfer function of its own, except for the sampler. Unfortunately, a sampler does not have a transfer function, which complicates the determination of the closed-loop transfer function of a sampled-data system. To get around this problem, we commute the sampling operator with the summer, and so reformulate the model of Fig. 9.48 in the equivalent form shown in Fig. 9.50, where the signals entering into the analysis are now all represented by their respective Laplace transforms. The usual approach in sampled-data systems analysis is to relate the sampled version of the input, $X_\delta(s)$, to a sampled version of the output, $Y_\delta(s)$. That is, we analyze the closed-loop transfer function, $T_\delta(s)$, contained in the dashed box in Fig. 9.50. This approach describes the behavior of the plant output $y(t)$ *at the sampling instants* but provides no information on how the output varies between the sampling instants.

In Fig. 9.50, the transfer function of the zero-order hold has been split into two parts. One part, represented by $(1 - e^{-\mathcal{T}s})$, has been integrated with the transfer function $D_\delta(s)$ of the digital controller. The other part, represented by $1/s$, has been integrated with the transfer function $G(s)$ of the plant. In so doing, we now have only two kinds of transforms to think about in the model of Fig. 9.50:

▶ Transforms of continuous-time quantities, represented by the Laplace transform $y(t) \overset{\mathcal{L}}{\longleftrightarrow} Y(s)$ and the transfer function

$$B(s) = \frac{G(s)}{s} \tag{9.116}$$

▶ Transforms of discrete-time quantities, represented by the Laplace transforms $x_\delta(t) \overset{\mathcal{L}}{\longleftrightarrow} X_\delta(s)$, $e_\delta(t) \overset{\mathcal{L}}{\longleftrightarrow} E_\delta(s)$, $y_\delta(t) \overset{\mathcal{L}}{\longleftrightarrow} Y_\delta(s)$, and the transfer function

$$A_\delta(s) = D_\delta(s)(1 - e^{-\mathcal{T}s}) \tag{9.117}$$

We are now ready to describe a straightforward procedure for the analysis of sampled-data systems:

1. Write cause-and-effect equations, using Laplace transforms to obtain the closed-loop transfer function $T_\delta(s)$.
2. Convert $T_\delta(s)$ to a discrete-time transfer function $T(z)$.
3. Use z-plane analysis tools, such as the root locus method, to assess system stability and system performance.

Although we have described this procedure in the context of the sampled-data system shown in Fig. 9.46 containing a single sampler, the procedure generalizes to a sampled-data system containing any number of samplers.

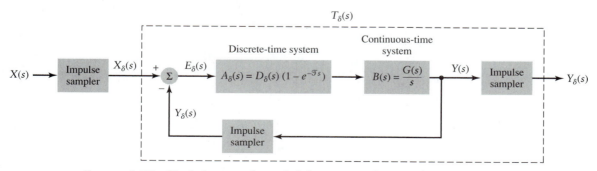

FIGURE 9.50 Block diagram of sampled-data system obtained by reformulating the model of Fig. 9.48.

Looking at Fig. 9.50, we may readily set up the following cause-and-effect equations:

$$E_\delta(s) = X_\delta(s) - Y_\delta(s) \tag{9.118}$$

$$Y(s) = A_\delta(s)B(s)E_\delta(s) \tag{9.119}$$

where $B(s)$ and $A_\delta(s)$ are defined by Eqs. (9.116) and (9.117), respectively. The impulse sampler applied to $y(t)$ has the same sampling period $\mathcal{T}$ and is synchronous with the impulse sampler at the front end of the system. Thus sampling $y(t)$ in this manner, we may rewrite Eq. (9.119) in the sampled form:

$$\begin{aligned}Y_\delta(s) &= A_\delta(s)B_\delta(s)E_\delta(s) \\ &= L_\delta(s)E_\delta(s)\end{aligned} \tag{9.120}$$

where $B_\delta(s)$ is the sampled form of $B(s)$, and $L_\delta(s)$ is defined by Eq. (9.115). Solving Eqs. (9.118) and (9.120) for the ratio $Y_\delta(s)/X_\delta(s)$, we may express the closed-loop transfer function of the sampled-data system of Fig. 9.48 as

$$\begin{aligned}T_\delta(s) &= \frac{Y_\delta(s)}{X_\delta(s)} \\ &= \frac{L_\delta(s)}{1 + L_\delta(s)}\end{aligned} \tag{9.121}$$

Finally, adapting Eq. (9.109) to our present situation, we may rewrite Eq. (9.121) in terms of the z-transform as follows:

$$T(z) = \frac{L(z)}{1 + L(z)} \tag{9.122}$$

where

$$L(z) = L_\delta(s)\big|_{e^{\mathcal{T}s}=z}$$

and

$$T(z) = T_\delta(s)\big|_{e^{\mathcal{T}s}=z}$$

As stated previously, Eq. (9.122) defines the transfer function $T(z)$ between the sampled input of the original sampled-data system in Fig. 9.46 and the plant output $y(t)$ measured only at the sampling instants.

EXAMPLE 9.16   In the sampled-data system of Fig. 9.46, the transfer function of the plant is

$$G(s) = \frac{a}{s + a}$$

and the z-transform of the digital controller (computer program) is

$$D(z) = \frac{K}{1 - z^{-1}}$$

Determine the closed-loop transfer function $T(z)$ of the system.

Solution: Consider first $B(s) = G(s)/s$ expressed in partial fractions as

$$\begin{aligned}B(s) &= \frac{a}{s(s + a)} \\ &= \frac{1}{s} - \frac{1}{s + a}\end{aligned}$$

The inverse Laplace transform of $B(s)$ is

$$b(t) = \mathcal{L}^{-1}[B(s)] = 1 - e^{-at}$$

Hence, adapting the definition of Eq. (9.108) for the problem at hand, we have

$$B_\delta(s) = \sum_{n=-\infty}^{\infty} b[n]e^{-sn\mathcal{T}}$$

$$= \sum_{n=-\infty}^{\infty} (1 - e^{-an\mathcal{T}})e^{-sn\mathcal{T}}$$

$$= \sum_{n=-\infty}^{\infty} e^{-sn\mathcal{T}} - \sum_{n=-\infty}^{\infty} e^{-(s+a)n\mathcal{T}}$$

$$= \frac{1}{1 - e^{-\mathcal{T}s}} - \frac{1}{1 - e^{-a\mathcal{T}}e^{-\mathcal{T}s}}$$

$$= \frac{(1 - e^{-a\mathcal{T}})e^{-\mathcal{T}s}}{(1 - e^{-\mathcal{T}s})(1 - e^{-a\mathcal{T}}e^{-\mathcal{T}s})}$$

For convergence, we have to restrict our analysis to values of s for which both $|e^{-\mathcal{T}s}|$ and $|e^{-\mathcal{T}(s+a)}|$ are less than one. Next, applying the formula of Eq. (9.110) to the given z-transform $D(z)$, we get

$$D_\delta(s) = \frac{K}{1 - e^{-\mathcal{T}s}}$$

the use of which in Eq. (9.117) yields

$$A_\delta(s) = K$$

Hence, using the results obtained for $A_\delta(s)$ and $B_\delta(s)$ in Eq. (9.115), we obtain

$$L_\delta(s) = \frac{K(1 - e^{-a\mathcal{T}})e^{-\mathcal{T}s}}{(1 - e^{-\mathcal{T}s})(1 - e^{-a\mathcal{T}}e^{-\mathcal{T}s})}$$

Finally, setting $e^{-\mathcal{T}s} = z^{-1}$, we get the z-transform

$$L(z) = \frac{K(1 - e^{-a\mathcal{T}})z^{-1}}{(1 - z^{-1})(1 - e^{-a\mathcal{T}}z^{-1})}$$

$$= \frac{K(1 - e^{-a\mathcal{T}})z}{(z - 1)(z - e^{-a\mathcal{T}})}$$

which has a zero at the origin, a pole at $z = e^{-a\mathcal{T}}$ inside the unit circle, and a pole at $z = 1$ on the unit circle in the z-plane.

▶ **Drill Problem 9.17** The transfer function of a plant is

$$G(s) = \frac{1}{(s + 1)(s + 2)}$$

Determine $(G(s)/s)_\delta$.

Answer: $\left(\dfrac{G(s)}{s}\right)_\delta = \dfrac{\frac{1}{2}}{1 - e^{-\mathcal{T}s}} - \dfrac{1}{1 - e^{-\mathcal{T}}e^{-\mathcal{T}s}} + \dfrac{\frac{1}{2}}{1 - e^{-2\mathcal{T}}e^{-\mathcal{T}s}}$ ◀

■ STABILITY

The stability problem in a sampled-data system is different from its continuous-time counterpart, because we are performing our analysis in the z-plane instead of the s-plane. The stability domain for a continuous-time system is represented by the left half of the s-plane. On the other hand, the stability domain for a sampled-data system is represented by the interior of the unit circle in the z-plane.

Referring to Eq. (9.122), we see that the stability of the sampled-data system of Fig. 9.46 is determined by the poles of the closed-loop transfer function $T(z)$ or, equivalently, the roots of the characteristic equation:

$$1 + L(z) = 0$$

or, equivalently,

$$L(z) = -1 \qquad (9.123)$$

The significant point to note about this equation is that it has the same mathematical form as the corresponding equation for the continuous-time feedback system described in Eq. (9.73). Accordingly, the mechanics of constructing the root locus in the z-plane are exactly the same as the mechanics of constructing the root locus in the s-plane. In other words, all the properties of the s-plane root locus described in Section 9.13 carry over to the z-plane root locus. The only point of difference is that in order for the sampled-data feedback system to be stable, all the roots of the characteristic equation (9.123) must be confined to the interior of the unit circle in the z-plane.

In a similar way, the principle of the argument used to derive the Nyquist criterion in Section 9.15 applies to the z-plane as well as the s-plane. This time, however, the imaginary axis of the s-plane is replaced by the unit circle in the z-plane, and all the poles of the closed-loop transfer function $T(z)$ are required to be inside the unit circle.

EXAMPLE 9.17   Continuing with Example 9.16, assume that $e^{-a\Im} = \frac{1}{2}$. The corresponding value of $L(z)$ is

$$L(z) = \frac{\frac{1}{2}Kz}{(z-1)(z-\frac{1}{2})}$$

Construct the z-plane root locus of the system.

Solution: The characteristic equation of the system is

$$(z-1)(z-\tfrac{1}{2}) + \tfrac{1}{2}Kz = 0$$

that is,

$$z^2 + \tfrac{1}{2}(K-3)z + \tfrac{1}{2} = 0$$

This is a quadratic equation in z; its two roots are given by

$$z = -\tfrac{1}{4}(K-3) \pm \tfrac{1}{4}\sqrt{K^2 - 6K + 1}$$

The root locus of the system is shown in Fig. 9.51, where we note the following:

▶ Starting with $K = 0$, the breakaway point of the root locus occurs at $z = 1/\sqrt{2} = 0.707$ for $K = 3 - 2\sqrt{2} = 0.172$.

▶ For $K = 3 + 2\sqrt{2} = 5.828$, the root locus again intersects the real axis of the z-plane, but this time at $z = -1/\sqrt{2} = -0.707$.

▶ For $0.172 \leq K \leq 5.828$, the roots of the characteristic equation trace a circle centered on the origin of the z-plane and with radius equal to $1/\sqrt{2}$.

▶ For $K > 5.828$, the two roots start separating from each other, with one root moving toward the zero at the origin and the other root moving toward infinity.

▶ For $K = 6$, the two roots of the characteristic equation move to $z = -\tfrac{1}{2}$ and $z = -1$. Thus for this value of K the system is on the verge of instability, and for $K > 6$ the system becomes unstable.

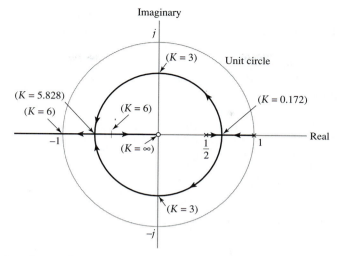

FIGURE 9.51   Root locus of sampled-data system with loop transfer function

$$L(z) = \frac{\frac{1}{2}Kz}{(z-1)(z-\frac{1}{2})}$$

9.18   *Design of Control Systems*

In the preceding sections of this chapter, we have described the benefits of feedback applied to a control system and presented a detailed treatment of the stability problem. It is therefore befitting that we now consider the design of a control system in light of the material covered herein.

The design of a feedback control system is concerned with three interrelated matters:

▶ Planning of a suitable structure for the system
▶ Selection of suitable sensors and controllers
▶ Determination of adjustable parameters of the controllers.

The design begins with a set of specifications that include the following issues that impinge on system performance in one way or another:

▶ Sensitivity to parameter variations and disturbances
▶ Steady-state errors, measured with respect to step, ramp, and parabolic inputs
▶ Transient response, specified in terms of rise time, peak time, percentage overshoot, and settling time for a step input
▶ Relative stability, measured in terms of gain and phase margins

The design can be carried out in the frequency domain or time domain. In the frequency-domain approach, the design is performed manually by graphical means. Specifically, the designer uses a compensator (controller) to alter or reshape the characteristics of the system using the Bode diagram so as to satisfy the design specifications. The design methodology is usually based on the notion of a dominant second-order approximation, and the relationship between the damping ratio and phase margin, as discussed in Section 9.16.

Alternatively, the designer uses a time-domain approach. In this second approach, the compensator is used to alter or reshape the root locus so that the closed-loop poles of the system include a dominant pair of complex-conjugate poles at desired locations in the s-plane. With the availability of high-powered and user-friendly computer software, the designer is able to arrive at a satisfactory solution by going through a large number of design runs in a relatively short period of time.

In this section we present a couple of design examples illustrative of both of these approaches.

EXAMPLE 9.18 *Design of a Phase-Lag Compensator Using the Bode Diagram*
Consider the use of a compensator in the form of a *phase-lag network* whose transfer function is defined by

$$H(s) = \frac{\alpha\tau s + 1}{\tau s + 1}, \quad \alpha < 1$$

You are given a plant with the following transfer function:

$$G(s) = \frac{K}{s(s + 1)}$$

The requirement is to design a phase-lag compensator to achieve the following performance specifications:

▶ Closed-loop step response with an overshoot less than 25%
▶ Steady-state error of less than 0.1, measured with respect to a ramp input

Having designed the compensator, construct an operational amplifier circuit for its implementation.

Solution: Figure 9.52 shows the pole–zero map of the compensator. With $\alpha < 1$, the pole at $s = -1/\tau$ lies closer to the origin than the zero at $s = -1/\alpha\tau$. Figure 9.53 shows the corresponding Bode diagram with two corner frequencies defined in accordance with the frequency response

$$H(j\omega) = \frac{j\alpha\tau\omega + 1}{j\tau\omega + 1}, \quad \alpha < 1$$

From Fig. 9.53(a) we note that for $\omega \geq 10/\alpha\tau$, the compensator is characterized by (1) *high-frequency attenuation* dependent on α and (2) *almost zero* phase lag. We will put this property to good use in our design strategy.

Recall that the percentage overshoot for a second-order system is defined in terms of the damping ratio ζ by (see Example 9.5)

$$\text{P.O.} = 100e^{-\pi\zeta/\sqrt{1-\zeta^2}}$$

With P.O. $\leq 25\%$, this relation is satisfied by choosing $\zeta \geq 0.4$.

The steady-state error for a ramp input is $1/K_v$, where K_v is the velocity error constant (see Section 9.8). With this error specified to be less than 0.1, we must have

$$K_v \geq \frac{1}{0.1} = 10$$

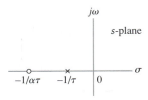

FIGURE 9.52   Pole–zero map of phase-lag compensation network, assuming that $\alpha < 1$.

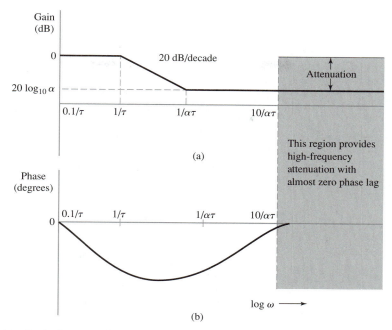

FIGURE 9.53   Bode diagram of phase-lag compensation network with $\alpha < 1$. (a) Gain response. (b) Phase response.

Figure 9.54 shows the block diagram of the feedback control system incorporating the phase-lag compensator. The loop transfer function of the system is

$$L(s) = G(s)H(s)$$

$$= \frac{K(\alpha\tau s + 1)}{s(\tau s + 1)(s + 1)}$$

This loop transfer function was considered in Example 9.4. Using the result obtained therein, we find that with $K_v = 10$ the desired loop gain K is

$$K = K_v = 10$$

The Bode diagram of the uncompensated system is plotted in Fig. 9.55, from which we find that the gain crossover frequency is

$$\omega_g = 3.2 \text{ rad/s}$$

and the phase margin is

$$\phi_m = 23°$$

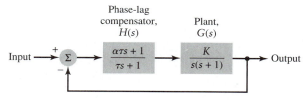

FIGURE 9.54   Block diagram of compensated feedback control system.

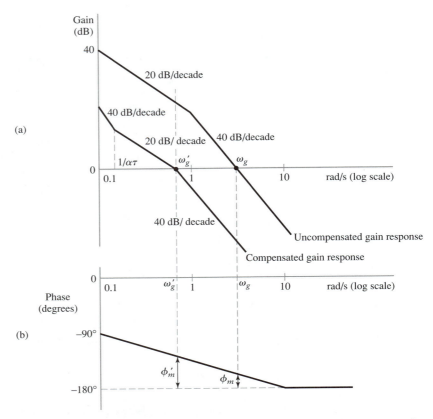

FIGURE 9.55 Bode diagram. (a) Gain responses of uncompensated and compensated systems. (b) Phase response of uncompensated system.

With a desired damping ratio $\zeta \geq 0.4$ for the compensated system, the corresponding phase margin is 45° (see Fig. 9.44). It follows therefore that the phase margin of the uncompensated system must be increased. We shall increase the phase margin by using the compensator to decrease the gain crossover frequency without appreciably changing the phase. Allowing 5° for the phase lag introduced by the compensator, the desired phase margin is

$$\phi_m' = 45° + 5° = 50°$$

We now locate the frequency ω on the phase response of the uncompensated system that corresponds to (see Fig. 9.55(b))

$$180° - \phi_m' = 180° - 50° = 130°$$

The frequency $\omega_g' = 0.69$ rad/s so located is the desired gain crossover frequency.

Earlier we pointed out that for $\omega \geq 10/\alpha\tau$ the compensator introduces attenuation with almost zero phase lag. We may therefore set

$$\frac{10}{\alpha\tau} = \omega_g' = 0.69 \text{ rad/s}$$

From the gain response of the uncompensated system, the corresponding gain is 23 dB at $\omega = \omega_g'$. The attenuation introduced by the compensator is therefore 23 dB; that is,

$$-20 \log_{10} \alpha = 23 \text{ dB}$$

or, equivalently,

$$\alpha = 0.07$$

Solving for the time constant τ, we obtain

$$\tau = \frac{10}{0.69\alpha} = 208$$

The design is now complete, with the compensator defined by

$$H(s) = \frac{0.07(s + 0.069)}{s + 0.0048}$$

For the implementation of the phase-lag compensator including the gain K, we may use the operational amplifier circuit shown in Fig. 9.56. The transfer function of this circuit is

$$\frac{V_2(s)}{V_1(s)} = -\frac{R_2}{R_3}\left(\frac{R_1C_1s + 1}{(R_1 + R_2)C_1s + 1}\right)$$

We may therefore make the following identifications:

$$K = \frac{R_2}{R_3}$$

$$\tau = (R_1 + R_2)C_1$$

$$\alpha\tau = R_1C_1$$

Pick $C_1 = 10\ \mu$F. Then with $K = 10$, $\alpha = 0.07$, and $\tau = 208$, we find that the remaining elements of the operational amplifier circuit in Fig. 9.56 have the following values:

$$R_1 = 1.46\ M\Omega$$

$$R_2 = 19.3\ M\Omega$$

$$R_3 = 1.93\ M\Omega$$

In closing this example, it is of interest to see how close the designed system performance is to the design specifications. The loop transfer function of the compensated system is

$$L(s) = G(s)H(s)$$

$$= \frac{145.6s + 10}{208s^3 + 209s^2 + s}$$

The actual gain and phase responses computed from this loop transfer function with $s = j\omega$ are plotted in Fig. 9.57. Here we see that the phase margin is $52.86°$ and the gain crossover

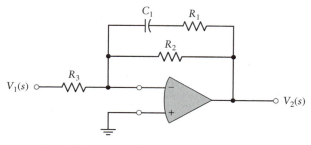

FIGURE 9.56   Operational amplifier circuit for implementing the phase-lag compensator in the control system of Fig. 9.54.

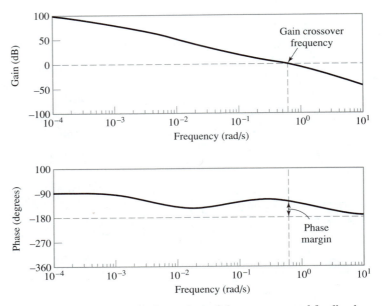

FIGURE 9.57 Actual closed-loop gain–phase plots of the compensated feedback control system designed using the frequency-domain approach.

frequency is 0.6032 rad/s. These are fairly close to the corresponding values that we aimed for in carrying out the design. The closed-loop transfer function of the system is

$$T(S) = \frac{L(s)}{1 + L(s)}$$

$$= \frac{145.6s + 10}{208s^3 + 209s^2 + 146.6s + 10}$$

Figure 9.58 shows the step response of this system. The percentage overshoot from this figure is about 20%, which is less than the specified value of 25%. So we can indeed say that we have successfully met the design specifications.

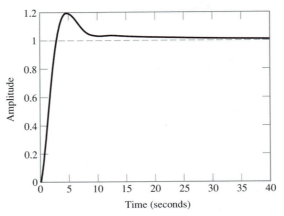

FIGURE 9.58 Step response of the compensated feedback control system designed using the frequency-domain approach.

EXAMPLE 9.19 *Design of Phase-Lag Compensator Using Root Locus*
   Repeat the design of the phase-lag compensator so as to meet the performance specifications described in Example 9.18, but this time use the root locus method.

Solution: Consider first the uncompensated system. The loop transfer function of the system is $K/s(s + 1)$. The uncompensated root locus is therefore as shown in Fig. 9.59(a). For an overshoot less than 25%, the damping ratio is $\zeta \geq 0.4$. To find the locations of the closed-loop poles on the root locus corresponding to $\zeta = 0.4$, we note that they are defined by $-\zeta\omega_n \pm j\omega_n\sqrt{1 - \zeta^2}$ (see Eq. (9.55)). Thus with $\zeta\omega_n = 0.5$, it follows that the natural frequency ω_n is

$$\omega_n = \frac{0.5}{0.4} = 1.25$$

The closed-loop poles are therefore at $-0.5 \pm j1.146$, as indicated in Fig. 9.59(a). The value of gain K for which the closed-loop poles are at these locations is $K = 1.56$. The velocity error constant for the uncompensated system is therefore

$$K_v = K = 1.56$$

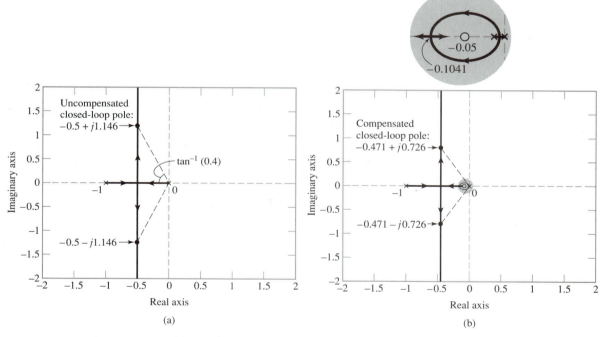

FIGURE 9.59   (a) Root locus of the uncompensated feedback control system defined by the loop transfer function

$$L(s) = \frac{K}{s(s + 1)}$$

(b) Root locus of the compensated feedback control system defined by the loop transfer function

$$L(s) = \frac{K'(s + 0.05)}{s(s + 1)(s + 0.004)}$$

The shaded insert presents details of the root locus in the neighborhood of the origin.

The resulting steady-state error for a ramp input is 0.64, which is much greater than the specified value of 0.1. To meet this specification, we therefore need to compensate the system.

For this purpose, we may use a phase-lag compensator. As before, the transfer function of the compensator is

$$H(s) = \frac{\alpha(s + 1/\alpha\tau)}{s + 1/\tau}, \quad \alpha < 1$$

The scaling factor α is absorbed in the system gain. The ratio of the zero to pole locations of the compensator is therefore

$$\left|\frac{-1/\alpha\tau}{-1/\tau}\right| = \frac{1}{\alpha} = \frac{K_{v,\text{comp}}}{K_{v,\text{uncomp}}} = \frac{10}{1.56}$$

Hence, $\alpha = 0.156$. Examining the uncompensated root locus of Fig. 9.59(a), we see that a reasonable choice for the zero location of the compensator is

$$-\frac{1}{\alpha\tau} = -0.05$$

With $\alpha = 0.156$, the corresponding pole location of the compensator is

$$-\frac{1}{\tau} = -0.05 \times 0.156 = -0.0078$$

The loop transfer function of the compensated system is therefore

$$L(s) = \frac{K'(s + 0.05)}{s(s + 1)(s + 0.0078)}$$

where $K' = \alpha K$. To realize a steady-state error of 0.1 in response to a ramp input, we require

$$K' = \frac{10 \times 1 \times 0.0078}{0.05} = 1.56$$

With $K' = 1.56$, the closed-loop transform function of the system is

$$T(s) = \frac{1.56s + 0.078}{s^3 + 1.0078s^2 + 1.568s + 0.078}$$

The resulting step response of the system is shown in Fig. 9.60(a). Here we see that the overshoot is 30%, which is slightly larger than the specified value of 25%.

In order to reduce the overshoot, we will keep the zero of the compensator at -0.05 but move the pole closer to the origin by locating it at $s = -0.004$. For our second try at the loop transfer function of the compensated system, we thus have

$$L(s) = \frac{K'(s + 0.05)}{s(s + 1)(s + 0.004)}$$

Then proceeding in a manner similar to that described for our first try at compensation, we may report the following results:

▶ The gain $K' = 0.8$ for a steady-state error of 0.1 in response to a ramp input.
▶ The closed-loop transfer function is

$$T(s) = \frac{0.8s + 0.04}{s^3 + 1.004s^2 + 0.804s + 0.04}$$

▶ A dominant pair of complex-conjugate poles exist at $s = -0.471 \pm j0.726$, which, as expected, have shifted from their uncompensated locations (see the root locus of Fig. 9.59(b)).

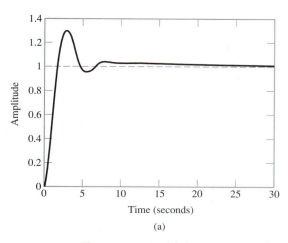
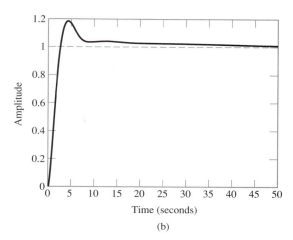

FIGURE 9.60   (a) Step response of the compensated feedback control system defined by the closed-loop transfer function

$$T(s) = \frac{1.56s + 0.078}{s^3 + 1.0078s^2 + 1.568s + 0.078}$$

(b) Step response of the second compensated feedback control system defined by the closed-loop transfer function

$$T(s) = \frac{0.8s + 0.04}{s^3 + 1.004s^2 + 0.804s + 0.04}$$

Both of these responses were obtained using the time-domain approach for designing the control system.

▶ The step response is shown in Fig. 9.60(b), which exhibits an overshoot of about 19%.
▶ The root locus has three breakaway points at $s = -0.002, -0.1041, -0.4708$.

The design specifications are thus met by using a phase-lag compensator with a zero at $s = -0.05$, a pole at $s = -0.004$, and a gain $K' = 0.8$. Note that the system gain K has been increased from the uncompensated value of 1.56 to $K'/\alpha = 10$.

9.19   *Exploring Concepts with MATLAB*

In this chapter, we studied the role of feedback in feedback amplifiers and control systems. The issue of stability is of paramount importance in the study of feedback systems. Specifically, we are given the (open) loop transfer function of the system, denoted by $L(s)$, and the requirement is to determine the closed-loop stability of the system. In the material presented in this chapter, two basic methods for the study of this problem were presented:

1. Root locus method.
2. Nyquist stability criterion.

The MATLAB Control System Toolbox is designed to explore these two methods in a computationally efficient manner.

■ CLOSED-LOOP POLES OF FEEDBACK SYSTEM

Let the loop transfer function $L(s)$ be expressed as the ratio of two polynomials in s, as shown by

$$L(s) = K \frac{P(s)}{Q(s)}$$

where K is a scaling factor. The characteristic equation of the feedback system is defined by

$$1 + L(s) = 0$$

or, equivalently,

$$Q(s) + KP(s) = 0$$

The roots of this equation define the poles of the closed-loop transfer function of the feedback system. To extract these roots, we use the command **roots** introduced in Section 6.8.

This command was used to compute the results presented in Table 9.3, detailing the roots of the characteristic equation of a third-order feedback system, namely,

$$s^3 + 3s^2 + 3s + K + 1 = 0$$

for $K = 0$, 5, and 10. For example, for $K = 10$ we have

```
≫ sys = [1, 3, 3, 11];
≫ roots(sys)
ans =
    -3.1544
     0.0772 + 1.8658i
     0.0772 - 1.8658i
```

A related issue of interest is that of calculating the natural frequencies and damping factors corresponding to the poles of the closed-loop transfer function of the system or the roots of the characteristic equation. This calculation is easily accomplished on MATLAB using the command **damp**. Specifically, the command

```
[Wn, z] = damp(sys)
```

returns vectors **Wn** and **z** containing the natural frequencies and damping factors of the feedback system, respectively.

Continuing with the above example, suppose we want to calculate the natural frequencies and damping factors pertaining to the closed-loop poles of the third-order feedback system for $K = 10$. For this system, we write and get

```
≫ sys = [1, 3, 3, 11];
≫ damp(sys)
   Eigenvalue           Damping       Freq. (rad/s)
   0.0772 + 1.8658i     -0.0414          1.8674
   0.0772 - 1.8658i     -0.0414          1.8674
  -3.1544                1.000           3.1544
```

The values returned in the first column are the roots of the characteristic equation. The term **Eigenvalue** is merely a reflection of the way in which this part of the calculation is performed.

■ ROOT LOCUS DIAGRAM

Construction of the root locus of a feedback system requires that we calculate and plot the locus of the roots of the characteristic equation

$$Q(s) + KP(s) = 0$$

for varying K. This construction is easily accomplished using the MATLAB command

```
rlocus(tf(num, den))
```

where num and den denote the coefficients of the numerator polynomial $P(s)$ and denominator polynomial $Q(s)$, respectively, in descending powers of s.

This command was used to generate the results plotted in Figs. 9.22, 9.26, and 9.28. For example, the root locus of Fig. 9.28 pertains to the loop transfer function

$$L(s) = \frac{0.5K(s + 2)}{s(s + 12)(s - 4)}$$

$$= \frac{K(0.5s + 1.0)}{s^3 + 8s^2 - 48s}$$

The root locus is computed and plotted by using the commands

```
≫ num = [.5, 1];
≫ den = [1, 8, -48, 0];
≫ rlocus(tf(num, den))
```

▶ **Drill Problem 9.18**   Use the command rlocus to plot the root locus of a feedback system having the following loop transfer function:

$$L(s) = \frac{K}{(s + 1)^3}$$ ◀

Answer: Breakaway point $= -1$. The system is on the verge of instability for $K = 8$, for which the closed loop poles of the system are at $s = -3$ and $s = \pm 1.7321i$. ◀

▶ **Drill Problem 9.19**   Use the command rlocus to plot the root locus of the following loop transfer function:

$$L(s) = \frac{K'(s + 0.05)}{s(s + 1)(s + 0.004)}$$

Hence, verify the result shown in Fig. 9.59(b). ◀

▶ **Drill Problem 9.20**   Use the command step to plot the closed-loop transient response pertaining to the feedback system specified in Drill Problem 9.19. Hence, verify the result plotted in Fig. 9.60(b). ◀

Another useful command is rlocfind, which finds the value of scaling factor K required to realize a specified set of roots on the root locus. To illustrate the use of this command, consider again the root locus of Fig. 9.28 and give the commands

```
≫ num = [.5, 1];
≫ den = [1, 8, -48, 0];
≫ rlocus(tf(num, den));
≫ K = rlocfind(num, den)
Select a point in the graphics window
```

We then respond by placing the cursor at point A, representing the location of the root in the top left-hand quadrant, say, a symbol "+" as indicated in Fig. 9.61. Upon entering this command, MATLAB responds as follows:

```
selected point =
    -1.6166 + 6.393i
K =
    213.68
```

■ REDUCED-ORDER MODELS

In Section 9.14, we discussed the use of a reduced-order model to approximately represent a higher-order feedback system. There are two basic steps to the derivation of a reduced-order model (see Fig. 9.29):

1. *Determine the dominant closed-loop poles of the system.* These poles are those roots of the system's characteristic equation that lie closest to the $j\omega$-axis of the s-plane. The remaining roots lie far to the left of the $j\omega$-axis and are discarded. We may identify the dominant closed-loop poles using the `roots` command.
2. *Determine the dominant closed-loop zeros of the system.* We may identify these zeros as those particular zeros of $L(s)$ that lie closest to the $j\omega$-axis of the s-plane, with the remaining zeros lying further away.

To assess the quality of the approximation, we may compare the step response of the reduced-order model to that of the original feedback system. Let the closed-loop transfer function of the feedback system be denoted by

$$T(s) = \frac{b_m s^m + b_{m-1} s^{m-1} + \cdots + b_1 s + b_0}{a_n s^n + a_{n-1} s^{n-1} + \cdots + a_1 s + a_0}$$

Let `numorig` and `denorig` denote the coefficients of the numerator and denominator polynomials of $T(s)$, respectively, arranged in descending powers of s. To compute the step response of this system, we use the command

```
step(tf(numorig, denorig))
```

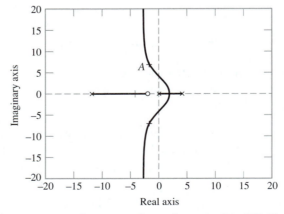

FIGURE 9.61   Root locus diagram illustrating the application of MATLAB command `rlocfind`.

Consider next the reduced-order model represented by the transfer function

$$T'(s) = \frac{b'_l s^l + b'_{l-1} s^{l-1} + \cdots + b'_1 s + b'_0}{a'_r s^r + a'_{r-1} s^{r-1} + \cdots + a'_1 s + a'_0}$$

where $l < m$ and $r < n$. For the two step responses to have the same value at $t = \infty$, we require that

$$\frac{b_0}{a_0} = \frac{b'_0}{a'_0}$$

Let `numapp` and `denapp` denote the numerator and denominator polynomials of reduced-order transfer function $T'(s)$, respectively. To compute the step response of the reduced-order model, we reapply the above MATLAB command, as shown here:

```
step(tf(numapp, denapp))
```

For the feedback system considered in Example 9.12, we have

$$T(s) = \frac{48}{s^3 + 6s^2 + 11s + 54}$$

and

$$T'(s) = \frac{8.3832}{s^2 + 0.2740s + 9.4308}$$

Hence we may write

```
≫ numorig = [48];
≫ denorig = [1, 6, 11, 54];
≫ numapp = [8.3832];
≫ denapp = [1, 0.2740, 9.4308];
≫ step(tf(numorig, denorig))
≫ hold
≫ step(tf(numapp, denapp))
```

The results of this computation are shown in Fig. 9.31.

■ *NYQUIST STABILITY CRITERION

Construction of the Nyquist diagram involves making a polar plot of the loop frequency response $L(j\omega)$, which is obtained from the loop transfer function $L(s)$ by putting $s = j\omega$. The frequency ω is varied over the range $-\infty < \omega < \infty$. To proceed with the construction, we first express $L(j\omega)$ as the ratio of two polynomials in descending powers of $j\omega$, as shown by

$$L(j\omega) = \frac{p'_m (j\omega)^m + p'_{m-1}(j\omega)^{m-1} + \cdots + p'_1(j\omega) + p'_0}{q_n(j\omega)^n + q_{n-1}(j\omega)^{n-1} + \cdots + q_1(j\omega) + q_0}$$

where $p'_i = K p_i$ for $i = m, m - 1, \ldots, 1, 0$. Let `num` and `den` denote the numerator and denominator coefficients of $L(j\omega)$, respectively. We may then construct the Nyquist diagram by using the MATLAB command

```
nyquist(tf(num, den))
```

The results displayed in Fig. 9.37 for Example 9.13 were obtained using this MATLAB command. For that example, we have

$$L(j\omega) = \frac{36}{(j\omega)^3 + 6(j\omega)^2 + 11(j\omega) + 6}$$

To compute the Nyquist diagram, we therefore write

```
≫ num = [36];
≫ den = [1, 6, 11, 6];
≫ nyquist(tf(num, den))
```

▶ **Drill Problem 9.21**   Using the command `nyquist`, plot the Nyquist diagram for the feedback system defined by

$$L(j\omega) = \frac{6}{(1 + j\omega)^3}$$

Hence, determine whether the system is stable or not.

Answer: The system is stable.   ◀

■ BODE DIAGRAM

The Bode diagram for a linear feedback system consists of two graphs. In one graph, the loop gain response $20 \log_{10}|L(j\omega)|$ is plotted versus the logarithm of ω. In the other graph, the loop phase response $\arg\{L(j\omega)\}$ is plotted versus the logarithm of ω. With the given loop frequency response expressed in the form

$$L(j\omega) = \frac{p'_m(j\omega)^m + p'_{m-1}(j\omega)^{m-1} + \cdots + p'_1(j\omega) + p'_0}{q_n(j\omega)^n + q_{n-1}(j\omega)^{n-1} + \cdots + q_1(j\omega) + q_0}$$

we first set up the vectors `num` and `den` to represent the coefficients of the numerator and denominator polynomials of $L(j\omega)$, respectively. The Bode diagram for $L(j\omega)$ may then be easily constructed by using the MATLAB command

```
margin(tf(num, den))
```

This command calculates the gain margin, phase margin, and associated crossover frequencies from frequency response data. The result also includes plots of both the loop gain and phase responses.

The above command was used to compute the results presented in Fig. 9.42 for Example 9.14. For that example, we have

$$L(j\omega) = \frac{36}{(j\omega)^3 + 6(j\omega)^2 + 11(j\omega) + 6}$$

The commands for computing the Bode diagram, including the stability margins, are as follows:

```
≫ num = [36];
≫ den = [1, 6, 11, 6];
≫ margin(tf(num, den))
```

▶ **Drill Problem 9.22**   Compute the Bode diagram, the stability margins, and the associated crossover frequencies for the loop frequency response

$$L(j\omega) = \frac{6}{(1 + j\omega)^3}$$

Answer:

Gain margin $= 2.499$ dB

Phase margin $= 10.17°$

Phase crossover frequency $= 1.7321$

Gain crossover frequency $= 1.5172$ ◀

9.20 *Summary*

In this chapter we discussed the concept of feedback, which is of fundamental importance to the study of feedback amplifiers and control systems. The application of feedback has beneficial effects:

- ▶ It reduces the sensitivity of the closed-loop gain of the system with respect to changes in the gain of a plant inside the loop.
- ▶ It reduces the effect of a disturbance generated inside the loop.
- ▶ It reduces nonlinear distortion due to deviation of the plant from a linear behavior.

Indeed, these improvements get better as the amount of feedback, measured by the return difference, is increased.

However, feedback is like a double-edged sword that can become harmful if it is used improperly. In particular, it is possible for a feedback system to become unstable, unless special precautions are taken. Stability features prominently in the study of feedback systems. There are two fundamentally different methods for assessing the stability of linear feedback systems:

1. Root locus, a transform-domain method, is related to the transient response of the closed-loop system.
2. Nyquist stability criterion, a frequency-domain method, is related to the open-loop frequency response of the system.

From an engineering perspective, it is not enough to ensure that a feedback system is stable. Rather, the design of the system must include an adequate margin of stability to guard against parameter variations due to external factors. The root locus technique and Nyquist stability criterion cater to this requirement in their own specific ways.

The Nyquist stability criterion may itself be pursued using one of two presentations:

- ▶ Nyquist diagram (locus). In this method of presentation, the open-loop frequency response of the system is plotted in polar form, with attention focusing on whether the critical point $(-1, 0)$ is encircled or not.
- ▶ Bode diagram. In this second method, the open-loop frequency response of the system is presented as a combination of two graphs. One graph plots the loop gain response and the other graph plots the loop phase response. In an unconditionally stable system, we should find that the gain crossover frequency is smaller than the phase crossover frequency.

FURTHER READING

1. In an engineering context, it appears that the term "feedback" was first used in 1920 by personnel of Bell Telephone Laboratories in the United States of America. Notable among them were:

   ▶ Harold S. Black (1898–1981) who invented the negative feedback amplifier.

   ▶ Harry Nyquist (1889–1976) who formulated the Nyquist criterion.

   ▶ Hendrick W. Bode (1905–1982) who devised the Bode diagram.

   For a short history of control systems, see

   ▶ Dorf, R. C., and R. H. Bishop, *Modern Control Systems*, Seventh Edition (Addison-Wesley, 1995)

   ▶ Phillips, C. L., and R. D. Harbor, *Feedback Control Systems*, Third Edition (Prentice Hall, 1996)

2. For a complete treatment of automatic control systems, see the books:

   ▶ Belanger, P. R., *Control Engineering: A Modern Approach* (Saunders, 1995)

   ▶ Dorf, R. C., and R. H. Bishop, *Modern Control Systems*, Seventh Edition (Addison-Wesley, 1995)

   ▶ Kuo, B. C., *Automatic Control Systems*, Seventh Edition (Prentice Hall, 1995)

   ▶ Palm, W. J. III, *Control Systems Engineering* (Wiley, 1986)

   ▶ Phillips, C. L., and R. D. Harbor, *Feedback Control Systems*, Third Edition (Prentice Hall, 1996)

   These books cover both continuous-time and discrete-time aspects of control systems. They also present detailed system design procedures.

3. Feedback amplifiers are discussed in the books:

   ▶ Siebert, W. McC., *Circuits, Signals, and Systems* (MIT Press, 1986)

   ▶ Waldhauer, F. D., *Feedback* (Wiley, 1982)

4. For a discussion of operational amplifiers and their applications, see the books:

   ▶ Kennedy, E. J., *Operational Amplifier Circuits* (Holt, Rinehart, and Winston, 1988)

   ▶ Wait, J. V., L. P. Huelsman, and G. A. Korn, *Introduction to Operational Amplifier Theory and Applications*, Second Edition (McGraw-Hill, 1992)

5. For a proof of Property 5, embodying Eqs. (9.82) and (9.83), see:

   ▶ Truxal, J. G., *Control System Synthesis* (McGraw-Hill, 1955), pp. 227–228.

6. For a discussion of the practical issues involved in the operation of D/A converters, see the article:

   ▶ Hendriks, P. "Specifying communication DACs," *IEEE Spectrum*, vol. 34, pp. 58–69, July 1997.

PROBLEMS

9.1 A transistor amplifier has a gain of 2500. Feedback is applied around the amplifier using a passive network that returns a fraction of the amplifier output to the input, namely, $\beta = 0.01$.

(a) Calculate the closed-loop gain of the feedback amplifier.

(b) The gain of the transistor amplifier changes by 10% due to external factors. Calculate the corresponding change in the closed-loop gain of the feedback amplifier.

9.2 Figure P9.2 shows the block diagram of a position-control system. The preamplifier has a gain G_a. The gain of the motor and load combination (i.e., plant) is G_p. The sensor in the feedback path returns a fraction H of the motor output to the input of the system.

(a) Determine the closed-loop gain T of the feedback system.

(b) Determine the sensitivity of T with respect to changes in G_p.

(c) Assuming that $H = 1$ and nominally $G_p = 1.5$, what is the value of G_a that would make the sensitivity of T with respect to changes in G_p equal to 1%?

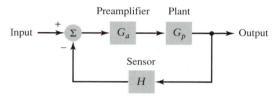

FIGURE P9.2

9.3 Figure P9.3 shows the simplified block diagram of a radar tracking system. The radar is represented by K_r, denoting some gain; θ_{in} and θ_{out} denote the angular positions of the target (being tracked) and the radar antenna, respectively. The controller has gain G_c, and the plant (made up of motor, gears, and antenna pedestal) is represented by G_p. To improve performance, "local" feedback via the sensor H is applied around the plant. In addition, the system uses unity feedback, as indicated in Fig. P9.3. The purpose of the system is to drive the antenna so as to track the target with sufficient accuracy. Determine the closed-loop gain of the system.

9.4 Figure P9.4 shows the circuit diagram of an inverting op amp circuit. The op amp is modeled as having infinite input impedance, zero output impedance, and infinite voltage gain. Evaluate the transfer function of this circuit, that is, $V_2(s)/V_1(s)$.

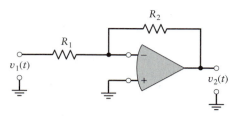

FIGURE P9.4

9.5 Figure P9.5 shows a practical differentiator using an op amp. Assume that the op amp is ideal, having infinite input impedance, zero output impedance, and infinite gain.

(a) Determine the transfer function of this circuit.

(b) What is the range of frequencies for which this circuit acts like an ideal differentiator?

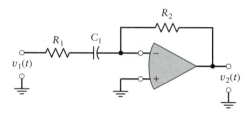

FIGURE P9.5

*9.6 Figure P9.6 shows the linearized block diagram of a phase-locked loop.

(a) Show that the closed-loop gain of the system is

$$\frac{V(s)}{\Phi_1(s)} = \frac{(s/K_v)L(s)}{1 + L(s)}$$

where K_v is a constant. The loop transfer function $L(s)$ is itself defined by

$$L(s) = K_0 \frac{H(s)}{s}$$

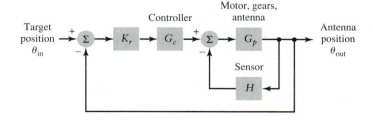

FIGURE P9.3

where K_0 is a gain factor and $H(s)$ is the transfer function of the loop filter.

(b) Specify the condition for which the phase-locked loop acts as an ideal differentiator. Under this condition, define the output voltage $v(t)$ in terms of the phase angle $\phi_1(t)$ acting as input.

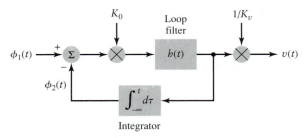

FIGURE P9.6

9.7 Figure P9.7 shows a control system with unity feedback. Determine the steady-state error of the system for (i) unit-step input, (ii) unit-ramp input, and (iii) unit-parabolic input. Do these calculations for each of the following cases:

(a) $G(s) = \dfrac{15}{(s + 1)(s + 3)}$

(b) $G(s) = \dfrac{5}{s(s + 1)(s + 4)}$

(c) $G(s) = \dfrac{5(s + 1)}{s^2(s + 3)}$

(d) $G(s) = \dfrac{5(s + 1)(s + 2)}{s^2(s + 3)}$

You may assume that, in each case, the closed-loop transfer function of the system is stable.

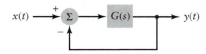

FIGURE P9.7

9.8 Using the Routh–Hurwitz criterion, demonstrate the stability of the closed-loop transfer function of the control system of Fig. P9.7 for all four cases specified in Problem 9.7.

9.9 Use the Routh–Hurwitz criterion to determine the number of roots in the left half, on the imaginary axis, and in the right half of the s-plane for each of the following characteristic equations:

(a) $s^4 + 2s^2 + 1 = 0$

(b) $s^4 + s^3 + s + 0.5 = 0$

(c) $s^4 + 2s^3 + 3s^2 + 2s + 4 = 0$

9.10 Using the Routh–Hurwitz criterion, find the range of parameter K for which the following characteristic equation represents a stable system:

$$s^3 + s^2 + s + K = 0$$

9.11 (a) The characteristic equation of a third-order feedback system is defined, in general, by

$$A(s) = a_3 s^3 + a_2 s^2 + a_1 s + a_0 = 0$$

Using the Routh–Hurwitz criterion, determine the conditions which the coefficients a_0, a_1, a_2, and a_3 must satisfy for the system to be stable.

(b) Revisit Problem 9.10 in light of the result obtained in part (a).

9.12 The loop transfer function of a feedback control system is defined by

$$L(s) = \frac{K}{s(s^2 + s + 2)}, \qquad K > 0$$

The system uses unity feedback.

(a) Sketch the root locus of the system for varying K.

(b) What is the value of K for which the system is on the verge of instability?

9.13 The block diagram of a feedback control system using a *proportional (P) controller* is shown in Fig. P9.13. This form of compensation is used when satisfactory performance is attainable merely by setting the constant K_P. For the plant specified in Fig. P9.13, determine the value of K_P needed to realize a natural (undamped) frequency $\omega_n = 2$. What are the corresponding values of (a) the damping factor and (b) the time constant of the system?

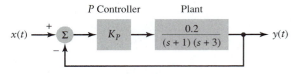

FIGURE P9.13

9.14 Figure P9.14 shows the block diagram of a feedback control system using a *proportional-plus-integral (PI) controller*. This form of controller, characterized by the parameters K_P and K_I, is used to improve the steady-state error of the system by increasing the system type by 1. You are given that $K_I/K_P = 0.1$. Plot the root locus of the system for varying K_P. Hence find the value of K_P needed to place a pole of the closed-loop transfer function for the system at $s = -5$. What is the steady-state error of the system for a unit-ramp input?

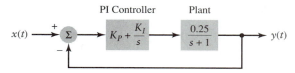

FIGURE P9.14

9.15 Figure P9.15 shows the block diagram of a feedback control system using a *proportional-plus-derivative (PD) controller*. This form of compensation, characterized by the parameters K_P and K_D, is used to improve the transient response of the system. You are given that $K_P/K_D = 4$. Plot the root locus of the system for varying K_D. Hence determine the value of K_D needed to place a pair of poles of the closed-loop transfer function of the system at $s = -2 \pm j2$.

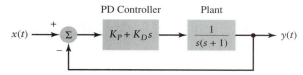

FIGURE P9.15

*9.16 Referring to the PI and PD controllers of Problems 9.14 and 9.15, we may make the following statements in the context of their frequency responses:

(a) The PI controller is *phase-lag*, in that it adds a negative contribution to the angle criterion of the root locus.

(b) The PD controller is *phase-lead*, in that it adds a positive contribution to the angle criterion of the root locus.

Justify the validity of these two statements.

9.17 Figure P9.17 shows the block diagram of a control system using a popular compensator known as the *proportional-plus-integral-plus-derivative (PID) controller*. The parameters K_P, K_I, and K_D of the controller are chosen to introduce a pair of complex-conjugate zeros at $s = -1 \pm j2$ into the loop transfer function of the system. Plot the root locus of the system for increasing

K_D. Hence determine the range of values of K_D for the system to remain stable.

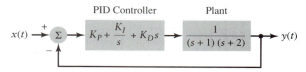

FIGURE P9.17

*9.18 Figure P9.18 shows an *inverted pendulum* that moves in a vertical plane on a cart. The cart itself moves along a horizontal axis under the influence of a force. The force is applied to keep the pendulum vertical. The transfer function of the inverted pendulum on a cart, viewed as the plant, is given by

$$G(s) = \frac{(s + 3.1)(s - 3.1)}{s^2(s + 4.4)(s - 4.4)}$$

Assuming the use of a proportional controller in a manner similar to that described in Problem 9.13, is the use of such a controller sufficient to stabilize the system? Justify your answer. How would you stabilize the system?

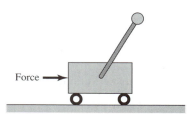

FIGURE P9.18

9.19 Consider a control system with unity feedback, whose loop transfer function is given by

$$L(s) = \frac{K}{s(s + 1)}$$

Plot the closed-loop transient response of the system for the following values of gain factor:

(a) $K = 0.1$

(b) $K = 0.25$

(c) $K = 2.5$

9.20 Figure P9.20 shows the block diagram of a feedback control system of type 1. Determine the damping ratio, natural frequency, damped frequency, and time constant of this system for $K = 20$.

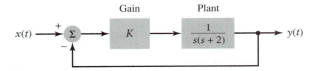

FIGURE P9.20

*9.21 Consider a two-stage feedback amplifier, for which the loop transfer function is

$$L(s) = \frac{K}{(s + 1)^2}$$

Using the Nyquist criterion, show that this system is stable for all $K > 0$.

*9.22 Consider a three-stage feedback amplifier, for which the loop transfer function is

$$L(s) = \frac{K}{(s + 1)^2(s + 5)}$$

(a) Using the Nyquist criterion, investigate the stability of this system for varying K.

(b) Find the value of the scaling factor K needed for a gain margin of 5 dB.

(c) For the value of K found in part (b), what is the phase margin?

*9.23 The loop transfer function of a unity-feedback control system of type 1 is defined by

$$L(s) = \frac{K}{s(s + 1)}$$

Using the Nyquist criterion, investigate the stability of this system for varying K. Hence show that the system is stable for all $K > 0$.

*9.24 A unity-feedback control system of type 2 has the transfer function

$$L(s) = \frac{K}{s^2(s + 1)}$$

Using the Nyquist criterion, show that the system is unstable for all gains $K > 0$. Also, verify your answer using the Routh–Hurwitz criterion.

*9.25 The loop transfer function of a unity-feedback system of type 1 is defined by

$$L(s) = \frac{K}{s(s + 1)(s + 2)}$$

(a) Using the Nyquist stability criterion, show that the system is stable for $0 < K < 6$. Also, verify your answer using the Routh–Hurwitz criterion.

(b) For $K = 2$, determine the gain margin in decibels and the phase margin in degrees.

(c) A phase margin of 20° is required. What is the value of K to attain this requirement? What is the corresponding value of the gain margin?

*9.26 Figure 9.43 illustrates the definitions of gain margin and phase margin using the Bode diagram. Illustrate the definitions of these two measures of relative stability using the Nyquist diagram.

9.27 (a) Construct the Bode diagram for the loop frequency response

$$L(j\omega) = \frac{K}{(j\omega + 1)(j\omega + 2)(j\omega + 3)}$$

for $K = 7, 8, 9, 10,$ and 11. Hence show that the three-stage feedback amplifier characterized by this loop frequency response is stable for $K = 7, 8,$ and 9; on the verge of instability for $K = 10$; and unstable for $K = 11$.

(b) Calculate the gain and phase margins of the feedback amplifier for $K = 7, 8,$ and 9.

9.28 In this problem we revisit Example 9.12, involving the approximation of a third-order feedback amplifier with a second-order model. In that example, we used the step response as the basis for assessing the quality of approximation. In this problem we use the Bode diagram as the basis for assessing the quality of approximation. Specifically, plot the Bode diagram for the second-order model

$$T'(s) = \frac{8.3832}{s^2 + 0.2740s + 9.4308}$$

and compare it to the Bode diagram for the original system

$$T(s) = \frac{48}{s^3 + 6s^2 + 11s + 54}$$

Comment on your results.

9.29 Sketch the Bode diagram for each of the following loop transfer functions:

(a) $L(s) = \dfrac{50}{(s + 1)(s + 2)}$

(b) $L(s) = \dfrac{10}{(s + 1)(s + 2)(s + 5)}$

(c) $L(s) = \dfrac{5}{(s + 1)^3}$

(d) $L(s) = \dfrac{10(s + 0.5)}{(s + 1)(s + 2)(s + 5)}$

9.30 Investigate the stability performance of the system described in Example 9.10 for varying gain K. This time, however, use the Bode diagram to do the investigation.

*9.31 Consider the sampled-data system shown in Fig. P9.31. Express the z-transform of the sampled output $y(t)$ as a function of the z-transform of the sampled input $x(t)$.

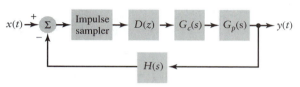

FIGURE P9.31

*9.32 Figure P9.32 shows the block diagram of a satellite control system using digital control. The transfer function of the digital controller is defined by

$$D(z) = K\left(1.5 + \frac{z-1}{z}\right)$$

Find the closed-loop transfer function of the system, assuming that the sampling period $\mathcal{T} = 0.1$ s.

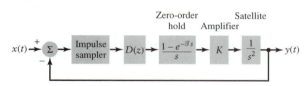

FIGURE P9.32

*9.33 Figure P9.33 shows the block diagram of a sampled-data system.

(a) Determine the closed-loop transfer function of the system for a sampling period $\mathcal{T} = 0.1$ s.

(b) Repeat the problem for $\mathcal{T} = 0.05$ s.

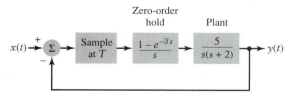

FIGURE P9.33

9.34 In the design procedures presented in this chapter we only discussed the use of 1-degree-of-freedom controller structures. Figure P9.34 presents two alternative 2-degree-of-freedom controller structures. Contrast the benefits of the two feed-forward compensation schemes shown in Fig. P9.34 in the contexts of (a) noise reduction and (b) pole–zero placement.

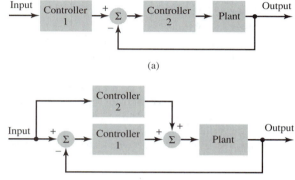

FIGURE P9.34

▶ *Computer Experiments*

9.35 Continuing with the third-order feedback system studied in Drill Problem 9.18, use the MATLAB command `rlocfind` to determine the value of scaling factor K in the loop transfer function

$$L(s) = \frac{K}{(s+1)^3}$$

that satisfies the following requirement: The complex-conjugate closed-loop poles of the feedback system have a damping factor equal to 0.5. What is the corresponding value of the natural frequency?

9.36 In Problem 9.12, we considered a unity-feedback system with the loop transfer function

$$L(s) = \frac{K}{s(s^2 + s + 2)}$$

Use the following MATLAB commands to evaluate stability performance of the system:

(a) `rlocus` for constructing the root locus diagram of the system.

(b) rlocfind for determining the value of K for which the complex-conjugate closed-loop poles of the system have a damping factor of about 0.707.

(c) margin for evaluating the gain and phase margins of the system for $K = 1.5$.

9.37 The loop transfer function of a feedback system is defined by

$$L(s) = \frac{K(s - 1)}{(s + 1)(s^2 + s + 1)}$$

This transfer function includes an allpass component represented by $(s - 1)/(s + 1)$, which is so called because it passes all frequencies with no amplitude distortion.

(a) Use the MATLAB command rlocus to construct the root locus of the system. Next, use the command rlocfind to determine the value of K for which the system is on the verge of instability. Check the value so obtained by using the Routh–Hurwitz criterion.

*(b) Use the command nyquist to plot the Nyquist diagram of the system for $K = 0.8$; hence confirm that the system is stable.

(c) For $K = 0.8$, use the command margin to assess the stability margins of the system.

9.38 The loop transfer function of a feedback system is defined by

$$L(s) = \frac{K(s + 1)}{s^4 + 5s^3 + 6s^2 + 2s - 8}$$

This system is stable only when K lies inside a certain range $K_{min} < K < K_{max}$.

(a) Use the MATLAB command rlocus to plot the root locus of the system.

(b) Find the critical limits of stability, K_{min} and K_{max}, by using the command rlocfind.

(c) For K lying midway between K_{min} and K_{max}, determine the stability margins of the system using the command margin.

*(d) For the value of K used in part (c), confirm the stability of the system by using the command nyquist to plot the Nyquist diagram.

9.39 In this problem, we study the design of a unity-feedback control system using a *phase-lead compensator* to improve the transient re-

sponse of the system. The transfer function of this compensator is defined by

$$G_c(s) = \frac{\alpha \tau s + 1}{\tau s + 1}, \qquad \alpha > 1$$

It is referred to as a phase-lead compensator because it introduces a phase advance into the loop frequency response of the system. The loop transfer function of the uncompensated system is defined by

$$L(s) = \frac{K}{s(s + 1)}$$

The lead compensator is connected in cascade with the open-loop system, resulting in the modified loop transfer function

$$L_c(s) = G_c(s)L(s)$$

(a) For $K = 1$, determine the damping factor and natural frequency of the closed-loop poles of the uncompensated system.

(b) Consider next the compensated system. Suppose that the transient specifications require that we have closed-loop poles with a damping factor $\zeta = 0.5$ and a natural frequency $\omega_n = 2$. Assuming that $\alpha = 10$, show that the angle criterion (pertaining to root locus construction) is satisfied by choosing the time constant $\tau = 0.027$ for the lead compensator.

(c) Use the MATLAB command rlocfind to confirm that the use of a phase-lead compensator with $\alpha = 10$ and $\tau = 0.027$ does indeed satisfy the desired transient response specifications.

9.40 Consider a unity-feedback control system using the cascade connection of a plant and phase-lag compensator, as shown in Fig. 9.54. The transfer function of the plant is

$$G(s) = \frac{10}{s(0.2s + 1)}$$

The transfer function of the compensator is defined by

$$H(s) = K\left(\frac{\alpha \tau s + 1}{\tau s + 1}\right), \qquad \alpha < 1$$

The compensator must be designed to achieve the following requirements:

(a) The steady-state error to a ramp input of unit slope should be 0.1.

(b) The overshoot of the step response should not exceed 10%.

(c) The 5% settling time of the step response should be less than 2 s.

Carry out this design using (a) the frequency-domain approach based on the Bode diagram, and (b) the time-domain approach using the root locus method. Having designed the compensation network by either approach, construct an operational amplifier circuit for its implementation.

9.41 Throughout this chapter, we have treated the adjustable scale factor K in the loop transfer function

$$L(s) = K \frac{P(s)}{Q(s)}$$

as a positive number. In this last problem, we use MATLAB to explore feedback systems for which K is negative.

(a) Use the `rlocus` command to plot the root locus of the loop transfer function

$$L(s) = \frac{K(s - 1)}{(s + 1)(s^2 + s + 1)}$$

where K is negative. Hence, using the `rlocfind` command, show that this feedback system is on the verge of instability for $K = -1.0$. Verify this result using the Routh–Hurwitz criterion.

(b) Use the `rlocus` command to show that a feedback system with loop transfer function

$$L(s) = \frac{K(s + 1)}{s^4 + 5s^3 + 6s^2 + 2s - 8}$$

is unstable for all $K < 0$.

10 | Epilogue

In this book we have presented an introductory treatment of signals and systems. Insofar as signals are concerned, we placed particular emphasis on Fourier analysis as a method for their representation. Fourier theory is an essential part of the signal-processing practitioner's kit of tools. Basically, it enables us to transform a signal described in the time domain into an equivalent representation in the frequency domain, subject to certain conditions imposed on the signal. Most importantly, the transformation is one-to-one, in that there is no loss of information as we go back and forth from one domain to the other. As for the analysis of systems, we restricted our attention primarily to a special class of systems that satisfies two distinct properties: linearity and time invariance. The motivation for invoking these two properties is to make system analysis mathematically tractable.

The purpose of this final chapter of the book is to place these tools in a proper perspective in relation to the physical realities of signals and systems as encountered in real-life situations. In clarifying these issues, we will briefly discuss other tools that overcome the limitations of Fourier analysis and linear time-invariant systems. In so doing, the reader is given a broad picture of the toolkit available to signal and system analysts/designers.

10.1 Physical Properties of Real-Life Signals

Many of the real-life signals encountered in practice exhibit a nonstationary behavior and some form of randomness. The ways in which these properties manifest themselves naturally depend on the type of signal being considered.

■ NONSTATIONARITY

A signal is said to be *nonstationary* if its intrinsic characteristics vary with time. For example, a speech signal is nonstationary, as explained here.

A simple model of the speech production process is given by a form of filtering, in which a sound source excites a vocal tract filter. The vocal tract is then modeled as a tube of nonuniform cross-sectional area, beginning at the glottis (i.e., the opening between the vocal cords) and ending at the lips, as outlined by the dashed lines in Fig. 10.1. This figure shows a sagittal plane x-ray photograph of a human vocal system. Depending on the mode

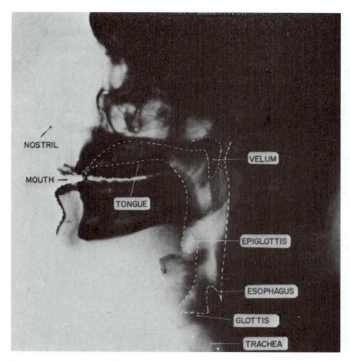

FIGURE 10.1 Saggital plane x-ray of the human vocal apparatus. (Reproduced from J. L. Flanagan et al., "Speech coding," *IEEE Transactions in Communications*, vol. COM-27, pp. 710–737, 1979; courtesy of the IEEE.)

of excitation provided by the source, the sounds constituting a speech signal may be classified into two distinct types:

► *Voiced sounds*, for which the source of excitation is pulse-like and periodic. In this case, the speech signal is produced by forcing air (from the lungs) through the glottis with the vocal cords vibrating in a relaxation oscillation. An example of voiced sounds is /e/ in *eve*; the symbol / / is commonly used to denote a phoneme, a basic linguistic unit.

► *Unvoiced sounds*, for which the source of excitation is noise-like (i.e., random). In this second case, the speech signal is produced by forming a constriction in the vocal tract toward the mouth end and forcing a continuous stream of air through the constriction at high velocity. An example of unvoiced sound is /f/ in *f*ish.

Voiced sounds during vowels are characterized by quasi-periodicity, low-frequency content, and large amplitude. In contrast, unvoiced sounds or fricatives are characterized by randomness, high-frequency content, and relatively low amplitude. The transition in time between voiced and unvoiced sounds is gradual (on the order of tens of milliseconds). Thus recognizing that a typical speech signal contains many voiced and unvoiced sounds strung together in a manner that depends on what is being spoken, we can readily appreciate that a speech signal is indeed a nonstationary signal.

Other examples of nonstationary signals include the time series representing the fluctuations in stock prices observed at the various capital markets around the world, the received signal of a radar system monitoring variations in prevalent weather conditions, and the received signal of a radio telescope listening to radio emissions from the galaxy of stars around us.

■ RANDOMNESS

Another property of real-life signals is the random-like appearance of their waveforms. This property may arise because the signal of interest is generated by a physical process that is *intrinsically random in nature*. A signal is said to be random if it is unpredictable. An ensemble of such waveforms, each having a certain probability of occurrence, constitutes a *random process* or *stochastic process*.

A speech signal, made up of voiced and unvoiced sounds, is an example of a random signal. This is well illustrated in the "filtered" version of a speech signal displayed in Fig. 8.21(b). This waveform is typical of speech signals transmitted over a telephone channel. Figure 8.21(b) clearly illustrates the random behavior of speech signals, exemplified by the fact that the amplitude of the speech signal fluctuates between positive and negative values in an apparently unpredictable fashion.

Another example of a random process is *thermal noise*, which refers to electrical noise arising from the random motion of electrons in a conductor. This form of noise is observed at the front end of every communication system receiver. For another example of a random process, we mention the *quantization noise* produced by the conversion of an analog signal into a digital signal. Thermal noise and quantization noise, in their own individual ways, impose limits on the capacity of a communication system to transmit information from a source to a user over a channel.

Although the precise prediction of a random signal is not possible, it may be described in terms of its *statistical properties*. These properties include the average power in the random signal, and the spectral distribution of this average power as a function of frequency. The mathematical discipline that deals with the statistical characterization of random signals is *probability theory*.

■ CHAOTIC BEHAVIOR

With predictability as the criterion of interest, we classified signals in Chapter 1 as either deterministic or random. In reality, however, there is another class of signals known as *chaotic signals*, which exhibit distinct properties of their own. A chaotic signal is an observation of a *chaotic system*, defined by a coupled system of nonlinear differential equations or nonlinear difference equations, whose parameters are all fixed. In light of what we have said so far, it would be tempting to say that the future behavior of a chaotic system is completely determined by its past. In actual fact, however, this is not so due to an intrinsic property of all chaotic systems, known as *sensitivity to initial conditions*. Specifically, any uncertainty in the initial conditions of the chaotic system, no matter how small, grows exponentially with time. Accordingly, a chaotic signal is only predictable in the short term.

Despite the fact that a chaotic signal is deterministic in nature, it can exhibit a waveform so complicated in appearance that it looks random. Indeed a chaotic signal may pass many of the statistical tests ordinarily associated with random signals, which makes it easy to confuse a chaotic signal for a random one.

In short, we may define a chaotic signal as a random-like signal generated by a coupled system of nonlinear differential/difference equations with fixed parameters, and which, unlike noise, is slightly predictable but not as predictable as a purely deterministic process. This definition lacks mathematical precision; nevertheless, it is adequate for our present discussion.

Chaos can arise in one of two ways:

1. *Known chaotic systems*, examples of which include the logistic and Lorenz systems, as described here.

   ▶ *Logistic system*, whose dynamics are governed by the nonlinear difference equation:

$$y[n] = 4y[n - 1](1 - y[n - 1]) \qquad (10.1)$$

   The previous sample $y[n - 1]$ determines the present sample $y[n]$. It is for this reason that the logistic system is referred to as a first-order nonlinear process. The logistic system is known to be chaotic on the interval [0, 1].

   ▶ *Lorenz system*, whose dynamics are governed by the coupled system of nonlinear differential equations:

$$\frac{dx(t)}{dt} = -\sigma x(t) + \sigma y(t)$$

$$\frac{dy(t)}{dt} = -x(t)z(t) + rx(t) - y(t) \qquad (10.2)$$

$$\frac{dz(t)}{dt} = x(t)y(t) - bz(t)$$

   where σ, r, and b are dimensionless parameters. Typical values for these parameters are $\sigma = 10$, $b = \frac{8}{3}$, and $r = 28$. The Lorenz system is named after the meteorologist Edward Lorenz. Figure 10.2 shows a trajectory of the point represented by $(x(t), y(t), z(t))$ for varying time t in three-dimensional space.

2. *Real-life phenomena*, where chaotic behavior has been confirmed experimentally. Examples of such phenomena include the following:

   ▶ *Turbulence* observed in fluids

   ▶ Radar backscatter from an ocean surface, commonly referred to as *sea clutter* or *sea echo*

The study of chaos is rooted in *nonlinear dynamical systems*.

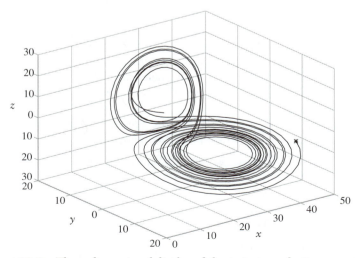

FIGURE 10.2   Three-dimensional display of the trajectory of a Lorenz attractor.

▌ 10.2 *Time–Frequency Analysis*

Fourier theory is valid only for stationary signals. For the analysis of a nonstationary signal, the preferred method is to use a description of the signal that involves both time and frequency. As the name implies, *time–frequency analysis* maps a signal (i.e., a one-dimensional function of time) onto an image (i.e., a two-dimensional function of time and frequency) that displays the signal's spectral components as a function of time. In conceptual terms, we may think of this mapping as a *time-varying spectral representation* of the signal. This representation is analogous to a musical score, with time and frequency representing the two principal axes. The values of the time–frequency representation of the signal provide an indication of the specific times at which certain spectral components of the signal are observed.

Basically, there are two classes of time–frequency representations of signals: *linear* and *quadratic*. In this section, we concern ourselves with linear representations only. Specifically, we present brief expositions of the short-time Fourier transform and the wavelet transform, in that order.

■ SHORT-TIME FOURIER TRANSFORM

The short-time Fourier transform (STFT) is a natural extension of the ordinary Fourier transform. Let $x(t)$ denote a signal assumed to be "stationary" when it is viewed through a temporal *window* $w(t)$ that is of limited extent; in general, $w(t)$ is complex valued. The STFT of $x(t)$, denoted by $X(\tau, j\omega)$, is defined as the Fourier transform of the windowed signal $x(t)w^*(t - \tau)$, where τ is the center position of the window and the asterisk denotes complex conjugation; that is, we write

$$X(\tau, j\omega) = \int_{-\infty}^{\infty} x(t)w^*(t - \tau)e^{-j\omega t}\, dt \tag{10.3}$$

$X(\tau, j\omega)$ is linear in the signal $x(t)$. The parameter ω plays a role similar to that of angular frequency in the ordinary Fourier transform. For a given $x(t)$, the result obtained by computing $X(\tau, j\omega)$ is dependent on the choice of the window $w(t)$. (In the literature on time–frequency analysis, the short-time Fourier transform is usually denoted by $X(\tau, \omega)$; we have used $X(\tau, j\omega)$ here to be consistent with the terminology used in this book.)

Many different window shapes are used in practice. Typically, they are symmetric, unimodal, and smooth; examples include a Gaussian window as illustrated in Fig. 10.3(a) or a single period of a Hanning window (i.e., raised-cosine window) as illustrated in Fig. 10.3(b); see Eq. (8.61) for definition of the Hanning window in discrete time.

In mathematical terms, the integral of Eq. (10.3) represents the inner (scalar) product of the signal $x(t)$ with a two-parameter family of basis functions, $w(t - \tau)e^{j\omega t}$, for varying τ and ω. It is important to note that, in general, these basis functions do not constitute an orthonormal set.

Many of the properties of the Fourier transform are carried over to the STFT. In particular, we may mention the following two signal-preserving properties:

1. *The STFT preserves time shifts, except for a linear modulation:* that is, if $X(\tau, j\omega)$ is the STFT of a signal $x(t)$, then the STFT of the time-shifted signal $x(t - t_0)$ is equal to $X(\tau - t_0, j\omega)e^{-j\omega t_0}$.

2. *The STFT preserves frequency shifts:* that is, if $X(\tau, j\omega)$ is the STFT of a signal $x(t)$, then the STFT of the modulated signal $x(t)e^{-j\omega_0 t}$ is equal to $X(\tau, j\omega - j\omega_0)$.

An issue of major concern in using the STFT is that of *time–frequency resolution.* To be specific, consider a pair of purely sinusoidal signals whose angular frequencies are

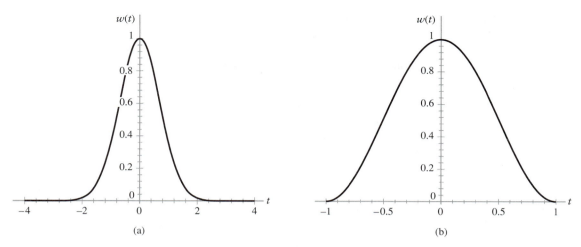

FIGURE 10.3 (a) Gaussian window. (b) Hanning window.

spaced $\Delta\omega$ rad/s apart. The smallest value of $\Delta\omega$ for which the two signals are resolvable is called the *frequency resolution*. The corresponding duration of the window $w(t)$ is called the *time resolution*, denoted by $\Delta\tau$. The frequency resolution $\Delta\omega$ and time resolution $\Delta\tau$ are inversely related, as shown by

$$\Delta\tau\,\Delta\omega \geq \tfrac{1}{2} \tag{10.4}$$

which is a manifestation of the duality property of the STFT, inherited from the Fourier transform. This relationship is referred to as the *uncertainty principle*, a term borrowed from analogy with statistical quantum mechanics; it was discussed in Section 3.6 in the context of time–bandwidth product. The best that we can do is to satisfy Eq. (10.4) with the equality sign. This special condition is satisfied by the use of a Gaussian window. Consequently, the time–frequency resolution capability of the STFT is *fixed* over the entire time–frequency plane. This point is illustrated in Fig. 10.4(a), where the time–frequency plane is partitioned into *tiles* of the same shape and size. Figure 10.4(b) displays the real

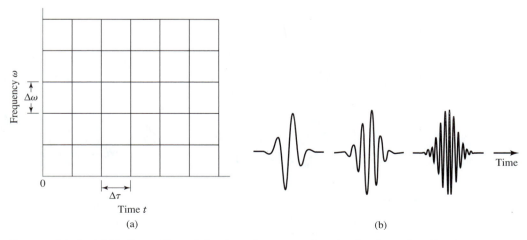

FIGURE 10.4 (a) Uniform tiling of the time–frequency plane by the short-time Fourier transform. (b) Real parts of associated basis functions.

parts of the associated basis functions of the STFT; they all have exactly the same duration, but different frequencies.

The squared magnitude of the STFT of a signal $x(t)$ is referred to as the *spectrogram* of the signal; it is defined by

$$|X(\tau, j\omega)|^2 = \left| \int_{-\infty}^{\infty} x(t)w^*(t - \tau)e^{-j\omega t} \, dt \right|^2 \tag{10.5}$$

The spectrogram represents a simple and yet powerful extension of classical Fourier theory. In physical terms, it provides a measure of the signal energy in the time–frequency plane.

■ SPECTROGRAMS OF SPEECH SIGNALS

Figure 10.5 shows the spectrograms of prefiltered and postfiltered versions of a speech signal that were displayed in Figs. 8.21(a) and (b). (Two versions of Fig. 10.5 are presented: black and white here, and color on the color plate.) These spectrograms were computed using a raised-cosine window, 256 samples long. The color of a particular pattern in the spectrograms of Fig. 10.5 is indicative of the signal energy in that pattern. The color code (in decreasing energy) is as follows: red is the highest, followed by yellow, and then blue.

The following observations pertaining to the characteristics of speech signals can be made from the spectrograms of Fig. 10.5:

▶ The resonant frequencies of the vocal tract are represented by the brightest (red) areas of the spectrograms.

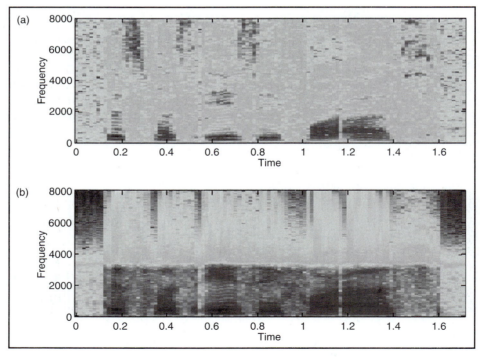

FIGURE 10.5 Spectrograms of speech signals. (a) Noisy version of the speech signal produced by a female speaker saying the phrase: "This was easy for us." (b) Filtered version of the speech signal. Color version of this figure appears on the color plate.

▶ In the voiced regions, the striations appear brighter as the energy is concentrated in narrow frequency bands.

▶ The unvoiced sounds have lower amplitudes because their energy is distributed over a broader band of frequencies.

The sharp horizontal boundary at about 3.1 kHz between regions of significant energy and low (almost zero) energy seen in the spectrogram of Fig. 10.5(b), representing the postfiltered speech signal, is due to the action of the FIR digital lowpass filter.

▨ WAVELET TRANSFORM

To overcome the time–frequency resolution limitation of the STFT, we need a form of mapping that has the ability to trade off time resolution for frequency resolution, and vice versa. One such method is known as the *wavelet transform* (WT). The WT acts like a "mathematical microscope," in the sense that different parts of the signal may be examined by adjusting the "focus."

Given a nonstationary signal $x(t)$, the WT is defined as the inner product of $x(t)$ with the two-parameter family of basis functions denoted by

$$\psi_{\tau,a}(t) = |a|^{-1/2}\psi\left(\frac{t-\tau}{a}\right) \tag{10.6}$$

where a is a *scale factor* (also referred to as a *dilation parameter*), and τ is a time delay. Thus in mathematical terms the WT of the signal $x(t)$ is defined by

$$W_x(\tau, a) = |a|^{-1/2}\int_{-\infty}^{\infty} x(t)\psi^*\left(\frac{t-\tau}{a}\right)dt \tag{10.7}$$

In wavelet analysis, the basis function $\psi_{\tau,a}(t)$ is an oscillating function; there is therefore no need to use sines and cosines (waves) as in Fourier analysis. More specifically, the basis function $e^{j\omega t}$ in Fourier analysis oscillates forever; in contrast, the basis function $\psi_{\tau,a}(t)$ in wavelet analysis is localized in time—it lasts for only a few cycles. Like the Fourier transform, the wavelet transform is invertible; if we are given $W_x(\tau, a)$, the original signal $x(t)$ can be reconstructed without loss of information.

The basis functions $\psi_{\tau,a}(t)$ are called *wavelets*; they constitute the building blocks of wavelet analysis. According to Eq. (10.6), the wavelets are scaled and translated versions of a prototype $\psi(t)$, called the *basic wavelet* or *mother wavelet*. The delay parameter τ gives the position of the wavelet $\psi_{\tau,a}(t)$, while the scale factor a governs its frequency content. For $a \ll 1$, the wavelet $\psi_{\tau,a}(t)$ is a very highly concentrated and shrunken version of the basic wavelet $\psi(t)$, with frequency content concentrated mostly in the high-frequency range. On the other hand, for $|a| \gg 1$, the wavelet $\psi_{\tau,a}(t)$ is very much spread out and has mostly low frequencies. Figure 10.6 shows two such examples of the wavelet $\psi(t)$. The *Haar wavelet*, shown in Fig. 10.6(a), is the simplest example of a wavelet. The *Daubechies wavelet*, shown in Fig. 10.6(b), is a more sophisticated example of a wavelet. Both of these wavelets have finite (compact) support in time. The Daubechies wavelet has length $L = 15$ and is therefore less local than the Haar wavelet that has length $L = 2$. However, the Daubechies wavelet is continuous and has better frequency resolution than the Haar wavelet.

In windowed Fourier analysis the goal is to measure the *local frequency content* of the signal. On the other hand, in wavelet analysis we measure *similarity* between the signal $x(t)$ and the wavelet $\psi_{\tau,a}(t)$ for varying τ and a. The dilations by $1/a$ result in several *magnifications* of the signal, with distinct *resolutions*.

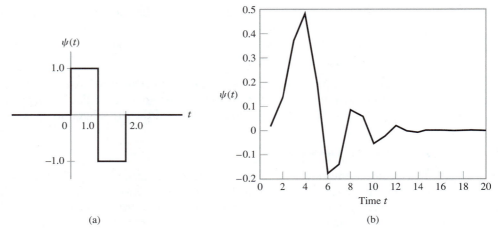

FIGURE 10.6   (a) Haar wavelet. (b) Daubechies wavelet.

Just as the STFT has signal-preserving properties of its own, so does the WT as described here:

1. *The WT preserves time shifts:* that is, if $W_x(\tau, a)$ is the WT of a signal $x(t)$, then $W_x(\tau - t_0, a)$ is the WT the time-shifted signal $x(t - t_0)$.
2. *The WT preserves time scaling:* that is, if $W_x(\tau, a)$ is the WT of a signal $x(t)$, then the WT of the time-scaled signal $|a_0|^{1/2} x(a_0 t)$ is equal to $W_x(a_0 \tau, a/a_0)$.

However, unlike the STFT, the WT does not preserve frequency shifts.

In general, the basic wavelet $\psi(t)$ can be any bandpass function. To establish a connection with the modulated window in the STFT, we choose

$$\psi(t) = w(t)e^{-j\omega_0 t} \tag{10.8}$$

The window $w(t)$ is typically a lowpass function. Thus Eq. (10.8) descibes the basic wavelet $\psi(t)$ as a complex, linearly modulated signal whose frequency content is concentrated essentially around its own carrier frequency ω_0. Let ω denote the carrier frequency of an analyzing wavelet $\psi_{\tau,a}(t)$. The scale factor a of $\psi_{\tau,a}(t)$ is inversely related to the carrier frequency ω, as shown by

$$a = \frac{\omega_0}{\omega} \tag{10.9}$$

Since, by definition, a wavelet is a scaled version of the same prototype, it follows that we may also write

$$\frac{\Delta\omega}{\omega} = Q \tag{10.10}$$

where $\Delta\omega$ is the frequency resolution of the analyzing wavelet $\psi_{\tau,a}(t)$ and Q is a constant. Choosing the window $w(t)$ to be a Gaussian function, and therefore using Eq. (10.4) with the equality sign, we may express the time resolution of the wavelet $\psi_{\tau,a}(t)$ as

$$\Delta\tau = \frac{1}{2\Delta\omega}$$

$$= \frac{1}{2Q\omega} \tag{10.11}$$

In light of Eqs. (10.10) and (10.11), we may now formally state the time–frequency resolution properties of the WT as follows:

▶ The time resolution $\Delta\tau$ varies inversely with the carrier frequency ω of the analyzing wavelet $\psi_{\tau,a}(t)$; hence it can be made arbitrarily small at high frequencies.

▶ The frequency resolution $\Delta\omega$ varies linearly with the carrier frequency ω of the analyzing wavelet $\psi_{\tau,a}(t)$; hence it can be made arbitrarily small at low frequencies.

Thus the WT is well suited for the analysis of nonstationary signals containing high-frequency transients superimposed on longer-lived low-frequency components.

In the previous subsection we stated that the STFT has a fixed resolution, as illustrated in Fig. 10.4(a). In contrast, the WT has a multiresolution capability, as illustrated in Fig. 10.7(a). In the latter figure we see that the WT partitions the time–frequency plane into tiles of the same area, but with varying widths and heights that depend on the carrier frequency ω of the analyzing wavelet $\psi_{\tau,a}(t)$. Thus, unlike the STFT, the WT provides a trade-off between time and frequency resolutions, which are represented by the widths and heights of the tiles, respectively (i.e., narrower widths/heights correspond to better resolution). Figure 10.7(b) displays the real parts of the basis functions of the WT. Here we see that every time the basis function is compressed by a factor of 2, say, its carrier frequency is increased by the same factor.

The WT performs a time-scale analysis. Thus its squared magnitude is called a *scalogram*, defined by

$$|W_x(\tau, a)|^2 = \frac{1}{a}\left|\int_{-\infty}^{\infty} x(t)\psi^*\left(\frac{t-\tau}{a}\right)dt\right|^2 \tag{10.12}$$

The scalogram represents a distribution of the signal energy in the time-scale plane.

In both the spectrogram and scalogram, phase information pertaining to the signal $x(t)$ is lost; neither one of them can therefore be inverted in general.

■ IMAGE COMPRESSION USING THE WAVELET TRANSFORM

The transmission of an image over a communication channel can be made more efficient by compressing the image at the transmitting end of the system and reconstructing the

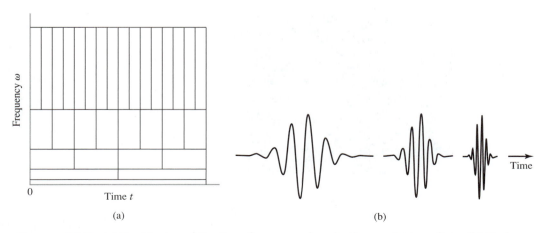

(a)   (b)

FIGURE 10.7   (a) Partitioning of the time–frequency plane by the wavelet transform. (b) Real parts of associated basis functions.

original image at the receiving end. This combination of signal-processing operations is called *image compression*. Basically, there are two types of image compression:

1. *Lossless compression*, which operates by removing the *redundant* information contained in the image. Lossless compression is completely reversible in that the original image can be reconstructed exactly.
2. *Lossy compression*, which involves the loss of some information and may therefore not be completely reversible. It is, however, capable of achieving a compression ratio higher than that attainable with lossless methods.

In many cases, lossy compression is the preferred method as it does not significantly alter the perceptual quality of the image. For most applications, this is acceptable, and in high-throughput applications such as the Internet it is a necessity.

Wavelets provide a powerful linear method for lossy image compression, because the coefficients of the wavelet transform are localized in both space and frequency. Figure 10.8(a) shows a *magnetic resonance image* (MRI). Figure 10.8(b) shows the same image after 20 : 1 compression and reconstruction. The wavelet transform used to perform the compression was based on the Daubechies wavelet of Fig. 10.6(b). Comparing the compressed MRI with the original MRI, we see that the visual loss resulting from the com-

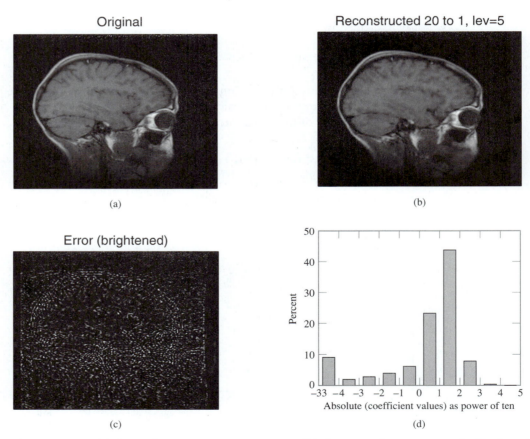

FIGURE 10.8   (a) Original magnetic resonance image. (b) Compressed image, using the Daubechies wavelet of Fig. 10.6(b). (c) Difference image between the original and compressed images. This image has been brightened up to make the differences more visible. (d) Distribution of the wavelet coefficients.

pression is barely discernible. In reality, however, there is some information loss, as shown in the difference image displayed in Fig. 10.8(c), which is obtained by subtracting the compressed MRI from the original MRI. The difference image can be reduced by using a more elaborate form of the Daubechies wavelet than that of Fig. 10.6(b), if so desired. It is also important to note that in some applications (e.g., a clinical setting), the diagnostic quality of the image must not be degraded on reconstruction.

Figure 10.8(d) displays the profile of the distribution of wavelet coefficients by highlighting the percentage of coefficients that have absolute values inside specific intervals. For the 20:1 compression ratio achieved in Fig. 10.8(b), only the coefficients represented by the rightmost of approximately 2.5 bars (i.e., the top 5%) are retained, and the remaining coefficients are reduced to zero. In so doing, we readily see the power and practical benefit of image compression techniques.

10.3 *Departures from the "Linear Time-Invariant System" Model*

We turn next to the idealized system model, namely, the linear time-invariant (LTI) system. This model rests on a well-developed body of mathematical theory, exemplified by the material presented in the previous chapters of this book. Indeed, many physical systems do admit the use of a LTI model. However, strictly speaking, a physical system may depart from this ideal model due to the presence of nonlinear components or time-varying parameters in the system or both, depending on the conditions under which the system is operated.

■ NONLINEAR SYSTEMS

For the linearity assumption to be satisfied the amplitudes of the signals encountered in a system (e.g., control system) would have to be restricted to lie inside a range small enough for all components of the system to operate in their "linear region." This would ensure that the principle of superposition is essentially satisfied, so that, for all practical purposes, the system can be viewed as linear. But when the amplitudes of the signals are permitted to lie outside the range of linear operation, the system can no longer be considered linear. For example, a transistor amplifier used in a control system exhibits an input–output characteristic that runs into saturation when the amplitude of the signal applied to the amplifier input is large. Other sources of nonlinearity in a control system include backlash between coupled gears and friction between moving parts. In any event, when the deviation from linearity is relatively small, some form of distortion is introduced into the characterization of the system. The effect of this distortion can be reduced by applying feedback around the system, as discussed in Chapter 9.

However, what if the operating amplitude range of a system is required to be large? The answer to this question depends on the intended application and how the system is designed. For example, in a control application, a linear control system is likely to perform poorly or to become unstable, because a linear design procedure is incapable of properly compensating the effects of large deviations from linearity. On the other hand, a nonlinear control system may perform better by directly incorporating nonlinearities into the system design. This point is demonstrated by studying the motion control of a robot, where many of the dynamic forces experienced vary as the speed squared. When a linear control system is used for such an application, nonlinear forces associated with the motion of robot links are neglected, with the result that the control accuracy degrades rapidly as the speed of motion is increased. Accordingly, in a robot task such as "pick-and-place," the speed of motion has to be maintained relatively low so as to realize a prescribed control accuracy. In contrast, a highly accurate control for a large range of robot speeds in a large workplace can be attained by employing a nonlinear control system that compensates the nonlinear forces experienced in robot motion. The benefit so gained is improved productivity.

Notwithstanding the mathematical difficulty of analyzing nonlinear systems, serious efforts have been made on several fronts to develop theoretical tools for the study of nonlinear systems. In this context, the following three approaches deserve special mention:

1. *Phase-space analysis*
The basic idea of this method is to use a graphical approach to study a nonlinear system of first-order differential equations written as follows:

$$\frac{d}{dt} x_j(t) = f_j(x_j(t)), \qquad j = 1, 2, \ldots, p \tag{10.13}$$

where the x_j define the elements of a p-dimensional *state vector* $\mathbf{x}(t)$, and the f_j are non-linear functions; p is referred to as the *order* of the system. Equation (10.13) may be viewed as describing the motion of a point in a p-dimensional space, commonly referred to as the *phase space* of the system; this terminology is borrowed from physics. The phase space is important because it provides us with a visual/conceptual tool for analyzing the dynamics of a nonlinear system described by Eq. (10.13). It does so by focusing our attention on the global characteristics of the motion rather than the detailed aspects of analytic or numeric solutions of the equations.

Starting from a set of initial conditions, Eq. (10.13) defines a solution represented by $x_1(t), x_2(t), \ldots, x_p(t)$. As time t is varied from zero to infinity, this solution traces a

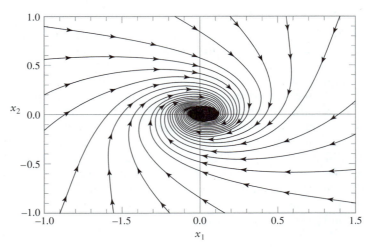

FIGURE 10.9 Phase portrait of a two-dimensional nonlinear dynamical system described by the pair of state equations

$$\frac{dx_1(t)}{dt} = x_2(t) - x_1(t)(x_1^2(t) + x_2^2(t) - c)$$

$$\frac{dx_2(t)}{dt} = -x_1(t) - x_2(t)(x_1^2(t) + x_2^2(t) - c)$$

for control parameter $c = -0.2$. (Reproduced from T. S. Parker and L. O. Chua, *Practical Numerical Algorithms for Chaotic Systems*, Springer-Verlag, 1989; courtesy of Springer-Verlag.)

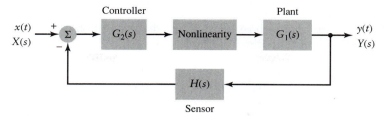

FIGURE 10.10   Feedback control system containing a nonlinear element in its feedback loop.

curve in the phase space. Such a curve is called a *trajectory*. A family of trajectories, corresponding to different initial conditions, is called a *phase portrait*. Figure 10.9 illustrates the phase portrait of a two-dimensional nonlinear dynamical system.

Largely due to the limitations of our visual capability, the graphical power of phase-space analysis is limited to second-order systems (i.e., $p = 2$) or systems that can be approximated by second-order dynamics.

2. *Describing-function analysis*

With a nonlinear element subjected to a sinusoidal input, the *describing function* of the element is defined as the complex ratio of the fundamental component of the output to the sinusoidal input. Thus the essence of the describing-function method is to approximate the nonlinear elements of a nonlinear system with quasilinear equivalents, and then exploit the power of frequency-domain techniques to analyze the approximating system.

Consider, for example, the nonlinear control system described in Fig. 10.10. The feedback loop of the system includes a nonlinearity and three linear elements represented by the transfer functions $G_1(s)$, $G_2(s)$, and $H(s)$. For example, the nonlinear element may be a relay. For a sinusoidal input $x(t)$, the input to the nonlinearity will not be sinusoidal due to harmonics of the input signal generated by the nonlinearity and fed back to the input via the sensor $H(s)$. If, however, the linear elements $G_1(s)$, $G_2(s)$, and $H(s)$ are of a lowpass type such that the harmonics so generated are attenuated to insignificant levels, we then would be justified to assume that the input to the nonlinearity is essentially sinusoidal. In a situation of this kind, application of the describing-function method would produce accurate results.

The describing-function method is used mainly to predict the occurrence of *limit cycles* in nonlinear feedback systems. By a "limit cycle" we mean a closed trajectory in the phase space onto which other trajectories converge asymptotically, from both inside and outside, as time approaches infinity. This form of convergence in phase space is illustrated in Fig. 10.11. A limit cycle is a periodic motion peculiar to nonlinear feedback systems.

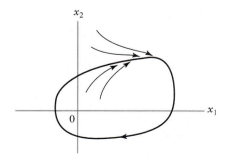

FIGURE 10.11   Limit cycle in two-dimensional phase space.

3. *Lyapunov theory*

This third approach to the stability analysis of nonlinear systems is built around Lyapunov's indirect and direct methods:

▶ The *indirect method* states that the stability properties of a nonlinear-feedback system in the neighborhood of an equilibrium point are essentially the same as those obtained through a linearized approximation to the system. By an "equilibrium point" we mean a point in phase space where the state vector of the system can reside forever.

▶ The *direct method* is a generalization of the concept of energy associated with a mechanical system. It states that the motion of a mechanical system is stable if the energy of the system is a decreasing function of time. To apply the method, we need to formulate a scalar energy-like function, called the *Lyapunov function*, and then see if that function decreases with time. Herein lie both the power and limitation of the method. The generality of Lyapunov's direct method gives it analytic power: it is applicable to every kind of nonlinear feedback system, be it time-varying or time-invariant, finite-dimensional or infinite-dimensional. The limitation of the method is that it is often difficult to find a Lyapunov function for a nonlinear system of interest. However, the inability to find a Lyapunov function does not prove instability of the system. Rather, the existence of a Lyapunov function is sufficient but not necessary for stability.

We thus see that the phase-space method, describing-function method, and Lyapunov theory have advantages and limitations of their own. Between them, they provide a powerful set of tools for the study of nonlinear dynamical systems.

■ ADAPTIVE FILTERS

Nonlinearity in a physical system is one cause of deviation from the idealized LTI system model. Another cause of deviation is variation of system parameters with time. Such variation may be due to a variety of physical factors. There may also be unforeseen changes in the statistical properties of external inputs and disturbances applied to the system, the effects of which may be viewed as changes in the environment in which the system operates. The tools of conventional LTI system theory, yielding a system design with fixed parameters, are usually inadequate for dealing with these real-life situations. To produce a satisfactory performance over the entire range of parameter variations, the preferred approach is to use an adaptive filter.

An *adaptive filter* is defined as a time-varying system that is provided with an iterative mechanism for adjusting its parameters in a step-by-step fashion so as to operate in an optimum manner according to some specified criterion. Applications of adaptive filters include adaptive equalization and system identification, described in the sequel.

In a telecommunications environment, the channel is usually time varying. For example, in a switched telephone network we find that two factors contribute to the distribution of signal distortion on different link connections:

▶ Differences in the transmission characteristics of the individual links that may be switched together

▶ Differences in the number of links used on a particular connection

The result is that the telephone channel is random in the sense of being one of an ensemble (group) of possible physical realizations available for use. Consequently, the performance of a fixed equalizer for the transmission of digital data over a telephone channel may not be adequate, when it is designed using LTI system theory based on average channel characteristics. To realize the full data transmission capability of a telephone channel, there is need for *adaptive equalization*. The device used to perform adaptive equalization is called an *adaptive equalizer*. The adaptive equalizer, made up of a FIR filter, is ordinarily placed at the front end of the receiver. Figure 10.12 shows the block diagram of an adaptive equalizer. This equalizer adjusts its own parameters (i.e., FIR filter coefficients) continuously and automatically by operating on a pair of signals:

- *Input signal*, $x[n]$, representing a distorted version of the signal transmitted over the channel
- *Desired response*, $d[n]$, representing a replica of the transmitted signal

One's first reaction to the availability of a replica of the transmitted signal is: If such a signal is available at the receiver, why do we need adaptive equalization? To answer this question, we first note that a typical telephone channel changes little during an average data call. Accordingly, prior to data transmission, the equalizer is adjusted with the guidance of a binary *training sequence* transmitted through the channel. A synchronized version of this training sequence is generated at the receiver, where (after a time shift equal to the transmission delay through the channel) it is applied to the equalizer as the desired response. A training sequence commonly used in practice is the *pseudonoise (PN) sequence*, which consists of a deterministic periodic sequence with noise-like characteristics. Two identical PN sequence generators are used, one at the transmitter and the other at the receiver. When the training process is completed, the adaptive equalizer is ready for normal operation. The training sequence is switched off and information-bearing binary data are then transmitted over the channel. The equalizer output is passed through a threshold device, and a decision is made on whether the transmitted binary data symbol is a "1" or a "0." In normal operation, the decisions made by the receiver are correct most of the time. This means that the sequence of symbols produced at the output of the threshold device represents a fairly reliable estimate of the transmitted data sequence and may therefore be used as a substitute for the desired response, as indicated in Fig. 10.13. This second mode of operation is called a *decision-directed mode*, the purpose of which is to *track*

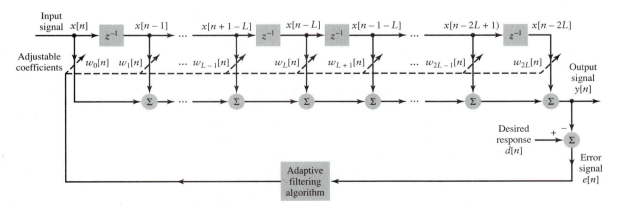

FIGURE 10.12   Adaptive equalizer built around a FIR digital filter.

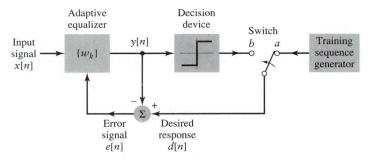

FIGURE 10.13   The two modes of operating an adaptive equalizer. (1) When the switch is in position *a*, the equalizer operates in its training mode. (2) When the switch is moved to position *b*, the equalizer operates in its decision-directed mode.

relatively slow variations in channel characteristics that may take place during the course of normal data transmission. The adjustment of filter coefficients in the equalizer is thus performed using an *adaptive filtering algorithm* that proceeds as follows:

1. *Training mode*
   (i) Given the FIR filter coefficients at iteration *n*, the corresponding value $y[n]$ of the actual equalizer output is computed in response to the input signal $x[n]$.
   (ii) The difference between the desired response $d[n]$ and equalizer output $y[n]$ is computed; this difference constitutes the *error signal*, denoted by $e[n]$.
   (iii) The error signal $e[n]$ is used to apply corrections to the FIR filter coefficients.
   (iv) Using the updated filter coefficients of the equalizer, steps (i) through (iii) are repeated until the equalizer reaches a steady-state condition, whereafter no noticeable changes in the filter coefficients are observed.

   To initiate this sequence of iterations, the filter coefficients are set equal to some suitable values (e.g., zero for all of them) at $n = 0$. The details of the corrections applied to the filter coefficients from one iteration to the next are determined by the type of adaptive filtering algorithm employed.

2. *Decision-directed mode*. This second mode of operation starts where the training mode finishes and uses the same set of steps except for two modifications:
   ▶ The output of the threshold device is substituted for the desired response.
   ▶ The adjustments to filter coefficients of the equalizer are continued throughout the data transmission.

Another useful application of adaptive filtering is in *system identification*. In this application, we are given an unknown dynamic plant, the operation of which cannot be interrupted, and the requirement is to build a model for it. Figure 10.14 shows the block diagram of how such a model, consisting of a FIR filter, may be computed. The input signal $x[n]$ is applied simultaneously to the plant and the model. Let $d[n]$ and $y[n]$ denote the corresponding values of the plant output and model output, respectively. The plant output $d[n]$ serves the purpose of desired response in this application. The difference between $d[n]$ and $y[n]$ thus defines the error signal $e[n]$, which is to be minimized according to some specified criterion. This minimization is attained by using an adaptive filtering algorithm that adjusts the model parameters (i.e., FIR filter coefficients) in a step-by-step

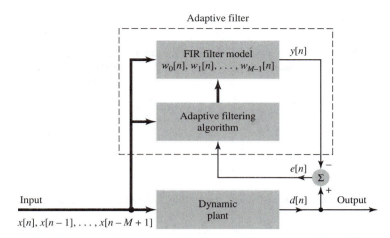

FIGURE 10.14   Block diagram of a FIR model whose coefficients are adjusted by an adaptive filtering algorithm for the identification of an unknown dynamic plant.

fashion, following a procedure similar to that described for the training mode of the adaptive equalizer.

Adaptive equalization and system identification are just two of the many applications of adaptive filters, which span such diverse areas as communications, control, radar, sonar, seismology, radio astronomy, and biomedical engineering.

10.4   *Concluding Remarks*

The material presented in the previous nine chapters provides a theoretical treatment of signals and systems, which paves the way for detailed studies of digital signal processing, communication systems, and control systems. The theory presented therein rests on the following idealizations:

▶ Stationary signals
▶ Linear time-invariant systems

In this chapter we highlighted limitations of that theory, viewed from the perspective of real-life signals and systems. In so doing, we also briefly touched on the topics of chaos, random processes, time–frequency analysis, nonlinear systems, and adaptive filters. These topics and those presented in previous chapters emphasize the enormous breadth of what is generally encompassed by the subject of Signals and Systems.

Another noteworthy point is that throughout this book we have focused on time as the independent variable. We may therefore refer to the material covered herein as *temporal processing*. In *spatial processing*, on the other hand, spatial coordinates play the role of independent variables. Examples of spatial processing are encountered in continuous aperture antennas, (discrete) antenna arrays, and processing of images. Much of the material presented in this book applies equally well to spatial processing, pointing further to its fundamental nature.

▌FURTHER READING

1. The classic approach to the characterization of speech signals is discussed in the following books:
   - ▶ Flanagan, J. L., *Speech Analysis: Synthesis and Perception* (Springer-Verlag, 1972)
   - ▶ Rabiner, L. R., and R. W. Schafer, *Digital Processing of Speech Signals* (Prentice Hall, 1978)

   For a more complete treatment of the subject, see the book:
   - ▶ Deller, J. R. Jr., J. G. Proakis, and J. H. L. Hanson, *Discrete-Time Processing of Speech Signals* (Prentice Hall, 1993)

2. For an introductory treatment of probability theory and random processes, see the following books:
   - ▶ Helstrom, C. W., *Probability and Stochastic Processes for Engineers*, Second Edition (Macmillan, 1990)
   - ▶ Leon-Garcia, A., *Probability and Random Processes for Electrical Engineering*, Second Edition, (Addison-Wesley, 1994)
   - ▶ Thomas, J. B., *Introduction to Probability* (Springer-Verlag, 1986)

3. For a highly readable account of nonlinear dynamics, including chaos, see the book:
   - ▶ Abraham, R. H., and C. D. Shaw, *Dynamics: The Geometry of Behavior* (Addison-Wesley, 1992)
   - ▶ The May 1998 issue of the *IEEE Signal Processing Magazine*, Volume 15, No. 3, presents two articles devoted to chaotic signal processing.

4. The subject of time–frequency analysis is covered in the book:
   - ▶ Cohen, L., *Time-Frequency Analysis* (Prentice Hall, 1995)

   For a discussion of wavelets, see:
   - ▶ Meyer, Y., *Wavelets: Algorithms and Applications* (SIAM, 1993)
   - ▶ Vetterli, M., and J. Kovacevic, *Wavelets and Subband Coding* (Prentice Hall, 1995)

   For an easy-to-read article on medical applications of the wavelet transform, see:
   - ▶ Akay, M., "Wavelet applications in medicine," *IEEE Spectrum*, vol. 34, no. 5, pp. 50–56, May 1997

5. For a study of nonlinear system analysis, see the books:
   - ▶ Atherton, D. P., *Nonlinear-Control Engineering* (Van Nostrand-Reinhold, 1975)
   - ▶ Slotine, J.-J. E., and W. Li, *Applied Nonlinear Control* (Prentice Hall, 1991)

6. The theory of adaptive filters and their applications are covered in the following books:
   - ▶ Haykin, S., *Adaptive Filter Theory*, Third Edition (Prentice Hall, 1996)
   - ▶ Widrow, B., and S. D. Stearns, *Adaptive Signal Processing* (Prentice Hall, 1985)

A — Selected Mathematical Identities

A.1 *Trigonometry*

Consider the right triangle depicted in Fig. A.1. The following relationships hold:

$$\sin \theta = \frac{y}{r}$$

$$\cos \theta = \frac{x}{r}$$

$$\tan \theta = \frac{y}{x} = \frac{\sin \theta}{\cos \theta}$$

$$\cos^2 \theta + \sin^2 \theta = 1$$

$$\cos^2 \theta = \tfrac{1}{2}(1 + \cos 2\theta)$$

$$\sin^2 \theta = \tfrac{1}{2}(1 - \cos 2\theta)$$

$$\cos 2\theta = 2\cos^2 \theta - 1$$

$$= 1 - 2\sin^2 \theta$$

Other identities include:

$$\sin(\theta \pm \phi) = \sin \theta \cos \phi \pm \cos \theta \sin \phi$$

$$\cos(\theta \pm \phi) = \cos \theta \cos \phi \mp \sin \theta \sin \phi$$

$$\sin \theta \sin \phi = \tfrac{1}{2}[\cos(\theta - \phi) - \cos(\theta + \phi)]$$

$$\cos \theta \cos \phi = \tfrac{1}{2}[\cos(\theta - \phi) + \cos(\theta + \phi)]$$

$$\sin \theta \cos \phi = \tfrac{1}{2}[\sin(\theta - \phi) + \sin(\theta + \phi)]$$

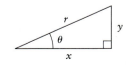

FIGURE A.1 Right triangle.

A.2   Complex Numbers

Let w be a complex number expressed in rectangular coordinates as $w = x + jy$, where $j = \sqrt{-1}$, $x = \text{Re}\{w\}$ is the real part of w, and $y = \text{Im}\{w\}$ is the imaginary part. We express w in polar coordinates as $w = re^{j\theta}$, where $r = |w|$ is the magnitude of w and $\theta = \arg\{w\}$ is the phase of w. The rectangular and polar representations for the number w are depicted in the complex plane of Fig. A.2.

■ CONVERTING FROM RECTANGULAR TO POLAR COORDINATES

$$r = \sqrt{x^2 + y^2}$$

$$\theta = \arctan\left(\frac{y}{x}\right)$$

■ CONVERTING FROM POLAR TO RECTANGULAR COORDINATES

$$x = r \cos \theta$$
$$y = r \sin \theta$$

■ COMPLEX CONJUGATE

If $w = x + jy = re^{j\theta}$, then using the asterisk to denote complex conjugation,

$$w^* = x - jy = re^{-j\theta}$$

■ EULER'S FORMULA

$$e^{j\theta} = \cos \theta + j \sin \theta$$

■ OTHER IDENTITIES

$$ww^* = r^2$$

$$x = \text{Re}(w) = \frac{w + w^*}{2}$$

$$y = \text{Im}(w) = \frac{w - w^*}{2j}$$

$$\cos \theta = \frac{e^{j\theta} + e^{-j\theta}}{2}$$

$$\sin \theta = \frac{e^{j\theta} - e^{-j\theta}}{2j}$$

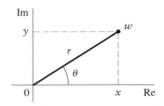

FIGURE A.2   The complex plane.

A.3 *Geometric Series*

If β is a complex number, then the following relationships hold:

$$\sum_{n=0}^{N-1} \beta^n = \begin{cases} \dfrac{1 - \beta^N}{1 - \beta}, & \beta \neq 1 \\ N, & \beta = 1 \end{cases}$$

$$\sum_{n=0}^{\infty} \beta^n = \frac{1}{1 - \beta}, \qquad |\beta| < 1$$

$$\sum_{n=k}^{\infty} \beta^n = \frac{\beta^k}{1 - \beta}, \qquad |\beta| < 1$$

$$\sum_{n=0}^{\infty} n\beta^n = \frac{\beta}{(1 - \beta)^2}, \qquad |\beta| < 1$$

A.4 *Definite Integrals*

$$\int_a^b x^n \, dx = \frac{1}{n + 1} x^{n+1} \Big|_a^b, \qquad n \neq -1$$

$$\int_a^b e^{cx} \, dx = \frac{1}{c} e^{cx} \Big|_a^b$$

$$\int_a^b x e^{cx} \, dx = \frac{1}{c^2} e^{cx}(cx - 1) \Big|_a^b$$

$$\int_a^b \cos(cx) \, dx = \frac{1}{c} \sin(cx) \Big|_a^b$$

$$\int_a^b \sin(cx) \, dx = -\frac{1}{c} \cos(cx) \Big|_a^b$$

$$\int_a^b x \cos(cx) \, dx = \frac{1}{c^2} \left(\cos(cx) + cx \sin(cx) \right) \Big|_a^b$$

$$\int_a^b x \sin(cx) \, dx = \frac{1}{c^2} \left(\sin(cx) - cx \cos(cx) \right) \Big|_a^b$$

$$\int_a^b e^{gx} \cos(cx) \, dx = \frac{e^{gx}}{g^2 + c^2} \left(g \cos(cx) + c \sin(cx) \right) \Big|_a^b$$

$$\int_a^b e^{gx} \sin(cx) \, dx = \frac{e^{gx}}{g^2 + c^2} \left(g \sin(cx) - c \cos(cx) \right) \Big|_a^b$$

■ **GAUSSIAN PULSES**

$$\int_{-\infty}^{\infty} e^{-x^2/2\sigma^2} \, dx = \sigma\sqrt{2\pi}, \qquad \sigma > 0$$

$$\int_{-\infty}^{\infty} x^2 e^{-x^2/2\sigma^2} \, dx = \sigma^3\sqrt{2\pi}, \qquad \sigma > 0$$

■ **INTEGRATION BY PARTS**

$$\int_a^b u(x)\ dv(x) = u(x)v(x)\ |_a^b - \int_a^b v(x)\ du(x)$$

A.5 *Matrices*

A *matrix* is a set of numbers arranged in a rectangular array. For example,

$$\mathbf{A} = \begin{bmatrix} 2 & 3 \\ -1 & 4 \end{bmatrix}$$

is a matrix with two columns and two rows. We thus say that $\mathbf{A}$ is a two by two matrix. The first and second rows of $\mathbf{A}$ are given by $[2\ \ 3]$ and $[-1\ \ 4]$, respectively. We index the elements of the matrix in terms of their location, which is measured by the row and column in which the element lies. For example, the element in the first row and second column of $\mathbf{A}$ is 3. Boldface uppercase symbols are used to denote matrix quantities.

A *vector* is a matrix containing a single column or a single row. A *column vector* consists of a single column. For example,

$$\mathbf{b} = \begin{bmatrix} 3 \\ -2 \end{bmatrix}$$

is a two-dimensional column vector. A *row vector* consists of a single row. For example,

$$\mathbf{c} = [2\ \ \ -1]$$

is a two-dimensional row vector. Vectors are denoted with lowercase boldface symbols.

■ **ADDITION**

If a_{ij} and b_{ij} are the elements in the ith row and jth column of matrices $\mathbf{A}$ and $\mathbf{B}$, respectively, then the matrix $\mathbf{C} = \mathbf{A} + \mathbf{B}$ has elements $c_{ij} = a_{ij} + b_{ij}$.

■ **MULTIPLICATION**

If a_{ik} is the element in the ith row and kth column of an M by N matrix $\mathbf{A}$ and b_{kj} is the element in the kth row and jth column of an N by L matrix $\mathbf{B}$, then the M by L matrix $\mathbf{C} = \mathbf{AB}$ has elements $c_{ij} = \Sigma_{k=1}^N a_{ik} b_{kj}$.

Two by Two Matrix Inverse

$$\begin{bmatrix} a & b \\ c & d \end{bmatrix}^{-1} = \frac{1}{ad - bc} \begin{bmatrix} d & -b \\ -c & a \end{bmatrix}$$

Inverse of Product of Matrices

If $\mathbf{A}$ and $\mathbf{B}$ are invertible, then

$$(\mathbf{AB})^{-1} = \mathbf{B}^{-1}\mathbf{A}^{-1}$$

B | Partial Fraction Expansions

Partial fraction expansions are used to express a ratio of polynomials as a sum of ratios of lower order polynomials. In essence, the partial fraction expansion is the inverse operation to placing a sum of fractions over a common denominator. The partial fraction expansion is used in signals and systems problems to determine inverse Fourier, Laplace, and z-transforms. In this context we use a partial fraction expansion to express an arbitrary ratio of polynomials as a sum of terms for which the inverse transform is known.

There are two different standard forms for ratios of polynomials that occur in our study of signals and systems. One arises in the context of representations for continuous-time signals and systems, while the other arises in the context of discrete-time signals and systems. We shall treat these separately, since the method for performing the partial fraction expansion differs slightly in each case.

B.1 *Partial Fraction Expansions for Continuous-Time Representations*

In the study of continuous-time signals and systems we generally encounter ratios of polynomials of the form

$$W(u) = \frac{B(u)}{A(u)}$$
$$= \frac{b_M u^M + b_{M-1} u^{M-1} + \cdots + b_1 u + b_0}{u^N + a_{N-1} u^{N-1} + \cdots + a_1 u + a_0} \tag{B.1}$$

In a Fourier transform problem the variable u represents $j\omega$ and in a Laplace transform problem u represents s. Note that the coefficient of u^N in $A(u)$ is unity. We assume that $W(u)$ is a proper rational function: that is, the order of $B(u)$ is less than that of $A(u)$ ($M < N$). If this condition is not satisfied, then long division of $A(u)$ into $B(u)$ is used to write $W(u)$ as the sum of a polynomial in u and a proper rational function representing the remainder of the division. The partial fraction expansion is then applied to the remainder.

The first step in performing a partial fraction expansion is to factor the denominator polynomial. If the N roots d_i are distinct, then we may rewrite $W(u)$ as shown by

$$W(u) = \frac{B(u)}{(u - d_1)(u - d_2) \cdots (u - d_N)}$$

In this case the partial fraction expansion for $W(u)$ takes the form

$$W(u) = \frac{C_1}{u - d_1} + \frac{C_2}{u - d_2} + \cdots + \frac{C_N}{u - d_N} \tag{B.2}$$

If a root $u = r$ occurs with multiplicity L, then $W(u)$ is of the form

$$W(u) = \frac{B(u)}{(u - r)^L (u - d_1)(u - d_2) \cdots (u - d_{N-L})}$$

and the partial fraction expansion for $W(u)$ is given as

$$
\begin{aligned}
W(u) = {} & \frac{C_1}{u - d_1} + \frac{C_2}{u - d_2} + \cdots + \frac{C_{N-L}}{u - d_{N-L}} + \frac{K_{L-1}}{u - r} \\
& + \frac{K_{L-2}}{(u - r)^2} + \cdots + \frac{K_0}{(u - r)^L}
\end{aligned}
\tag{B.3}
$$

Note that as the power to which the denominator terms $(u - r)$ are raised increases, the indices i of the corresponding coefficients K_i decrease.

The constants C_i and K_i are called *residues*. We may obtain the residues using two different approaches. In the method of linear equations we place all the terms in the partial fraction expansion for $W(u)$ over a common denominator and equate the coefficient of each power of u to the corresponding coefficient in $B(u)$. This gives a system of N linear equations that may be solved to obtain the residues, as illustrated in the following example. For hand calculations this approach is generally limited to $N = 2$ or $N = 3$.

EXAMPLE B.1   Determine the partial fraction expansion for the function

$$W(u) = \frac{3u + 5}{u^3 + 4u^2 + 5u + 2}$$

Solution: The roots of the denominator polynomial are $u = -2$ and $u = -1$ with multiplicity two. Hence the partial fraction expansion for $W(u)$ is of the form

$$W(u) = \frac{K_1}{u + 1} + \frac{K_0}{(u + 1)^2} + \frac{C_1}{u + 2}$$

The residues K_1, K_0, and C_1 may be determined by placing the terms in the partial fraction expansion over a common denominator, as shown by

$$
\begin{aligned}
W(u) &= \frac{K_1(u + 1)(u + 2)}{(u + 1)^2(u + 2)} + \frac{K_0(u + 2)}{(u + 1)^2(u + 2)} + \frac{C_1(u + 1)^2}{(u + 1)^2(u + 2)} \\
&= \frac{(K_1 + C_1)u^2 + (3K_1 + K_0 + 2C_1)u + (2K_1 + 2K_0 + C_1)}{u^3 + 4u^2 + 5u + 2}
\end{aligned}
$$

Equating the coefficient of each power of u in the numerator on the right-hand side of this equation to those in $B(u)$ gives the system of three equations in the three unknowns K_1, K_0, and C_1 shown by

$$
\begin{aligned}
0 &= K_1 + C_1 \\
3 &= 3K_1 + K_0 + 2C_1 \\
5 &= 2K_1 + 2K_0 + C_1
\end{aligned}
$$

Solving these equations we obtain $K_1 = 1$, $K_0 = 2$, and $C_1 = -1$ so the partial fraction expansion of $W(u)$ is given by

$$W(u) = \frac{1}{u + 1} + \frac{2}{(u + 1)^2} - \frac{1}{u + 2}$$

The *method of residues* is based on manipulating the partial fraction expansion so as to isolate each residue. Hence this method is usually easier to use than solving linear equations. Consider multiplying each side of Eq. (B.3) by $(u - d_i)$, as shown by

$$(u - d_i)W(u) = \frac{C_1(u - d_i)}{u - d_1} + \frac{C_2(u - d_i)}{u - d_2} + \cdots + C_i + \cdots + \frac{C_{N-L}(u - d_i)}{u - d_{N-L}}$$
$$+ \frac{K_{L-1}(u - d_i)}{u - r} + \frac{K_{L-2}(u - d_i)}{(u - r)^2} + \cdots + \frac{K_0(u - d_i)}{(u - r)^L}$$

On the left-hand side the multiplication by $(u - d_i)$ cancels the $(u - d_i)$ term in the denominator of $W(u)$. If we now evaluate this expression at $u = d_i$, then all the terms on the right-hand side are zero except for C_i and we obtain the expression for C_i given by

$$C_i = (u - d_i)W(u)\big|_{u=d_i} \tag{B.4}$$

Isolation of the residues associated with the repeated root $u = r$ requires multiplying both sides of Eq. (B.3) by $(u - r)^L$ and differentiating. We have

$$K_i = \frac{1}{i!}\frac{d^i}{du^i}\{(u - r)^L W(u)\}\bigg|_{u=r} \tag{B.5}$$

The following example uses Eqs. (B.4) and (B.5) to obtain the residues.

EXAMPLE B.2   Find the partial fraction expansion for

$$W(u) = \frac{3u^3 + 15u^2 + 29u + 21}{(u + 1)^2(u + 2)(u + 3)}$$

Solution: Here we have a root of multiplicity two at $u = -1$ and distinct roots at $u = -2$ and $u = -3$. Hence the partial fraction expansion for $W(u)$ is of the form

$$W(u) = \frac{K_1}{u + 1} + \frac{K_0}{(u + 1)^2} + \frac{C_1}{u + 2} + \frac{C_2}{u + 3}$$

We obtain C_1 and C_2 using Eq. (B.4) as shown by

$$C_1 = (u + 2)\frac{3u^3 + 15u^2 + 29u + 21}{(u + 1)^2(u + 2)(u + 3)}\bigg|_{u=-2}$$
$$= -1$$

$$C_2 = (u + 3)\frac{3s^3 + 15u^2 + 29u + 21}{(u + 1)^2(u + 2)(u + 3)}\bigg|_{u=-3}$$
$$= 3$$

Now we may obtain K_1 and K_0 using Eq. (B.5) as follows:

$$K_0 = (u + 1)^2\frac{3u^3 + 15u^2 + 29u + 21}{(u + 1)^2(u + 2)(u + 3)}\bigg|_{u=-1}$$
$$= 2$$

$$K_1 = \frac{1}{1!}\frac{d}{du}\left\{(u + 1)^2\frac{3u^3 + 15u^2 + 29u + 21}{(u + 1)^2(u + 2)(u + 3)}\right\}\bigg|_{u=-1}$$
$$= \frac{(9u^2 + 30u + 29)(u^2 + 5u + 6) - (3u^3 + 15u^2 + 29u + 21)(2u + 5)}{(u^2 + 5u + 6)^2}\bigg|_{u=-1}$$
$$= 1$$

Hence the partial fraction expansion for $W(u)$ is given by

$$W(u) = \frac{1}{u + 1} + \frac{2}{(u + 1)^2} - \frac{1}{u + 2} + \frac{3}{u + 3}$$

We may draw the following conclusions about the residues from Eqs. (B.4) and (B.5):

▶ The residue associated with a real root is real.

▶ The residues associated with a pair of complex-conjugate roots are the complex conjugate of each other, and thus only one of them needs to be computed.

B.2 *Partial Fraction Expansions for Discrete-Time Representations*

In the study of discrete-time signals and systems we frequently encounter ratios of polynomials having the form shown by

$$W(u) = \frac{B(u)}{A(u)}$$

$$= \frac{b_M u^M + b_{M-1} u^{M-1} + \cdots + b_1 u + b_0}{a_N u^N + a_{N-1} u^{N-1} + \cdots + a_1 u + 1} \tag{B.6}$$

In a discrete-time Fourier transform problem the variable u represents $e^{-j\Omega}$, while in a z-transform problem u represents z^{-1}. Note that the coefficient of the zeroth power of u in $A(u)$ is unity here. We again assume that $W(u)$ is a proper rational function: that is, the order of $B(u)$ is less than that of $A(u)$ ($M < N$). If this condition is not satisfied, then long division of $A(u)$ into $B(u)$ is used to write $W(u)$ as the sum of a polynomial in u and a proper rational function representing the remainder of the division. The partial fraction expansion is then applied to the remainder.

Here we write the denominator polynomial as a product of first-order terms as shown by

$$A(u) = (1 - d_1 u)(1 - d_2 u) \cdots (1 - d_N u) \tag{B.7}$$

where d_i^{-1} is a root of $A(u)$. Equivalently, d_i is a root of the polynomial $\tilde{A}(u)$ constructed by reversing the order of coefficients in $A(u)$. That is, d_i is a root of

$$\tilde{A}(u) = u^N + a_1 u^{N-1} + \cdots + a_{N-1} u + a_N$$

If all the d_i are distinct, then the partial fraction expansion is given by

$$W(u) = \frac{C_1}{1 - d_1 u} + \frac{C_2}{1 - d_2 u} + \cdots + \frac{C_N}{1 - d_N u} \tag{B.8}$$

If a term $1 - ru$ occurs with multiplicity L in Eq. (B.7), then the partial fraction expansion has the form shown by

$$W(u) = \frac{C_1}{1 - d_1 u} + \frac{C_2}{1 - d_2 u} + \cdots + \frac{C_{N-L}}{1 - d_{N-L} u}$$

$$+ \frac{K_{L-1}}{(1 - ru)} + \frac{K_{L-2}}{(1 - ru)^2} + \cdots + \frac{K_0}{(1 - ru)^L} \tag{B.9}$$

The residues C_i and K_i may be determined analogously to the continuous-time case. We may place the right-hand side of Eq. (B.8) or (B.9) over a common denominator and obtain a system of N linear equations by equating coefficients of like powers of u in the numerator polynomials. Alternatively, we may solve for the residues directly by manipulating the partial fraction expansion in such a way as to isolate each coefficient. This yields the following two relationships:

$$C_i = (1 - d_i u)W(u)|_{u=d_i^{-1}} \tag{B.10}$$

$$K_i = \frac{1}{i!}(-r^{-1})^i \frac{d^i}{du^i}\{(1 - ru)^L W(u)\}\Big|_{u=r^{-1}} \tag{B.11}$$

EXAMPLE B.3   Find the partial fraction expansion for the discrete-time function

$$W(u) = \frac{-14u - 4}{8u^3 - 6u - 2}$$

Solution: The constant term is not unity in the denominator, and so we first divide the denominator and numerator by -2 to express $W(u)$ in standard form. We may then write

$$W(u) = \frac{7u + 2}{-4u^3 + 3u + 1}$$

The denominator polynomial $A(u)$ is factored by rooting the related polynomial

$$\tilde{A}(u) = u^3 + 3u^2 - 4$$

This polynomial has a single root at $u = 1$, and a root of multiplicity two at $u = -2$. Hence $W(u)$ can be expressed as

$$W(u) = \frac{7u + 2}{(1 - u)(1 + 2u)^2}$$

and the partial fraction expansion has the form given by

$$W(u) = \frac{C_1}{1 - u} + \frac{K_1}{1 + 2u} + \frac{K_0}{(1 + 2u)^2}$$

The residues are evaluated using Eqs. (B.10) and (B.11) as follows:

$$C_1 = (1 - u)W(u)|_{u=1}$$
$$= 1$$
$$K_0 = (1 + 2u)^2 W(u)|_{u=-1/2}$$
$$= -1$$
$$K_1 = \frac{1}{1!}\left(\frac{1}{2}\right)\frac{d}{du}\{(1 + 2u)^2 W(u)\}\Big|_{u=-1/2}$$
$$= \frac{7(1 - u) + (7u + 2)}{2(1 - u)^2}\Big|_{u=-1/2}$$
$$= 2$$

We conclude that the partial fraction expansion is shown by

$$W(u) = \frac{1}{1 - u} + \frac{2}{1 + 2u} - \frac{1}{(1 + 2u)^2}$$

C Tables of Fourier Representations and Properties

C.1 Basic Discrete-Time Fourier Series Pairs

Time Domain	Frequency Domain
$x[n] = \sum_{k=\langle N \rangle} X[k]e^{jkn\Omega_o}$ Period $= N$	$X[k] = \dfrac{1}{N} \sum_{n=\langle N \rangle} x[n]e^{-jkn\Omega_o}$ $\Omega_o = \dfrac{2\pi}{N}$
$x[n] = \begin{cases} 1, & \lvert n \rvert \le M \\ 0, & M < \lvert n \rvert \le N/2 \end{cases}$ $x[n] = x[n+N]$	$X[k] = \dfrac{\sin\left(k\,\dfrac{\Omega_o}{2}\,(2M+1)\right)}{N\sin\left(k\,\dfrac{\Omega_o}{2}\right)}$
$x[n] = e^{jp\Omega_o n}$	$X[k] = \begin{cases} 1, & k = p,\, p \pm N,\, p \pm 2N,\, \ldots \\ 0, & \text{otherwise} \end{cases}$
$x[n] = \cos(p\Omega_o n)$	$X[k] = \begin{cases} \frac{1}{2}, & k = \pm p,\, \pm p \pm N,\, \pm p \pm 2N,\, \ldots \\ 0, & \text{otherwise} \end{cases}$
$x[n] = \sin(p\Omega_o n)$	$X[k] = \begin{cases} \dfrac{1}{2j}, & k = p,\, p \pm N,\, p \pm 2N,\, \ldots \\ \dfrac{-1}{2j}, & k = -p,\, -p \pm N,\, -p \pm 2N,\, \ldots \\ 0, & \text{otherwise} \end{cases}$
$x[n] = 1$	$X[k] = \begin{cases} 1, & k = 0,\, \pm N,\, \pm 2N,\, \ldots \\ 0, & \text{otherwise} \end{cases}$
$x[n] = \sum_{p=-\infty}^{\infty} \delta[n - pN]$	$X[k] = \dfrac{1}{N}$

C.2   Basic Fourier Series Pairs

Time Domain	Frequency Domain
$x(t) = \sum\limits_{k=-\infty}^{\infty} X[k]e^{jk\omega_o t}$ $Period = T$	$X[k] = \dfrac{1}{T} \int_{\langle T \rangle} x(t)e^{-jk\omega_o t}\, dt$ $\omega_o = \dfrac{2\pi}{T}$
$x(t) = \begin{cases} 1, & \lvert t \rvert \le T_s \\ 0, & T_s < \lvert t \rvert \le T/2 \end{cases}$	$X[k] = \dfrac{\sin(k\omega_o T_s)}{k\pi}$
$x(t) = e^{jp\omega_o t}$	$X[k] = \delta[k - p]$
$x(t) = \cos(p\omega_o t)$	$X[k] = \tfrac{1}{2}\delta[k - p] + \tfrac{1}{2}\delta[k + p]$
$x(t) = \sin(p\omega_o t)$	$X[k] = \dfrac{1}{2j}\delta[k - p] - \dfrac{1}{2j}\delta[k + p]$
$x(t) = \sum\limits_{p=-\infty}^{\infty} \delta(t - pT)$	$X[k] = \dfrac{1}{T}$

C.3   Basic Discrete-Time Fourier Transform Pairs

Time Domain	Frequency Domain
$x[n] = \dfrac{1}{2\pi} \int_{\langle 2\pi \rangle} X(e^{j\Omega})e^{j\Omega n}\, d\Omega$	$X(e^{j\Omega}) = \sum\limits_{n=-\infty}^{\infty} x[n]e^{-j\Omega n}$
$x[n] = \begin{cases} 1, & \lvert n \rvert \le M \\ 0, & \text{otherwise} \end{cases}$	$X(e^{j\Omega}) = \dfrac{\sin\left[\Omega\left(\dfrac{2M + 1}{2}\right)\right]}{\sin\left(\dfrac{\Omega}{2}\right)}$
$x[n] = \alpha^n u[n], \qquad \lvert \alpha \rvert < 1$	$X(e^{j\Omega}) = \dfrac{1}{1 - \alpha e^{-j\Omega}}$
$x[n] = \delta[n]$	$X(e^{j\Omega}) = 1$
$x[n] = u[n]$	$X(e^{j\Omega}) = \dfrac{1}{1 - e^{-j\Omega}} + \pi \sum\limits_{p=-\infty}^{\infty} \delta(\Omega - 2\pi p)$
$x[n] = \dfrac{1}{\pi n} \sin(Wn), \qquad 0 < W \le \pi$	$X(e^{j\Omega}) = \begin{cases} 1, & \lvert \Omega \rvert \le W \\ 0, & W < \lvert \Omega \rvert \le \pi, \end{cases} \quad X(e^{j\Omega}) \text{ is } 2\pi \text{ periodic}$
$x[n] = (n + 1)\alpha^n u[n]$	$X(e^{j\Omega}) = \dfrac{1}{(1 - \alpha e^{-j\Omega})^2}$

C.4 *Basic Fourier Transform Pairs*

Time Domain $x(t) = \dfrac{1}{2\pi} \displaystyle\int_{-\infty}^{\infty} X(j\omega)e^{j\omega t}\, d\omega$	*Frequency Domain* $X(j\omega) = \displaystyle\int_{-\infty}^{\infty} x(t)e^{-j\omega t}\, dt$
$x(t) = \begin{cases} 1, & |t| \le T \\ 0, & \text{otherwise} \end{cases}$	$X(j\omega) = \dfrac{2\sin(\omega T)}{\omega}$
$x(t) = \dfrac{1}{\pi t}\sin(Wt)$	$X(j\omega) = \begin{cases} 1, & |\omega| \le W \\ 0, & \text{otherwise} \end{cases}$
$x(t) = \delta(t)$	$X(j\omega) = 1$
$x(t) = 1$	$X(j\omega) = 2\pi\delta(\omega)$
$x(t) = u(t)$	$X(j\omega) = \dfrac{1}{j\omega} + \pi\delta(\omega)$
$x(t) = e^{-at}u(t), \qquad \text{Re}\{a\} > 0$	$X(j\omega) = \dfrac{1}{a + j\omega}$
$x(t) = te^{-at}u(t), \qquad \text{Re}\{a\} > 0$	$X(j\omega) = \dfrac{1}{(a + j\omega)^2}$
$x(t) = e^{-a|t|}, \qquad a > 0$	$X(j\omega) = \dfrac{2a}{a^2 + \omega^2}$
$x(t) = \dfrac{1}{\sqrt{2\pi}}\, e^{-t^2/2}$	$X(j\omega) = e^{-\omega^2/2}$

C.5 *Fourier Transform Pairs for Periodic Signals*

Periodic Time-Domain Signal	*Fourier Transform*
$x(t) = \displaystyle\sum_{k=-\infty}^{\infty} X[k]e^{jk\omega_o t}$	$X(j\omega) = 2\pi \displaystyle\sum_{k=-\infty}^{\infty} X[k]\delta(\omega - k\omega_o)$
$x(t) = \cos(\omega_o t)$	$X(j\omega) = \pi\delta(\omega - \omega_o) + \pi\delta(\omega + \omega_o)$
$x(t) = \sin(\omega_o t)$	$X(j\omega) = \dfrac{\pi}{j}\delta(\omega - \omega_o) - \dfrac{\pi}{j}\delta(\omega + \omega_o)$
$x(t) = e^{j\omega_o t}$	$X(j\omega) = 2\pi\delta(\omega - \omega_o)$
$x(t) = \sum_{n=-\infty}^{\infty}\delta(t - n\mathcal{T})$	$X(j\omega) = \dfrac{2\pi}{\mathcal{T}} \displaystyle\sum_{k=-\infty}^{\infty} \delta\left(\omega - k\dfrac{2\pi}{\mathcal{T}}\right)$
$x(t) = \begin{cases} 1, & |t| \le T_s \\ 0, & T_s < |t| < T/2 \end{cases}$ $x(t + T) = x(t)$	$X(j\omega) = \displaystyle\sum_{k=-\infty}^{\infty} \dfrac{2\sin(k\omega_o T_s)}{k}\delta(\omega - k\omega_o)$

C.6 *Discrete-Time Fourier Transform Pairs for Periodic Signals*

Periodic Time-Domain Signal	*Discrete-Time Fourier Transform*
$x[n] = \sum\limits_{k=\langle N \rangle} X[k] e^{jk\Omega_o n}$	$X(e^{j\Omega}) = 2\pi \sum\limits_{k=-\infty}^{\infty} X[k]\delta(\Omega - k\Omega_o)$
$x[n] = \cos(\Omega_1 n)$	$X(e^{j\Omega}) = \pi \sum\limits_{k=-\infty}^{\infty} \delta(\Omega - \Omega_1 - k2\pi) + \delta(\Omega + \Omega_1 - k2\pi)$
$x[n] = \sin(\Omega_1 n)$	$X(e^{j\Omega}) = \dfrac{\pi}{j} \sum\limits_{k=-\infty}^{\infty} \delta(\Omega - \Omega_1 - k2\pi) - \delta(\Omega + \Omega_1 - k2\pi)$
$x[n] = e^{j\Omega_1 n}$	$X(e^{j\Omega}) = 2\pi \sum\limits_{k=-\infty}^{\infty} \delta(\Omega - \Omega_1 - k2\pi)$
$x[n] = \sum\limits_{k=-\infty}^{\infty} \delta(n - kN)$	$X(e^{j\Omega}) = \dfrac{2\pi}{N} \sum\limits_{k=-\infty}^{\infty} \delta\left(\Omega - \dfrac{k2\pi}{N}\right)$

C.7 Properties of Fourier Representations

Property	Fourier Transform $x(t) \xleftrightarrow{FT} X(j\omega)$ $y(t) \xleftrightarrow{FT} Y(j\omega)$	Fourier Series $x(t) \xleftrightarrow{FS;\ \omega_o} X[k]$ $y(t) \xleftrightarrow{FS;\ \omega_o} Y[k]$ Period $= T$
Linearity	$ax(t) + by(t) \xleftrightarrow{FT} aX(j\omega) + bY(j\omega)$	$ax(t) + by(t) \xleftrightarrow{FS;\ \omega_o} aX[k] + bY[k]$
Time shift	$x(t - t_o) \xleftrightarrow{FT} e^{-j\omega t_o}X(j\omega)$	$x(t - t_o) \xleftrightarrow{FS;\ \omega_o} e^{-jk\omega_o t_o}X[k]$
Frequency shift	$e^{j\gamma t}x(t) \xleftrightarrow{FT} X(j(\omega - \gamma))$	$e^{jk_o\omega_o t}x(t) \xleftrightarrow{FS;\ \omega_o} X[k - k_o]$
Scaling	$x(at) \xleftrightarrow{FT} \dfrac{1}{\lvert a \rvert} X\!\left(\dfrac{j\omega}{a}\right)$	$x(at) \xleftrightarrow{FS;\ a\omega_o} X[k]$
Differentiation-time	$\dfrac{d}{dt}x(t) \xleftrightarrow{FT} j\omega X(j\omega)$	$\dfrac{d}{dt}x(t) \xleftrightarrow{FS;\ \omega_o} jk\omega_o X[k]$
Differentiation-frequency	$-jtx(t) \xleftrightarrow{FT} \dfrac{d}{d\omega} X(j\omega)$	—
Integration/Summation	$\displaystyle\int_{-\infty}^{t} x(\tau)\, d\tau \xleftrightarrow{FT} \dfrac{X(j\omega)}{j\omega} + \pi X(j0)\delta(\omega)$	—
Convolution	$\displaystyle\int_{-\infty}^{\infty} x(\tau)y(t - \tau)\, d\tau \xleftrightarrow{FT} X(j\omega)Y(j\omega)$	$\displaystyle\int_{\langle T\rangle} x(\tau)y(t - \tau)\, d\tau \xleftrightarrow{FS;\ \omega_o} TX[k]Y[k]$
Modulation	$x(t)y(t) \xleftrightarrow{FT} \dfrac{1}{2\pi} \displaystyle\int_{-\infty}^{\infty} X(j\nu)Y(j(\omega - \nu))\, d\nu$	$x(t)y(t) \xleftrightarrow{FS;\ \omega_o} \displaystyle\sum_{l=-\infty}^{\infty} X[l]Y[k - l]$
Parseval's Theorem	$\displaystyle\int_{-\infty}^{\infty} \lvert x(t)\rvert^2\, dt = \dfrac{1}{2\pi}\int_{-\infty}^{\infty} \lvert X(j\omega)\rvert^2\, d\omega$	$\dfrac{1}{T}\displaystyle\int_{\langle T\rangle} \lvert x(t)\rvert^2\, dt = \sum_{k=-\infty}^{\infty} \lvert X[k]\rvert^2$
Duality	$X(jt) \xleftrightarrow{FT} 2\pi x(-\omega)$	$x[n] \xleftrightarrow{DTFT} X(e^{j\Omega})$ $X(e^{jt}) \xleftrightarrow{FS;1} x[-k]$
Symmetry	$x(t)$ real $\xleftrightarrow{FT} X^*(j\omega) = X(-j\omega)$ $x(t)$ imaginary $\xleftrightarrow{FT} X^*(j\omega) = -X(-j\omega)$ $x(t)$ real and even $\xleftrightarrow{FT} \text{Im}\{X(j\omega)\} = 0$ $x(t)$ real and odd $\xleftrightarrow{FT} \text{Re}\{X(j\omega)\} = 0$	$x(t)$ real $\xleftrightarrow{FS;\ \omega_o} X^*[k] = X[-k]$ $x(t)$ imaginary $\xleftrightarrow{FS;\ \omega_o} X^*[k] = -X[-k]$ $x(t)$ real and even $\xleftrightarrow{FS;\ \omega_o} \text{Im}\{X[k]\} = 0$ $x(t)$ real and odd $\xleftrightarrow{FS;\ \omega_o} \text{Re}\{X[k]\} = 0$

Discrete-Time FT	Discrete-Time FS
$x[n] \xleftrightarrow{\text{DTFT}} X(e^{j\Omega})$ $y[n] \xleftrightarrow{\text{DTFT}} Y(e^{j\Omega})$	$x[n] \xleftrightarrow{\text{DTFS; }\Omega_o} X[k]$ $y[n] \xleftrightarrow{\text{DTFS; }\Omega_o} Y[k]$ Period $= N$
$ax[n] + by[n] \xleftrightarrow{\text{DTFT}} aX(e^{j\Omega}) + bY(e^{j\Omega})$	$ax[n] + by[n] \xleftrightarrow{\text{DTFS; }\Omega_o} aX[k] + bY[k]$
$x[n - n_o] \xleftrightarrow{\text{DTFT}} e^{-j\Omega n_o}X(e^{j\Omega})$	$x[n - n_o] \xleftrightarrow{\text{DTFS; }\Omega_o} e^{-jk\Omega_o n_o}X[k]$
$e^{j\Gamma n}x[n] \xleftrightarrow{\text{DTFT}} X(e^{j(\Omega - \Gamma)})$	$e^{jk_o\Omega_o n}x[n] \xleftrightarrow{\text{DTFS; }\Omega_o} X[k - k_o]$
$x_z[n] = 0, \qquad n \neq lp$ $x_z[pn] \xleftrightarrow{\text{DTFT}} X_z(e^{j\Omega/p})$	$x_z[n] = 0, \qquad n \neq lp$ $x_z[pn] \xleftrightarrow{\text{DTFS; }p\Omega_o} pX_z[k]$
—	—
$-jnx[n] \xleftrightarrow{\text{DTFT}} \dfrac{d}{d\Omega} X(e^{j\Omega})$	—
$\displaystyle\sum_{k=-\infty}^{n} x[k] \xleftrightarrow{\text{DTFT}} \dfrac{X(e^{j\Omega})}{1 - e^{-j\Omega}} + \pi X(e^{j0}) \sum_{k=-\infty}^{\infty} \delta(\Omega - k2\pi)$	—
$\displaystyle\sum_{l=-\infty}^{\infty} x[l]y[n - l] \xleftrightarrow{\text{DTFT}} X(e^{j\Omega})Y(e^{j\Omega})$	$\displaystyle\sum_{l=\langle N \rangle} x[l]y[n - l] \xleftrightarrow{\text{DTFS; }\Omega_o} NX[k]Y[k]$
$x[n]y[n] \xleftrightarrow{\text{DTFT}} \dfrac{1}{2\pi} \displaystyle\int_{\langle 2\pi\rangle} X(e^{j\Gamma})Y(e^{j(\Omega - \Gamma)})\, d\Gamma$	$x[n]y[n] \xleftrightarrow{\text{DTFS; }\Omega_o} \displaystyle\sum_{l=\langle N\rangle} X[l]Y[k - l]$
$\displaystyle\sum_{n=-\infty}^{\infty} |x[n]|^2 = \dfrac{1}{2\pi} \int_{\langle 2\pi\rangle} |X(e^{j\Omega})|^2\, d\Omega$	$\dfrac{1}{N} \displaystyle\sum_{n=\langle N\rangle} |x[n]|^2 = \sum_{k=\langle N\rangle} |X[k]|^2$
$x[n] \xleftrightarrow{\text{DTFT}} X(e^{j\Omega})$ $X(e^{jt}) \xleftrightarrow{\text{FS; }1} x[-k]$	$X[n] \xleftrightarrow{\text{DTFS; }\Omega_o} \dfrac{1}{N} x[-k]$
$x[n]$ real $\xleftrightarrow{\text{DTFT}} X^*(e^{j\Omega}) = X(e^{-j\Omega})$ $x[n]$ imaginary $\xleftrightarrow{\text{DTFT}} X^*(e^{j\Omega}) = -X(e^{-j\Omega})$ $x[n]$ real and even $\xleftrightarrow{\text{DTFT}} \text{Im}\{X(e^{j\Omega})\} = 0$ $x[n]$ real and odd $\xleftrightarrow{\text{DTFT}} \text{Re}\{X(e^{j\Omega})\} = 0$	$x[n]$ real $\xleftrightarrow{\text{DTFS; }\Omega_o} X^*[k] = X[-k]$ $x[n]$ imaginary $\xleftrightarrow{\text{DTFS; }\Omega_o} X^*[k] = -X[-k]$ $x[n]$ real and even $\xleftrightarrow{\text{DTFS; }\Omega_o} \text{Im}\{X[k]\} = 0$ $x[n]$ real and odd $\xleftrightarrow{\text{DTFS; }\Omega_o} \text{Re}\{X[k]\} = 0$

C.8 Relating the Four Fourier Representations

Let

$$g(t) \xleftrightarrow{\;FS;\; \omega_o = 2\pi/T\;} G[k]$$

$$v[n] \xleftrightarrow{\;DTFT\;} V(e^{j\Omega})$$

$$w[n] \xleftrightarrow{\;DTFS;\; \Omega_o = 2\pi/N\;} W[k]$$

■ FT REPRESENTATION FOR A CONTINUOUS-TIME PERIODIC SIGNAL

$$g(t) \xleftrightarrow{\;FT\;} G(j\omega) = 2\pi \sum_{k=-\infty}^{\infty} G[k]\delta(\omega - k\omega_o)$$

■ DTFT REPRESENTATION FOR A DISCRETE-TIME PERIODIC-SIGNAL

$$w[n] \xleftrightarrow{\;DTFT\;} W(e^{j\Omega}) = 2\pi \sum_{k=-\infty}^{\infty} W[k]\delta(\Omega - k\Omega_o)$$

■ FT REPRESENTATION FOR A DISCRETE-TIME APERIODIC SIGNAL

$$v_\delta(t) = \sum_{n=-\infty}^{\infty} v[n]\delta(t - n\mathcal{T}) \xleftrightarrow{\;FT\;} V_\delta(j\omega) = V(e^{j\Omega})\Big|_{\Omega=\omega\mathcal{T}}$$

■ FT REPRESENTATION FOR A DISCRETE-TIME PERIODIC SIGNAL

$$w_\delta(t) = \sum_{n=-\infty}^{\infty} w[n]\delta(t - n\mathcal{T}) \xleftrightarrow{\;FT\;} W_\delta(j\omega) = \frac{2\pi}{\mathcal{T}} \sum_{k=-\infty}^{\infty} W[k]\delta\left(\omega - \frac{k\Omega_o}{\mathcal{T}}\right)$$

C.9 Sampling and Aliasing Relationships

Let

$$x(t) \xleftrightarrow{\;FT\;} X(j\omega)$$

$$v[n] \xleftrightarrow{\;DTFT\;} V(e^{j\Omega})$$

■ IMPULSE SAMPLING FOR CONTINUOUS-TIME SIGNALS

$$x_\delta(t) = \sum_{n=-\infty}^{\infty} x(n\mathcal{T})\delta(t - n\mathcal{T}) \xleftrightarrow{\;FT\;} X_\delta(j\omega) = \frac{1}{\mathcal{T}} \sum_{k=-\infty}^{\infty} X\left(j\left(\omega - k\frac{2\pi}{\mathcal{T}}\right)\right)$$

$X_\delta(j\omega)$ is $2\pi/\mathcal{T}$ periodic.

■ Sampling a Discrete-Time Signal

$$y[n] = v[qn] \xrightarrow{\quad DTFT \quad} Y(e^{j\Omega}) = \frac{1}{q} \sum_{m=0}^{q-1} V(e^{j(\Omega - m2\pi)/q})$$

$Y(e^{j\Omega})$ is 2π periodic.

■ Sampling the DTFT in Frequency

$$w[n] = \sum_{m=-\infty}^{\infty} v[n + mN] \xleftarrow{\quad DTFS; \; \Omega_o = 2\pi/N \quad} W[k] = \frac{1}{N} V(e^{jk\Omega_o})$$

$w[n]$ is N periodic.

■ Sampling the FT in Frequency

$$g(t) = \sum_{m=-\infty}^{\infty} x(t + mT) \xleftarrow{\quad FS; \; \omega_o = 2\pi/T \quad} G[k] = \frac{1}{T} X(jk\omega_o)$$

$g(t)$ is T periodic.

D | Tables of Laplace Transforms and Properties

D.1 Basic Laplace Transforms

Signal	Transform	ROC
$u(t)$	$\dfrac{1}{s}$	$\text{Re}\{s\} > 0$
$tu(t)$	$\dfrac{1}{s^2}$	$\text{Re}\{s\} > 0$
$\delta(t - \tau), \quad \tau > 0$	$e^{-s\tau}$	for all s
$e^{-at}u(t)$	$\dfrac{1}{s + a}$	$\text{Re}\{s\} > -a$
$te^{-at}u(t)$	$\dfrac{1}{(s + a)^2}$	$\text{Re}\{s\} > -a$
$[\cos(\omega_1 t)]u(t)$	$\dfrac{s}{s^2 + \omega_1^2}$	$\text{Re}\{s\} > 0$
$[\sin(\omega_1 t)]u(t)$	$\dfrac{\omega_1}{s^2 + \omega_1^2}$	$\text{Re}\{s\} > 0$
$[e^{-at}\cos(\omega_1 t)]u(t)$	$\dfrac{s + a}{(s + a)^2 + \omega_1^2}$	$\text{Re}\{s\} > -a$
$[e^{-at}\sin(\omega_1 t)]u(t)$	$\dfrac{\omega_1}{(s + a)^2 + \omega_1^2}$	$\text{Re}\{s\} > -a$

■ **BILATERAL LAPLACE TRANSFORMS FOR SIGNALS THAT ARE NONZERO FOR** $t \le 0$

Signal	Bilateral Transform	ROC
$\delta(t - \tau),\ \tau \le 0$	$e^{-s\tau}$	for all s
$-u(-t)$	$\dfrac{1}{s}$	$\text{Re}\{s\} < 0$
$-tu(-t)$	$\dfrac{1}{s^2}$	$\text{Re}\{s\} < 0$
$-e^{-at}u(-t)$	$\dfrac{1}{s + a}$	$\text{Re}\{s\} < -a$
$-te^{-at}u(-t)$	$\dfrac{1}{(s + a)^2}$	$\text{Re}\{s\} < -a$

D.2 *Laplace Transform Properties*

Signal	Unilateral Transform	Bilateral Transform	ROC
$x(t)$	$X(s)$	$X(s)$	R_x
$y(t)$	$Y(s)$	$Y(s)$	R_y
$ax(t) + by(t)$	$aX(s) + bY(s)$	$aX(s) + bY(s)$	At least $R_x \cap R_y$
$x(t - \tau)$	$e^{-s\tau}X(s)$ if $x(t - \tau)u(t) = x(t - \tau)u(t - \tau)$	$e^{-s\tau}X(s)$	R_x
$e^{s_o t}x(t)$	$X(s - s_o)$	$X(s - s_o)$	$R_x - \text{Re}\{s_o\}$
$x(at)$	$\dfrac{1}{|a|} X\left(\dfrac{s}{a}\right)$	$\dfrac{1}{|a|} X\left(\dfrac{s}{a}\right)$	$\dfrac{R_x}{|a|}$
$x(t) * y(t)$	$X(s)Y(s)$	$X(s)Y(s)$	At least $R_x \cap R_y$
$-tx(t)$	$\dfrac{d}{ds} X(s)$	$\dfrac{d}{ds} X(s)$	R_x
$\dfrac{d}{dt} x(t)$	$sX(s) - x(0^+)$	$sX(s)$	At least R_x
$\displaystyle\int_{-\infty}^{t} x(\tau)\, d\tau$	$\dfrac{1}{s} \displaystyle\int_{-\infty}^{0^+} x(\tau)\, d\tau + \dfrac{X(s)}{s}$	$\dfrac{X(s)}{s}$	At least $R_x \cap \{\text{Re}\{s\} > 0\}$

■ INITIAL VALUE THEOREM

$$\lim_{s \to \infty} sX(s) = x(0^+)$$

This result does not apply to rational functions $X(s)$ in which the order of the numerator polynomial is equal to or greater than the order of the denominator polynomial, since this implies $X(s)$ contains terms of the form cs^k, $k \geq 0$. Such terms correspond to the impulses and their derivatives located at time $t = 0$ and are excluded from the unilateral Laplace transform.

■ FINAL VALUE THEOREM

$$\lim_{s \to 0} sX(s) = \lim_{t \to \infty} x(t)$$

This result requires that all the poles of $sX(s)$ be in the left half of the s-plane.

■ UNILATERAL DIFFERENTIATION PROPERTY—GENERAL FORM

$$\frac{d^n}{dt^n} x(t) \xleftrightarrow{\mathcal{L}_u} s^n X(s) - \left.\frac{d^{n-1}}{dt^{n-1}} x(t)\right|_{t=0^+} - s\left.\frac{d^{n-2}}{dt^{n-2}} x(t)\right|_{t=0^+} - \cdots - s^{n-2}\left.\frac{d}{dt} x(t)\right|_{t=0^+} - s^{n-1}x(0^+)$$

E Tables of z-Transforms and Properties

E.1 Basic z-Transforms

Signal	Transform	ROC
$\delta[n]$	1	All z
$u[n]$	$\dfrac{1}{1 - z^{-1}}$	$\lvert z \rvert > 1$
$\alpha^n u[n]$	$\dfrac{1}{1 - \alpha z^{-1}}$	$\lvert z \rvert > \lvert \alpha \rvert$
$n\alpha^n u[n]$	$\dfrac{\alpha z^{-1}}{(1 - \alpha z^{-1})^2}$	$\lvert z \rvert > \lvert \alpha \rvert$
$[\cos(\Omega_1 n)]u[n]$	$\dfrac{1 - z^{-1}\cos\Omega_1}{1 - z^{-1}2\cos\Omega_1 + z^{-2}}$	$\lvert z \rvert > 1$
$[\sin(\Omega_1 n)]u[n]$	$\dfrac{z^{-1}\sin\Omega_1}{1 - z^{-1}2\cos\Omega_1 + z^{-2}}$	$\lvert z \rvert > 1$
$[r^n\cos(\Omega_1 n)]u[n]$	$\dfrac{1 - z^{-1}r\cos\Omega_1}{1 - z^{-1}2r\cos\Omega_1 + r^2 z^{-2}}$	$\lvert z \rvert > r$
$[r^n\sin(\Omega_1 n)]u[n]$	$\dfrac{z^{-1}r\sin\Omega_1}{1 - z^{-1}2r\cos\Omega_1 + r^2 z^{-2}}$	$\lvert z \rvert > r$

▪ **BILATERAL TRANSFORMS FOR SIGNALS THAT ARE NONZERO FOR $n < 0$**

Signal	Bilateral Transform	ROC
$u[-n - 1]$	$\dfrac{1}{1 - z^{-1}}$	$|z| < 1$
$-\alpha^n u[-n - 1]$	$\dfrac{1}{1 - \alpha z^{-1}}$	$|z| < |\alpha|$
$-n\alpha^n u[-n - 1]$	$\dfrac{\alpha z^{-1}}{(1 - \alpha z^{-1})^2}$	$|z| < |\alpha|$

E.2 z-Transform Properties

Signal	Unilateral Transform	Bilateral Transform	ROC
$x[n]$	$X(z)$	$X(z)$	R_x
$y[n]$	$Y(z)$	$Y(z)$	R_y
$ax[n] + by[n]$	$aX(z) + bY(z)$	$aX(z) + bY(z)$	At least $R_x \cap R_y$
$x[n - k]$	See below	$z^{-k}X(z)$	R_x except possibly $|z| = 0, \infty$
$\alpha^n x[n]$	$X\left(\dfrac{z}{\alpha}\right)$	$X\left(\dfrac{z}{\alpha}\right)$	$|\alpha| R_x$
$x[-n]$	—	$X\left(\dfrac{1}{z}\right)$	$\dfrac{1}{R_x}$
$x[n] * y[n]$	$X(z)Y(z)$	$X(z)Y(z)$	At least $R_x \cap R_y$
$nx[n]$	$-z\dfrac{d}{dz}X(z)$	$-z\dfrac{d}{dz}X(z)$	R_x except possibly addition or deletion of $z = 0$

▪ **UNILATERAL z-TRANSFORM TIME-SHIFT PROPERTY**

$$x[n - k] \xleftrightarrow{z_u} x[-k] + x[-k + 1]z^{-1} + \cdots + x[-1]z^{-k+1} + z^{-k}X(z) \quad \text{for } k > 0$$

$$x[n + k] \xleftrightarrow{z_u} -x[0]z^k - x[1]z^{k-1} - \cdots - x[k - 1]z + z^k X(z) \quad \text{for } k > 0$$

Index